Springer Series in
SOLID-STATE SCIENCES 164

Springer Series in

# SOLID-STATE SCIENCES



The Springer Series in Solid-State Sciences consists of fundamental scientific books prepared by leading researchers in the field. They strive to communicate, in a systematic and comprehensive way, the basic principles as well as new developments in theoretical and experimental solid-state physics.

Please view available titles in *Springer Series in Solid-State Sciences* on series homepage http://www.springer.com/series/682

Claudine Lacroix
Philippe Mendels
Frédéric Mila
Editors

# Introduction to Frustrated Magnetism

## Materials, Experiments, Theory

With 244 Figures

Springer

*Editors:*

Claudine Lacroix
Institut Néel
CNRS-UJF
38042 Grenoble Cedex 9
France
E-mail: claudine.lacroix@grenoble.cnrs.fr

Frédéric Mila
Institute of Theoretical Physics
Ecole Polytechnique Fédérale de Lausanne
1015 Lausanne
Switzerland
E-mail: frederic.mila@epfl.ch

Philippe Mendels
Laboratoire de Physique des Solides
Université Paris Sud
91405 Orsay Cedex
France
E-mail: mendels@lps.u-psud.fr



Springer Series in Solid-State Sciences ISSN 0171-1873
ISBN 978-3-642-10588-3 e-ISBN 978-3-642-10589-0
DOI 10.1007/978-3-642-10589-0
Springer Heidelberg Dordrecht London New York

Library of Congress Control Number: 2010938611



*Cover design*: eStudio Calamar Steinen

Printed on acid-free paper

Springer is part of Springer Science+Business Media (www.springer.com)

# Preface

Frustrated magnetism has become an extremely active field of research. After undergoing a revival in the 1990s in the context of applying Anderson's resonating-valence-bond (RVB) theory to high-temperature superconductors, the subject has experienced a tremendous burst of theoretical and experimental activity in the last decade. Since its first edition in 2000, the major conference in the field, "Highly Frustrated Magnetism," now takes place every 2 years and has seen very rapid growth in participant number. Broad research networks on frustration and related topics have been established recently both in Europe and in Japan.

Within the context of the European Network on Highly Frustrated Magnetism, which gathers 14 countries and is supported by the European Science Foundation (2005–2010), we made the decision to edit a book that would cover all of the important aspects of the field. The subject matter would span each of its three pillars: materials, experiment, and theory. The summer school which was held in Trieste in summer 2007 as an activity of the ESF network presented the ideal opportunity for the definitive launch of this project. The response was very positive, and the 400-page volume planned at the outset has increased in size over the intervening months due to the enthusiasm of the authors, who have produced beautiful reviews both significantly longer and broader in scope than initially foreseen.

The driving force behind the considerable current activity is the conviction, now demonstrated by numerous examples, that frustrated magnetism presents an excellent proving ground in which to discover new states and new properties of matter. For theorists, this conviction comes from the simple observation that long-range magnetic order, the standard low-temperature instability, cannot be achieved due to the proliferation of possible ground states. This is true both for classical systems, where averaging over the various ground states often leads to decaying correlations at large distances, and for quantum systems, where this proliferation translates into a very soft spectrum and diverging fluctuations.

The list of proposed alternatives to long-range order is already impressive (residual entropy, algebraic or dipolar correlations, gapped or gapless spin liquids, spin nematics, ...), and one may anticipate further increases over the coming years. A good example is the role played by frustration in many multiferroic systems. While the proposals are theoretical, at the root of many of these developments have been materials scientists. They have invented new families of compounds, rediscovered some existing minerals, found synthesis routes to produce them in a form purer than the natural species, and even discovered some variants of these. Artificial frustrated lattices have begun to pave the way towards further new challenges. However, despite all of these efforts, most of the possible states of matter predicted by theory remain largely unexplored at the experimental level, and the ultra-clean realizations of materials required to probe critical behavior are still lacking. There is no doubt that materials science has a leading role to play in the future of frustrated magnetism.

Experimentally, the absence of a phase transition at a temperature of the same order as the typical coupling constant signals the possibility of unconventional low-temperature physics. This is often quantified by the ratio $\Theta_{CW}/T_c$, sometimes called the Ramirez frustration ratio, where $\Theta_{CW}$ is the Curie–Weiss temperature and $T_c$ the critical temperature (if any). Strong frustration always produces a large ratio. Values around 50 are not uncommon, and there is no upper limit: in a number of systems, no phase transition to any type of order has been detected to date.

The concept of frustration is now ubiquitous in physics. In the context of magnetism, the name first appeared for spin glasses to describe the difficulty experienced by these systems in reaching their true ground state when both disorder and frustration act together. While disorder certainly plays a role in the properties of the materials to be discussed in this volume, the main focus is on the intrinsic effects of "geometrical" frustration, the general term for the competition between interaction pathways which arises in clean and periodic, but frustrated, systems.

Frustration and one-dimensionality share many common features: they both lead to diverging fluctuations, to exotic excitations, and to reduced critical temperatures due to small additional interactions. However, in frustrated systems, the presence of several competing states leads to a very large number of low-lying excitations, or more generally to a redistribution of spectral weight into narrow bands. This can be manifest as an anomalously large specific heat at low temperatures (and even as a residual entropy in examples such as spin ice), or as narrow, compressible phases between incompressible magnetization plateaus.

The field of frustrated magnetism is vast, and dichotomies have emerged naturally over the years: classical vs. quantum, 2D vs. 3D, rare-earth vs. transition-metal ions, corner-sharing vs. edge-sharing lattices, Ising vs. Heisenberg interactions, etc. It would be dangerous to neglect these differences, because this is a field where details matter. Indeed, small interactions are often responsible for the ultimate selection from among several candidate ground states. This is why, with the exception of a small number of more general chapters, a significant fraction of the contributions to this volume concentrate on specific aspects or materials rather than on broader principles.

The goal of this book is two-fold: the first is to provide a solid introduction to Highly Frustrated Magnetism for researchers and PhD students beginning their activities in the field; the second is to review the more advanced topics of current interest which, in some cases, remain under development. The volume is divided into six sections. The first section contains two chapters which provide a general introduction to classical and quantum frustrated magnetism. The second section reviews the primary spectroscopic approaches (neutron scattering, resonance techniques, light scattering) upon which, in large part, our current experimental understanding of frustrated magnets is based. The third section is devoted to the synthesis and crystal growth of frustrated magnets, and to experimental reviews of the leading families of compounds with frustrated geometries (triangular, kagomé, spinel, pyrochlore). The fourth section deals with physical effects characteristic of frustration, such as magnetization plateaus, the spin-Jahn–Teller transition, spin ice, and spin nematics. The fifth section is devoted entirely to theory, presenting the primary approaches which have been deployed in trying to cope with the extremely difficult problem raised by the interplay between quantum fluctuations and frustration. Finally, the last section addresses the effects of frustration in systems with further degrees of freedom, such as mobile carriers or orbital degeneracy.

What are the leading open issues and challenges in the field? Each reader will probably have her or his own priority list among the outstanding problems to be found in this volume. The experimental realization of an RVB spin liquid and the solution of the Heisenberg model on the kagomé or on the pyrochlore lattice are long-standing ones. We hope that this book may serve both as a reference and as a springboard which can contribute to the solution of these fascinating problems.

We would like to express our sincere thanks to Bruce Normand, who proof-read the chapters written by non-native speakers of English and, in the course of this ungrateful task, also provided many insightful comments and suggestions.

In conclusion, we would like to thank the International Center for Theoretical Physics in Trieste for hosting the 2007 Summer School, as well as for supporting part of its costs, and the European Science Foundation for its constant support of the HFM network.

France — *Claudine Lacroix*
France — *Philippe Mendels*
Switzerland — *Frédéric Mila*
August 2010

# Contents

## Part II Probing Frustrated Magnets

# Contributors

**Daniel P. Arovas** Department of Physics, University of California at San Diego, La Jolla, CA 92093, USA, darovas@ucsd.edu

**Assa Auerbach** Physics Department, Technion, Haifa 32000, Israel, assa@physics.technion.ac.il

**Federico Becca** CNR-IOM-Democritos and International School for Advanced Studies (SISSA), Via Beirut 2-4, 34014 Trieste, Italy, becca@sissa.it

**Steven T. Bramwell** London Centre for Nanotechnology, University College London, 17-19 Gordon Street, London WC1H 0AH, UK, s.t.bramwell@ucl.ac.uk

**Luca Capriotti** Global Modelling and Analytics Group, Investment Banking Division, Credit Suisse Group, Eleven Madison Avenue, New York, NY 10010-3629, USA, luca.capriotti@gmail.com

**Pietro Carretta** Dipartimento di Fisica "A. Volta", University of Pavia, Via Bassi, 6, 27100 Pavia, Italy, pietro.carretta@unipv.it

**Robert J. Cava** Department of Chemistry, Princeton University, Princeton, NJ 08540, USA, rcava@Princeton.EDU

**John T. Chalker** Theoretical Physics, Oxford University, 1, Keble Road, Oxford OX1 3NP, UK, j.chalker1@physics.ox.ac.uk

**Gia-Wei Chern** Department of Physics, University of Wisconsin, 1150 University Avenue, Madison, WI 53706, USA, gchern@physics.wisc.edu

**Joachim Deisenhofer** Experimentalphysik V, Center for Electronic Correlations and Magnetism, Institute for Physics, Augsburg University, 86135 Augsburg, Germany, joachim.deisenhofer@physik.uni-augsburg.de

**Jason S. Gardner** Department of Physics, Indiana University, Bloomington, IN 47405, USA
and
NIST Centre of Neutron Research, Gaithersburg, MD 20899-6102, USA, jsg@nist.gov

**Bruce D. Gaulin** Department of Physics and Astronomy, McMaster University, Hamilton, ON, Canada L8S 4C6
and
Canadian Institute for Advanced Research, 180 Dundas Street W, Toronto, ON, Canada, gaulin@mcmaster.ca

**Michel J.P. Gingras** Department of Physics and Astronomy, University of Waterloo, Waterloo, ON, Canada N2L 3G1
and
Canadian Institute for Advanced Research/Quantum Materials Program, 180 Dundas Street West, Toronto, ON, Canada M5G 1Z8, gingras@gandalf.uwaterloo.ca

**Zenji Hiroi** Institute for Solid State Physics, University of Tokyo, 5-1-5 Kashiwa, Chiba 277-8581, Japan, hiroi@issp.u-tokyo.ac.jp

**Katharine L. Holman** Department of Chemistry, Princeton University, Princeton, NJ 08540, USA, katiehughes@gmail.com

**Amit Keren** Physics Department, Technion-Israel Institute of Technology, Haifa 32000, Israel, keren@physics.technion.ac.il

**Claudine Lacroix** Institut Néel, 25, avenue des Martyrs, Bâtiment K, BP 166, 38042 Grenoble cedex 9, France, Claudine.Lacroix@grenoble.cnrs.fr

**Andreas M. Läuchli** Max Planck Institut fur Physik komplexer Systeme, Nothnitzer Strasse 38, 01187 Dresden, Germany, laeuchli@comp-phys.org

**Peter Lemmens** Institute for Condensed Matter Physics, Braunschweig University of Technology, 38106 Braunschweig, Germany, p.lemmens@tu-bs.de

**Claire Lhuillier** Laboratoire de Physique Théorique de la Matière Condensée, Université Pierre et Marie Curie, case courrier 121, 4 place Jussieu, 75252 Paris cedex 05, France, claire.lhuillier@umpc.fr

**Tyrel McQueen** Department of Chemistry and Department of Physics and Astronomy, Johns Hopkins University, 3400 N. Charles St, Baltimore, MD 21218, USA, mcqueen@jhu.edu

**Philippe Mendels** Laboratoire de Physique des Solides, Université Paris Sud, 91405 Orsay Cedex, France, mendels@lps.u-psud.fr

**Frédéric Mila** Institute of Theoretical Physics, Ecole Polytechnique Fédérale de Lausanne, 1015 Lausanne, Switzerland, frederic.mila@epfl.ch

**Grégorie Misguich** Institut de Physique Théorique, Orme des Merisiers batiment 774, Point courrier 136, CEA/DSM/IPhT, CEA/Saclay, 91191 Gif-sur-Yvette Cedex, France, gregoire.misguich@cea.fr

**Shin Miyahara** Multiferroics Project (MF), ERATO, Japan Science and Technology Agency (JST), c/o Department of Applied Physics, University of Tokyo, 7-3-1 Hongo, Tokyo 113-8656, Japan, miyahara@erato-mf.t.u-tokyo.ac.jp

**Roderich Moessner** Max-Planck-Institut für Physik komplexer Systeme, 01187 Dresden, Germany, moessner@pks.mpg.de

**Seiji Niitaka** RIKEN (The Institute of Physical and Chemical Research), Wako 351-0198, Japan, niitaka@riken.jp

**Zohar Nussinov** Department of Physics, Washington University, St. Louis, MO 63160, USA, zohar@wuphys.wustl.edu

**Masao Ogata** Department of Physics, University of Tokyo, 7-3-1 Hongo, Bunkyo-ku, Tokyo, 113-0033 Japan, ogata@phys.s.u-tokyo.ac.jp

**Andrzej M. Oleś** Marian Smoluchowski Institute of Physics, Jagellonian University, Reymonta 4, 30059 Krakow, Poland
and
Max-Planck-Institut für Festkörperforschung, Heisenbergstrasse 1, 70569 Stuttgart, Germany, a.m.oles@fkf.mpg.de

**Alberto Parola** Dipartimento di Fisica e Matematica, Università dell'Insubria, Via Valleggio 11, 22100 Como, Italy, parola@mi.infm.it

**Karlo Penc** Research Institute for Solid State Physics and Optics, P.O. Box 49, 1525 Budapest, Hungary, penc@szfki.hu

**Didier Poilblanc** Laboratoire de Physique Théorique, C.N.R.S. and Université de Toulouse III, 118, route de Narbonne, 31062 Toulouse cedex, France, didier.poilblanc@irsamc.ups-tlse.fr

**Kumar S. Raman** University of California at Riverside, Riverside, CA 92521, USA, kumar.raman@ucr.edu

**Kai Phillip Schmidt** Lehrstuhl für Theoretische Physik I, TU Dortmund, 44221 Dortmund, Germany, schmidt@fkt.physik.uni-dortmund.de

**Sandro Sorella** CNR-IOM-Democritos and International School for Advanced Studies (SISSA), Via Beirut 2-4, 34014 Trieste, Italy, sorella@sissa.it

**Hidenori Takagi** Department of Advanced Materials, University of Tokyo, Kashiwa 277-8651, Japan, htakagi@k.u-tokyo.ac.jp

**Masashi Takigawa** Institute for Solid State Physics, The University of Tokyo, Kashiwa, Chiba 277-8581, Japan, masashi@issp.u-tokyo.ac.jp

**Oleg Tchernyshyov** Department of Physics and Astronomy, Johns Hopkins University, 3400 N. Charles St., Baltimore, MD 21218, USA, olegt@jhu.edu

**Hirokazu Tsunetsugu** Institute for Solid State Physics, University of Tokyo, Kashiwanoha 5-1-5, Chiba 277-8581, Japan, tsune@issp.u-tokyo.ac.jp

**Jeroen van den Brink** Institute Lorentz for Theoretical Physics, Leiden University, P.O. Box 9506, 2300 RA Leiden, The Netherlands, j.van.den.brink@ifw-dresden.de

**Eric J. Welsh** Department of Chemistry, Princeton University, Princeton, NJ 08540, USA, ericjwelch@gmail.com

**D. Vincent West** Department of Materials Science, University of Pennsylvania, 3231 Walnut St., Philadelphia, PA 19104, USA, barleytone@gmail.com

**Anthony Williams** Department of Chemistry, Princeton University, Princeton, NJ 08540, USA, da_tonester@hotmail.com

**Andrew S. Wills** Department of Chemistry, University College London, 20 Gordon Street, London, WC1H 0AJ, UK, a.s.wills@ucl.ac.uk

# Part I
# Basic Concepts in Frustrated Magnetism

# Chapter 1
# Geometrically Frustrated Antiferromagnets: Statistical Mechanics and Dynamics

**John T. Chalker**

**Abstract** These lecture notes provide a simple overview of the physics of geometrically frustrated magnets. The emphasis is on classical and semiclassical treatments of the statistical mechanics and dynamics of frustrated Heisenberg models, and on the ways in which the results provide an understanding of some of the main observed properties of these systems.

## 1.1 Introduction

This chapter is intended to give an introduction to the theory of thermal fluctuations and their consequences for static and dynamic correlations in geometrically frustrated antiferromagnets, focusing on the semiclassical limit, and to discuss how our theoretical understanding leads to an explanation of some of the main observed properties of these systems. A central theme will be the fact that simple, classical models for highly frustrated magnets have a ground state degeneracy which is *macroscopic*, though *accidental* rather than a consequence of symmetries. We will be concerned in particular with (a) the origin of this degeneracy and the possibility that it is lifted by thermal or quantum fluctuations; (b) correlations within ground states; and (c) low-temperature dynamics. We concentrate on Heisenberg models with large spin $S$, referring to the Chap. 16 by G. Misguich for a discussion of quantum spin liquids, and to the Chap. 12 by M. Gingras for an overview of geometrically frustrated Ising models in the context of spin ice materials. Several earlier reviews provide useful further reading, including [1–3] for experimental background, and [4, 5] for an alternative perspective on theory.

To provide a comparison, it is useful to begin by recalling the behaviour of an *unfrustrated* antiferromagnet. To be definite, consider the Heisenberg model with nearest neighbour exchange $J$ on a simple cubic lattice. As the lattice is bipartite – it can be separated into two interpenetrating sublattices, in such a way that sites of one sublattice have as their nearest neighbours only sites from the other sublattice – the classical ground states are two-sublattice Néel states, in which spins on one sublattice all have the same orientation, and those on the other sublattice have the opposite orientation. These states are unique up to global spin rotations, which are a symmetry of the model. Their only low energy excitations are long wavelength spinwaves. These are Goldstone modes – a consequence of the symmetry breaking in ground states – and have a frequency $\omega(\mathbf{k})$ that is linear in wavevector $\mathbf{k}$ at small $k$. This classical picture carries over to the quantum system, and for $S \gg 1$ it is sufficient to take account of fluctuations using harmonic spinwave theory. In particular, within this approximation the sublattice magnetisation at low temperature is reduced from its classical ground state value $S$ by an amount

$$\delta S = \frac{1}{\Omega} \int_{\mathrm{BZ}} \frac{zJS}{\hbar\omega(\mathbf{k})} [\langle n(\mathbf{k})\rangle + 1/2]\,\mathrm{d}^3\mathbf{k} - \frac{1}{2}, \tag{1.1}$$

where $\langle n(\mathbf{k})\rangle$ is a Bose factor giving the number of thermally excited spin waves at wavevector $\mathbf{k}$ and $1/2$ represents the zero-point contribution, with the integral running over the Brillouin zone of volume $\Omega$. Fluctuations increase with temperature and the sublattice magnetisation falls to zero at the Néel temperature $T_{\mathrm{N}}$. Within mean field theory $k_{\mathrm{B}} T_{\mathrm{N}} = zJS^2$, where $z$ is the number of nearest neighbour sites (six on the simple cubic lattice).

A central reason for the interest in geometrically frustrated magnets is that they hold out the possibility of evading Néel order. At the simplest level, there is a tendency for frustrated systems to have many low-frequency modes, which means both that excitations are effective in reducing the ordered moment, because of the factor of $zJS/\hbar\omega(\mathbf{k})$ in (1.1), and that they are thermally populated even at low temperature. More fundamentally, we will see that frustration may lead to classical ground states of a quite different kind and suppress $T_{\mathrm{N}}$ to zero.

Since the term *frustration* is used in several different contexts, it is worthwhile to set out some distinctions before going further. In general terms, classical frustrated systems have Hamiltonians with competing interactions which make contributions to the energy that cannot simultaneously be minimised. The concept was originally discussed in relation to spin glasses, but these are set apart from the systems we are considering here by the fact that quenched disorder has a controlling influence on their properties. Frustration as a way of destabilising Néel order has been studied extensively in models with competing nearest neighbour and further neighbour interactions, notably the $J_1 - J_2$ model on the square lattice [6], illustrated in Fig. 1.1. The classical ground state of this model depends on the ratio $J_1/J_2$. For $J_1 > 2J_2 > 0$ neighbouring spins are antiparallel, enforcing ferromagnetic alignment of second neighbours and frustration of the interaction $J_2$. In the other regime, $2J_2 > J_1 > 0$, second neighbours are antiferromagnetically aligned at the expense of frustration of half of the $J_1$ interactions. Interest focuses on the point $J_1 = 2J_2$ where these alternative classical states are degenerate, and the consequences of frustration are likely to be largest. While models of this kind provide an attractive starting point for theoretical work, there are likely to be difficulties in finding experimental realisations with interaction strengths that place them close to the degeneracy point. From this perspective, the long-appreciated [7, 8] attraction of geometrically

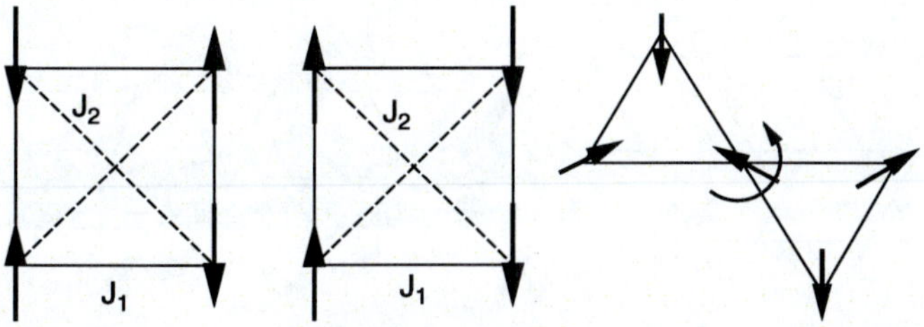


**Fig. 1.1** *Left and centre*: $J_1 - J_2$ model, showing ground state spin configurations for: $2J_1 > J_2$ (*left*); and $J_2 > 2J_1$ (*centre*). *Right*: ground states of classical Heisenberg spins at vertices of two corner-sharing triangles, with degeneracy arising from rotations about the common spin, as indicated

frustrated magnets is that they are systems in which structure alone may destabilise Néel order, with only nearest neighbour interactions. To illustrate this at an elementary level, consider Heisenberg spins at the vertices of two corner-sharing triangles with nearest-neighbour antiferromagnetic interactions, also shown in Fig. 1.1. The ground states are configurations in which spins within each triangle are coplanar and at relative angles of $2\pi/3$. They have an accidental degeneracy (in addition to that arising from symmetry) under relative rotations of the spin planes for the two triangles about the axis defined by the orientation of their common spin.

## 1.2 Models

The models we are concerned with extend some features present in the simple system of two corner-sharing triangles to a periodic lattice. In general, we will consider non-bipartite lattices constructed from corner-sharing arrangements of frustrated clusters, with local magnetic moments at the vertices of each cluster and exchange interactions of equal strength between all moments in each cluster (other arrangements are also of interest, but typically show less dramatic consequences of frustration) [9, 10]. An important example in two dimensions is the kagomé lattice, formed from corner-sharing triangles; a three-dimensional analogue is the pyrochlore lattice, built from corner-sharing tetrahedra: see Fig. 1.2 for illustrations of both.

The Hamiltonian for these models, written in terms of the exchange energy $J$ and the spin operators $\mathbf{S}_i$ at sites $i$, has the form

$$\mathcal{H} = J\sum_{\langle ij\rangle} \mathbf{S}_i \cdot \mathbf{S}_j \equiv \frac{J}{2}\sum_{\alpha} |\mathbf{L}_\alpha|^2 + c$$

$$\text{where} \quad \mathbf{L}_\alpha = \sum_{i\in\alpha} \mathbf{S}_i\,. \qquad (1.2)$$

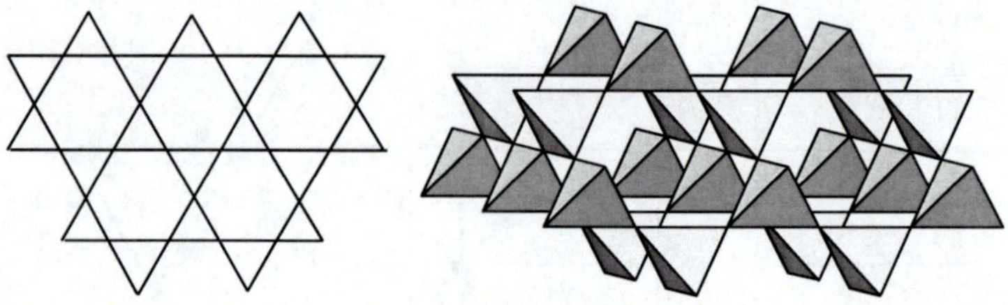

**Fig. 1.2** *Left*: kagomé lattice. *Right*: pyrochlore lattice

Here, the first sum runs over neighbouring pairs $ij$ of sites, while the second sum is over clusters $\alpha$. To recognise that this second expression is a correct rewriting of $\mathcal{H}$ in terms of the total spin $\mathbf{L}_\alpha$ of each cluster $\alpha$, it is necessary only to note that expansion of $|\mathbf{L}_\alpha|^2$ generates the required off-diagonal terms $\mathbf{S}_i \cdot \mathbf{S}_j$, together with diagonal terms $\mathbf{S}_i^2$ that contribute to the constant $c$. The apparent simplicity of this second form is of course deceptive, since the operators $\mathbf{L}_\alpha$ and $\mathbf{L}_\beta$ associated with two clusters $\alpha$ and $\beta$ that share a site are not independent.

For future reference it is useful to introduce some terminology. The frustrated clusters are in general simplices, and their centres occupy sites of a second lattice, called the simplex lattice. Spins in our models are located at the mid-points of nearest-neighbour links of the simplex lattice. For the kagomé magnet the simplex lattice is the honeycomb lattice, and for the pyrochlore magnet it is the diamond lattice.

While the Hamiltonian of (1.2) provides a useful basis for understanding the properties of a range of geometrically frustrated magnetic materials, various additional physical contributions to a realistic model may also be important. These include single-ion anisotropy [11], further neighbour exchange [12, 13], dipolar interactions [14, 15], Dzyaloshinskii–Moriya interactions [16], magnetoelastic coupling [17], site dilution [18, 19] and exchange randomness [20]. In many cases, the associated energy scales are small. They set a temperature scale much smaller than nearest-neighbour exchange, below which they may induce magnetic order or spin freezing, but they can be neglected at higher temperatures. We omit all these perturbations and restrict our discussion to models with only nearest neighbour exchange.

## 1.3 Some Experimental Facts

The single most revealing property of a geometrically frustrated magnet is arguably the dependence on temperature $T$ of its magnetic susceptibility $\chi$. It is convenient to consider plots of $\chi^{-1}$ vs $T$, which at high temperature have the linear form

$$\chi^{-1} \propto T - \Theta_{\rm CW}, \qquad (1.3)$$

where the Curie–Weiss constant $\Theta_{CW}$ characterises the sign and strength of interactions. In an antiferromagnet $\Theta_{CW}$ is negative, and for the model of (1.2) one has $k_B\Theta_{CW} = -zJS^2$. Without frustration, magnetic order, signalled by a cusp in $\chi$, appears below the Néel temperature, $T_N \sim |\Theta_{CW}|$. By contrast, in geometrically frustrated systems nothing sharp is observed at the temperature scale set by interaction strength: instead, the paramagnetic phase extends to temperatures $T \ll \Theta_{CW}$. Ordering or spin freezing may appear at a lower temperature $T_c$, but a large value for the ratio $f \equiv |\Theta_{CW}|/T_c$ is a signature of frustration [1]. This behaviour is illustrated schematically in Fig. 1.3; references to experimental papers are given in Table 1.1.

More detailed information on low temperature behaviour is provided by magnetic neutron scattering (see the Chap. 3 by S.T. Bramwell). Again, we sketch typical observations in Fig. 1.3, and give references in Table 1.1. The dynamical structure factor $S(Q,\omega)$ has a broad peak at finite wavevector $Q$, showing that spin correlations are predominantly short-range and antiferromagnetic. The width of this peak indicates a correlation length of order the lattice spacing, while the small value of the elastic scattering cross-section for $Q \to 0$ shows that correlations suppress long wavelength fluctuations in magnetisation density. This form stands in contrast both to that in unfrustrated antiferromagnets, where Néel order leads to magnetic Bragg peaks, and to that in systems with short-range ferromagnetic correlations, where the structure factor is peaked at $Q = 0$. Inelastic scattering has a width in frequency $\omega$ that decreases with decreasing temperature, and in materials that show spin freezing, scattering weight is transferred from the inelastic to the elastic response with little change in $Q$-dependence on cooling through $T_c$.

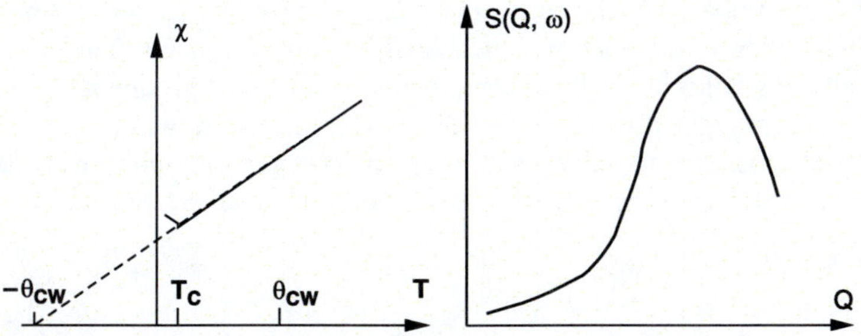


**Fig. 1.3** Characteristic behaviour of a geometrically frustrated antiferromagnet. *Left*: sketch of $\chi^{-1}$ vs $T$. *Right*: sketch of $S(Q,\omega)$ vs $Q$

**Table 1.1** Three geometrically frustrated antiferromagnets

| Material | Structure | $\lvert\Theta_{CW}\rvert$ | $T_c$ | References |
|---|---|---|---|---|
| $SrGa_3Cr_9O_{19}$ | Pyrochlore slabs | 515 K | 4 K | [21–24] |
| Hydronium iron jarosite | Kagomé | 700 K | 14 K | [25] |
| $Y_2Mo_2O_7$ | Pyrochlore | 200 K | 22 K | [26, 27] |

Properties of three well-studied geometrically frustrated antiferromagnets are set out in Table 1.1.[1] Two basic theoretical questions arise. Why is there no magnetic ordering at $T \sim |\Theta_{\mathrm{CW}}|$? And what is the nature of correlations in the strongly interacting regime $T \ll \Theta_{\mathrm{CW}}$?

## 1.4 Classical Ground State Degeneracy

To get insight into the answers to these questions, we start by considering ground states of models defined by (1.2) in the classical limit, in which the $\mathbf{S}_i$ are not operators but three-component vectors of magnitude $S$. As a first step, it is useful to examine a single tetrahedral cluster of four spins, with the Hamiltonian

$$\mathcal{H} = \frac{J}{2}|\mathbf{L}|^2 + c \qquad \text{where} \qquad \mathbf{L} = \mathbf{S}_1 + \mathbf{S}_2 + \mathbf{S}_3 + \mathbf{S}_4. \tag{1.4}$$

By writing the Hamiltonian in terms of the cluster spin $\mathbf{L}$ we see at once that ground states are those with $\mathbf{L} = \mathbf{0}$. Such an arrangement of four vectors, each having three components, with resultant zero is shown in Fig. 1.4: these ground states have two internal degrees of freedom, indicated in Fig. 1.4 by the angles $\theta$ and $\phi$, in addition to the degeneracies under global rotations which are expected from the symmetry of $\mathcal{H}$.

We should next understand how this accidental ground state degeneracy extends from a single cluster to a periodic lattice. We can do so using a counting argument [9,10], which compares $F$, the number of degrees of freedom in the system with $K$, the number of constraints that must be satisfied in ground states. The central point is that if all constraints are independent, then the number of ground state degrees of freedom is given by the difference $F - K$. Such an argument was used by Maxwell in 1864 to discuss the stability of mechanical systems of jointed rods [28], and is sometimes referred to as a *Maxwellian counting argument*. For a system of $N_s$

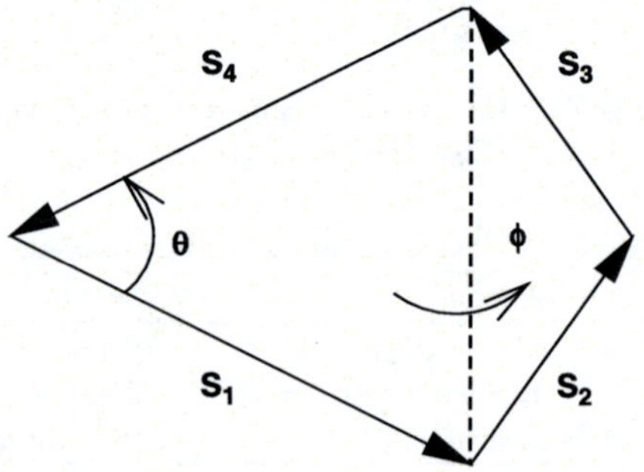


**Fig. 1.4** A ground state configuration for a frustrated cluster of four classical Heisenberg spins

[1] All three examples show spin freezing below $T_c$.

classical Heisenberg spins, $F = 2N_s$, since two angles are required to specify the orientation of each spin. And in a system with the Hamiltonian of (1.2) consisting of $N_c$ clusters, $K = 3N_c$, since in ground states all three components of $\mathbf{L}_\alpha$ must be zero for every cluster $\alpha$. Under the assumptions that all constraints can be satisfied simultaneously, and that they are all linearly independent, we arrive at an estimate for $D$, the number of ground-state degrees of freedom: $D = F - K$. Taking the example of the pyrochlore lattice, we have $N_s = 2N_c$ (since four spins are associated with each tetrahedron, but every spin is shared between two tetrahedra) and hence $D = N_c$, an extensive quantity.

This is a striking conclusion: it suggests that there are local degrees of freedom which can fluctuate independently without the system leaving its ground state manifold. The argument has two implications for our understanding of the experimental results summarised in Sect. 1.3. First, macroscopic degeneracy may prevent long range order at the temperature scale set by interaction strength, since there are many low-energy configurations that lack order. Second, since the magnetisation of each cluster is zero in *all* ground states, the amplitude of long wavelength fluctuations in the magnetisation density is small at low temperature, and so the dynamical structure factor $S(Q, \omega)$ is small at low $Q$.

At this point it is worth pausing to consider possible limitations to the counting argument that has been presented. As noted, it rests on an assumption that all ground state constraints are linearly independent. If this is not the case, we underestimate $D$. In our context, corrections are important if they make an extensive contribution to $D$. This occurs in the kagomé lattice Heisenberg antiferromagnet: in this case our estimate yields $D = 0$ (since, for a lattice built from corner-sharing triangles, $N_s = 3N_c/2$), but by explicit construction one finds sets of states with special spin arrangements [29, 30] for which $D = N_s/9$. Such an arrangement is illustrated in Fig. 1.5. By contrast, for the pyrochlore Heisenberg antiferromagnet, it is known [9, 10] that corrections to the estimate for $D$ are at most sub-extensive.

The view of classical geometrically frustrated Heisenberg antiferromagnets that emerges at this stage is summarised by the cartoon of phase space given in Fig. 1.6:

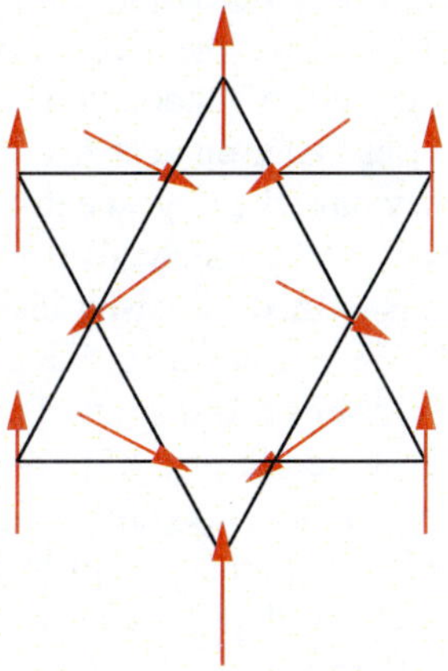

**Fig. 1.5** Illustration of how ground state degrees of freedom arise for the Heisenberg model on the kagomé lattice: spins on the central hexagon may be rotated together through any angle about the axis defined by the outer spins, without leaving the ground state

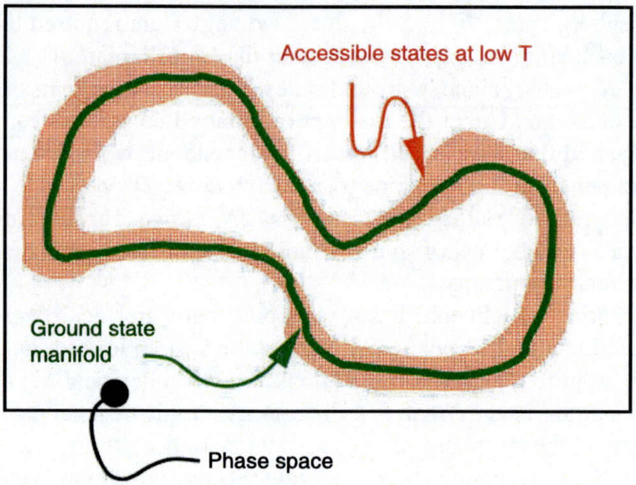


**Fig. 1.6** Schematic view of phase space for a geometrically frustrated magnet

within (high-dimensional) phase space for the system as a whole, the ground states form a manifold with a dimension that is much smaller but nevertheless extensive. At temperatures small compared to the Curie–Weiss constant ($k_B T \ll JS^2$), the system is confined to a region of phase space that forms a thin layer around the ground state manifold. Quantum effects can be neglected provided $JS \ll k_B T$, and so a strongly correlated, classical window, $JS \ll k_B T \ll JS^2$, opens for large $S$.

## 1.5 Order by Disorder

The fact that extensive ground state degeneracy in classical, geometrically frustrated antiferromagnets is, in the technical sense, accidental prompts us to ask whether it has robust consequences in the presence of thermal or quantum fluctuations. Specifically, since the degeneracy is not a consequence of symmetry, one expects the spectrum of fluctuations around each ground state to be different: the possibility arises that ground states with the lowest excitation frequencies are selected, because they have the largest entropy and the smallest zero-point energy. Such an apparently paradoxical mechanism, by which fluctuations enhance order instead of suppressing it, is termed 'order-by-disorder' [31, 32].

We will consider first the effects of thermal fluctuations, and begin by discussing a cluster of four spins. Two ground states with fluctuations of contrasting types are illustrated in Fig. 1.7. For the configuration shown on the left, the total spin of the cluster has a magnitude $|\mathbf{L}|$ that varies with the departure $\delta\theta$ from the ground state as $|\mathbf{L}| \propto \delta\theta$. Since the excitation energy is proportional to $|\mathbf{L}|^2$, it has a conventional, quadratic dependence on $\delta\theta$. By contrast, for the excitation from a collinear ground state shown on the right, $|\mathbf{L}| \propto (\delta\theta)^2$: this mode is therefore soft, with an energy

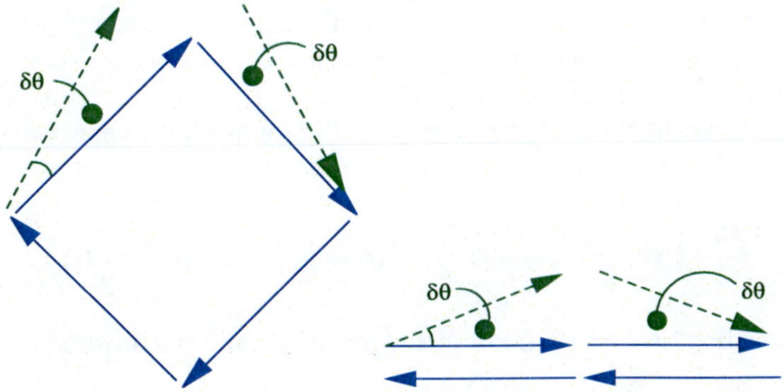


**Fig. 1.7** Fluctuations away from ground state configurations for a cluster of four spins. *Left*: a conventional fluctuation; *right*: a soft mode

**Fig. 1.8** Coordinate system used for configurations of four spins

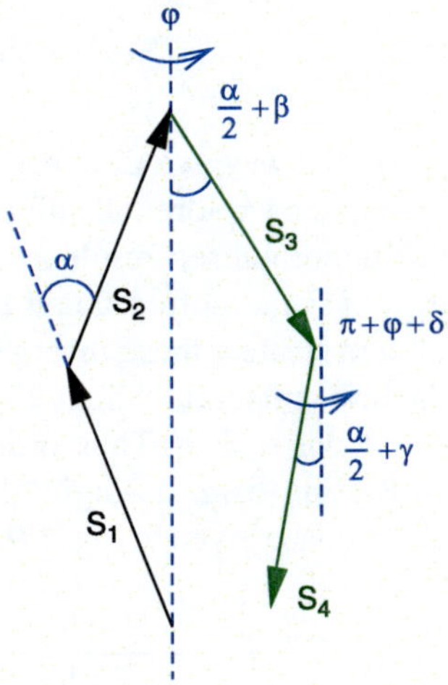


proportional to $(\delta\theta)^4$. We wish to understand whether the presence of this soft mode leads almost collinear configurations to dominate at low temperature.

Analysis for a cluster of four spins is simple enough that it can be followed through in full. To illustrate the range of possible outcomes, we will consider spins with $n$ components, comparing behaviour for $n = 3$ and $n = 2$. We use the coordinate system shown in Fig. 1.8. Our aim is to evaluate the thermal probability distribution $P_n(\alpha)$ of the angle $\alpha$ between the pair of spins $\mathbf{S}_1$ and $\mathbf{S}_2$. The distribution $P_n(\alpha)\mathrm{d}\alpha$ is a product of two factors. One stems from the measure for $\mathbf{S}_2$, and is $\sin(\alpha)\mathrm{d}\alpha$ or $\mathrm{d}\alpha$, for $n = 3$ or $n = 2$ respectively. The other comes from integrating over orientations of $\mathbf{S}_3$ and $\mathbf{S}_4$: it is

$$\mathcal{Z}_n(\alpha) \propto \int \mathrm{d}\mathbf{S}_3 \int \mathrm{d}\mathbf{S}_4 \exp\left(-\frac{J}{2T}\,|\mathbf{S}_3 + \mathbf{S}_4 - 2S\cos(\alpha/2)\hat{z}|^2\right). \tag{1.5}$$

In the low temperature limit, this can be evaluated by expanding the energy to quadratic order in deviations from a ground state. For Heisenberg spins ($n = 3$) the low energy configurations have $|\beta|$, $|\gamma|$, $|\delta| \ll 1$; for $n = 2$, spins are coplanar and two coordinates are fixed: $\varphi = \delta = 0$. In a quadratic approximation, the energy is

$$\frac{JS^2}{2}\left\{(\beta-\gamma)^2\cos^2(\alpha/2)+\left[(\beta+\gamma)^2+\delta^2\right]\sin^2(\alpha/2)\right\}, \tag{1.6}$$

so that (including for $n = 3$ a factor of $\sin^2(\alpha/2)$ arising from $d\mathbf{S}_3 d\mathbf{S}_4$)

$$\mathcal{Z}_3(\alpha) \propto [\cos(\alpha/2)]^{-1} \qquad \text{and} \qquad \mathcal{Z}_2 \propto [\cos(\alpha/2)\sin(\alpha/2)]^{-1}. \tag{1.7}$$

Combining contributions, we have

$$P_3(\alpha) \propto \sin(\alpha/2) \qquad \text{and} \qquad P_2(\alpha) \propto \frac{1}{\sin(\alpha)}. \tag{1.8}$$

In this way, we discover contrasting behaviour for the two cases. With $n = 3$, the system explores all values of $\alpha$ even in the low temperature limit. But for $n = 2$ our unnormalised result for $Z_2(\alpha)$ has non-integrable divergences at $\alpha = 0$ and $\alpha = \pi$: in a more detailed treatment, retaining contributions to the energy quartic in coordinates, these divergences are cut off on a scale set by temperature, but in the low-temperature limit $P_2(\alpha)$ approaches a sum of two delta functions, located at $\alpha = 0$ and $\alpha = \pi$. Thus, order by disorder is absent for $n = 3$ but perfect for $n = 2$.

Passing from a single cluster to an extended system, consider the sketch of phase space given in Fig. 1.9. Here, repeating the convention of Fig. 1.6, the shading

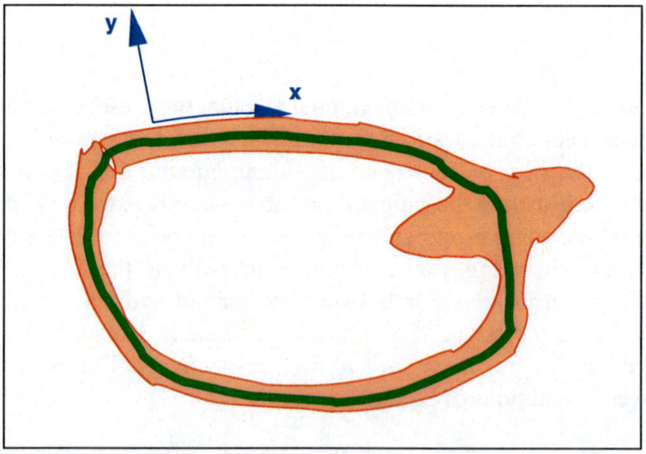


**Fig. 1.9** Schematic view of phase space. The full curve represents the ground state manifold. Coordinates $x$ and $y$ are respectively parallel and perpendicular to it

indicates the region accessible at low temperature. One part of this region is concentrated near points on the ground state manifold at which there are soft modes, as represented by in Fig. 1.9 by a bulge, while another part is distributed in the neighbourhood of the remainder of the ground state manifold. To decide whether the system displays order by disorder, we need to understand which of these two parts dominates. Introducing coordinates $\mathbf{x}$ and $\mathbf{y}$, respectively parallel and perpendicular to the ground state manifold, in the low temperature limit we obtain a measure $P(\mathbf{x})$ on the ground state manifold by integrating over transverse fluctuations [9, 10]. Characterising these fluctuations by dynamical frequencies $\omega_l(\mathbf{x})$, we obtain

$$P(\mathbf{x}) \propto \prod_l \left( \frac{k_B T}{\hbar \omega_l(\mathbf{x})} \right). \tag{1.9}$$

The system has extra soft modes (in addition to those associated with the ground state coordinates $\mathbf{x}$) at points $\mathbf{x_0}$ where one or more of the harmonic frequencies $\omega_l(\mathbf{x_0})$ vanishes. At these points $P(\mathbf{x_0})$ is divergent. As for a cluster of four spins, behaviour depends on whether any such divergence is integrable. If it is, the system explores the whole of the ground state manifold in the low temperature limit, but if it is not, then those ground states with soft modes are selected by thermal fluctuations. Detailed considerations, tested using Monte Carlo simulations, show for the Heisenberg antiferromagnet on the kagomé lattice that there is coplanar spin order in the low temperature limit [29], while on the pyrochlore lattice there is no order by disorder [9, 10].

The possibility of ground state selection due to quantum fluctuations can be discussed using an approach similar in spirit to the one we have taken for thermal fluctuations, although the outcome has significant differences. Referring again to Fig. 1.9, one can treat excitations around a particular point $\mathbf{x}$ on the ground state manifold using harmonic spin wave theory. Excitations involving the coordinates $\mathbf{y}$ locally orthogonal to the ground state manifold are conventional modes with non-zero frequencies $\omega_l(\mathbf{x})$, which have already made an appearance in (1.9). By contrast, fluctuations involving the coordinates $\mathbf{x}$ are, within a harmonic approximation, zero modes. The zero-point energy of the conventional, finite-frequency modes provides an effective Hamiltonian for these remaining degrees of freedom, the classical ground state coordinates. This Hamiltonian takes the form

$$\mathcal{H}_{\text{eff}}(\mathbf{x}) = \frac{1}{2} \sum_l \hbar \omega_l(\mathbf{x}). \tag{1.10}$$

The components of $\mathbf{x}$ consist of pairs that are, within the approximations of harmonic spin wave theory (see (1.14)), canonically conjugate. Treating them as classical commuting variables, the ground state is the set of points $\mathbf{x}_G$ on which $\mathcal{H}_{\text{eff}}(\mathbf{x})$ is minimised. More accurately, the ground state wavefunction for large $S$ is peaked at $\mathbf{x}_G$, but has zero-point fluctuations in an effective potential defined by $\mathcal{H}_{\text{eff}}(\mathbf{x})$.

It is not straightforward to anticipate what features of a classical ground state spin configuration will minimise $\mathcal{H}_{\text{eff}}(\mathbf{x})$: since all $\omega_l(\mathbf{x})$ contribute, one could equally imagine focusing on either the highest frequencies or the lowest ones. In the examples that have been studied in detail, however, it seems that minima lie at points $\mathbf{x}_G$ where some $\omega_l(\mathbf{x})$ vanish, which we have seen are also the states favoured by thermal fluctuations. In particular, for Heisenberg antiferromagnets at large $S$ the selected spin configurations are coplanar on the kagomé lattice [33] and collinear on the pyrochlore lattice [34]. In both examples, one-third of the $\omega_l(\mathbf{x})$ become soft modes at the corresponding points $\mathbf{x}_G$: the coplanar or collinear configurations, respectively.

The principal difference between the ordering effects of thermal and quantum fluctuations is that in the first case, as we have seen, order may or may not arise on the limit $J \gg T$, depending on the nature of the thermal ground state distribution $P(\mathbf{x})$, while in the second case we always expect order for $S \gg 1$, because by taking $S$ sufficiently large, one can ensure that quantum fluctuations around the minimum of $\mathcal{H}_{\text{eff}}(\mathbf{x})$ are arbitrarily small. Within this framework, the scenario by which one arrives at a spin liquid on reducing $S$ is clear, at least in principle. For smaller $S$, the quantum fluctuations are larger and the ground state wavefunction is less well localised around the minimum of $\mathcal{H}_{\text{eff}}(\mathbf{x})$, while below a critical value of $S$, the quantum ground state wavefunction becomes delocalised over the entire classical ground state manifold and the system loses magnetic order. At large, fixed $S$ long range order induced by quantum fluctuations is suppressed thermally above a critical temperature $T_c \sim JS$. While the expectation that spin liquids are favoured at small $S$ is common to our discussions of both (1.1) and (1.10), one should of course remember that the two equations embody different physics: harmonic and anharmonic fluctuations, respectively.

Efforts to identify experimental examples of order by disorder must face the problem of establishing that fluctuations, rather than additional interaction terms in the Hamiltonian, are the cause of what is observed. For the garnet $Ca_3Fe_2Ge_3O_{12}$, a material with two interpenetrating magnetic lattices coupled via zero-point fluctuations, it has been shown that a spinwave gap in the Néel ordered state indeed arises mainly in this way, by independent determination of the size of single ion anisotropy (the other possible origin for the gap) [35], and via the characteristic temperature dependence of the gap [36].

## 1.6 Ground State Correlations

As we have seen, in some circumstance a model geometrically frustrated magnet (for example, the classical Heisenberg model on the pyrochlore lattice) explores its entire ground state manifold at low temperature, and the experimental evidence from elastic and inelastic neutron scattering suggests that this is a reasonable picture for the behaviour of a range of frustrated magnetic materials. We are led to ask in this section whether there are any important correlations within the ground state manifold. We will find (for a large class of models: those in which the simplex lattice is bipartite) that there are indeed long-range correlations within ground states [10], and that these can be characterised in terms of fluctuations of a Gaussian, divergenceless field [37,38]. For this reason, the set of ground states is said to constitute a Coulomb phase.

The possibility that spin correlations, averaged over ground states, have a long-range component, is not self-evident. Indeed, one might expect the fact that there are a macroscopic number of ground state degrees of freedom to signal the opposite, since their existence implies that the set of ground states includes local degrees of freedom that can fluctuate independently. In turns out, however, that some ground state correlations are impervious to all local fluctuations: in this sense they can be said to be topological.

The simplest way to appreciate the existence of long-range correlations within ground states is to start from (1.2) and the fact that the total spin $\mathbf{L}_\alpha$ of each frustrated cluster $\alpha$ vanishes within all ground states. A consequence of this on the pyrochlore lattice can be visualised with reference to Fig. 1.2. In particular, consider for any ground state the total magnetisation $\mathbf{m}(z)$ of the lowest plane of sites in this figure. Its value (which may be small – for example, of order the square root of the number of sites within the plane, if spins are randomly orientated within the layer) is perfectly correlated with the magnetisation of other parallel planes making up the lattice. Indeed, let $\mathbf{m}(z+1)$ be the magnetisation of the plane neighbouring the lowest one. Since the sites in both planes taken together make up a layer of complete tetrahedra, the overall magnetisation of the layer is zero, and so $\mathbf{m}(z+1) = -\mathbf{m}(z)$. By extension, $\mathbf{m}(z+n) = (-1)^n \times \mathbf{m}(z)$ for any $n$, a signal of long range correlations. The correlations give rise to sharp features, termed *pinch points* or *bow ties* in the Fourier transform of the two-point correlation function, averaged over ground states, as obtained from simulations [39], large-$n$ calculations [40], and diffuse neutron scattering measurements [41]. These singularities distinguish the diffraction pattern of the frustrated system from that of a paramagnet, but are weaker than those of Bragg peaks arising from Néel order. While their structure can be understood by building on our discussion of $\mathbf{m}(z)$ [10], a more complete approach uses a long-wavelength description of ground states.

This continuum description of the Coulomb phase is obtained by mapping spin configurations onto configurations of vector fields in such a way that the ground state condition $\mathbf{L}_\alpha = \mathbf{0}$, involving the specifics of the lattice structure, is translated into the requirement that the vector fields have lattice divergence zero. This second version of the constraint has the advantage that it can be implemented in the continuum [37, 38]. To describe the mapping in detail, it is necessary first to discuss some features of the simplex lattice, introduced in Sect. 1.2. We require the simplex lattice to be bipartite. This is the case, for example, for the diamond lattice, the simplex lattice associated with the pyrochlore lattice. (In models without a bipartite simplex lattice, the correlations discussed in this section are absent [42].) For a bipartite simplex lattice, one can adopt an orientation convention for links, taking them to be directed from the simplices of one sublattice to those of the other. In this way one can define unit vectors $\hat{e}_i$ oriented according to the convention and placed at the mid-points of links of the simplex lattice, which are also the locations of spins.

Considering in the first instance Ising spins, a spin configuration is represented as a vector field $\mathbf{B}(\mathbf{r})$ on the lattice via

$$\mathbf{B}(\mathbf{r}_i) = S_i \hat{e}_i . \tag{1.11}$$

The condition satisfied in ground states, $\sum_{i\in\alpha} S_i = 0$, fixes the lattice divergence of $\mathbf{B}(\mathbf{r})$ to be zero. More generally, for $n$-component spins $\mathbf{S}_i$ we require $n$ flavours of vector field $\mathbf{B}^l(\mathbf{r})$, with $l = 1, 2 \ldots n$. These are related to spin components $S_i^l$ via $\mathbf{B}^l(\mathbf{r}) = S_i^l \hat{e}_i$, so that in ground states each flavour has divergence zero. Note that the fields $\mathbf{B}^l(\mathbf{r})$ are defined in real space, and that the global O($n$) symmetry of the spin Hamiltonian is implemented as a transformation within the space of flavours, $l$.

Continuum versions of these vector fields $\mathbf{B}^l(\mathbf{r})$ result from coarse-graining and the restriction to ground state configurations is imposed exactly by requiring the continuum fields to be solenoidal. Each coarse-grained state represents many microscopic configurations and should have an entropic weight that reflects this. It is plausible that small continuum field strengths will arise from many different microscopic configurations, and that large field strengths will be generated by fewer microscopic states. This suggests [37, 42] the weight

$$P[\mathbf{B}^l(\mathbf{r})] \propto \exp\left(-\frac{\kappa}{2}\int \mathrm{d}^d\mathbf{r} \sum_l |\mathbf{B}^l(\mathbf{r})|^2\right). \tag{1.12}$$

This theory has a single parameter, the stiffness $\kappa$, whose value affects the amplitude but not the form of correlations and is determined microscopically. In $d = 3$ dimensions all other terms consistent with symmetry that might be added to the effective action are irrelevant in the scaling sense, and so (1.12) is expected to have a universal validity. The resulting correlation function,

$$\langle B_i^l(\mathbf{0}) B_j^m(\mathbf{r})\rangle = \frac{\delta_{lm}}{4\pi\kappa}\left(\frac{3r_ir_j - r^2\delta_{ij}}{r^5}\right), \tag{1.13}$$

falls off with a fixed, integer power of distance, and has a characteristic, dipolar angle dependence.

The fixed, integer power appearing in (1.13) stands in contrast to behaviour in two other situations in statistical mechanics for which power-law correlations appear: those of a system undergoing a continuous phase transition, and of the low-temperature phase in the xy model. The form of correlations in (1.13) is instead similar to those generated by Goldstone modes in the ordered phase of a system with a spontaneously broken continuous symmetry, and an equivalence between that and the Coulomb phase can be developed by passing to a dual description of the frustrated magnet [43].

At finite temperature thermal fluctuations out of the ground state manifold generate a finite correlation length $\xi$ which acts as a cut-off for the power-law in (1.13).

The scale $\xi$ diverges at low temperature, as $\xi \sim T^{-1/2}$ in a Heisenberg model, and exponentially in an Ising model.

## 1.7 Dynamics

As we have seen in some detail, geometrically frustrated magnets in the temperature window $T_c \ll T \ll T_{CW}$ are strongly correlated, yet lack long range order. Their dynamics in this regime has novel features which we summarise in this section.

An obvious first step to understanding low temperature dynamics is to apply harmonic spinwave theory, starting from one of the ground states. The results one expects are summarised in terms of the density of states $\rho(\omega)$ in frequency $\omega$ in Fig. 1.10. Within the harmonic approximation, excitations are of two types. One type, similar to those in conventional magnets, forms a band of finite-frequency states, with a maximum frequency $\sim \mathcal{O}(JS/\hbar)$. The other type (those associated with the ground state coordinates **x**, in the discussion of (1.10)) are zero modes. For example, for excitations from a generic ground state of the Heisenberg antiferromagnet on the pyrochlore lattice, one quarter are zero modes.

There is a clear interest in understanding in more detail the nature of these zero modes. In the most cases, however there is a obstacle to a simple, analytical treatment, which stems from the fact that spin configurations in representative classical ground states do not have long range order. This means that, even though the lattice itself is periodic, the equations of motion cannot be diagonalised by Fourier transform, and results of the kind sketched in Fig. 1.10 can be obtained only numerically.

To circumvent this difficulty and illustrate in a simple fashion how a dispersionless band of modes can arise, it is interesting to consider a geometrically frustrated Heisenberg antiferromagnet in the presence of a magnetic field $h$ strong enough that

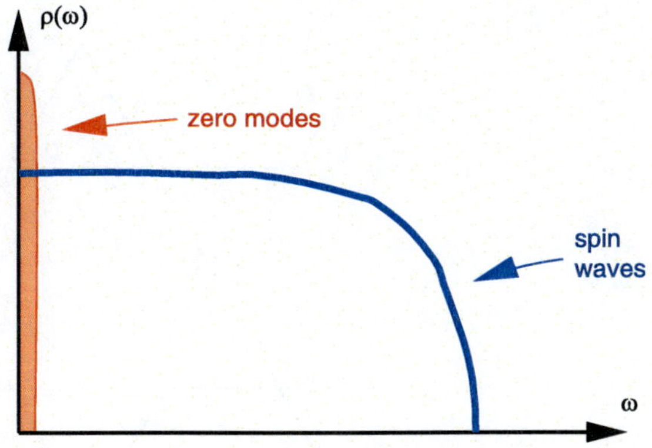


**Fig. 1.10** Density of states in frequency for harmonic excitations in a geometrically frustrated antiferromagnet

the ground state is fully polarised. Using the standard Holstein–Primakoff transformation to write operators for spin components $S_i^l$ in terms of boson creation and annihilation operators $a_i^\dagger$ and $a_i$, with

$$\begin{aligned} S_i^z &= S - a_i^\dagger a_i \\ S_i^+ &= (2S)^{1/2} a_i + \dots \\ S_i^- &= (2S)^{1/2} a_i^\dagger + \dots, \end{aligned} \tag{1.14}$$

we have

$$\mathcal{H} = J \sum_{ij} \mathbf{S}_i \cdot \mathbf{S}_j - h \sum_i S_i^z = JS \sum_{ij} \left[ a_i^\dagger a_j + a_j^\dagger a_i \right] - \mu \sum_i a_i^\dagger a_i + \mathcal{O}(S^0). \tag{1.15}$$

The right-hand form of (1.15) is a tight-binding model for bosons moving on the lattice with a nearest-neighbour hopping amplitude $JS$ and a chemical potential $\mu \equiv zJS - h$ that is linear in the magnetic field $h$. It is a characteristic of the lattices we are concerned with that such a tight-binding model has a dispersionless band with eigenvalue $-2JS$, which lies at the bottom of the spectrum for $J > 0$.

Eigenvectors of the tight-binding Hamiltonian from the dispersionless band are straightforward to picture. One for the kagomé lattice is represented in Fig. 1.11. Here, eigenvector amplitudes are zero at all sites except for those around one hexagon of the lattice, on which they have equal magnitude and alternating signs. The state is an eigenvector because there is destructive interference between hopping processes that move the boson off the occupied hexagon. It belongs to a dispersionless band since it is degenerate with many other, equivalent states, based on the other hexagons of the lattice. The condition for an arbitrary vector, with site amplitudes $\psi_i$, to be a linear superposition of such states is that $\sum_{i\in\alpha} \psi_i = 0$ for each triangle $\alpha$. Both the extension of this condition to other lattices constructed from corner-sharing simplices and its parallel with the ground state condition in

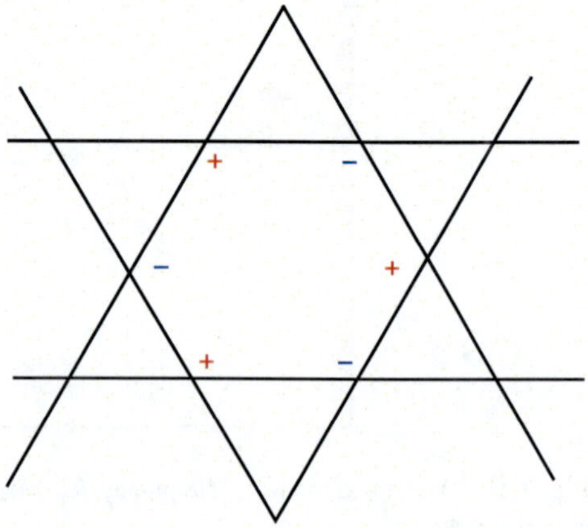


**Fig. 1.11** A magnon mode from the dispersionless band

spin models, $\mathbf{L}_\alpha = \mathbf{0}$ for all $\alpha$, are obvious. There is a gap to excitations for large $h$, when $\mu$ is large and negative. The gap falls to zero at the critical field strength $h_\mathrm{c} = JS(z+2)$ at which $\mu$ crosses the energy of the dispersionless magnon band. For $h < h_\mathrm{c}$ these states are populated, and the magnetisation deviates from its saturated value. In this field range we recover the classical ground state degeneracy of the zero-field problem.

We now turn to a discussion of dynamics beyond the harmonic approximation. As a starting point we should consider the full equation of motion

$$\hbar \frac{\mathrm{d}\mathbf{S}_i}{\mathrm{d}t} = \mathbf{S}_i \times \mathbf{H}_i = \frac{J}{2} \left\{ \left[ \mathbf{L}_\alpha + \mathbf{L}_\beta \right] \times \mathbf{S}_i - \mathbf{S}_i \times \left[ \mathbf{L}_\alpha + \mathbf{L}_\beta \right] \right\}. \tag{1.16}$$

Here $\mathbf{H}_i$ is the exchange field acting at site $i$. It is parallel to $\mathbf{S}_i$ in a classical ground state, but in excited states has transverse components. These can be expressed as shown, in terms of the total magnetisations $\mathbf{L}_\alpha$ and $\mathbf{L}_\beta$ of the clusters $\alpha$ and $\beta$ that share the site $i$. For large $S$ we can treat this equation of motion classically. Within the harmonic approximation, obtained by linearising the right-hand side, the exchange field $\mathbf{H}_i$ is a superposition of contributions from finite frequency modes, which maintain phase coherence indefinitely, and therefore average to zero over long times. Anharmonic interactions have two consequences, which are distinct but turn out to be closely linked [9, 10]. One is to generate a lifetime for the finite frequency modes. The other is to introduce coupling between the ground state coordinates $\mathbf{x}$ and the coordinates $\mathbf{y}$ orthogonal to the ground state manifold. Over long timescales this coupling drives the system around the ground state manifold. Within the framework of (1.16), this long-time component to the dynamics arises because, once spinwaves have a finite lifetime, the exchange field is no longer a superposition of exactly harmonic contributions. Instead, on timescales longer than the lifetime, it is better thought of as a stochastic quantity. In turn, the long-time motion of the system around the ground state manifold is itself a source of dephasing for finite frequency excitations. Specifically, since the harmonic Hamiltonian is time-dependent, an adiabatic approximation is not exact, and modes are mixed at long times. There is a separation of timescales, since typical spinwave periods are fixed, while spinwave lifetimes diverge as $T^{-1/2}$ and the timescale for motion between groundstates diverges faster, as $T^{-1}$ [10].

These ideas suggest a much simpler approach to calculating the spin autocorrelation function, in which we treat the exchange field as a stochastic quantity, using the equation of motion

$$\frac{\mathrm{d}\mathbf{S}(t)}{\mathrm{d}t} = \mathbf{S}(t) \times \mathbf{H}(t) \tag{1.17}$$

with the correlation function

$$\langle H^l(t) H^m(t') \rangle = \Gamma \delta_{lm} \delta(t - t') \tag{1.18}$$

for components $H^l(t)$ of $\mathbf{H}(t)$. The noise intensity $\Gamma$ can be estimated using equipartition. We have

$$\Gamma = \int_{-\infty}^{\infty} \langle \mathbf{H}_i(0) \cdot \mathbf{H}_i(t) \rangle \, \mathrm{d}t \sim \int_{-\infty}^{\infty} \langle \mathbf{L}_\alpha(0) \cdot \mathbf{L}_\alpha(t) \rangle \, \mathrm{d}t. \tag{1.19}$$

In addition, $\mathbf{L}_\alpha(t)$ is a superposition of contributions with amplitudes $A_\omega$ from thermally excited spinwaves:

$$\mathbf{L}_\alpha(t) = \sum_l A_{\omega_l} \mathrm{e}^{-\mathrm{i}\omega_l t}. \tag{1.20}$$

Combining these assumptions, we find [10]

$$\Gamma \sim \langle |A_\omega|^2 \rangle \rho(\omega) \big|_{\omega \to 0} \sim \frac{k_\mathrm{B} T}{JS}. \tag{1.21}$$

The Langevin equation (1.18) itself is straightforward to solve, and yields

$$\langle \mathbf{S}(0) \cdot \mathbf{S}(t) \rangle = S(S+1) \exp(-c k_\mathrm{B} T t / \hbar S), \tag{1.22}$$

where $c \sim \mathcal{O}(1)$ is an undetermined numerical constant. This result is notable for the fact that temperature alone sets the time scale: $J$ drops out of the long-time dynamics in the low temperature regime. In this sense, behaviour matches that expected at a quantum critical point, although the underlying physics is quite different.

The predictions of (1.22) have been tested both in simulations and in experiment. Molecular dynamics simulations proceed by direct integration of the equations of motion, (1.16), with an initial configuration drawn from a thermal distribution and generated via Monte Carlo simulation. Results for the classical pyrochlore Heisenberg antiferromagnet [9, 10] in the temperature range $k_\mathrm{B} T \ll JS^2$ reproduce both the functional form of (1.22) for the time dependence of the autocorrelation function and the scaling of relaxation rate with temperature. In experiment, inelastic neutron scattering offers direct access to spin dynamics. Early measurements of the energy width of quasielastic scattering in $CsNiFe_6$ [44] were fitted to a Lorentzian, the transform of the time-dependence given in (1.22), yielding a relaxation rate for $T < |\Theta_\mathrm{CW}|$ that is strongly temperature dependent, although without specific evidence for the (subsequently proposed) linear variation with $T$. More detailed data for SCGO ([45]; Lee et al. unpublished) confirms a relaxation rate of order $k_\mathrm{B} T/\hbar$, and very recent measurements on $Y_2Ru_2O_7$ [46] display rather clearly a relaxation rate proportional to temperature.

Recent theoretical work [47] has mapped out the low temperature dynamics as a function of wavevector throughout the Brillouin zone.

## 1.8 Final Remarks

In conclusion, we have seen how geometrical frustration in classical magnets can lead to macroscopic ground state degeneracy and the suppression of long range order. Low temperature states in model systems, although disordered, are very different from those of a non-interacting paramagnet: correlations are power-law in space, and decay in time at a rate set by temperature alone. Many experimental systems display these features within the temperature window $T_{\mathrm{c}} < T < |\Theta_{\mathrm{CW}}|$ where behaviour is dominated by nearest neighbour exchange. Behaviour in this regime is well summed up in the term coined by Jacques Villain [8], a pioneer in the field: *Cooperative Paramagnetism.*

**Acknowledgements**

I am very grateful to my collaborators in the research on geometrically frustrated antiferromagnets that I have been involved with, especially E.F. Shender, P.C.W. Holdsworth and R. Moessner, and also J.F.G. Eastmond, S.E. Palmer, P. Hogan, M.Y. Veillette, R. Coldea, T.E. Saunders, M.J. Bhaseen and T.S. Pickles. In addition, I have benefitted from discussions with many other colleagues. I thank EPSRC for supporting this work.

## References

1. A.P. Ramirez, Annu. Rev. Mater. Sci. **24**, 453 (1994)
2. P. Schiffer, A.P. Ramirez, Comments Cond. Mat. Phys. **18**, 21 (1996)
3. M.J. Harris, M.P. Zinkin, Mod. Phys. Lett. B**10**, 417 (1996)
4. R. Moessner, Can. J. Phys. **79**, 1283 (2001)
5. C.L. Henley, Can. J. Phys. **79**, 1307 (2001)
6. P. Chandra, B. Doucot, Phys. Rev. B **38**, 9335 (1988)
7. P.W. Anderson, Phys. Rev. **102**, 1008 (1956)
8. J. Villain, Z. Phys. B **33**, 31 (1979)
9. R. Moessner, J.T. Chalker, Phys. Rev. Lett. **80**, 2929 (1998)
10. R. Moessner, J.T. Chalker, Phys. Rev. B **58**, 12049 (1998)
11. M. Harris, S.T. Bramwell, D.F. McMorrow, T. Zeiske, K.W. Godfrey, Phys. Rev. Lett. **79**, 2554 (1997)
12. J.N. Reimers, A.J. Berlinsky, A.-C. Shi, Phys. Rev. B **43**, 865 (1991)
13. J.N. Reimers, J.E. Greedan, M. Björgvinsson, Phys. Rev. B **45**, 7295 (1991)
14. N. Raju, M. Dion, M.J.P. Gingras, T.E. Mason, J.E. Greedon, Phys. Rev. B **59**, 14489 (1999)
15. S.E. Palmer, J.T. Chalker, Phys. Rev. B **62**, 488 (2000)
16. M. Elhajal, B. Canals, R. Sunyer, C. Lacriox, Phys. Rev. B **71**, 94420 (2005)
17. O. Tchernyshyov, R. Moessner, S.L. Sondhi, Phys. Rev. B **66**, 64403 (2002)
18. E.F. Shender, V.B. Cherepanov, P.C.W. Holdsworth, A.J. Berlinsky, Phys. Rev. Lett. **70**, 3812 (1993)
19. R. Moessner, A.J. Berlinsky, Phys. Rev. Lett. **83**, 3293 (1999)
20. T.E. Saunders, J.T. Chalker, Phys. Rev. Lett. **98**, 157201 (2007)
21. A.P. Ramirez, G.P. Espinoza, A.S. Cooper, Phys. Rev. Lett. **64**, 2070 (1990)

22. C. Broholm, G. Aeppli, G.P. Espinosa, A.S. Cooper, Phys. Rev. Lett. **65**, 3173 (1990)
23. B. Martinez, F. Sandiumenge, A. Rouco, A. Labarta, J. Rodriguez-Cavajal, M. Tovar, M.T. Causa, S. Gali, X. Obradors, Phys. Rev. B **46**, 10786 (1992)
24. S.H. Lee, C. Brohlom, G. Aeppli, A. Ramirez, T.G. Perring, C.J. Carlile, M. Adams, T.L. Jones, B. Hessen, Europhys. Lett. **35**, 127 (1996)
25. A.S. Wills, A. Harrison, S.A.M. Menthink, T.E. Mason, Z. Tun, Europhys. Lett. **42**, 325 (1998)
26. M.J.P. Gingras, C.V. Stager, N.P. Raju, B.D. Gaulin, J.E. Greedan, Phys. Rev. Lett. **78**, 947 (1997)
27. J.S. Gardner, B.D. Gaulin, S.-H. Lee, C. Broholm, N.P. Raju, J.E. Greedan, Phys. Rev. Lett. **83**, 211 (1999)
28. J.C. Maxwell, Philos. Mag. **27**, 294 (1864)
29. J.T. Chalker, P.C.W. Holdsworth, E.F. Shender, Phys. Rev. Lett. **68**, 855 (1992)
30. I. Ritchey, P. Chandra, P. Coleman, Phys. Rev. **47**, 15342 (1993)
31. J. Villain, R. Bidaux, J.P. Carton, R.J. Conte, J. Phys. Paris **41**, 1263 (1980)
32. E.F. Shender, Sov. Phys. JETP **56**, 178 (1982)
33. A.V. Chubukov, Phys. Rev. Lett. **69**, 832 (1992)
34. C.L. Henley, Phys. Rev. Lett. **96**, 47201 (2006)
35. A.G. Gukasov, T. Brückel, B. Dorner, V.P. Plakhty, W. Prandtl, E.F. Shender, O.P. Smirnov, Europhys. Lett. **7**, 83 (1988)
36. T. Brückel, B. Dorner, A.G. Gukasov, V.P. Plakhty, Phys. Lett. A **162**, 357 (1992)
37. S.V. Isakov, K. Gregor, R. Moessner, S.L. Sondhi, Phys. Rev. Lett. **93**, 167204 (2004)
38. C.L. Henley, Phys. Rev. B **71**, 014424 (2005)
39. M. Zinkin, M.J. Harris, T. Zeiske, Phys. Rev. B **56**, 11786 (1997)
40. B. Canals, D. Garanin, Can. J. Phys. **79**, 1323 (2001)
41. T. Fennell, S.T. Bramwell, D.F. McMorrow, P. Manuel, Nat. Phys. **3**, 566 (2007)
42. D.A. Huse, W. Krauth, R. Moessner, S.L Sondhi, Phys. Rev. Lett. **91**, 167004 (2003)
43. L. Jaubert, J.T. Chalker, P.C.W. Holdsworth, R. Moessner, Phys. Rev. Lett. **100**, 067207 (2008)
44. M.J. Harris, M.P. Zinkin, T. Zeiske, Phys. Rev. B **52**, R707 (1995)
45. S.-H. Lee, Ph.D. thesis, Johns Hopkins University (1996)
46. J. van Duijn, N. Hur, J.W. Taylor, Y. Qiu, Q.Z. Huang, S.-W. Cheong, C. Broholm, T.G. Perring, arXiv:0707.2678 (2007)
47. P. Conlon, J.T. Chalker, Phys. Rev. Lett. **102**, 237106 (2009)

# Chapter 2
# Introduction to Quantum Spin Liquids

**Claire Lhuillier and Grégoire Misguich**

**Abstract** In SU(2)-invariant spin models with frustrated interactions and low spin quantum number, long-ranged magnetic order and breaking of the SU(2) symmetry is not the most general situation, especially in low dimensions. Many such systems are, loosely speaking, "quantum paramagnets" down to zero temperature, states conveniently represented in terms of spins paired into rotationally invariant singlets, or "valence bonds" (VBs). In this large family of states, at least two very different physical phases should be distinguished, valence bond crystals (or solids) and resonating-valence-bond (RVB) liquids.

This chapter provides a basic introduction to the concept of short-range resonating-valence-bond (srRVB) states, and emphasizes the qualitative differences between VB *crystals* and RVB *liquids*. We explain in simple terms why the former sustain only integer-spin excitations while the latter possess spin-$\frac{1}{2}$ excitations (the property of fractionalization). We then elaborate qualitatively on the notion of *macroscopic quantum resonances*, which are behind the 'R' of the RVB spin liquid, to motivate the idea that spin liquids are not disordered systems but possess instead hidden quantum order parameters. After giving a brief list of models and materials of current interest as candidate spin liquids, we conclude by mentioning the role of the parity of the net spin in the unit cell (half-odd-integer or integer) in spin-liquid formation, by recalling the contrasting results for kagomé and pyrochlore lattices.

## 2.1 Introduction

In isotropic (SU(2)-invariant) Heisenberg spin systems, frustration of individual bond energies, arising due to competing interactions and/or lattice topology, and extreme quantum fluctuations, due to low spin values and low coordination numbers, can prevent $T = 0$ magnetic ordering (defined as the existence of a non-zero on-site magnetization, $\langle \mathbf{s}_i \rangle > 0$). These factors lead instead to a large variety of quantum phases, sometimes given the misleading name "spin liquids." We begin the task of categorizing this large zoo of phases by excluding those which do possess an order parameter, but merely one more complex than an on-site magnetization, such as a quadrupolar moment or a spin current [1–3]). These exotic forms of complex order break SU(2) symmetry and support Goldstone modes, and hence can be understood by semiclassical (spin-wave-like) approximations.

However, quantum effects may generate more radical situations where SU(2) symmetry is not broken in the ground state. One of the most common ways of achieving this is that the spins are paired in rotationally invariant singlets, or valence bonds (VBs). Such states are by definition non-magnetic, and thus qualify as "quantum paramagnets".[1] This requirement on the ground-state wave function is insufficient to characterize the physics of an extended system, its low-energy excitations and the long-distance behavior of its correlation functions. In this large family of quantum paramagnets, a crucial distinction should be introduced between spin systems possessing *long-ranged order in the VB arrangements*,[2] which are called valence-bond crystals (VBCs), and spin systems that *do not develop long-range order (LRO) in any local order parameter at any temperature*. The latter are the true spin liquids.

With this foundation we will also show, following Feynmann's statement, that an understanding of the structure of the ground-state wave function of these phases gives important indications concerning the nature of their excitations: specifically, the excitations of the VBC have integer spins, whereas those of the RVB liquids may be fractionalized. In the second case, integer-spin excitations of such systems will decay immediately into more or less complex states of spin-$\frac{1}{2}$ entities, known as spinons, which further are *deconfined*. This property of the true spin liquid was recognized very early in their investigation [6–11], and in Sects. 2.2 and 2.3 we provide a simple explanation in the framework of a *short-range* VB basis. The restriction to short-ranged VBs is a very serious one, and it implies that the short-range RVB spin liquids satisfying our schematic representations are, except at quantum critical points, gapped in the triplet sector: a finite energy is required to break a microscopic constitutive singlet block and promote it to a triplet state.[3] In fact there are some reasons to believe that spin liquids with gapless excitations and algebraically decaying spin correlations also exist in 2D, as argued in [14–17]. In algebraic spin liquids, the concept of spinons remains valid but the schematic discussion of Sect. 2.2 would fail, and spinons may be strongly interacting particles over their whole range of

[1] We use at this stage this word with a fuzzy acceptance. We will subsequently precise the various types of "quantum paramagnets" and abandon this terminology which has been used in the past, with different meanings, by theoreticians and experimentalists.

[2] The actual size of the repeated singlet pattern (unit cell) maybe larger than a single valence-bond: quadrumer (four-site plaquettes), etc. It can even reach 36 sites, as suggested by recent numerics on the ground-state of the Heisenberg model on the kagomé lattice [4, 5].

[3] Whether such systems may also be gapless in the singlet sector seems unlikely: it would be necessary to invoke a theory of low-lying excitations with massive spinons ("matter fields") and compact gapless gauge fields in two spatial dimensions (2D). Such a theory is believed not to exist [12]. It would correspond to quantum electrodynamics with massive electrons and interactions mediated by massless photons, but due to lattice effects the gauge field is compact and acquires a mass. The spectrum of all excitations would then be fully gapped in 2D. On the other hand, a phase with such properties, the "Coulomb phase," exists in 3D [13].

energy. We refer to [12, 18] for an introduction to the more elaborate techniques required to address this class of systems.

On lattices where all nearest neighbor bonds are equivalent, all nearest-neighbor VB configurations have the same (variational) energy at zeroth order. For the ground state to be a VBC, it is necessary that an ordered (periodic) pattern of VBs offers a competitive energy advantage compared to disordered configurations: this is trivially true if the model has different bond strengths, as in the spin-Peierls Hamiltonian on the 1D chain, and also the case for the 2D Shastry–Sutherland model [19]. For models where long-ranged dimer order results from a spontaneous breaking of the original lattice symmetries, the competitive advantage of a given VBC pattern may be rather delicate to ascertain, and results from a mechanism known as *"order by disorder"* where fluctuations around the zeroth-order pattern are particularly favorable.[4] We discuss the physics of order by disorder and present some examples in Sect. 2.3.

For the ground state to be an RVB liquid, it is necessary that macroscopic, large-scale resonances between different configurations lead through interference effects to a noticeable energy gain: in other words, macroscopically large tunneling effects should dominate the low temperature physics.[5] This property, which is not contained in the discussion of Sects. 2.2 and 2.3, is essential. Thus, in Sect. 2.4, we elaborate on this crucial concept of *macroscopic resonances*[6] in the ground states and low-lying excitations of RVB spin liquids, and present some arguments which help to explain how the specific physics of the spin-liquid state is hidden in the specificity of the superpositions of VB configurations. In the case of $\mathbb{Z}_2$ spin liquids, these properties are also closely related to $S = 0$ vison excitations. More generally, the question of the hidden symmetries of RVB superpositions – for any VB lengths – can be considered as the question of a hidden, non-local order parameter characterizing the spin liquid.[7] We have defined spin liquids as spin systems that *do not develop LRO in any local order parameter at any temperature*. Despite this property, spin liquids are not

[4] Some authors restrict the use of "VBC" to the situations with a spontaneous symmetry breaking, and call "quantum paramagnets" the others [20]. Here, we will use "type (i)" and "type (ii)" VBC (see Sect. 2.3), to underline the close similarity between the wave functions and their excitations.

[5] A simpler and already well understood problem sheds additional light on the mechanism we wish to describe: what is the quantum mechanical transition which drives (at $T = 0$) solid $^4$He to liquid (superfluid) $^4$He when decreasing pressure at approximately 25 bar? The solid under pressure is stabilized by dominant potential effects, but the balance between kinetic energy and potential energy is broken in favor of the quantum zero-point kinetic energy as the pressure is lowered. The ground state of the system changes in a first-order transition from a regular array of atoms to a macroscopic superposition of all permuted atomic configurations [21]. This is the origin of the hidden order parameter (LRO in the off-diagonal one-body density matrix) and of the superfluid transition [22, 23]. The macroscopic tunneling processes between different spatial configurations are the essential ingredient of this quantum regime.

[6] A concept with a deep connection to the problem of macroscopic quantum entanglement [24].

[7] For comparison, this question in spin liquids is the conceptual equivalent of the question solved by Penrose and Onsager for $^4$He when they showed that the off-diagonal LRO in the one-body density matrix is the hidden order parameter intimately related to the superfluid fraction of this liquid [25].

"disordered systems," and most of them are expected to possess a definite "hidden order," which shapes both their ground states and their low-energy collective excitations. In this sense, spin liquids are much more akin to superconductors or quantum Hall liquids than to spin glasses. The question of hidden symmetries and the classification of "quantum orders" has been addressed by Wen [26]: using the concept of the *projective symmetry group*, he classified hundreds of spin-liquid candidates on the square lattice and approximately one hundred on the triangular lattice. A method to detect whether a given wave function corresponds to a fractionalized liquid (such as an RVB liquid) was proposed recently [27,28]. Although probably difficult to use for realistic quantum spin models, it was shown to work for simple quantum dimer models [29].

Understanding RVB spin-liquid physics as the result of macroscopic interference led us to investigate the effects of *multiple-spin* exchange interactions, which are in essence direct realizations of collective quantum tunneling processes [30]. As a first example, four-spin cyclic exchange on a rhombus maps directly one of the nearest-neighbor VB coverings to the other, whereas the same effect is obtained only at third order in the perturbative expansion of the Heisenberg interaction (which is a two-body exchange term). A study of the phase diagram of this Hamiltonian on the triangular lattice indeed led to the first indications in favor of an extended, fully gapped RVB spin-liquid phase in an SU(2)-symmetric antiferromagnet [31,33, 32,34].

Rokhsar and Kivelson (RK) pioneered an apparently different route to RVB liquid phases, by focusing directly on the VB in the form of *quantum dimers* [35]. In fact the four-spin cyclic exchange term projected in the nearest-neighbor VB subspace reduces to the kinetic term of the RK model. Unfortunately, RK developed their study of the quantum dimer model only on the square lattice, where they found that, except for a critical point, this model exhibits only VBC phases.[8] However, Moessner and Sondhi [36] later showed that the quantum dimer model on the triangular lattice has indeed a fully gapped, or topological/$\mathbb{Z}_2$, liquid phase, further details of which may be found in R. Moessners and K.S. Raman's Chap. 17.

Besides this rather simple approach of forcing tunneling effects, another popular approach to spin-liquid behavior is to focus on the regions of phase diagrams where frustrating interactions lead to competition between different classical phases. The frustrated $J_1$-$J_2$ models on 2D lattices (in the parameter range between different Néel phases) have thus been the object of a large number of such studies. Here, we do not discuss spin-$\frac{1}{2}$ $J_1$-$J_2$ models on the square lattice, for which the interested reader may consult [37] and the Chap. 15 by F. Becca et al. Spin-$\frac{1}{2}$ $J_1$-$J_2$-$J_3$ models on the hexagonal lattice have been studied by Fouet et al. [38]. In this latter class of model, there are some VBC phases and possibly also gapped spin-liquid phases, which to date are not fully characterized.

[8] Here, the nature of the lattice, bipartite or otherwise, is crucial: the multiple-spin exchange model on the square lattice exhibits a large number of phases, including magnetically ordered, nematic, and VB phases, but no spin-liquid phase [1,3].

A third approach somewhat related to the previous one has been to look for systems in which at the classical level there is an extreme degeneracy between different ground states which prevents the classical system from ordering at $T = 0$. The Heisenberg model both on the pyrochlore and on the kagomé lattice presents the archetype of such a situation. In the final section of this review, we present a critical account of the current understanding of these systems.

Many spin systems in real materials display behavior that may be indicative of spin-liquid properties, including spin-$\frac{1}{2}$ magnets on the anisotropic triangular lattices $Cs_2CuCl_4$ [39], $\kappa(ET)_2X$ [40], and $EtMe_3Sb[Pd(dmit)_2]_2$ [41], and on the kagomé lattice as found in the spin-$\frac{1}{2}$ compounds volborthite [42] and herbertsmithite [74–79] and the spin-$\frac{3}{2}$ system SCGO [43,44].

## 2.2 Basic Building Blocks of VBC and RVB Physics: The Valence Bonds

A valence bond is a singlet state which connects two $S = \frac{1}{2}$ spins at sites $a$ and $b$ of a lattice, and will be denoted hereafter as $|ab\rangle$. This is the state which minimizes the energy of the antiferromagnetic Heisenberg coupling between the two spins, $\langle ab|\mathbf{S}_a \cdot \mathbf{S}_b|ab\rangle = -3/4$.

A VB *configuration* is a direct (tensor) product of VBs encompassing all the sites of the lattice: on an $N$-site lattice, a VB configuration contains $N/2$ VBs (also referred to as dimers). It is a spin configuration with total spin $\mathbf{S}^2_{\text{tot}} = S_{\text{tot}}(S_{\text{tot}} + 1) = 0$. If we consider bonds of any length, the full set of VB configurations is overcomplete, but does span the whole $S_{\text{tot}} = 0$ sector of the spin system. In particular, any non-ferromagnetic or -ferrimagnetic state can be described in the $S_{\text{tot}} = 0$ sector as a superposition of VB configurations [45,46].

*Classical Néel order versus purely quantum ground states.*

Let us begin for simplicity with the nearest-neighbor Heisenberg Hamiltonian on a lattice with coordination number $z$, and consider the variational energy per spin of the purely classical Néel state (denoted hereafter as $e_{cl}$) compared to that of a VB configuration ($e_{vb}$). On a bipartite lattice, these energies are $e_{cl} = -\frac{z}{2}\frac{1}{4}$ and $e_{vb} = -\frac{3}{8}$, while on a 2D tripartite lattice they are $e_{cl} = -\frac{z}{2}\frac{1}{8}$ and $e_{vb} = -\frac{3}{8}$. There are some lattices for which the nearest-neighbor VB configuration wins at this crude level of approximation: the Heisenberg model on the chain, on the kagomé lattice, and on the 2D[9] and 3D pyrochlore lattices. The two variational energies are equal for the triangular and honeycomb lattices [38,47], while on the square lattice the classical solution wins.

Rather generally, in those cases where this zeroth-order comparison favors the classical picture, more sophisticated approaches, such as exact diagonalization with finite-size scaling, do confirm Néel order at $T = 0$ and in dimensions $D > 1$. The

[9] Also called the checkerboard lattice.

classical result is renormalized by long-wavelength fluctuations (spin waves), but the physics may be understood semi-quantitatively on the basis of (semi-)classical approaches such as spin-wave calculations and the non-linear sigma model. The ground states of these models have LRO in their spin correlations, break SU(2) symmetry at $T = 0$, and their lowest-lying excitations are the Goldstone modes of the broken symmetry: gapless magnons.

In those cases where the comparison is clearly in favor of a VB configuration (namely Heisenberg models on the lattices mentioned above, and also examples including the $J_1$-$J_2$ model on the honeycomb lattice, whose discussion we defer to Fig. 2.2), the quantum system does not develop long-ranged Néel order at $T = 0$. Any VB configuration has in fact two features which explain these results: it optimizes the energy of the bonds involved in the singlets, and it gives zero energy for all other bonds. The second property implies that a lattice with a large coordination number will favor Néel order, while a combination of the first and second properties explains why a VB configuration can minimize frustration effects due to competing interactions. However, only in very rare cases is a VB configuration an exact eigenstate of the Hamiltonian [48], so the effect of fluctuations around these VB configurations will be essential in shaping the nature of the true ground states.

In some cases, the system is able to lower its energy still further by adopting a dominant regular VB arrangement "dressed" by local fluctuations. The simplest way to have a regular VB ground state is to consider an explicitly dimerized model, such as the $J$-$J'$ Heisenberg model shown in Fig. 2.1a, where pairs of sites are strongly coupled by an antiferromagnetic interaction $J$, and pairs are coupled by a weaker interaction $J'$. For $J' \ll J$, the ground state is unique and can be viewed as a product of spin singlets on the $J$ bonds (Fig. 2.1b), dressed perturbatively by fluctuations induced by $J'$ (Fig. 2.1c). In Sect. 2.3 we discuss the more subtle case where there is no preferred, pre-existing regular dimer configuration (for example, $J' = J$) and the selection of an ordered pattern for the ground state, if one exists, occurs through the fluctuation-driven mechanism of "order by disorder".

However, in some cases this mechanism does not apply and the system gains energy through resonances between many "disordered" VB configurations. The ground state is then a highly intricate superposition where no spatial symmetry is broken. It contains significant weights for a very large number of VB configurations

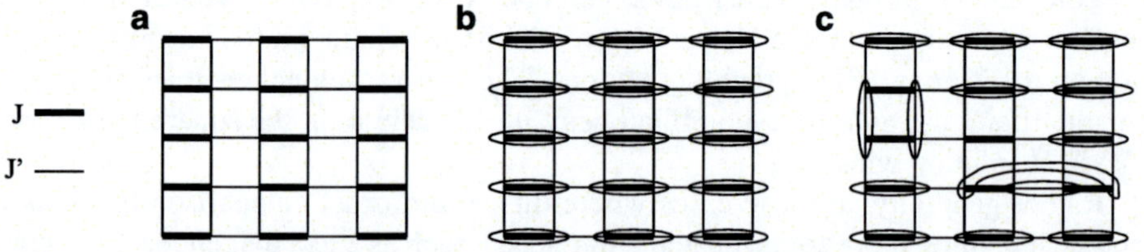


**Fig. 2.1** (**a**) an explicitly dimerized $J$-$J'$ Heisenberg model. (**b**) in the limit $J' = 0$, the ground state is (trivially) a product of spin singlets (indicated by the *ellipses*) on the $J$ bonds. (**c**) for small but non-zero $J'$, other VB configurations contribute to the ground-state wave function. These "fluctuations" dress the zeroth-order state shown in (**b**)

which are *very* different from each other, and this fully resonant situation is the paradigm for the RVB spin liquid. In order to ascertain the differences between RVB spin liquids and VBC order, we describe these phases in detail in the next two sections.

## 2.3 Valence-Bond Crystals

More general than the columnar dimer state of Fig. 2.1, a VBC can be viewed as any state in which the spins group themselves spontaneously into small clusters (which have an even number of sites) arranged in a spatially regular pattern. Within each cluster, the spins form a singlet state.[10]

Two different types of VBC should be distinguished. In the first family (i), the energetically optimal VB pattern is *unique*, imposed by the Hamiltonian (for example through inequivalent bonds, as in Fig. 2.1), and respects all the lattice symmetries in that each cluster fits into the unit cell of the lattice. In a second family (ii), each singlet cluster involves spins from a number (at least two) of unit cells, and *several* VB patterns, which are related by symmetry, are *degenerate*. In case (ii), VBC formation spontaneously breaks some discrete lattice symmetries, and from this point of view, it is truly an ordered state (hence the use of the word "crystal"). A system of weakly coupled dimers, as in Fig. 2.1, is an example of class (i), while the frustrated Heisenberg model of Fig. 2.2 produces a VBC of class (ii). Although the wave functions of the two cases appear similar, they possess some slightly different symmetry (and correlation-function) properties, and the mechanisms leading to the formation of a type-(ii) VBC are manifestly more complex than for type (i). Concerning their magnetic excitations (Sect. 2.3.3), the two types of VBC are very similar.

VBCs do not break any continuous symmetries and do not have gapless Golstone modes. For this reason, they do not "melt" at an infinitesimal temperatures (in contrast to conventional solids in 1D or 2D), they are protected by a gap in their excitation spectrum. For type-(ii) VBC, sufficiently strong thermal fluctuations are expected to restore the broken symmetry, and give rise to a (or several) finite-temperature phase transition. Type-(i) VBC, instead, can evolve continuously (no phase transition) up to high temperatures. Because magnetoelastic coupling between the spin and lattice degrees of freedom are often non-negligible in real systems, a lattice distortion and spontaneous enlargement of the unit cell is expected to occur when the spins enter a type-(ii) VBC phase upon cooling (spin-Peierls transition), and in fact would cause a type-(ii) VBC to transform into a type-(i) VBC. Many type-(i) VBC have been discovered and studied experimentally, but the triangular-lattice organic material studied in [49] may be the only example of

[10] The clusters may have more than two sites (plaquettes, etc.), in which case the name VBC may sound like an abuse of notation. Notice, however, that the singlet ground state of each cluster can always be decomposed over VB coverings.

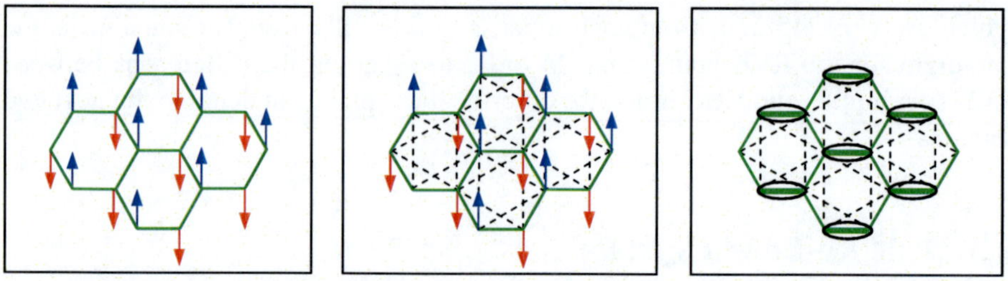

**Fig. 2.2** *Left*: classical arrangement of spins in the ground state of the nearest-neighbor antiferromagnetic Heisenberg Hamiltonian on the honeycomb lattice. *Center*: a second-neighbor antiferromagnetic coupling $J_2$ (*dashed lines*) frustrates the collinear Néel order. *Right*: when the frustrating coupling is sufficiently strong ($J_2/J_1 \sim 0.4$), the formation of a columnar VBC best combats the frustration. The ellipses denote the VBs, while *fat lines* represent bonds which are (at lowest order) fully satisfied and *thin lines* are used for all the couplings "ignored" at this level. In the VBC phase, the ground state is three-fold degenerate in the thermodynamic limit, because there are three symmetry-equivalent VB patterns related by $2\pi/3$ rotations of the hexagonal lattice

VBC-(ii) known to date in $D > 1$ (spontaneous dimerization and opening of a spin gap at low temperature).

Unless explicitly stated, the term VBC in the following refers to type-(ii) VBCs.

### *2.3.1 Zeroth-Order VBC Wave Function*

In the most elementary approximation, the VBC wave function is just the tensor product of these singlet states (one for each cluster) and the spin–spin correlation function is zero for distances larger than the unit-cell size, but the system has LRO in its cluster–cluster correlations. This is best illustrated for dimer correlations, where $\langle (\mathbf{S}_1 \cdot \mathbf{S}_{1'}).(\mathbf{S}_2 \cdot \mathbf{S}_{2'}) \rangle$, with $(1, 1')$ in a given cluster and $(2, 2')$ in a second, will be a finite quantity even when the clusters are separated by macroscopic distances. In this sense, these systems are indeed crystals, and *LRO in the dimer–dimer correlation function in the absence of spin LRO will be our definition of a VBC*.

Examples of VBCs in 1D are well known: both the dimerized phase of the $J_1$-$J_2$ model (for $J_2/J_1 > 0.24$, as discussed by S. Miyahara in Chap. 19) and the spin-Peierls phase are VBCs. Because these phases do not break SU(2) spin symmetry and have no long-range spin correlations (and therefore no magnetic Bragg peaks), they are sometimes defined as "spin liquids". However, in view of their long-ranged dimer correlations, which are observed in Raman and in X-ray experiments, this extension of the nomenclature is rather awkward.

There are numerous examples of spin models with VBC ground states in dimensions $D > 1$. The evidence is often provided by numerical exact diagonalization studies. These include the honeycomb [38], square [50, 51], checkerboard [52], and Shastry–Sutherland [53] lattices. Of particular importance in this context is [48], which provides one of the few Hamiltonians where exact VBC ground-state wave functions are known. Although the issue has not yet been resolved definitively, the "columnar phase" represented in Fig. 2.3a might be realized in the $J_1$-$J_2$ model on the square lattice close to $J_2 = J_1/2$ (we refer the reader to Sect. 2 of [37] and references therein).

### 2.3.2 *Quantum Fluctuations in VBCs*

The spatially regular VB configuration is usually not an eigenstate of the Hamiltonian. It is thus interesting to consider in more detail the action of a Heisenberg coupling term,[11] as shown in Figs. 2.3a,b. The spin–spin interaction can be rewritten as

$$2\mathbf{S}_i \cdot \mathbf{S}_j = P_{ij} - \frac{1}{2}, \tag{2.1}$$

where $P_{ij}$ is the permutation operator of the spins on sites $i$ and $j$, i.e. up to a constant term, $\mathbf{S}_i \cdot \mathbf{S}_j$ permutes the states of sites $i$ and $j$. This may lead to a simple change of sign of the wave function if the sites $i$ and $j$ belong to the same VB ($P_{ab}|ab\rangle = -|ab\rangle$), or to the creation of two longer singlet bonds in any other situation. From a regular pattern, this creates new configurations which differ slightly from the "parent" VB configuration. The effect of a small number of products of transpositions is represented in panel (b) of Fig. 2.3. These fluctuations can be described graphically by the loops drawn on the lattice by superposition of the initial and perturbed VB configurations. This set of loops is called a transition graph. In a VBC wave function, the typical transition graphs obtained by taking snapshots of the wave function in the VB basis consist of a collection of *small* loops.[12] These perturbations around the zeroth-order wave function contribute to a dressing of the bare VB configuration and act to lower its energy. Although (many) different VB configurations may have the same bare (zeroth-order) energy, they do not in general receive the same energy corrections from fluctuations. The VBC pattern which is selected is then the one which optimizes the energy gain, and this is the microscopic mechanism for "order by disorder" – one ordered VBC configuration emerges from the disorder of the different fluctuating coverings. As one example, the columnar pattern of Fig. 2.3 is the one which allows the largest number of four-site resonance loops (the smallest possible loop, as in the upper right plaquette of Fig. 2.3b). If the spin interactions are such that these resonances are the most important ones (which is not usually obvious a priori), such a VBC pattern is likely to be that stabilized by the fluctuations. These fluctuations also confer some finite range to the spin correlations in the exact ground state, and renormalize the crystal order parameter (the dimer structure factor) downwards.

---

[11] Other SU(2)-symmetric interactions between the $S = \frac{1}{2}$ spins can be written as more complicated permutations, and have similar effects.

[12] If there were many large loops, the regular pattern would have melted under the action of the spin interactions, rendering the system a spin liquid.

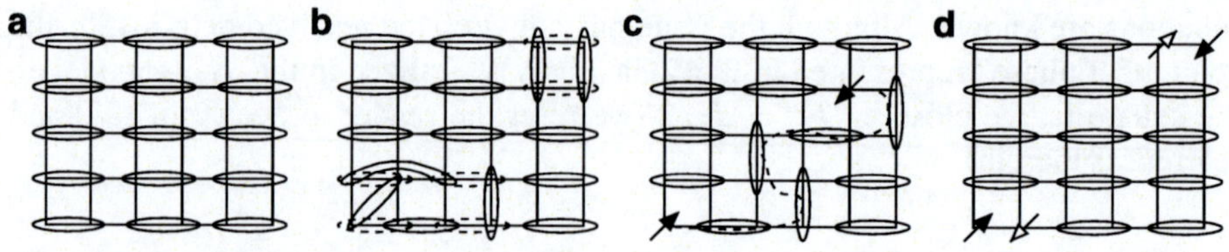


**Fig. 2.3** (**a**) schematic representation of a columnar VBC on the square lattice. (**b**) a realistic VBC wave function contains fluctuations, viewed here as VB configurations which differ from the parent configuration (**a**) by *small loops* in their transition graph. (**c**) representation of the confinement potential experienced at intermediate distances by two test spinons in a VBC. The ordered background is perturbed ("shifted") along a path connecting the spinons, providing an effective potential which grows linearly with distance. (**d**) for larger separations, where the energy becomes similar to the spin gap, it is advantageous to break one more singlet bond and create two more unpaired spins (*empty arrows*). These extra spin-$\frac{1}{2}$ objects "screen" the test spinons, and restore the regular pattern of the VBC. The energy of the system therefore becomes constant on further increasing the separation of the test spinons due to dressing by the extra spins. This makes it impossible to isolate spatially a single spin-$\frac{1}{2}$ excitation in a VBC

### *2.3.3 VBC Excitations*

The simplest magnetic excitations of a VBC (of type (i) or type (ii)) involve one basic cluster, and are $\Delta S = 1$ transitions. *These are gapped, carry integer spin, and appear as sharp modes in an inelastic neutron scattering experiment.* Such excitations may be more or less mobile, depending on the strength and geometry of the couplings between the basic cluster unit and its environment. When the kinetic energy of the excitations is sufficiently low compared to their mutual repulsion, they may crystallize at certain rational densities, giving rise to magnetization plateaus (as discussed by M. Takigawa and F. Mila in Chap. 10).[13]

An important property of RVB liquids (discussed in the next section) is the fact that they have spin-$\frac{1}{2}$ excitations, called spinons. To understand the nature of these (rather exotic) excitations, it is first useful to understand why spinons *cannot* exist in a VBC. The two excited states shown in panels (c) and (d) of Fig. 2.3 make clear why it is energetically unfavorable to disturb the dimer order of a VBC in an attempt to liberate two free spin-$\frac{1}{2}$ entities. Such a trial state has a string of misaligned dimers between the two unpaired spin-$\frac{1}{2}$ objects, and these interfere with the pattern selection in the ground state: the fluctuations (resonance loops) which stabilize the VBC pattern in the ground state are not available in the vicinity of the string, leading to an energy cost for the trial state due to fluctuation processes. Specifically, the string has an energetic cost which is approximately proportional to its length, and this creates

---

[13] If the basic cluster contains more than two sites (i.e. it does not reduce to a dimer), different excited modes with integer spin (0, 1, 2 . . . ) may exist [52]. These have the same generic properties as the triplets discussed here, but cannot be detected as easily by neutron spectroscopy; Raman scattering and nuclear magnetic resonance (NMR) are more appropriate probes for studying these additional modes.

an effective restoring force which forbids an infinite separation of the initial two spin-$\frac{1}{2}$ defects.[14] This long-ranged effective force is associated with the disruption of the regular arrangement of local singlet clusters when a pair of distant spinons is forced into the system. The two spinons are said to be confined. More precisely, Fig. 2.3d illustrates the physical mechanism contributing to the screening of spinons in a VBC at large distances, and thus the impossibility of observing free spin-$\frac{1}{2}$ objects in the bulk of such phases. A further discussion may be found in [18, 37].

## 2.4 Resonating-Valence-Bond Spin Liquids

In our discussion of the VBC, the central idea is that a small number of spatially ordered VB configurations (one in the case of a type-(i) VBC without lattice symmetry-breaking, a number $\mathcal{O}(1)$ in a type (ii) VBC with spontaneous symmetry-breaking) are favored at $T = 0$. In some geometries, however, this is far from evident, and the nearest-neighbor Heisenberg model on the kagomé lattice provides one of the most straightforward examples which nevertheless contains some of the most complex physics.

If there is no preferred ordering pattern, the ground state will have weights of the same order of magnitude over an extensive fraction of the manifold of VB configurations.[15] From this picture of an RVB liquid, one may anticipate the following properties:

- Such a system has no LRO in any spin, dimer, or (finite) higher-order correlation functions. It is a true liquid.
- The absence of an "ordered parent VB configuration" implies that there is no evident mechanism to confine spinon excitations.[16] These spin-$\frac{1}{2}$ objects are thus essential to a full understanding of the magnetic excitations of spin liquids. In an inelastic neutron scattering experiment, spinons are necessarily created in (at least) pairs, because a neutron changes $S^z_{\mathrm{tot}}$ by $\pm 1$. The two spinons of a given pair can then propagate independently, each one carrying a fraction of the energy

[14] This conclusion is indeed also true for a VBC of type (i), even without invoking fluctuations.

[15] One may be more precise in defining the subspace generated by VB configurations: this could be restricted to *nearest-neighbor* VB states, or to VB configurations including singlets between sites with some maximum separation $r_{\mathrm{max}}$. One may also require that the probability to find a VB between two sites of separation $r$ decay exponentially with $r$, or include long VB configurations provided that their weight in the wave function decay "sufficiently rapidly" with distance (in the spirit of [45]). One further systematic refinement is to adopt a Hilbert space where the basis states are nearest-neighbor VB configurations dressed by one application of the Heisenberg interaction (in the spirit of [54]).

[16] These spinons occupy a "sea" of VBs whose fluctuations can mediate short-distance interactions, and the deconfinement of spinons in some cases may be only asymptotic (meaning that bound states may occur at short distances). Here also the word "sea" is used as an image, and the state does not necessarily bear any resemblance to a Fermi sea.

and momentum. *The signature of such deconfined spinons will thus be a continuum in energy and momentum.* The unfortunate consequence of this property is that the spin spectral weight may be distributed over a broad range of energy and thus the spinons may be difficult to observe in a neutron scattering experiment. It is nevertheless expected that the spectral weight is maximal at the low-energy threshold of the continuum. Incoherent fluctuations down to temperatures equal to the gap or to $T = 0$ in case of gapless spin liquid, provide an essential experimental signature of a spin liquid which would be revealed by local probes such as NMR or muon spectroscopy.

- The absolute ground-state wave function is the most stable superposition of VB configurations. By varying the relative phases of each VB configuration, one may construct a *macroscopic* number of states which are orthogonal to the ground state. These states are true $S = 0$ excitations which may be either gapped with respect to the ground state or gapless [55].[17]
- An RVB liquid ground-state wave function contains long exchange cycles between different VB configurations. More precisely, one may choose an initial VB configuration, $c_1$, and consider the transition graph (a set of closed loops) obtained by superimposing a new configuration $c_2$ onto $c_1$. By "long exchange cycles," we mean that the transition graphs obtained by sampling such pairs of configurations $(c_1, c_2)$ (according to their weights in the wave function) will lead statistically to graphs containing some long loops, whose length diverges in the thermodynamic limit. This is to be contrasted with VBC wave functions, where typical pairs $(c_1, c_2)$ give only a set of small loops. The presence of these long resonance loops in the ground-state wave function, and in the lowest excitations, is reminiscent of the quantum properties of superfluid $^4$He, where it has been understood [21–23] that it is the building of these long, *coherent, macroscopic exchange cycles* upon cooling which lies at the origin of the $\lambda$ transition and of the superfluidity. In liquid $^4$He, these exchange loops have *definite relative phase factors*, which shape the dynamics of the low-temperature system; in this case they are all positive due to Bose statistics. From the same point of view, an RVB liquid wave function may be considered as a kind of "VB condensate", but in such spin systems the exchange loops are not constrained to have positive phase factors.[18]

[17] For the more theoretically inclined reader, we note that this freedom to change certain phases in a VB superposition may permit the construction, in gapped liquids and in the thermodynamic limit, of several degenerate ground states if the lattice has periodic boundary conditions (i.e. a non-trivial topology). This property is known as topological degeneracy [9,56].

[18] The presence of long exchange loops in 2D electronic wave functions has also been discussed in the context of the fractional quantum Hall effect [57,58], where the loops correspond to processes (in an imaginary-time, path-integral formulation) in which large numbers of electrons exchange their positions in a cyclic way. These processes can be made coherent due to Aharonov–Bohm phases, which originate from the (physical, external) magnetic flux piercing each exchange loop. This leads to a direct analogy with RVB spin liquids, and in particular with chiral spin liquids, where a (fictitious) magnetic field may emerge and where the role of the electrons is played by spinons; more details may be found in [18]).

In fact there is a rich variety of possible RVB spin liquids, including but not limited to $\mathbb{Z}_2$ spin liquids, U(1) spin liquids, and SU(2) spin liquids (the reader is referred to [12, 18, 26, 59, 60] and references therein for an extensive treatment). In very broad terms, this classification refers to the properties of the gauge fields associated with the structure and symmetries of the phase factors in front of the resonance loops. These gauge fields also mediate the interaction between the spinons introduced above.

- $\mathbb{Z}_2$ spin liquids are the simplest and best understood spin liquids: they are gapped, with (at least) two kinds of well defined excitations, spin-$\frac{1}{2}$ spinons and spin-0 visons [9, 11]. The wave functions of the spin-0 excitations can be approximated by taking the ground-state wave function and applying phase factors of $-1$ to selected VB configurations. These excitations can also be approached by (approximate) mappings to Ising gauge theories, where they appear as the vortices of this dual gauge theory (whence their name, 'v-isons') [61–65]. Of particular interest for pedagogical purposes are an exactly solvable spin model on the square lattice [65] and a dimer model on the kagomé lattice where visons appear as free particles [63].
- When the spinons and singlet excitations are gapless, the liquid is in a critical state and many correlation functions (including $\langle \mathbf{S}_i \cdot \mathbf{S}_j \rangle$) decay algebraically with distance [14, 66, 67]. In such cases, the spinon and singlet excitations cannot be considered as quasi-free quasiparticles, because they remain strongly interacting down to the lowest energies/temperatures.

Where can one look for RVB spin-liquid physics? Some spin-$\frac{1}{2}$ triangular-lattice models are good theoretical candidates [31, 32, 68]. The nearest-neighbor Heisenberg model on the triangular lattice is thought to be Néel ordered [47], but cyclic four-spin exchange terms on rhombic plaquette units can destabilize this order and lead to one (or possibly more) gapped phase(s) [33].[19] For a field-theoretic approach to one of these possible spin-liquid phases on the triangular lattice, the reader is referred to [20] and to the references therein. As indicated in the introduction, there are now several triangular-lattice systems which are under active experimental investigation as candidate spin liquids [39–41, 71–73].

At the same time, spin-$\frac{1}{2}$ models on the kagomé lattice remain perhaps the most promising candidates for the observation of RVB spin liquids in nature – either with Heisenberg (below) or with Ising or XY spin interactions [16, 69, 70]. The recent discovery of herbertsmithite ($ZnCu_3(OH)_6Cl_2$), a "structurally perfect" $S = \frac{1}{2}$ kagomé material [74–79], represents an important step in this direction. The fact that herbertsmithite shows neither spin freezing nor the opening of a spin gap down to temperatures lower than 1/4000 of the coupling constant tends to support the theoretical speculations of the past 20 years, and demands further efforts by theorists on this fundamental and complex problem. More recently, improved samples of

[19] More generally, numerous theoretical models which have been shown to demonstrate spin-liquid physics contain these multiple-spin (or ring-)exchange terms [69, 70].

volborthite also appear to exhibit spin-liquid behavior over a large range of temperature and magnetic field [42].

## 2.5 VBCs or RVB Spin Liquids on Kagomé and Pyrochlore Lattices?

As emphasized in the introduction, the classical Heisenberg model both on the kagomé and on the pyrochlore lattice has an extensive $T = 0$ entropy. The question of the quantum ground state of these systems arose very early in the study of frustrated quantum magnetism [80–82], but the answers remain at present only partial. The spin-$\frac{1}{2}$ Heisenberg model on the kagomé and pyrochlore lattices has been the object of considerable theoretical attention since the late 1980s. Numerical approaches [54, 83–86] were quick to show that such spin-$\frac{1}{2}$ systems are indeed quite unusual. In the spirit of this chapter, a description of the singlet sector in a srRVB basis [54, 87–89] was developed: this picture captures the short distance physics of the kagomé magnet rather nicely but the low energy, long distance physics remains a mystery.

In this short discussion, we will not present a review of the many studies performed to investigate these questions, but focus on the most recent results and understanding; we refer the reader to [37] for a brief guide to the earlier literature. In the following, we list some open questions, underline those points which now seem to be reasonably well established, and highlight those which remain under debate.

To within the accuracy of modern numerical approaches, it is generally acknowledged that the spin-$\frac{1}{2}$ Heisenberg models on the pyrochlore [86] and kagomé lattices have no LRO in their spin correlations [82, 83, 90]. On the kagomé lattice, it has also been found that dimer–dimer, chiral–chiral, and nematic–nematic correlation functions decrease rapidly at short distances. However, the available system sizes remain too small for a systematic determination of their true asymptotic behavior (specifically, to distinguish between exponential and algebraic decay, or to differentiate either form from very weak VBC order).

From numerical studies (exact diagonalization, the contractor renormalization method or the density-matrix renormalization group), one may list the following points as appearing either well established or at minimum highly probable:

- The spin-$\frac{1}{2}$ Heisenberg magnet on the 2D pyrochlore (checkerboard) lattice is a VBC with a large gap both to spin excitations and in the singlet sector [52, 91]. From the numerical real-space renormalization approach [92] and exact diagonalization on a $N = 32$ system [52], it seems that the same situation is true for the 3D pyrochlore lattice. Still, more analysis of the pyrochlore case is essential to verify this point, and to give more information about the excitation spectrum.[20]

[20] Such a structure is also very sensitive to spin-lattice couplings, which, as noted above, act to favor VBC formation [93].

- The spin-1 Heisenberg magnet on the kagomé lattice has a large gap [94], it may be in the Affleck–Kennedy–Lieb–Tasaki class [95, 96], and have spin-$\frac{1}{2}$ edge excitations. These kind of phases were qualified in [95, 96] as Valence Bond Solid (VBS). The VBS states are closely related to the type (i) VBC.
- In the small systems studied, the spectra of the spin-$\frac{1}{2}$ Heisenberg model on the kagomé and pyrochlore lattices are quite different. A clear gap is present in the singlet sector for the latter (its value is $\sim 0.21\,J$ on a $N = 32$ sample [52], where $J$ is the coupling constant), whereas for the former the singlet sector seems to be gapless: the average interval between consecutive levels of this continuum on a $N = 36$ cluster lies below $J/50$ [97].
- A study of the spin-$\frac{3}{2}$ Heisenberg model on the kagomé lattice [98] points to a finite-size spin gap decreasing rapidly with the size (opening the possibility of a gapless system), and to a structure of the low-lying singlet sector notably different from that of the spin-$\frac{1}{2}$ model. Given the very small cluster sizes, these results could even be consistent with a semi-classical picture – something clearly not the case for the spin-$\frac{1}{2}$ system and absolutely not for the spin-1 system.

These four points illustrate the role of the value (integer or half-odd-integer) of the total spin in the unit cell. With the present nulmerical data, we can hypothesize that systems with integer spins in the unit cell, spin-1 kagomé, spin-$\frac{1}{2}$ 2D pyrochlore, and apparently spin-$\frac{1}{2}$ 3D pyrochlore, have a rather large gap in the singlet and triplet sectors and are likely to be VBCs. Spin-$\frac{1}{2}$ and spin-$\frac{3}{2}$ systems appear rather different, they may have a very small or vanishing gap. A discussion of the Lieb–Schultz–Mattis theorem in this context may be found in [18, 60] and provides the first steps to understand theoretically the role played by the total spin per unit cell (integer or not).

For the spin-$\frac{1}{2}$ kagomé antiferromagnet, different competing hypotheses remain open. Either the system is in a VBC phase with a large unit cell or it may be a gapped RVB spin liquid (with a very small gap) or critical gapless spin liquid.

- The first hypothesis, already studied by a number of authors [99–101] has been substantiated recently by a state-of-the-art series expansion starting from an ordered VB pattern [4, 5]. In this approach, a large unit cell of 36 spins is proposed as the basic cluster emerging through the spontaneous formation of a [type-(ii)] VBC. This spontaneous symmetry breaking is an illustration of the mechanism of "order by disorder" invoked in Sect. 2.3. This calculation shows that a VBC phase is at minimum very close in energy to the absolute ground-state, and does not melt under relatively small loop exchange cycles (the series expansion is performed up to 5th order).[21] Nevertheless, the importance of the bias introduced by the choice of an initial ordered VB configuration remains quite unclear. Recent DMRG calculation [90] for systems up to 108 spins, which evaluate in the thermodynamic limit a finite spin gap of order $J/20$ exclude a

[21] In other approaches, with their own different limitations, the crystal is shown to melt under the effects of fluctuations. [54, 92].

long-ranged dimer–dimer order, which is at variance with the VBC hypothesis, whatever the size of the unit cell of the crystal.

- The second hypothesis of a "short" range RVB $Z_2$ spin liquid has been advanced very early by Sachdev [102] within a large-$N$ Schwinger bosons approach. The behavior of the finite size gap and spin–spin correlations in samples with all the symmetries of the infinite lattice [37, 83, 90] does not contradict such an hypothesis, but the analysis of the 36-sites dynamical spin–spin susceptibility [103] shows that the maximum of this susceptibility is not at the corner of the Brillouin zone as would have been expected in this hypothesis.
- The last hypothesis is that of gapless spin liquid. Ran et al. have proposed a gapless algebraic spin liquid as a possible state to describe the low-energy physics of this compound [17, 67], whereas Ryu et al. consider another kind of critical spin liquid, the "algebraic vortex liquid," which is possibly relevant in describing the spin-$\frac{1}{2}$ XY magnet on the same lattice [16]. A critical spin liquid would explain more simply (than the explanation of [4, 5]) the quasi-continuum observed in exact diagonalization studies for all spin sectors ($S = 0, \frac{1}{2}, 1, \ldots$). It is also fully consistent with the dynamical mean-field theory results of [104]. A discussion of the spin gap issue extracted from finite size calculations (exact diagonalizations) will be soon available to complement this discussion [97].

For a discussion of the possible experimental realizations of such a model we refer to P. Mendels chapter in this book.

## 2.6 Conclusion

VBC and srRVB liquids are states of matter based on the same building blocks. Both types of wave function can be viewed as linear combinations of VB configurations, which are energetically favorable states in the presence of frustrated interactions. The properties of these two types of state at long distances are completely different: a VB crystal wave function is "localized" in the vicinity of one (or a small number of) simple "parent" VB configuration(s) which is (are) spatially regular; the wave function of an RVB liquid spreads over macroscopically many VB configurations which are very different, and distant from each other in configuration space (i.e. with long loops in their mutual transition graphs). This difference is connected at a fundamental level to the nature of their magnetic excitations: integer-spin excitations in VBCs (a broken singlet bond which propagates and becomes "dressed" by small resonance loops) contrast sharply with deconfined spin-$\frac{1}{2}$ spinons in RVB liquids (an unpaired spin in a sea of VBs).

It is usually a difficult, but nevertheless crucial, question to decide whether a given spin model has a VBC or a short range RVB-liquid ground state or, in fact, something else again. These states differ in their long-distance properties: confinement versus deconfinement of spinons and long- versus short-ranged bond–bond correlation functions. Analytical and numerical approaches often need to be combined to reach a conclusion. However, the kagomé and pyrochlore cases discussed in Sect. 2.5 are examples where the definitive answer is not yet known. To discuss these issues in a more formal manner – something we have avoided here – it is instructive to formulate the spin model in terms of spinons interacting with gauge fields. This approach is discussed in [18], including its application to other types of spin-liquid wave function based on states with long-ranged VBs, and possibly also possessing gapless spinons.

**Acknowledgements**

We thank Bruce Normand and Frédéric Mila for helpful contributions to both the content and the presentation of this contribution.

## References

1. A. Läuchli, J.C. Domenge, C. Lhuillier, P. Sindzingre, M. Troyer, Phys. Rev. Lett. **95**, 137206 (2005)
2. T. Momoi, P. Sindzingre, N. Shannon, Phys. Rev. Lett. **97**, 257204 (2006)
3. N. Shannon, T. Momoi, P. Sindzingre, Phys. Rev. Lett. **96**, 027213 (2006)
4. R.R.P. Singh, D.A. Huse, Phys. Rev. B **76**, 180407 (2007)
5. R.R.P. Singh, D.A. Huse, Phys. Rev. B **77**, 144415 (2008)
6. G. Baskaran, Z. Zou, P.W. Anderson, Solid State Comm. **88**, 853 (1987)
7. S.A. Kivelson, D.S. Rokhsar, J.P. Sethna, Phys. Rev. B **35**, 8865 (1987)
8. B. Sutherland, Phys. Rev. B **37**, 3786 (1988); Phys. Rev. B **38**, 7192 (1988)
9. N. Read, B. Chakraborty, Phys. Rev. B **40**, 7133 (1989)
10. F.D.M. Haldane, H. Levine, Phys. Rev. B **40**, 7340 (1989)
11. N. Read, S. Sachdev, Phys. Rev. Lett. **66**, 1773 (1991)
12. P.A. Lee, N. Nagaosa, X.-G. Wen, Rev. Mod. Phys. **78**, 17 (2006)
13. M. Levin, X.-G. Wen, Rev. Mod. Phys. **77**, 871 (2005)
14. W. Rantner, X.-G. Wen, Phys. Rev. Lett. **86**, 3871 (2001)
15. M. Hermele, T. Senthil, M.P.A. Fisher, Phys. Rev. B **72**, 104404 (2005)
16. S. Ryu, O.I. Motrunich, J. Alicea, M.P.A. Fisher, Phys. Rev. B **75**, 184406 (2007)
17. M. Hermele, Y. Ran, P.A. Lee, X.-G. Wen, Phys. Rev. B **77**, 224413 (2008)
18. G. Misguich, Quantum Spin Liquids and Fractionalization. in *Introduction to Frustrated Magnetism*, ed. by C. Lacroix (Springer, Heidelberg, 2010)
19. B.S. Shastry, B. Sutherland, Phys. B (Amsterdam) **108**, 1069 (1981)
20. S. Sachdev, arXiv:0901.4103
21. R.P. Feynman, Phys. Rev. **91**, 1301 (1953)
22. D.M. Ceperley, Rev. Mod. Phys. **67**, 279 (1995)
23. P. Grueter, D. Ceperley, F. Laloë, Phys. Rev. Lett. **79**, 3549 (1997)
24. X.-G. Wen, Phys. Lett. A **300**, 175 (2002)
25. O. Penrose, L. Onsager, Phys. Rev. **104**, 576–584 (1956)
26. X.-G. Wen, Phys. Rev. B **65**, 165113 (2002)
27. A.Y. Kitaev, J. Preskill, Phys. Rev. Lett. **96**, 110404 (2006)
28. M. Levin, X.-G. Wen, Phys. Rev. Lett. **96**, 110405 (2006)
29. S. Furukawa, G. Misguich, Phys. Rev. B **75**, 214407 (2007)
30. D.J. Thouless, Proc. Phys. Soc. **86**, 893 (1965)
31. G. Misguich, B. Bernu, C. Lhuillier, C. Waldtmann, Phys. Rev. Lett. **81**, 1098 (1998)
32. G. Misguich, C. Lhuillier, B. Bernu, C. Waldtmann, Phys. Rev. B **60**, 1064 (1999)

33. W. LiMing et al., Phys. Rev. B **62**, 6372 (2000)
34. G. Misguich, C. Lhuillier, M. Mambrini, P. Sindzingre, Eur. Phys. J. B **26**, 167 (2002)
35. D.S. Rokhsar, S.A. Kivelson, Phys. Rev. Lett. **61**, 2376 (1988)
36. R. Moessner, S. Sondhi, Phys. Rev. Lett. **86**, 1881 (2001)
37. G. Misguich, C. Lhuillier, in *Frustrated Spin Systems*, ed. by H.T. Diep (World Scientific, Singapore, 2005), arXiv:cond-mat/0310405
38. J.B. Fouet, P. Sindzingre, C. Lhuillier, Eur. Phys. J. B **20**, 241 (2001)
39. J.O. Fjærestad, W. Zheng, R.R.P. Singh, R.H. McKenzie, R. Coldea, Phys. Rev. B **75**, 174447 (2007)
40. M. Yamashita et al., Nat. Phys. **5**, 44 (2008)
41. T. Itou, A. Oyamada, S. Maegawa, M. Tamura, R. Kato, Phys. Rev. B **77**, 104413 (2008)
42. H. Yoshida et al., J. Phys. Soc. Jpn. **78**, 043704 (2009) ; Phys. Rev. Lett. **103**, 077207 (2009)
43. Y.J. Uemura et al., Phys. Rev. Lett. **73**, 3306 (1994)
44. D. Bono et al., Phys. Rev. Lett. **93**, 187201 (2004)
45. S. Liang, B. Douçot, P.W. Anderson, Phys. Rev. Lett. **61**, 365 (1988)
46. F. Becca et al., Variational Wave Functions for Frustrated Magnetic Models. In *Introduction to Frustrated Magnetism*, ed. by C. Lacroix (Springer, Heidelberg, 2010)
47. B. Bernu, P. Lecheminant, C. Lhuillier, L. Pierre, Phys. Rev. B **50**, 10048 (1994)
48. A. Gellé, A. Läuchli, B. Kumar, F. Mila, Phys. Rev. B **77**, 014419 (2008)
49. M. Tamura, A. Nakao, R. Kato, J. Phys. Soc. Jpn. **75**, 093701 (2006)
50. A. Läuchli et al., Phys. Rev. Lett. **95**, 137206 (2005)
51. M. Mambrini, A. Läuchli, D. Poilblanc, F. Mila, Phys. Rev. B **74**, 144422 (2006)
52. J.-B. Fouet, M. Mambrini, P. Sindzingre, C. Lhuillier, Phys. Rev. B **67**, 054411 (2003)
53. A. Läuchli, S. Wessel, and M. Sigrist, Phys. Rev. B **66**, 014401 (2002)
54. C. Zeng, V. Elser, Phys. Rev. B **51**, 8318 (1995)
55. P.W. Anderson, Mat. Res. Bull. **8**, 153 (1973)
56. X.-G. Wen, Phys. Rev. B **44**, 2664 (1991)
57. S. Kivelson, C. Kallin, D.P. Arovas, J.R. Schrieffer Phys. Rev. Lett. **56**, 873 (1986)
58. D.H. Lee, G. Baskaran, S. Kivelson, Phys. Rev. Lett. **59**, 2467 (1987)
59. X.-G. Wen, *Quantum Field Theory of Many-body Systems: From the Origin of Sound to an Origin of Light and Electrons*, (Oxford University Press, Oxford, 2004)
60. G. Misguich, Lectures notes of the Les Houches summer school on "Exact Methods in Low-dimensional Statistical Physics and Quantum Computing", July 2008, arXiv:0809.2257
61. A. Kitaev, Ann. Phys. **303**, 2 (2003)
62. T. Senthil, M.P.A. Fisher, Phys. Rev. B **62**, 7850 (2000)
63. G. Misguich, D. Serban, V. Pasquier, Phys. Rev. Lett. **89**, 137202 (2002)
64. R. Moessner, S.L. Sondhi, E. Fradkin, Phys. Rev. B **65**, 024504 (2002)
65. X.-G. Wen, Phys. Rev. Lett. **90**, 016803 (2003)
66. M. Hermele et al., Phys. Rev. B **70**, 214437 (2004)
67. Y. Ran, M. Hermele, P.A. Lee, X.-G. Wen, Phys. Rev. Lett. **98**, 117205 (2007)
68. O.I. Motrunich, Phys. Rev. B **72**, 045105 (2005)
69. L. Balents, M.P.A. Fisher, S.M. Girvin, Phys. Rev. B **65**, 224412 (2002)
70. D.N. Sheng, L. Balents, Phys. Rev. Lett. **94**, 146805 (2005)
71. E. Collin, S. Triqueneaux, R. Harakaly, M. Roger, C. Bäuerle, Yu. Bunkov, H. Godfrin, Phys. Rev. Lett. **86**, 2447 (2001)
72. R. Masutomi, Y. Karaki, H. Ishimoto, Phys. Rev. Lett. **92**, 025301 (2004)
73. H. Akisato, S. Murakawa, Y. Mastumoto, D. Tsuji, H. Kambara, H. Fukuyama J. Low Temp. Phys. **138**, 265 (2005)
74. M.P. Shores, E.A Nytko, B.M. Bartlett, D.G. Nocera, J. Am. Chem. Soc. **127**, 13462 (2005)
75. J.S. Helton et al., Phys. Rev. Lett. **98**, 107204 (2007)
76. P. Mendels et al., Phys. Rev. Lett. **98**, 077204 (2007)
77. F. Bert et al., Phys. Rev. B **76**, 132411 (2007)
78. A. Olariu et al., Phys. Rev. Lett. **100**, 087202 (2008)
79. T. Imai et al., Phys. Rev. Lett. **100**, 077203 (2008)
80. V. Elser, Phys. Rev. Lett. **62**, 2405 (1989)

81. P. Chandra, P. Coleman, I. Ritchey, Int. J. Mod. Phys. B, **5**, 171 (1991)
82. J.T. Chalker, J.F.G. Eastmond, Phys. Rev. B **46**, 14201 (1992)
83. P.W. Leung, V. Elser, Phys. Rev. B **47**, 5459 (1993)
84. P. Lecheminant et al., Phys. Rev. B **56**, 2521 (1997)
85. C. Waldtmann et al., Eur. Phys. J. B **2**, 501 (1998)
86. B. Canals, C. Lacroix, Phys. Rev. Lett. **80**, 2933 (1998)
87. F. Mila, Phys. Rev. Lett. **81**, 2356 (1998)
88. M. Mambrini, F. Mila, Eur. Phys. J. B **17**, 651 (2001)
89. S. Dommange, M. Mambrini, B. Normand, F. Mila, Phys. Rev. B **68**, 224416 (2003)
90. H.C. Jiang, Z.Y. Weng, D.N. Sheng, Phys. Rev. Lett. **101**, 117203 (2008)
91. B. Canals, Phys. Rev. B **65**, 184408 (2002)
92. E. Berg, E. Altman, A. Auerbach, Phys. Rev. Lett. **90**, 147204 (2003)
93. O. Tchernyshyov, R. Moessner, S.L. Sondhi, Phys. Rev. Lett. **88**, 067203 (2002)
94. K. Hida, J. Phys. Soc. Jpn. **69**, 4003 (2000)
95. I. Affleck, T. Kennedy, E. Lieb, H. Tasaki, Phys. Rev. Lett. **59**, 799 (1987)
96. I. Affleck, T. Kennedy, E. Lieb, H. Tasaki, Commun. Math. Phys. **115**, 477 (1988)
97. P. Sindzingre, C. Lhuillier, Europhys. Lett. **88**, 27009 (2009)
98. A. Läuchli, S. Dommange, B. Normand, F. Mila, Phys. Rev. B **76**, 144413 (2007)
99. J.B. Marston, C. Zeng, J. Appl. Phys. **69**, 5962 (1991)
100. A.V. Syromyatnikov, S.V. Maleyev, Phys. Rev. B **66**, 132408 (2002)
101. P. Nikolic, T. Senthil, Phys. Rev. B **68**, 214415 (2003)
102. S. Sachdev, Phys. Rev. B **45**, 12377 (1992)
103. A. Läuchli, C. Lhuillier, arXiv:0901.1065
104. A. Georges, R. Siddhartan, S. Florens, Phys. Rev. Lett. **87**, 277203 (2001)

# Part II
# Probing Frustrated Magnets

# Chapter 3
# Neutron Scattering and Highly Frustrated Magnetism

**Steven T. Bramwell**

**Abstract** Neutron scattering is a paradigm technique for the determination of spin correlation functions. It is the method of choice for the study of magnetic order and magnetic interactions, and, in favourable cases, it may provide an unambiguous distinction between conventional paramagnetism and cooperative paramagnetism. This chapter is an introduction to neutron scattering with an emphasis on its interpretation in terms of spin correlation functions and generalised susceptibilities. The main aim is to illustrate how neutron scattering is used to interrogate the nature of the disordered and highly degenerate states that occur in highly frustrated magnets.

## 3.1 Introduction

The study of the phases of matter might be seen, at a fundamental level, as the study of broken symmetries, of which there are many kinds [1]. Magnetic neutron scattering essentially measures the spin–spin correlation function, so most clearly distinguishes magnetic phases that have correlation functions of differing symmetry. The simplest example is the distinction of a paramagnet and a spin ordered phase (ferromagnet or antiferromagnet). In the former the spin–spin correlation is a decreasing function of distance between the spins, which gives rise to continuous or 'diffuse' neutron scattering in reciprocal space; in the latter the correlation function acquires a component with periodic symmetry that gives rise to discrete scattering in the form of Bragg peaks. Thus, given sufficient instrumental resolution, there is no ambiguity in distinguishing ordered and disordered spin components as their signatures in neutron scattering are radically different (Fig. 3.1).

The game of highly frustrated magnetism is to find new and exotic phases, that may generally be described as 'cooperative paramagnets' [2]. Bulk measurements may infer the existence of such phases, but they do not generally indicate a distinctive difference between these and the conventional paramagnetic phase, so we should ask, is it even correct to imagine a basic distinction between the two? If the answer were no, then highly frustrated magnetism would be a very boring subject! Fortunately, there is such a basic distinction, at least for one broad class of highly

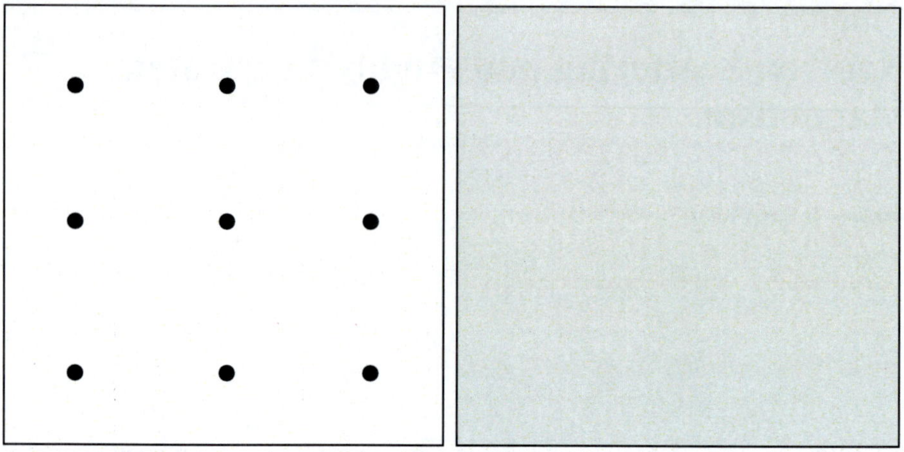

**Fig. 3.1** Schematic neutron scattering map of an ordered phase with sharp Bragg spots (*left*) and an ideal paramagnetic phase, with broad diffuse scattering (*right*)

frustrated magnets, and this is clearly revealed by neutron scattering. The clarity of the distinction afforded by neutron scattering again reflects symmetry differences in the underlying spin correlation functions of a conventional and a cooperative paramagnet, even though neither phase exhibits periodic spin order.

To emphasise this, Fig. 3.2 illustrates reciprocal space maps of the correlation function $S(\mathbf{Q})$ (defined below), a quantity directly estimable by neutron scattering. Assuming a cubic system, the maps may be parameterised in polar coordinates $\{q, \theta\}$ with respect to a Brillouin zone centre. We consider four possibilities: (a) that S depends neither on $q$ nor $\theta$, (b) that it depends on $q$ but not $\theta$, (c) that it depends on $\theta$ but not $q$, and (d) that it depends on both. In fact, this classification turns out to correspond respectively to the case of (a) an ideal paramagnet, (b) a conventional magnet in its paramagnetic phase, (c) an ideal highly frustrated magnet with 'emergent gauge' structure, and (d) a more general highly frustrated magnet. This correspondence is a vivid demonstration of how the most straightforward classification of neutron maps naturally identifies distinct paramagnetic phases that fit neatly with our concepts of frustration.

In this article, I want to elaborate upon these remarks in order to explain how neutron scattering may be used (and has been used so far) to interrogate the nature of cooperative paramagnetic phases. This interrogation may be direct – by measuring spin–spin correlations – or indirect: by establishing the spin–spin interactions, typically through characterisation of magnetic structures, spin waves or crystal field excitations. My additional aim is to give the reader a basic introduction to neutron scattering and its relation to statistical mechanics.

The article is divided into four sections. The first Sect. 3.2 describes what neutron scattering measures (within the kinematic theory, stemming from the first Born approximation) and how it relates to spin correlation functions, generalised susceptibilities and magnetic moments. The second Sect. 3.3 contrasts the signatures,

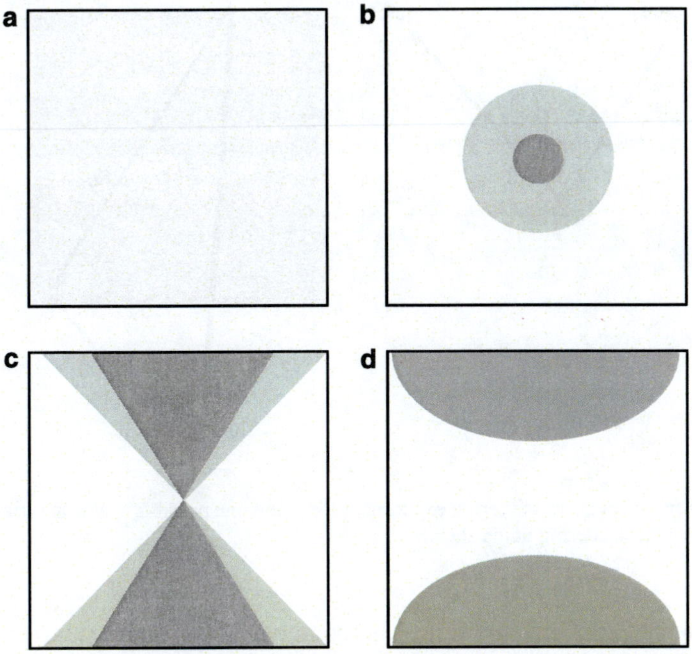


Fig. 3.2 Schematic neutron scattering maps classified by their radial and angular dependence near to a Brillouin zone centre: see text for the definition of (a)–(d)

in neutron scattering, of conventional and highly frustrated magnetism. The third Sect. 3.4 surveys some experimental results: these are chosen to illustrate general principles, rather than to exhaustively review the subject. The final Sect. 3.5 draws some conclusions. I do not discuss the generation of neutron beams or any aspects of neutron instrumentation and I make no attempt to describe the full array of neutron scattering techniques.

## 3.2 What Neutron Scattering Measures

### *3.2.1 Scattering Triangle*

The basic neutron scattering experiment is illustrated in Fig. 3.3. An incident neutron of known wavevector $\mathbf{k}_i$ is scattered by the sample into an outgoing (final) neutron of wavevector $\mathbf{k}_f$.[1] It is possible to measure the change in both the direction and magnitude of the neutron wavevector, or equivalently, the changes in the

[1] More precisely, one should conceive of a probability current of incoming and scattered neutrons.

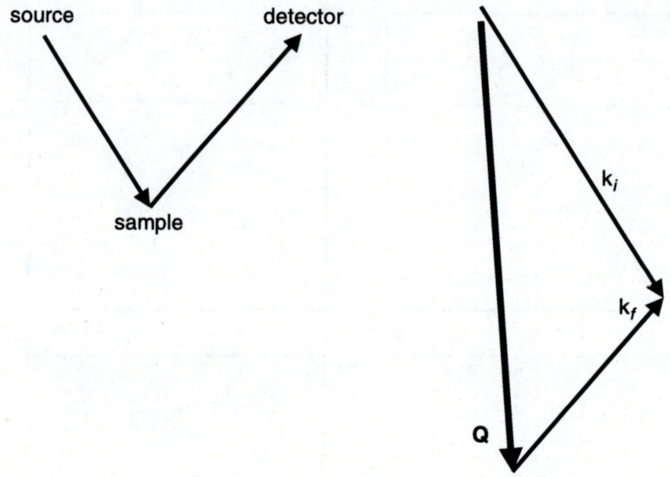


**Fig. 3.3** Schematic neutron scattering experiment (*left*) and one possible scattering triangle that is consistent with the scattering angle shown

neutron's momentum $\Delta\mathbf{p} = \hbar\Delta\mathbf{k}$. The conservation of momentum leads to the vector relationship:

$$\Delta\mathbf{p}/\hbar = \mathbf{k}_i - \mathbf{k}_f = \mathbf{Q}, \tag{3.1}$$

which is summarised in the scattering triangle, Fig. 3.3. The wavevector $\mathbf{Q}$ is called the *scattering vector*. If $|\mathbf{k}_i| \neq |\mathbf{k}_f|$ then the triangle is non-isoscelese and there is a change in neutron energy $\Delta E = \hbar^2 \Delta(k^2)/2m$, where $m$ is the neutron mass. This may be expressed as a frequency $\omega$:

$$\Delta E = \frac{\hbar^2}{2m}\left(k_i^2 - k_f^2\right) = \hbar\omega, \tag{3.2}$$

where $\Delta E = \hbar\omega$ is called the *energy transfer*. The scattering vector and energy transfer are the two key parameters in neutron scattering that can be selected by the experimenter.

The usual neutron experiment delivers a beam of unpolarised neutrons and detection does not discriminate between scattered spin states. However, neutrons may also be delivered in a polarised state with analysis of the scattered neutron spin states. In this way, changes in spin angular momentum may also be characterised.

### *3.2.2 Partial Differential Cross Section*

Like any triangle, the scattering triangle can be fully specified in terms of three variables, which we can chose in different ways. For example, we might specify the incoming wavelength, scattering angle and change in energy. In order to detect neutrons we need to collect the scattered neutrons over a range of angles in three dimensions, i.e., a range of solid angles $\Delta\Omega$, as well as a range of scattered neutron energies $\Delta E$. We can conceive of an ideal high resolution experiment in which these ranges are small and can be taken as differential quantities, $\mathrm{d}\Omega$ and $\mathrm{d}E$. The number of neutrons scattered per second per unit incident flux, into a range of solid angle $\mathrm{d}\Omega$ and with range of energies between $E$ and $E + \mathrm{d}E$, is called the partial differential scattering cross section, $\mathrm{d}^2\sigma/\mathrm{d}\Omega\mathrm{d}E$. We can expect this quantity to be the product of several terms. First, a characteristic length squared to give it the right dimensions – call it $r_0^2$; second, a term $(k_f/k_i)$ because it depends on scattered/incident number of neutrons per unit time, quantities which are proportional to the velocity of the neutrons and hence to $k$; and third, some function (say $\phi$) of $\mathbf{Q}$ and $\omega$ as well as the possible spin states of the neutrons:

$$\frac{\mathrm{d}^2\sigma}{\mathrm{d}\Omega\mathrm{d}E} = r_0^2\frac{k_f}{k_i}\phi\left(\mathbf{Q}, \omega, \mathrm{spin}\right). \tag{3.3}$$

### 3.2.3 *Relation to Sample Properties*

Neutrons are scattered by atomic nuclei ('nuclear scattering') and magnetic moments or spins ('magnetic scattering'). The key feature of neutron scattering is that because momentum and energy are conserved quantities, then the changes in neutron linear momentum ($\hbar\mathbf{Q}$), neutron angular momentum ($\mathbf{s}$) and neutron energy ($\hbar\omega$) are exactly mirrored by momentum and energy changes in the sample. Two examples from nuclear scattering illustrate this. The first is phonon scattering where the neutron excites or de-excites a phonon of wavevector $\mathbf{q} = \mathbf{Q}$ and frequency $\omega_0 = \omega$. Thus, scattering is only observed at those values of $\{\mathbf{Q}, \omega\}$ that coincide with the phonon 'dispersion' curve $\omega_0(\mathbf{q})$. The second example is Bragg reflection. Here, the scattering is *elastic* ($\omega = 0$ so $|\mathbf{k}_i| = |\mathbf{k}_f|$), and *coherent*,[2] so produces diffraction: because the angle of incidence is equal to the angle of 'reflection', the momentum transfer $\hbar\mathbf{Q}$ is normal to the set of crystal planes $(hkl)$ doing the diffracting. By applying Bragg's law it is easily shown that diffraction may only occur when $\mathbf{Q} = \mathbf{G}$ where $\mathbf{G}$ is a reciprocal lattice vector with $|\mathbf{G}| = 2\pi/d_{hkl}$ (here $d$ is the interplanar spacing).

One can see from these examples that any physical property of the sample that can be represented in terms of wavevectors and frequencies may be directly observable to neutron scattering. Thus, neutron scattering directly looks at reciprocal space (the set of all wavevectors) and frequency space (the set of all frequencies) and most clearly reveals sample properties like phonons and crystal planes that are naturally represented in these spaces.

[2] Bragg scattering is coherent in the absolute sense, which means that the scattered beam interferes with the incoming beam. Diffuse scattering, treated below, may only be coherent in the relative sense which means that only the scattered waves produce interference effects [3].

### 3.2.4 Scattering from Atomic Magnetic Moments

Here, we are interested in magnetic neutron scattering from the electrons in a sample and I will assume that the nuclear scattering and other unwanted scattering (from say nuclear spins or the sample environment) can always be removed to reveal only this component (in reality this may be tricky in a powder experiment but it is usually straightforward in a single crystal experiment). Also, when I speak of 'spins' it is to be understood that these could be effective spin operators that describe the combined spin and orbital angular momentum.[3]

The neutrons are scattered by the spatially varying magnetic field $\mathbf{B}(\mathbf{r})$ that arises from the spins in the sample. Neutron scattering sees the Fourier transformed field $\mathbf{B}(\mathbf{q})$ which is proportional (within the constraint discussed below) to the Fourier transformed spin field $\mathbf{S}(\mathbf{q})$. Interference effects are produced by the relative coherence in the scattering from different atoms and bearing in mind that the spin field can vary in time as well as space, second order dynamical perturbation theory shows that the scattering is essentially proportional to the time ($t$) and spatial Fourier transform of the two-spin correlation function of the sample:[4]

$$S^{\alpha\beta}(\mathbf{q},\omega) = \frac{1}{2\pi}\sum_{\mathbf{R}}\int_{-\infty}^{+\infty} e^{i\mathbf{q}\cdot\mathbf{R}-\omega t}\langle S_{\mathbf{0}}^{\alpha}(0)S_{\mathbf{R}}^{\beta}(t)\rangle dt, \tag{3.4}$$

where $\mathbf{R}$ defines the spin positions, $\langle\ldots\rangle$ denotes a thermal average and $\alpha,\beta = x,y,z$ are cartesian components. $S^{\alpha\beta}(\mathbf{q},\omega)$ (called, in the context of neutron scattering the 'dynamic scattering function') is the sample property that the neutron scattering experiment seeks to estimate. However, as explained below, only some of the components of this tensor can be observed at any particular setting of the neutron instrument.

### 3.2.5 Orientation Factor and Form Factor

The observed scattering is modified by several extra factors, of which two are particularly important. The first of these is a direct consequence of Maxwell's Law $\nabla\cdot\mathbf{B} = 0$ that implies that $\mathbf{B}(\mathbf{q})$ has no component parallel to $\mathbf{q}$.[5] This means that neutrons only 'see' spin components perpendicular to the scattering vector $\mathbf{Q} = \mathbf{q}$. Consequently, the scattering function $S^{\alpha\beta}(\mathbf{Q},\omega)$ must be multiplied by

---

[3] Atomic magnetic moments may always be treated as proportional to effective angular momentum operators: the so-called *magnetomechnical equivalence*.

[4] Small $\mathbf{q}$ will be used to represent sample properties: it is always equal to the vector $\mathbf{Q}$ of the scattering experiment, but there is a logical distinction.

[5] The Fourier transform of $\nabla\cdot\mathbf{X} = 0$ is $\mathbf{q}\cdot\mathbf{X}(\mathbf{q}) = 0$, a result that we shall use again below (here $\mathbf{X}$ is a vector field).

an orientation factor that projects out its parallel components. This factor may be written $\delta_{\alpha\beta} - \hat{Q}_\alpha \hat{Q}_\beta$ where $\hat{Q}_{\alpha,\beta}$ are components of the unit vector $\hat{\mathbf{Q}} = \mathbf{Q}/|\mathbf{Q}|$.

The second modification arises from the fact that the unpaired spins actually inhabit spatially extended valence orbitals, so neutrons are really scattered by spin density rather than spins localised on lattice points. Fortunately, there is a rigorous theorem – the Wigner–Eckart theorem – that shows that to within a factor the spin density on a given atom or ion may be replaced with an effective spin operator localised on a lattice point [4]. $S^{\alpha\beta}(\mathbf{Q}, \omega)$ is derived from these localised operators so must be corrected to take into account the spatially extended spin density. The correction factor is $f(\mathbf{Q})^2$ where $f$ is the magnetic form factor, which is essentially the Fourier transform of the normalised spin density on a single atom or ion (and hence depends on atom or ion *type*). Because ions are approximately spherical, it is typically a simple decreasing function of $|\mathbf{Q}|$ which (at least for our purposes) may be treated as a experimental correction to be looked up in tables [5].

### 3.2.6 General Expression for the Neutron Scattering

Collecting together the factors discussed above and recalling (3.3), the scattered intensity may be written

$$\frac{d^2\sigma}{d\Omega dE} = r_0^2 \frac{k_f}{k_i} f(\mathbf{Q})^2 \sum_{\alpha\beta} \left(\delta_{\alpha\beta} - \hat{q}_\alpha \hat{q}_\beta\right) S^{\alpha\beta}(\mathbf{Q}, \omega) \tag{3.5}$$

where the summation over $\alpha, \beta$ is specific to a an unpolarised neutron experiment (see below). The factor $r_0$, called the magnetic scattering length, depends on a number of fundamental constants that define the interaction of the neutron with the atomic magnetic moment: specifically, $r_0 = \gamma_n e^2/m_e c^2 = -0.5391 \times 10^{-14}$ m, where $\gamma_n$ is the neutron gyromagnetic ratio [6]. Equation (3.5) should also be corrected by a Debye–Waller type thermal factor that describes the blurring of the lattice due to thermal vibrations. The equation as written may be applied directly to the scattering from a single crystal sample: with a polycrystalline sample one further needs to average over crystallite orientations, which obscures much useful information.

If necessary, the components of $S^{\alpha\beta}(\mathbf{Q}, \omega)$ can be separated by the technique of polarisation analysis [7], thus removing the summation in (3.5). As the neutron has spin $s = 1/2$, the magnetic scattering might either flip this spin or not, with spin angular momentum conserved by a corresponding change in the spin state of the sample. Polarisation analysis measures spin flip and non spin flip cross sections as a function of incident neutron polarisation, from which the components of $S^{\alpha\beta}$ can be reconstructed [8].

### 3.2.7 *Real Experiments*

In a real experiment, in addition to applying (3.5), one needs to account for instrumental resolution [9, 10], sample imperfection, absorption by the sample and counting statistics. It is usual to assume a parameterised form of $S^{\alpha\beta}(\mathbf{Q},\omega)$, then to convolve this with a parameterised resolution function (that includes the 'mosaic spread' of the sample), and finally to least-squares fit to the (possibly absorption-corrected) neutron scattering data. The experimental statistics in neutron scattering are strictly Poissonian, i.e., an error of $\sqrt{n}$ on count $n$, which renders this sort of approach to data analysis particularly robust. It should also be noted that (3.5) is only valid for a sufficiently small or 'ideally imperfect' sample, which naturally avoids the problem of multiple scattering and consequent extinction effects. The typical mosaic structure of real crystals is an advantage here as the coherent scattering volume is limited to a few microns. Generally speaking, the maximum crystal size should be a few $\mathrm{mm}^3$ for Bragg scattering and a few $\mathrm{cm}^3$ for diffuse scattering. Within this constraint the crystal volume should be maximised, to optimize counting statistics.

### 3.2.8 *Powder Averaging*

For most types of neutron experiment single crystal samples are preferable to powders. When a powder sample is used, there is averaging over the orientations of **Q** which obscures a great deal of information. However, there are certain situations in which powder samples may be preferred even if single crystals are available. For example, in magnetic crystallography, powder samples are often used, in order to cover more reciprocal space and to avoid extinction effects, which can affect Bragg intensities. Nevertheless, probably the most common reason for measuring powder samples is simply to maximise sample volume in cases where sufficiently large single crystals are not available, or neutron beam time is limited.

### 3.2.9 *Static Approximation*

In the case that energy transfers to and from the sample are negligible, the dynamic scattering function approaches a delta function in $\omega$ at the origin. Integration over frequencies then gives the so-called 'static approximation':

$$S^{\alpha\beta}(\mathbf{Q}) = \sum_{\mathbf{R}} \langle S_0^{\alpha}(0) S_{\mathbf{R}}^{\beta}(0)\rangle \mathrm{e}^{\mathrm{i}\mathbf{Q}\cdot\mathbf{R}} = \int_{-\infty}^{+\infty} S^{\alpha\beta}(\mathbf{Q},\omega)\mathrm{d}\omega. \quad (3.6)$$

The quantity on the left hand side of (3.6) is an instantaneous correlation function (i.e., one realised at time $t = 0$), but in a real experiment, such instantaneous correlations are sampled at a sequence of times that are widely separated on the time scale of spin fluctuations. This process is akin to taking an ergodic average, which should be well represented by an ensemble average in equilibrium statistical mechanics. The assumed fact that energy transfers are negligible means that the equilibrium state of the sample is not perturbed by the neutrons, so $S(\mathbf{q})$ measures a genuine equilibrium correlation function.

In this case, the differential cross section, which no longer depends on energy, becomes:

$$\frac{\mathrm{d}\sigma}{\mathrm{d}\Omega} = r_0^2 \frac{k_f}{k_i} f(\mathbf{Q})^2 \sum_{\alpha\beta} \left(\delta_{\alpha\beta} - \hat{Q}_\alpha \hat{Q}_\beta\right) S^{\alpha\beta}(\mathbf{Q}) \tag{3.7}$$

One may thus regard the total scattering at $\mathbf{Q}$ as an estimate of $S(\mathbf{Q})$.

### 3.2.10 Wavevector Dependent Magnetic Moment and Susceptibility

A very useful alternative formulation of (3.7) exploits the general relationship between correlation function, susceptibility $\chi$ and average magnetic moment $M$. For a sample with $N$ spins located on a Bravais lattice of position vectors $\mathbf{R}$, we define the (intensive) wavevector dependent magnetic moment,

$$\mathbf{M}(\mathbf{q}) = N^{-1} \sum_{\mathbf{R}} S_{\mathbf{R}} \mathrm{e}^{\mathrm{i}\mathbf{q}\cdot\mathbf{R}} \tag{3.8}$$

from which follows

$$S^{\alpha\beta}(\mathbf{q}) = N \left\langle M^\alpha(\mathbf{q}) M^\beta(-\mathbf{q}) \right\rangle, \tag{3.9}$$

which may be verified from (3.8), by invoking the translational invariance of the sample (so $\sum_{\mathbf{R}_0} \sum_{\mathbf{R}} \to N \sum_{\mathbf{R}-\mathbf{R}_0}$). Defining the (intensive) wavevector dependent susceptibility:

$$\chi^{\alpha\beta}(\mathbf{q}) = \frac{\mathrm{d}M^\alpha}{\mathrm{d}H^\beta} = \frac{N}{kT}\left[\langle M^\alpha(\mathbf{q}) M^\beta(-\mathbf{q})\rangle - \langle M^\alpha(\mathbf{q})\rangle \langle M^\beta(-\mathbf{q})\rangle\right], \tag{3.10}$$

we see that

$$S^{\alpha\beta}(\mathbf{q}) = kT\chi^{\alpha\beta}(\mathbf{q}) + N\langle M^\alpha(\mathbf{q})\rangle \langle M^\beta(-\mathbf{q})\rangle. \tag{3.11}$$

Neutron scattering thus measures both wavevector dependent magnetic moments and susceptibilities.[6]

[6] This susceptibilty is really a reduced susceptibility that should be multiplied the form factor squared to convert it to the 'real' susceptibility.

### 3.2.11 Fully Ordered Magnet

In a fully ordered magnet, the susceptibility components are all zero[7] and all scattering is due to the wavevector dependent magnetic moment. Assuming that the crystal structure has the symmetry of a Bravais lattice (as is nearly always the case for geometrically frustrated magnets), then the spin positions $\mathbf{R}$ may be represented as $\mathbf{R} = \mathbf{T}+\mathbf{r}$ where $\mathbf{T}$ is a translation vector that defines a particular magnetic unit cell and $\mathbf{r}$ is the position of a magnetic atom or spin within a unit cell. If the wavevectors of the *magnetic* reciprocal lattice are $\mathbf{G}$ then $\langle \mathbf{M}(\mathbf{q})\rangle$ can only be finite for $\mathbf{q} = \mathbf{G}$. At these wavevectors $\langle \mathbf{M}(\mathbf{q})\rangle$ becomes $N^{-1}\sum_{\mathbf{T}}\sum_{\mathbf{r}}\langle \mathbf{S}(\mathbf{T}+\mathbf{r})\rangle \mathrm{e}^{\mathrm{i}\mathbf{G}\cdot(\mathbf{T}+\mathbf{r})} = \sum_{\mathbf{r}}\langle \mathbf{S}(\mathbf{r})\rangle \mathrm{e}^{\mathrm{i}\mathbf{G}\cdot\mathbf{r}}$, since $\mathrm{e}^{\mathrm{i}\mathbf{G}\cdot\mathbf{T}} = 1$ [3]. Plugging this result into (3.11), we find:

$$\frac{\mathrm{d}\sigma}{\mathrm{d}\Omega} = r_0^2\frac{k_f}{k_i} f(\mathbf{Q})^2 \left|\sum_{\mathbf{r}}\langle \mathbf{S}_\perp(\mathbf{r})\rangle \mathrm{e}^{\mathrm{i}\mathbf{Q}\cdot\mathbf{r}}\right|^2 \delta(\mathbf{Q}-\mathbf{G}), \tag{3.12}$$

where the orientation factor (which projects out parallel spin components) has been absorbed into the '$\perp$' symbol, $\mathbf{S}_\perp$ being the component, perpendicular to $\mathbf{Q}$, of a particular spin in the unit cell of the magnetic structure. The scattering is seen to occur as delta functions, or Bragg peaks, at the magnetic reciprocal lattice vectors, with intensity proportional to the square of the ordered component. In the above equation, the summation is over the unit cell rather then the whole sample and the term $|\ldots|$ is referred to as the *magnetic structure factor* [11].

Magnetic Bragg peaks do not necessarily occur at all the magnetic reciprocal lattice points. There are two very general reasons for this. First, the orientation factor (the '$\perp$' in (3.12) might cause certain Bragg reflections to have zero intensity. Second, the phase factors $\mathrm{e}^{\mathrm{i}\mathbf{Q}\cdot\mathbf{r}}$ might cause the scattering from different spins in the unit cell to interfere destructively, leading to systematic absences. This always occurs if the magnetic and nuclear unit cells are non-coincident, causing the magnetic Bragg peaks to occur at different positions in reciprocal space to the nuclear Bragg peaks, as illustrated in Fig. 3.4. In the example, the magnetic unit cell is doubled with respect to the nuclear unit cell along one direction. The nuclear reciprocal lattice $\{\mathbf{G}_{\text{nuclear}}\}$ is a subset of the magnetic reciprocal lattice $\{\mathbf{G}_{\text{magnetic}}\}$, but the set $\{\mathbf{G}_{\text{nuclear}}\}$ are systematic absences for magnetic diffraction, leading to a neat separation of nuclear and magnetic Bragg peaks.

If we define $\{\mathbf{G}_{\text{magnetic Bragg}}\}$ as the set of magnetic Bragg peak positions then we can write

$$\mathbf{G}_{\text{magnetic Bragg}} = \mathbf{G}_{\text{nuclear}} + \sum_i \mathbf{k}_i \tag{3.13}$$

where the $\mathbf{k}_i$s are a finite set of propagation vectors of the magnetic structure [11]. In the example (Fig. 3.4), the set $\{\mathbf{k}_i\}$ comprises only the single vector $\mathbf{k} = (0, 0, \frac{1}{2})$, but more generally, so-called multi-$k$ structures are also possible [11]. As described

[7] At least in the hypothetical case where the magnet has no possible fluctuations.

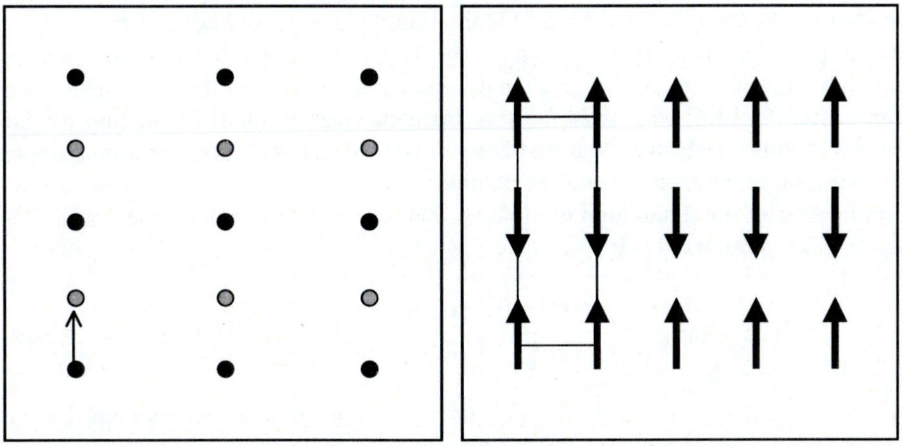

**Fig. 3.4** The magnetic structure on the right (box is the magnetic unit cell which is twice the nuclear unit cell) produces to a diffraction pattern in reciprocal space like that shown on the left. Magnetic Bragg peaks (*grey circles*) appear intercalated between nuclear Bragg peaks (*black circles*) and are related to them by a propagation vector **k** (*arrow*)

later (Sect. 3.4), examples of a single-$k$ and a four-$k$ structure are provided by the rare earth pyrochlores $Gd_2Sn_2O_7$ [12] and $Gd_2Ti_2O_7$ [13] respectively.

In practice, magnetic structures are solved by fitting experimental data, via (3.12) to candidate models: group theoretic techniques (representational analysis) may be used to identify the possibilities.

### 3.2.12 *Magnet with Full or Partial Disorder*

Magnetic disorder is characterised by finite susceptibilities. In general, the wavevector dependent susceptibility is a smoothly varying function of wavevector so the scattering arising from the magnetic disorder is diffuse, occurring throughout reciprocal space. For this reason, it is easily distinguished from the diffraction that arises from any ordered component, that is localised into sharp Bragg peaks. The diffuse scattering may nevertheless exhibit considerable detail, which can be analysed to reveal the structural properties of the magnetic disorder, as shown below.

### 3.2.13 *Validity of the Static Approximation*

The preceding sections show how equilibrium thermodynamic properties (which are necessarily time independent) may be directly obtained by neutron scattering under the assumption that energy transfers to and from the sample are negligible.

In the case that they are not negligible the sample is generally not in a true state of equilibrium, but rather in a uniform steady state, close to equilibrium, in which the neutron current continuously exchanges energy with the sample. Integration over frequencies in (3.6) still gives the true instantaneous correlation function for the sample in this steady state, but that is no longer the desired correlation function of the equilibrium state that would be obtained in the absence of the neutrons current. Application of the fluctuation dissipation theorem and detailed balance establishes the following relation [14]:

$$S^{\alpha\beta}(\mathbf{Q},\omega)_{\text{diffuse}} = \frac{kT\chi^{\alpha\beta}(\mathbf{Q})}{g^2\mu_B^2}\frac{\hbar\omega/kT}{1-\exp(-\hbar\omega/kT)}F^{\alpha\beta}(\mathbf{Q},\omega), \tag{3.14}$$

where $F$, called the spectral shape function, is normalised to unit area when integrated over $\omega$. If $\alpha = \beta$ or if $\mathbf{q}$ is equivalent to $-\mathbf{q}$ it is an even function of $\omega$.

Thus, to get a true equilibrium correlation function or susceptibility the integral in (3.6) should be corrected by a frequency and temperature dependent factor:

$$kT\chi^{\alpha\beta}(\mathbf{Q}) = \int_{-\infty}^{+\infty}\frac{1-\exp(-\hbar\omega/kT)}{\hbar\omega/kT}S^{\alpha\beta}(\mathbf{Q},\omega)\mathrm{d}\omega. \tag{3.15}$$

The actual integration can either be performed explicitly on the estimated $S(\mathbf{Q},\omega)$ or automatically, by using a large enough incident wavevector so that the spread of scattered wavevectors is within the instrumental resolution.[8] The latter method is usually preferable.

The diffuse scattering at zero energy transfer measures $\sim T\chi(\mathbf{Q})F(\mathbf{Q},0)$ which is not equal to the equilibrium $T\chi(\mathbf{Q})$. While it may occasionally be expedient to consider this measured quantity (or more generally $\sim T\chi(\mathbf{Q})F(\mathbf{Q},\omega_0)$ where $\omega_0$ is some fixed frequency), it should be born in mind that it is only an approximation to the wavevector-dependent susceptibility. In general, the validity of the static approximation should always be carefully considered when discussing experimental data.

### 3.2.14 *Generalised Susceptibility*

In order to compare the results of neutron scattering with those of other experimental techniques (particularly *ac*-susceptibility), it is useful to adopt the 'common language' of generalised susceptibility. The dynamic scattering function $S^{\alpha\beta}(\mathbf{Q},\omega)_{\text{diffuse}}$ may be alternatively expressed in terms of an complex frequency-dependent susceptibility $\chi(\mathbf{Q},\omega) \equiv \mathrm{Re}[\chi(\mathbf{Q},\omega)] + i\,\mathrm{Im}[\chi(\mathbf{Q},\omega)]$. Here, the real

[8] Note that $\mathbf{Q}$ scales linearly with $k$, while energy scales quadratically, which means that the spectrum effectively narrows with increasing $\mathbf{k}_i$.

and imaginary parts mediate, respectively, the in phase and out of phase response of the system to a time dependent magnetic field. Bragg scattering, as it reflects a broken ergodicty in the system, does not depend on time or frequency. The diffuse scattering, in contrast, usually will depend on time, as it reflects the disordered part of the spin system. The fluctuation-dissipation theorem connects the dissipation of neutron energy to the spectrum of spontaneous fluctuations of the sample bathed in the neutron beam. The following relation can be established [14]:

$$S^{\alpha\alpha}(\mathbf{Q}, \pm|\omega|)_{\text{diffuse}} \propto \left(n_{|\omega|} + \frac{1}{2} \pm \frac{1}{2}\right) \text{Im}[\chi(\mathbf{Q}, |\omega|)], \tag{3.16}$$

where $n_{|\omega|} = (\mathrm{e}^{\hbar|\omega|/kT} - 1)^{-1}$ may be interpreted as a harmonic oscillator occupation number. If neutrons accept energy from the sample (neutron energy gain), what is measured is essentially the thermal spectrum, while if neutrons give energy to the sample (neutron energy loss), the scattering measures states excited by the neutrons, which need not be thermally populated. The two cross sections differ by a factor of $\mathrm{e}^{-\hbar|\omega|/kT}$, an expression of the condition of 'detailed balance'.

### *3.2.15 Neutron Spectroscopy*

In the case that the sample contains discrete excitations, these correspond to poles in the dynamical susceptibility and it is appropriate to adopt a more 'spectroscopic' description, involving matrix elements and selection rules. Perhaps the most basic excitations in ionic magnetic solids are so-called "magnetic excitons": dispersed single ion excitations with (possibly gapped) dispersion $\omega^i(\mathbf{q})$ [15]. Here, $i = 0, 1, 2\ldots$ represent the ladder of corresponding single ion excitations. The lowest energy excitations ($i = 0$) are the magnons while higher excitons originate in the single ion crystal field states. If exchange is small compared to the crystal field splitting, as is typical in rare earth salts, then the dispersion of the magnetic excitons may be neglected and neutron scattering can be used to estimate the single ion crystal field spectrum. The unpolarised cross section for a particular transition $i \to j$ is

$$\frac{\mathrm{d}^2\sigma}{\mathrm{d}\Omega\mathrm{d}E} = r_0^2 \frac{k_f}{k_i} F(|\mathbf{Q}|)^2 n_i \left|\langle j|\hat{\mu}_\perp|i\rangle\right|^2 \delta(E_i - E_j + E) \tag{3.17}$$

(neglecting Debye–Waller factor) where $n_i$ is the number of ions in the $i$th state, of energy $E_i$, and $\hat{\mu}_\perp$ is the component of the magnetic moment operator perpendicular to the scattering vector. It can be seen from this expression that the intensities of the transitions depend in a very sensitive way on the single ion wavefunctions, so neutron scattering is a direct and powerful method of estimating these functions and deriving the crystal field parameters. However, it should be noted that powder averaging gives potential ambiguity in the assignment of crystal field transitions, so results obtained from powder samples should always be checked for consistency with other measurements, for example, bulk susceptibility or heat capacity.

## 3.3 Typical Neutron Scattering Patterns

In the previous section, I have tried to illustrate some of the different interpretations of neutron scattering. Thus, in the static approximation neutron scattering may be regarded as a thermodynamic probe that reveals properties the sample's equilibrium state, independent of the neutron flux. Beyond the static approximation it can either be regarded as probing the sample's relaxation towards that equilibrium state, following the weak disturbance of the neutron flux, or as a spectroscopy that reveals the sample's energy level structure, within the usual caveat of selection rules.

In this section, I aim to illustrate how these various probes of the sample may be used to classify magnetic states, and in particular, to discriminate conventional magnets from highly frustrated magnets. To this end I consider some typical neutron scattering patterns.

### *3.3.1 Scattering Plane*

In a typical neutron scattering experiment, the scattering triangle is in a horizontal plane called the *scattering plane*. As illustrated in Fig. 3.5, we may adopt a set of cartesian axes in which $z$ is normal to the scattering plane (and so perpendicular to $\mathbf{q}$) and $x$ and $y$ are within the scattering plane. The unpolarised experiment (3.7 and 3.11) measures

$$\chi^{zz}(\mathbf{q}) + (1-\hat{q}_x^2)\chi^{xx}(\mathbf{q}) + (1-\hat{q}_y^2)\chi^{yy}(\mathbf{q}) - 2\hat{q}_x\hat{q}_y\chi^{xy}(\mathbf{q}). \quad (3.18)$$

where $\hat{q}_x^2 + \hat{q}_y^2 = 1$. Consider, for example, a conventional cubic paramagnet with $z = [001]$, in which case $\chi^{xx}(\mathbf{q}) = \chi^{yy}(\mathbf{q}) = \chi^{zz}(\mathbf{q}) = \chi(\mathbf{q}); \chi^{xy} = 0$. The above expression simply becomes $2\chi(\mathbf{q})$ so the wavevector dependent susceptibility is directly measured.

A large number of neutron scattering experiments on cubic frustrated magnets (spinels or pyrochlores) have typically investigated the $[1\bar{1}0]$ scattering plane. This plane, with two-fold symmetry, contains the major symmetry directions [001], [110] and [111]. Its reciprocal lattice, for face centred cubic symmetry, is illustrated in Fig. 3.6 [16].

### *3.3.2 Free Energy*

In order to evaluate the wavevector dependent susceptibility, we consider the Helmholtz free energy, which is the thermodynamic potential to be minimised when the magnetic moment $M$ is constrained to a certain value: although this can't be achieved in practice we can use it conceptually to discuss the susceptibility. At a fixed temperature, the magnetic field $H$ is the first derivative of the free energy with

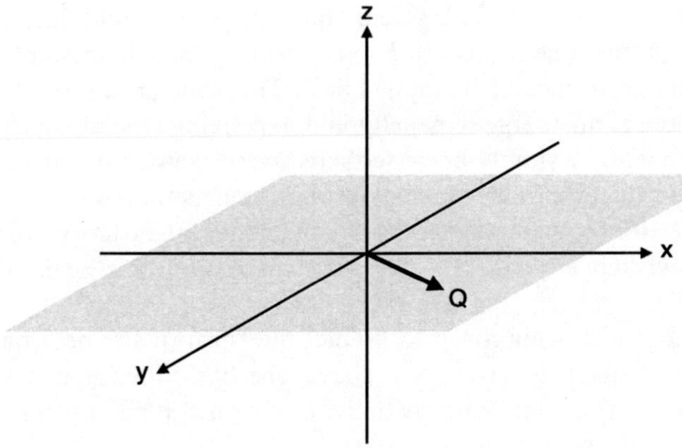


**Fig. 3.5** Illustration of a horizontal scattering plane (*grey*) and axis system in a conventional single crystal scattering experiment

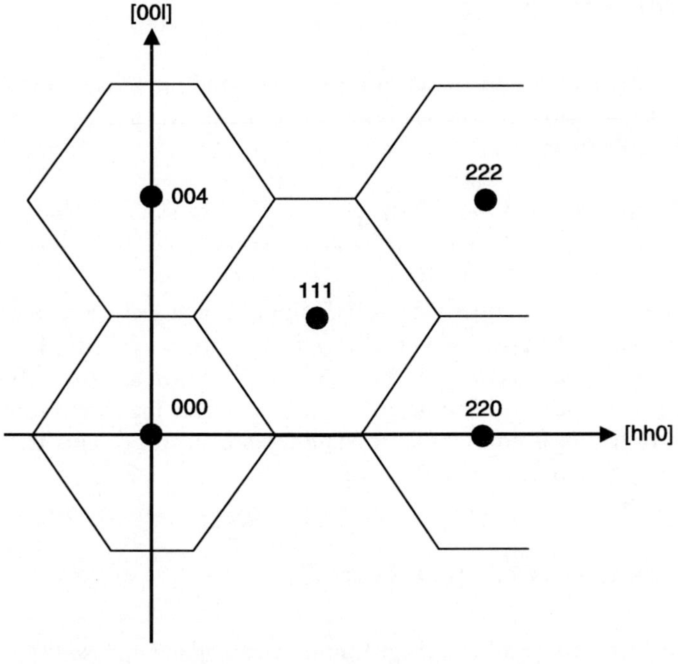


**Fig. 3.6** Reciprocal lattice for the *hhl* plane in face centred cubic symmetry

respect to magnetic moment, $H = (\partial F/\partial M)_T$, and the inverse susceptibility is the second derivative: $\chi^{-1} = (\partial^2 F/\partial M^2)_T$ [17]. As an example, consider the case of an ideal paramagnet. If we write $F(M) = F_0 + \frac{1}{2}\chi_0^{-1}M^2 - MH$ (which assumes that the magnetic moment and field vectors are parallel) then a first differentiation shows that the free energy is minimised at $M = \chi_0 H$, which is indeed a property of a paramagnet, and a second differentiation confirms that the susceptibility

$\chi_0$ is consistently defined (here $F_0$ is the component of $F$ that does not depend on $M$). The quadratic dependence of $F$ on $M$ shows that the paramagnet resists magnetization in the absence of an applied field. This resistance is purely statistical or entropic as there is no energetic penalty in magnetizing a paramagnet: it doesn't magnetise spontaneously simply because there are so many more states with zero magnetic moment than there are states with finite magnetic moment. Because the internal energy $U$ for the paramagnet is zero, the above free energy expression can alternatively be written $F = -TS - MH$. The susceptibility is given by the Curie Law: $\chi_0 = C/T$.

We can generalise this approach to predict the form of the neutron scattering in various types of magnets as we now show. The basic idea is to write down a similar free energy expression for a spatially varying magnetic moment field $\mathbf{M(r)}$ transformed into reciprocal space.

### 3.3.3 Ideal Paramagnet

We first consider the case of the ideal paramagnet and note that, as well as being resistant to magnetization, it is equally resistant to ordering at any wavevector $\mathbf{q}$ so we should be able to write:

$$F - F_0 = \frac{1}{2}\chi_0^{-1} \sum_{\mathbf{q}} \mathbf{M(q)} \cdot \mathbf{M(-q)} - M_x(\mathbf{q}) H_x(-\mathbf{q}) \tag{3.19}$$

where the field direction defines the $x$-axis. Minimising this expression with respect to each component of $\mathbf{M(q)}$ individually gives $\chi^{\alpha\alpha}(\mathbf{q}) = H^{\alpha}(\mathbf{q})/M^{\alpha}(\mathbf{q}) = \chi_0$, with terms in $\alpha\beta$ equal to zero. Then from (3.7, 3.11) we see that the diffuse neutron scattering does not vary with temperature and, apart from the slow variation due to magnetic form factor, is constant throughout reciprocal space. This is a scattering pattern of the type anticipated in Fig. 3.1a.

### 3.3.4 Conventional Magnet Above $T_C$

We next consider a conventional magnet in its paramagnetic phase [14]. There is a tendency to order at some wavevector $\mathbf{k}$ (see Sect. 3.2.11), but the statistical resistance to ordering is sufficient to overcome that tendency. For simplicity we consider the case of a ferromagnet with $\mathbf{k} = 0$. We have to add a term to our free energy expression that represents the internal energy arising from the magnetic exchange coupling. We imagine a mean magnetic moment field surrounding a given lattice site $\mathbf{r}$ that is just slightly smaller than the magnetic moment field at the lattice site itself. The mean field may be represented as a Taylor expansion: $\mathbf{M}_{MF} \propto \mathbf{M(r)} + (a^2/2)\nabla^2\mathbf{M(r)}$ where $a$ is the lattice constant. Summing over $z$ nearest neighbours the total exchange energy is $-zJ \int \mathbf{M(r)} \cdot \big(\mathbf{M(r)} + (a^2/2)\nabla^2\mathbf{M(r)}\big)\, d\mathbf{r}$, where $J$ is the exchange constant. By substituting (3.8) into this expression, and integrating, we find that the internal energy of the magnet, can be represented in reciprocal space by the expression $U = zJ\sum_{\mathbf{q}} \mathbf{M}(\mathbf{q}) \cdot \mathbf{M}(-\mathbf{q})((a^2/2)q^2 - 1)$. Including the entropic factor and field energy (3.19), the free energy becomes:

$$F - F_0 = \frac{1}{2}\sum_{\mathbf{q}} \mathbf{M}(\mathbf{q}) \cdot \mathbf{M}(-\mathbf{q})(\chi_0^{-1} - zJ + zJa^2q^2) - \mathbf{H}(\mathbf{q}) \cdot \mathbf{M}(-\mathbf{q}). \quad (3.20)$$

Differentiating, and forming $\chi^{\alpha\beta} = \mathrm{d}M^{\alpha}/\mathrm{d}H^{\beta}$ we find the following expression for the wavevector dependent susceptibility:

$$\chi^{\alpha\alpha}(\mathbf{q}) = \chi(q) = \frac{\chi(0)\kappa^2}{q^2 + \kappa^2} \quad (3.21)$$

with $\chi^{\alpha\beta} = 0$ for $\alpha \neq \beta$. In (3.21) the parameters $\chi(0)$ and $\kappa$ are defined by the two relationships $\chi(0) = C/(T - zJC)$, which we recognise as the Curie–Weiss expression, and $\chi(0) = (a\kappa)^{-2}$, a scaling law.[9] Here, we have assumed a cubic sample (in other cases both $\chi$ and $\kappa$ would depend on $\alpha = x, y, z$), and used $\chi_0 = C/T$. Equation (3.21) leads to the famous Ornstein–Zernike form of the scattering function $S(\mathbf{Q}) \sim T\chi(\mathbf{q})$. It may be Fourier transformed to give a correlation function $g(r) = r^{-1}\exp(-\kappa r)$, so it can be seen that $\xi = 1/\kappa$ measures the size of the correlated regions, i.e., $\xi$ is the correlation length. In reciprocal space, the scattering forms diffuse blobs of spherical symmetry centred on the ferromagnetic Brillouin zone centre with width $\sim \kappa$, the form anticipated in Fig. 3.2b (in a non cubic magnet these blobs will generally have ellipsoidal symmetry).

### 3.3.5 Conventional Magnet Below $T_C$

In this case, following (3.11) the neutron scattering is a superposition of (3.21) (Fig. 3.1b) with a sharp Bragg peak at the Brillouin zone centre. Just below $T_C$, the intensity of the Bragg peak grows as the square of the magnetization, i.e., as $\sim(T_C - T)^{2\beta}$, where $\beta$ is the order parameter critical exponent. The diffuse scattering is much weaker than the Bragg scattering apart from very close to the critical temperature, where it is approximately described by (3.21), with $\chi(0) \sim |T - T_C|^{-\gamma}$ and $\kappa \sim |T - T_C|^{\nu}$: here $\gamma$ an $\nu$ are the critical exponents for susceptibility and correlation length respectively.

---

[9] This is a mean field expression: in critical point theory the susceptibility acquires an 'anomalous' dimension $\chi(0) \sim \kappa^{-2+\eta}$ [18].

### 3.3.6 Cooperative Paramagnet

As a prototypical case, we consider spin ice, the highly frustrated pyrochlore ferromagnet [16, 19]. Its low temperature cooperative paramagnetic phase is governed by the ice rule constraint 'two spins in, two spins out' per tetrahedron of the pyrochlore structure. The magnetic moment field $M(\mathbf{r})$ is therefore solenoidal (i.e., a solenoid has as many magnetic flux lines flowing into it as out of it) and obeys the local 'divergence free' condition:

$$\nabla \cdot \mathbf{M}(\mathbf{r}) = 0, \tag{3.22}$$

from which follows

$$\mathbf{q} \cdot \mathbf{M}(\mathbf{q}) = 0 \tag{3.23}$$

and the system is thus completely described by Fourier components for which magnetic moment is perpendicular to its wavevector $\mathbf{q}$. As an approximation, which ignores the details of the local spin structure, we may consider spin ice to be a paramagnet in which the free energy (3.19) is minimised subject to this constraint, so as before, we seek to calculate $\chi(\mathbf{q})$ by differentiating the free energy. In three dimensions, the algebra is laborious, but in two dimensions it is very easy [20, 21]. The constraint (3.23) may be written

$$q_x M_x(\mathbf{q}) + q_y M_y(\mathbf{q}) = 0, \tag{3.24}$$

so the magnetization components are not independent. First, consider the response of $M_x$ to a field along $x$, so $\mathbf{H} = (H_x, 0, 0)$. The free energy is:

$$F = \sum_{\mathbf{q}} \frac{1}{2} \chi_0^{-1} \left( M_x^2(\mathbf{q}) + M_y^2(\mathbf{q}) \right) - M_x(\mathbf{q}) H_x(\mathbf{q}), \tag{3.25}$$

so by eliminating $M_y$, using (3.24), and by minimising with respect to $M_x$ as above, we find:

$$\chi^{xx} = \chi_0 \frac{q_y^2}{q_x^2 + q_y^2}. \tag{3.26}$$

We could have alternatively considered the response of $M_x$ to a field along $y$, with the result:

$$\chi^{xy} = -\chi_0 \frac{q_x q_y}{q_x^2 + q_y^2}. \tag{3.27}$$

The neutron scattering function $S^{\alpha\alpha}(\mathbf{q}) \sim \chi^{\alpha\alpha}(\mathbf{q})T$, is given by (3.26). This is an unusual functional form that contains a singularity as $q \to 0$: if the origin is approached along $y$ at $q_x = 0$ then $\chi(0) = \chi_0$ whereas if it is approached along $x$ at $q_y = 0$ then $\chi(0) = 0$. The scattering function is therefore singular at the origin: a so-called pinch point singularity or 'bow tie' of scattering, as anticipated in Fig. 3.2c. Recalling that we ignored the details of the local spin structure, this form of the scattering would not be expected to apply at large wavevector in the Brillouin zone, but it should be a good approximation near the zone centre.

The pinch point scattering ($\sim\chi T$) may be re-expressed in polar coordinates $q = \sqrt{q_x^2 + q_y^2}$, $\theta = \arcsin(q_x/q)$, showing that the diffuse scattering near the zone centre depends only on angle $\theta$, not on $q$ (see Fig. 3.2c). Alternatively one can write (3.26), (3.27) as

$$\chi^{\alpha\beta} = \chi_0 \left(\delta_{\alpha,\beta} - \hat{q}_\alpha \hat{q}_\beta\right), \tag{3.28}$$

where $\hat{q}_{\alpha,\beta}$ are unit vector components.

This analysis is easily extended to three dimensions. Taking into account the summation over cartesian components in (3.7) pinch points may not be visible to unpolarised neutrons in a four-fold scattering plane like [100], but should be visible in a two fold scattering plane like [1$\bar{1}$0]; in either case it is clear that they would be better studied by polarised neutron scattering which can isolate the components of $\chi^{\alpha\beta}$.

Pinch point scattering is not only a property of spin ice but also applies in principal to any highly frustrated magnet that obeys a local ice-rule type constraint [22, 23]. The scattering is seen to be fundamentally different to that of a simple paramagnet or a disordered ferro- or antiferromagnet. When (3.28) is Fourier transformed, the spin correlations in direct space are seen to take the form of a dipolar interaction, with a complicated angular dependence and a slow algebraic decay as the inverse cube of distance [20]. This contrasts the decay of correlations in a conventional magnet above its ordering temperature, where the decay of correlations is like a screened Coulomb interaction (3.21). This relationship of correlation functions to charge interactions was first pointed out by Youngblood, Axe and McCoy [21] for the analogous case of ice-rule paraelectrics. The pinch point scattering functions of cubic magnetic systems have been analysed in detail by Isakov et al. [22] and by Henley [23]. Their relevance to frustrated magnets was first realised by Zinkin et al., on the basis of numerical simulations [24], and the first clear experimental observation was recently reported by Fennell et al. [25].

Comparing Figs. 3.2a and 3.1b we finally see how neutron scattering reveals a symmetry difference, in the spin correlation function, between a cooperative paramagnet and a conventional paramagnet. The cooperative paramagnetic phase may be called a 'Coulomb phase', by analogy with field theories, because its basic excitations can be shown to be effective charges or magnetic monopoles [26].

### *3.3.7 Absent Pinch Points*

In any real frustrated magnet, it is certain that the pinch point singularity will be removed by one mechanism or another, even though this effect may be very small. A first possibility is that the zero divergence constraint (3.22) is relaxed at a long length scale $\xi_{\text{ice}}$ by adding a term in $(\nabla \cdot \mathbf{M}(\mathbf{r}))^2$ to the free energy [20]. This will cause a broadening of the pinch point to give it a width $\xi_{\text{ice}}^{-1}$ (see Fig. 3.7, *left*), and the susceptibility takes the form:

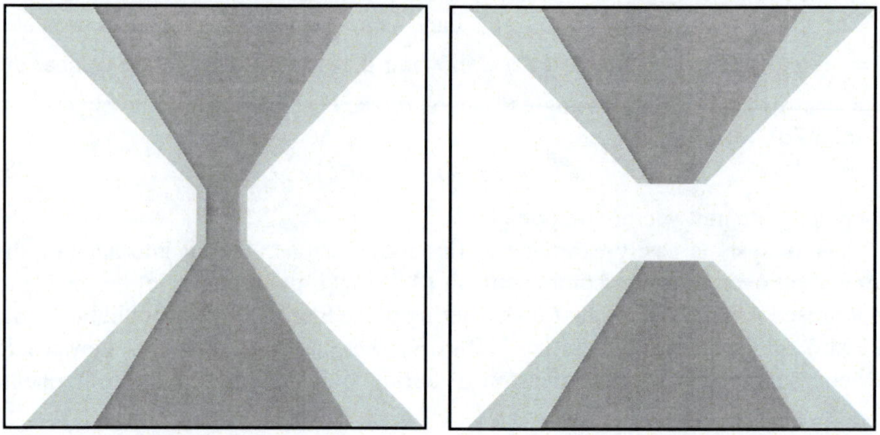

**Fig. 3.7** Absent pinch points (schematic). *Left*: the pinch point is replaced by a ridge of width $\xi_{\text{ice}}^{-1}$. *Right*: the scattering near the zone centre is completely suppressed, indicating finite spin clusters (see text)

$$\chi^{xx} = \chi_0 \frac{q_y^2 + \xi_{\text{ice}}^{-2}}{q_x^2 + q_y^2 + \xi_{\text{ice}}^{-2}}. \tag{3.29}$$

A second possibility arises when there is an additional length scale in the problem that removes the power law correlations associated with the pinch point. This is most easily seen in the case of spin ice, where the spin ice state can be constructed from spin loops of all sizes, from the smallest possible, a hexagon, to loops that span the entire sample. Each of these loops is closed and carries zero magnetization: the algebraic nature of the correlations arises from the fact that there are loops on all scales. It is possible, however, to construct a state that consists *only* of hexagonal loops [27]. It is impossible to construct a long wavelength magnetization fluctuation from these finite loops so the scattering function is completely suppressed in the Brillouin zone centre (Fig. 3.7, *right*). In reality, unless there is some breaking of spin symmetry (typically signaled by a phase transition), then a complete suppression of the pinch point is unlikely.

A similar argument may be applied to frustrated antiferromagnets: a pattern like Fig. 3.7b would be indicative of a tendency to finite size clusters of some sort. In fact, experiments on systems such as Y(Sc)$Mn_2$ [28], $ZnCr_2O_4$ [29] and $MgCr_2O_4$ [30] reveal that this is the common situation in pyrochlore lattice spin liquids (see Sect. 3.4).

### 3.3.8 *Dynamical Signature of Cooperative Paramagnetism*

It would be wrong to speak of a single dynamical signature of cooperative paramagnetism, as the spin dynamics depend in great detail on the actual interactions in the system. For example, in Ising-like systems such as spin ice, the dynamics are thermally activated [26, 31] while in systems with continuous spin symmetry, they may persist to the limit of low temperature. Nevertheless, two aspects of spin dynamics in highly frustrated magnets could be regarded as generic.

First, for the classical Heisenberg antiferromagnet on the pyrochlore lattice, Moessner and Chalker derived the following form for the time dependence of the spin autocorrelation function [32]:

$$\langle \mathbf{S}_{\mathbf{r}}(0)\mathbf{S}_{\mathbf{r}}(t)\rangle = \exp(-cTt), \tag{3.30}$$

where $c$ is a constant. This Fourier transforms to a Lorentzian scattering function in frequency space with width $\sim T$. A relaxation of the type (3.30) might be expected to be quite typical for highly frustrated magnets as it reflects a very large degeneracy of ground states and low lying excited states, that include dynamical modes of zero energy. Thus, even the smallest temperatures mediate an extremely rapid relaxation of any disturbance, and it is only at zero temperature that all motions completely disappear.

The second aspect of spin dynamics expected to be generic for highly frustrated magnets is the elementary thermal excitation of the Coulomb phase, such as the magnetic monopoles of spin ice [26]. These cost energy but do not couple directly to neutrons as their creation or destruction does not involve a single unit of spin angular momentum. Thus they could only be detected indirectly by their effect on the thermal or spatial evolution of the spin correlation functions.

## 3.4 Experimental Results

In this section, I describe various results of key neutron scattering experiments on highly frustrated magnets. This is not a comprehensive review of the literature, but rather, examples are chosen to illustrate the general ways by which neutron scattering may contribute to the field of highly frustrated magnetism.

### *3.4.1 Cooperative Paramagnet States*

#### 3.4.1.1 Spin Ice

Spin ice is the Ising-like highly frustrated pyrochlore lattice ferromagnet, in which spin configurations map onto proton configurations in Pauling's model of water ice [16, 19]. $Ho_2Ti_2O_7$, the canonical spin ice material [16, 33], is an ideal subject for neutron scattering, as it freezes into a disordered magnetic state below $\sim 2$ K, in which the static approximation is almost ideally obeyed. Thus, mapping of the diffuse scattering in reciprocal space gives a direct measure of $\chi^{\alpha\beta}(\mathbf{q})$. Furthermore, the moments on $Ho^{3+}$ are about the largest of any magnetic substance ($10\mu_B$), and since the neutron cross section scales as the square of the moment, its magnetic scattering is orders of magnitude stronger than that of a typical magnet. This is also true of another spin ice material, $Dy_2Ti_2O_7$ [34], but in other respects latter is less suited to neutron scattering as dysprosium is strongly neutron absorbing. To compensate for this it has been studied as the isotopically enriched $^{162}Dy_2Ti_2O_7$ [35].

The (unpolarised) scattering pattern of spin ice materials has been extensively mapped in several planes of reciprocal space in both zero and applied magnetic field [16,25,33,35–38]. In zero applied field the narrowing of the diffuse unpolarised scattering towards the Brillouin zone centre suggests the pinch points that are characteristic of pseudo-dipolar correlations, but it is only very recently that these have been clearly revealed by polarised neutron scattering [39]. Application of a magnetic field on [111] gives the reduced dimensionality kagomé ice phase [40]: this shows very clear pinch points (Fig. 3.8) [25]. The existence of pinch points shows that there is no characteristic length scale for the disorder in kagomé ice, so it is very different to a conventional paramagnet.

Away from the Brillouin zone centre the scattering pattern reflects the fine details of the spin–spin interactions in spin ice. The theory is well developed and a particularly accurate spin Hamiltonian has been established for $Dy_2Ti_2O_7$ by fitting the experimental $S(\mathbf{Q})$ to a parameterised theoretical form [27].

At temperatures above about 0.3 K, the spin relaxation of spin ice comes into the experimental window. It has been studied by the technique of neutron spin echo,

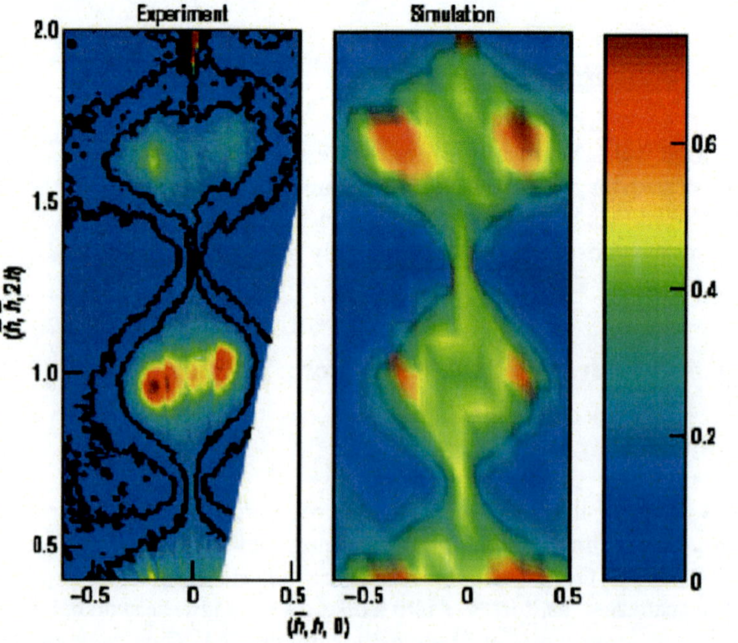


**Fig. 3.8** Pinch point scattering in kagomé ice: experiment (*left*) versus theory (*right*) [25]

which measures $S(\mathbf{Q}, t)$ (called the intermediate scattering function), the Fourier transform of $S(\mathbf{Q}, \omega)$ [41, 42]. This revealed an athermal regime of constant relaxation, stretching from 2 to 15 K. Such behaviour is consistent with the diffusion of monopole defects [31]. At higher temperatures the same study revealed that relaxation occurs via single ion excitations to crystal field states. It should be noted that the spatial characteristics of neutron scattering were key to establishing this fact [41].

#### 3.4.1.2 Pyrochlore and Spinel Antiferromagnets

The classical Heisenberg antiferromagnet on the pyrochlore lattice is one of the most basic spin models to exhibit a cooperative paramagnetic, or spin liquid state [32]. The system has a local ordering principle of zero magnetization per tetrahedron of the pyrochlore structure, that leads, as in the case of the ice rules, to pinch point singularities at Brillouin zone centres [22, 23]. However, as discussed in [43], the pure Heisenberg model is unphysical in the sense that it implies a spin symmetry that cannot exist in real pyrochlore materials. The inclusion of a realistic Ising or XY spin symmetry eliminates the spin liquid state in favour of ordered states, and, surprisingly, it is only the Ising *ferromagnet* (spin ice), rather than the antiferromagnet, that exhibits a macroscopic ground state degeneracy.

Nevertheless, there are several pyrochlore antiferromagnets with peculiarities that stabilise cooperative paramagnetic states. The one that perhaps comes closest to being a true Heisenberg antiferromagnet is the spinel $ZnCr_2O_4$: at low temperatures (12 K) this system has an ordering transition with a lattice distortion, but above that temperature it approximates the Heisenberg antiferromagnet on an undistorted pyrochlore lattice. Lee et al. reported neutron scattering maps of the $hhl$ plane at finite energy transfer: essentially maps of $S(\mathbf{Q}, \omega_0)$ at a finite $\omega_0$ [29]. This function can be interpreted as roughly approximating the equilibrium spin correlations, but with some weighting towards those configurations preferentially excited by the neutron flux.[10] Pinch points were not clearly observed: rather the data appeared to show diffuse blobs of scattering, reminiscent of the expected pattern, but separated by broad saddle points rather than by sharp pinch points. From the analysis of Sect. 3.3, this can be interpreted as evidence for a characteristic length scale in the disordered state. Indeed, Lee et al. showed that the scattering pattern can be well fitted by a model of hexagonal, non interacting clusters ('emergent excitations') (Fig. 3.9). The neutron scattering pattern bears a similarity to that predicted for the quantum Heisenberg antiferromagnet [44]. More recently, similar scattering has been observed in the related spinel $MgCr_2O_4$ [30].

Another material of great interest is the metallic alloy $(Y_{0.97}Sc_{0.03})Mn_2$, in which the Mn atoms occupy a pyrochlore lattice. The related phase $YMn_2$ is

[10] Measuring at finite energy transfer typically 'cleans up' the pattern, by removing nuclear elastic scattering.

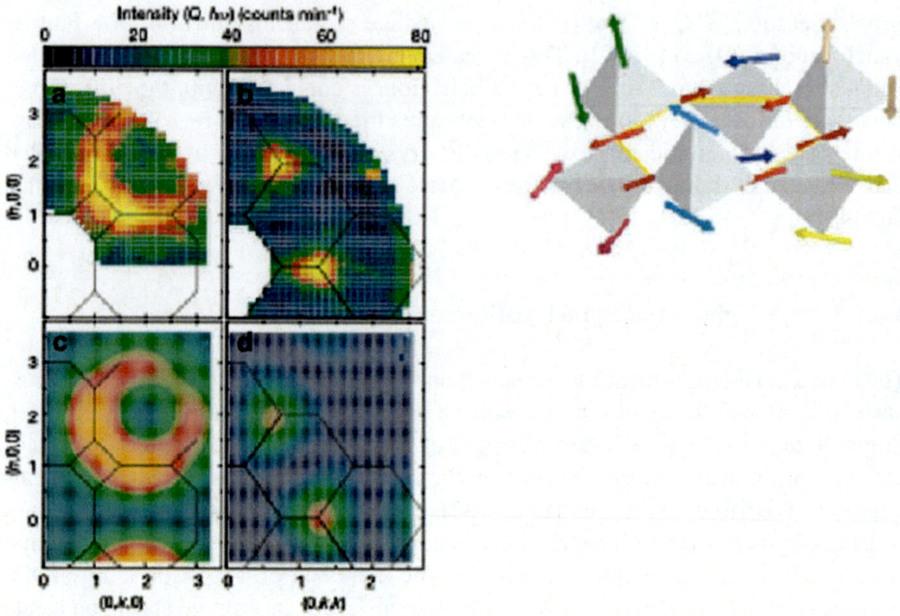


**Fig. 3.9** Cluster like scattering in $ZnCr_2O_4$ [29]. *Left*: experiment (*top*) versus theory (*bottom*); *right*: a six spin cluster

antiferromagnetically ordered below 100 K, but the application of pressure or 'chemical pressure' (the substitution of some Sc) leads to an itinerant state with heavy Fermion behaviour. Neutron inelastic scattering shows that the magnetism in this phase is almost entirely dynamic in nature: localised spin fluctuations out of an itinerant state. Ballou et al. mapped the neutron scattering in the $hhl$ plane at finite energy transfer. The scattering is strikingly similar to that of $ZnCr_2O_4$, described above, as well as the predicted form for the quantum Heisenberg antiferromagnet [44].

A third Heisenberg-like antiferromagnet to be studied in detail is $CsMnCrF_6$, in which Cr and Mn are distributed over a pyrochlore lattice [45]. Here, the $Cr^{3+}$ and $Mn^{2+}$ ions have only weak anisotropy, which is further frustrated by the positional disorder of the ions (Banks et al., unpublished). Neutron scattering is consistent with the expected pattern for the classical Heisenberg model, and broadened pinch points have recently been identified. The temperature evolution of the dynamic scattering function is broadly consistent with the expression (3.30) (see Fig. 3.10).

Finally, in this section, we mention the case of $Tb_2Ti_2O_7$ which has a spin liquid state at Kelvin temperatures. The diffuse scattering again shows broad features with no clearly resolved pinch points [46]. The degeneracy of the spin liquid can be lifted by the application of pressure, which induces a magnetic ordering transition [47].

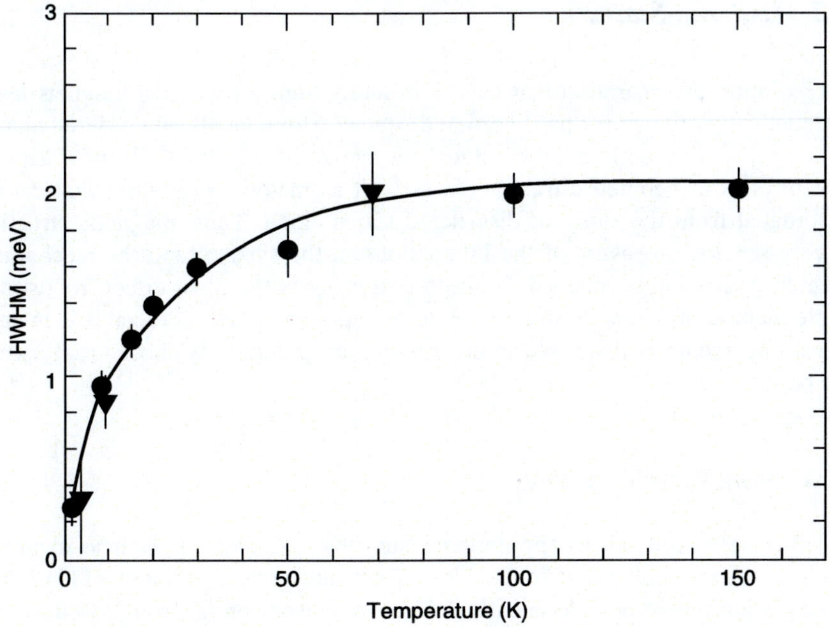


**Fig. 3.10** Width of the nelastic scattering peak in $CsMnCrF_6$ as a function of temperature, showing it to narrow as $T \to 0$ [45]

#### 3.4.1.3 Powder Experiments

The effect of powder averaging is to reduce the four-dimensional scattering function $S(\mathbf{Q}, \omega)$ into a two-dimensional one, $S(Q, \omega)$. This is a major problem in the study of cooperative paramagnets which are strongly characterised by the angular dependence of the scattering function (see Sect. 3.3). Nevertheless, the powder averaged scattering function does give some spatial information about the spin correlations. In the static approximation, powder averaging of $S(\mathbf{Q})$ gives

$$S(Q) = \sum_r \langle \mathbf{S}_0 \cdot \mathbf{S}_r \rangle z_r \sin(Qr)/(Qr), \tag{3.31}$$

where $z_r$ is the number of spins at distance $r$. Using this expression, models for the spin correlation function can be tested against neutron scattering data. This technique has been applied to the study of spin correlations in many spin disordered phases: for example, the pyrochlores $Y_2Mo_2O_7$, $Tb_2Ti_2O_7$ and $Tb_2Mo_2O_7$ [48], the kagomé system deuteronium iron jarosite [49], gadolinium gallium garnet [50], and $\beta$-$Mn_{1-x}Co_x$ [51], as well as partially ordered $Gd_2Ti_2O_7$ [13] (see below).

## *3.4.2 Ordered States*

The resolution of frustration that occurs in many highly frustrated magnets leads to magnetically ordered (or partly ordered) states. Although all spin ordered phases are, in a sense, conventional, their study may nevertheless reveal essential aspects of the physics of geometric frustration, as well as material properties that cannot be estimated from the study of disordered states alone. Thus, magnetic structure is a very sensitive measure of the interactions in the system, and the mechanism by which frustration is relieved. Neutron scattering is the most direct, robust and flexible method of magnetic structure determination, so plays a crucial role in such studies, The examples discussed in this section are particularly chosen to illustrate these points.

### 3.4.2.1 Ising Pyrochlore: $FeF_3$

$FeF_3$ may be considered to approximate the most basic of all realistic magnetic models on the pyrochlore lattice: the Heisenberg antiferromagnet with $\langle 111 \rangle$ Ising-like single ion anisotropy. As mentioned above, frustration is strongly relieved to give an ordered state with spins on alternate tetrahedra of the pyrochlore structure all pointing ‘in’ or ‘out’. This magnetic structure was confirmed by powder neutron scattering [52]. Despite being ordered, the low temperature state shows an interesting property that has been associated with the geometric frustration of the lattice. The temperature dependence of the magnetic Bragg peaks, and hence the magnetic order parameter, is unconventional. The magnetic order parameter exponent $\beta$ has been estimated to take the value $\beta = 0.18$, as distinct to the usual $\beta \approx 1/3$ for a typical three-dimensional magnet. It would be interesting to measure the exponent with greater precision by neutron diffraction on a single crystal sample of $FeF_3$. Unconventional exponents are also observed in triangular lattice frustrated systems [53], but it should be noted that the example of $Er_2Ti_2O_7$, discussed below, rules out the possibility that they are universal in ordered frustrated magnets.

### 3.4.2.2 XY Pyrochlore: $Er_2Ti_2O_7$

Erbium titanate approximates the Heisenberg pyrochlore antiferromagnet with local $\langle 111 \rangle$ XY-like single ion anisotropy [54]. The classical model relieves frustration with order induced by thermal fluctuations (‘order by disorder’) [54,55]. The magnetic structure formed has a single tetrahedron basis with every spin lying along a local $\langle 211 \rangle$ axis, in the ‘XY’ plane. The real material orders at $T_N = 1.2\,K$ to form exactly this structure. There is no mechanism other than order by disorder known to stabilise it: in particular, the leading perturbation to the model Hamiltonian – dipolar interactions – favour a different structure, as discussed below. Thus, it was suggested in [54] that the magnetic structure is stabilised by quantum order by disorder that mimics the classical thermal order by disorder. Two caveats on this argument have recently been removed. First, the existence of (previously unobserved) Goldstone mode spin waves has been demonstrated by neutron ineleastic scattering [56], lending support to a Hamiltonian with continuous symmetry, like that proposed. Second, 6-fold crystal field anisotropy has been shown to be too weak to be the sole cause of the observed ordering pattern, again lending support to the simplified Hamiltonian [57]. In this context, further theoretical studies of the quantum XY pyrochlore will be of great interest to confirm if $Er_2Ti_2O_7$ really does have a ground state stabilised through the order by disorder mechanism.

$Er_2Ti_2O_7$ exhibits two other points of interest [54]. First, as mentioned above, the order parameter exponent $\beta$ has been measured by single crystal neutron diffraction. It takes a conventional value, $\beta = 0.33$, which is close to the theoretical value for the three-dimensional XY antiferromagnet, thus ruling out the idea that unconventional exponents are *universal* for frustrated magnets. Second, neutron diffraction leads to an ambiguity in the magnetic structure as two different (symmetry related) candidate structures give identical Bragg intensities, when magnetic domains are accounted for. The resolution of this difficulty illustrates some general neutron diffraction methodology. In the earliest study a magnetic field was applied to the single crystal sample to stabilise a magnetic monodomain: changes in diffracted intensity clearly discriminated the candidate structures, but it could always be argued that the application of a field to a frustrated magnet might alter the magnetic state. Thus, a more recent study [58] used a sophisticated zero field neutron polarimetry method to distinguish the structures and estimate domain populations, which finally confirmed the magnetic structure proposed in [54].

### 3.4.2.3 Dipolar Ordering: $Gd_2Sn_2O_7$

Dipole–dipole interactions are inescapable in any real magnet, but are rarely sufficiently strong to dictate the ordering pattern. Heavy rare earth salts have particularly strong dipolar interactions and relatively weak exchange. Salts of $Gd^{3+}$ furthermore have relatively weak (though non-negligible) crystal field anisotropy, so gadolinium pyrochlores may be taken as a prototypical Heisenberg-dipolar frustrated magnets. Early ideas that $Gd_2Ti_2O_7$ should be a model system of this sort were quickly dispelled by neutron diffraction of the isotopically enriched $^{160}Gd_2Ti_2O_7$ [59],[11] but more recently it has been shown by powder neutron diffraction [12] that its close relative (enriched) $Gd_2Sn_2O_7$ does exhibit the magnetic structure expected for dipole–dipole interactions on the pyrochlore lattice [60]. This structure has spins in local 'XY' planes along $\langle 110 \rangle$ axes. It is probable that the single ion anisotropy of $Gd^{3+}$[12] favours local XY symmetry, so does not compete with the dipolar ordering pattern.

[11] The isotope was used because natural Gd is strongly absorbing.

[12] Although $Gd^{3+}$ is nominally an $S$-state ion, it normally exhibits single ion $g$- factor anisotropy of magnitude several tenths of a Kelvin.

#### 3.4.2.4 Partial Order: $Gd_2Ti_2O_7$

Originally considered a good candidate for the Heisenberg-dipolar pyrochlore antiferromagnet [60, 61], there have been many conjectures about the properties of $Gd_2Ti_2O_7$, but these have been mainly ruled out by neutron diffraction evidence. At the time of writing the magnetic structures it exhibits remain only poorly understood, but are, nevertheless, of considerable interest.

In zero applied field $Gd_2Ti_2O_7$ shows two ordering transitions at 1.1 K and 0.7 K. In the first transition at 1.1 K only three quarters of the spins of the pyrochlore structure order magnetically [13]. The structure is partially ordered, consisting of one disordered tetrahedron pre unit cell, as illustrated in Fig. 3.11. It is instructive to note that this is a '4-$k$' structure that looks totally different to the corresponding single $k$ variant (illustrated in [59]). As described in Sect. 3.2, simple neutron diffraction cannot distinguish the single $k$ and multi-$k$ variants, but a study of the diffuse scattering arising from the disordered spin set, clearly favours the 4-$k$ solution (see Fig. 3.12) [13]. The 0.7 K ordering transition then corresponds to the ordering of these spins, although they show strong quantum mechanical fluctuation in the ground state.

The actual powder pattern for $Gd_2Ti_2O_7$ is illustrated in Fig. 3.13 For symmetry reasons the $(1/2, 1/2, 1/2)$ magnetic peak arises solely from the ordering of the spin set that is remains disordered between 0.7 and 1.1 K: the rapid growth of this peak below 0.7 K proves the ordering of this set and gives a strong experimental handle on the ground state moment.

#### 3.4.2.5 Kagomé Lattice: Jarosite Materials

The Jarosite family, in which $M^{3+}$ spins (M = V, Cr, Fe) occupy a kagomé lattice in well-separated layers, affords many chemical variations which generally show

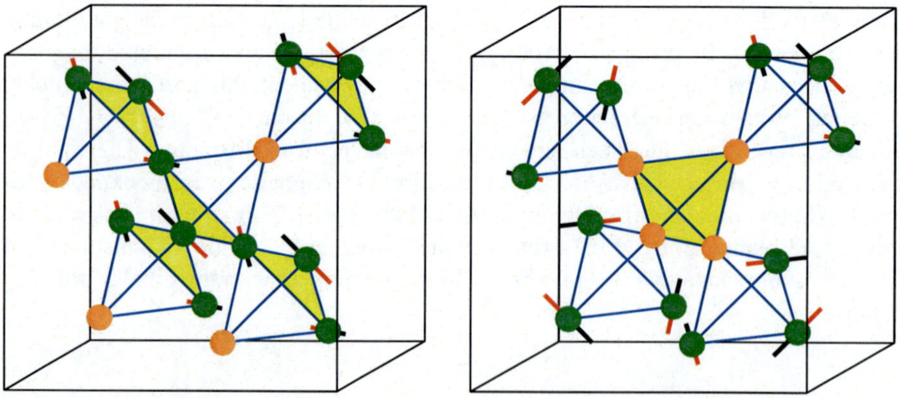

**Fig. 3.11** Partially ordered magnetic structures of $Gd_2Ti_2O_7$ [13]. *Left*: the 1-$k$ structure [59] and right the 4-$k$ structure [13]. The 4-$k$ structure is that found in experiment (see Fig. 3.12). The orange circles represent spins that are disordered above 0.7 K

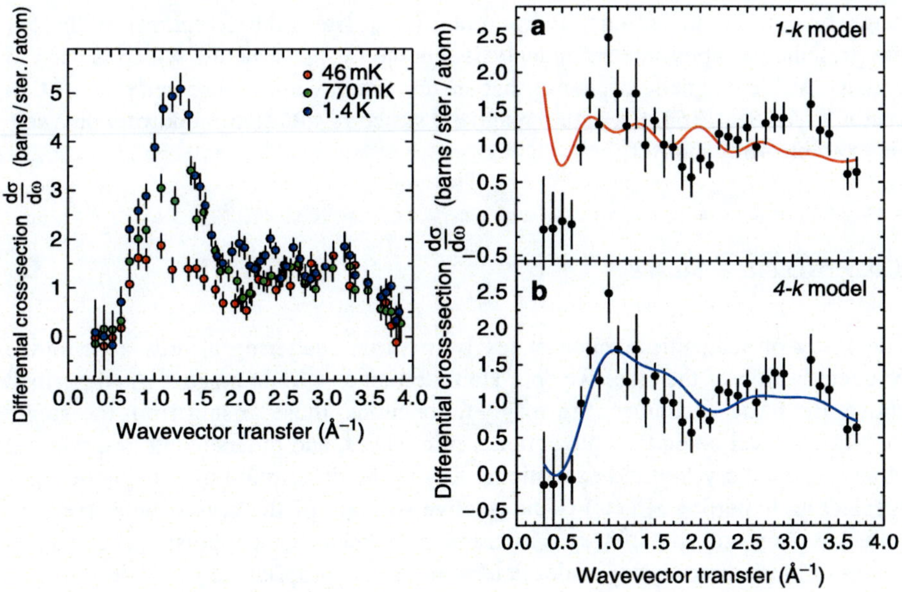


**Fig. 3.12** Diffuse scattering from partly ordered $Gd_2Ti_2O_7$ [13]. The fits show that the disordered spins of the 4-$k$ model describe the data much better than do those of the 1-$k$ model (see Fig. 3.11)

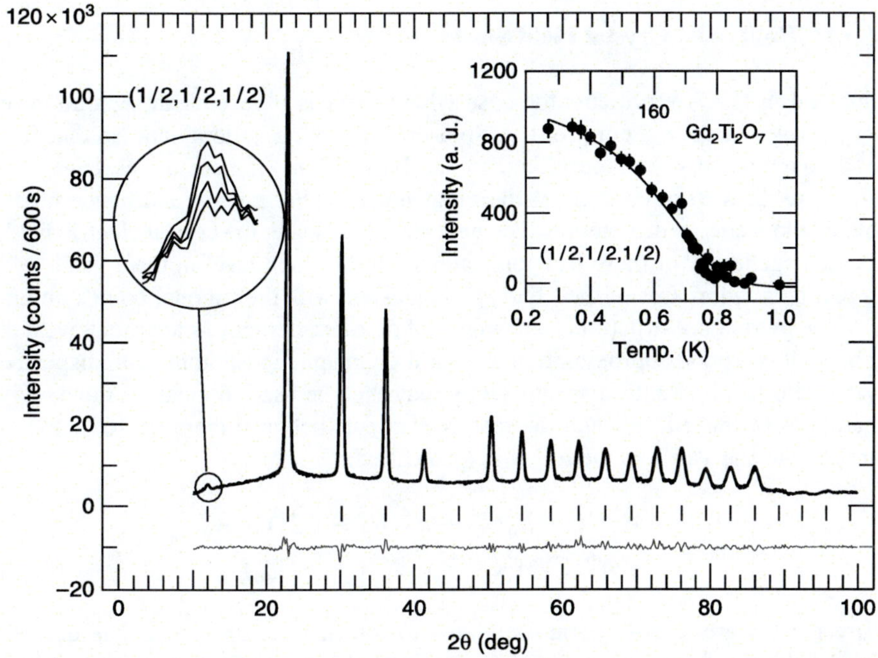


**Fig. 3.13** Powder diffraction pattern of $Gd_2Ti_2O_7$ [13]. Lower line is observed data minus that calculated using the model of Fig. 3.11 and the insets show the growth of the small peak that arises exclusively from the spin set that is disordered above 0.7 K

magnetic ordering transitions (hydronium iron jarosite is the exception). Wills [62] has presented a representational analysis scheme for the structure which is used to derive possible magnetic structures that are consistent with an observed propagation vector **k** (see Sect. 3.2). For actual magnetic structure solutions on powder data see, for example, [62] and [63].

### 3.4.3 Excited States

The study of magnetic excited states by neutron scattering affords a definitive characterisation of the effective spin Hamiltonian of a ionic magnet. As described above, the basic excitations are magnetic excitons, those arising from the single ion ground level being the conventional spin waves, and others being best thought of as dispersed crystal field excitations. Key to the description of a magnetic solid (neglecting hyperfine effects) is the relative strength of the crystal field, the spin orbit coupling and the spin–spin interactions (exchange or dipolar). The relative values of these three energy scales determines the appropriate physical description. At a practical level the neutron scattering energy resolution is a fourth energy scale that, in comparison with the other three, determines what can be measured. The examples in this section are chosen to represent the various possibilities.

#### 3.4.3.1 $Ho_2Ti_2O_7$: Crystal Field States

Spin ice $Ho_2Ti_2O_7$ represents the case where there is a very clean separation of energy scales in the system [16, 19]. Spin-orbit coupling is dominant, so that free $Ho^{3+}$ ions are very well described in the basis $|J, m_J\rangle$: here $J = 8, m_J = 8, 7, 6 \cdots -8$, with other $J$-levels being at much higher energy and hence negligible in a low energy description. The next strongest scale, the crystal electric field, lifts the degeneracy of these free ion states such that each level in the crystal field is a particular admixture of the $|J, m_J\rangle$'s consistent with the trigonal point symmetry of the pyrochlore structure. Exchange and dipolar interactions are much weaker again, so it is a good approximation to regard the magnetic excitons as undispersed crystal field levels. Neutron spectroscopy may then be used to measure the energy separations of these levels and the matrix elements between them. In [64], a spin Hamiltonian equivalent to the following was fitted:

$$
\begin{aligned}
H_{CF} = {} & B_0^2 C_0^2 + B_0^4 C_0^4 + B_3^4 (C_3^4 - C_{-3}^4) + B_0^6 C_0^6 \\
& + B_3^6 (C_3^6 + C_{-3}^6) + B_6^6 (C_6^6 + C_{-6}^6),
\end{aligned}
\tag{3.32}
$$

where the $C$s are tensor operators related to the Stevens operators. The derived crystal field parameters are given in [64]. Crystal field wavefunctions are a further product of the analysis: these show that the ground term is a doublet consisting almost entirely of $|8, \pm 8\rangle$, with the next excited state at $\sim$300 K above the ground state. As the dipolar coupling in $Ho_2Ti_2O_7$ is only of order $\sim$1.5 K (exchange being weaker still) it is thus an extremely good approximation to regard the $Ho^{3+}$ effective spins as Ising-like, the basis of spin ice physics.

Although the separation of energy scales in $Ho_2Ti_2O_7$ allows an extremely accurate characterisation of the spin Hamiltonian, this comes at a price. First, the weak coupling is not far off the limit of neutron resolution, meaning that excitations arising from the magnetic coupling cannot be easily resolved. Second, the Ising-like 'spins' do not have a true $S = 1/2$, so cannot be flipped by a neutron; thus the basic monopole excitations in spin ice cannot be excited or de-excited by the neutron flux, so are not directly visible to neutron spectroscopy.

#### 3.4.3.2 $Er_2Ti_2O_7$: Weakly Dispersed Crystal Field States

$Er_2Ti_2O_7$ makes an interesting comparison to $Ho_2Ti_2O_7$ as the dispersion of the magnetic excitions is clearly visible to neutron spectroscopy (see Fig. 3.14), indicating magnetic exchange of a few Kelvin [54]. Ignoring this dispersion, the crystal field scheme has been solved by neutron scattering [65]. In its ground state $Er^{3+}$ is roughly the 'opposite' of $Ho^{3+}$ having dominant XY anisotropy. The spin wave excitations arising from the exchange coupling can just be resolved by neutron scattering, but a model has not yet been fitted to the data [56].

It is a fairly common practice to 'extrapolate' crystal field parameters from one compound to another. However, the comparison of the true crystal field parameters of $Er_2Ti_2O_7$ with those of $Ho_2Ti_2O_7$ shows that this method is quite inaccurate [65]. Thus, considering the parameter $B_2^0$, for example, the value for $Ho^{3+}$ is $-0.08$ and the exprapolated value for $Er^{3+}$ is $+0.08$, but the real value is $>0.13$ [65]. Whenever possible, neutron scattering should be applied directly to determine the crystal fields.

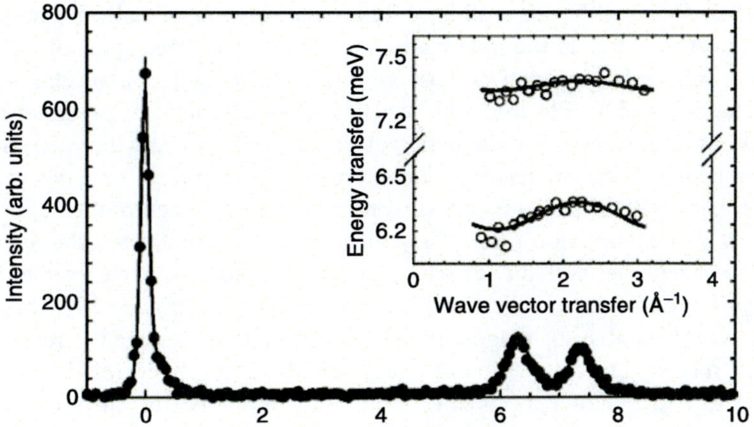


**Fig. 3.14** Magnetic excitons in $Er_2Ti_2O_7$ [54]. The main figure shows the powder spectrum and the inset the dispersion of the magnetic excitons at $\sim$6 and 7 meV

#### 3.4.3.3 Potassium Iron Jarosite: Spin Wave Dispersions

If a magnet is periodically ordered then neutron scattering affords the most direct way of confirming a spin Hamiltonian and determining its parameters, by measurement of the spin wave dispersion. As regards geometrically frustrated magnets, it is generally necessary that the magnet is ordered, as otherwise the spin waves (if they exist at all) are not classified by the wavevector **Q**, so their dispersion cannot be resolved. This means that the technique can only be applied to those systems where the frustration is ultimately resolved into magnetic order at low temperature. A good example of this is potassium iron jarosite, for which very detailed spin wave measurements on a single crystal sample [66] and an accompanying theory [67] have been reported.

## 3.5 Conclusions

In this article, I have tried to illustrate what different magnetic states look like to neutron scattering, and I have emphasised how neutron scattering provides a direct measurement of spin correlation functions, or equivalently, wavevector dependent susceptibilities and magnetic moments. The key figures are Figs. 3.2 and 3.7 which are essentially look-up tables for different magnetic states. A real frustrated magnet might exhibit any of these patterns, either in a pure form or else as a mixture. For example, a frustration-induced suppression of the ordering temperature would generally result in a broad blob of type Fig. 3.2b (broad because the correlations are short) but additional hints of a pattern like Fig. 3.2c would imply that an ice-type rule is operating (but note that not all frustrated magnets obey an ice rule).

A fairly common experience in practical neutron scattering is to prove that some particular material, previously touted as exceptional on the basis of bulk measurements, is in fact highly conventional! The interpretation of bulk measurements inevitably leaves much to the imagination, and neutron scattering, as often as not, provides a sobering correction to those who let their imagination get the better of their scientific caution. For this reason it is always advisable, if possible, to perform neutron scattering on a magnetic substance under investigation. Even if the scattering pattern is poorly resolved, there are usually sufficient statistics to check for consistency with whatever model is being tested. Indeed, neutron scatterers have a saying that 'neutrons don't lie'. They mean that the technique is the ideal passive probe of material systems for which the interpretation of experimental results is particularly transparent, robust and, in most cases, unambiguous. Neutron scattering may not be able to image all sorts of magnetic phases, but it remains the most important and direct method of examining the two spin correlation function, so provides the paradigm experiment for the study of highly frustrated magnets.

## References

1. P.W. Anderson, *Basic Notions of Condensed Matter Physics* (Perseus Books Group, 1997)
2. J. Villain, Z. Phys. B **33**, 31 (1979)
3. V.F. Sears, *Neutron Optics: An Introduction to the Theory of Neutron Optical Phenomena and Their Applications* (Oxford University Press, 1989)
4. R.M. White, *Quantum Theory of Magnetism*, 2nd edn. (Springer, Berlin Heidelberg New York, 1983)
5. P.J. Brown, Magnetic form factors, Chap. 4.4.5, in *International Tables for Crystallography*, vol. C, ed. by A.J.C. Wilson (D. Reidel Publishing, Dordrecht, Holland, 1983–1993), pp. 391–399
6. S.W. Lovesey, *Theory of Neutron scattering from Condensed Matter*, vols. 1, 2 (Oxford University Press, 1984)
7. R.M. Moon, T. Riste, W.C. Koehler, Phys. Rev. **181**, 920 (1969)
8. W.G. Williams, *Polarized Neutrons* (Oxford University Press, 1988)
9. M.J. Cooper, R. Nathans, Acta Cryst. **23**, 357 (1967)
10. B. Dorner, Acta Cryst. **A28**, 319 (1972)
11. J. Rossat-Mignod, in *Neutron Scattering*, ed. by K. Skjold, D.L. Price, *Methods in Experimental Physics*, vol. 23C (Academic, New York, 1987)
12. A.S. Wills et al., J. Phys. Condens. Matter **18**, L37 (2006)
13. J.R. Stewart et al., J. Phys. Condens. Matter **16**, L321 (2004)
14. W. Marshall, R.D. Lowde, Rep. Prog. Phys. **31**, 705 (1968)
15. B. Grover Phys. Rev. **140** A1944 (1965)
16. M.J. Harris et al., Phys. Rev. Lett. **79**, 2554 (1997)
17. H.E. Stanley, *Introduction to Phase Transitions and Critical Phenomena* (Oxford University Press, 1971)
18. N. Goldenfeld, *Lectures on Phase Transitions and the Renormalization Group* (Addison-Wesley, Reading, MA, 1992)
19. S.T. Bramwell, M.J.P. Gingras, Science **294**, 1495 (2001)
20. R.W. Youngblood, J.D. Axe, Phys. Rev. B **23**, 232 (1981)
21. R.W. Youngblood, J.D. Axe, B.M. McCoy Phys. Rev. B **21**, 5212 (1980)
22. S.V. Isakov et al., Phys. Rev. Lett. **95**, 217201 (2005)
23. C. Henley, Phys. Rev. B **71**, 014424 (2005)
24. M.P. Zinkin, M.J. Harris, T. Zeiske, Phys. Rev. B **56**, 11786 (1997)
25. T. Fennell et al., Nat. Phys. **3**, 566 (2007)
26. C. Castelnovo, R. Moessner, S.L. Sondhi, Nature **451**, 42 (2008)
27. T. Yavors'kii et al., Phys. Rev. Lett. **101**, 037204 (2008)
28. R. Ballou, E. Lelièvre-Berna, B. Fak, Phys. Rev. Lett. **76**, 2125 (1996)
29. S.-H. Lee et al., Nature **418**, 856 (2002)
30. K. Tomyasu et al., Phys. Rev. Lett. **101**, 177401 (2008)
31. L.D.C. Jaubert, P.C.W. Holdsworth, Nat. Phys. **5**, 258 (2009)
32. R. Moessner, J.T. Chalker, Phys. Rev. Lett. **80**, 2929 (1998)
33. S.T. Bramwell et al., Phys. Rev. Lett. **87**, 047205 (2001)
34. A.P. Ramirez et al., Nature **399**, 333 (1999)
35. T. Fennell et al., Phys. Rev. B **70**, 134408 (2004)
36. M. Kanada et al., J. Phys. Soc. Jpn. **71**, 313 (2002)
37. J.P. Clancy et al., Phys. Rev. B **79**, 014408 (2009)
38. T. Fennell et al., Phys. Rev. B **72**, 224411 (2005)
39. T. Fennell et al., Science **326**, 415 (2009)
40. K. Matsuhira et al., J. Phys. Condens. Matter **14**, L559 (2002)
41. G. Ehlers et al., J. Phys. Condens. Matter **15**, L9 (2003)
42. G Ehlers et al., J. Phys. Condens. Matter **16**, S635 (2004)
43. S.T. Bramwell, M.J. Harris, J. Phys. Condens. Matter **10**, L215 (1998)
44. B. Canals, C. Lacroix, Phys. Rev. Lett. **80**, 2933 (1998)

45. M.P. Zinkin, M.J. Harris, T. Zeiske, Phys. Rev. B **56**, 11786 (1997)
46. J. Gardner et al., Phys. Rev. B **68**, 180401 (2003)
47. I. Mirabeau et al., Nature **420**, 54 (2002)
48. B.D. Gaulin et al., Physica B **241–243**, 511 (1998)
49. B. Fåk et al., Europhys. Lett. **81**, 17006 (2008)
50. O.A. Petrenko et al., Phys. Rev. Lett. **80**, 4570 (1998)
51. J.R. Stewart, R. Cywinski, J. Phys. Condens. Matter **21**, 124216 (2009)
52. J.N. Reimers et al., Phys. Rev. B **45**, 7295 (1991)
53. H. Kawamura J. Phys. Soc. Jpn. **55**, 2095 (1986)
54. J.D.M. Champion et al., Phys. Rev. B **68**, 020401 (2003)
55. S.T. Bramwell, M.J.P. Gingras, J.N. Reimers, J. Appl. Phys. **75**, 5523 (1994)
56. J.P. Ruff et al., Phys. Rev. Lett. **101**, 147205 (2008)
57. P.A. McClarty, S.H. Curnoe, M.J.P. Gingras, J. Phys. Conf. Ser. **145** 012032 (2009)
58. A. Poole, A.S. Wills, E. Lelièvre-Berna, J. Phys. Condens. Matter **19**, 452201 (2007)
59. J.D. Champion et al., Phys. Rev. B **64**, 140407 (2001)
60. S.E. Palmer, J.T. Chalker, Phys. Rev. B **62**, 488 (2000)
61. N.P. Raju et al., Phys. Rev. B **59**, 14489 (1999)
62. A.S. Wills, Phys. Rev. B **63**, 064430 (1981)
63. T. Inami et al., Phys. Rev. B **61**, 12181 (2000)
64. S. Rosenkranz et al., J. Appl. Phys. **87**, 5914 (2000)
65. M. Shirai, Ph.D. Thesis, University College London (2007)
66. K. Matan et al., Phys. Rev. Lett. **96**, 247201 (2006)
67. T. Yildirim, A.B., Harris, Phys. Rev. B **73**, 214446 (2006)

# Chapter 4
# NMR and $\mu$SR in Highly Frustrated Magnets

**Pietro Carretta and Amit Keren**

**Abstract** In this chapter, we present a brief overview of some of the most significant achievements of the nuclear magnetic resonance (NMR) and muon spin resonance ($\mu$SR) techniques in the study of highly frustrated magnets. We begin by discussing the physical quantities measured by the two techniques and their connection to the microscopic static and dynamical spin susceptibilities of a magnetic system. Our presentation of the physical insights obtained from NMR and $\mu$SR experiments is structured by starting from the most simple frustrated units, molecular nanomagnets, and continuing through artificially constructed frustrated systems, such as $^3$He on graphite, to magnets with a macroscopically degenerate ground state of the type familiar from the pyrochlore and kagomé lattices.

## 4.1 Basic Aspects of NMR and $\mu$SR Techniques

NMR and $\mu$SR are very powerful experimental techniques which allow the microscopic properties of spin systems to be investigated through the study of the time-evolution of the nuclear magnetization, $\boldsymbol{M}(t)$, and of the muon spin polarization, $\boldsymbol{P}(t)$, respectively [1–5]. Each technique has its advantages and disadvantages. In NMR, one knows the crystallographic position of the nuclei under investigation, and therefore NMR results can be compared quite readily to theoretical analysis. However, NMR experiments cannot be performed in compounds where only low-sensitivity nuclei are present, or where rapid nuclear relaxation processes prevent the observation of an NMR signal. When polarized muons are injected into the sample to probe the local microscopic properties of the system, it is possible to detect relaxation times shorter than 0.1 $\mu$s, approximately two orders of magnitude shorter than the shortest relaxation time accessible by NMR. Because the nuclear magnetization is the quantity detected in NMR experiments, in general a magnetic field must be applied to generate it. By contrast, the muon beam is already polarized before entering the sample, so that the system can be studied in zero field. This aspect is particularly important if one wishes to investigate the intrinsic properties of a certain system without the perturbation of a magnetic field. Nevertheless, novel ground states can be induced by the application of high magnetic fields (typically above 10 Tesla), where $\mu$SR experiments cannot be performed while NMR experiments can. Hence, although both techniques appear to measure similar quantities, as discussed in the next section, in fact they are often complementary and it is their combination which represents a rather powerful investigative tool. Here we apply this combination to probe the local microscopic properties of frustrated magnets.

### 4.1.1 Line Shift and Line Width

The time-evolution of $\boldsymbol{M}(t)$ and $\boldsymbol{P}(t)$ is determined by the hyperfine interactions, which can be summarized in the form

$$\mathcal{H} = \mathcal{H}_z + \mathcal{H}_{n-n} + \mathcal{H}_{n-e} + \mathcal{H}_{EFG}. \tag{4.1}$$

The effects of the four terms on the hyperfine levels and on the NMR spectra for a nuclear spin $I = 3/2$ are depicted in Fig. 4.1. We remind the reader that the intensity of the whole spectrum is proportional to the nuclear magnetization. The first term describes the Zeeman interaction with an external field. The dipole–dipole interaction $\mathcal{H}_{n-n}$ among the nuclear spins, or between the muon and the nuclear spins,

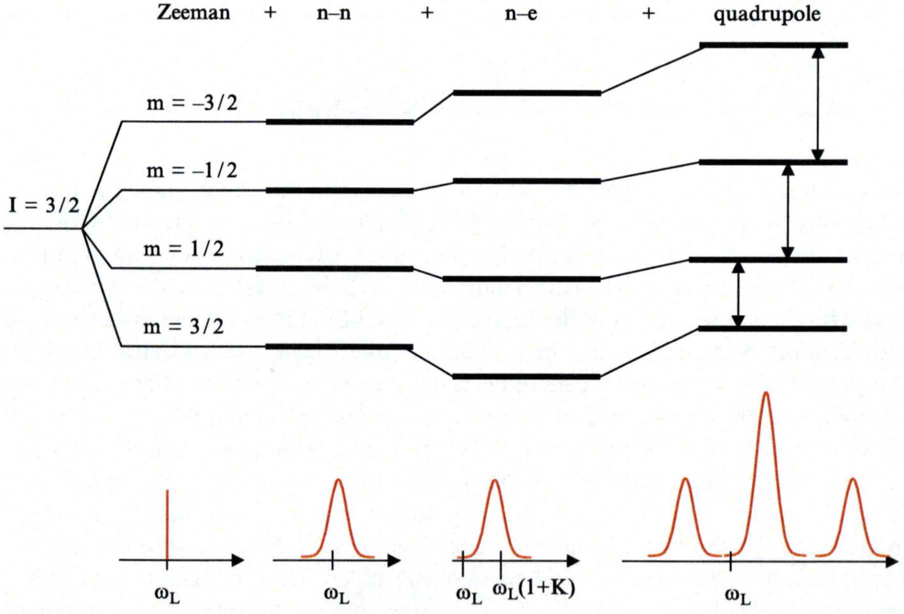


**Fig. 4.1** Schematic illustration of the modifications in the hyperfine levels of $I = 3/2$ nuclei due to the different terms in the nuclear hyperfine Hamiltonian of (4.1). The corresponding modifications of the NMR spectra are shown in the lower panel

yields a broadening of the NMR [1,2] and $\mu$SR [3–5] spectra. In certain compounds, including the cuprates, the nuclear dipole–dipole interaction is mediated by the electron spins, and information about the static electron spin susceptibility $\chi'(\boldsymbol{q})$ is obtained from the dipolar broadening [6]. The final term, $\mathcal{H}_{EFG}$, describes the interaction between the nuclear electric-quadrupole moment, $Q$, and the electric-field gradient (EFG) generated by the charge distribution around the nucleus. This term is finite for nuclear spins with $I > 1/2$ and thus is not present in the muon interaction Hamiltonian. The quadrupole interaction is very sensitive to modifications of the local configuration, and thus can be used to probe distortions induced by the spin-lattice coupling [7] or the presence of an inhomogeneous charge distribution induced by charge ordering [8].

The third term is the one most relevant to the investigation of frustrated magnetism, as it describes the hyperfine interaction with the electron spins $\boldsymbol{S}$. Because most of the systems we discuss here are insulators, we write

$$\mathcal{H}_{n-e} = -\gamma\hbar \sum_{i,k} \boldsymbol{I}_i \tilde{A}_{ik} \boldsymbol{S}_k, \tag{4.2}$$

with $\tilde{A}_{ik}$ the hyperfine coupling tensor, $I$ and $S$, respectively, the nuclear or $\mu^+$ and electronic spin operators, $i$ and $k$ respectively the nuclear or $\mu^+$ and electronic spin indexes, and $\gamma$ the nuclear or $\mu^+$ gyromagnetic ratio. The hyperfine field at the $i$th nucleus or muon is then given by $\boldsymbol{h}_i = \sum_k \tilde{A}_{ik}\boldsymbol{S}_k$, and in the presence of a non-zero average polarization $\langle \boldsymbol{S} \rangle$, $\boldsymbol{h}_i = \sum_k \tilde{A}_{ik}\langle \boldsymbol{S}_k \rangle$. Thus $\langle \boldsymbol{S} \rangle$ can be estimated directly from the precessional frequency $\omega = \gamma \sum_k \tilde{A}_k \langle \boldsymbol{S}_k \rangle$ of the nuclei or of the muons around the local field.

When an external field $\boldsymbol{H}_0 \parallel \hat{z}$ is applied, the local magnetic field becomes

$$\boldsymbol{B} = \boldsymbol{H}_0 + \sum_k \tilde{A}_k \langle \boldsymbol{S}_k \rangle \tag{4.3}$$

and the resonance frequency is shifted to

$$\omega = \omega_{\mathrm{L}}(1 + K), \tag{4.4}$$

where $\omega_{\mathrm{L}} = \gamma H_0$ is the Larmor frequency and, for $H_0 \gg |\sum_k \tilde{A}_k \langle \boldsymbol{S}_k \rangle|$,

$$K = \frac{\left(\sum_k \tilde{A}_k \langle \boldsymbol{S}_k \rangle\right)_z}{H_0}. \tag{4.5}$$

In general, $K$ is a tensor and $\tilde{K} = \sum_k \tilde{A}_k \tilde{\chi}(\boldsymbol{q} = 0, \omega = 0)$, whence the shift in the precessional frequency of the nuclei (or of the $\mu^+$) can be used to derive the static uniform susceptibility associated only with those electron spins which are coupled to the nuclei under investigation (Fig. 4.2) or to the muon.

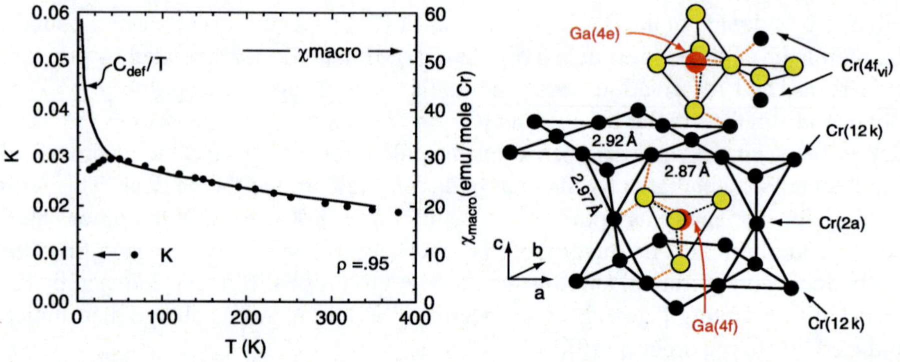


**Fig. 4.2** Temperature-dependence of $^{71}$Ga(4f) NMR shift, $K$, in $SrCr_{9p}Ga_{12-9p}O_{19}$ ($p = 0.95$) compared to the macroscopic susceptibility derived with a magnetometer. Because the $^{71}$Ga(4f) nuclei are coupled strongly only to chromium ions within the kagomé layers, one may isolate only the intrinsic susceptibility of the latter, whereas the macroscopic susceptibility also detects the contributions from defects [9]

If, for any reason, the separation between the nuclei or muon and the electronic spins is random, or there are missing spins in the sample, the hyperfine coupling $\tilde{A}$ will be a random or multi-valued variable, leading to a distribution of precessional frequencies and to an increase in the relaxation of the magnetization or polarization. For an external field applied perpendicular to the initial polarization of the muons, one finds

$$P_{\perp}(t) = \exp\left(-\left[\frac{t}{T_2^*}\right]^2\right)\cos(\omega t); \tag{4.6}$$

$1/T_2^*$ could also represent the decay rate of the NMR signal after an RF pulse. Assuming a distribution of hyperfine fields in the $\hat{z}$ direction, $\tilde{A}_k$ can be expressed as the sum of a mean value $\overline{A_k}$ plus a fluctuating component $\delta A_k$. For the distribution

$$\rho(\delta A_k) = \frac{1}{\sqrt{2\pi}\sigma_k}\exp\left(-\frac{\delta A_k^2}{2\sigma_k^2}\right),$$

one finds that the width of the spectrum is

$$\frac{1}{T_2^*} = \gamma\chi H_0\left(\sum_k \sigma_k^2\right)^{1/2}, \tag{4.7}$$

and the average frequency shift is

$$K = \frac{\omega - \gamma H_0}{\gamma H_0} = \chi\sum_k \overline{A_k}. \tag{4.8}$$

If $\sigma_k$ and $\overline{A}$ are temperature-independent, we expect

$$1/T_2^* \propto K, \tag{4.9}$$

where the temperature is an implicit parameter. We emphasize that this proportionality is valid for any functional form of the relaxation (4.6). Breakdown of the validity of (4.9) would indicate a modification of the hyperfine couplings, which is usually expected when lattice distortions occur (Sects. 4.2.2 and 4.2.6). On the other hand, there are certain types of system in which the hyperfine coupling is constant but the spin polarization can be site-dependent. In this case, (4.9) is no longer applicable and both the line-broadening and line shape reflect the distribution of the local spin polarization [10].

### 4.1.2 Nuclear and Muon Spin-Lattice Relaxation Rate $1/T_1$

The transitions among the hyperfine levels, driven by the time-dependent part of the hyperfine Hamiltonian [not shown in (4.1)], modify the nuclear spin population on each level, and thus the longitudinal component of the nuclear magnetization. In a magnetic system, it is then possible to derive information about the spin dynamics, which drives the fluctuations of the hyperfine field, from the time-evolution of the nuclear magnetization. If the nuclear spin ensemble is brought out of equilibrium by an arbitrary radio-frequency (RF) pulse sequence, the recovery of the longitudinal component of the nuclear magnetization is described by a characteristic relaxation rate, $1/T_1$. For relaxation mechanisms driven by fluctuations $\boldsymbol{h}(t)$ of the hyperfine field ($h$ is small with respect to $H_0$), time-dependent perturbation theory combined with the assumption that the frequency of the field fluctuations is much greater than the relaxation rate, $\omega = 2\pi\nu \gg 1/T_1$, yields

$$\frac{1}{T_1} = \frac{\gamma^2}{2}\int_{-\infty}^{+\infty} \mathrm{e}^{\mathrm{i}\omega_{\mathrm{L}}t}\langle h_+(t)h_-(0)\rangle \mathrm{d}t. \tag{4.10}$$

In this fundamental expression, $1/T_1$ is driven by the transverse components of the fluctuating field at the nucleus, which is a consequence of the magnetic-dipole selection rules, and is proportional to the Fourier transform of the correlation function at the resonance frequency (energy-conservation criterion). Alternatively stated, $1/T_1$ probes the spectral density at the resonance frequency, $\omega_{\mathrm{L}}$, a quantity typically in the MHz range, and thus orders of magnitude lower than the spectral range accessed by inelastic neutron scattering experiments. It should be noted that this does not imply that relevant energy scales much greater than $\hbar\omega_{\mathrm{L}}$ cannot be estimated from $1/T_1$: when sum-rules apply, as is often the case in spin systems, the amplitude of the low-frequency spectral density is determined by the characteristic frequency of the fluctuations $\omega \gg \omega_{\mathrm{L}}$.

In general, when collective spin excitations are present, one may write

$$\boldsymbol{h}(t) = \frac{1}{\sqrt{N}} \sum_{\boldsymbol{q}} \sum_{k} \mathrm{e}^{\mathrm{i}\boldsymbol{q}\boldsymbol{r}_k} \tilde{A}_k \boldsymbol{S}(\boldsymbol{q}, t), \tag{4.11}$$

and by substituting the transverse components of $\boldsymbol{h}(t)$ in the expression (4.10) one obtains

$$\frac{1}{T_1} = \frac{\gamma^2}{2} \frac{1}{N} \sum_{\boldsymbol{q}, \alpha = x,y,z} \left( |A_{\boldsymbol{q}}|^2 S_{\alpha\alpha}(\boldsymbol{q}, \omega_{\mathrm{L}}) \right)_{\perp}. \tag{4.12}$$

Being the nucleus a truly local probe, $1/T_1$ is related to the integral over the Brillouin zone of the component $S_{\alpha\alpha}(\boldsymbol{q}, \omega_{\mathrm{L}})$ of the dynamical structure factor, i.e. at the Larmor frequency. In (4.12), $|A_{\boldsymbol{q}}|^2$ is the form factor, which gives the hyperfine coupling of the nuclei with the spin excitations at wave vector $\boldsymbol{q}$, as illustrated in Fig. 4.3. The subscript $\perp$ indicates that contributions are given only by the products $|A_{\boldsymbol{q}}|^2 S_{\alpha\alpha}(\boldsymbol{q}, \omega_0)$ associated with the perpendicular components of the hyperfine field at the nucleus. By recalling that typical Larmor frequencies are such that $k_B T \gg \hbar\omega_{\mathrm{L}}$, one may use the fluctuation-dissipation theorem to write

$$\frac{1}{T_1} = \frac{\gamma^2}{2} \frac{k_B T}{\hbar} \frac{1}{N} \sum_{\boldsymbol{q}, \alpha = x,y,z} \left( |A_{\boldsymbol{q}}|^2 \frac{\chi''_{\alpha\alpha}(\boldsymbol{q}, \omega_{\mathrm{L}})}{\omega_{\mathrm{L}}} \right)_{\perp}, \tag{4.13}$$

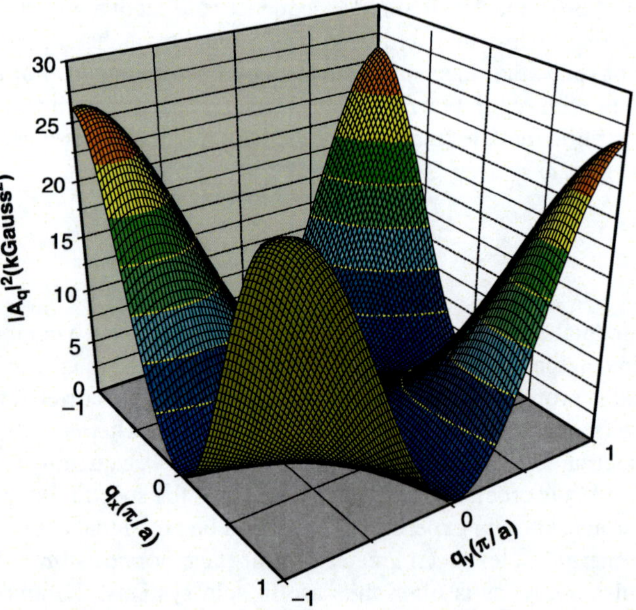


**Fig. 4.3** $^{29}$Si form factor in the first Brillouin zone of the two-dimensional frustrated antiferromagnet $Li_2VOSiO_4$. Excitations at wave vectors $(\pm\pi/a, 0)$ or $(0, \pm\pi/a)$ are filtered out, i.e. $^{29}$Si $1/T_1$ is not sensitive to these modes

which is a quite general expression of the relationship between the nuclear or muon spin-lattice relaxation and the spectrum of excitations characteristic of any magnetic system.

These equations for $1/T_1$ apply also to the $\mu$SR spin-lattice relaxation rate when a large magnetic field is applied. However, for $\mu$SR experiments at zero field, standard perturbative methods of the type used in (4.10) are no longer valid, because no transverse direction is defined and the internal field is not small compared to $H$. Accordingly, different methods are required to analyze data for the muon relaxation function in zero and small external fields. Further, usually the muon stopping site is not known, and the discussion must be phrased in terms of the field $\mathbf{B}$ at the muon site rather than the hyperfine coupling $A$. The treatment of the muon $T_1$ in zero or small fields requires two steps: the first is the static case, where $T_1 \to \infty$, and in the second the dynamic fluctuations are added so that $T_1$ becomes finite.

### 4.1.3 $\mu$SR: The Static Case

After entering the sample, a fully polarized muon comes to rest in a magnetic environment. Because the mechanism which stops the muon is much stronger than any magnetic interaction, the muon maintains its polarization while losing its kinetic energy. Only when the muon is at its stopping site does its spin begin to evolve in the local field $\mathbf{B}$. The muon polarization $P_z$ along the $\hat{\mathbf{z}}$-direction is given by the double-projection expression

$$P_z(\mathbf{B}, t) = \cos^2\theta + \sin^2\theta \cos(\gamma_\mu |\mathbf{B}|\, t), \tag{4.14}$$

where $\theta$ is the angle between the initial muon spin and the local field direction (Fig. 4.4). This angle is related to the field values by

$$\cos^2\theta = \frac{B_z^2}{\mathbf{B}^2}, \quad \sin^2\theta = \frac{B_x^2 + B_y^2}{\mathbf{B}^2}.$$

In a real sample, however, there is a distribution of internal fields and the averaged polarization is

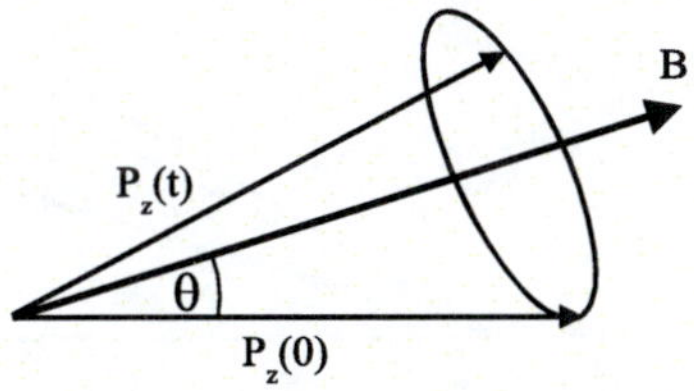


**Fig. 4.4** Muon spin polarization rotating around a magnetic field of arbitrary orientation

$$\overline{P}_z(t) = \int \rho(\mathbf{B}) \left[ \frac{B_z^2}{\mathbf{B}^2} + \frac{B_x^2 + B_y^2}{\mathbf{B}^2} \cos(\gamma_\mu \, |\mathbf{B}| \, t) \right] \mathrm{d}^3\mathbf{B}, \qquad (4.15)$$

where $\overline{P}_z(t)$ is the sample-averaged polarization and $\rho(\mathbf{B})$ is the field distribution, which is normalized according to $\int \rho(\mathbf{B})\mathrm{d}\mathbf{B}^3 = 1$. If the distribution of internal fields is a function only of $|\mathbf{B}|$, then one may write

$$\overline{P}_z(t) = \int \rho(|\mathbf{B}|) \left[\cos^2\theta + \sin^2\theta \cos(\gamma_\mu \, |\mathbf{B}| \, t)\right] B^2 \mathrm{d}B \mathrm{d}\Omega.$$

It is convenient to define $\rho'(|\mathbf{B}|) = 4\pi\rho(|\mathbf{B}|)$, so that $\int \rho'(|\mathbf{B}|)\mathbf{B}^2\mathrm{d}B = 1$ and the integral over the angular dependence can be performed to yield

$$\overline{P}_z(t) = \frac{1}{3} + \frac{2}{3} \int \rho'(|\mathbf{B}|) \cos(\gamma_\mu \, |\mathbf{B}| \, t) B^2 \mathrm{d}B.$$

If, for example, the system has long-range order, the field at the muon site is centered around a specific value $\omega_0/\gamma_\mu$ and thus

$$\rho'(|\mathbf{B}|) = \frac{\gamma_\mu}{\sqrt{2\pi}\Delta\mathbf{B}^2} \exp\left[ -\gamma_\mu^2 \left( |\mathbf{B}| - \frac{\omega_0}{\gamma_\mu^2} \right)^2 / 2\Delta^2 \right],$$

whence

$$\overline{P}_z(\omega_0, \Delta, t) = \frac{1}{3} + \frac{2}{3} \exp\left( -\frac{\Delta^2 t^2}{2} \right) \cos(\omega_0 t)$$

and oscillations will be observed in the data. At long times, the polarization will relax to $1/3$, because this is the fraction of muons experiencing a field parallel to their initial spin direction and therefore not relaxing.

When a longitudinal field $H$ is applied in the direction of the initial muon spin, as in Fig. 4.5, the situation becomes more complicated, and there is no closed-form expression. However, some simplifications can be made to reduce the dimension of the integrals for the purposes of a numerical calculation. For example, if the local field is completely random with a Gaussian distribution, then

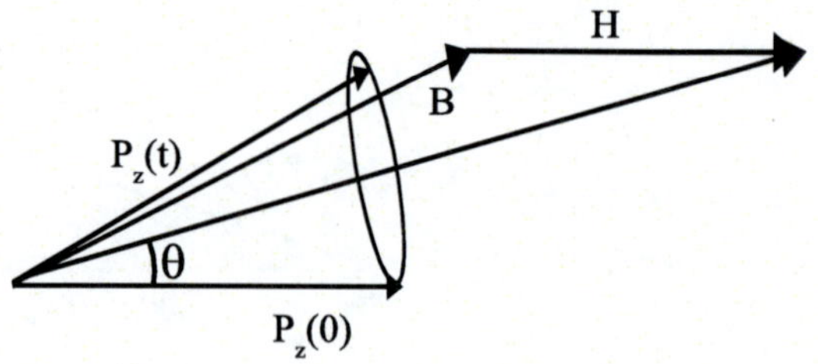


**Fig. 4.5** Muon spin polarization rotating around the vector sum of an arbitrary internal field and an external magnetic field in the initial polarization direction of the muon

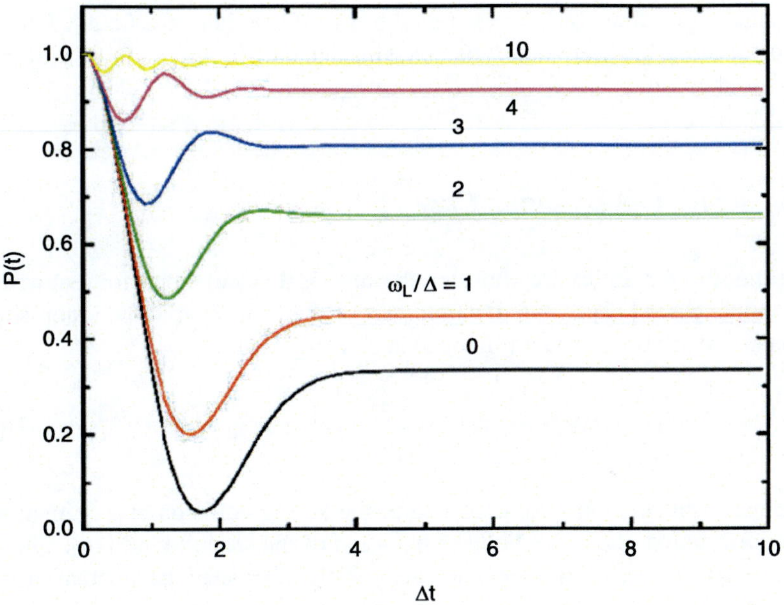


**Fig. 4.6** Muon polarization function in a Gaussian internal-field distribution and with different values of an external field $H$ oriented in the direction of the initial muon spin

$$\rho(\mathbf{B}) = \frac{\gamma_\mu^3}{(2\pi)^{3/2}\Delta^3}\exp\left(-\frac{\gamma_\mu^2[\mathbf{B} - H_0\hat{\mathbf{z}}]^2}{2\Delta^2}\right), \tag{4.16}$$

and (4.15) can be simplified to [11]

$$\overline{P}_z(\omega_\mathrm{L}, \Delta, t) = 1 - \frac{2\Delta^2}{(\omega_\mathrm{L})^2}\left[1 - \exp\left(-\frac{1}{2}\Delta^2 t^2\right)\cos(\omega_\mathrm{L} t)\right] + \frac{2\Delta^4}{(\omega_\mathrm{L})^3}\int_0^t \exp\left(-\frac{1}{2}\Delta^2\tau^2\right)\sin(\omega_\mathrm{L}\tau)\mathrm{d}\tau. \tag{4.17}$$

This is known as the static-Gaussian-longitudinal-field Kubo–Toyabe (KT) function. Figure 4.6 shows $\overline{P}_z(\omega_\mathrm{L}, \Delta, t)$ for a variety of values of $\omega_\mathrm{L}$. Despite the fact that the external field is in the muon spin direction, oscillations are observed in the polarization, and their frequency is given by $\omega_\mathrm{L}$. When $\omega_\mathrm{L} \gg \Delta$, the muon no longer relaxes, because the field at the muon site is nearly parallel to the initial muon spin direction. In this situation, the external field is said to decouple the muon spin from the internal field. Finally, in the zero-field case ($\omega_\mathrm{L} = 0$), (4.17) reduces to [11]

$$\overline{P}_z(0, \Delta, t) = \frac{1}{3} + \frac{2}{3}(1 - \Delta^2 t^2)\exp\left(-\frac{1}{2}\Delta^2 t^2\right). \tag{4.18}$$

This polarization function is known as the static-Gaussian-zero-field KT function. At short times it has Gaussian-type behavior, reaching a minimum on a time scale set by $\Delta$, after which it recovers and saturates again to $1/3$.

### 4.1.4 $\mu$SR: The Dynamic Case

When the spin dynamics are considered, numerical calculations are required apart from simple cases [12]. If the dynamic part of the local field at the muon site, $\delta\mathbf{B}$, fluctuates both in time and amplitude in such a way that

$$\langle \delta\mathbf{B}(t)\delta\mathbf{B}(0)\rangle = \frac{3\Delta^2}{\gamma_\mu^2}\exp(-2\nu t), \tag{4.19}$$

where $\nu$ is a fluctuation rate, then within the strong-collision approximation, the muon polarization obeys the Volterra equation of the second kind [16]. The polarization, $\overline{P}_z(\nu, \omega_\mathrm{L}, \Delta, t)$, now depends also on the characteristic fluctuation rate $\nu$, and obeys

$$\overline{P}_z(\nu, \omega_\mathrm{L}, \Delta, t) = \mathrm{e}^{-\nu t}\overline{P}_z(0, \omega_\mathrm{L}, \Delta, t) + \nu \int_0^t \mathrm{d}t' \overline{P}_z(\nu, \omega_\mathrm{L}, \Delta, t-t')\mathrm{e}^{-\nu t'}\overline{P}_z(0, \omega_\mathrm{L}, \Delta, t'), \tag{4.20}$$

where $\overline{P}_z(0, \omega_\mathrm{L}, \Delta, t)$ is the static relaxation function, namely the polarization if the local field were frozen in time. The factor $\mathrm{e}^{-\nu t}$ is the probability that the field remains unchanged until time $t$. The factor $\mathrm{e}^{-\nu t'}\nu \mathrm{d}t'$ is the probability density for the muon to experience a field change only between $t'$ and $t' + \mathrm{d}t'$. The first term on the right-hand side of (4.20) is the polarization at time $t$ due to muons that have not experienced any changes in their local field. The second is the contribution from those muons which have experienced their first change in field at time $t'$. The factor $\mathrm{e}^{-\nu t'}\overline{P}_z(0, \omega_\mathrm{L}, \Delta, t')\nu \mathrm{d}t'$ is the amplitude for the polarization function which evolves from time $t'$ to $t$, and can contain further field changes recursively. This equation can be solved numerically [13].

There are three ways of using (4.20) to obtain dynamical information. The first is in simple cases where $\overline{P}_z(0, \omega_\mathrm{L}, \Delta, t)$ is known analytically, as in the analysis of the F-$\mu$-F bond performed by Brewer et al. [14]. The second is when $\overline{P}_z(0, \omega_\mathrm{L}, \Delta, t)$ must be obtained numerically, as in the cases of Gaussian [11] or Lorenzian [15] field-fluctuation distributions and with an external longitudinal field. The third way is to measure $\overline{P}_z(0, \omega_\mathrm{L}, \Delta, t)$ by cooling the system to temperatures sufficiently low that dynamic fluctuations are no longer present, and to use the measured $\overline{P}_z(0, \omega_\mathrm{L}, \Delta, t)$ as input in the Volterra equation [16].

Taking the polarizations generated by the static-field distribution of (4.17) with $\omega_\mathrm{L} = 0$, and using this as input in the Volterra equation, gives the dynamic

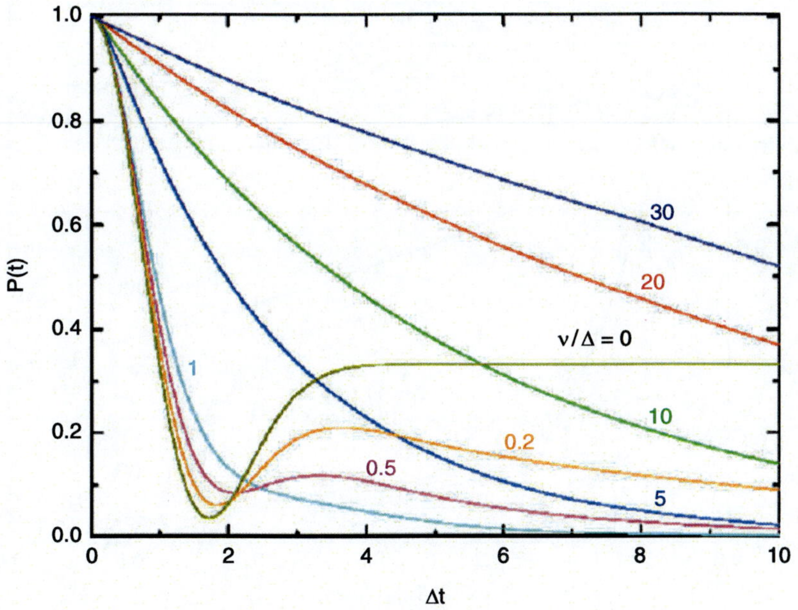


**Fig. 4.7** Expected muon relaxation in a dynamical field with Gaussian instantaneous distribution and no external field. Different values of the fluctuation rate $\nu$ are illustrated

polarizations shown in Fig. 4.7. This figure demonstrates that the $1/3$ recovery is lost when dynamic fluctuations are present. As the fluctuation rate $\nu$ increases, the Gaussian-type relaxation at $t \to 0$ is also lost, and when $\nu > \Delta$ the relaxation becomes exponential.

Finally, we present in Fig. 4.8 (solid lines) an example of the most complicated relaxation function, which combines a Gaussian field distribution, fluctuations, and a longitudinal external field. This is known as the dynamic-Gaussian-longitudinal-field KT relaxation function. In this figure, we have chosen special values of the parameters $\Delta = 11.8$ MHz and $\nu = 12.2$ MHz, for reasons which will become clear in Sect. 4.2.5, to illustrate the modification of the polarization as $H$ is varied.

We mention here that, for the case $\nu \geq \Delta$, there is an approximate expression for the dynamic-Gaussian-longitudinal-field KT relaxation function, given by [17]

$$\overline{P}_z(t) = \exp(-\Gamma(t)t) \tag{4.21}$$

with

$$\Gamma(t)t = \frac{2\Delta^2\{[\omega_L^2 + \nu^2]\nu t + [\omega_L^2 - \nu^2][1 - e^{-\nu t}\cos(\omega_L t)] - 2\nu\omega_L e^{-\nu t}\sin(\omega_L t)\}}{(\omega_L^2 + \nu^2)^2}.$$

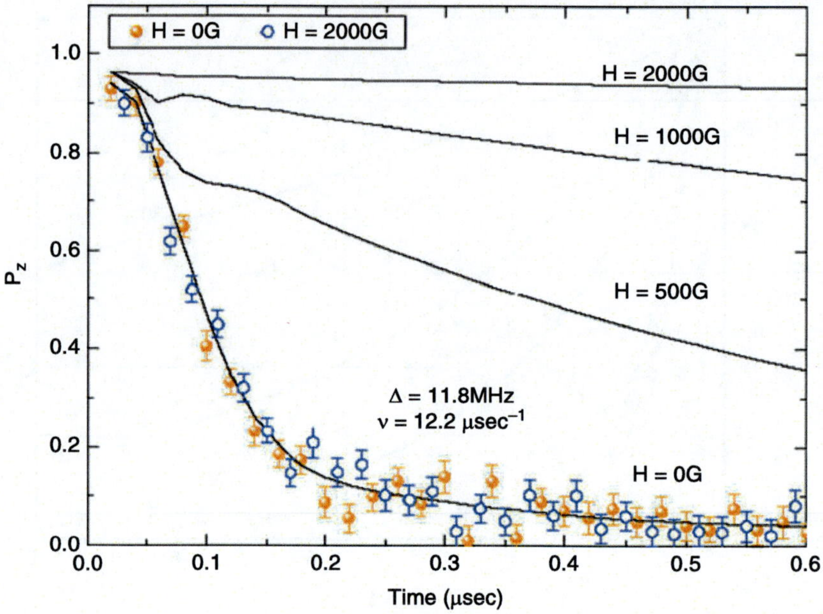


**Fig. 4.8** *Solid lines*: expected muon relaxation in a combination of internal field fluctuations with Gaussian instantaneous distribution and longitudinal external field $H$. *Symbols*: relaxation data from $SrCr_{9p}Ga_{12-9p}O_{19}$ from [52] for two different values of $H$. The discrepancy between the data and the model indicates unconventional behavior and is discussed in Sect. 4.2.6

In the long-time limit ($\nu t \gg 1$), one finds that $\lim_{t\to\infty} \overline{P}_z(t) = P_0 \exp(-t/T_1)$, leading to the standard expression for $T_1$,

$$\frac{1}{T_1} = \frac{2\Delta^2\nu}{(\omega_L^2 + \nu^2)}. \tag{4.22}$$

This result demonstrates that, when the external field is small ($\omega_L \ll \nu$), $T_1$ has no field-dependence. By contrast, when the field is large ($\omega_L \gg \nu$), $T_1$ increases with increasing field. Field-dependent measurements can therefore be used to provide information about $\nu$, and to distinguish between static and dynamic cases.

Equation (4.22) can be related to (4.13) for the value of $1/T_1$ deduced by NMR. In systems characterized by a spin-spin correlation function which decays in time according to $\exp(-\nu t)$, if the $q$-dependence is neglected one may express the spin susceptibility in the form

$$\chi''_{\alpha\alpha}(0,\omega) = \frac{(g\mu_B)^2}{\hbar V}\frac{\hbar\omega}{k_B T} N \left\langle S_\alpha^2 \right\rangle \frac{\nu}{\nu^2 + \omega^2},$$

and thus relate $\Delta^2$ in (22) to the amplitude of the spin fluctuations,

$$\Delta^2 = \frac{\gamma^2}{12} S(S+1) \sum_{q} \left( |A_q|^2 \right)_\perp .$$

When the applied field is small or the fluctuations are fast, (4.22) reduces to

$$\frac{1}{T_1} = \frac{2\Delta^2}{\nu}. \tag{4.23}$$

Typical values are $T_1 \sim 0.1\ \mu$s, $\Delta \sim 10$ MHz, and $\nu \sim 10\ \mu s^{-1}$.

## 4.2 From Zero- to Three-Dimensional Frustrated Magnets

### *4.2.1 Molecular Magnets*

In recent years, systems of molecular crystals containing molecules of magnetic ions have attracted considerable attention. These magnetic ions have significant intramolecular exchange couplings but negligible intermolecular couplings, so that each molecule in the crystal can be considered as an independent nanomagnet [18].

Some vanadium nanomagnets contain $V^{4+}$ ions, which have $S = 1/2$, in a triangular lattice. The best example of this is $V_{15}$, where at low temperatures there are 12 $V^{4+}$ spins coupled into singlets while the remaining three ions form a triangle. From a study of the $^{51}$V NMR line shift below 100 mK, Furukawa et al. [19] provided a preliminary estimate of the hyperfine coupling, and then deduced the expectation values of the $V^{4+}$ magnetic moments at zero field. The resonance frequency of the $i$th nucleus is given by $\omega_i = \gamma(\boldsymbol{H}_0 + \tilde{A}\langle \boldsymbol{S}_i \rangle)$, where the internal fields $\tilde{A}\langle \boldsymbol{S}_i \rangle$ have different orientations and magnitudes on the different sites. In order to derive $\langle \boldsymbol{S}_i \rangle$, the authors studied the magnetic field-dependence of $\omega_i$, as shown in Fig. 4.9. The expectation values were found to be consistent with a doubly degenerate ground state of the form

$$\begin{aligned}\psi_a &= \frac{1}{\sqrt{2}}(|\downarrow\downarrow\uparrow\rangle - |\downarrow\uparrow\downarrow\rangle)\\ \psi_b &= \frac{1}{\sqrt{6}}(2|\uparrow\downarrow\downarrow\rangle - |\downarrow\uparrow\downarrow\rangle - |\downarrow\downarrow\uparrow\rangle),\end{aligned} \tag{4.24}$$

where the symbol $\uparrow$ or $\downarrow$ represents the orientation of each one of the three spins. Measurements on the same compound by $\mu$SR [20] indicate a nearly temperature-independent longitudinal relaxation rate at low temperatures, which is thought to be associated with the transitions between these two degenerate states. If one considers that the frequency of the spin fluctuations satisfies $\omega \gg \omega_L$, one may derive the characteristic tunneling rate between the two states. We note here that a nearly

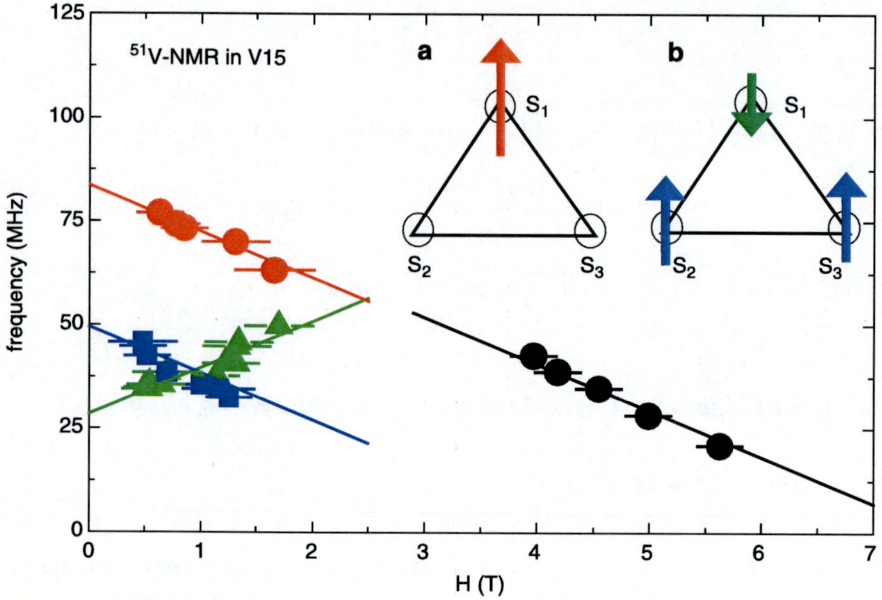


**Fig. 4.9** Magnetic field-dependence of the resonance frequency for different $^{51}V$ sites in $V_{15}$, used to extract the spin polarization of each $V^{4+}$ ion [19]. In the inset is shown a schematic representation of the twofold-degenerate ground state

temperature-independent relaxation is observed also in frustrated magnets with a macroscopically degenerate ground state (Sect. 4.2.6).

## 4.2.2 *Antiferromagnets on a Square Lattice with Competing Interactions: The $J_1$-$J_2$ Model*

$V^{4+}$ ions can also form other magnetic structures characterized by a strong frustration of the magnetic moments. Certain vanadates form some of the best known prototypes of frustrated magnets on a square lattice, where the frustration arises from the competition between nearest-neighbour ($J_1$) and next-nearest-neighbour ($J_2$) exchange couplings (see Chap. 15 by F. Becca et al.). The first NMR studies of such a system were performed on $Li_2VOSiO_4$. The $^7$Li-NMR spectra were observed to split into three different lines for $T < T_c \simeq 2.9\,K$ [21, 22], one unshifted and two shifted symmetrically with respect to the central one. This splitting of the NMR line was the first evidence that this compound is characterized by a collinear magnetic ground state, as confirmed some years later by neutron scattering experiments [23]. A careful study of the order parameter was conducted by means of zero-field $\mu$SR measurements (Fig. 4.10), where the $\mu^+$ polarization was shown to be characterized by oscillations at a frequency directly proportional to the $V^{4+}$ magnetic moment [24] (Sect. 4.1). The continuous increase of the order parameter for $T \to T_c$

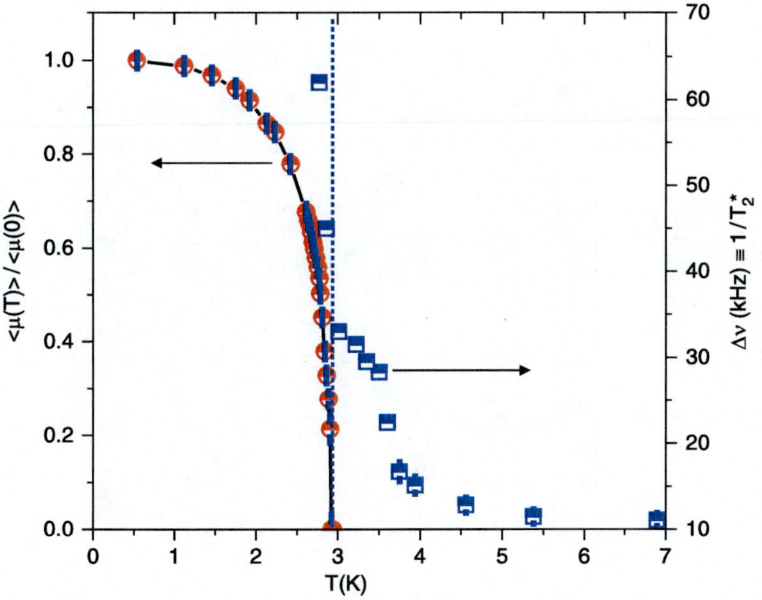


**Fig. 4.10** Temperature-dependence of the order parameter in the collinear phase of $Li_2VOSiO_4$, derived from zero-field $\mu$SR (*circles*), and of the $^7$Li-NMR line width for $\boldsymbol{H} \parallel ab$ (*squares*). Note that a broadening sets in well above the transition temperature (*vertical dotted line*) and is possibly caused by a frustration-driven lattice distortion

was found to be described by a critical exponent close to that expected for the 2D XY universality class. Above $T_c$, where no internal field is present, the NMR shift $K$ is expected to be proportional to the static, uniform spin susceptibility, $\chi_s$ (5), with a slope given by the hyperfine coupling. Remarkably, a change in the slope of $K(^7\text{Li})/\chi_s$ is apparent for $T_c < T < J_1 + J_2$ [25], suggesting a modification of the average hyperfine coupling (Sect. 4.1). Moreover, the broadening of the $^7$Li-NMR linewidth in the same temperature range can be explained only if the distribution of the hyperfine couplings is also broadening. We remark that in this temperature range, the average shift decreases on cooling below 5 K, while the linewidth increases, marking a breakdown of (9). A similar scenario is observed for $^{95}$Mo NMR in $MoVO_5$ [26]. These modifications of the hyperfine coupling are believed to be associated with a lattice distortion driven by the spin-lattice coupling, which lifts the degeneracy of the magnetic ground state.

The $T$-dependence of the in-plane correlation length, $\xi$, in $Li_2VOSiO_4$ has been estimated from measurements of $^7$Li $1/T_1$ [27] (Fig. 4.11). By scaling arguments, one may rewrite the dynamical structure factor (4.12) in terms of powers of $\xi$, and thus deduce that

$$\frac{1}{T_1} \sim \xi^z \sim \exp(2\pi z \rho_s / T), \tag{4.25}$$

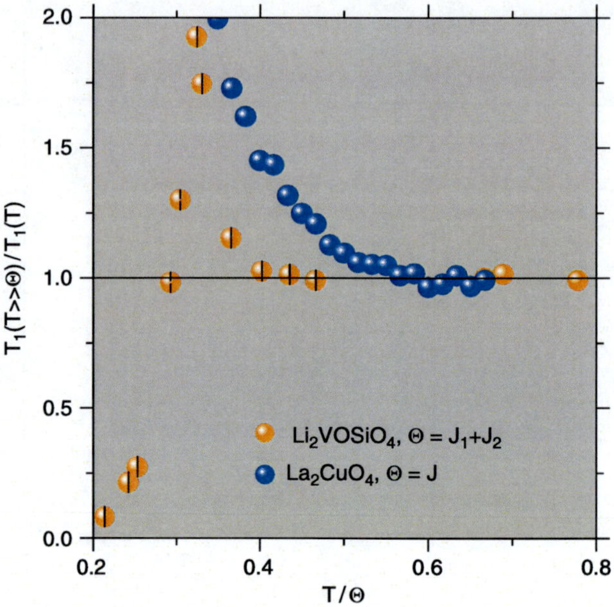


**Fig. 4.11** Comparison of the temperature-dependence of $^7$Li $1/T_1$ in $Li_2VOSiO_4$ with $^{63}$Cu $1/T_1$ in $La_2CuO_4$, where any frustration is negligible. The temperature is normalized to the Curie–Weiss temperature $\Theta$, while $1/T_1$ is normalized to its value at $T \gg \Theta$

where $\rho_s$ is the spin stiffness and $z$ the dynamical scaling exponent, which is estimated on the basis of additional physical considerations to be approximately $z = 1$ [27]. The conclusion from this type of analysis is that frustration in a two-dimensional antiferromagnet on a square lattice causes a less pronounced increase of $\xi$ on cooling, i.e. a decrease in the spin stiffness.

The decay of the longitudinal muon polarization in $Li_2VOSiO_4$ and $Li_2VOGeO_4$ is found to be described accurately by (4.21) and provides evidence of dynamical processes at frequencies well below the Heisenberg exchange frequency [24]. This behavior has been ascribed to fluctuations within an effectively twofold-degenerate ground state whose levels are separated by a barrier [28]

$$E(T) = \left(\frac{J_1^2 S^2}{2J_2}\right)\left[0.26\left(\frac{1}{S}\right) + 0.318\left(\frac{T}{J_2 S^2}\right)\right]\xi^2(T).$$

Recently, the spin dynamics and the ground-state of vanadates characterized by a ferromagnetic nearest neighbour exchange coupling has been studied by means of $\mu$SR and NMR [29, 30].

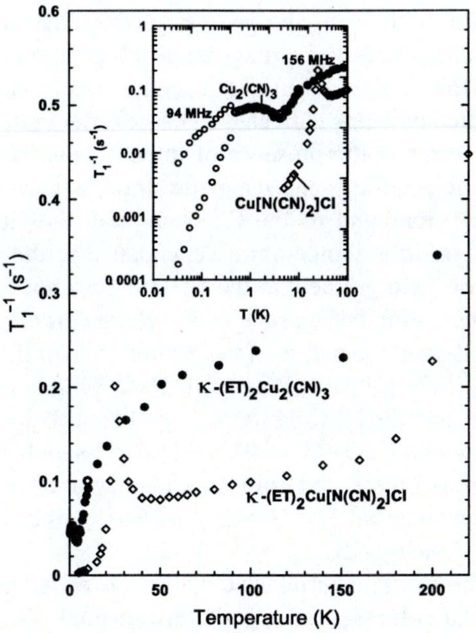


**Fig. 4.12** $^1$H nuclear spin-lattice relaxation rate $1/T_1$ for a single crystal (*open circles*) and a polycrystalline sample (*closed circles*) of $\kappa$-$(ET)_2Cu_2(CN)_3$ and a single crystal of $\kappa$-$(ET)_2Cu[N(CN)_2]Cl$ (*open diamonds*) [35]. The inset shows the low-temperature regime on logarithmic scales. Figure reprinted from Y. Shimizu, K. Miyagawa, K. Kanoda, M. Maesato, and G. Saito, Phys. Rev. Lett. **91**, 107001 (2003). Copyright (2003) by the American Physical Society

### *4.2.3 Magnetic Frustration on a Triangular Lattice*

The simplest two-dimensional lattice with purely geometrical frustration of the nearest-neighbor interactions is triangular. Several systems presenting magnetic moments on a triangular lattice, both insulating [31, 32] and metallic [33, 34], have been investigated in the past decade. Some of these compounds display a rather rich phase diagram as a function of the magnetic field intensity, as discussed in Chap. 10 by M. Takigawa and F. Mila in this volume. The molecular crystal $\kappa$-$(ET)_2Cu_2(CN)_3$, which is insulating at ambient pressure, also has $S = 1/2$ spins arranged on a slightly distorted triangular lattice, and here it has been observed (Fig. 4.12) that $^1$H $1/T_1$ decreases abruptly at low temperature [35]. This decrease can be associated with the onset of a gap $\Delta$ between collective singlet and triplet states, which causes the behavior $1/T_1 \propto \exp(-\Delta/T)$ for $T \ll \Delta$, and has been suggested as the first evidence for a spin-liquid phase in a triangular antiferromagnet. This may even be experimental evidence for the long-sought RVB state suggested by Anderson [36]. Yet more provocative is that the application of hydrostatic pressure is observed to drive this system into a superconducting ground state, which can also be explained in terms of an RVB description.

Another spin system with triangular geometry which has attracted much attention is $Na_xCoO_2$, where a rich phase diagram develops upon Na doping [37] (see also Chap. 22 by Z. Hiroi and M. Ogata). An accurate study of the $^{23}$Na and $^{59}$Co NMR spectra in oriented powders indicated a charge order in the range $x \simeq 0.7$ [8]. The signature of this order is the presence of three distinct Na sites characterized by different quadrupole couplings and magnetic shifts, which imply a well-defined order both of the $Na^+$ ions and of the Co charges in the $CoO_2$ planes. On the other hand, for $x = 0.5$, the temperature-dependence of the $^{59}$Co NMR spectra reveals that the electric-field gradient at the Co site does not change at the metal–insulator transition, indicating the absence of any charge ordering [38]. These NMR measurements have clarified the nature of the ground state of this system in the doping range $0.5 \leq x \leq 0.75$ [39, 40]. When full Na-doping is achieved, the system becomes nonmagnetic and the NMR shift vanishes [41, 42]. Somewhat remarkably, the isostructural compound $CoO_2$ ($x = 0$) is found to be metallic, with a crossover from a strongly correlated metal to Fermi-liquid behavior at low temperature [43]: the temperature-dependence of $1/T_1T$ starts to flatten, the hallmark of a Fermi liquid [1, 2], at approximately 4 K.

The recently synthesized material $NaCrO_2$ is an excellent realization of an isotropic $S = 3/2$ triangular Heisenberg antiferromagnet. The remarkable feature of this compound is that, while specific-heat and magnetization measurements indicate the onset of a transition at $T_c \simeq 40$ K, both muon spin-rotation and NMR data reveal a fluctuating crossover regime extending well below $T_c$, with a peak in $1/T_1$ at approximately 25 K (Fig. 4.13) [44]. This apparent discrepancy may suggest

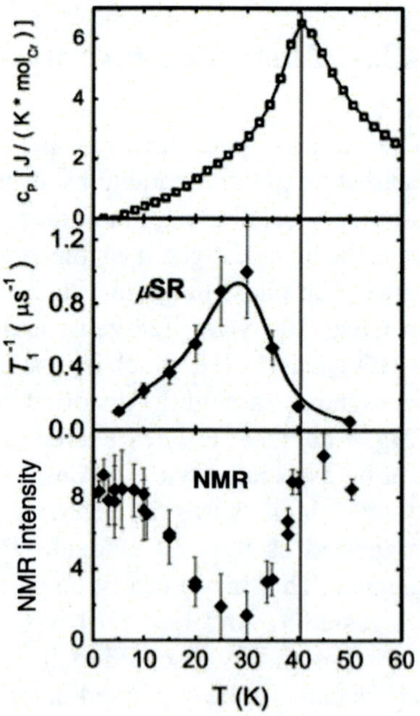


**Fig. 4.13** Specific heat (*top*), muon spin-lattice relaxation rate (*center*), and $^{23}$Na NMR signal intensity (*bottom*) in $NaCrO_2$ as functions of temperature. Note that the peaks in the specific heat and in $1/T_1$ occur at different temperatures. The loss of $^{23}$Na signal is possibly associated with an increase of the NMR relaxation rate

the occurrence of vortex–antivortex decoupling [45] yielding a spin dynamic in the MHz range around 25 K.

Magnetic frustration can be associated not only with the geometry of the electronic spins and their interactions, but also with the question of indistinguishable nuclear spins, as it is the case in $^3$He. The triangular lattice topology can be achieved by evaporating a single layer of $^3$He on a graphite substrate. Then it is convenient to use the intensity of the NMR signal, which is proportional to the nuclear magnetization, to track the $T$-dependence of the nuclear spin susceptibility [46]. $^3$He-NMR measurements have been performed down temperatures of some tens of $\mu$K, more than an order of magnitude below the effective exchange coupling. The low-temperature increase in the nuclear spin susceptibility demonstrates the relevance of higher-order multiple-spin-exchange interactions which, together with the triangular geometry of the nuclear spins, makes the system strongly frustrated and possibly characterized by a spin-liquid ground state [46–48].

### 4.2.4 $\mu$SR and NMR in the Spin-1/2 Kagomé Lattice $ZnCu_3(OH)_6Cl_2$

One of the most promising model compounds for the study of frustrated magnetism is herbertsmithite, $ZnCu_3(OH)_6Cl_2$, which has the kagomé geometry. The moments in this material originate from $Cu^{2+}$ ions, and have spin $S = 1/2$, making the system ideal for the investigation of quantum ground states. Unfortunately, different probes such as muons [49] and NMR on the O [50], Cu, and Cl ions [51] all suggest a different behavior of the line shifts below temperatures of order 50 K, and the origin of these variations is not yet clear. Here we present only the $\mu$SR results for $ZnCu_3(OH)_6Cl_2$, beginning with the muon shift and $T_2^*$, considered both independently (Fig. 4.14) and for their relative scaling (inset Fig. 4.14). More can be found in the Chap. 9 by Mendels and Wills. The muon shift, and hence the susceptibility, increase continuously upon cooling and saturate below 200 mK. This indicates that the spin-1/2 kagomé system in $ZnCu_3(OH)_6Cl_2$ does not freeze or form singlets. The ground state is paramagnetic. In addition, $K$ and $1/T_2^*$ track each other as the temperature is lowered, as expected from (4.9). Although $1/T_2^*$ is not a perfectly linear function of $K$, there is no reason to expect a modification in the hyperfine coupling upon cooling i.e. that a lattice deformation may occur in this case.

We next examine whether the ground state is separated by a gap from the excited states. If such a gap were to exist, a finite temperature would be required to generate excitations and achieve a non-zero value of $\chi''_{\alpha\alpha}(\boldsymbol{q}, \omega_0)$. Thus from (4.13) one would expect $1/(T_1 T)$ to extrapolate to zero. Such behavior has not been observed experimentally [50, 51] (Ofer et al., unpublished): the relaxation rate $^{37}$Cl $1/T_1$ shown in Fig. 4.15 remains proportional to $T$, suggesting that $\chi''_{\alpha\alpha}(\boldsymbol{q}, \omega_0)$ is finite in the limit $T \rightarrow 0$. With no evidence for a gap, the kagomé lattice seems to be an exotic magnet with no broken continuous symmetry but gapless excitations.

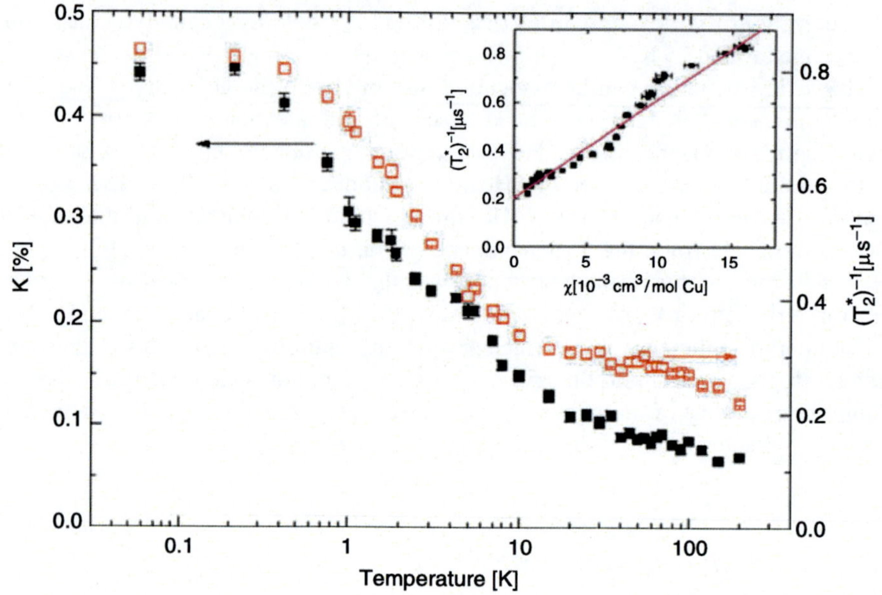


**Fig. 4.14** Muon shift $K$ and transverse relaxation time $1/T_2^*$ as functions of temperature in $ZnCu_3(OH)_6Cl_2$. Inset: scaling of $1/T_2^*$ with susceptibility

### 4.2.5 *The Problem of $\mu^+$ Relaxation in Some Kagomé Lattices*

At the very beginnings of research in the field of frustrated magnetism, it was noticed that the muon-relaxation function is unusual [52]. The symbols in Fig. 4.8 show the polarization at a temperature of 100 mK in the kagomé system $SrCr_{9p}Ga_{12-9p}O_{19}$ (SCGO), at zero field and in a longitudinal field of 2 kG [52]. First, no oscillations are found, meaning that the internal field is random with either static or dynamic nature. Second, the relaxation at short times is Gaussian, with a time scale of 0.1 $\mu$sec, so $\Delta$ must be on the order of 10 MHz. Third, there is no recovery, so there must be some dynamical field fluctuations as in Fig. 4.7, but with $\nu \sim \Delta$: if $\nu$ were larger than $\Delta$, the initial relaxation would be Lorenzian ((4.21)–(4.22) and Fig. 4.7); if $\nu$ were much smaller than $\Delta$, the polarization would have recovered at least partially. In these circumstances, a field of 2 kG, which is equivalent to $\omega_L = 170$ MHz should "decouple" the relaxation, but this does not happen: the solid lines in Fig. 4.8 represent the expected decoupling, which is very different from the measured data. A model has been proposed to explain this problem [52], which has received the name sporadic dynamics (SD) [53, 54].

In the SD model, the spin fluctuations in the sample do not relax the muon spin at all times but only for a fraction $f$ of the time, as illustrated in Fig. 4.16. In zero field, it is clear that such a form of the relaxation would lead to a sporadic dynamic polarization $\overline{P}_z^{\,sd}(\nu, 0, \Delta, t) = \overline{P}_z(\nu, 0, \Delta, ft)$. However, even when a field is applied, the polarization changes only when the internal field relaxes the muon spin, i.e. the

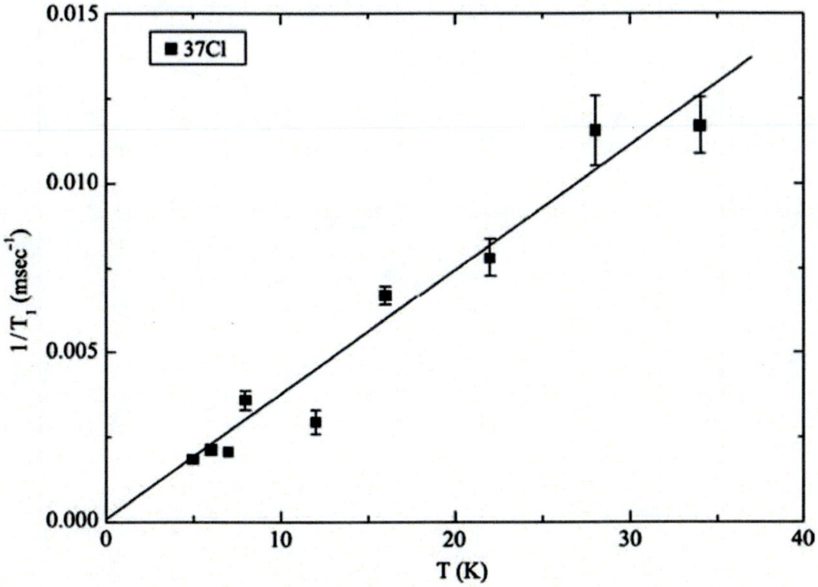


**Fig. 4.15** $^{37}$Cl $1/T_1$ in herbertsmithite ($ZnCu_3(OH)_6Cl_2$) at low temperature

flat parts of the curve in Fig. 4.16 remain flat even when $H$ is finite. One therefore expects that $\overline{P}_z^{\,sd}(\nu, \omega_L, \Delta, t) = \overline{P}_z(\nu, \omega_L, \Delta, ft)$ for all values of $\omega_L$. The fact that $\nu$, $\omega_L$, and $\Delta$ always enter the relaxation function as a product with $t$ [see for example (4.21)] mandates the result $\overline{P}_z^{\,sd}(\nu, \omega_L, \Delta, t) = \overline{P}_z(f\nu, f\omega_L, f\Delta, t)$.

When analyzing the data within this framework, one is in effect estimating that $f\Delta$ and $f\nu$ are 10 MHz, whence $\overline{P}_z^{\,sd} = \overline{P}_z(10, f\omega_L, 10, t)$. The effect of the field is therefore reduced by a factor $f$. In the fraction of the time for which the field is finite, $\Delta$ and $\nu$ are actually much higher than 10 MHz, and this is the reason that no decoupling is observed.

The SD model is very successful in explaining $\mu$SR data in frustrated systems. It has two possible physical interpretations. One is that the field at the muon site is zero for most of the time due to singlet formation of the spins in the sample, but occasional breaking of a singlet causes a fluctuating field and relaxation [52]. The problem with this scenario is that the muon relaxation is temperature-independent below 2–3 K, similar to the case presented in Fig. 4.17. This should happen when the system is in its ground state, but then there can be no time evolution, meaning that the internal field cannot fluctuate from zero to finite and the muon spin cannot relax. If the system is not in its ground state at $T \simeq 2$ K, then a lower energy scale, below the 100 mK limit of the experiment, must exist which separates the ground from the first excited state. Accordingly, the constant relaxation observed in the range 100 mK $\leq T \leq$ 2 K would arise from quantum fluctuations within a manifold of weakly coupled, nearly degenerate states. The other possibility is that the muon is hopping, although not regularly, between two sites with different relaxation rates.

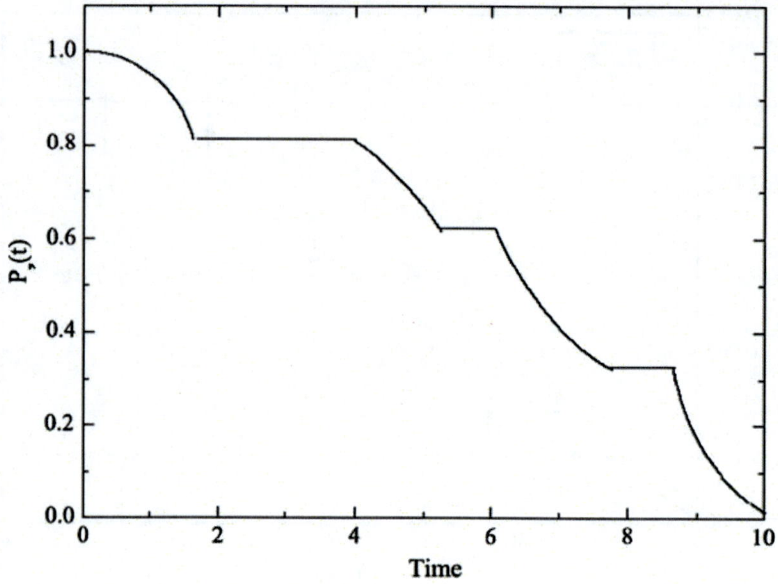


**Fig. 4.16** Schematic muon relaxation function for a situation in which the muons relax only sporadically, during certain time intervals

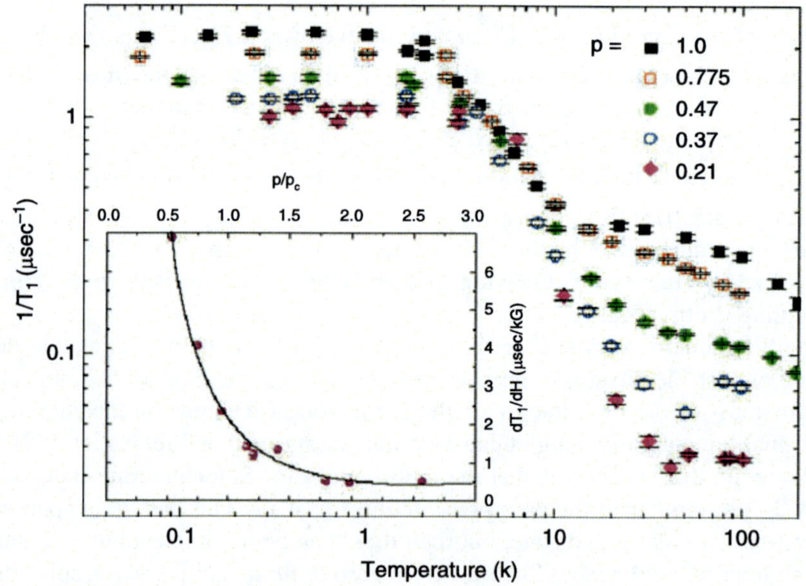


**Fig. 4.17** Temperature-dependence of the muon relaxation rate $1/T_1$ at $H = 50\,\mathrm{G}$ in $(\mathrm{Tb}_p\mathrm{Y}_{1-p})_2\mathrm{Ti}_2\mathrm{O}_7$ for different values of the magnetic ion concentration $p$. Inset: rate of change of $T_1$ with field as a function of $p$ normalized to the percolation threshold $p_c$

This, however, would be unusual at such low temperatures. A full description of the muon spin relaxation in SCGO and similar systems is still lacking. More information on SCGO and related materials can be found in Chap. 9 by P. Mendels and A.S. Wills.

### 4.2.6 Persistent Dynamics and Lattice Distortions in the Pyrochlore Lattice

If one common message can be drawn from the study of frustrated magnets by $\mu$SR, it is that these systems maintain a fluctuating component of their moments even when the temperature is lowered far below the coupling energy scale. This is manifest in the saturation of $1/T_1$ upon cooling. At zero field, $1/T_1$ is given by (4.23), and thus if the relaxation rate remains finite, it is because at least part of the moment continues to fluctuate and $\Delta$ cannot be zero. While both $T_1$ and $\Delta$ vary with the physical system, the typical scale for $\nu$ is $\mu$ sec$^{-1}$.

In Fig. 4.17, we present what may be the canonical example of persisting fluctuations to the lowest temperatures, $Tb_2Ti_2O_7$ [55]. The solid, black symbols denote $1/T_1$ over a wide temperature range. The increase in $1/T_1$ upon cooling indicates that $\nu$ decreases, meaning that the spin fluctuations slow down. However, in conventional magnets, long-range order or spin-freezing sets in at some temperature, and the amplitude of the fluctuations, $\delta B$ (4.19), decreases upon cooling, which is reflected in a decrease in $\Delta^2$. The net result is a peak in $1/T_1$ at, or close to, the critical temperature. The $1/T_1$ peak is missing in $Tb_2Ti_2O_7$ suggesting that no static magnetic moment develops in this system. Similar results have been obtained in other pyrochlore lattices, such as $Tb_2Sn_2O_7$ [56] and $Gd_2Ti_2O_7$ [54, 57].

This conclusion leads to yet another open question in this area of research, namely which type of excitation will dominate at low temperatures: candidates include spin-wave, spinless, and spinon excitations. This question can be investigated by measuring $1/T_1$ as a function of the magnetic ion concentration, $p$, both above and below the percolation threshold, $p_c$. A strong dependence of $1/T_1$ on $p$ close to $p_c$ would suggest that the fluctuations emerge from a collective phenomenon. By contrast, if $1/T_1$ varies smoothly across $p_c$, it would suggest that the excitations are local in nature and not sensitive to the overall coverage of the lattice. Figure 4.17 shows $1/T_1$ data for $(Tb_pY_{1-p})_2Ti_2O_7$ samples in which the magnetic Tb ion is replaced by nonmagnetic Y [58]. It is clear that the fluctuations have similar behavior both above and below the percolation threshold, which here is at $p_c = 0.39$ [59].

Moreover, the muon relaxation rate $T_1$ is found to be a linear function of $H$ [58]. Such behavior is very different from (4.22) and suggests that the internal-field correlations do not decay exponentially, but rather with a power-law form. In this case, the dynamical properties of the system would be contained in the quantity $\mathrm{d}T_1/\mathrm{d}H$. The inset of Fig. 4.17 shows $\mathrm{d}T_1/\mathrm{d}H$ as a function of $p$: there is no anomaly at $p_c$, suggesting that $T_1$ is indeed controlled by local excitations.

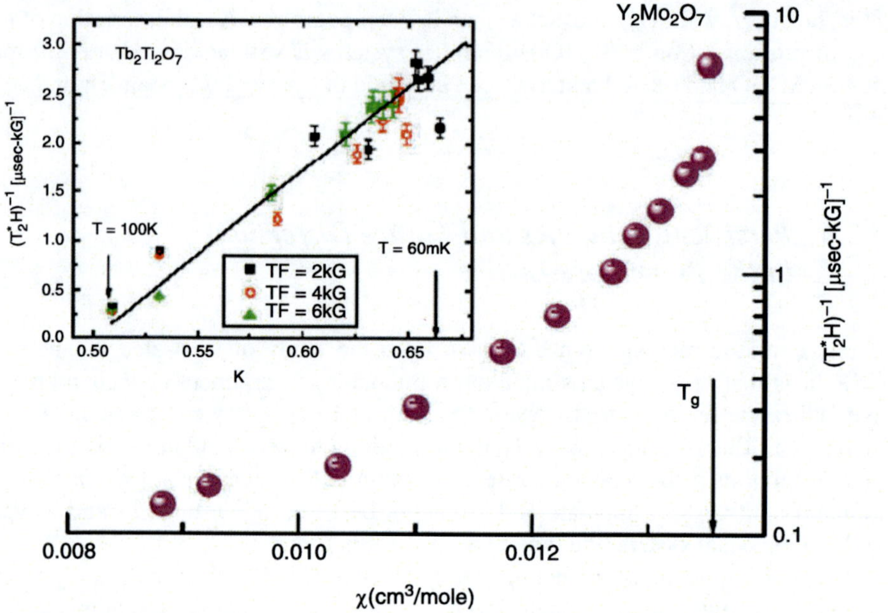


**Fig. 4.18** Field-normalized relaxation rate $1/T_2^*$ for $Y_2Mo_2O_7$, shown scaled with the susceptibility $\chi$ on a semi-log scale, with temperature an implicit parameter. Inset: field-normalized $1/T_2^*$ for $Tb_2Ti_2O_7$ on a linear scale

Finally, as in the case of $Li_2VOSiO_4$ (Sect. 4.2.2), magnetoelastic coupling [60–67] (Wang and Vishwanath, unpublished) is also very important in the pyrochlore system (as discussed in Chap. 11 of O. Tchernyshyov and G.-W. Chern), often leading to a lattice deformation in order to relieve the frustration. In fact, this interaction will cause some lattice deformation for arbitrarily small interaction strengths. Alternatively stated, there is no such thing as a perfect Heisenberg pyrochlore lattice, because it must distort. In practice, however, very small magnetoelastic couplings may give an undetectably small distortion. Nevertheless, the investigation of the ground state in the presence of magnetoelastic coupling is a growing theoretical sub-field which is accompanied by an intensifying experimental search for this effect.

As explained above, a comparison between the muon relaxation rate and the muon shift or macroscopic susceptibility can provide evidence for lattice deformations. In the example of the pyrochlore $Y_2Mo_2O_7$, $1/T_2^*$ depends exponentially on the susceptibility, as demonstrated in the semi-log graph of Fig. 4.18. The fact that the lattice parameters in $Y_2Mo_2O_7$ vary upon cooling has received confirmation from NMR [67] experiments. In contrast, the lattice of $Tb_2Ti_2O_7$ does not distort [68, 69], and indeed $1/T_2^*$ depends linearly on the susceptibility, as shown in the inset of Fig. 4.18.

In conclusion, in this chapter we reviewed the main aspects of NMR and $\mu$SR which are relevant for the study of frustrated magnets. We showed how the shift, splitting, and broadening of the spectra provides significant information on the ground-state and on the structural deformations induced by the spin-lattice coupling. We demonstrated how the nuclear and muon relaxation rates are sensitive to the peculiar low-energy excitations associated with the nearly degenerate ground-state which characterizes frustrated magnets. We provided examples of the application of these tools to the investigation of several model systems as, for instance: to triangular molecular magnets with a two-fold degenerate ground state; to $J_1$–$J_2$ and to pyrochlores systems with lattice distortions and low $T$ fluctuations, to triangular antiferromagnets with an energy gap in the low-energy excitations and to the kagomé lattice where there is no evidence of such gap. Many more subtle aspects of variety of frustrated systems are also presented.

**Acknowledgements**

The authors would like to acknowledge Y. Furukawa and K. Kanoda for contributing to this manuscript with figures reporting their results. A. Keren acknowledges support by the Israel-U.S.A. binational science foundation.

## References

1. C.P. Slichter, in *Principles of Magnetic Resonance*, 3rd ed. (Springer, Berlin, 1990)
2. A. Abragam, in *Principles of Nuclear Magnetism* (Oxford University Press, New York, 1961)
3. S.L. Lee, S.H. Kilcoyne, R. Cywinski (eds.), *Muon science* (IOP Publishing, Bristol and Philadelphia, 1999)
4. A. Schenck, *Muon Spin Rotation: Principles and Applications in Solid State Physics* (Adam Hilger, Bristol, 1985)
5. S. Blundell, Contemp. Phys. **40**, 175 (1999)
6. C.H. Pennington, C.P. Slichter, Phys. Rev. Lett. **66**, 381 (1991)
7. R. Ofer, S. Levy, A. Kanigel, A. Keren, Phys. Rev. B **73**, 012503, (2006)
8. I.R. Mukhamedshin, H. Alloul, G. Collin, N. Blanchard, Phys. Rev. Lett. 93, 167601 (2004)
9. L. Limot, P. Mendels, G. Collin, C. Mondelli, B. Ouladdiaf, H. Mutka, N. Blanchard, M. Mekata, Phys. Rev. B **65**, 144447 (2002)
10. H. Alloul, J. Bobroff, M. Gabay, P.J. Hirschfeld, Rev. Mod. Phys. **81**, 45 (2009) and references therein
11. R.S. Hayano, Y.J. Uemura, J. Imazato, N. Nishida, T. Yamazaki, R. Kubo, Phys. Rev. B **20**, 850 (1979)
12. A. Keren, L.P. Le, G.M. Luke, B.J. Sternlied, W.D. Wu, Y.J. Uemura, S. Tajima, S. Uchida, Phys. Rev. B **48**, 12926 (1993)
13. W.H. Press, B.P. Flannery, A.A. Teukolsky, W.T. Vetterling, *Numerical Recipes* (Cambridge University Press, Cambridge, 1989)
14. J.H. Brewer, D.R. Harshman, R. Keitel, S.R. Kreitzman, G.M. Luke, D.R. Noakes, R.E. Turner, Hyperfine Interact. **32**, 677 (1986)
15. Y.J. Uemura, T. Yamazaki, D.R. Harshman, M. Senab, E.J. Ansaldo, Phys. Rev. B **31** (1985)
16. A. Keren, J. Phys. Condens. Matter **16**, 1 (2004)
17. A. Keren, Phys. Rev. B **50**, 10039 (1994)
18. D. Gatteschi, A. Caneschi, L. Pardi, R. Sessoli, Science **265**, 54 (1994)
19. Y. Furukawa, Y. Nishisaka, K. Kumagai, P. Kögerler, F. Borsa, Phys. Rev. B **75**, 220402 (2007)

20. D. Procissi, A. Lascialfari, E. Micotti, M. Bertassi, P. Carretta, Y. Furukawa, P. Kögerler, Phys. Rev. B **73**, 184417 (2006)
21. R. Melzi, P. Carretta, A. Lascialfari, M. Mambrini, M. Troyer, P. Millet, F. Mila, Phys. Rev. Lett. **85**, 1318 (2000)
22. R. Melzi, S. Aldrovandi, F. Tedoldi, P. Carretta, P. Millet, F. Mila, Phys. Rev. B **64**, 024409 (2001)
23. A. Bombardi, F. de Bergevin, S. Di Matteo, L. Paolasini, P.Carretta, J. Rodriguez-Carvajal, P. Millet, R. Caciuffo, Phys. Rev. Lett. **93**, 027202 (2004)
24. P. Carretta, R. Melzi, N. Papinutto, P. Millet, Phys. Rev. Lett. **88**, 047601 (2002)
25. P. Carretta, N. Papinutto, R. Melzi, P. Millet, S. Gouthier, P. Mendels, P. Wzietek, J. Phys. Condens. Matter **16**, S849 (2004)
26. P. Carretta, N. Papinutto, C.B. Azzoni, M.C. Mozzati, E. Pavarini, S. Gonthier, P. Millet, Phys. Rev. B **66**, 094420 (2002)
27. N. Papinutto, P. Carretta, S. Gonthier, P. Millet, Phys. Rev. B **71**, 174425 (2005)
28. P. Chandra, P. Coleman, A.I. Larkin, Phys. Rev. Lett. **64**, 88 (1990)
29. P. Carretta, M. Filibian, R. Nath, C. Geibel, P.J.C. King, Phys. Rev. B **79**, 224432 (2009)
30. R. Nath, Y. Furukawa, F. Borsa, E.E. Kaul, M. Baenitz, C. Geibel, D.C. Johnston, Phys. Rev. B **80**, 214430 (2009)
31. L.K. Alexander, N. Büttgen, R. Nath, A.V. Mahajan, A. Loidl, Phys. Rev. B **76**, 064429 (2007)
32. A. Olariu, P. Mendels, F. Bert, B.G. Ueland, P. Schiffer, R.F. Berger, R.J. Cava, Phys. Rev. Lett. **97**, 167203 (2006)
33. Y. Kurosaki, Y. Shimizu, K. Miyagawa, K. Kanoda, G. Saito, Phys. Rev. Lett. **95**, 177001 (2005)
34. A. Kawamoto, Y. Honma, K. Kumagai, Phys. Rev. B **70**, 060510 (2004)
35. Y. Shimizu, K. Miyagawa, K. Kanoda, M. Maesato, G. Saito, Phys. Rev. Lett. **91**, 107001 (2003)
36. P.W. Anderson, Mater. Res. Bull. **8**, 153 (1973)
37. M.L. Foo, Y. Wang, S. Watauchi, H.W. Zandbergen, T. He, R.J. Cava, N.P. Ong, Phys. Rev. Lett. **92**, 247001 (2004)
38. J. Bobroff, G. Lang, H. Alloul, N. Blanchard, G. Collin, Phys. Rev. Lett. **96**, 107201 (2006)
39. J.L. Gavilano, B. Pedrini, K. Magishi, J. Hinderer, M. Weller, H.R. Ott, S.M. Kazakov, J. Karpinski, Phys. Rev. B **74**, 064410 (2006)
40. M.-H. Julien, C. de Vaulx, H. Mayaffre, C. Berthier, M. Horvatic, V. Simonet, J. Wooldridge, G. Balakrishnan, M.R. Lees, D.P. Chen, C.T. Lin, P. Lejay, Phys. Rev. Lett. **100**, 096405 (2008)
41. G. Lang, J. Bobroff, H. Alloul, P. Mendels, N. Blanchard, G. Collin, Phys. Rev. B **72**, 094404 (2005)
42. C. de Vaulx, M.-H. Julien, C. Berthier, M. Horvatic, P. Bordet, V. Simonet, D. P. Chen, C.T. Lin, Phys. Rev. Lett. **95**, 186405 (2005)
43. C. de Vaulx, M.-H. Julien, C. Berthier, S. Hübert, V. Pralong, A. Maignan, Phys. Rev. Lett. **98**, 246402 (2007)
44. A. Olariu, P. Mendels, F. Bert, B.G. Ueland, P. Schiffer, R.F. Berger, R.J. Cava, Phys. Rev. Lett. **97**, 167203 (2006)
45. H. Kawamura, J. Phys. Condens. Matter 10, 4707 (1998)
46. H. Ikegami, R. Masutomi, K. Obara, H. Ishimoto, Phys. Rev. Lett. **85**, 5146 (2000)
47. M. Roger, C. Bäuerle, Yu. M. Bunkov, A.-S. Chen, H. Godfrin, Phys. Rev. Lett. **80**, 1308 (1998)
48. H. Ishimoto, R. Masutomi, H. Ikegami, Y. Karaki, A. Yamaguchi, J. Phys. Chem. Solids **66**, 1417 (2005)
49. O. Ofer, A. Keren, Phys. Rev. B **79** 134424 (2009)
50. A. Olariu, P. Mendels, F. Bert, F. Duc, J.C. Trombe, M.A. de Vries, A. Harrison, Phys. Rev. Lett. **100** 087202 (2008)
51. T. Imai, E.A. Nytko, B.M. Bartlett, M.P. Shores, D.G. Nocera, Phys. Rev. Lett. **100** 077203, (2008)
52. Y.J. Uemura, A. Keren, K. Kojima, L.P. Le, G.M. Luke, W.D. Wu, Y. Ajiro, T. Asano, Y. Kuriyama, M. Mekata, H. Kikuchi, K. Kakurai, Phys. Rev. Lett. **73**, 3306 (1994)

53. D. Bono, P. Mendels, G. Collin, N. Blanchard, F. Bert, A. Amato, C. Baines, A.D. Hillier, Phys. Rev. Lett. **93**, 187201 (2004)
54. A. Yaouanc, Physica B **374–375**, 145 (2006)
55. S.R. Dunsiger, R.F. Kiefl, K.H. Chow, B.D. Gaulin, M.J.P. Gingras. J.E. Greedan, A. Keren, K. Kojima, G.M. Luke, W.A. MacFarlane, N.P. Raju, J.E. Sonnier, Y.J. Uemura, W.D. Wu, Phys. Rev. B **54**, 9019 (1996)
56. P. Dalmas de Réotier, A. Yaouanc, L. Keller, A. Cervellino, B. Roessli, C. Baines, A. Forget, C. Vaju, P.C.M. Gubbens, A. Amato, P.J.C. King, Phys. Rev. Lett. **96**, 127202 (2006)
57. A. Yaouanc, P. Dalmas de Réotier, V. Glazkov, C. Marin, P. Bonville, J.A. Hodges, P.C.M. Gubbens, S. Sakarya, C. Baines, Phys. Rev. Lett. **95**, 047203 (2005)
58. A. Keren, S. Gardner, G. Ehlers, A. Fukaya, E. Segal, Y.J. Uemura, Phys. Rev. Lett. **92**, 107204 (2004)
59. C.L. Henley, Can. J. Phys. **79**, 1307 (2001)
60. K. Terao, J. Phys. Soc. Jpn. **65**, 1413 (1996)
61. Y. Yamashita, K. Ueda, Phys. Rev. Lett. **85**, 4960 (2000)
62. S.-H. Lee, C. Broholm, T.H. Kim, W. Ratcliff II, S.-W. Cheong, Phys. Rev. Lett. **84**, 3718 (2000)
63. O. Tchernyshyov, R. Moessner, S.L. Sondhi, Phys. Rev. Lett. **88**, 067203 (2002)
64. J. Richter, O. Derzhko, J Schulenburg, Phys. Rev. Lett. **93**, 107206 (2004)
65. D.L. Bergman, R. Shindou, G.A. Fiete, L. Balents, Phys. Rev. B **74**, 134409 (2006)
66. K. Penc, N. Shannon, H. Shiba, Phys. Rev. Lett. **93**, 197203 (2004)
67. A. Keren, J.S. Gardner, Phys. Rev. Lett. **87**, 177201 (2001)
68. S.-W. Han, J.S. Gardner, C.H. Booth, Phys. Rev. B **69**, 024416 (2004)
69. O. Ofer, A. Keren, C. Baines, J. Phys. Condens. Matter **19**, 145270 (2007)

# Chapter 5
# Optical Techniques for Systems with Competing Interactions

**Joachim Deisenhofer and Peter Lemmens**

**Abstract** Optical spectroscopy and Raman scattering can be used to probe all the relevant degrees of freedom in a solid. In this chapter, we present an overview of the elementary and collective excitations accessible in compounds where spin, orbital, and lattice degrees of freedom compete strongly. For illustration, we will focus in particular on the magnetic response of the weakly coupled spin-tetrahedron system $Cu_2Te_2O_5(Br,Cl)_2$, on optical phonons in the highly frustrated pyrochlore antiferromagnets $CdCr_2O_4$ and $ZnCr_2O_4$, and on exciton-magnon excitations in the quasi-one-dimensional spin-1/2 Heisenberg antiferromagnet $KCuF_3$.

## 5.1 Introduction

In an inelastic light-scattering process, monochromatic light is scattered by variations in the electronic polarizability. The involvement of virtual electron-hole pairs leads to a strong sensitivity to local lattice distortions, as exemplified by local orbital states in manganites, or to local magnetic exchange processes in Heisenberg exchange-coupled systems. While magnetic light scattering in long-range-ordered systems leads to two-magnon signals whose shape is reminiscent of a density of states, frustrated spin topologies and topological ordering may lead to sharper, quasiparticle-like modes. Indeed, inelastic light-scattering has a very high sensitivity to singlet excitations, for example in weakly coupled spin tetrahedra. However, depending on the exchange geometry of the system and the temperature of the experiment, a multitude of overlapping modes, or quasielastic scattering, has also been observed [1, 2]. Inelastic scattering by electronic transitions is allowed at low energies if the scattering rate is large and screening can be prohibited. A number of specialized reviews have been written which cover correlated electron systems [3], high-temperature superconductors [4, 5], two-leg spin ladders [6, 7], and the materials aspects of perovskites [8].

In infra-red (IR) optical spectroscopy, the properties of a system are probed through the intensity and the polarization state of the reflected, absorbed, or transmitted light over an extended frequency range. The dominant contribution to the excitation processes observable by IR spectroscopy are transitions which are electric-dipole-allowed (ED-allowed, e.g. IR-active phonons), and hence the spin degrees of freedom are generally considered not to be directly accessible by IR spectroscopy. Magnetic-dipole (MD) transitions, however, are optically allowed and can be observed at far-infra-red (FIR) frequencies in magnetically ordered systems. This one-magnon excitation process is called antiferromagnetic resonance (AFMR) and can provide valuable information on the symmetry of a magnetically ordered phase. Because MD transitions are several orders of magnitude weaker than ED transitions, their detection is more probable in transmission geometry than in reflection. Recently, the prototypical antiferromagnetic insulator CoO, which has the rock-salt structure, was investigated in the FIR regime in reflection geometry and ED-active phonons, MD transitions, and electric quadrupole transitions were detected [9]. In addition to these single-ion excitations, cooperative excitations can occur which involve simultaneous transitions on two neighboring exchange-coupled magnetic ions: these include two-magnon absorptions, multi-magnon-plus-phonon processes, exciton–magnon, and exciton–exciton transitions [10].

In the following we will first describe general aspects of and typical set-ups for light-scattering experiments and IR spectroscopy. Then we will discuss examples of Raman and IR experiments where the spin degrees of freedom are clearly detectable in the excitation spectra.

## 5.2 Inelastic Light-Scattering

In an inelastic light-scattering experiment, monochromatic light in the visible range, for example with the wavelength $\lambda = 532.1$ nm, and with well-defined polarization is focused on the sample surface. Inelastically scattered light with a "Raman shift" (frequency-shift) of 5–3,000 $cm^{-1}$ is emitted, which has dipole characteristics and is sampled by a photo-lens of sufficiently large aperture. The Raman shift is an energy-proportional scale based on the inverse wavelength, on which the incident light given above corresponds to 18,794 $cm^{-1}$ and numbers on the order of 300 $cm^{-1}$ correspond to a typical phonon energy scale in a transition-metal oxide.

A second lens with an adapted focal length is used to focus the beam of parallel light on the entrance slit of the Raman spectrometer. The two lenses between the sample chamber or cryostat and the spectrometer allow for flexibility with respect to the distance between the experimental components. Figure 5.1 shows a typical set-up for Raman scattering.

Neither sensitivity nor frequency resolution are limiting factors in present-day Raman-scattering experiments. CCD arrays (back-illuminated and liquid-nitrogen-cooled) with long sampling times (20 s–20 min.) and multiple accumulations to remove spurious signals from cosmic muons allow the simultaneous acquisition of frequency shifts in a broad window corresponding to excitations of quasi-particles in the solid. The spectral resolution, which is of order several $cm^{-1}$ ($\equiv$1.44 K$\equiv$0.125 meV), depends on geometrical factors including the focal length,

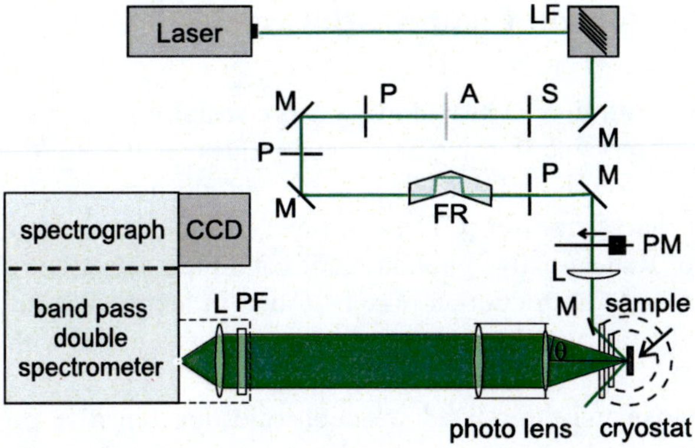


**Fig. 5.1** Schematic representation of a typical Raman set-up with mirrors (M), pinholes (P), holographic laser filter (LF), lenses (L), power meter (PM), fresnel rhombus (FR), polarization filter (PF), variable slit (S), and attenuator (A). The triple spectrometer includes a subtractive double monochromator that is used as a band-pass filter and a single monochromator that disperses the frequency band of inelastically scattered light on the CCD detector. The *thick line* denotes the light path from the laser to the sample and from the sample to the spectrometer

the slit widths, the number of grooves on the grating, and the density of pixels on the CCD array.

A limiting factor, however, is the set-up- and sample-related background level of fluorescence and stray light. Its origin may be surface contaminations and defects. As a consequence, freshly cleaved single-crystal surfaces are desirable. Further sources of background contributions include the glass of the collecting lenses, other optical components, and the laser itself. Intrinsic fluorescence processes are observed only rather rarely in extended solid states due to their quenching by nonradiative decay processes.

Taking into account that the scattering cross-section of magnetic and electronic excitations is one to two orders of magnitude smaller than the scattering intensity of phonons, and that these contributions have both low energy and a large bandwidth, the suppression of scattering with a similarly broad energy distribution is essential. For the laser, a pre-monochromator or holographic filter is used. Optical pinholes and a long beam path from the laser to the focussing lens further decrease background radiation. The optical beam path and the optical elements are separated for the exciting and for the scattered light.

Finally a controllable suppression of the elastically scattered light from the sample is gained by using a subtractive double monochromator in front of the dispersing spectrometer. Including the dispersing ("spectrograph") stage, the spectrometer is then referred to as a triple spectrometer. Single monochromator stages with supernotch filters, as for example in a high-aperture Raman microscope, do not generally comply with these requirements and are more suitable for material analysis and phonon studies.

## 5.3 Inelastic Phonon Light-Scattering

Optical phonons with $q \approx 0$ provide very large scattering cross-sections. Further, they are easily observable in transition-metal oxides, as the exciting light and the corresponding electron-hole pairs are generally in close proximity to a resonance to interband or charge-transfer (CT) excitations. The factor-group analysis, i.e. the determination of Raman-active phonon modes and their selection rules, can follow different schemes. An introduction may be found in text books on light scattering [11, 12]. The Bilbao Crystallograhic server provides web pages with corresponding services [13, 14].[1]

Optical phonons and the related selection rules are sensitive probes for lattice instabilities or anomalies whose origin lies in spin–phonon or electron–phonon coupling to fluctuating electronic or magnetic degrees of freedom. In such systems, anomalies of the phonon line width, such as asymmetric phonon lines (Fano line shape) and changes in the slopes of the temperature-dependent frequency or intensities, are very common observations. Quite generally, nature prevents energetic degeneracies and competing interactions of electronic or magnetic origin by using phonon degrees of freedom in a dynamic or a static way. The results are frequency shifts and intensity anomalies as a function of temperature, as represented in Fig. 5.2. Dynamic phonon anomalies spanning large temperature ranges are obviously of most interest [15].

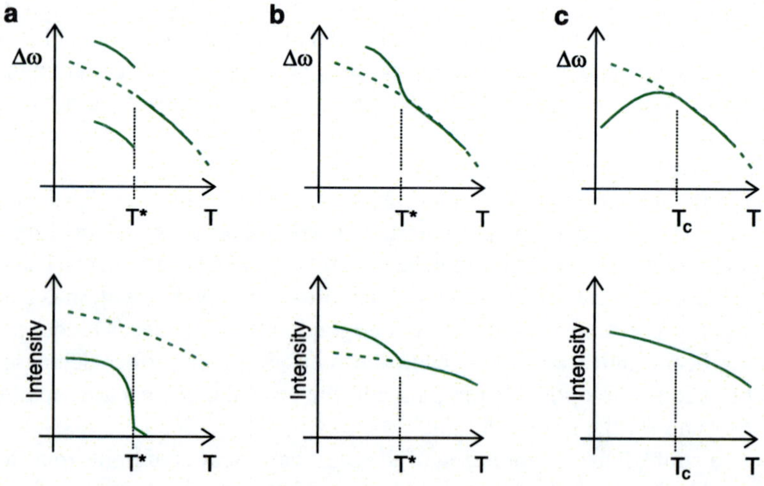


**Fig. 5.2** Phonon effects in Raman shift (*upper graph*) and intensity (*lower graph*) which may be observed in different compounds. Dashed lines correspond to the unperturbed situation. Full lines correspond to observations in a system with **(a)** a structural instability, **(b)** a crossover due to the mutual coupling of spin and phonon systems, and (c) orbital dynamics or other fluctuating states. In **(c)**, $T_c$ represents the Curie temperature, for example of a weakly doped manganite such as (La,Sr)$MnO_3$, which precedes the formation of orbital polarons

[1] http://www.cryst.ehu.es/rep/sam.html.

**Table 5.1** Phonon anomalies in Raman scattering observed in systems featuring mutual interplay of structural, electronic, and magnetic degrees of freedom, with respective examples

| Effect | Example |
|---|---|
| Spin-Peierls transition | $CuGeO_3$ [19,20] |
| Structural ordering | $SrCu_2(BO_3)_2$ [21,22] |
| Structural and charge ordering | $NaV_2O_5$ [23] |
| Fluctuation-induced phonon softening | TiOCl, TiOBr [24–26] |
| Crossover behavior | $\alpha$-$TeVO_4$ [27] |
| | $KCuCl_3$ [28], $TlCuCl_3$ [29,30] |
| Random interactions | $KCu_5V_3O_{13}$ [31] |
| Spin-state fluctuations | $[Fe(pmd)(H_2O)\{Au(CN)_2\}_2]\cdot H_2O$ [32] |
| Orbital-polaron formation | $(La_{1-x}Sr_x)MnO_3$ [16–18] |
| Canted orbital polarons | $(La_{0.5}Sr_{0.5})MnO_4$ [33] |
| Charge/spin stripes | $(La,Sr)_2NiO_4$ [34,35] |
| Local distortions | $CdCr_2S_4$ (Gnezdilov et al., unpublished) |

Compounds with incommensurate magnetic ordering often show related phonon anomalies close to the transition temperature. In contrast, doped manganites show a continuous softening of phonon frequencies without a characteristic temperature, which is due to the formation of orbital polarons [16–18]. In the sense of a local electronic defect, this type of anomaly shares a number of aspects with the effect of local polar distortions in a system such as $CdCr_2S_4$ (Gnezdilov et al., unpublished), which shows magnetocapacitive effects.

Different again is multi-phonon scattering, which is enhanced in the compounds under discussion and is usually non-monotonic in intensity but constant in frequency as function of temperature. This higher-energy scattering very often overlaps with magnetic scattering, reflecting the mixing of degrees of freedom under such conditions. In Table 5.1, several examples are listed which demonstrate the use of phonon modes to characterize the electronic or spin degrees of freedom in systems with competing interactions.

## 5.4 Inelastic Magnetic, Quasielastic, and Electronic Light Scattering

The study of magnetic excitations is helpful because they probe the typical energy scales of the magnetic system. Reviews of both the basic and the advanced aspects of magnetic light scattering in low-dimensional systems may be found in [1, 15], and for three-dimensional classical spin systems in [36].

The magnetic excitation spectra of compounds with a frustrated exchange geometry of quantum spins show a tendency towards spectral downshift compared to the two-magnon density of states of a long-range-ordered system. If the ground state is nonmagnetic, due for example to the formation of a spin gap in a $S = 1/2$ system, asymmetric, low-energy singlet modes may be observed. Further triplet modes are observed at similar energy scales. Singlet and triplet modes differ with respect to their behavior in a magnetic field. Similar to two-magnon scattering in gapless systems, singlet modes do not shift in frequency in a magnetic field. Longitudinal magnons (triplet excitations) occupy an intermediate position: they may be observed in long-range-ordered states close to a quantum critical point and thus where the ordered moment is strongly reduced. These modes show the temperature dependence expected for the order parameter, but no shift in a magnetic field.

Systems with helical or incommensurate order often show a multitude of sharp excitations which may overlap partly with the background, which in turn is a broader scattering continuum. This complex line shape is difficult to describe and depends on details of the magnetic ordering. The ordered moments of these systems are nevertheless larger when compared to the types of system discussed above, whence spin fluctuations are less pronounced. In Fig. 5.3, we illustrate schematically the approximate line shapes which may be observed in the parameter space given by the degree of competition between the interactions and the connectivity of the spin system. Examples of specific systems are listed in Table 5.2.

Systems of weakly coupled spin tetrahedra form a well established class of materials in which one may study the effect of competing interactions on magnetic excitations by light-scattering [37]. In Fig. 5.4, we show the crystal structure together with the low-energy excitation spectra of several compounds whose stoichiometry is derived from $Cu_2Te_2O_5(Br,Cl)_2$ [38, 39]. The lone-pair ion $Te^{4+}$, in combination with halogen and oxygen ions, leads to open structures which form the framework for magnetic exchange with restricted connectivity [1].

While the largest exchange constants are those within the tetrahedra, the most important exchange paths in the spin-tetrahedron compounds consist of halogen–oxygen bonds which couple the tetrahedra of $Cu^{2+}$ ions within the crystallographic *ab* plane. These interactions can be modified by replacing Cl by Br, leading to a weakening of the in-plane with respect to the out-of-plane coupling. The arrangement of the tetrahedra can also be modified as in $Cu_4Te_5O_{12}Cl_4$ [39], leading to a three-dimensionally coupled system more amenable (than the coupled-chain and planar materials) to a mean-field description.

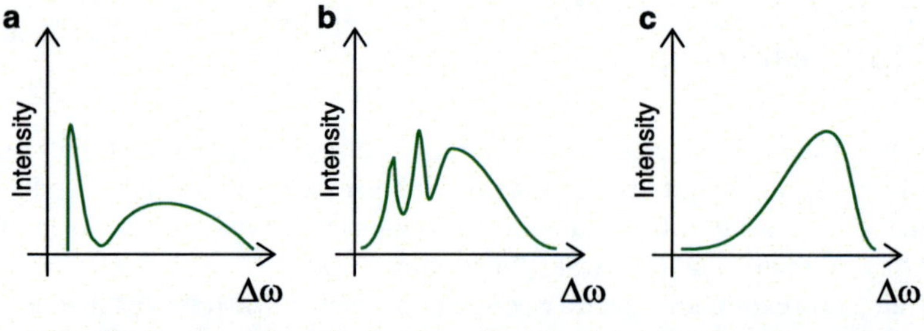


**Fig. 5.3** Representation of magnetic light-scattering intensity in different exchange geometries as a function of energy. The spectral-weight distribution depends on the presence of a spin gap at low energy and on the extent of spin frustration. Cases shown are (**a**) frustrated, dimerized spin chains, (**b**) coupled spin tetrahedra, and (**c**) a two-dimensional antiferromagnet

**Table 5.2** Examples of magnetic and electronic Raman scattering giving the observed phenomenon, the basic exchange geometry, and example systems. Mode labels are taken primarily from the references given. Mode line widths are determined by an interplay of the system connectivity with the degree of competition between spin exchange constants. This may be characterized by considering the ratio of intra- to inter-chain/dimer/tetrahedron coupling in the spin system. Fluctuations are enhanced as anisotropies increase, making the systems increasingly low-dimensional

| Phenomenon | Exchange system | Example |
|---|---|---|
| Singlet bound state | Dimerized, frustrated chain | $CuGeO_3$ [19] |
| Longitudinal magnon | Weakly coupled tetrahedra | $Cu_2Te_2O_5Br_2$ [37,38] |
| Sharp triplets | Coupled tetrahedra | $Cu_2Te_2O_5Cl_2$ [39] |
| Defect-induced bound state | Dimerized, frustrated chain | (Cu,Zn)$GeO_3$ [50,51] |
| Sharp triplets | Dimerized, coupled chains | $(VO_2)P_2O_7$ [52] |
| Sharp singlets and triplets | Frustrated dimers | $SrCu_2(BO_3)_2$ [53–55] |
| | 2D coupled dimers | $NaV_2O_5$ [56–59] |
| Three-magnon scattering | Chain, spin-Peierls | $CuGeO_3$ [60] |
| | 3D coupled dimers | $KCuCl_3$ [28] |
| Two-magnon continuum | 2D AF with stripes | $La_2NiO_4$ [61] |
| Two-magnon continuum | 3D AF | $KNiF_3$ [36,62,63] |
| Continuum, smaller line width | 3D coupled dimers | (Tl,K)$CuCl_3$ [28–30] |
| Resonance RS, gap formation | Spin ladder | $Sr_{14}Cu_{24}O_{41}$ [7,18,64] |
| Resonance RS | Square lattice | undoped HTSC [65,66] |
| Quasi-elastic Lorentzian | Alternating chain | $(VO_2)P_2O_7$ [52] |
| | Chain, spin-Peierls | $CuGeO_3$ [67] |
| | Chain with helical order | $NaCu_2O_2$ [68] |
| Electronic scattering | 1D collision-dominated | $BaVS_3$ [69] |
| | 2D square lattice | HTSC [3] |
| | 2D triangular lattice | $Na_xCoO_2$ [49] |
| | 3D polaronic | EuO, (Eu,La)$B_6$ [70] |

These systems of weakly coupled spin tetrahedra are found to be at the boundary between a localized product ground state of spin singlets and helical ground states with large unit cells and reduced ordered moments. It has been suggested that this crossover can be parameterized by considering the ratio of inter-tetrahedron coupling and a Dzyaloshinskii–Moriya interaction which is antisymmetric with respect to intra-tetrahedron exchange [38].

All known spin-tetrahedron compounds realize long-range ordering with critical temperatures $T_c < 20$ K [39]. However, the magnitudes of the ordered moments, the ordering vectors, and the energy scales in the inelastic excitation spectra differ strongly. As shown in Fig. 5.4a–c, the low-energy spectrum consists essentially of magnetic excitations with a well-defined and distinctive temperature-dependence. The frequency window of magnetic excitations is approximately 30–80 $cm^{-1}$ for all systems. In addition, a low-frequency "collective mode" and quasielastic scattering are observed for frequencies below 25 $cm^{-1}$. In the compound with the smallest $T_c$, and presumably the smallest in-plane inter-tetrahedron interaction, a low-energy mode has been identified as a longitudinal magnon. This excitation follows the

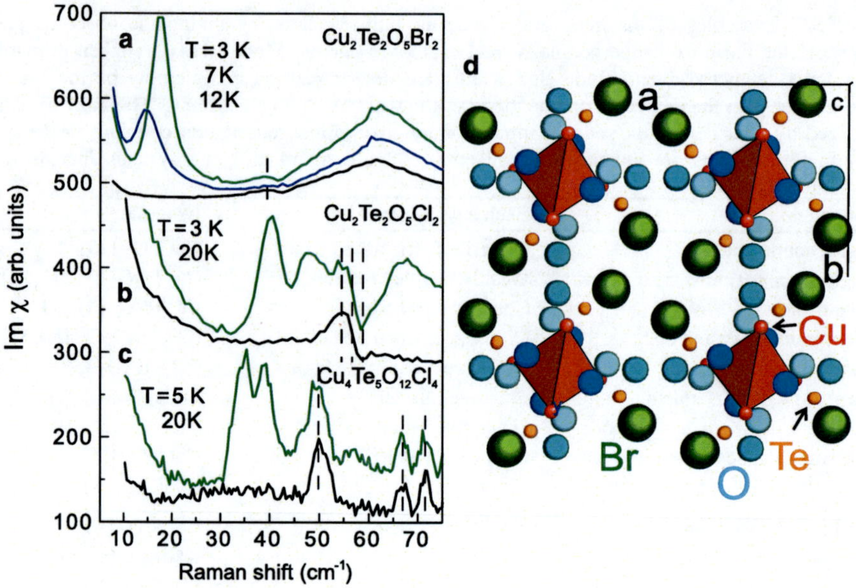


**Fig. 5.4** Low-energy Raman scattering in three different compounds with coupled spin tetrahedra at temperatures above and below an ordering temperature $T_c$. Data are shown for (**a**) $Cu_2Te_2O_5Br_2$ ($T_c = 11.4\,K$), (**b**) $Cu_2Te_2O_5Cl_2$ (18.2 K), and (**c**) $Cu_4Te_5O_{12}Cl_4$ (13.6 K); the curves are shifted for clarity. Dashed lines denote scattering intensity due to phonons. (**d**) Representation of the crystal structure in *ab*-plane projection. The tetrahedra formed from Cu ions coordinated by oxygen ions on two different sites are filled. Exchange paths between the tetrahedra are formed by Br (*large circles*) and O ions (*intermediate-sized circles*)

temperature-dependence of the order parameter, i.e. of the sublattice magnetization. By contrast, the higher-energy continuum shows no shift and only a depression of its intensity with increasing temperature (Fig. 5.4a). Increasing the in-plane coupling suppresses this mode completely. Instead, a larger number of excitations with smaller line width appears. The more three-dimensional system $Cu_4Te_5O_{12}Cl_4$ shows only two very sharp modes which have no temperature-dependence of their energy (Fig. 5.4c). This effect is a consequence of the suppression of spin fluctuations in the latter system.

For two- or three-dimensional compounds with larger connectivity and ground-state degeneracy, a multitude of singlet states leading to a continuum of scattering is expected. With the exception of Raman scattering, these singlet excitations can only be observed through their contribution to the specific heat [40]. Examples where such excitations have been detected definitively in Raman studies remain rather rare, although good candidates are compounds with square planar lattices featuring additional diagonal coupling ($J_1$–$J_2$ systems [41]) and kagomé materials [42, 43]. The theoretical relevance of Raman scattering in kagomé systems is worth noting because it can distinguish spin-liquid from symmetry-broken ground states by observing the anisotropy of the magnetic scattering [44] and its line shape at low energies (Läuchli and Lhuillier, unpublished).

We include in this section a discussion of quasielastic and electronic scattering. Because all Raman-scattering processes are by definition electronic, we follow the pragmatic approach of defining as quasielastic scattering any very low-energy scattering which arises due to energy-density fluctuations of the magnetic system. These processes are mediated by spin–phonon coupling and lead to a Lorentzian line width and a temperature-dependence related to the magnetic specific heat [45, 46]. This type of scattering is observed with different intensities in all magnetic compounds where competing interactions are present.

Electronic Raman scattering with a single scattering time has the same Lorentzian line shape and is screened at finite frequencies [5]. However, if the fluctuations consist of variations of the effective mass within an anisotropic band or if sources of anomalous scattering exist, then finite-energy scattering is observable. While the first case was demonstrated first for doped semiconductors under the application of anisotropic pressure [47], the second is relevant in systems such as dirty or correlated metals [48]. In non-Fermi-liquid systems, an anomalous energy- and frequency-dependent scattering rate, arising from polaronic effects or from nesting, leads to a frequency-independent scattering continuum that extends to rather high energies. The best-known candidates for such scattering are high temperature superconductors (HTSC), where the electronic Raman continuum extends to several thousand wavenumbers [3], and the hydrated cobaltates, $Na_xCoO_2{\cdot}yH_2O$ [49], where this contribution is observed in Raman shifts up to $550\,cm^{-1}$ (Table 5.2).

## 5.5 The IR Experiment

A typical experimental set-up for performing IR measurements is shown in Fig. 5.5. Before impinging on the sample, the light is passed through a Michelson interferometer unit where the beam is split by a beamsplitter (BS) and reflected by a fixed and a scanning mirror (FM and SM). The result is an interferogram which provides the desired spectral information after Fourier transformation.

The shape of the resulting spectrum depends on the emission of the source, and is influenced by all of the optical elements in the beam path. The reflectivity of a sample must thus be obtained as the ratio of its spectrum in comparison to the spectrum of a reference material such as gold, silver, or aluminum, which are assumed to be ideally reflective in a certain frequency range. Extrapolating the measured reflectivity or transmission data to zero and to infinite frequency allows one to perform a Kramers–Kronig transformation to obtain the real and imaginary parts of the dielectric function or the optical conductivity. A direct experimental determination of both imaginary and real parts of the dielectric function becomes possible by performing ellipsometry measurements, i.e. by measuring the exact polarization state of the elliptically polarized light reflected from the sample [71].

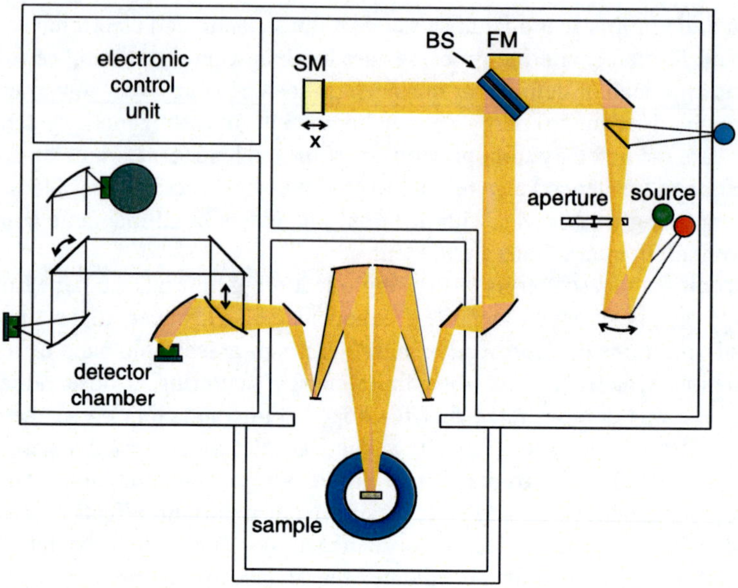


**Fig. 5.5** Representation of an IR Fourier spectrometer (Bruker IFS66v) with a Michelson interferometer unit consisting of a beam-splitter (BS), a scanning (SM), and a fixed mirror (FM) [73]

## 5.6 Spins, Phonons, and Light

To illustrate how electromagnetic radiation with electric- and magnetic-field vectors **E** and **H** couples to the lattice and spin degrees of freedom, we follow the discussion of Mizuno and Koide [72] and start with the Hamiltonian in the form

$$H = \sum_i \frac{\mathbf{p_i}^2}{2m_i} + \sum_i D_i(\mathbf{E}) + \sum_i \mathbf{f_i}(\mathbf{E}) \cdot \mathbf{r_i} + \sum_{i \leq j} \mathbf{r_i} \cdot k_{ij}(\mathbf{E}) \cdot \mathbf{r_j} \\ + \sum_{i<j} \mathbf{S_i} \cdot J_{ij}(\mathbf{r}, \mathbf{E}) \cdot \mathbf{S_j} + \sum_i K_i(\mathbf{S_i}, \mathbf{E}, \mathbf{r}) + \beta \sum_i \mathbf{H} \cdot g_i(\mathbf{r}) \cdot \mathbf{S_i}, \quad (5.1)$$

where $m_i$ is the mass, $\mathbf{p_i}$ the momentum, $\mathbf{r_i}$ the displacement from its equilibrium position, and $\mathbf{S_i}$ the spin of the $i$th ion. With no external electric field, the first line of the Hamiltonian (5.1) reduces to the conventional form for a lattice in the harmonic approximation and the second line to the effective spin Hamiltonian for an exchange-coupled system with exchange constants $J_{ij}$, single-ion anisotropy $K_i$, and a Zeeman term to account for MD transitions. By neglecting the latter two (single-ion) terms, assuming $J_{ij}$ to be a scalar (isotropic, diagonal interactions), and expanding the Hamiltonian with respect to **r** and **E**, one obtains at zeroth order the well-known lattice Hamiltonian, a Heisenberg term, and a set of terms linear in **E**,

$$H_E = \sum_i \mathbf{E}\cdot\left(\frac{\partial D_i}{\partial \mathbf{E}}\right)_0 + \sum_i \mathbf{E}\cdot\left(\frac{\partial \mathbf{f}_i}{\partial \mathbf{E}}\right)_0 \cdot \mathbf{r_i} + \sum_{i\leq j} \mathbf{r_i}\cdot\mathbf{E}\cdot\left(\frac{\partial k_{ij}}{\partial \mathbf{E}}\right)_0 \cdot \mathbf{r_j}$$
$$+ \sum_{i<j} \mathbf{E}\cdot\left(\frac{\partial J_{ij}}{\partial \mathbf{E}}\right)_0 \cdot (\mathbf{S_i}\cdot\mathbf{S_j}) + \sum_{i<j} (\mathbf{S_i}\cdot\mathbf{S_j}) \sum_k \mathbf{E}\cdot\left(\frac{\partial^2 J_{ij}}{\partial \mathbf{E}\partial \mathbf{r_k}}\right)_0 \cdot \mathbf{r_k}. \quad (5.2)$$

Here the second and third terms account respectively for the one- and two-phonon processes which dominate the FIR spectra in dielectric crystals. The fourth and the fifth terms contain spin-dependent effective dipole moments, and describe, respectively, dipole-active two-magnon excitation processes and two-magnon-plus-one-phonon absorptions. The former were first observed in $FeF_2$ [74] and $MnF_2$ [75] and were described theoretically by Tanabe and coworkers [10, 76]. Microscopically, the effective dipole moment of the two-magnon excitation arises from off-diagonal terms in the superexchange coupling involving two spin-flips (because of spin conservation) on two neighboring sites in the ground state [77]. In a lattice with inversion symmetry, such as the NaCl-type structure of NiO, the two-magnon excitation is not allowed. In order to break the inversion symmetry, the additional excitation of an odd-symmetry phonon is necessary, leading to two-magnon-plus-one-phonon processes which are ED-allowed [72, 78]. Observation of these processes in the parent compound of the cuprates, $La_2CuO_4$, led to a generalized description of multi-magnon-plus-phonon excitations [79, 80]. As in the two-magnon process, spin-flips can also occur accompanied by the excitation of one or both of the neighboring ions into a higher orbital state, giving rise to exciton–magnon or exciton–exciton transitions, respectively [77, 81, 82]. Very recently, a hybridized magnon–phonon excitation, dubbed the electromagnon, has been discovered experimentally [83]. The electromagnon becomes observable in the presence of strong spin-lattice coupling and appears to be a feature of multiferroic systems with spiral spin structure, but the exact mechanism of the coupling is still the subject of active discussion [83, 84]. In Table 5.3, we present a short list of excitations observable by optical measurements, distinguishing between excitations related to either lattice or spin degrees of freedom and excitations arising due to coupling between the subsystems. The examples and references were chosen to include both early and recent studies and systems of differing effective magnetic dimensionality. Comprehensive review articles are indicated where available.

In the expansion of the Hamiltonian (5.1), a further term appears which does not depend on **E**, but is linear in the ion displacement,

$$H_{SP} = \sum_{i<j} (\mathbf{S_i}\cdot\mathbf{S_j}) \sum_k \left(\frac{\partial J_{ij}}{\partial \mathbf{r_k}}\right)_0 \cdot \mathbf{r_k}. \quad (5.3)$$

This term describes a magnetoelastic coupling between the spins and the lattice, reflecting the influence of exchange coupling on the lattice vibrations probed by

**Table 5.3** Elementary excitations detectable by optical spectroscopy in crystals with lattice, orbital, and spin degrees of freedom

| Excitation | Degree of Freedom | Examples | Monographs/ Reviews |
|---|---|---|---|
| Phonon | Lattice | $CdCr_2S_4$ [85] | [86] |
| One-magnon | Spins | $FeF_2$ [87], $LuMnO_3$ [84] | |
| Two-magnon | Spins | $FeF_2$ [74], $MnF_2$ [75], $SmTiO_3$ [88] | [10] |
| Exciton-magnon | Spins | $MnF_2$ [89], $KCuF_3$ [90] | |
| Exciton-exciton | Spins | $MnF_2$,$RbMnF_3$ [91] | |
| | Coupling | | |
| Vibron | Electron–phonon | $CuGeO_3$ [92], $LaTiO_3$ [93] | [94] |
| Polaron | Electron–phonon | $La_{1-x}(Ca,Sr)_xMnO_3$ [95] | [96] |
| Phonon-assisted | Magnon–phonon | NiO [72,78] | [97] |
| Multi-magnon | | $La_2CuO_4$ [79,80] | |
| | | $Sr_2CuO_3$ [98] | |
| electromagnon | Magnon–phonon | $TbMnO_3$ [83], $TbMn_2O_5$ [84] | |

IR spectroscopy. This term is discussed in detail for the case of the magnetically frustrated pyrochlore lattice in Chap. 11 by Tchernyshyov and Chern. In the following, we discuss experimental results for the FIR phonon spectra in Cr spinels, compounds which are the paradigm for frustrated antiferromagnets displaying this type of physics.

## 5.7 Spin–Phonon Interaction in Cr Spinels

The quest for realizations of highly frustrated magnets leads naturally to the spinel systems $ACr_2X_4$ ($X = $ O,S,Se). As described in Chap. 7 by Takagi and Niitaka, in one spinel family the $A$ site is occupied by nonmagnetic ions such as Zn or Cd [99]. The $Cr^{3+}$ ions have spin $S = 3/2$ and are situated on corner-sharing tetrahedra which form a pyrochlore lattice (inset Fig. 5.6). Examples such as $CdCr_2O_4$ (with frustration parameter $f = \Theta_{CW}/T_N = 32$) and $ZnCr_2O_4$ ($f = 8.7$) can be regarded as paradigms for systems where the antiferromagnetic nearest-neighbor Heisenberg exchange leads to inherent frustration and considerable degeneracy of the magnetic ground state [100]. The phase diagram in Fig. 5.6 shows that the strong frustration in $ZnCr_2O_4$, where $\Theta_{CW} = -3,98$ K, is diminished when the lattice parameter is enlarged by substituting Cd and Hg for Zn or S and Se for O. Weakening of the direct Cr–Cr coupling, which is antiferromagnetic, is accompanied by a strengthening of ferromagnetic next-nearest neighbor interactions, which lead to bond frustration with $\Theta_{CW} = T_N \approx 10$ K for $ZnCr_2S_4$, and finally to ferromagnetism in cases such as $CdCr_2S_4$. In reality, the degeneracy of the magnetic ground state is lifted at finite temperatures by the coupling to other degrees of freedom. The spin-Jahn–Teller effect is an example of this phenomenon, and

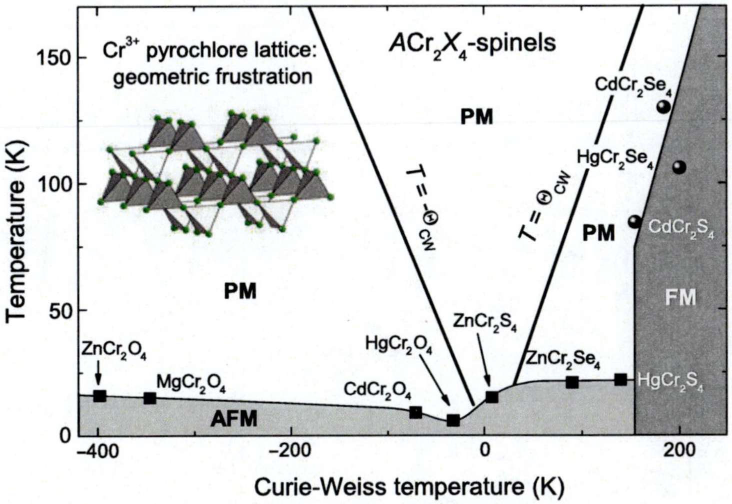


**Fig. 5.6** Phase diagram for $ACr_2X_4$ ($A$ = Zn,Mg,Cd, $X$ = O,S,Se) spinels. The Cr spins occupy a pyrochlore lattice, leading to strong geometric frustration of the dominant antiferromagnetic interactions. For $X$ = S,Se, competition with ferromagnetic interactions leads to a change in sign of the Curie–Weiss temperature and thus to a ferromagnetic ground state

both the single-tetrahedron case and the collective case of the pyrochlore lattice (specifically $CdCr_2O_4$ and $ZnCr_2O_4$) are presented in Chap. 11 by Tchernyshyov and Chern. This spin–Jahn–Teller effect in the Cr spinels leads to a tetragonal or an orthorhombic distortion of the lattice and to corresponding splitting of the IR-active phonon modes.

Experimentally, four phonon excitations are observed in the IR spectra of these spinel systems, which have cubic symmetry and space group $Fd\bar{3}m$ [101]. The corresponding reflectivity spectrum for $CdCr_2O_4$ is shown in Fig. 5.7a for the paramagnetic state at 15 K. The modes observed are consistent with the irreducible representation of the optically active phonons which can be obtained by normal mode analysis [11],

$$\Gamma = A_{1g} + E_{1g} + 3T_{2g} + 4T_{1u}. \tag{5.4}$$

In systems with inversion symmetry, the Raman and IR modes are mutually exclusive [102], i.e. modes even with respect to inversion can only be observed in a Raman experiment [103, 104] while only odd modes generate a dipole moment which allows for IR activity. Hence, the observed IR-active phonons correspond to the four triply degenerate $T_{1u}(i)$ ($i = 1, 2, 3, 4$) modes.

However, the reflectivity data at 5 K, below the magnetic ordering temperature $T_N$ = 8 K, shows clearly that mode $T_{1u}(2)$ is split in the low-temperature phase (Fig. 5.7a). The temperature-dependence of its phonon frequency is shown in Fig. 5.7c and reveals a clear splitting of the mode by $\Delta\omega = 10\,\text{cm}^{-1}$ [100, 105].

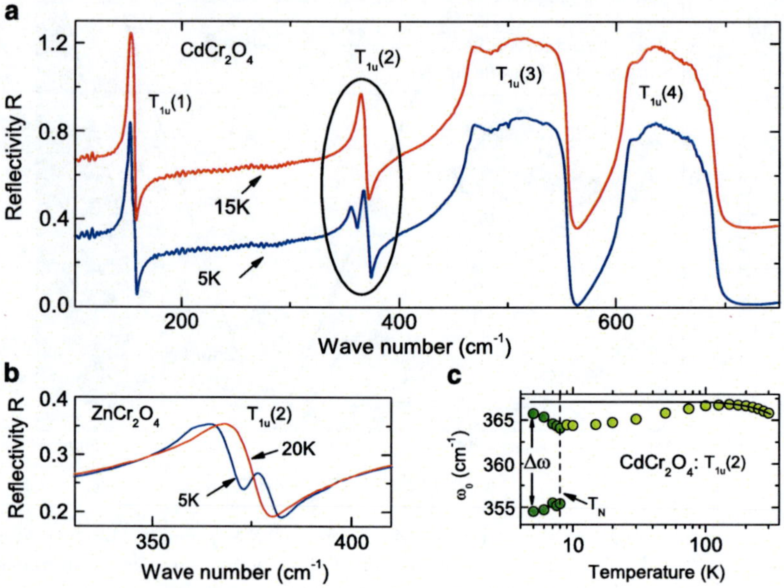


**Fig. 5.7** (**a**) Reflectivity spectra of $CdCr_2O_4$ at 15 K in the paramagnetic state (shifted upwards for clarity) and at 5 K, below the antiferromagnetic ordering temperature $T_N = 8$ K. The $T_{1u}(2)$ mode is clearly split at 5 K [100]. (**b**) Splitting of the $T_{1u}(2)$ phonon in $ZnCr_2O_4$. (**c**) Temperature-dependence of the phonon eigenfrequency of the $T_{1u}(2)$ mode, revealing the splitting at $T_N = 8$ K in $CdCr_2O_4$

Similar behavior has been found for $ZnCr_2O_4$ (Fig. 5.7b) at the magnetostructural transition at $T_c = 12.5$ K, with a splitting $\Delta\omega = 11\,\text{cm}^{-1}$ [106], and for several other Cr spinel systems [100, 107].

The magnetoelastic coupling is already manifest above the magnetostructural phase transition, because it influences the phonon frequency as soon as finite spin correlations appear. In the temperature-dependence of the phonon frequency of mode $T_{1u}(2)$ in $CdCr_2O_4$ (Fig. 5.7c), a deviation from the expected behavior, which would be a stiffening of the lattice vibrations due to anharmonic effects (solid line in Fig. 5.7c), is clearly visible in the softening of the phonon below 100 K, which is the order of the Curie–Weiss temperature in this compound.

By considering a nearest-neighbor Heisenberg spin system, Baltensberger and Helman related the frequency shift of lattice vibrations in magnetic crystals to the magnetic energy of the system, $\Delta\omega \approx \lambda\langle S_i \cdot S_j\rangle$: these authors introduced the spin–phonon coupling parameter $\lambda$, which depends on the derivatives of the exchange constants with respect to the coordinates of the magnetic ions [108, 109]. Information concerning the spin–phonon coupling can thus be extracted from optical measurements of phononic excitations by IR spectroscopy, if the spin–spin correlation functions can be obtained independently. By estimating the magnetic contribution from a nearest-neighbor Heisenberg term with coupling $J$ to the specific heat $C_m$, the temperature-dependence of the nearest-neighbor spin–spin correlations was extracted via

$$\langle S_i \cdot S_j \rangle = const + \frac{1}{6N_A J}\int_{T_N}^{T} C_m(T')\mathrm{d}T', \tag{5.5}$$

where $6N_A$ is the number of Cr–Cr bonds per mole. Another approach is to obtain the nearest-neighbor spin–spin correlations directly from the magnetic susceptibility via

$$\langle S_i \cdot S_j \rangle = \frac{k_B T \chi(T)}{N_A g^2 \mu_B} - \frac{s(s+1)}{3}. \tag{5.6}$$

The values for $ZnCr_2O_4$ were found to be in the range $\lambda \approx 4.0 - 6.2\,\mathrm{cm}^{-1}$, and $\lambda \approx 3.3–3.9\,\mathrm{cm}^{-1}$ was derived in the paramagnetic regime for $CdCr_2O_4$ [104,106].

The spin–phonon coupling derived from the phonon shift in the paramagnetic state should also describe the phonon splitting upon lowering the symmetry from, for example, $O_h$ to $D_{4h}$ at the magnetic phase transition. As a result of this symmetry-lowering, the $T_{1u}$ triplets should split into an $E_u$ doublet and an $A_{2u}$ singlet [110]. The magnetostructural transition can be characterized by the spin-Peierls order parameter $n_{sp} = \langle S_1 \cdot S_2 - S_1 \cdot S_4 \rangle$ for the triply degenerate $T_{1u}$ modes, where the spins $S_i$ reside on the corners of each tetrahedron. The order parameter accounts for the difference in tetrahedral bonds with ferromagnetic or antiferromagnetic coupling and vanishes accordingly in the undistorted paramagnetic state. The induced phonon anisotropy for $ZnCr_2O_4$ in the magnetically ordered phase has been calculated from first principles [110] and the spin–phonon coupling involved in the splitting of the $T_{1u}(2)$ mode,

$$\Delta\omega \approx \lambda \langle S_1 \cdot S_2 - S_1 \cdot S_4 \rangle \tag{5.7}$$

was calculated to be $\lambda \approx 11\mathrm{cm}^{-1}$. This value is a about twice the one derived in the paramagnetic state with $\lambda \approx 4.0–6.2\,\mathrm{cm}^{-1}$ [104, 106]. In the case of $ZnCr_2O_4$, the $T_{1u}(2)$ mode reportedly splits into a lower-lying doublet and a higher singlet state [106], which is consistent with a contraction of the lattice in $c$-direction as illustrated in Chap. 11 by Tchernyshyov and Chern. For this distortion, which has $E_g$ symmetry, averaging over the corresponding bonds leads to $n_{sp} = 9/2$, and the experimental splitting $\Delta\omega = 11\,\mathrm{cm}^{-1}$ yields a spin–phonon coupling of $\lambda = 2.4\,\mathrm{cm}^{-1}$. For $CdCr_2O_4$, the doublet lies higher than the singlet [105], corresponding to an elongation along the $c$-axis with $E_u$ symmetry (depicted in Chap. 11 by Tchernyshyov and Chern). The spin-Peierls order parameter is then $n_{sp} = -2.25$, whence one can estimate $\lambda \approx 4\,\mathrm{cm}^{-1}$ from the experimentally measured splitting $\Delta\omega \approx 9\,\mathrm{cm}^{-1}$. This values compares nicely to the one from the paramagnetic regime, justifying the assumption of a nearly collinear configuration of the spins in the ground state. Moreover, the observation of optical phonons which become both Raman and IR active [104] at the magneto-structural transition provides evidence for the loss of inversion symmetry in $CdCr_2O_4$ in agreement with the results of Tchernyshyov and Chern.

The reason for the discrepancy between the values derived from the splitting, $\lambda_{AFM} = 2.4\,cm^{-1}$, and from the phonon shift above $T_N$, $\lambda_{PM} = 4.0$–$6.2\,cm^{-1}$, in $ZnCr_2O_4$ may originate from the oversimplified assumption of such a collinear antiferromagnetic ground state: experimental studies indicate that the true low-temperature spin configuration of $ZnCr_2O_4$ is considerably more complex [111–113]. Hence, a value much lower than $n_{sp} = 9/2$ can be expected, which is in accordance with the fact that a complete splitting of some of the triply degenerate $T_{1u}$ modes was reported for both $ZnCr_2O_4$, confirming a lower symmetry in the magnetically ordered state [100, 104].

## 5.8 Exciton–Magnon Absorption in $KCuF_3$

The perovskite $KCuF_3$ is considered as a prototypical system for a cooperative Jahn–Teller effect, orbital ordering, and a quasi-one-dimensional antiferromagnetic Heisenberg chain. This is related to the particular type of orbital order in $KCuF_3$, in which a single hole alternately occupies $3d_{x^2-z^2}$ and $3d_{y^2-z^2}$ orbital states of the $Cu^{2+}$ ions ($3d^9$ electronic configuration) [114]. The cooperative Jahn–Teller distortion is characterized by $CuF_6$ octahedra elongated along the $a$ and $b$ axes and arranged in an antiferrodistortive pattern in the $ab$-plane. The system orders antiferromagnetically at $T_N = 39\,K$, but the spinon excitation continuum, one characteristic property of a 1D spin chain, remains observable below $T_N$ and persists up to approximately 200 K [115].

In Fig. 5.8a, the absorption coefficient $\alpha$ for light polarized parallel ($\mathbf{E} \parallel c$) and perpendicular ($\mathbf{E} \perp c$) to the $c$-axis, the direction of the dominant antiferromagnetic coupling, is shown at 8 K. For both orientations, four broad excitation bands $A_1$–$A_4$ are clearly visible. These absorption bands correspond to the local $d$-$d$ CF excitations of the $Cu^{2+}$ ions in a distorted octahedral environment (the level-splitting is shown as an inset in Fig. 5.8a). ED transitions between $d$-levels are parity-forbidden, but they can become allowed as a consequence of a perturbation of the CF potential by phonons (vibronic interaction). Here, we focus on the sharp features which can be seen at the onset of the $A_2$ and $A_3$ bands. Two sets of sidebands become visible on the enlarged scale shown in Fig. 5.8b, the respective onset energies being 8,508 $cm^{-1}$ and 9,775 $cm^{-1}$. These vibronic $A_2$ and $A_3$ transitions were ascribed to MD transitions between the CF levels, while the first three sidebands appear at separations of 88(86), 114(112), 174(175), and 200(203) $cm^{-1}$ from the transitions, as indicated in Fig. 5.8b. Further absorptions related to the $A_2$ transition can be identified at 287, 356, 432, and 553 $cm^{-1}$.

The appearance of sidebands related to CF transitions was first reported in 1965 for $MnF_2$ and explained in terms of magnon sidebands arising due to an exchange-induced dipole mechanism [116, 117]. The separation in energy of the magnon sideband from the magnetic dipole transition corresponds to the magnon energy at the Brillouin zone boundary. In $KCuF_3$, neutron scattering investigations reported zone-boundary energies of magnon excitations at $m_1 = 88.7$, $m_2 = 117$, and

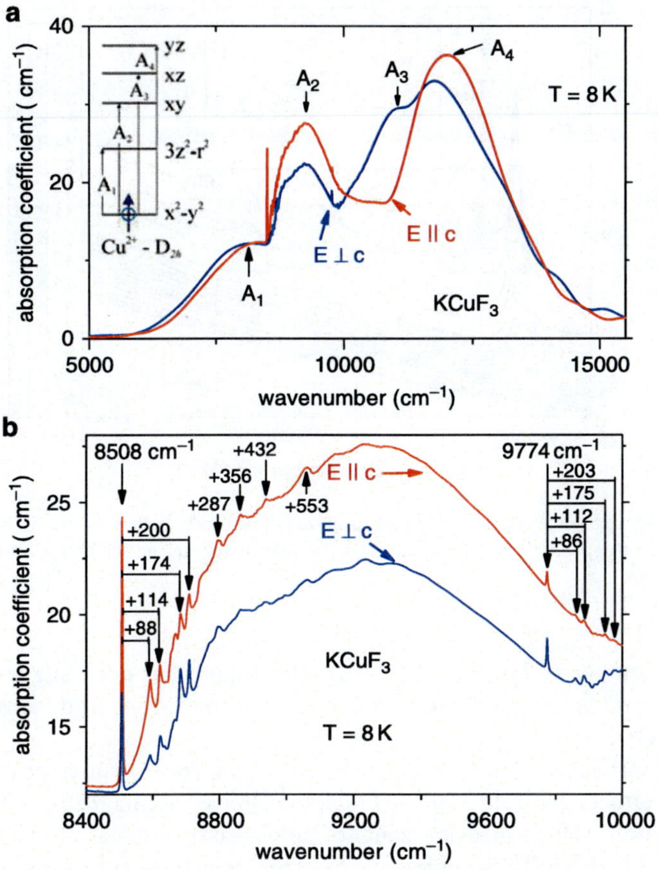


**Fig. 5.8** (**a**) Absorption spectra of $KCuF_3$ at 8 K for light polarized perpendicular and parallel to the antiferromagnetic $c$-axis. Inset: crystal-field (CF) level-splitting of $Cu^{2+}$ in $D_{2h}$ symmetry. (**b**) Fine structures related to the CF transitions $A_2$ and $A_3$ for $E \parallel c$ and $E \perp c$ at 8 K. Frequency shifts with respect to the MD lines at 8,508 and 9,774 cm$^{-1}$ are indicated

$m_3 = 221\,\text{cm}^{-1}$ [118, 119]. The first and second sideband peaks are in good agreement with the first two magnon energies, the third sideband can be assigned to a two-magnon process involving the 88.7 cm$^{-1}$ magnon, and the fourth corresponds to a two-magnon process involving the 88.7 and the 117 cm$^{-1}$ magnons. The further sidebands related to $A_2$ also agree excellently with $2m_1 + m_2 = 294.4\,\text{cm}^{-1}$, $4m_1 = 354.8\,\text{cm}^{-1}$, $m_1 + 3m_2 = 439.7\,\text{cm}^{-1}$, and $m_1 + 4m_2 = 556.7\,\text{cm}^{-1}$. The fact that five or more magnon sidebands can be observed with such clarity is not yet understood, but it may be connected with the coexistence of spinon and magnon excitations in this compound [115]. The temperature-dependence of the MD transition at 8,508 cm$^{-1}$ and of the magnon sidebands is displayed in Fig. 5.9 for **E** ∥ $c$. The fine structure emerges below a temperature $T_S = 50$ K, while the integrated intensity of the MD transition, shown as an inset in Fig. 5.9, displays

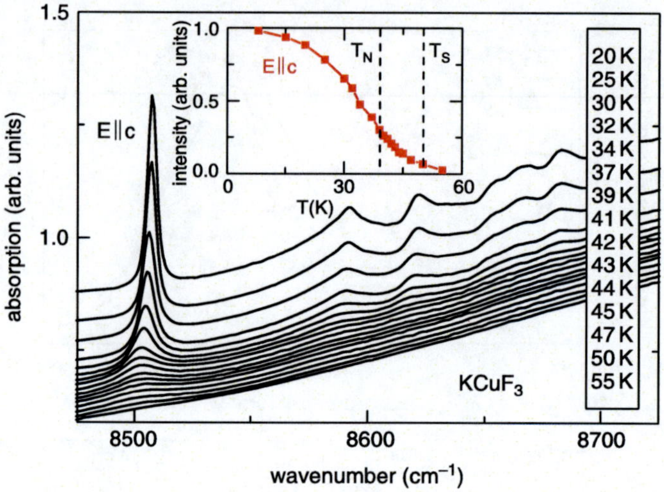


**Fig. 5.9** Temperature-dependence of the exciton–magnon bands below $T_S = 50\,\mathrm{K}$ for $E \parallel c$. Inset: temperature-dependence of the intensity of the zero-magnon line for $E \parallel c$. The solid line is a guide to the eye

no visible anomaly at $T_N = 39\,\mathrm{K}$. This temperature has been suggested as the onset of short-range order arising as the result of a freezing-in of dynamical lattice distortions [120].

This example shows that optical spectroscopy is a very sensitive probe of short-range-order effects and finite spin correlations above the magnetic ordering temperature. Hence, optical measurements of spin-forbidden local orbital transitions can provide important information on the spin-excitation spectrum both at low temperatures and in the correlated paramagnetic phase of frustrated magnets.

We thank Gernot Güntherodt and Alois Loidl for inspiring discussions and assistance both before and during the preparation of this chapter.

## References

1. P. Lemmens, G. Güntherodt, C. Gros, Phys. Rep. **375**, 1 (2003)
2. P. Lemmens, K.Y. Choi, in *Scattering: Inelastic Scattering Technique - Raman, "Encyclopaedia of Condensed Matter Physics"*, eds. by G. Bassani, G. Liedl, P. Wyder (Elsevier, 2005)
3. T.P. Devereaux, R. Hackl, Rev. Mod. Phys. **79**, 175 (2007)
4. S. Sugai, in *Magneto-Optics* (Springer, New York, 2000), vol. 128, Chap. 3, *Raman spectroscopy of Magnetic Compounds with Strong Electron-Correlation*, pp. 75–106
5. E.Y. Sherman, O.V. Misochko, P. Lemmens, *Spectroscopy of High Temperature Superconductors*, Ed. N.M. Plakida (Taylor & Francis Inc., London and New York, 2003), *What can one learn from Raman Spectra of High Temperature Superconductors?*, pp. 97–157

6. P. Lemmens, M. Fischer, M. Grove, P.H.M. v. Loosdrecht, G. Els, E. Sherman, C. Pinettes, G. Güntherodt, in *Advances in Solid State Physics 39*, ed. by B. Kramer (Vieweg Verlag, Braunschweig, 1999), pp. 281–290
7. A. Gozar, G. Blumberg, in *Frontiers in Magnetic Materials* (Springer, Heidelberg, 2005), *Collective Spin and Charge Excitations in* $(Sr,La)_{14-x}Ca_xCu_{24}O_{41}$ *Quantum Spin Ladders*, pp. 653–695
8. M. Johnsson, P. Lemmens, *Crystallography and Chemistry of Perovskites* (Wiley, Chichester, UK, 2006), pp. 2098–2106
9. Ch. Kant, T. Rudolf, F. Schrettle, F. Mayr, J. Deisenhofer, P. Lunkenheimer, M.V. Eremin, A. Loidl, Phys. Rev. B **78**, 245103 (2008)
10. N. Kojima, *Magnetooptics*, (Springer, Berlin, 2002)
11. D.L. Rousseau, R.P. Bauman, S.P.S. Porto, J. Raman Spectrosc. **10**, 253 (1981)
12. W. Hayes R. Loudon, *Scattering of Light by Crystals* (Willey, New York, 1987)
13. M.I. Aroyo, J.M. Perez-Mato, C. Capillas, E. Kroumova, S. Ivantchev, G. Madariaga, A. Kirov, H. Wondratschek, Zeit. f. Kristallogr. **221**, 15 (2006)
14. M.I. Aroyo, A. Kirov, C. Capillas, J.M. Perez-Mato, H. Wondratschek, Acta Cryst. **A 62**, 115 (2006)
15. P. Lemmens, P. Millet, *Spin – Orbit – Topology, a triptych* in *Quantum Magnetism*, eds. by U. Schollwöck, J. Richter, D.J.J. Farrell, R.F. Bishop, (Lecture Notes in Physics, Springer, Heidelberg, 2004)
16. K.Y. Choi, P. Lemmens, G. Güntherodt, Y.G. Pashkevich, V.P. Gnezdilov, P. Reutler, L. Pinsard-Gaudart, B. Büchner, A. Revcolevschi, Phys. Rev. B **72**, 024301 (2005)
17. K.Y. Choi, P. Lemmens, G. Güntherodt, Y.G. Pashkevich, V.P. Gnezdilov, P. Reutler, L. Pinsard-Gaudart, B. Büchner, A. Revcolevschi, Phys. Rev. B **71**, 174402 (2005)
18. K.Y. Choi, Y.G. Pashkevich, V.P. Gnezdilov, G. Güntherodt, A.V. Yeremenko, D. A. Nabok, V.I. Kamenev, S.N. Barilo, S.V. Shiryaev, A.G. Soldatov, P. Lemmens, Phys. Rev. B **74**, 064406 (2006)
19. P.H.M. van Loosdrecht, J.P. Boucher, G. Martinez, G. Dhalenne, A. Revcolevschi, Phys. Rev. Lett. **76**(2), 311 (1996)
20. M. Weiden, R. Hauptmann, W. Richter, C. Geibel, P. Hellmann, M. Köppen, F. Steglich, M. Fischer, P. Lemmens, G. Güntherodt, A. Krimmel, G. Nieva, Phys. Rev. B **55**, 15067 (1997)
21. K. Sparta, G.J. Redhammer, P. Roussel, G. Heger, G. Roth, P. Lemmens, A. Ionescu, M. Grove, G. Güntherodt, F. Hüning, H. Lueken, H. Kageyama, K. Onizuka, Y. Ueda, Eur. Phys. J. B **19**, 507 (2001)
22. K.Y. Choi, Y.G. Pashkevich, K.V. Lamonova, H. Kageyama, Y. Ueda, P. Lemmens, Phys. Rev. B **68**, 104418 (2003)
23. E. Ya. Sherman, M. Fischer, P. Lemmens, P.H.M. van Loosdrecht, G. Güntherodt, Europhys. Lett. **48**, 648 (1999)
24. P. Lemmens, K.Y. Choi, G. Caimi, L. Degiorgi, N.N. Kovaleva, A. Seidel, F.C. Chou, Phys. Rev. B **70**, 134429 (2004)
25. G. Caimi, L. Degiorgi, P. Lemmens, F.C. Chou, J. Phys. Cond. Mat. **16**, 5583 (2004)
26. P. Lemmens, K.Y. Choi, R. Valentí, T. Saha-Dasgupta, E. Abel, Y.S. Lee, F.C. Chou, J. Phys. **7**, 74 (2005)
27. V. Gnezdilov, P. Lemmens, A.A. Zvyagin, V.O. Cheranovskii, K. Lamonova, Y.G. Pashkevich, R.K. Kremer, H. Berger, Phys. Rev. B **78**, 184407 (2008)
28. K.Y. Choi, A. Oosawa, H. Tanaka, P. Lemmens, Phys. Rev. B **72**, 024451 (2005)
29. K.Y. Choi, G. Güntherodt, A. Oosawa, H. Tanaka, P. Lemmens, Phys. Rev. B **68**, 174412 (2003)
30. K.Y. Choi, A. Oosawa, H. Tanaka, P. Lemmens, Prog. Theor. Phys. **159**, 195 (2005)
31. K.Y. Choi, P. Lemmens, J. Pommer, A. Ionescu, G. Güntherodt, S. Hiroya, H. Sakurai, K. Yoshimura, A. Matsuo, Phys. Rev. B **70**, 174417 (2004)
32. V. Gnezdilov, P. Lemmens, P. Scheib, M. Ghosh, Y. G. Pashkevich, H. Paulsen, V. Schünemann, J.A. Wolny, G. Agustí, J.A. Real, Phys. Rev. B **79**, 045122 (2009)

33. K.Y. Choi, P. Lemmens, D. Heydhausen, G. Güntherodt, C. Baumann, R. Klingeler, P. Reutler, B. Büchner, Phys. Rev. B **77**, 064415 (2008)
34. Y.G. Pashkevich, V.A. Blinkin, V.P. Gnezdilov, V.V. Tsapenko, V.V. Eremenko, P. Lemmens, M. Fischer, M. Grove, G. Güntherodt, L. Degiorgi, P. Wachter, J.M. Tranquada, D.J. Buttrey, Phys. Rev. Lett. **84**, 3919 (2000)
35. V.P. Gnezdilov, A.V. Yeremenko, Y.G. Pashkevich, G. Güntherodt, P. Lemmens, J.M. Tranquada, D.J. Buttrey, K. Nakajima, Ukr. J. Low. Temp. Phys. **28**, 510 (2002)
36. M.G. Cottam, D.J. Lockwood, *Light scattering in Magnetic Solids* (Wiley-Interscience, New York, 1986)
37. P. Lemmens, K.Y. Choi, E.E. Kaul, C. Geibel, K. Becker, W. Brenig, R. Valentí, C. Gros, M. Johnsson, P. Millet, F. Mila, Phys. Rev. Lett. **87**, 227201 (2001)
38. C. Gros, P. Lemmens, M. Vojta, R. Valentí, K.Y. Choi, H. Kageyama, Z. Hiroi, N. Mushnikov, T. Goto, M. Johnsson, P. Millet, Phys. Rev. B **67**, 174405 (2003)
39. R. Takagi, M. Johnsson, V. Gnezdilov, R.K. Kremer, W. Brenig, P. Lemmens, Phys. Rev. B **74**, 014413 (2006)
40. V.N. Kotov, M.E. Zhitomirsky, O.P. Sushkov, Phys. Rev. B **63**, 064412 (2001)
41. V.N. Kotov, J. Oitmaa, O.P. Sushkov, W.H. Zheng, Phys. Rev. B **60**, 14613 (1999)
42. J.S. Helton, K. Matan, M.P. Shores, E.A. Nytko, B.M. Bartlett, Y. Yoshida, Y. Takano, A. Suslov, Y. Qiu, J.H. Chung, D.G. Nocera, Y.S. Lee, Phys. Rev. Lett. **98**, 107204 (2007)
43. P. Mendels, F. Bert, M.A.D. Vries, A. Olariu, A. Harrison, F. Duc, J.C. Trombe, J.S. Lord, A. Amato, C. Baines, Phys. Rev. Lett. **98**, 077204 (2007)
44. O. Cepas, J.O. Haerter, C. Lhuillier, Phys. Rev. B **77**, 172406 (2008)
45. J.W. Halley, Phys. Rev. Lett. **41**, 1605 (1978)
46. G.F. Reiter, Phys. Rev. **13**, 169 (1976)
47. M. Chandrasekhar, M. Cardona, E.O. Kane, Phys. Rev. B **16**, 3579 (1977)
48. A. Zawadowski, M. Cardona, Phys. Rev. B **42**, 10732 (1990)
49. P. Lemmens, K.Y. Choi, V. Gnezdilov, E.Y. Sherman, D.P. Chen, C.T. Lin, F.C. Chou, B. Keimer, Phys. Rev. Lett. **96**, 167204 (2006)
50. G. Els, G.S. Uhrig, P. Lemmens, H. Vonberg, P.H.M. van Loosdrecht, G. Güntherodt, O. Fujita, J. Akimitsu, G. Dhalenne, A. Revcolevschi, Europhys. Lett. **43**, 463 (1998)
51. M. Fischer, P.H.M. van Loosdrecht, P. Lemmens, G. Güntherodt, B. Büchner, T. Lorenz, M. Breuer, J. Zeman, G. Martinez, G. Dhalenne, A. Revcolevschi, Phys. Rev. B **57**(13), 7749 (1998)
52. M. Grove, P. Lemmens, G. Els, G. Güntherodt, B.C. Sales, F. Büllesfeld, W. Assmus, Phys. Rev. B **61**, 6126 (2000)
53. P. Lemmens, M. Grove, M. Fischer, G. Güntherodt, V.N. Kotov, H. Kageyama, K. Onizuka, Y. Ueda, Phys. Rev. Lett. **85**, 2605 (2000)
54. A. Gozar, B.S. Dennis, H. Kageyama, G. Blumberg, Phys. Rev. B **72**, 064405 (2005)
55. A. Gozar, G. Blumberg, in *Frontiers in Magnetic Materials* (Springer, Heidelberg, 2005), *Collective Magnetic Excitations in* $SrCu_2(BO_3)_2$, pp. 735–754
56. M. Fischer, P. Lemmens, G. Els, G. Güntherodt, E. Ya. Sherman, E. Morré, C. Geibel, F. Steglich, Phys. Rev. B **60**, 7284 (1999)
57. P. Lemmens, M. Fischer, G. Güntherodt, A. Mishchenko, M. Weiden, R. Hauptmann, C. Geibel, F. Steglich, Phys. Rev. B **58**, 14159 (1998)
58. E.Ya. Sherman, C. Ambrosch-Draxl, P. Lemmens, G. Güntherodt, P.H.M. van Loosdrecht, Phys. Rev. B **63**, 224305 (2001)
59. A. Gozar, G. Blumberg, in *Frontiers in Magnetic Materials* (Springer, Heidelberg, 2005), *Electronic Properties of* $\alpha' - NaV_2O_5$, pp. 697–734
60. G. Els, P.H.M. van Loosdrecht, P. Lemmens, H. Vonberg, G. Güntherodt, G. Uhrig, O. Fujita, J. Akimitsu, G. Dhalenne, A. Revcolevschi, Phys. Rev. Lett. **79**, 5138 (1997)
61. K. Yamamoto, T. Katsufuji, T. Tanaba, Y. Tokura, Phys. Rev. Lett. **80**, 1493 (1998)
62. P.A. Fleury, R. Loudon, Phys. Rev. **166**(2), 514 (1968)
63. P.A. Fleury, H.J. Guggenheim, Phys. Rev. Lett. **24**, 1346 (1970)
64. A. Gozar, B. Blumberg, B.S. Dennis, B.S. Shastry, N. Motoyama, H. Eisaki, S. Uchida, Phys. Rev. Lett. **87**, 197202 (2001)

65. S. Sugai, S. Shamoto, M. Sato, Phys. Rev. B **38**, 6436 (1988)
66. A.V. Chubukov, D.M. Frenkel, Phys. Rev. B **52**, 9760 (1995)
67. H. Kuroe, J.I. Sasaki, T. Sekine, N. Koide, Y. Sasago, K. Uchinokura, M. Hase, Phys. Rev. B **55**, 409 (1997)
68. K.Y. Choi, V.P. Gnezdilov, P. Lemmens, L. Capogna, M.R. Johnson, M. Sofin, A. Maljuk, M. Jansen, B. Keimer, Phys. Rev. B **73**, 094409 (2006)
69. K.-Y. Choi, D. Wulferding, H. Berger, P. Lemmens, Phys. Rev. B **80**, 245108 (2009)
70. C.S. Snow, S.L. Cooper, D.P. Young, Z. Fisk, A. Comment, J.P. Ansermet, Phys. Rev. B **64**, 174412 (2001)
71. H.G. Tompkins, E.A. Irene, *Handbook of Ellipsometry*, (Springer, Berlin, 2005)
72. Y. Mizuno, S. Koide, Phys. Kondens. Materie **2**, 179 (1964)
73. F. Mayr, Dissertation, Universität Augsburg, 2002
74. J.W. Halley, I.F. Silvera, Phys. Rev. Lett. **15**, 654 (1965)
75. S.J. Allen, Jr., R. Loudon, P.L. Richards. Phys. Rev. Lett. **16**, 463 (1966)
76. Y. Tanabe, T. Moriya, S. Sugano, Phys. Rev. Lett. **15**, 1023 (1965)
77. Y. Tanabe, K. Gondaira, J. Phys. Soc. Jpn. **21**, 1527 (1966)
78. R. Newman, R.M. Chrenko, Phys. Rev. **114**, 1507 (1959)
79. J.D. Perkins, J.M. Graybeal, M.A. Kastner, R.J. Birgeneau, J.P. Falck, M. Greven, Phys. Rev. Lett. **71**, 1621 (1993)
80. J. Lorenzana, G.A. Sawatzky, Phys. Rev. Lett. **74**, 1867 (1995)
81. K. Shinagawa, Y. Tanabe, J. Phys. Soc. Jpn. **30**, 1280 (1971)
82. T. Fujiwara, W. Gebhardt, K. Petanides, Y. Tanabe, J. Phys. Soc. Jpn. **33**, 39 (1972)
83. A. Pimenov, A.A. Mukhin, V.Y. Ivanov, V.D. Travkin, A. Balbashov, A. Loidl, Nature Phys. **2**, 97 (2006)
84. A.B. Sushkov, R. Valdés Aguilar, S. Park, S.-W. Cheong, H.D. Drew, Phys. Rev. Lett. **98**, 027202 (2007)
85. T. Rudolf, Ch. Kant, F. Mayr, J. Hemberger, V. Tsurkan, A. Loidl, Phys. Rev. B **76**, 174307 (2007)
86. P. Brüesch, *Phonons, Theory and Experiments II*, (Springer, Berlin Heidelberg New York, 1987)
87. R.C. Ohlmann, M. Tinkham, Phys. Rev. **123**, 425 (1961)
88. M.G. Hildebrand, A. Slepkov, M. Reedyk, G. Amow, J.E. Greedan, D.A. Crandles, Phys. Rev. B **59**, 6938 (1999)
89. D.D. Sell, R.L. Greene, R.M. White, Phys. Rev. **158**, 489 (1967)
90. J. Deisenhofer, I. Leonov, M.V. Eremin, Ch. Kant, P. Ghigna, F. Mayr, V.V. Iglamov, V.I. Anisimov, D. van der Marel, Phys. Rev. Lett. **101**, 157406 (2008)
91. S.E. Stokowski, D.D. Sell, H.J. Guggenheim, Phys. Rev. B **4**, 3141 (1971)
92. M. Bassi, P. Camagni, R. Rolli, G. Samoggia, F. Parmigiani, G. Dhalenne, A. Revcolevschi, Phys. Rev. B **54**, R11030 (1996)
93. R. Rückamp, E. Benckiser, M.W. Haverkort, H. Roth, T. Lorenz, A. Freimuth, L. Jongen, A. Möller, G. Meyer, P. Reutler, B. Büchner, A. Revcolevschi, S.-W. Cheong, C. Sekar, G. Krabbes, M. Grüninger, New J. Phys. **7**, 144 (2005)
94. C.J. Ballhausen, *Introduction to Ligand Field Theory* (McGraw-Hill, New York, 1962)
95. Ch. Hartinger, F. Mayr, J. Deisenhofer, A. Loidl, T. Kopp, Phys. Rev. B **69**, 100403(R) (2004)
96. P. Calvani, *Optical Properties of Polarons* (La Rivista del Nuovo Cimento, Bologna, 2001)
97. J. Lorenzana, G.A. Sawatzky, Phys. Rev. B **52**, 9576 (1995)
98. H. Suzuura, H. Yasuhara, A. Furusaki, N. Nagaosa, Y. Tokura, Phys. Rev. Lett. **76**, 2579 (1996)
99. S.-H. Lee, C. Broholm, W. Ratcliff, G. Gasparovic, Q. Huang, T.H. Kim, S.-W. Cheong, Nature **418**, 856 (2002)
100. T. Rudolf, Ch. Kant, F. Mayr, J. Hemberger, V. Tsurkan, A. Loidl, New J. Phys. **9**, 76 (2007)
101. E.J.W. Verwey, E.L. Heilmann, J. Chem. Phys. **15**, 174 (1947)
102. H. Poulet, J.P. Mathieu, *Vibration Spectra and Symmetry of Crystals* (Gordon & Breach, New York, 1976)

103. J. Himmrich, H.D. Lutz, Solid State Commun. **79**, 447 (1991)
104. Ch. Kant, J. Deisenhofer, T. Rudolf, F. Mayr, F. Schrettle, A. Loidl, V. Gnezdilov, D. Wulferding, P. Lemmens, V. Tsurkan, Phys. Rev. B **80**, 214417 (2009)
105. R. Valdés Aguilar, A.B. Sushkov, Y.J. Choi, S.-W. Cheong, H.D. Drew, Phys. Rev. B **77**, 092412 (2008)
106. A.B. Sushkov, O. Tchernyshyov, W. Ratcliff II, S.-W. Cheong, H.D. Drew, Phys. Rev. Lett. **94**, 137202 (2005)
107. T. Rudolf, Ch. Kant, F. Mayr, M. Schmidt, V. Tsurkan, J. Deisenhofer, A. Loidl, Eur. Phys. J. B **68**, 153 (2009)
108. W. Baltensperger, J.S. Helman, Helv. Phys. Acta **41**, 668 (1968)
109. W. Baltensperger, J. Appl. Phys. **41**, 1052 (1970)
110. C.J. Fennie, K.M. Rabe, Phys. Rev. Lett. **96**, 205505 (2006)
111. S.-H. Lee, G. Gasparovic, C. Broholm, M. Matsuda, J.-H. Chung, Y.J. Kim, H. Ueda, G. Xu, P. Zschak, K. Kakurai, H. Takagi, W. Ratcliff, T.H. Kim, S.-W. Cheong, J. Phys. Condens. Matter **19**, 145259 (2007)
112. I. Kagomiya, Y. Hata, D. Eto, H. Yanagihara1, E. Kita, K. Nakajima, K. Kakurai, M. Nishi, K. Ohoyama, J. Phys. Soc. Jpn. **76**, 064710 (2007)
113. S. Ji, S.-H. Lee, C. Broholm, T.Y. Koo, W. Ratcliff, S.-W. Cheong, P. Zschack, Phys. Rev. Lett. **103**, 037201 (2009)
114. K.I. Kugel, D.I. Khomskii, Sov. Phys. Usp. **25**, 231 (1982)
115. B. Lake, D.A. Tennant, C.D. Frost, S.E. Nagler, Nat. Mater. **4**, 329 (2005)
116. R.L. Greene, D.D. Sell, W.M. Yen, A.L. Schawlow, R.M. White, Phys. Rev. Lett. **15**, 656 (1965)
117. Y. Tanabe, K. Gondaira, J. Phys. Soc. Jpn. **22**, 573 (1967)
118. S.K. Satija, J.D. Axe, G. Shirane, H. Yoshizawa, K. Hirakawa, Phys. Rev. B **21**, 2001 (1980)
119. B. Lake, D.A. Tennant, S.E. Nagler, Phys. Rev. Lett. **85**, 832 (2000)
120. M.V. Eremin, D.V. Zakharov, H.-A. Krug von Nidda, A. Shuvaev, A. Pimenov, P. Ghigna, J. Deisenhofer, A. Loidl, Phys. Rev. Lett. **101**, 147601 (2008)

# Part III
# Frustrated Systems

# Chapter 6
# The Geometries of Triangular Magnetic Lattices

**Robert J. Cava, Katharine L. Holman, Tyrel McQueen, Eric J. Welsh, D. Vincent West, and Anthony J. Williams**

**Abstract** A survey of the crystal structures of materials that are currently being studied in the context of geometric magnetic frustration in triangular lattices is presented, based on a purely geometrical picture of their magnetic lattices. Two-dimensional materials are first described, followed by three dimensional materials of increasing structural complexity. When known, both transition metal and rare earth based examples within a lattice type are noted. The ideal cases are described, as well as the geometries that are found in some materials for which ideal triangular symmetry is broken in different ways. A brief description of materials preparation and crystal growth is also presented.

## 6.1 Introduction

For solids where the geometry of the magnetic lattice is based on triangles, magnetic ordering is often suppressed to temperatures below what is expected from near neighbor magnetic interactions. Such compounds continue to be of substantial interest. The ordered state that finally occurs can often be a consequence of a subtle balance among different factors. Deviation from perfect triangular symmetry in the materials often plays an important role, for example, as do interactions between second or higher order nearest neighbors. There are many excellent review articles on this topic, written from chemical, experimental, and theoretical perspectives, describing the properties of geometrically frustrated magnetic materials and their theoretical understanding (see, for instance, [1–7]). Current theoretical and experimental reviews can be found in this volume. What this brief review adds to this body of information is to present a survey of a wide range of triangular lattice materials based on a purely geometrical picture of their magnetic lattices.

The magnetic lattices of interest are often imbedded in complex arrays of other atoms that frequently influence the manner in which the magnetic system picks its ultimate ground state. For the purposes of this review, so that the magnetic lattice geometry will not be obscured, the non-magnetic atoms of the compounds of interest are not included in the figures (with one exception). Factors beyond pure geometry, such as magnetic ion anisotropy, exchange paths and spin-orbit coupling, impact the effective magnetic geometry. These have to be considered on a case-by-case basis and will generally not be addressed in this review. Arranged by lattice type, we first describe the ideal cases and then some variants in which the triangular symmetry is broken in different ways. First, the two-dimensional materials are described, and then, the three dimensional materials. When known, both transition metal and rare earth based examples within a lattice type are noted. The magnetic lattices are presented in the figures, and some of the interatomic distances relevant to the magnetism are enumerated in the tables. Due to the large number of triangular magnetic lattices encountered in materials, some selectivity has been employed by the authors in this compilation.

## 6.2 Two-Dimensional Structures

### *6.2.1 Planes of Edge-Sharing Triangles*

The most commonly found triangular lattices in magnetically frustrated materials are based on planes of edge-sharing equilateral triangles. In spite of the simple, ideal geometry of such a lattice, various types of magnetic ordering are found within this lattice type. These ordering schemes depend on factors beyond the lattice geometry, such as the magnitude of the spin, the anisotropy of the orbital occupancy, the relative intraplane and interplane distances between magnetic ions, and the influence of the ligands on superexchange and crystal field. The stacking geometry of neighboring planes is critical in determining the ordering in the third dimension. The chemistry of magnetic compounds based on planes of edge-shared equilateral triangles is quite diverse, and many different types of transition-metal compounds with this lattice are known. Surprisingly, there are very few published examples of purely planar edge-sharing equilateral lattices based on rare earths ions.

"Eclipsed" triangular plane lattices stack directly on top of each other (Fig. 6.1), while for the "staggered" lattices the layers are shifted out of registry. In the latter case, the atoms in the neighboring layers often cap the triangles in the central plane to form elongated tetrahedra (Fig. 6.2). For the eclipsed stacking in this lattice type, there are two magnetic out-of-plane neighbors, one in the plane above and one below; while for staggered stacking, there are typically six out-of-plane neighbors, three in each neighboring plane. The interplane interactions in staggered lattices, because they involve three ions in the neighboring planes, can frustrate three-dimensional ordering. In planar lattices based on edge-sharing equilateral triangles, the ratio of out-of-plane magnetic atom spacing to in-plane atom spacing varies over a very wide range, from a maximum of about 3.31 in compounds like $NiGa_2S_4$ [8] to a minimum of 0.41 in compounds like $CsNiCl_3$ [9]. The former compound is therefore two-dimensional in its properties [10] while the latter are more like triangular arrangements of magnetic chains (Fig. 6.1) [11].

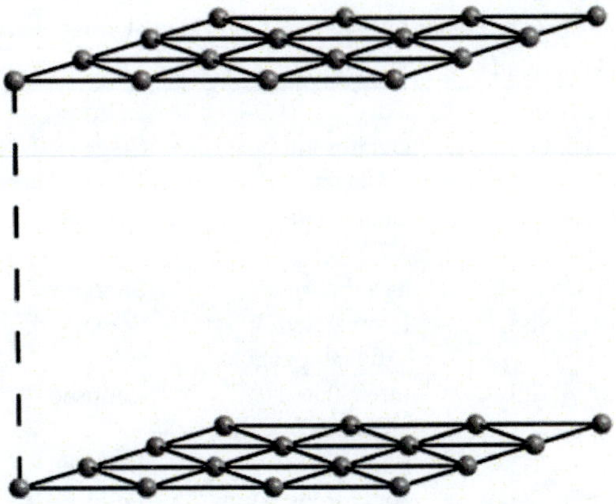

**Fig. 6.1** The eclipsed equilateral triangle layers in $NiGa_2S_4$ (*upper*) and $CsNiCl_3$ (*lower*). The distances between the layers are dramatically different

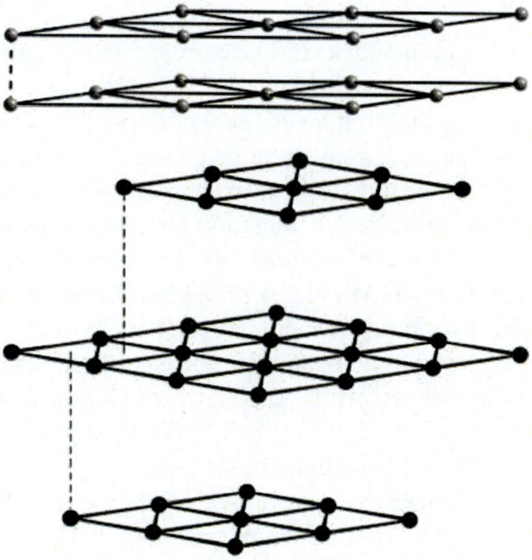

**Fig. 6.2** The layers of edge-sharing equilateral triangles in $\alpha$-$NaFeO_2$. The three layers per cell are stacked in a staggered fashion

Examples of materials that display planes of equilateral triangles are presented in Table 6.1 Some of simplest examples are classic layered compounds like $VCl_2$, where the [Cl-V-Cl]-[Cl-V-Cl] sequence yields weak Van der Waals bonding between the layers and weak interplane magnetic coupling [12]. Another classic example is $Fe_{1/3}NbS_2$, where the transition metals are in a triangular lattice that 1/3 fills the sites in the Van der Waals layer in the $NbS_2$ host, i.e. the layer sequence is [S-Nb-S]-[$Cr_{1/3}$]-[S-Nb-S]; compounds in which the Van der Waals layer is 1/4 filled also yield triangular layers of magnetic ions, such as is found in $Cr_{1/4}NbS_2$ [13, 14]. In the latter case the relatively large in-plane spacing of the Cr makes the intraplane spacing shorter than the interplane spacing. Somewhat more frequently studied planar lattices based on equilateral triangles are found in the Delafossite family. In this crystal structure, layers of edge-sharing $MO_6$ or $LnO_6$ octahedra are

**Table 6.1** Planar lattices based on Edge-sharing Equilateral triangles

| Compound | Structure type | Stacking type | M–M in-plane | M–M next plane |
|---|---|---|---|---|
| $VCl_2$ | $CdI_2$ | Eclipsed | 3.60 | 5.84 |
| $CuCoO_2$ | Delafossite 3-layer | Staggered | 2.85 | 5.95 |
| $CuFeO_2$ | Delafossite 2-layer | Eclipsed | 3.04 | 5.73 |
| $NaCrO_2$ | $\alpha$-$NaFeO_2$ | Staggered | 2.98 | 5.59 |
| $Fe_{1/3}NbS_2$ | Intercalated dichalcogenide | Staggered | 5.76 | 6.95 |
| $Cr_{1/4}NbS_2$ | Intercalated Dichalcogenide | Eclipsed | 6.63 | 5.99 |
| $CsNiCl_3$ | $BaNiO_3$ | Eclipsed | 7.17 | 2.97 |
| $RbFe(SO_4)_2$ | Anhydrous Alum | Eclipsed | 4.82 | 8.22 |
| $NiGa_2S_4$ | $NiGa_2S_4$ | Eclipsed | 3.63 | 12.00 |

the building blocks for the equilateral triangular magnetic plane, with non-magnetic, stick geometry $Cu^{1+}$ or $Ag^{1+}$ (which are $d^{10}$ and non-magnetic) layers acting as spacers, yielding layer sequences of, for example, [O-Nd-O]-Cu-[O-Nd-O]. There are two-layer and three-layer variants of Delafossites, differences that result in eclipsed or staggered near-neighbor planes, respectively. This structure type is also found for magnetic transition elements in compounds such as $CuFeO_2$ [15–17].

Much more common are equilateral triangle geometry layered phases of the $\alpha$-$NaFeO_2$ type (Fig. 6.2). Again, these are based on a triangular plane of edge-sharing transition metal oxygen octahedra. Unlike Delafossites, the spacing layer in $\alpha$-$NaFeO_2$-type materials consists of alkali-or alkaline earth-oxygen octahedra or triangular prisms: H, Li, Na, K, or Ca spacer layers are known. The magnetic atom can be Ti, V, Cr, Fe, Co, or Ni (see e.g. [18]). Many different magnetic ordering schemes are encountered for this family.

Another large family of compounds based on edge-sharing equilateral triangles has the $CsNiCl_3$ structure type (Fig. 6.1). In these compounds, the basic structural features are columns of face-sharing $MCl_6$ octahedra. The columns are arranged in a triangular array and are spaced by the large Cs ions so that interchain coupling has an M-Cl-Cl-M superexchange pathway. These compounds are much more chain-like than triangle-like in their magnetic properties, though the triangular arrangement of chains can frustrate inter-chain ordering.

The anhydrous alums [19], typified by $KFe(SO_4)_2$, are relatively poorly studied magnetically, even though they represent a very large family in the class of edge-sharing triangular planar lattices. They are made from the stacking of triangular layers of alkalis and triangular layers of magnetic metal ion octahedra that share no common ligands, and layers of sulfate or molybdate tetrahedra. The magnetic ions are typically trivalent transition elements such as Fe, with a few variants containing rare earths. The anhydrous alums have a substantial number of variants for which the edge-sharing triangles are isoceles rather than equilateral (Fig. 6.4). The final member of this class that we will consider is $NiGa_2S_4$. $NiGa_2S_4$ doesn't have a

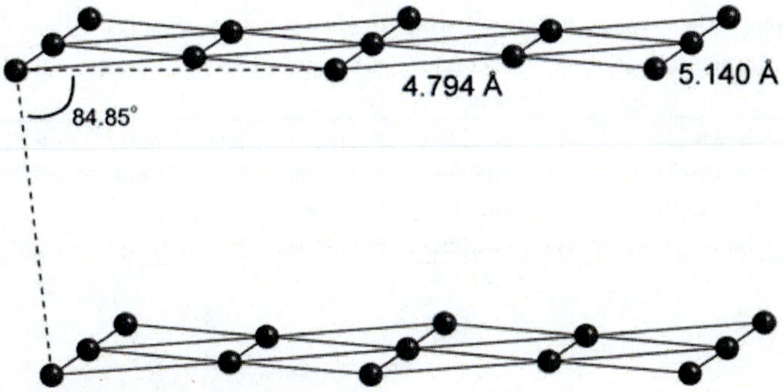


**Fig. 6.3** The planes of edge-sharing isosceles triangles in the anhydrous alum $KFe(SO_4)_2$. The layers are well separated, and slightly shifted

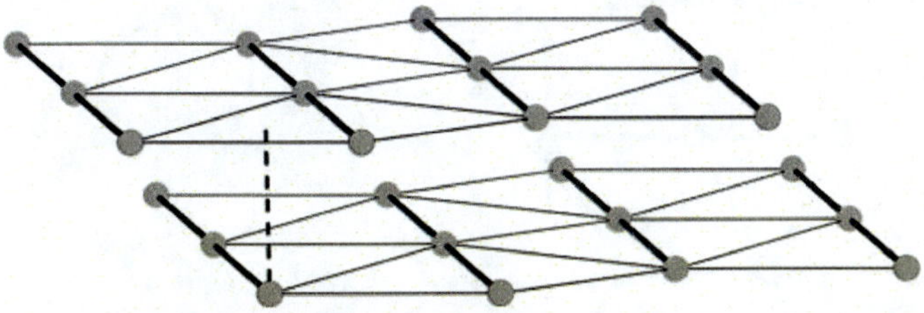

**Fig. 6.4** The layers of edge-sharing isosceles triangles in $Cs_2CuCl_4$. The layers are staggered and slightly buckled. The strongest interactions are shown as *bold lines* in the triangular planes

common crystal structure or analogous variants, but, because it contains triangular Ni layers quite far apart, it is one of the most two-dimensional, magnetic compounds in this class (Fig. 6.1).

There are fewer examples of magnetic compounds based on planes of edge-sharing, non-equilateral triangles. Typical examples are based on $CuO_6$ octahedra sharing edges, which, due to the Jahn–Teller distortion, have two long bonds and four short bonds; the elongated octahedra result in anisotropic triangular planes. The symmetry of magnetic interactions in such cases is lowered from the ideal case, but the properties are quite interesting since $Cu^{2+}$ is a spin $1/2$ ion. Examples of these materials in Table 6.2 are $Cs_2CuCl_4$ [20] and $Cu_2(OH)_3Cl$ [21]. The former (Fig. 6.5) has chain-like interactions, with in-plane coupling between chains and interlayer interactions that are all finely balanced. The latter compound, shown in Fig. 6.3, due to the arrangement of long and short bonds and occupied orbitals, has an interesting type of symmetry breaking in the triangular plane in which two slightly different Cu chains are coupled by weaker, crosslinking interactions. $NaNiO_2$ is a distorted variant of the $\alpha$-$NaFeO_2$ structure type due to the presence of Jahn–Teller distortions of the $NiO_6$ octahedra [22]. $NaTiO_2$ (a very difficult compound to synthesize!) is similarly distorted due to ordering of the occupied and unoccupied Ti $t_{2g}$ orbitals [23].

**Table 6.2** Planar lattices based on Edge-sharing of non-equilateral triangles

| Compound | Structure type | Triangle type | Stacking type | M–M in-plane | M–M next plane |
|---|---|---|---|---|---|
| $Cs_2CuCl_4$ | $Cs_2MCl_4$ | Isosceles | Staggered | 7.60, 7.26 | 6.81 |
| $Cu_2(OH)_3Cl$ | Botallackite | Scalene | Eclipsed | 3.06, 3.18, 3.23 | 5.72 |
| $KFe(SO_4)_2$ | Anhydrous Alum | Isosceles | Eclipsed | 4.79, 5.14 | 7.88 |
| $NaNiO_2$ | Distorted $\alpha$-$NaFeO_2$ | isosceles | Staggered | 2.84, 3.02 | 5.47, 5.58 |

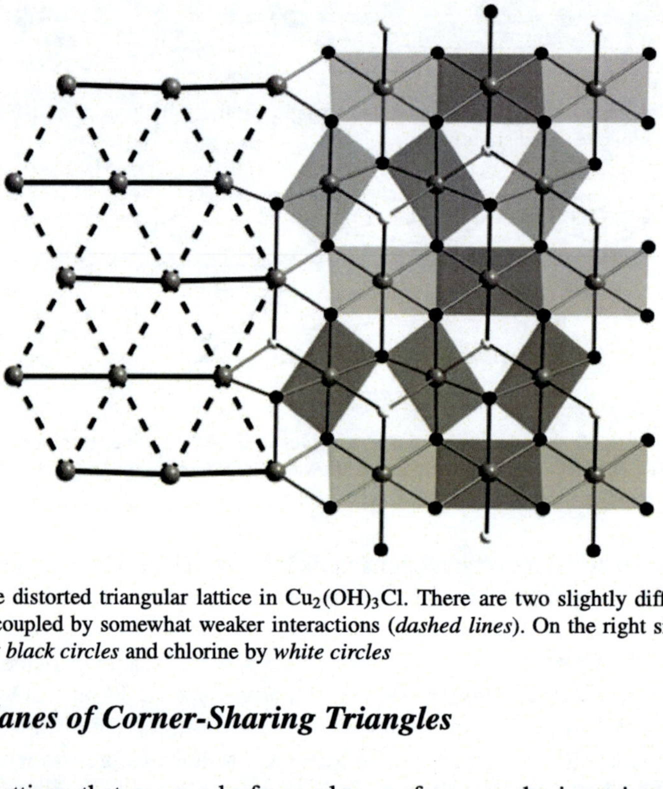

**Fig. 6.5** The distorted triangular lattice in $Cu_2(OH)_3Cl$. There are two slightly different chains (*bold lines*) coupled by somewhat weaker interactions (*dashed lines*). On the right side, oxygens are shown by *black circles* and chlorine by *white circles*

## *6.2.2 Planes of Corner-Sharing Triangles*

Magnetic lattices that are made from planes of corner-sharing triangles are of particular interest from the viewpoint of magnetic frustration because the systems are more under constrained than those based on edge-sharing. Most common and famous among these are kagomé lattice compounds. The kagomé lattice is a corner-sharing array of equilateral triangles that is derived from the edge-shared triangular lattice by periodic removal of 1/4 of the magnetic sites. The equilateral triangles share corners to form six-triangle rings. The most numerous kagomé compounds are found in the Jarosite family [24], a group of hydroxide-sulfates, typified by $KFe_3(OH)_6(SO_4)_2$, that can be made with different alkalis, ammonium, and magnetic Fe or Cr. The transition metals are found in $MO_6$ octahedra that share edges in the kagomé plane, so the intraplane magnetic coupling is quite strong. The kagomé layers in Jarosite are stacked in a staggered fashion (Fig. 6.6). Spin 1/2 kagomé lattices are of particular interest from a theoretical perspective, and

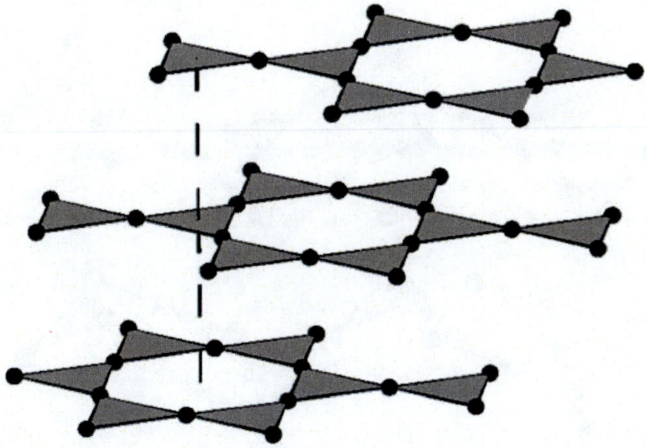

**Fig. 6.6** The kagomé lattice made from equilateral triangles in Jarosite. The stacking of the layers is staggered

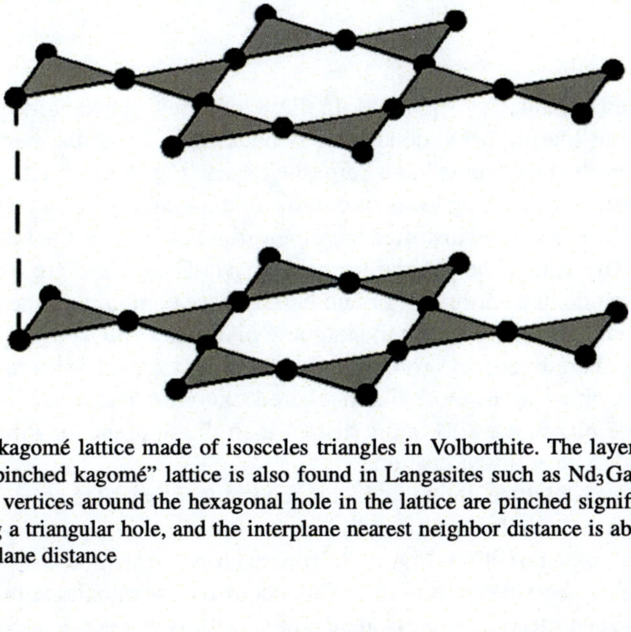

**Fig. 6.7** The kagomé lattice made of isosceles triangles in Volborthite. The layers are eclipsed. An eclipsed "pinched kagomé" lattice is also found in Langasites such as $Nd_3Ga_5SiO_{14}$; in that case, alternate vertices around the hexagonal hole in the lattice are pinched significantly inward, nearly forming a triangular hole, and the interplane nearest neighbor distance is about 20% larger than the intraplane distance

two inorganic materials based on $Cu^{2+}$ have been proposed as examples of this class: $Cu_3V_2O_7(OH)_2 \cdot 2H_2O$, Volborthite (Fig. 6.7, [25]); and $ZnCu_3(OH)_6Cl_2$, Paratacamite [26]. Due to the Jahn–teller distortions, the lattices are made from non-equilateral triangles. As the formulas imply, the Cu-based and jarosite-class materials described here are prepared hydrothermally, making them inappropriate for study by the synthetically-challenged.

There are two lower-symmetry derivatives of the kagomé lattice that have been of interest from a magnetic perspective. One of these is the "kagomé-staircase" family (Fig. 6.8), consisting of $M_3V_2O_8$, with $M = Mn^{2+}$, $Co^{2+}$, $Ni^{2+}$ and, in a

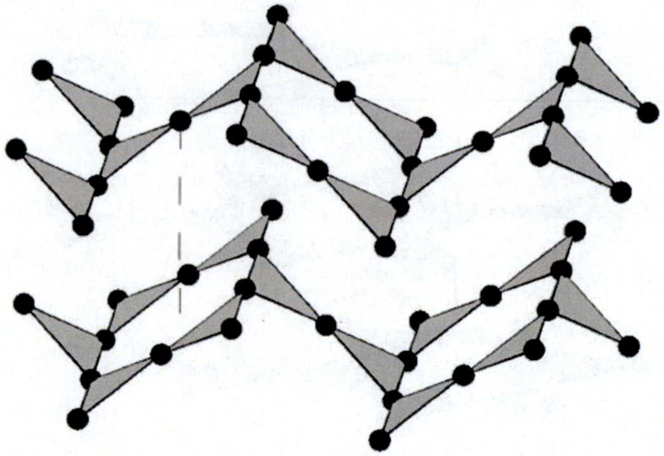

**Fig. 6.8** The buckled kagomé lattice in "kagomé staircase" compounds such as $Ni_3V_2O_8$. The triangles are isosceles, with stronger interactions along the spines

highly distorted variant, $Cu^{2+}$ [27,28]. In these materials, the kagomé plane, which consists of edge-sharing $MO_6$ octahedra, is buckled to break the perfect symmetry, resulting in distinct "spine" and "crosstie" sites (Fig. 6.8). In all cases known, the subtle symmetry breaking leads to complex temperature-applied field magnetic phase diagrams. The second distorted kagomé variant is found in the Langasite compounds [29]. These have the general formula $Ln_3Ga_5SiO_{14}$, and are found only for the larger magnetic rare earths e.g. Pr and Nd. In these compounds, the array of corner sharing triangles that make up the kagomé planes is strongly "pinched" inward on alternating shared vertices around the normally hexagonal holes in the kagomé planes to make those holes more like puckered triangles. The rare earths in neighboring kagomé planes are 20% more distant than the in-plane neighbors, with the planes stacked in an eclipsed fashion.

The Olivine [30] lattice, though it is not generally appreciated as such, is derived from the kagomé lattice by removing magnetic sites in periodic lines perpendicular to the kagomé spine direction (Fig. 6.9). The result is a saw tooth lattice of corner-sharing triangles (isosceles in real materials but also conceptually equilateral). The saw tooth has a central spine of equivalent atoms, with symmetry-independent atoms on the saw tooth tips. The olivines are a large crystal chemical family, with $Mn_2SiO_4$ and $Ni_2SiO_4$ the most investigated from a geometric magnetic frustration viewpoint. There are also cases where the spine sites are populated by magnetic ions but the saw tooth tips are occupied by non-magnetic ions resulting in magnetic chain compounds; $LiCoPO_4$ is one example [31]. Less well known, and only briefly surveyed, are the rare earth olivines, $ZnLn_2S_4$, which are known for the cases of Ln = Er, Tm and Yb [32]. Olivines may prove fertile ground for future investigation of magnetic frustration.

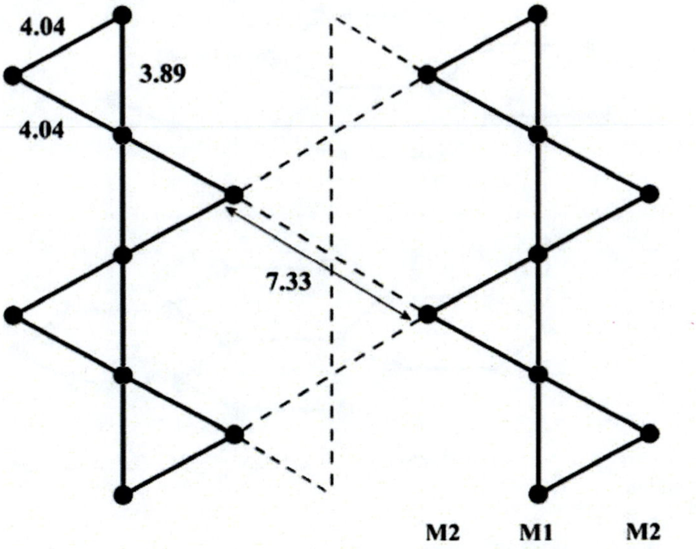


**Fig. 6.9** The incomplete kagomé plane in the rare earth olivine $ZnTm_2S_4$. These incomplete planes are stacked in a staggered fashion. M1 and M2 atoms are not equivalent. In some olivines only the M1 sites are occupied by magnetic atoms

Finally, honeycomb lattices, though usually not considered geometrically frustrating, are derived from the edge-sharing triangular planar lattice by removing 1/3 of the magnetic sites in an ordered fashion. Though less densely packed than the kagomé lattice at 2/3 rather than 3/4 filling, the arrangement of neighbors in honeycombs makes straightforward ordering schemes possible. Honeycomb lattice layers made from edge-sharing $MO_6$ octahedra are found as building blocks in important non-magnetic compounds such as sapphire, where the overall structure is highly three-dimensional. Layered honeycomb magnetic materials are found commonly in two families. Most studied are those in the structural family typified by $BaNi_2V_2O_8$ (Fig. 6.10), which has staggered, honeycomb layers of $NiO_6$ octahedra sharing edges, spaced by non-magnetic Ba and $V^{5+}$ oxide layers [32]. Magnetic Ni, Co, and Fe variants of this structure are known, and non-magnetic P or As can be present instead of V. Particularly interesting for this family is that the magnetic ordering observed is far from simple, with competition between near neighbor and next nearest neighbor magnetic interactions across the hexagonal hole in the honeycomb leading to frustration of simple magnetic ground states. These competing interactions must be surprisingly closely balanced, because the magnetic ordering is quite different for $BaNi_2V_2O_8$, $BaNi_2As_2O_8$, and $BaNi_2P_2O_8$, for example, which differ only slightly in crystal structure dimension [33] (Table 6.3).

Another honeycomb magnetic lattice is found in a different structural family, typified by members $Na_2Co_2TeO_6$ and $Na_3Co_2SbO_6$ [34]. For these materials, the triangular layer, instead of having 1/3 of the sites vacant to form the magnetic honeycomb, has a non-magnetic atom filling the vacant sites. The net magnetic lattice is the same. The layers are not as well separated as they are for the compounds in the $BaNi_2V_2O_8$ family. In one of the structural variants, the honeycombs are eclipsed

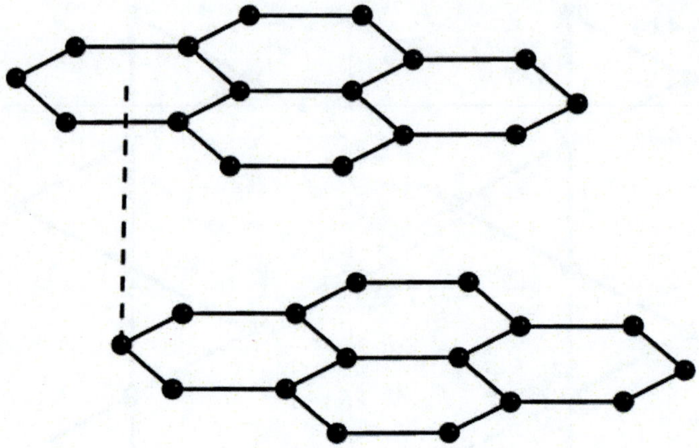

**Fig. 6.10** The regular honeycomb lattice of Ni in $BaNi_2V_2O_8$. The layers are stacked in a staggered fashion

**Table 6.3** Planar lattices based on corner-sharing triangles

| Compound | Structure type | Layer type | Stacking type | M–M in plane | M–M next plane |
|---|---|---|---|---|---|
| $NH_4Fe_3(SO_4)_2(OH)_6$ | Jarosite | Kagomé | Eclipsed | 3.65 | 6.13 |
| $ZnCu_3(OH)_6Cl_2$ | Paratacamite | Kagomé | Staggered | 3.39, 3.43 | 5.05, 5.15 |
| $Cu_3V_2O_7(OH)_2 \cdot 2H_2O$ | Volborthite | Kagomé | Eclipsed | 2.93, 3.03 | 7.21 |
| $Ni_3V_2O_8$ | $Co_3V_2O_8$ | Kagomé staircase | Staggered | 2.94, 2.96 | 5.71 |
| $Nd_3Ga_5SiO_{14}$ | Langasite | Pinched Kagomé | Eclipsed | 4.19, 5.83 | 5.03 |
| $Mn_2SiO_4$ | Olivine | Isosceles saw tooth | Staggered | 3.12, 3.35 | 3.69, 3.76, 4.02 |
| $ZnTm_2S_4$ | Olivine | Isosceles saw tooth | Staggered | 3.89, 4.04 | 4.81, 4.85, 5.02 |

while in the other they are displaced, though not symmetrically, with respect to the lower layer. The magnetic structures of these compounds have not been studied in any detail, so it is not known how the ordering compares to what is seen for other honeycombs. This family is interesting, however, because it includes the $Cu^{2+}$ spin 1/2 variants $Na_2Cu_2TeO_6$ and $Na_3Cu_2SbO_6$ [35,36]. For these latter two compounds, two nearest neighboring coppers are displaced towards each other to form dimers and therefore spin singlets, in a repeating pattern (Fig. 6.11). The net result of dimerizing the honeycomb lattice is to create a lattice of edge-sharing equilateral triangles of dimers. The very strong intradimer coupling apparently dominates the interdimer coupling. Relatively little detailed characterization is available for the $Na_2Co_2TeO_6$ family when compared to the $BaNi_2V_2O_8$ family, possibly due to the current lack of large single crystals for these compounds.

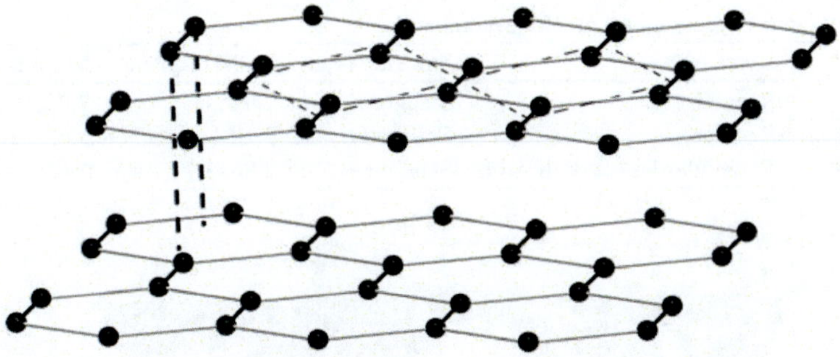

**Fig. 6.11** The honeycomb Cu planes in $Na_2Cu_2TeO_6$. The neighboring Cu in the honeycombs form dimers. The result is a triangular lattice of dimers. The stacking of the layers is staggered

## 6.3 Three-Dimensional Structures

When the magnetic lattice extends into the third dimension, a more diverse group of frustrating lattices is found. There is, however, a limit to the degree of complexity that can be modeled theoretically and understood conceptually, and so, for the most part, the frustrated three-dimensional magnetic lattices that have been studied in detail are relatively few. Most often studied are the rare earth oxide pyrochlores, no doubt due to their simple magnetic geometry and the fact that the floating zone crystal growth method can be used to grow large single crystals. After the pyrochlores, the transition metal spinels and the ferrites, which are derived from spinels, are also commonly studied, as are rare earth garnets, which have a more complex structure (Table 6.4).

Tetrahedra are the basic frustrating unit in many of the most commonly studied three-dimensional frustrated magnets. Interestingly, the number of known 3D lattices based on repeating magnetic tetrahedra is relatively small. The simplest such lattice is the FCC lattice, which is simply an array of edge-sharing tetrahedra (Fig. 6.12). Most importantly from a magnetic frustration perspective, the ubiquitous double perovskites, with formula $A_2BB'O_6$, display this kind of magnetic sublattice when only one of the B-type sites is filled with magnetic ions. The B site ions are in octahedral local coordination with oxygen. $Sr_2MgReO_6$ and $Ba_2HoNbO_6$, for example, have FCC lattices of magnetic Re or Ho, i.e. lattices of magnetic edge-sharing tetrahedra in three dimensions [37–40]. Many rare earth-based double perovskite variants are known, with Ba or Sr on the A site, and Nb, Ta or Sb on the nonmagnetic B site. Relatively little is known about the magnetic properties of these phases. As with all phases that involve a superlattice ordering of metals within a substructure, there is significant potential for intermixing of metals on the B and B′ sites in $A_2BB'O_6$, a situation that would add frustration due to disorder to the effects of geometric frustration, which is not a desired situation. Therefore, care must be taken by the chemist in both selection of B and B′, which should be as chemically dissimilar (e.g. in size, charge and electronegativity) as possible (e.g. BB′ = HoNb or HoSb should be well ordered but BB′ = NiTi or

**Table 6.4** Planar Lattices based on Honeycombs

| Compound | Layer type | Stacking type | M–M in-plane | M–M next-plane |
|---|---|---|---|---|
| $BaNi_2V_2O_8$ | Regular | Staggered | 2.91 | 7.99, 7.87 |
| $Na_3Co_2SbO_6$ | Regular | Eclipsed | 3.10 | 5.65 |
| $Na_2Cu_2TeO_6$ | Triangular lattice of dimers | Staggered | 2.86, 3.21 | 5.65 |

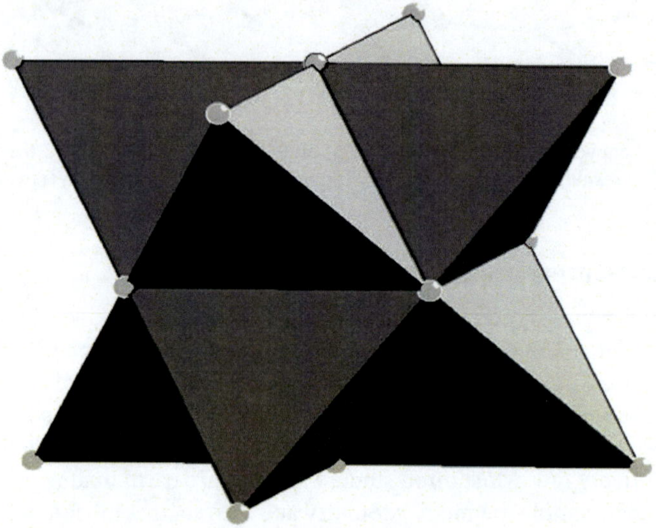

**Fig. 6.12** The FCC lattice is made from edge sharing of regular tetrahedra. This magnetic lattice is found in double perovskites such as $Ba_2HoNbO_6$, $Sr_2MgIrO_6$, and $Sr_2MgReO_6$

FeMo may display disorder). Detailed structural characterization is needed when chemically similar atoms are present to be sure the degree of intermixing is small.

Less constrained are three-dimensional lattices derived from magnetic tetrahedra that share corners. Surprisingly, spinels and the pyrochlores have the same tetrahedron-based geometry for certain subsets of their component atoms, though they are, on the surface, quite different. For $A_2B_2O_7$ pyrochlores, both the A and B sublattices (Fig. 6.13) are comprised of corner-sharing tetrahedra, which interpenetrate in the structure [41]. Magnetic frustration has most often been studied for pyrochlores with magnetic atoms on the A sites, e.g. $Ho_2Ti_2O_7$, $Dy_2Ti_2O_7$, and $Ho_2Sn_2O_7$. The chemistry and synthesis are straightforward enough for these materials. Crystal growth is relatively easy, and chemical manipulation, such as dilution of the magnetic lattice, e.g. in solid solutions such as $(Ho_{2-x}Y_x)Ti_2O_7$, is relatively easily achieved. This kind of magnetic site dilution allows single ion effects to be distinguished from collective effects, and also, as in all frustrated magnetic systems, allows some relief of the frustration through removal of a competing magnetic interaction on a nearest neighbor site. The magnetic properties of a small number of pyrochlores with transition metals on the A site have been reported, i.e. $Mn_2Sb_2O_7$ [42]. Relatively low temperatures are required for synthesis of these interesting

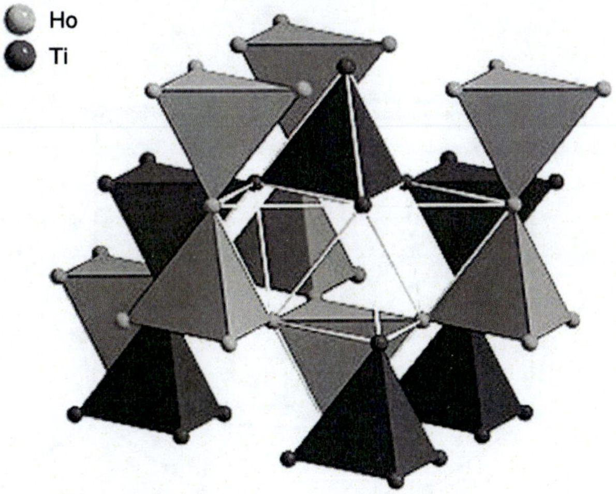


**Fig. 6.13** The interpenetrating A (Ho) and B (Ti) sublattices (*different shades of gray*) made from corner sharing tetrahedra in $Ho_2Ti_2O_7$ pyrochlore. Filling both A and B sites with the same atom yields an FCC lattice

phases, and as of this writing no single crystals have been reported. A handful of B-site magnetic pyrochlores, most notably $Y_2Mo_2O_7$ [43], have also been studied; spin 1 $Mo^{4+}$ occupies the B site. The $Ln_2Ir_2O_7$ pyrochlores, with spin $1/2$ $Ir^{4+}$ on the pyrochlore B sites, show various magnetic phases [44]. The interpenetrating A and B sublattices of the pyrochlore structure type are tetrahedra that are equal in size and spaced from each other by distances that are exactly equal to the distances between the vertices of the tetrahedra themselves. The result of this unique geometry is that, if both sublattices were to be populated by the same magnetic ion, the resulting magnetic lattice would be edge-sharing tetrahedra. Thus, by having a magnetic rare earth fully occupying the pyrochlore A-site, and then slowly substituting the same rare earth on the B-site, one could continuously titrate the magnetic system from a lattice of corner-sharing tetrahedra to one of edge-sharing tetrahedra (Fig. 6.12). This has recently been shown [45] to be possible in the "stuffed spin ices" $Ho_2(Ti_{2-x}Ho_x)O_{7-x/2}$ and $Dy_2(Ti_{2-x}Dy_x)O_{7-x/2}$. The study of these materials is just now beginning. The B-site lattice in the $AB_2X_4$ spinels has the same kind of corner-sharing tetrahedron lattice as is found for both A and B sublattices in pyrochlores.

A much more chemically and structurally varied family based on the same kind of corner-sharing tetrahedral magnetic lattice are the spinels and the ferrites that are derived from the spinels. The $AB_2X_4$ spinel structure is found very commonly among transition metal oxides, sulfides and selenides. Based on a cubic close packed X array with A atoms occupying 1/8 of the available tetrahedral interstices and B atoms occupying half of the available octahedral interstices, the number and kinds of chemical systems in this structural family is very large. As described above, the B atom sublattice, which in this case is made of $BX_6$ octahedra sharing edges in

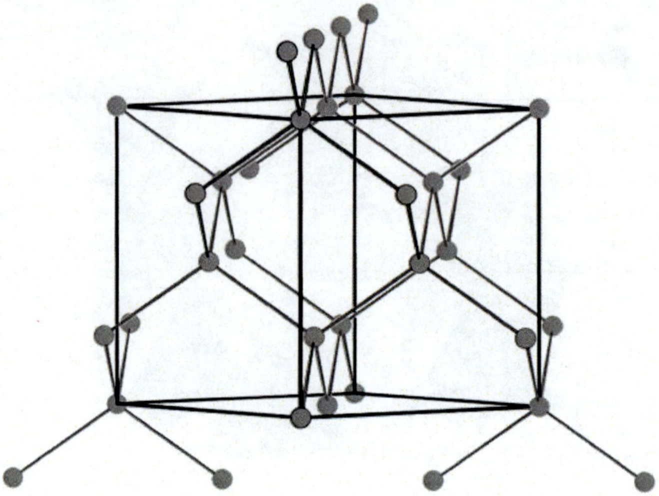

**Fig. 6.14** The A sites in spinel have the diamond structure. This is the Mn array for $MnSc_2S_4$

[111] chains that intersect in three dimensions, is made of corner-sharing tetrahedra. Most well known among the frustrated spinels are simple compounds such as $ZnCr_2O_4$ [46]. Much less frequently studied are the rare earth spinels of the type $CdLn_2(S, Se)_4$, stable for the smaller rare earths, with examples such as $CdDy_2Se_4$ and $CdHo_2Se_4$ [47]. The A site of the $AB_2X_4$ spinels forms a diamond lattice (Fig. 6.14), where each magnetic atom has four magnetic neighbors [48]. This lattice is larger than the B site lattice. If the A site is occupied by a magnetic atom, and the B site is occupied by a non-magnetic atom, then geometrically frustrated magnetism is reportedly present in this case as well. This has recently been reported [49] for the sulfide spinels $MnSc_2S_4$ and $FeSc_2S_4$. A recent review of frustrated magnetism in oxide spinels may be of interest [50].

The spinel structure is of particular interest as the parent structure of a very large family of compounds known as ferrites (Fig. 6.15). Considering the geometry of the magnetic part of the lattice only, the spinel structure can be disassembled into a sequence of kagomé (K) layers (the bases of magnetic tetrahedra) and triangular (T) layers (the tips of the magnetic tetrahedra) that stack alternately without interruption, i.e. in the infinite sequence [K-T-K-T-$\cdots$], to build a three-dimensional lattice. These layers are perpendicular to the 111 direction in the spinel. Ferrites are derived from spinels by extracting a [K-T-K] layer sandwich and separating it from the next [K-T-K] sandwich by intermediate non-spinel spacing layers. These spacing layers also frequently contain magnetic atoms, so that the spinel sandwiches are, in most cases, magnetically coupled in the third dimension as well. The M-type ferrite (magnetoplumbite), $SrGa_4Cr_8O_{19}$ ("SCGO"), is the most frequently studied in this family [51] and has significant coupling between spinel sandwich layers through the apices of hexagonal pyramids [HP] whose base is the hexagonal ring in the kagomé layer (Fig. 6.15). M-type ferrites such as SCGO can therefore be described

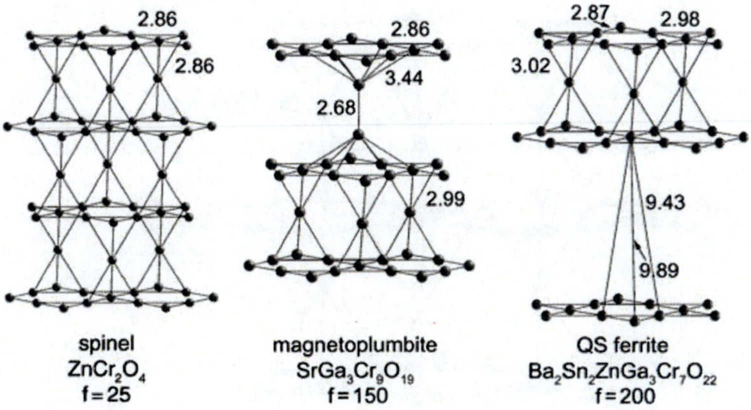


**Fig. 6.15** A comparison of representative Spinel, Magnetoplumbite, and QS ferrite magnetic lattices. kagomé layers are seen in all compounds and there are different kinds of connections between layers

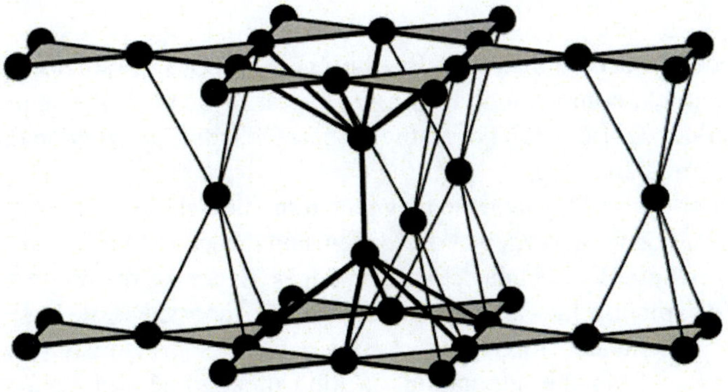

**Fig. 6.16** The magnetic lattice in the R-type ferrite $SrCo_6O_{11}$. The connecting layers between the eclipsed kagomé nets are a hybrid of the two types of connecting layers in the M-type ferrite; they are occupied by magnetic Co ions. In the Cr-based R-type ferrite, $SrSn_2Ga_{1.3}Cr_{2.7}O_{11}$, only the kagomé nets are occupied by magnetic ions; the intermediary layers contain only non-magnetic ions

by the sequence [K-T-K]-[HP]-[K-T-K]. The R-type ferrite, typified by $SrCo_6O_{11}$ [52] is a hybrid between a spinel and an M-type ferrite, with a mixed geometry coupling layer that includes both tetrahedra and hexagonal pyramids (Fig. 6.16). This type of ferrite can be described by the sequence [K-T-K]-[T + HP]-[K-T-K]. This structure type has recently been reported for the Cr-ferrite family for the compound $SrSn_2Ga_{1.3}Cr_{2.7}O_{11}$ [53]. In this Cr-based variant, magnetic Cr is found only in the kagomé (K) layers, and thus the material is highly two-dimensional. The QS ferrite [54] is a highly two-dimensional of the ferrite family. In one variant, the spinel-derived magnetic sandwiches are separated by both a large distance and the complete absence of magnetic atoms in the intermediate layer (Fig. 6.15). The

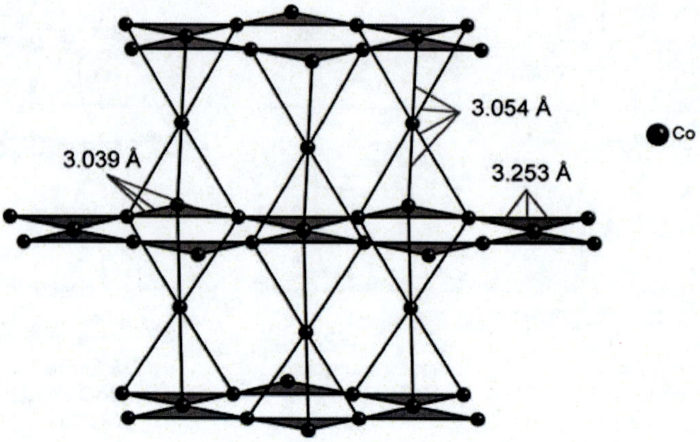


**Fig. 6.17** The spinel–like magnetic lattice in $YBa_2Co_4O_7$. In this case, there are corner plus face sharing tetrahedra connecting the kagomé planes, rather than corner sharing tetrahedra found in spinels

formula needed to accomplish this, $Ba_2Sn_2ZnGa_3Cr_7O_{22}$ is rather complex [55]. Denoting the thick, nonmagnetic $Ba_2Sn_2O_6$ spacer layer as Q, the sequence in QS ferrite is [K-T-K]–Q–[K-T-K]. For all the Cr-based ferrites, disorder on the magnetic lattices is a significant issue.

In the spinel-derived ferrites, the magnetic transition elements are found in octahedral coordination with oxygen or chalcogen, and the kagomé and triangular layers are made from these $BX_6$ octahedra sharing edges. Recently, the properties of a different type of spinel-related compound have been reported [56], based on $Co^{2+}$ tetrahedrally coordinated to oxygen, $YBa_2Co_4O_7$. As previously described, the spinel is derived from the alternating stacking of kagomé and triangular layers, resulting in a three-dimensionally connected lattice of corner-sharing tetrahedra. As seen in Fig. 6.15, the corner-sharing between tetrahedra occurs through the tips of the tetrahedra. (These tips lie in the triangular plane we have designated as T.) The kagomé layers consist of rings of 6 triangles sharing corners. In the spinel, three of these kagomé layer triangles are capped by magnetic atoms in the T layer above, and three are capped by magnetic atoms in the T layer below, in an alternating sequence. In $YBa_2Co_4O_7$, there are again three capping magnetic atoms above and three below the kagomé layer, but in this case they are on the same three triangles. The alternate triangles in the kagomé layer in this structure type are not capped. Thus, the three-dimensional magnetic lattice in $YBa_2Co_4O_7$ is unique – consisting of both corner-sharing and face-sharing tetrahedra (Fig. 6.17).

After the pyrochlores, spinels, and ferrites, the three-dimensional frustrated lattices get rather complicated. The others whose properties have been studied are based on edge- and corner-sharing triangles snaking through the crystal structure in three dimensions, rather than being based on a more complex arrangement of tetrahedra than is found in spinels and ferrites. Best known of these systems are the rare earth garnets, of the formula $Ln_3Ga_5O_{12}$ [57]. For the Garnets, the lanthanide Gd

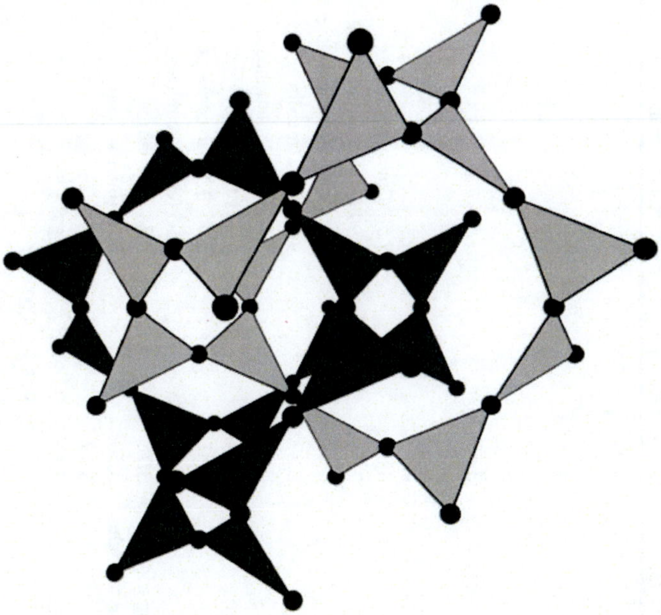

**Fig. 6.18** The two interpenetrating 10-triangle bifurcated ring systems in Garnets such as $Gd_3Ga_5O_{12}$ and $Mn_3Al_2Si_3O_{12}$

variant ("GGG") has been particularly well studied. The crystal structure of garnet is based on lanthanide in a seven-fold coordination with oxygen, and tetrahedral and octahedral coordination with oxygen for the smaller nonmagnetic ions, in a 2:3 ratio. The rare earth ions fall on two interpenetrating ring lattices, made from corner-sharing triangles that are equivalent by symmetry. Unlike the kagomé net, where the corner-sharing triangles form in six-membered rings, the rings in garnet (Fig. 6.18) are made of ten corner-sharing triangles. Each ten-membered ring is actually bifurcated. If triangles 1 and 5 in the 10-membered ring are denoted as first and mid-ring triangles, then there are 4 triangles between the first and mid-ring triangles, and, in addition, another 4-triangle ring connected to the free vertices of the first and mid-ring triangles. There are *two* of these bifurcated 10-triangle ring systems in the garnet structure. These ring systems are interpenetrating. The lanthanides in the two interpenetrating equivalent ring systems at their closest approach, in $Tb_3Ga_5O_{12}$ for instance, are 5.78 distant, compared to 3.78 angstroms within the triangles, so the magnetic interactions can be considered to be strongest within the rings, and weaker between rings. There are also a small number of garnets with transition metal ions on the triangular lattice sites and non-magnetic ions on the other sites, notably $Mn_3Al_2Si_3O_{12}$ (spessartine) and $Fe_3Al_2Si_3O_{12}$ (almandine) [58, 59]. The garnets that are based on magnetic rare earths on the large atom sites and iron on the tetrahedral and/or octahedral sites, compounds such as $Gd_3Fe_5O_{12}$, are well known high temperature insulating ferromagnets [60] (Table 6.5).

**Table 6.5** Three-dimensional materials with triangle-based lattices

| Compound | Structure type | Lattice type | M–M in-triangle | M–M next nearest |
|---|---|---|---|---|
| $Ba_2HoSbO_6$ | Double Perovskite | Edge-sharing tetrahedra (FCC) | 5.93 | |
| $Sr_2MgIrO_6$ | Double Perovskite | Edge-sharing tetrahedra (FCC) | 5.58 | |
| $Dy_2Ti_2O_7$ | Pyrochlore A site | Corner-sharing tetrahedral | 3.58 | |
| $Mn_2Sb_2O_7$ | Pyrochlore A site | Corner-sharing tetrahedra | 3.59 | |
| $Y_2Mo_2O_7$ | Pyrochlore B site | Corner-sharing tetrahedra | 3.62 | |
| $Dy_{2+x}Ti_{2-x}O_{7-x/2}$ | Stuffed Pyrochlore | Corner to edge-sharing tetrahedra | 3.58–3.66 | |
| $ZnCr_2O_4$ | Spinel (B site) | Corner-sharing tetrahedra | 2.86 | |
| $CdDy_2S_4$ | Spinel (B site) | Corner-sharing tetrahedra | 3.98 | |
| $MnSc_2S_4$ | Spinel (A site) | Diamond | 4.59 | 7.50 (2nd neighbor) |
| $SrGa_4Cr_8O_{19}$ | M-type Ferrite | Kagomé+tetrahedra+pyramid connectors | 2.86 | 2.99 (cap of tetrahedron)<br>3.44, 2.68 (to connectors) |
| $Ba_2Sn_2Ga_3ZnCr_7O_{22}$ | QS ferrite | Kagomé+tetrahedra | 2.87, 2.98 | 9.49, 9.86 (interlayer)<br>3.05 (cap of tetrahedron) |
| $SrSn_2Ga_{1.3}Cr_{2.7}O_{11}$ | R-type Ferrite | Kagomé | 2.94 | 6.72 (interlayer) |
| $SrCo_6O_{11}$ | R-type Ferrite | Kagomé+connector | 2.81 | 3.36, 3.54 (to connectors) |
| $YBaCo_4O_7$ | spinel-like | Kagomé+tetrahedra | 3.03, 3.25 | 3.05 (cap of tetrahedron) |
| $Tb_3Ga_5O_{12}$ | Garnet | Interpenetrating lattices of corner-sharing triangles | 3.78 | 5.78 (interlattice) |
| $Mn_3Al_2Si_3O_{12}$ | Garnet | Interpenetrating lattices of corner-sharing triangles | 3.50 | 5.35 (interlattice) |
| $Na_4Ir_3O_8$ | half-Garnet | Corner-sharing triangles | 3.19 | |
| $Er_2Ge_2O_7$ | Tetrag. pyro- germanate | Edge-sharing +corner-sharing triangles | 3.57, 3.59 | 5.57–5.85 |
| $SrHo_2O_4$ | $NaFe_2O_4$ | Honeycombs of edge-sharing triangles | 3.41, 3.49, 3.87, 4.02 | |
| $\beta$-$CaCr_2O_4$ | $NaFe_2O_4$ | Honeycombs of edge-sharing triangles | 2.96, 3.02, 3.57, 3.63 | |

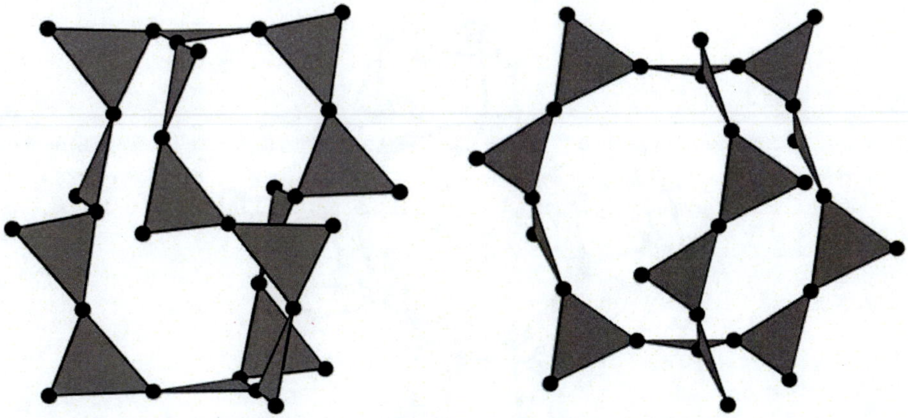

**Fig. 6.19** Comparison of the magnetic ring of triangles in $Na_4Ir_3O_8$ (*right*) with one of the rings of magnetic triangles in Garnet (*left*)

Recently, magnetic frustration has been reported in $Na_4Ir_3O_8$ [61], a compound that has the $Na_4Sn_3O_8$ structure type [62]. The magnetic $Ir^{4+}$ spin $1/2$ ions in this compound are in $IrO_6$ octahedra that share corner oxygens, with Na in the interstitial positions. This compound has been described as being a "hyper-kagomé". This name does not describe its magnetic lattice in detail, however, due to the type of interconnectivity of the triangles present. The lattice of corner-sharing, magnetic triangles in $Na_4Ir_3O_8$, in fact, has the same character and connectivity as one of the two symmetrically equivalent ten-fold bifurcated ring lattices in garnet, not the ring system seen in the kagomé net. Given that the character of the magnetic connectivity is an important parameter in geometrically frustrated systems, it may be productive to think about the magnetic lattice in $Na_4rIr_3O_8$ as a "half-garnet" rather than a "hyper-kagomé" (Fig. 6.19).

The two final three-dimensional magnetic lattices that will be described are those in $Ln_2Ge_2O_7$ and $SrLn_2O_4$. In spite of the similarity of the formula to that of the pyrochlore, the crystal structure of $Ho_2Ge_2O_7$, known as the tetragonal pyrogermanate structure, involves rare earth ions in seven-fold coordination with oxygen, and Ge in tetrahedral coordination with oxygen. The magnetic lattice consists of dimers of edge-shared triangles that then share corners with each other to form chains [63]. These chains form chiral spirals along the unique crystallographic axis (Fig. 6.20). The magnetic properties of some of these compounds have been studied [64]. Finally, though the magnetic lattice in $SrHo_2O_4$ [65] is conceptually simple, it can be confusing on first sight. Simply stated, it consists of three-dimensional honeycomb columns. The walls of these honeycomb columns consist of edge-sharing triangles – more fully, the walls are planes of corner-sharing triangles wrapped up to create six-sided columns with walls two edge-sharing triangles wide in a honeycomb array. The symmetry of the structure is such that there are four different wall types that differ slightly from each other, though this may have only a secondary effect on the magnetic properties (Fig. 6.21). Almost nothing is known about the

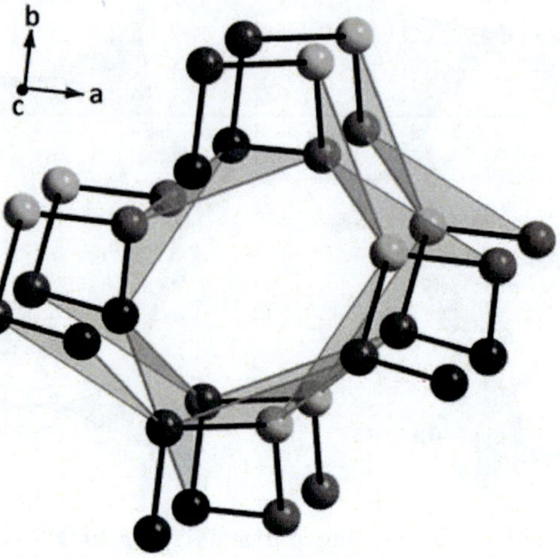


**Fig. 6.20** The magnetic lattice in the $Ho_2Ge_2O_7$ tetragonal pyrogermanate consists of spirals of corner and edge-sharing triangles

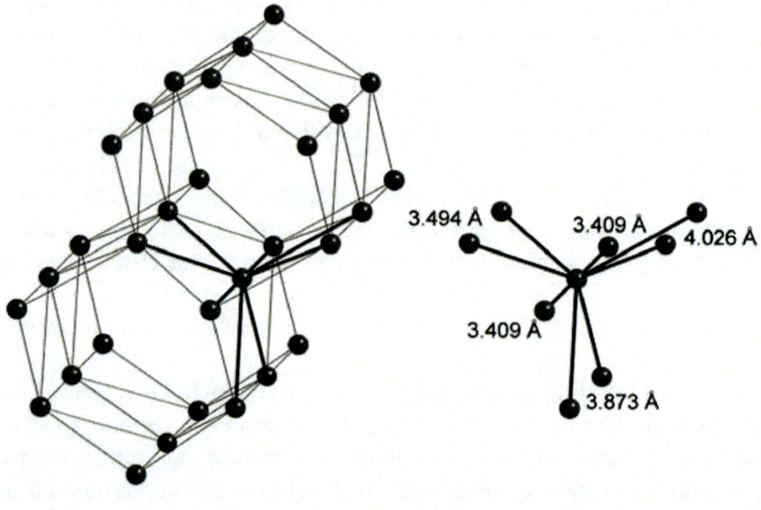


**Fig. 6.21** (*Left*) The honeycomb columns of edge sharing Ho triangles in $SrHo_2O_4$. The interatomic spacings are shown on the *right*. The same magnetic lattice geometry is found for β-$CaCr_2O_4$

magnetism in these phases. The beta form of the compound $CaCr_2O_4$ displays the same magnetic lattice (the general name for the structure is the $NaFe_2O_4$-type) as these rare earth compounds. Recent studies have shown that the magnetic phase diagram for this phase is very complex due to the highly frustrated three-dimensional magnetic lattice, which contains sublattices of lower effective dimensionality [66].

## 6.4 Note on Synthesis of the Compounds

For the most part, judged by state-of-the-art synthesis in solid state chemistry, the geometrically frustrated magnetic materials described in this summary are easy to make. In a real sense, ease of synthesis is one of the most critical characteristics that a material must possess to become widely studied, as is chemical stability in normal laboratory air. Chemical stability is a significant problem for $VCl_2$ and related chloride compounds, for example, limiting their extensive study. Fortunately, some of the materials described here can easily be made as large single crystals (e.g. the rare earth pyrochlores, oxide spinels, and some jarosites). This has been an especially important factor in the explosion of work on rare earth pyrochlores in recent years – they are straightforwardly grown as large crystals by the floating zone method. Crystals of other materials described here, such as the tetragonal pyrogermanates, garnets, and the kagomé staircase compounds are reasonably easily grown from fluxes. Very nice crystals of jarosite have been grown by a hydrothermal flux method. Others of the materials of interest are relatively easy to synthesize in polycrystalline form but, so far, are impossible to grow as crystals. One very important material for which this is a problem is the M-type ferrite $Sr(Ga,Cr)_{12}O_{19}$. The QS Cr-based ferrite is relatively easily made in polycrystalline form but the R-type ferrite is very difficult to make in any form. Special synthesis methods that exclude air are, of course, needed for making sulfides and selenides, and relatively large crystals of compounds such as $Cr_{1/3}NbS_2$ can be grown by vapor transport in an evacuated quartz tube. Some of the chalcogenides, such as the rare earth sulfide spinels and olivines, are extremely difficult to make in any form due to the high volatility of some components coupled with the poor reactivity of others. Thus though many frustrated magnetic materials are available and easily studied in polycrystalline form, the same cannot be said for the availability of single crystals. For all systems, particularly the ferrites, and even some conceptually "simple" ones like the spinels and double perovskites, disorder due to mixing of other atoms in the magnetic lattice can be a vexing problem. Detailed structural characterization is needed to resolve this issue in all the relevant cases, as is the development of synthesis methods to minimize the disorder. More work will be required to find simple systems such as the rare earth pyrochlores, where structural disorder is minimized, the properties are interesting, and crystals are easy to grow. Crystal growth of more complex frustrated materials should be considered a very important area for future research in this field.

## 6.5 Conclusion

Materials with their magnetic ions placed on triangular lattices have been recognized for many years as being of interest in probing the properties of frustrated spin systems, but it is only in the past 10 or 15 years that solid state chemists have joined the field. We have specifically been looking to expand this class of compounds by finding new structure types with spins on triangles. The efforts of the solid state community are represented in some of the tables in this compilation. The recent exploitation of the floating zone crystal growth method to grow many different rare earth pyrochlores has led to a substantial increase in the overall activity in this field on the experimental physics side. Geometric magnetic frustration is an area in particular where chemists and physicists have to work together to find new materials: in their search for interesting structures containing magnetic triangles, chemists may frequently come up with materials whose lattices are simply too complex to be reasonably modeled by theory or reasonably characterized by experiments (e.g. $SrHo_2O_4$). Conversely, no solid state chemist can be content to grow crystals of rare earth pyrochlores for very long – they are just too simple to hold a chemist's attention. The optimal new magnetically frustrated materials that will challenge chemists and physicists alike will only be found through continuing frequent interactions between scientists in the two disciplines.

## References

1. M.F. Collins, O.A. Petrenko, Can. J. Phys. **75**, 605 (1997)
2. S.T. Bramwell, M.J.P. Gingras, Science **294**, 1495 (2001)
3. J.E. Greedan, J. Mater. Chem. **11**, 37 (2001)
4. R. Moessner, Can. J. Phys. **79**, 1283 (2001)
5. A.P. Ramirez, Annu. Rev. Mat. Sci. **24**, 453 (1994)
6. P. Mendels, A. Olariu, F. Bert, D. Bono, L. Limot, G. Collin, B. Ueland, P. Schiffer, R.J. Cava, N. Blanchard, F. Duc, J. C. Trombe, J. Phys. Cond. Mat. **19**, 145224 (2007)
7. H.T. Diep (ed.), *Frustrated Spin Systems* (World Scientific Publishing Co. Inc., Hackensack NJ, 2004)
8. H.D. Lutz, W. Buchmeier, H. Siwert, Z. Anorg. Allg. Chem. **533**, 118 (1986)
9. M. Sassmannshausen, H.D. Lutz, Acta Crystallogr. C **54**, 704 (1998)
10. S. Nakatsuji, Y. Nambu, H. Tonomura, O. Sakai, S. Jonas, C. Broholm, H. Tsunetsugu, Y.M. Qiu, Y. Maeno, Science **309**, 1697 (2005)
11. H.B. Weber, T. Werner, J. Wosnitza, H. VonLohneysen, U. Schotte, Phys. Rev. B **54**, 15924 (1996)
12. C. Cros, M. Niel, G.L. Flem, M. Pouchard, P. Hagenmuller, Mat. Res. Bull. **10**, 461 (1975)
13. S.S.P. Parkin, R.H. Friend, Phil. Mag. B **41**, 65 (1980)
14. S.S.P. Parkin, R.H. Friend, Phil. Mag. B **41**, 95 (1980)
15. H. Haas, E. Kordes, Zeit. Krist. **129**, 250 (1969)
16. H. Effenberger, Acta Crystallogr. **C47**, 2644 (1991)
17. E.F. Bertaut, C. Delorme, Comptes Rendus Hebdomadaires des Seances de l'Academie des Sciences **238**, 1829 (1954)
18. W. Scheld, R. Hoppe, Z. Anorg. Allg. Chem. **568**, 151 (1989)
19. S.T. Bramwell, S.G. Carling, C.J. Harding, K.D.M. Harris, B.M. Kariuk, L. Nixon, I.P. Parkin, J. Phys. Cond. Mat. **8**, L123 (1996)
20. S. Bailleul, D. Svoronos, P. Porcher, A. Tomas, Comptes Rendus Hebdomadaires des Seances de l'Academie des Sciences, Serie **2** (313), 1149 (1991)
21. F.C. Hawthorne, Mineral. Mag. **49**, 87 (1985)
22. M. Sofin, M. Jansen, Zeitschrift fuer Naturforschung, Inorganische und Organische Chemie **B60**, 701 (2005)
23. S.J. Clarke, A.J. Fowkes, A. Harrison, R.M. Ibberson, M.J. Rosseinsky, Chem. Mat. **10**, 372 (1998)

24. S. Menchetti, C. Sabelli, Neues Jahnbuch fuer Mineralogie 406 (1976)
25. R. Basso, A. Palenzona, L. Zefiro, Neues Jahnbuch fuer Mineralogie 385 (1988)
26. J.S. Helton, K. Matan, M.P. Shores, E.A. Nytko, B.M. Bartlett, Y. Yoshida, Y. Takano, A. Suslov, Y. Qiu, J.H. Chung, D.C. Nocera, Y.S. Lee, Phys. Rev. Lett. **98**, 107204 (2007)
27. H. Fuess, E.F. Bertaut, R. Pauthenet, A. Durif, Acta Cryst. B **26**, 2036 (1970)
28. G. Lawes, M. Kenzelmann, N. Rogado, K.H. Kim, G.A. Jorge, R.J. Cava, A. Aharony, O. Entin-Wohlman, A.B. Harris, T. Yildirim, Q. Z. Huang, S. Park, C. Broholm, A.P. Ramirez, Phys. Rev. Lett. **93**, 247201 (2004)
29. H.D. Zhou, B.W. Vogt, J.A. Janik, Y.-J. Jo, L. Balicas, Y. Qiu, J.R.D. Copley, J.S. Gardner, C.R. Wiebe, Phys. Rev. Lett. **99**, 236401 (2007)
30. O. Ballet, H. Fuess, W. Wacker, E. Untersteller, W. Treutmann, E. Hellner, S. Hoyosa, J. Phys. Cond. Mat. **1**, 4955 (1989)
31. A. Goni, L. Lezama, G.E. Barberis, J.L. Pizarro, M.I. Arriortua, T. Rojo, J. Mag. Mag. Mat. **164**, 251 (1996)
32. G.C. Lau, B.G. Ueland, R.S. Freitas, M.L. Dahlberg, P. Schiffer, R.J. Cava, Phys Rev B **73**, 012413 (2006)
33. N. Rogado, Q. Huang, J.W. Lynn, A.P. Ramirez, D. Huse, R.J. Cava, Phys. Rev. B **65**, 14443 (2002)
34. L.P. Regnault, J. Rossat-Mignod, in *Magnetic Properties of Layered Transition Metal Compounds*, ed. by L.J. de Jongh, (Kluwer Academic, Dortrecht, 1989), pp. 271–320
35. L. Viciu, Q. Huang, E. Morosan, H.W. Zandbergen, N.I. Greenbaum, T.M. McQueen, R.J. Cava, J. Solid State Chem. **180**, 1060 (2007)
36. Y. Miura, R. Hirai, Y. Kobayashi, M. Sato, J. Phys. Soc. Jpn. **75**, 084707 (2006)
37. D.-Y. Jung, G. Demazeau, J. Solid State Chem. **115**, 447 (1995)
38. J.A. Alonso, C. Cascales, P. Casado-Garcia, J. Rasines, J. Solid State Chem. **128**, 247 (1997)
39. H. Karunadasa, Q. Huang, B.G. Ueland, P. Schiffer, R.J. Cava Proc. Nat. Acad. Sci. USA **100**, 8097 (2003)
40. C.R. Wiebe, J.E. Greedan, P.P. Kyriakou, G.M. Luke, J.S. Gardner, A. Fukaya, I.M. Gat-Malureanu, P.L. Russo, A. T. Savici, Y.J. Uemura, Phys. Rev. B **68**, 134410 (2003)
41. A.F. Fuentes, K. Boulallya, M. Maczka, J. Hanuza, U. Amador, Solid State Sci. **7**, 343 (2005)
42. H.D Zhou, C.R. Wiebe, A. Harter, N.S. Dalal, J.S. Gardner, J. Phys. Condens. Matter **20**, 325210 (2008)
43. J.N. Reimers, J.E. Greedan, M. Sato, J. Solid State Chem. **72**, 390 (1988)
44. D. Yanagishima, Y. Maeno, J. Phys. Soc. Jpn. **70**, 2880 (2001)
45. G.C. Lau, B.D. Muegge, T.M. McQueen, E.L. Duncan, R.J. Cava, J. Solid State Chem. **129**, 3126 (2006)
46. H. Sawata, Mat. Res. Bull. **32**, 873 (1997)
47. A.B. Agaev, P.G. Rustamov, O.M. Aliev, R.A. Azadaliev, Neorganicheskie Materialy **25**, 294 (1989)
48. L. Pawlak, M. Duczmal, Int. J. Mod. Phys. **7**, 1020 (1993)
49. V. Fritsch, J. Hemberger, N. Buttgen, E.W. Scheidt, H.A.K. von Nidda, A. Loidl, V.A. Tsurkan, Phys. Rev. Lett. **92**, 116401 (2004)
50. S.-H. Lee, H. Takagi, D. Louca, M. Matsuda, S. Ji, H. Ueda, Y. Ueda, T. Katsufuji, J.-H Chung, S. Park, S.-W. Cheong, C. Broholm, J. Phys. Soc. Jpn. **79**, 011004 (2010)
51. X. Boradors, A. Labarta, A. Isalgue, J. Tejada, J. Rodriguez, A. Pernet, Solid State Comm. **65**, 189 (1988)
52. S. Ishiwata, D. Wang, T. Saito, M. Takano, Chem. Mater. **17**, 2789 (2005)
53. I.D. Posen, T.M. McQueen, A.J. Williams, D.V. West, Q. Huang, R.J. Cava, Phys. Rev. B **81**, 134413 (2010)
54. M.C. Cadee, H.J.M. de Groot, L.J. de Jongh, D.W.J. Ijdo, J. Mag. Mag. Mat. **62**, 367 (1986)
55. I.S. Hagemann, Q. Huang, X.P.A. Gao, A.P. Ramirez, R.J. Cava, Phys. Rev. Lett. **86**, 894 (2001)
56. M. Valldor, M. Andersson, Solid State Sci. **4**, 923 (2002)
57. H. Sawada, J. Solid State Chem. **132**, 300 (1997)

58. G.C. Lau, T. Klimczuk, F. Ronning, T.M. McQueen, R.J. Cava, Phys. Rev. B **80**, 214414 (2009)
59. T. Armbruster, C.A. Geiger, G.A. Lager, Am. Mineral. **77**, 512 (1992)
60. S. Geller, Zeit. Fur Kristallgraphie, Kristallgeometrie, Kristallphysik, Kristallchemie **125**, 1 (1967)
61. Y. Okamoto, M. Nohara, H. Aruga-Katori, H. Takagi, Phys. Rev. Lett. **99**, 137207 (2007)
62. M. Iwasaki, H. Takizawa, K. Uheda, T. Endo, J. Mater. Chem. **12**, 1068 (2002)
63. Yu. I Smolin, Kristallographia **15**, 47 (1970)
64. E. Morosan, J.A. Fleitman, Q. Huang, J.W. Lynn, Y. Chen, X. Ke, M.L. Dahlberg, P. Schiffer, C.R. Craley, R.J. Cava, Phys. Rev. B **77**, 224423 (2008)
65. H. Karunadasa, Q. Huang, B.G. Ueland, J.W. Lynn, P. Schiffer, K.A. Regan, R.J. Cava, Phys. Rev. B **71**, 144414 (2005)
66. F. Damay, C. Martin, V. Hardy, A. Maignan, G. Andre, K. Knight, S.R. Giblin, L.C. Chapon, arXiv:0906.3378 (2009)

# Chapter 7
# Highly Frustrated Magnetism in Spinels

**Hidenori Takagi and Seiji Niitaka**

**Abstract** This chapter presents a review of highly frustrated magnetism in spinels and in related compounds. We begin by emphasizing the uniqueness of the spinel structure as a testing ground for the physics of frustration and provide an overview of the materials aspects. We then discuss selected topics from the physics of spinels, including the formation of the spin-liquid state, novel spin–orbit–lattice coupling in the presence of frustration, magnetic-field-induced magnetization-plateau states and charge frustration in mixed-valent spinels.

## 7.1 Introduction

The spinel structure, which has chemical formula $AB_2X_4$ (X = O, S, Se), is one of the most frequently stabilized among the wide variety of structural categories in complex transition-metal oxides and chalcogenides. Materials with the spinel structure have long provided physicists with a surprisingly rich variety of phenomena, including ferromagnetism, ferrimagnetism, and Jahn-Teller transitions [1]. Recently, spinels have attracted increasing interest as a playground for the physics of frustration.

In this review, we focus on frustrated magnetism and on the related phenomena which are produced by the simple but unique geometry of the spinel structure. Because many antiferromagnetic spinels exist as oxides rather than as chalcogenides, we will restrict our considerations essentially to oxides, with the exception of a small number of particularly interesting sulfides. We first describe the structure and electronic properties of spinel oxides, and discuss the uniqueness of the spinel structure for studying the physics of geometrical frustration. We then provide an overview of different spinel materials, emphasizing systems with spin and charge frustration on the B sites and with spin frustration on the A sites, to demonstrate the rich variety of physics amenable to systematic investigation in this class of compounds.

## 7.2 Spinel Structure

The spinel structure, shown in Fig. 7.1a, consists of two basic structural units, $AO_4$ tetrahedra and $BO_6$ octahedra [2]. The $BO_6$ octahedra are connected by their edges, which gives rise to a rutile-like array of edge-sharing chains along the $\langle 110\rangle$, $\langle 1\bar{1}0\rangle$, $\langle 101\rangle$, $\langle 10\bar{1}\rangle$, $\langle 011\rangle$, and $\langle 01\bar{1}\rangle$ directions. The A ions are located in the vacant spaces and are coordinated tetrahedrally by O ions, forming the $AO_4$ units. By considering six chains of B ions, it becomes clear that these form a three-dimensional network of corner-sharing tetrahedra. The B ions are located at every intersection of the chains, as shown in Fig. 7.1b, and this network is known as the pyrochlore lattice: it is realized in both the A and B sublattices of the $A_2B_2O_7$ "pyrochlore" structure, as well as by the B-sublattice sites of the $AB_2$ C15 Laves phase. The pyrochlore lattice is known to give rise to very strong geometrical frustration effects. It is useful to regard the pyrochlore structure from the $\langle 111\rangle$ direction, as shown in Fig. 7.1c, which reveals that there are two types of plane with alternate stacking: one is a two-dimensional triangular lattice, consisting of the apical B ions of the tetrahedra, while the other is a two-dimensional kagomé lattice consisting of the B triangles at the bases of the tetrahedra. The A sublattice of the spinel structure forms a diamond lattice, which is shown in Fig. 7.1d.

In many spinel oxides, one finds cubic symmetry at high temperatures, but this does not necessarily mean that $BO_6$ octahedra are undistorted. The octahedra can be either compressed or elongated along the $\langle 111\rangle$ direction, which is perpendicular to the two triangles facing each other and points towards the centers of the $B_4$ tetrahedra of the pyrochlore lattice. This gives rise to a local trigonal symmetry, which is one of the most important factors determining the electronic configuration of the B ions. The trigonal distortion is parameterized by the "$u$ parameter", which represents the position of the O ions within the cubic cell. For ideal O octahedra, $u = 3/8$, while $u < 0.375$ represents compression of the octahedra and $u > 0.375$ corresponds to elongation.

The most common spinel oxide is known as the "2-3 spinel," $A'^{2+}B'^{3+}_2O_4$, where there is a 1:2 ratio between divalent $A'^{2+}$ cations and trivalent $B'^{3+}$ cations. In the *normal* spinel, the $A'^{2+}$ cations occupy the A sites and the $B'^{3+}$ cations the B sites, while in the *inverse* spinel, $B'^{3+}$ cations occupy the A sites and the B sites are occupied by $A'^{2+}$ and $B'^{3+}$ cations with a 1:1 ratio, as in $B'^{3+}(A'^{2+}_{1/2}B'^{3+}_{1/2})_2O_4$. This can be viewed as a full site exchange between $A'^{2+}$ and $B'^{3+}$. In reality, an intermediate level of site exchange often occurs. Even in seemingly normal spinel materials, it is not easy to achieve complete elimination of such site exchange, which provides a major source of disorder in spinel oxides. There also exist "4-2 spinels" with tetravalent $A'^{4+}$ and divalent $B'^{2+}$ ions. Spinels with monovalent ($A'^{+}$) or trivalent ($A'^{3+}$) A cations form a special class: here the B ions are forced to have a mixed-valent situation with a 1:1 ratio of $B'^{3+}$ and $B'^{4+}$ or $B'^{2+}$ and $B'^{3+}$, respectively.

## 7.3 Basic Electronic Configuration

Most spinel oxides with transition-metal ions, if these are neither in the $d^0$ nor the $d^{10}$ configuration, are magnetic insulators because of the strong Coulomb repulsion within the narrow $d$-orbitals. In some exceptional cases, a spin singlet state may be formed and the system is non-magnetic. Mixed-valent spinels are sometimes, but not often, metallic. Spinel sulfides tend to be much more conducting than oxides due to their increased bandwidth, which comes about in part because of an enhanced $p$-$d$ hybridization.

The degeneracy of the $d$-orbitals in spinel oxides is lifted by the crystal field. When a transition-metal ion is located at the B site of a regular $BO_6$ octahedron, the fivefold degeneracy of the $d$-orbitals is split into high-lying, doubly degenerate $e_g$ and low-lying, triply degenerate $t_{2g}$ orbitals due to the cubic crystal field. In addition, a local trigonal distortion of the crystal field further splits the triply degenerate $t_{2g}$ orbitals into one $a_{1g}$ orbital, $\frac{1}{\sqrt{3}}(|xy\rangle + |yz\rangle + |zx\rangle)$, which is oriented towards the center of the B-lattice tetrahedron, and two degenerate $e'_g$ orbitals (different from the other $e_g$ doublet produced by a cubic crystal field) extending perpendicular to the $a_{1g}$ orbital [Fig. 7.1e]. Alternatively, a tetragonal distortion, which often occurs at low temperatures, acts to split the triply degenerate $t_{2g}$ orbitals into $xy$ and the doubly degenerate $yz/zx$ pair, and the doubly degenerate $e_g$ orbitals into nondegenerate $x^2 - y^2$ and $3z^2 - r^2$ orbitals.

If a transition-metal ion is placed in the tetrahedral A site, the splitting of the fivefold degenerate $d$-orbitals is into *high-lying*, triply degenerate $t_{2g}$ and *low-lying* doubly degenerate $e_g$ orbitals, i.e. opposite in sign to that on the octahedral B-site. The magnetic coupling in spinels usually varies dramatically with the number of $d$-electrons, reflecting changes in orbital character.

## 7.4 Uniqueness of the Spinel as a Frustrated Magnet

When antiferromagnetically coupled spins are placed on a pyrochlore lattice, very strong geometrical frustration is anticipated because of the triangle-based tetrahedral geometry. Within the broad spectrum of frustrated lattices, the pyrochlore lattice, realized as a sublattice of spinel oxides, occupies a very special position in that it is three-dimensional and that the frustration can be the strongest known [3]. To demonstrate the strength of pyrochlore frustration, let us reproduce the arguments articulated by Reimers et al. [4] and later by Moessner and Chalker [5]. Consider an $n$-component, classical spin system consisting of $N$ clusters with $q$ spin sites ($n = 3$ and $q = 4$ for a Heisenberg pyrochlore), the Hamiltonian for which may be expressed as

$$H = J \sum_{\langle i,j \rangle} \mathbf{S}_i \cdot \mathbf{S}_j \equiv \frac{1}{2} J \sum_{\alpha} |\mathbf{L}_\alpha|^2 - \frac{1}{2} J N q, \tag{7.1}$$

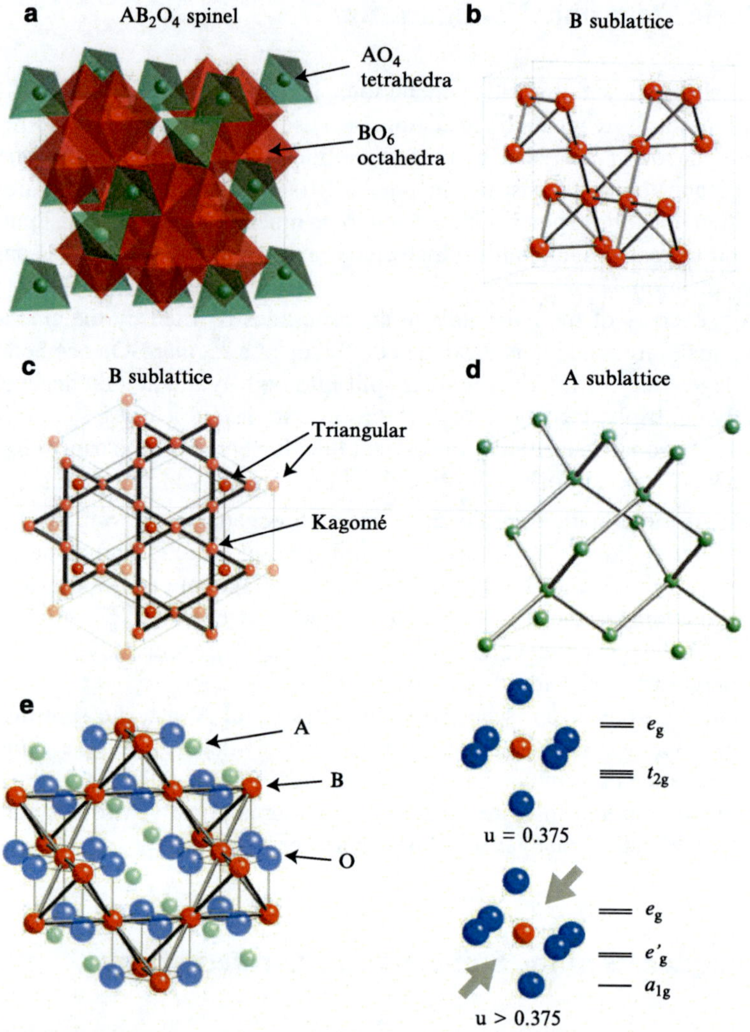


**Fig. 7.1** (**a**) Spinel structure, emphasizing the two basic structural units of $AO_4$ tetrahedra and $BO_6$ octahedra. (**b**) B sublattice of spinel structure, which defines a pyrochlore lattice. (**c**) B-sublattice structure (pyrochlore lattice) viewed from the $\langle 111 \rangle$ direction. (**d**) A sublattice (diamond lattice). (**e**) Spinel structure, shown to emphasize the coordination of B ions with surrounding oxygen ions. The inset shows the trigonal distortion of the $BO_6$ octahedra, which is characterized by the parameter $u$

where $\mathbf{S}_i$ is the spin on site $i$ with $|\mathbf{S}_i| = 1$, $J$ is the strength of the antiferromagnetic coupling and $\mathbf{L}_\alpha$ is the total spin in unit $\alpha$.

The ground state is the state with $L_\alpha = 0$ for all $\alpha$, namely with total spin per cluster (per tetrahedron in the pyrochlore case) equal to zero. There are in total $K = N(n-1)q/2$ degrees of freedom and the constraint $L_\alpha = 0$ imposes $K = Nn$. The remaining degrees of freedom are $F - K = N[n(q-2)-q]/2$, which is positive only for $q = 4$ and $n = 3$ among the possible sets of values $(q, n)$ with $q \leq 4$ and $n \leq 3$. The special nature of the case $q = 4$ and $n = 3$ demonstrates the uniqueness of the pyrochlore lattice. Closely linked to this large number of remaining degrees of freedom, it has been demonstrated that a classical spin liquid is stabilized in the classical pyrochlore Heisenberg model [5]. More recent work for $S = 1/2$ quantum Heisenberg spins on a pyrochlore lattice suggests that the ground state may be a spin liquid with a spin gap [6, 7].

## 7.5 Materials Overview of Spinels

Table 7.1 shows the structural and magnetic properties of some selected spinel oxides, most of which display frustrated magnetism. Following tradition, we represent the strength of frustration by the ratio between the Curie-Weiss temperature and the magnetic ordering temperature, $f = |\Theta_{CW}|/T_N$. It is clear that strongly frustrated magnetism, with an $f$-parameter of order 10 or greater, can be found in normal spinel oxides with nonmagnetic A ions and magnetic B ions, where the dominant antiferromagnetic interactions are between B ions ($\Theta_{CW} < 0$). Particularly noteworthy are the 2-3 spinels $ACr_2O_4$ (A = Mg, Zn, Cd) and $AV_2O_4$ (A = Mg, Zn, Cd). These systems indeed represent highly frustrated pyrochlore antiferromagnets, and will be described in detail in the following section. Those compounds with magnetic ions on both A and B sites can be frustrated, but this is not as obvious as in cases with nonmagnetic A sites, partly because of the coupling to magnetic A ions. In those materials with nonmagnetic, tetravalent Ge ions occupying the A sites, such as $GeFe_2O_4$ and $GeCo_2O_4$, the dominant interactions are seemingly ferromagnetic but are, in fact, frustrated because of competition from higher-order, antiferromagnetc interactions. $GeCu_2O_4$ forms a one-dimensional, $S = 1/2$ antiferromagnet due to Jahn-Teller distortion around the $Cu^{2+}$ ions.

The table also shows normal spinels in which magnetic A ions form a diamond lattice and the B ions are non-magnetic. One cannot expect frustration if only nearest-neighbor, antiferromagnetic interactions are considered. Nevertheless, some such compounds do show evidence of strong frustration: this phenomenon is known as A-site frustration and arises from competition between antiferromagnetic nearest- and next-nearest-neighbor couplings. $FeSc_2S_4$, $MnSc_2S_4$, $CoAl_2O_4$, $CoRh_2O_4$, and $FeAl_2O_4$ all fall into this category. A very different class of frustrated lattice can emerge from the spinel structure through cation ordering: $Na_4Ir_3O_8$ is a typical example where Ir and Na ions order on a pyrochlore lattice and form a "hyper-kagomé" geometry composed of a network of corner-sharing triangles. Another interesting class of spinel is the mixed-valent system introduced in Sect. 2, where $A'^{+}$ or $A'^{3+}$ ions occupy the A site and one may observe a charge analog of spin frustration. The most classical example for this is magnetite, $Fe_3O_4$ ($Fe^{3+}(Fe^{2.5+})_2O_4$), where $Fe^{3+}$ ions occupy the A sites and a 1:1 ratio of $Fe^{2+}$ and $Fe^{3+}$ ions the B sites. $LiV_2O_4$, $LiMn_2O_4$, $LiCo_2O_4$, $AlV_2O_4$, and $CuIr_2O_4$ fall into this category.

**Table 7.1** Frustrated spinel oxides

| 2-3 Normal spinel (A nonmagnetic, B magnetic) | | | | | | |
|---|---|---|---|---|---|---|
| | $S$ | $\Theta_{CW}$(K) | $T_N$(K) | $f$ | $T_S$(K) | Reference |
| $MgV_2O_4$ | 1 | −600 | 42 (CL) | 14.3 | 65 (C-T) | [8–10] |
| $ZnV_2O_4$ | 1 | −850 | 40 (CL) | 21.3 | 50 (C-T) | [11–14] |
| $CdV_2O_4$ | 1 | −400 | 35 (CL) | 11.4 | 97 (C-T) | [15,16] |
| $MgCr_2O_4$ | 3/2 | −400 | 12.5 (spiral) | 32 | 12.5 (C-T) | [17–19] |
| $ZnCr_2O_4$ | 3/2 | −390 | 12.5 (CL) | 31.2 | 12.5 (C-T) | [13,20–25] |
| $CdCr_2O_4$ | 3/2 | −70 | 7.8 (spiral) | 9.0 | 7.8 (C-T) | [23–28] |
| $HgCr_2O_4$ | 3/2 | −32.0 | 5.8 (CL) | 5.5 | 5.8 (C-O) | [24,29,30] |
| $MgMn_2O_4$ | 2 | −500 | 49 | 10.2 | 850 (C-T) | [31–33] |
| $ZnMn_2O_4$ | 2 | −537 | 21.5 | 25.0 | 1050 (C-T) | [33,34] |
| $CdMn_2O_4$ | 2 | −470 | 70 | 6.7 | 600 (C-T) | [35,36] |
| $ZnFe_2O_4$ | 5/2 | +120 | 13 | 9.2 | – | [37] |
| $CdFe_2O_4$ | 5/2 | −53 | 13 | 4.1 | – | [38] |
| $MgTi_2O_4$ | 1/2 | | VBS | | 260 (C-T) | [39,40] |
| 2-3 Normal spinel (A magnetic, B magnetic) | | | | | | |
| | $S_A, S_B$ | $\Theta_{CW}$(K) | $T_N$(K) | $f$ | $T_S$(K) | Reference |
| $MnV_2O_4$ | 5/2,1 | – | 61 (CL) 58 (NCL) | – | 58 (C-T) | [41–43] |
| $FeV_2O_4$ | 2,1 | +95 | 110 | 0.86 | 140,110,70 (C-T-O-T) | [44] |
| $MnCr_2O_4$ | 5/2,3/2 | – | 51 (ferri) 14(conical) | – | – | [45,46] |
| $FeCr_2O_4$ | 2,3/2 | −400 | 80 (CL) 35(NCL) | 5 | 135 (C-T) | [47] |
| $CoCr_2O_4$ | 3/2,3/2 | −650 | 93 (ferri) 25(conical) | 6.99 | – | [46,48] |
| $CuCr_2O_4$ | 1/2,3/2 | −600 | 150 (ferri) (NCL) | 4 | 865 (C-T) | [49,50] |
| 2-3 Normal spinel (A magnetic, B nonmagnetic) | | | | | | |
| | $S$ | $\Theta_{CW}$(K) | $T_N$(K) | $f$ | $T_S$(K) | Reference |
| $MnAl_2O_4$ | 5/2 | −143 | 40 | 3.6 | – | [62] |
| $FeAl_2O_4$ | 2 | −130 | 12 | 11 | – | [62] |
| $CoAl_2O_4$ | 3/2 | −89 | 4.8 | 19 | – | [62,63] |
| $CoRh_2O_4$ | 3/2 | −31 | 25 | 1.2 | – | [63] |

*(Continued)*

**Table 7.1** (Continued)

| | | | | | | |
|---|---|---|---|---|---|---|
| 2-3 Normal spinel (A magnetic, B nonmagnetic) | | | | | | |
| | $S$ | $\Theta_{CW}$(K) | $T_N$(K) | $f$ | $T_S$(K) | Reference |
| $MnSc_2S_4$ | 5/2 | −23 | 2.3,1.9 (spiral) | 10 | – | [57,58] |
| $FeSc_2S_4$ | 2 | −45 | < 2 | > 22.5 | – | [57,60,61] |
| 4-2 Normal spinel (A nonmagnetic, B magnetic) | | | | | | |
| $GeFe_2O_4$ | 2 | −15 | 10 | 1.5 | – | [88,89] |
| $GeCo_2O_4$ | 3/2 | +80 | 23.5 | 3.4 | – | [90–92] |
| $GeNi_2O_4$ | 1 | −15 | 12.1,11.5 | 1.2 | – | [93–95] |
| $GeCu_2O_4$ | 1/2 | −89 | 33 | 2.7 | 300 K (due to strong JT) | [87] |
| Cation-ordered spinel | | | | | | |
| $Na_4Ir_3O_8$ | 1/2 | −650 | < 2 | > 325 | – | [64] |
| Mixed-valent spinel | | | | | | |
| | | Ions | Ground state | | Remark | Reference |
| $LiTi_2O_4$ | | $Ti^{3+}$,$Ti^{4+}$ | SC | | $T_{SC} \sim 13$ K | [96] |
| $LiV_2O_4$ | | $V^{3+}$,$V^{4+}$ | Heavy fermion | | | [72,73,77,83,84] |
| $LiMn_2O_4$ | | $Mn^{3+}$,$Mn^{4+}$ | CO insulator | | $T_{CO} \sim 290$ K | [80] |
| $AlV_2O_4$ | | $V^{2+}$,$V^{3+}$ | CO insulator (heptamer VBS) | | $T_{CO} \sim 700$ K | [81,97] |
| $CuIr_2S_4$ | | $Ir^{3+}$,$Ir^{4+}$ | CO insulator (octamer VBS) | | $T_{CO} \sim 230$ K | [82] |
| $Fe_3O_4$ | | $Fe^{2+}$,$Fe^{3+}$ | CO insulator | | $T_{CO} \sim 120$ K | [71,78,79] |

Definition of symbols: $S$: spin quantum number, $T_N$: Néel ordering temperature, $\Theta_{CW}$: Curie-Weiss temperature from $\chi(T)$ (negative for antiferromagnetism), $f$: frustration factor $|\Theta_{CW}|/T_N$, $T_S$: structural phase transition temperature, $T_{SC}$: superconducting transition temperature, $T_{CO}$: charge-ordering temperature, $C$: cubic, $T$: tetragonal, $O$: orthorombic, $CL$: collinear, $NCL$: non-collinear, $SC$: superconducting, $CO$: charge-ordered, $VBS$: valence-bond solid, $JT$: Jahn-Teller coupling

## 7.6 Frustration in Selected Spinels

### *7.6.1 Pyrochlore Antiferromagnets in Spinel Oxides – B-site Frustration*

Let us first review selected normal spinels with magnetic B ions and nonmagnetic A ions. These spinels embody beautifully the physics of the strongly frustrated pyrochlore antiferromagnet, demonstrating in different systems a spin-liquid state, cooperative coupling with the lattice and the orbitals, quantum magnetization plateaus and the valence-bond-solid state.

#### 7.6.1.1 $ACr_2O_4$ (A = Zn, Cd, and Hg)

Cr spinel oxides, $ACr_2O_4$ (A = Zn, Cd, and Hg), exhibit the most dramatic signature of geometrical frustration associated with the pyrochlore geometry of the

B-sublattice. Cr in this spinel is trivalent. Because three electrons are accommodated in the triply degenerate $t_{2g}$ orbitals of $Cr^{3+}$, an $S = 3/2$ Mott insulator with no orbital degrees of freedom is realized. The presence of strong frustration is clear from the temperature-dependence of the magnetic susceptibility $\chi(T)$, shown in Fig. 7.2 for $ZnCr_2O_4$ and $CdCr_2O_4$. Curie-Weiss behavior is observed at high temperatures and an estimate of the effective moment gives $p_{\mathrm{eff}} \sim 3.7\ \mu_B$, consistent with the value expected for $S = 3/2$. The antiferromagnetic Curie-Weiss temperatures obtained from fits to data are $\Theta_{\mathrm{CW}} = -390$ K for $ZnCr_2O_4$ and $-70$ K for $CdCr_2O_4$. Despite the large value $|\Theta_{\mathrm{CW}}| \geq 100$ K, no evidence for ordering was observed down to $T \sim 10$ K. A discontinuous decrease of $\chi(T)$ can be recognized in Fig. 7.3 at $T_N = 12.5$ K for $ZnCr_2O_4$ and at 7.8 K for $CdCr_2O_4$, which corresponds to long-ranged antiferromagnetic order, but occurring at a temperature one order of magnitude lower than $|\Theta_{\mathrm{CW}}|$ [20, 23].

The paramagnetic phase well below the Curie-Weiss temperature and above $T_N$ can be viewed as a spin-liquid phase. As shown in Fig. 7.3a, the magnetic entropy $S_m(T)$ for $ZnCr_2O_4$, estimated from the magnetic specific heat $C_m(T)$, recovers to a value as large as nearly 40% of the total spin entropy $S = R \ln 4$ even at only 30 K, one order of magnitude lower in temperature than $|\Theta_{\mathrm{CW}}|$ [13, 21]. This large entropy should signal the presence of low-lying, strongly degenerate spin excitations produced by geometrical frustration. Such low-lying excitations were indeed detected directly by inelastic neutron scattering measurements, as a very intense signal around $\omega = 0$ above $T_N = 12.5$ K [20] [Fig. 7.3b]. Inelastic neutron scattering at a low energy of 1.5 meV and at $T = 15$ K ($T > T_N$) revealed the presence of hexagonal antiferromagnetic spin clusters [22]. As shown in Fig. 7.4, the pyrochlore lattice contains hexagons in its kagomé planes. Magnetic fluctuations select, from the ground-state manifold of states with zero spin on each tetrahedron, configurations in which the spins have local antiferromagnetic order around these hexagons. This gives rise to a set of local excitations linked to the orientation of the spins,

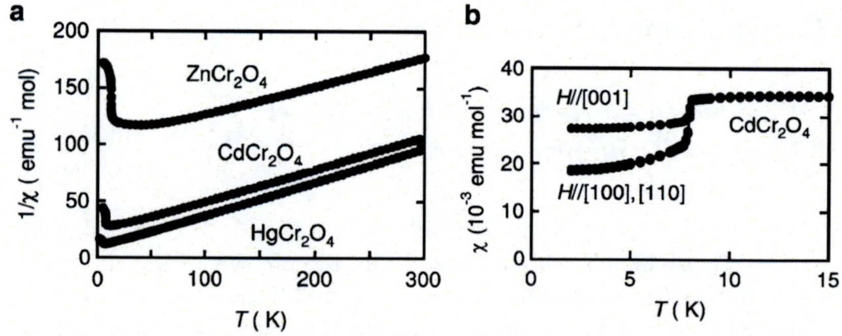


**Fig. 7.2** (**a**) Magnetic susceptibility $\chi(T)$ for the $S = 3/2$ pyrochlore antiferromagnet $ACr_2O_4$ (A = Zn, Cd, and Hg) from [24], shown as $1/\chi(T)$ vs. $T$. (**b**) $\chi(T)$ at low temperatures for $CdCr_2O_4$, from [23], indicating the presence of magnetic ordering at 7.8 K

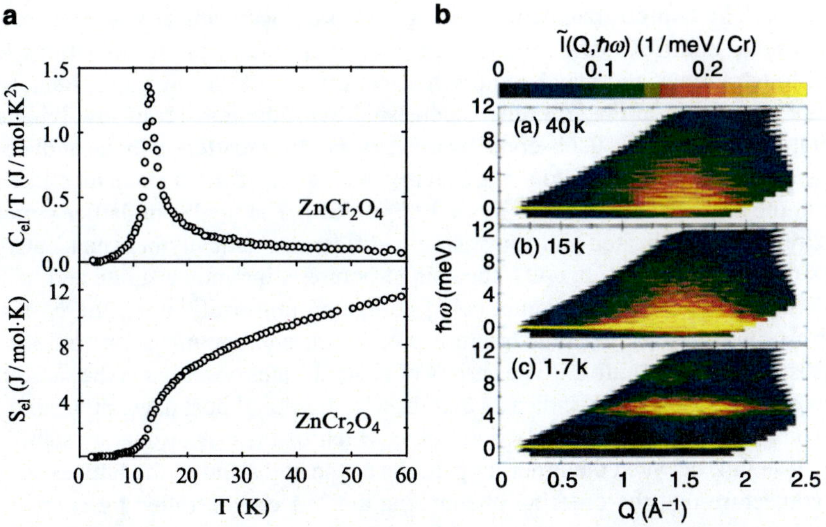


**Fig. 7.3** (**a**) Magnetic specific heat $C_m(T)$ and magnetic entropy $S_m(T)$, estimated from $C_m(T)$, for $ZnCr_2O_4$ from [13]. Note the large entropy recovered at temperatures much lower than $|\Theta_{CW}| = 350$ K. (**b**) Contour plot of magnetic neutron scattering intensity in the plane of energy transfer $\hbar\omega$ and wave number $Q$ for $ZnCr_2O_4$, taken from [20]

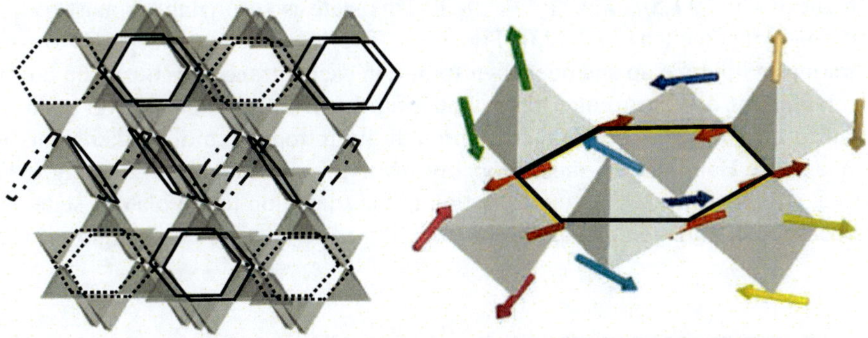

**Fig. 7.4** Antiferromagnetic hexagonal spin cluster formed in the spin-liquid phase of $ZnCr_2O_4$, from [22]

one for each antiferromagnetic hexagon, which very likely represents one of the zero-energy modes expected in pyrochlore antiferromagnets.

How can long-ranged magnetic order emerge from such a spin-liquid state? It is widely accepted that a spin–lattice coupling is responsible for the appearance of magnetic ordering. Distortion of the lattice in general makes magnetic coupling non-uniform, which inherently suppresses the geometrical frustration and lifts the spin degeneracy. Such degeneracy lifting by lattice distortion may happen cooperatively and may bring about a long-ranged magnetic order, which can be considered as a spin Jahn-Teller transition or as a three-dimensional analog of the spin-Peierls transition. The antiferromagnetic ordering processes occurring at low temperatures in $ZnCr_2O_4$ and $CdCr_2O_4$ are accompanied by structural phase transitions from cubic to tetragonal, and may be understood naturally in terms of a spin Jahn-Teller picture. As shown in Fig. 7.3, this is consistent with the transfer of low-lying spin excitations around $\omega = 0$, observed by inelastic neutron scattering in the undistorted phase ($T > T_N = 12.5$ K), to a high energy, where they form a resonance-like feature in the distorted phase ($T < T_N = 12.5$ K) [20]. The final spin–lattice structure is extremely complicated and, surprisingly, differs dramatically between $ZnCr_2O_4$ and $CdCr_2O_4$ [25]. In $CdCr_2O_4$, the spin structure is incommensurate and helical, while the lattice is rather simple, being elongated uniformly along the *c*-axis. In contrast, in $ZnCr_2O_4$ the spin structure is collinear and commensurate but the lattice distortion is not uniform. The variety of ground states observed in the same class of compounds most likely indicates that these states are almost degenerate, and that materials-dependent details may determine which of them is ultimately stabilized. The close link between the ordering patterns of the spins and of the lattices support the conjecture that the ordering phenomena are indeed controlled by *cooperative* interplay of spin and lattice degrees of freedom.

Tuning the frustration by applying a magnetic field may provide a unique opportunity to create a new phase of the spin–lattice complex in spinel materials. In the magnetization curve of $CdCr_2O_4$, there is indeed a first-order transition to a magnetization plateau, which arises due to the stabilization of a state with half the full moment of $Cr^{3+}$ ($1.5\mu_B$) at $H_c = 28$ T. This state is surprisingly robust over a field range of some tens of T [23] [Fig. 7.5a]. The 1/2 plateau corresponds to configurations with three up and one down spin(s) in each tetrahedron. Based on model calculations, it has been noted that stabilizing the 1/2-plateau state over such an extended field range is possible only through the introduction of further terms in the magnetic Hamiltonian, most probably involving a spin–lattice coupling [51]. Indeed, a relatively large magnetostriction at the transition to the plateau state has been observed, although the spin–orbit coupling is believed to be negligibly small

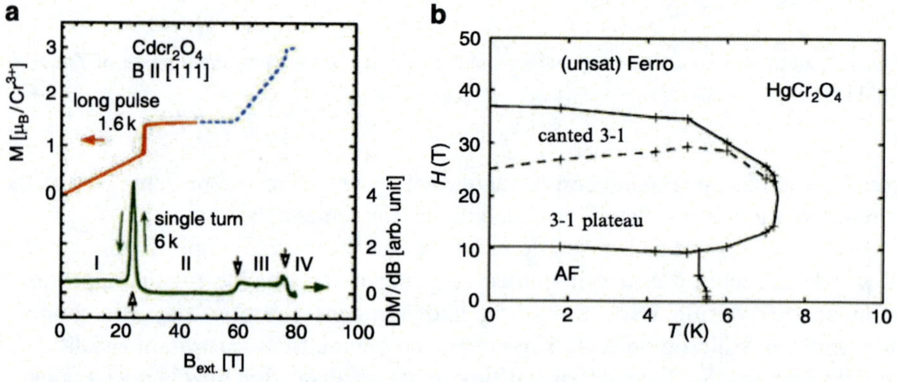


**Fig. 7.5** (**a**) Magnetiztion curve of the spinel $CdCr_2O_4$, from [23, 28]. (**b**) Phase diagram of $HgCr_2O_4$ in the *H*-*T* plane, from [29]

here. Later magnetization measurements, conducted up to higher magnetic fields, discovered two further transitions in addition to that into the 1/2-plateau state [28]. The first is from the 1/2-plateau state ("3-up, 1-down" state) to a canted state at 61 T, and the second is a first-order transition to a state with the fully polarized moment of $3\mu_B$ at 77 T.

The compound $HgCr_2O_4$ has a much smaller Curie-Weiss temperature, $|\Theta_{CW}| =$ 32 K, than do $ZnCr_2O_4$ and $CdCr_2O_4$ [29]. Essentially the same behavior of the magnetization curve was observed as in $CdCr_2O_4$. Taking advantage of the low energy scale of the magnetic interactions in this system, the full phase diagram in the $H$-$T$ plane can be measured, and is shown in Fig. 7.5b [29]. Further, the spin and lattice structure in the 1/2-plateau state has recently been determined successfully. It was found that the spin structure in the 1/2-plateau state has a rather complicated ordering pattern of "3-up, 1-down" tetrahedra with symmetry $P4_332$, which is reflected in the structural distortion [30].

#### 7.6.1.2 $AV_2O_4$ [A = Mg, Zn, and Cd]

$MgV_2O_4$, $ZnV_2O_4$, and $CdV_2O_4$ are $S = 1$ Mott insulators with a pyrochlore lattice. Because $V^{3+}$ accommodates two electrons in the triply degenerate $t_{2g}$ orbitals, these systems have, in contrast to Cr spinels, an orbital degree of freedom which provides a unique opportunity to study the interplay of frustrated spins and orbitals. At first glance, the magnetic properties of the V spinels are similar to those of the Cr spinels. Curie-Weiss fitting of $\chi(T)$ for $ZnV_2O_4$ yields a large, antiferromagnetic Curie-Weiss temperature $|\Theta_{CW}| \sim 850$ K [11], while a long-range magnetic ordering appears only at a temperature as low as $T_N = 40$ K [12], indicating the presence of very strong frustration. Very intense inelastic neutron scattering, a consequence of the low-lying and highly degenerate spin excitations, was observed around $\omega = 0$ [14].

The V spinels possess, however, many features which are quite distinct from those of the Cr spinels. As can be seen in Fig. 7.6a, two anomalies are observed clearly in $\chi(T)$ at low temperatures. On cooling, a discontinuous drop occurs at $T_S = 50$ K, which represents a structural phase transition from cubic to tetragonal [12]. No long-range magnetic ordering is detected either in neutron diffraction or by nuclear magnetic resonance (NMR) at this transition. With a further lowering of the temperature, a kink is observed at 40 K, which represents the onset of antiferromagnetic order. The structural phase transition, which occurs cooperatively with the magnetic transition in the Cr spinels, happens separately at temperatures higher than the onset of magnetic order in $ZnV_2O_4$. The separation between the structural and the magnetic transition is even more pronounced in $CdV_2O_4$ [15] (see also Table 7.1). The magnetic structure below $T_N$ is much simpler than in Cr spinels, consisting of collinear ($\langle 001 \rangle$ orientation), antiferromagnetic spin chains arranged along the $\langle 110 \rangle$ and $\langle 1\bar{1}0 \rangle$ directions. The ordered moment was found to be approximately $0.6\mu_B$.

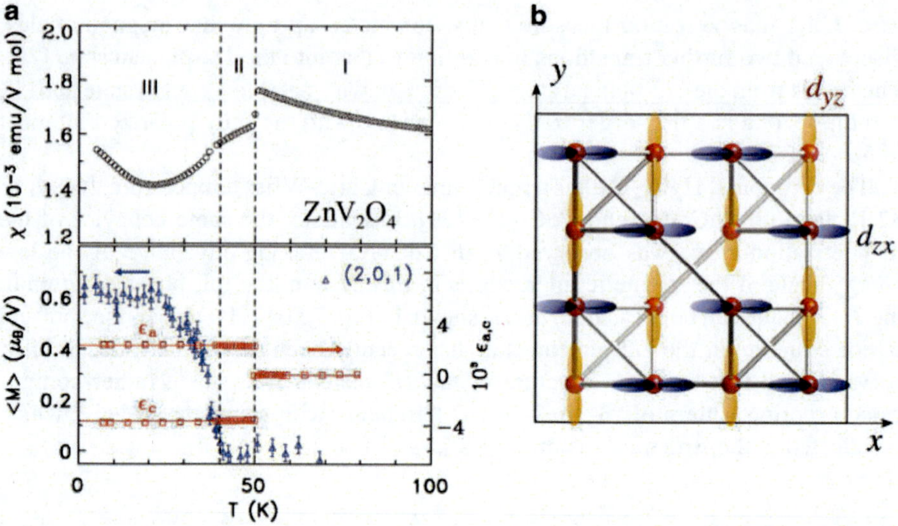


**Fig. 7.6** (**a**) Upper panel: temperature-dependent magnetic susceptibility $\chi(T)$ of $ZnV_2O_4$ (*upper panel*). Two anomalies are seen clearly. Lower panel: temperature-dependence of the ordered moment $M$ and the magnitude of the tetragonal splitting; both panels from [14]. (**b**) Ordering pattern proposed in [52] for $yz$ and $zx$ orbitals in $ZnV_2O_4$ below $T_S = 50$ K, as viewed from the $z$-direction. $xy$ orbitals are always occupied for all V sites

It has been argued that the different type of behavior in the V spinels can be understood in terms of orbital physics [53, 54]. The structural phase transition at $T_S = 50\,\mathrm{K}$ ($T_S > T_N$) takes the form of a lattice contraction in the $c$-direction, which should favor the $xy$ orbital and destabilize states with occupied $yz$ and $zx$ orbitals. When two electrons are introduced, one occupies $xy$ and the other the degenerate $yz/zx$ orbitals. It was pointed out theoretically [52, 53] that the orbital ordering in fact occurs at $T_S = 50$ K and that the degeneracy of $yz$ and $zx$ orbitals is lifted below $T_S$. The most likely orbital ordering pattern suggested by theoretical analysis [52, 53] is an alternate stacking along the $c$-axis of $yz$–$xy$ ferro-orbital planes with $\langle 110\rangle$-oriented V chains and $zx$–$xy$ ferro-orbital planes with $\langle 1\bar{1}0\rangle$-oriented V chains. This gives rise to an enhanced antiferromagnetic coupling along the $\langle 110\rangle$ and $\langle 1\bar{1}0\rangle$ chains compared to the other chain directions, as shown in Fig. 7.6b. This strongly anisotropic spin coupling, which is one-dimensional in nature, can lift the spin degeneracy due to frustration and therefore acts to promote magnetic ordering [52, 53]. Within this scenario, the magnetic ordering in V spinels might be viewed as being driven by orbital ordering. It should be emphasized that different forms of complex orbital order, as opposed to this alternating $yz/zx$ order, have also been proposed [54], and that it remains to clarify the orbital state below $T_S$ experimentally. The presence of orbital degrees of freedom may change not only the physics of degeneracy lifting but also the physics of the spin-liquid state above $T_S$. In the temperature range $T > T_S$, the characteristic wave vector $q_c$ of the low-lying spin excitations was found to be different from that of hexagonal clusters, indicating that the spin correlations in the liquid state are not the same as in $ZnCr_2O_4$ [14].

#### 7.6.1.3 $MgTi_2O_4$

In contrast to the other normal 2-4 spinels, $MgTi_2O_4$ is weakly metallic at room temperature, presumably because of the reduced correlation effect expected for early $3d$ transition-metal ions. The system undergoes a metal-insulator transition at 260 K on cooling, at which an almost discontinuous decrease of the susceptibility is observed, suggesting strongly the formation of a spin singlet ground state [39] [Figs. 7.7a and 7.7b]. This 260 K feature is accompanied by a structural phase transition [39], and the subsequent refinement of the crystal structure revealed the formation of Ti–Ti dimers aligned helically along the $c$-axis below the transition temperature, as shown in Fig. 7.7c [40]. The ground state of $MgTi_2O_4$ is therefore a spin singlet insulator rather than a $S = 1/2$ antiferromagnet. The mechanism behind the formation of this spin singlet, often called the valence-bond-solid state, is argued to be the orbital degeneracy [55, 56]. Similar orbital-driven spin singlet states have been found in a number of mixed-valent spinels, as will be discussed below.

### *7.6.2 Frustrated Spins on Spinel A Sites*

We turn now to spinels with magnetic A ions and nonmagnetic B ions. As noted in Sect. 2, in this case the magnetic A sublattice is a diamond lattice. Because the diamond lattice is bipartite, the magnetism of A ions would be free from frustration if only nearest-neighbor coupling were important. However, in real materials of this type, the next-nearest-neighbor coupling is often found not to be small by comparison, because the magnetic couplings between the A ions are mediated by rather complicated exchange paths including O-B-O links, rather than by purely direct exchange processes (Fig. 7.8). When the next-neighbor interaction $J_2$ is comparable to the nearest-neighbor term $J_1$, very strong geometrical frustration may be anticipated for A-site spins. An intuitive way of understanding this is as follows: the diamond lattice consists of two interpenetrating, face-centered-cubic (fcc) sublattices displaced by (1/4, 1/4, 1/4), as shown in Fig. 7.8a. The next-neighbor couplings of the diamond lattice are then simply the nearest-neighbor couplings within the fcc sublattice, and hence give rise to frustration if the coupling is antiferromagnetic. The nearest-neighbor couplings of the diamond lattice correspond to the couplings between the fcc sublattices, which, if antiferromagnetic, further enhance the frustration within the fcc sublattices.

The thiospinels $FeSc_2S_4$ and $MnSc_2S_4$ can be prototypical systems for such A-site frustration, although they are not oxide materials. $MnSc_2S_4$, with $S = 5/2$, shows an antiferromagnetic $|\Theta_{CW}|$ of 23 K, but orders to a spiral magnetic structure only at a temperature of 2 K [57, 58]. One theoretical analysis noted that this type of long-range order can be achieved by an order-by-disorder process [59], in which case the driving mechanism is weak. $FeSc_2S_4$, with $S = 2$, is an extreme case: no evidence for long-range order can be observed in the specific heat down to 50 mK, in spite of a sizable antiferromagnetic $|\Theta_{CW}| \sim 45$ K, suggesting a spin-liquid ground

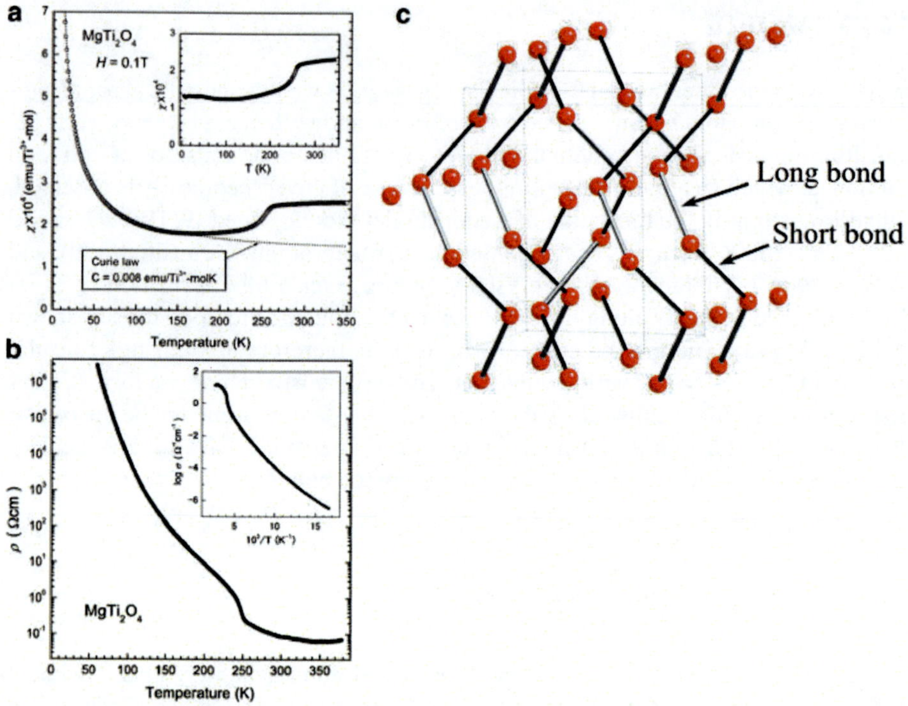


**Fig. 7.7** Temperature-dependence of (**a**) magnetic susceptibility and (**b**) resistivity for a polycrystalline sample of $MgTi_2O_4$. (**c**) Helical dimerization pattern observed for the $S = 1/2$ pyrochlore system $MgTi_2O_4$, from [40]

state. Neutron [60] and Sc-NMR [61] measurements indicated the presence of a finite but small spin gap of 0.2 meV. $Fe^{2+}$ ions in the tetrahedral geometry have an orbital degeneracy associated with the $e_g$ orbital and are Jahn-Teller-active. Nevertheless, no clear evidence for orbital ordering is observed in $FeSc_2S_4$. The strong orbital fluctuation suggested by Sc-NMR may be linked to the stability of spin-liquid state, and $FeSc_2S_4$ might in fact be best viewed as a spin–orbital complex liquid. The same A-site frustration appears to be at work in oxides, including $CoAl_2O_4$, $CoRh_2O_4$, and $FeAl_2O_4$ [62, 63], although in these materials it is not as dramatic as in the thiospinels. $CoAl_2O_4$ and $FeAl_2O_4$ show only spin-glass-like transitions at temperatures much lower than their Curie-Weiss temperatures [62].

### *7.6.3 Frustrated Magnets based on Cation-ordered Spinels: The Hyper-Kagomé Lattice of $Na_4Ir_3O_8$*

$Na_4Ir_3O_8$ [64] can be viewed as a cation-ordered spinel, a statement which becomes clear on rewriting the chemical formula as 2 $Na_{1.5}(Ir_{3/4}Na_{1/4})_2O_4$. The spinel B

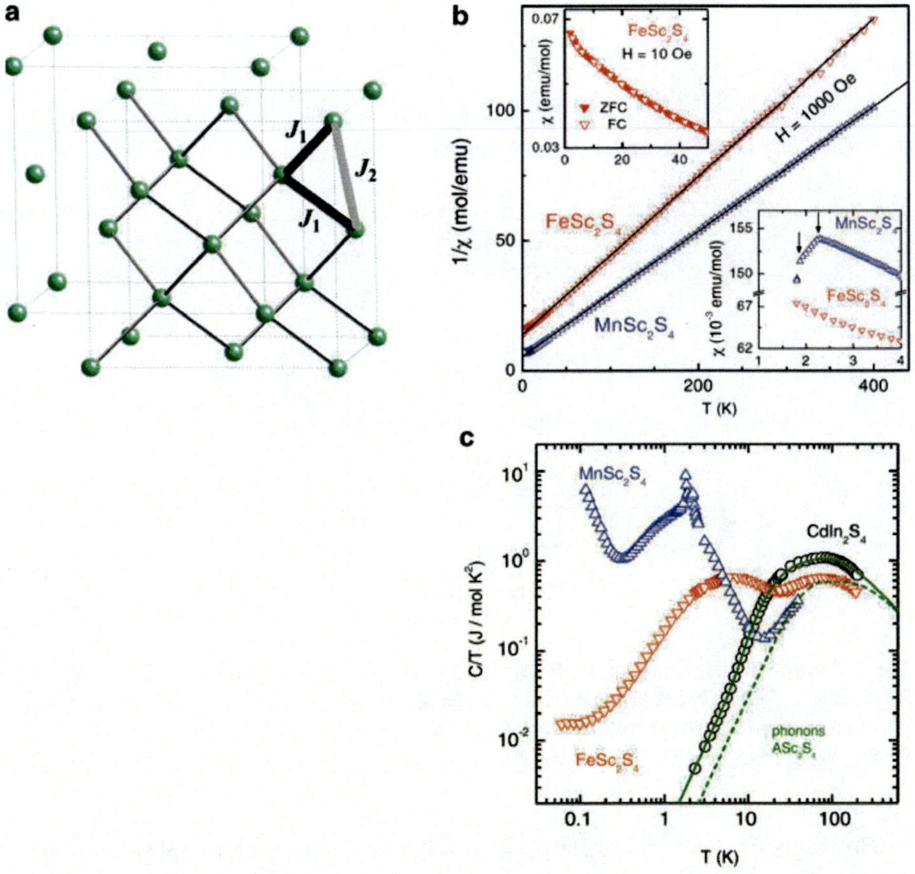


**Fig. 7.8** (**a**) Nearest-neighbor coupling $J_1$ and next-nearest-neighbor coupling $J_2$ in the diamond lattice. *Cubes* marked with *dotted lines* indicate the fcc sublattice. (**b**) Temperature-dependent magnetic susceptibility $\chi(T)$ and (**c**) specific heat displayed as $C(T)/T$ for the A-site-frustrated spinels $MnSc_2S_4$ and $FeSc_2S_4$; data from [57]

sites are occupied by both $S = 1/2$ $Ir^{+4}$ and nonmagnetic $Na^+$ ions, and each tetrahedron consists of three $Ir^{4+}$ ions and one $Na^+$. Because of the ordering of $Ir^{4+}$ and $Na^+$ on the B sites, the $Ir^{4+}$ ions form a network of corner-sharing triangles. This lattice is a twisted version of the two-dimensional kagomé lattice and may thus be called the "hyper-kagomé" lattice [Fig. 7.9a, b]. Creating a different class of frustrated lattice from the spinel structure by utilizing cation order is one of the most promising approaches in the chemistry of frustrated compounds. For example, the spinel B sublattice, the pyrochlore lattice, consists of stacks of triangular planes and kagomé planes alternating along the $\langle 111 \rangle$ direction, which opens the interesting possibility of producing not only hyper-kagomé structures but also kagomé or triangular magnets by introducing 3:1 ratios of two different cations at the B sites.

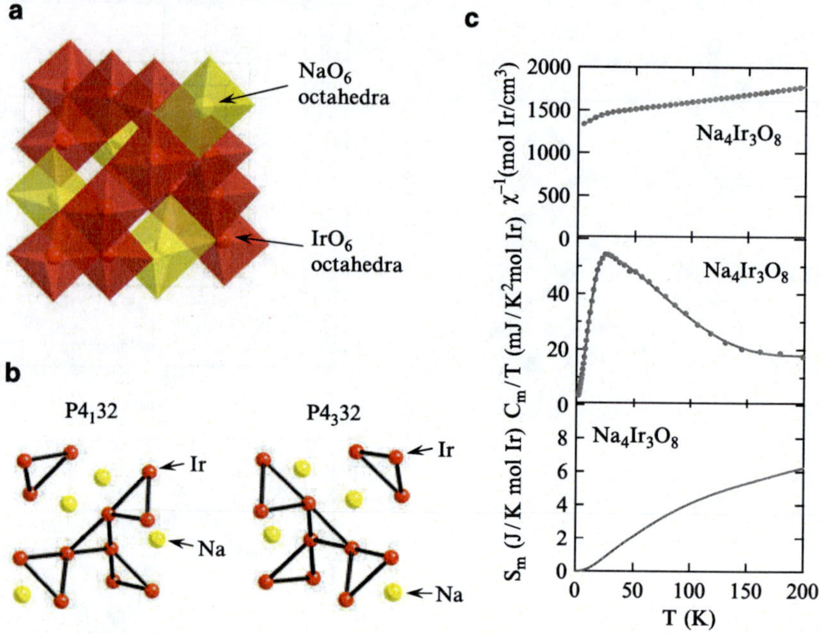


**Fig. 7.9** (**a**) Crystal structure of $Na_4Ir_3O_8$. (**b**) Two types of $Ir_3Na$ sublattice with different chirality. Ir ions form a hyper-kagomé lattice. (**c**) Magnetic susceptibility $\chi(T)$, magnetic specific heat $C_m(T)$, and magnetic entropy estimated from $C_m(T)$ for the hyper-kagomé system $Na_4Ir_3O_8$; panels (**b**) and (**c**) from [64]

No magnetic order has so far been found in the magnetic susceptibility [64] or the specific heat [64], or observed in neutron diffraction or Na-NMR [65] studies of $Na_4Ir_3O_8$, at least down to 2 K, despite the relatively large Curie-Weiss temperature, $|\Theta_{CW}| = 650$ K, estimated from $\chi(T)$. These observations suggest strongly that the ground state of this three-dimensional, $S = 1/2$ hyper-kagomé system is a quantum spin liquid. The presence of a spin-liquid ground state in the hyper-kagomé antiferromagnet has found theoretical support for classical spins [66] and more recently for quantum spins [67].

The behavior of the low-temperature specific heat is close to a pure $T^2$ form [64], similar to other frustrated systems with possible spin-liquid ground states [68, 69]. It is, however, not obvious why $T^2$ behavior is observed in this three-dimensional system, in common with the other candidate spin-liquid systems, which are two-dimensional [68, 69]. Only a very small $T$-linear term, with a $\gamma$ value of a few mJ/molK$^2$ and which might even be extrinsic, can be observed below 1 K, suggesting the complete absence of or at most an extremely low density of states in the ground state. The magnetic susceptibility at low temperatures tends nevertheless to approach a finite value [64]. These seemingly contradictory results were argued to be a consequence of spin–orbit coupling [70]. Much stronger spin–orbit coupling is present in heavy 5$d$ elements, such as Ir, than in 3$d$ and 4$d$ elements. A further interesting feature of $Na_4Ir_3O_8$, possibly related to this, is the presence of chirality in the lattice. As shown in Fig. 7.9c, there exist two structures ($P4_132$ and $P4_332$) with different chiralities, which depend on the local coordination of the $Ir^{4+}$ triangles. A novel interplay between the lattice chirality and the spin liquid through spin–orbit coupling might be anticipated, including a hidden chirality ordering. A full exploration of these possibilities is awaited.

### *7.6.4 Charge Frustration in Mixed-valent Spinels*

When the spinel A sites are occupied by either monovalent or trivalent ions, the 1:2 ratio of A to B ions enforces a half-integer net B-ion valence, or a 1:1 mixture of two sequential valences. For example, magnetite ($Fe_3O_4$ or $Fe^{3+}(Fe^{2.5+})_2O_4$) displays a 1:1 ratio of $Fe^{2+}$ and $Fe^{3+}$. In order to reduce the Coulomb energy, the system would prefer $Fe^{2+}$ and $Fe^{3+}$ ions to occupy neighboring sites, by analogy with the spin case where up and down spins should occupy neighboring sites. Because of the inherent geometrical frustration of the pyrochlore lattice, the system can have only two $Fe^{2+}$ and two $Fe^{3+}$ ions for each tetrahedron, which is known as the Anderson condition (the constant total charge per tetrahedron is equivalent to the condition of total spin zero per tetrahedron). This gives rise to a macroscopic degeneracy of the lowest-energy charge distribution and prevents long-range ordering of the charges [71].

An extreme case of such charge frustration is found in $LiV_2O_4$, which has a mixed-valent configuration of $V^{3+}$ and $V^{4+}$. Neither charge order nor magnetic order has been observed down to temperatures well below 1 K, due in all probability to the strong frustration and to the appearance of a heavy-fermion ground state with a specific-heat coefficient $\gamma \sim 400$ mJ/mol $K^2$, comparable to canonical Kondo heavy-fermion systems such as Ce, Yb, and U intermetallics [72, 73] (Fig. 7.10). It has been argued that this system exemplifies a new route to the formation of heavy-fermion states [74–76]. Particularly attractive in connection to frustrated magnetism is the possibility of viewing this system as the charge analog of a spin liquid [77].

In the other known mixed-valent spinels, charge ordering in fact occurs by coupling with further available degrees of freedom, such as those of the lattices and the orbitals, leading to nontrivial and complicated charge-ordering patterns in the ground states (a situation already familiar from the spin case). In $Fe_3O_4$, the most classic example of charge frustration, the ordering pattern below the Verwey transition [78] has been the subject of debate for many decades. Only very recently has a complicated charge-ordering pattern been revealed [79], which does not obey the Anderson condition. A very complicated charge-ordered state, also violating the Anderson condition, has been identified in $LiMn_2O_4$ [80]. In other mixed-valent spinels, such as $AlV_2O_4$ [81] and $CuIr_2S_4$ [82], the charge-ordering was found to occur through the formation of a valence-bond solid with spin singlet multimers. These valence-bond-solid states appear to be accompanied by orbital orderings and might therefore be viewed as a charge analog of the orbitally induced magnetic order

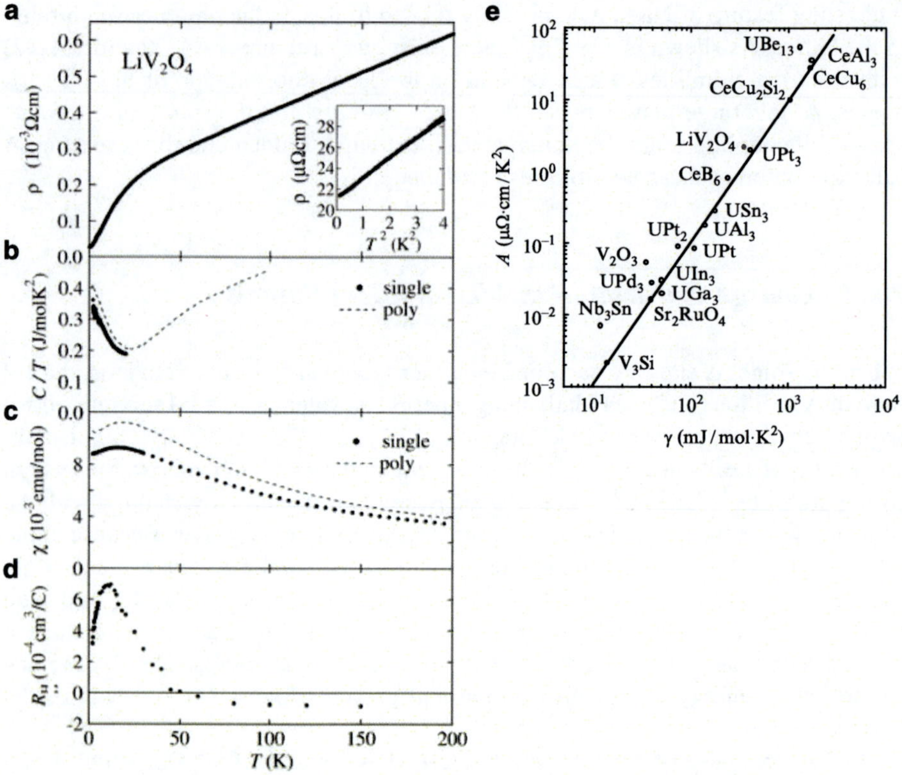


**Fig. 7.10** Heavy-fermion behavior in the frustrated, mixed-valent spinel $LiV_2O_4$. From top to bottom, resistivity $\rho(T)$ (**a**), specific heat $C(T)$ (**b**), magnetic susceptibility $\chi(T)$ (**c**), and Hall coefficient $R_H(T)$ (**d**). The inset in panel (**a**) demonstrates the $T^2$ dependence of the resistivity, indicative of a Fermi liquid. (**e**) The coefficient of the $T^2$-resistivity, $A$, and the square of the linear specific-heat coefficient, $\gamma^2$, shown for a variety of correlated metals (Kadowaki–Woods plot). Note that $LiV_2O_4$ appears in almost the same position as $UPt_3$

in $ZnV_2O_4$ [56]. The heavy-fermion oxide $LiV_2O_4$ under pressure shows a transition to a charge-ordered state [83], which appears to fall into the same category as $AlV_2O_4$ and $CuIr_2S_4$ [84].

## 7.7 Summary

Compared with other frustrated systems, an extremely strong geometrical frustration is realized for the pyrochlore B-sublattice of the spinel structure. This makes spinel oxides one of the most important avenues for research into the physics of quantum spin liquids, a topic which has attracted intense interest since Anderson's original proposal [85]. Indeed, some of the exotic properties expected for a spin liquid, including a zero-energy mode and a spin-liquid state seemingly stable down to the $T = 0$ limit, have recently been identified experimentally in spinel oxides. Thus the physics of the quantum spin liquid is definitively no longer merely a theoretical toy.

In many cases, the strong degeneracy of the spin excitations is lifted by coupling with the lattice and orbital degrees of freedom, and a non-trivial, self-organized, spin-charge-orbital-lattice complex state is formed. The exploration of such new states, and of the physics behind them, is definitely a new paradigm for quantum magnetism. The issue of multiferroics, pursued recently and enthusiastically in examples such as $CoCr_2O_4$ [48], falls into this category, where the interplay between the nontrivial spin structure, produced by frustration, and its coupling to the lattice plays a key role [86].

## References


1. J.B. Goodenough, Magnetism and the Chemical Bond (Interscience, NY, USA, 1963)
2. A.F. Wells, Structural Inorganic Chemistry, 5th edn. (Oxford University Press, Oxford, 1984)
3. P. Lacorre, J. Phys. C: Solid State Phys. 20, L775 (1987)
4. J.N. Reimers, A.J. Berlinsky, A.-C. Shi, Phys. Rev. B. 43, 865 (1991)
5. R. Moessner, J.T. Chalker, Phys. Rev. Lett. 80, 2929 (1998); see also contribution by J.T. Chalker in this volume
6. B. Canals, C. Lacroix, Phys. Rev. Lett. 80, 2933 (1998)
7. B. Canals, C. Lacroix, Phys. Rev. B. 61, 1149 (2000)
8. R. Plumier, A. Tardieu, C.R. Acad. Sci. 257, 3858 (1963)
9. H. Mamiya, M. Onoda, Solid State Commun. 95, 217 (1995)
10. H. Mamiya, M. Onoda, T. Furubayashi, J. Tang, I. Nakatani, J. Appl. Phys. 81, 5289 (1997)
11. S. Niziol, Phys. Stat. Sol. A 18, K11 (1973)
12. Y. Ueda, N. Fujiwara, H. Yasuoka, J. Phys. Soc. Jpn. 66, 778 (1997)
13. S. Kondo, C. Urano, Y. Kurihara, M. Nohara, H. Takagi, J. Phys. Soc. Jpn. 69 Suppl. B, 139 (2000)
14. S.-H. Lee, D. Louca, H. Ueda, S. Park, T.J. Sato, M. Isobe, Y. Ueda, S. Rosenkranz, P. Zschack, J. Íñiguez, Y. Qiu, R. Osborn, Phys. Rev. Lett. 93, 156407 (2004)
15. N. Nishiguchi, M. Onoda, J. Phys.: Condens. Matter 14, L551 (2002)
16. Z. Zhang, D. Louca, A. Visinoiu, S.-H. Lee, J.D. Thompson, T. Proffen, A. Llobet, Y. Qiu, S. Park, Y. Ueda, Phys. Rev. B 74, 014108 (2006)
17. R. Plumier, C. R. Acad. Sci. Paris B 267, 98 (1968)
18. M.T. Rovers, P.P. Kyriakou, H.A. Dabkowska, G.M. Luke, M.I. Larkin, A.T. Savici, Phys. Rev. B 66, 174434 (2002)
19. L. Ortega-San Martín, A.J. Williams, C.D. Gordon, S. Klemme, J.P. Attfield, J. Phys.: Condens. Matter 20, 104238 (2008)
20. S.-H. Lee, C. Broholm, T.H. Kim, W. Ratcliff II, S.-W. Cheong, Phys. Rev. Lett. 84, 3718 (2000)
21. H. Martinho, N.O. Moreno, J.A. Sanjurjo, C. Rettori, A.J. García-Adeva, D.L. Huber, S.B. Oseroff, W. Ratcliff II, S.-W. Cheong, P.G. Pagliuso, J.L. Sarrao, G.B. Martins, Phys. Rev. B 64, 024408 (2001)
22. S.-H. Lee, C. Broholm, W. Ratcliff, G. Gasparovic, Q. Huang, T.H. Kim, S.-W. Cheong, Nature 418, 856 (2002)
23. H. Ueda, H. Aruga-Katori, H. Mitamura, T. Goto, H. Takagi, Phys. Rev. Lett. 94, 047202 (2005)
24. H. Ueda, J. Yamaura, H. Mitamura, T. Goto, H. Aruga-Katori, H. Takagi, Y. Ueda, J. Mag. Mag. Mat. 310, 1275 (2007)

25. S.-H. Lee, G. Gasparovic, C. Broholm, M. Matsuda, J.-H. Chung, Y.J. Kim, H. Ueda, G. Xu, P. Zschack, K. Kakurai, H. Takagi, W. Ratcliff, T.H. Kim, S.-W. Cheong, J. Phys.: Condens. Matter 19, 145259 (2007)
26. M.T. Rovers, P.P. Kyriakou, H.A. Dabkowska, G.M. Luke, M.I. Larkin, A.T. Savici, Phys. Rev. B 66, 174434 (2002)
27. J.-H. Chung, M. Matsuda, S.-H. Lee, K. Kakurai, H. Ueda, T.J. Sato, H. Takagi, K.-P. Hong, S. Park, Phys. Rev. Lett. 95, 247204 (2005)
28. H. Mitamura, H. Ueda, H. Aruga-Katori, S. Takeyama, T. Sakakibara, Y. Ueda, H. Takagi, J. Phys. Soc. Jpn. 76, 085001 (2007)
29. H. Ueda, H. Mitamura, Y. Ueda, Phys. Rev. B 73, 094415 (2006)
30. M. Matsuda, H. Ueda, A. Kikkawa, Y. Tanaka, K. Katsumata, Y. Narumi, T. Inami, Y. Ueda, S.-H. Lee, Nat. Phys. 3, 397 (2007)
31. P. Ghigna, R. De Renzi, M.C. Mozzati, A. Lascialfari, G. Allodi, M. Bimbi, C. Mazzoli, L. Malavasi, C.B. Azzoni, Phys. Rev. B 73, 184402 (2006)
32. K. Muramori, S. Miyahara, J. Phys. Soc. Jpn. 15, 1906 (1960)
33. S.T. Kshirsagar, A.B. Biswas, J. Phys. Chem. Solids 28, 1493 (1967)
34. S. Åsbrink, A. Waśkowska, L. Gerward, J. Staun Olsen, E. Talik, Phys. Rev. B 60, 12651 (1999)
35. S.K. Dey, J.C. Anderson, Phil. Mag. 12, 975 (1965)
36. A.P.B. Sinha, N.R. Sanjana, A.B. Biswas, Z. Kristallogr 109, S410 (1957)
37. K. Kamazawa, Y. Tsunoda, H. Kadowaki, K. Kohn, Phys. Rev. B 68, 024412 (2003)
38. K. Kamazawa, S. Park, S.-H. Lee, T.J. Sato, Y. Tsunoda, Phys. Rev. B 70, 024418 (2004)
39. M. Isobe, Y. Ueda, J. Phys. Soc. Jpn. 71, 1848 (2002)
40. M. Schmidt, W. Ratcliff II, P.G. Radaelli, K. Refson, N.M. Harrison, S.-W. Cheong, Phys. Rev. Lett. 92, 056402 (2004)
41. K. Adachi, T. Suzuki, K. Kato, K. Osaka, M. Takata, T. Katsufuji, Phys. Rev. Lett. 95, 197202 (2005)
42. J.-H. Chung, J.-H. Kim, S.-H. Lee, T.J. Sato, T. Suzuki, M. Katsumata, T. Katsufuji, Phys. Rev. B 77, 054412 (2008)
43. V.O. Garlea, R. Jin, D. Mandrus, B. Roessli, Q. Huang, M. Miller, A.J. Schultz, S.E. Nagler, Phys. Rev. Lett. 100, 066404 (2008)
44. T. Katsufuji, T. Suzuki, H. Takei, M. Shingu, K. Kato, K. Osaka, M. Takata, H. Sagayama, T. Arima, J. Phys. Soc. Jpn. 77, 053708 (2008)
45. P. Holba, M. Nevriva, E. Pollert, Mat. Res. Bull. 10, 853 (1975)
46. K. Tomiyasu, J. Fukunaga, H. Suzuki, Phys. Rev. B 70, 214434 (2004)
47. G. Shirane, D.E. Cox, S.J. Pickart, J. Appl. Phys. 35, 954 (1964)
48. Y. Yamasaki, S. Miyasaka, Y. Kanako, J.-P. He, T. Arima, Y. Tokura, Phys. Rev. Lett. 96, 207204 (2006)
49. M. Tovar, R. Torabi, C. Welker, F. Fleischer, Physica B 385–386, 196 (2006)
50. E. Prince, Acta. Cryst 10, 554 (1957)
51. K. Penc, N. Shannon, H. Shiba, Phys. Rev. Lett. 93, 197203 (2004)
52. Y. Motome, H. Tsunetsugu, Phys. Rev. B. 70, 184427 (2004)
53. H. Tsunetsugu, Y. Motome, Phys. Rev. B. 68, 060405(R) (2003)
54. O. Tchernyshyov, Phys. Rev. Lett. 93, 157206 (2004)
55. S. Di Matteo, G. Jackeli, C. Lacroix, N.B. Perkins, Phys. Rev. Lett. 93, 077208 (2004)
56. D.I. Khomskii, T. Mizokawa, Phys. Rev. Lett. 94, 156402 (2005)
57. V. Fritsch, J. Hemberger, N. Buttgen, E.-W. Scheidt, H.-A. Krug von Nidda, A. Loidl, V. Tsurkan, Phys. Rev. Lett. 92, 116401 (2004)
58. A. Krimmel, M. Mücksch, V. Tsurkan, M.M. Koza, H. Mutka, C. Ritter, D.V. Sheptyakov, S. Horn, A. Loidl, Phys. Rev. B. 73, 014413 (2006)
59. D. Bergman, J. Alicea, E. Gull, S. Trebst, L. Balents, Nat. Phys. 3, 487 (2007)
60. A. Krimmel, M. Mücksch, V. Tsurkan, M.M. Koza, H. Mutka, A. Loidl, Phys. Rev. Lett. 94, 237402 (2005)
61. N. Büttgen, A. Zymara, C. Kegler, V. Tsurkan, A. Loidl, Phys. Rev. B. 73, 132409 (2006)

62. N. Tristan, J. Hemberger, A. Krimmel, H.-A. Krug von Nidda, V. Tsurkan, A. Loidl, Phys. Rev. B 72, 174404 (2005)
63. T. Suzuki, M. Nohara, H. Takagi, J. Phys.: Condens. Matter 19, 145265 (2007)
64. Y. Okamoto, M. Nohara, H. Aruga-Katori, H. Takagi, Phys. Rev. Lett. 99, 137207 (2007)
65. Fujiyama, private communication.
66. J.M. Hopkinson, S.V. Isakov, H.-Y. Kee, Y.B. Kim, Phys. Rev. Lett. 99, 037201 (2007)
67. M.J. Lawler, H.-Y. Kee, Y.B. Kim, A. Vishwanath, Phys. Rev. Lett. 100, 227201 (2008)
68. A.P. Ramirez, B. Hessen, M. Winklemann, Phys. Rev. Lett. 84, 2957 (2000)
69. S. Nakatsuji, Y. Nambu, H. Tonomura, O. Sakai, S. Jonas, C. Broholm, H. Tsunetsugu, Y. Qiu, Y. Maeno, Science 309, 1697 (2005)
70. G. Chen, L. Balents, Phys. Rev. B 78, 094403 (2008)
71. P.W. Anderson, Phys. Rev. 102, 1008 (1956)
72. S. Kondo, D.C. Johnston, C.A. Swenson, F. Borsa, A.V. Mahajan, L.L. Miller, T. Gu, A.I. Goldman, M.B. Maple, D.A. Gajewski, E.J. Freeman, N.R. Dilley, R.P. Dickey, J. Merrin, K. Kojima, G.M. Luke, Y.J. Uemura, O. Chmaissem, J.D. Jorgensen, Phys. Rev. Lett. 78, 3729 (1997)
73. C. Urano, M. Nohara, S. Kondo, F. Sakai, H. Takagi, T. Shiraki, T. Okubo, Phys. Rev. Lett. 85, 1052 (2000)
74. P. Fulde, A.N. Yaresko, A.A. Zvyagin, Y. Grin, Europhys. Lett. 56, 779 (2001)
75. J. Hopkinson, P. Coleman, Phys. Rev. Lett. 89, 267201 (2002)
76. R. Arita, K. Held, A.V. Lukoyanov, V.I. Anisimov, Phys. Rev. Lett. 98, 166402 (2007)
77. P.E. Jönsson, K. Takenaka, S. Niitaka, T. Sasagawa, S. Sugai, H. Takagi, Phys. Rev. Lett. 99, 167402 (2007)
78. E.J.W. Verwey, P.W. Haayman, Physica 8, 979 (1941)
79. J.P. Wright, J.P. Attfield, P.G. Radaelli, Phys. Rev. B. 66, 214422 (2002)
80. J. Rodríguez-Carvajal, G. Rousse, C. Masquelier, M. Hervieu, Phys. Rev. Lett. 81, 4660 (1998)
81. Y. Horibe, M. Shingu, K. Kurushima, H. Ishibashi, N. Ikeda, K. Kato, Y. Motome, N. Furukawa, S. Mori, T. Katsufuji, Phys. Rev. Lett. 96, 086406 (2006)
82. P.G. Radaelli, Y. Horibe, M.J. Gutmann, H. Ishibashi, C.H. Chen, R.M. Ibberson, Y. Koyama, Y.S. Hor, V. Kiryukhin, S.-W. Cheong, Nature 416, 155 (2002)
83. S. Niitaka et al., unpublished.
84. L. Pinsard-Gaudart, N. Dragoe, P. Lagarde, A.M. Flank, J.P. Itie, A. Congeduti, P. Roy, S. Niitaka, H. Takagi, Phys. Rev. B 76, 045119 (2007)
85. P.W. Anderson, Math. Res. Bull. 8, 153 (1973)
86. T. Kimura, Ann. Rev. Mater. Res. 37, 387 (2007)
87. T. Yamada, Z. Hiroi, M. Takano, M. Nohara, H. Takagi, J. Phys. Soc. Jpn. 69, 1477 (2000)
88. G. Blasse, J.F. Fast, Philips Res. Rept. 18, 393 (1963)
89. R. Plumier, C. R. Acad. Sci. Paris 263, 173 (1966)
90. S. Diaz, S. de Brion, G. Chouteau, B. Canals, V. Simonet, P. Strobel, Phys. Rev. B 74, 092404 (2006)
91. T. Hoshi, H. Aruga-Katori, M. Kosaka, H. Takagi, J. Mag. Mag. Mater. 310, 448 (2007)
92. S. Diaz, S. de Brion, G. Chouteau, P. Strobel, B. Canals, J.R. Carvajal, H. Rakoto, J.M. Broto, J. Appl. Phys. 97, 10A512 (2005)
93. M.K. Crawford, R.L. Harlow, P.L. Lee, Y. Zhang, J. Hormadaly, R. Flippen, Q. Huang, J.W. Lynn, R. Stevens, B.F. Woodfield, J. Boerio-Goates, R.A. Fisher, Phys. Rev. B 68, 220408 (2003)
94. S. Diaz, S. de Brion, G. Chouteau, B. Canals, V. Simonet, P. Strobel, Phys. Rev. B 74, 092404 (2006)
95. M. Matsuda, J.-H.Chung, S. Park, T.J. Sato, K. Matsuno, H. Aruga-Katori, H. Takagi, K. Kakurai, K. Kamazawa, Y. Tsunoda, I. Kagomiya, C.L. Henley, S.-H. Lee, Europhys. Lett. 82, 37006 (2008)
96. D.C. Johnston, J. Low Temp. Phys. 25, 145 (1976)
97. K. Matsuno, T. Katsufuji, S. Mori, Y. Moritomo, A. Machida, E. Nishibori, M. Takata, M. Sakata, N. Yamamoto, H. Takagi, J. Phys. Soc. Jpn. 70, 1456 (2001)

# Chapter 8
# Experimental Studies of Pyrochlore Antiferromagnets

**Bruce D. Gaulin and Jason S. Gardner**

**Abstract** Antiferromagnetically-coupled spins on the apex of corner sharing tetrahedra have generated significant interest over the last 20 years. Here we will discuss the experimental studies, mostly neutron scattering, of the gadolinium and terbium based pyrochlore magnets, for the purpose of illustrating the diverse nature of ground states which can occur.

## 8.1 Introduction

Geometrically-frustrated magnetic materials [1–3] have been a longstanding interest in materials physics and chemistry. Early theoretical work, such as that by Wannier [4], appreciated the nature of the problem associated with antiferromagnetically-coupled spins on triangular lattices, and that the problem was much more general, being associated with the incompatibility of local geometry and the satisfaction of pairwise interactions in cooperative systems. This problem finds great expression within magnetic materials as many crystalline architechtures are based on close-packing geometries, and triangular stackings, as occur in hexagonal and face centred cubic lattices, are very common in nature. In relation to geometrical frustration, the triangle is to two dimensions as the tetrahedron is to three. Antiferromagnetically-coupled spins on a single tetrahedron can possess a similar set of degenerate ground states as does the analogous case with a single triangle.

Magnetic materials display great variety in the nature of the magnetic moments which decorate the crystalline lattice – that is the nature of anisotropies which they display, as well as the nature of the interactions between neighbouring magnetic moments. As a result of much experimental work in the last 10–15 years, we now have many good examples of geometrically-frustrated magnetic materials which themselves display varied ground states at low temperatures. Many of these are described as exotic, in that they do not correspond to a conventional long range ordered magnetic state at temperatures below $\Theta_{CW}$, the Curie–Weiss temperature, which characterizes the average energy associated with magnetic interactions in the material. These exotic ground states tend to fall into three related categories: spin liquids, spin glasses [5], and spin ice [6]; although the complexity associated with real materials is such that a complete description of their ground state properties defies strict classification.

Because the leading order terms in the magnetic Hamiltonian are the ones which compete directly against each other, the actual ground state in frustrated materials is often selected by sub-leading terms in the Hamiltonian. These can be relatively weak terms which might be ignored in the study of unfrustrated magnetic materials. Examples of such interactions are dipolar interactions and exchange interactions beyond near-neighbour. For this reason, it is important to establish as comprehensive an understanding of the microscopic magnetic Hamiltonian as possible, in order to quantitatively understand the nature of the exotic magnetic ground state in these materials. Additionally, and for the same reasons, external perturbations, such as magnetic field and pressure, can have large effects in selecting a ground state, and indeed this will be a major theme of this review.

In this chapter we will focus on experimental studies of a class of rare-earth ion-based, pyrochlore magnets, and in particular the gadolinium and terbium based magnets, for the purpose of illustrating the diverse nature of ground states which can occur. It is, however, not intended to be a comprehensive review of the experimental side of geometrically-frustrated magnetism. Of course, the complete experimental characterization of these materials requires many different techniques be brought to bear on the problem. We will focus on scattering studies as they inform directly on the magnetic and chemical structure of the exotic ground state, and can spectroscopically probe the crystal field and cooperative Hamiltonian of the magnetic system. As a result such scattering studies play a central role in experimental studies of magnetic materials.

## 8.2 The Cubic Pyrochlores

Rare-earth-based cubic pyrochlores can be thought of in terms of magnetic moments decorating a three dimensional network of corner-sharing tetrahedra. In general, crystalline lattices which support geometrical frustration can be thought of as being made up of either edge or corner-sharing triangles, in two dimensions, or edge or corner-sharing tetrahedra in three dimensions. Of course all real materials are three dimensional, so geometrically-frustrated magnets based on triangular nets – either edge or corner-sharing – must be made up of a three dimensional stacking of such two dimensional nets. A two dimensional net comprised of edge sharing triangles gives the familiar two dimensional tringular lattice, which make up the basal plane of all hexagonal crystals. A similar net of corner-sharing triangles gives the two dimensional kagomé lattice, which has been the focus of intense theoretical and experimental interest. More complicated, but related, networks of triangles are observed in kagomé-staircase structures [7], hyper-kagomé lattices [8, 9], and some garnets [10, 11]. A three dimensional network of edge-sharing tetrahedra results in a face-centred cubic structure – which can also be throught of as a close packing of triangular planes along the [111] cube diagonal direction. Finally, a three dimensional network of corner-sharing tetrahedra is found in the cubic pyrochlores.

This architectural motif is not unique to the pyrochlores – it also occurs in Laves phase and Spinel materials. Indeed, as will be described, it is quite common in nature. The family we will focus on are the $A_2B_2O_7$ pyrochlores, where the A site is a trivalent rare-earth ion with eightfold oxygen coordination, and the B site is a tetravalent transition metal ion with sixfold oxygen coordination. Both the A site and the B site, independently, reside of a network of corner-sharing tetrahedra – as shown in Fig. 8.1. These networks are displaced from each other by a translation along the cube diagonal direction [111].

Such stoichiometries provide many good examples of pyrochlore magnets and hence geometrically frustrated magnets. Cases exist where the A site is magnetic (such as $Tb_2Ti_2O_7$ [12], $Tb_2Sn_2O_7$ [13, 14], $Ho_2Ti_2O_7$ [15]), where the B site is magnetic (such as $Y_2Mo_2O_7$ [16]), where both A and B sites are magnetic (such as $Tb_2Mo_2O_7$ [17, 18]), and where neither is magnetic (such as $Y_2Ti_2O_7$ [19]). The insulating rare-earth titanates have been well studied, as they are relatively easy to grow in single crystal form [20,21]. The differing nature of the magnetic anisotropy and magnetic coupling across the rare earth series has given rise to spectacular examples of exotic magnetic ground states arising from geometrical frustration. The related rare-earth stanates, $A_2Sn_2O_7$, have been much more difficult to produce in single crystal form, due to the very high vapour pressure of Sn. Nonetheless, members of this family have been investigated in polycrystal form and have provided intriguing counterpoints to the ground states exhibited by the titanates.

The $A_2Mo_2O_7$ family [22, 23] of molybdates is very interesting, as the conductivity evolves from insulating to metallic across the series. While our focus has been on insulating magnets wherein the exchange interactions are expected to be of shorter range and therefore to be more susceptable to effects related to the local

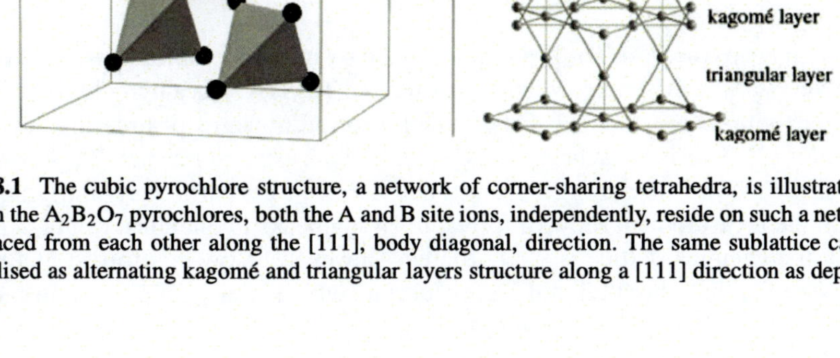


**Fig. 8.1** The cubic pyrochlore structure, a network of corner-sharing tetrahedra, is illustrated in (**a**). In the $A_2B_2O_7$ pyrochlores, both the A and B site ions, independently, reside on such a network displaced from each other along the [111], body diagonal, direction. The same sublattice can be visualised as alternating kagomé and triangular layers structure along a [111] direction as depicted in (**b**)

environment, there has been considerable interest in conducting pyrochlores and related materials, such as $Cd_2Re_2O_7$ [24] and the $\beta$-pyrochlores such as $KOs_2O_6$ [25] and $RbOs_2O_6$ [26], which are relatively low temperature superconductors ($T_C$'s of ∼1.3 K, 9.6 K, and 6.3 K, respectively).

It can be useful to decompose the pyrochlore lattice in at least two different ways, depending on the nature of the phenomena that is of interest in a particular material, or the nature of a perturbation which is applied to it. It can be thought of as an alternation of kagomé planes and triangular planes, stacked along the cube diagonal – the [111] direction (see Fig. 8.1b). There are pyrochlores which exhibit quasi-two dimensional spin correlations at low temperature, such as the recently studied ferromagnetic pyrochlore $Yb_2Ti_2O_7$ [27] as well as the spinel $Y_2Mn_2O_4$ [28], and this can be a convenient way to understand this phenomena. The pyrochlore lattice can also be decomposed into two sets of orthogonal chains, the $\alpha$ chains and the $\beta$ chains, both of which lie along orthogonal cube face diagonal directions – that is [110] directions. Half of all the spins lie along the $\alpha$ chains, while half lie along $\beta$ chains. A magnetic field applied along a [110] direction, for example, can polarize the spins on the $\alpha$ chains, leaving the spins on the $\beta$ chains decoupled, as occurs in certain spin ice ground states [29].

Finally, it is worth remembering that the real materials in the $A_2B_2O_7$ family which nominally crystallize within the cubic $Fd\bar{3}m$ space group are not perfect materials. They display weak but observable Bragg peaks of structural origin which violate the systematic absences of the $Fd\bar{3}m$ space group. For example, the (002) Bragg reflection is forbidden within $Fd\bar{3}m$, yet we have observed such a peak in a host of $A_2B_2O_7$ single crystals [29, 30]. This is a weak effect, typically 0.001 or less of the allowed (004) intensity. When first observed with triple axis, Bragg scattering techniques, it was ascribed to harmonic ($\lambda$/n) contamination in the beam. However, later experiments observed the same scattering in time-of-flight neutron experiments [30], which do not suffer from harmonic contamination. While this is somewhat of a detail, it is worth bearing in mind that a perfect pyrochlore lattice *may* not exist in nature, and that the lattice itself may well be a player, as opposed to a passive bystander, in the physics of geometrical frustration. We will return to this point later in this review.

## 8.3 The Spin Liquid Ground State in $Tb_2Ti_2O_7$

Interest in the pyrochlore $Tb_2Ti_2O_7$ was piqued by initial measurements, performed on polycrystalline materials, which characterized $Tb_2Ti_2O_7$ as a pyrochlore antiferromagnet with a $\Theta_{CW} \sim -13$ K, but which showed no signs of ordering down to ∼17 mK [12]. This is the achetypical signature of a frustrated magnet. Early powder neutron diffraction measurements, taken at $T = 2$ K, are particularly intriguing as this simple measurement shows a forrest of Bragg peaks of chemical origin, telling us that the chemical lattice is well ordered. However, below this forrest of Bragg peaks (see Fig. 8.2), broad liquid-like scattering, which is magnetic in origin can be

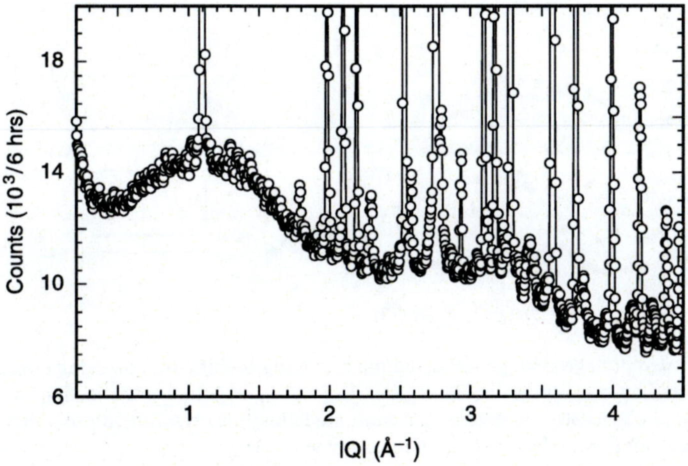


**Fig. 8.2** The raw powder neutron diffraction data taken at $T = 2$ K on $Tb_2Ti_2O_7$. Both the forest of sharp Bragg peaks, indicating well developed chemical order, and the liquid-like magnetic structure factor characteristic of the spin liquid state are seen [12]

observed, which tells us that the spins are correlated over about a single tetrahedron at these low temperatures (2 K).

Early neutron spectroscopic measurements [31] were also carried out on polycrystalline samples of $Tb_2Ti_2O_7$, with the intention of understanding the nature of the $Tb^{3+}$ magnetic moment. $Tb^{3+}$ possesses 8 f-electrons, an even number. By Kramer's theorm, it is not required to possess an, at least, two-fold degenerate ground state – which is to say that it need not be magnetic at all, but could exhibit a singlet ground state, depending on the local geometry of the crystalline environment around the $Tb^{3+}$ site. The $Tb^{3+}$ environment in $Tb_2Ti_2O_7$ has eightfold oxygen coordination. However, it does not lie in the middle of a perfect cube of oxygen – the "cube" is distorted such that six oxygen are relatively close to the $Tb^{3+}$ site and two are farther away.

The inelastic neutron measurements characterizing these crystal field states of $Tb^{3+}$ in this environment, are shown in Fig. 8.3. This data shows transitions from the ground state to excited states at $\sim$ 20 K, 120 K, and 175 K above the ground state. Modelling this inelastic data, and combining it with susceptibility measurements on $Tb_2Ti_2O_7$ and $(Tb_{0.02}Y_{0.98})_2Ti_2O_7$, where the $Tb^{3+}$ moments are sufficiently dilute to minimize interaction effects, results in a description for the six lowest lying states of the $(2J+1) = 13$, which make up the $J = 6$ Hund's rule ground state multiplet appropriate to $Tb^{3+}$. As shown in Fig. 8.3, this analysis gives a ground state doublet described by $\sim |\pm4>$, an excited state doublet at $\Delta E \sim 20$ K above the ground state described by $\sim |\pm5>$, followed by two higher energy singlet excited states described by $\sim |+3> \ \ +|-3>$ and $\sim -|+3> \ \ +|-3>$. Related measurements and analysis have recently been carried out which have revisited the nature of the crystal field levels in $Tb_2Ti_2O_7$ [32]. While the details of the eigenstates in this new study are not in agreement with the earlier work discussed here [31], the overall

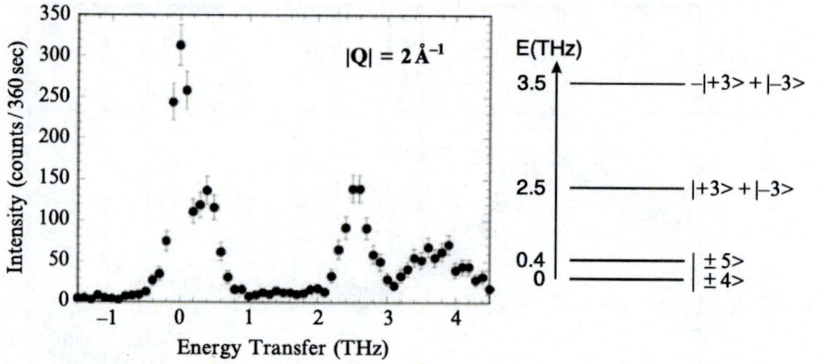


**Fig. 8.3** The *left panel* shows inelastic neutron scattering probing the low lying crystal field excitations of $Tb^{3+}$ in polycrystalline $Tb_2Ti_2O_7$. The *right panel* shows the leading order terms in the crystal field eigenstates, as determined from modelling the crystal field excitations along with magnetic susceptibility and heat capacity measurements [31]

nature of the ground state and expected Ising anisotropy of the $Tb^{3+}$ moments are reproduced in this newer analysis.

As far as the low temperature physics appropriate to geometrical frustration is concerned, only the lowest energy states will enter the discussion. The relatively low energy excited doublet is sufficiently close to the ground state doublet, that we will need to include all four of these states (the ground state doublet plus the low lying excited state doublet) in our description of the low temperature behaviour of $Tb_2Ti_2O_7$. These measurements and calculations also give a ground state moment of $\sim 5\ \mu_B$, distinct from the paramagnetic moment derived from the high temperature susceptibility, to which all the 13 states which make up the $J = 6$ multiplet contribute. They also tell us that, to the extent that the ground state doublet alone determines the anisotropy of the $Tb^{3+}$ moments, the Tb moments will display an Ising character, pointing directly into or out of the tetrahedra on which they reside. That is two, of eight, oxygens neighbouring the $Tb^{3+}$ site determine an easy axis for the moments, such that they would point into or out of their own tetrahedra.

The availability of floating zone image furnace techniques for single crystal growth resulted in the production of large and pristine single crystals of $Tb_2Ti_2O_7$ [20, 21] which would enable single crystal neutron scattering measurements. Such experiments can be much more informative than the corresponding measurements on polycrystalline materials, as the vector-**Q** dependence to the scattering can be investigated. Such a single crystal magnetic diffraction pattern [12, 33], taken in the [HHL] plane of reciprocal space in $Tb_2Ti_2O_7$, is shown in Fig. 8.4. We now see the **Q** pattern of magnetic diffuse scattering, in the form of a checkerboard. It is most intense around the (002) Bragg position, but extends across large regions of reciprocal space, consistent with earlier powder diffraction data.

This pattern to the diffuse scattering is very interesting. One can determine that the polarization dependence of the unpolarized neutron scattering cross section – the $\Sigma_{\alpha\beta}(\delta_{\alpha,\beta} - \hat{Q}_\alpha \hat{Q}_\beta) S^{\alpha\beta}(\mathbf{Q}, \omega)$ term in (1), would imply that no magnetic scattering should exist at the (002) position in reciprocal space for most models in which the

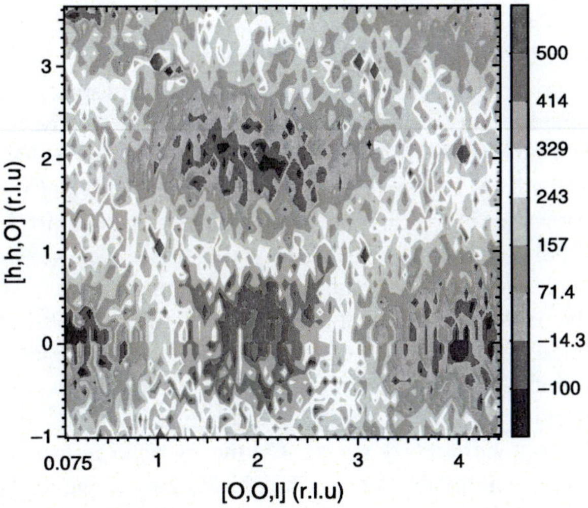


**Fig. 8.4** A difference map in reciprocal space from single crystal $Tb_2Ti_2O_7$, isolating the diffuse magnetic scattering characteristic of the spin liquid state. This data shows neutron diffraction data sets at $T = 9$ K and high temperatures subtracted from each other. Clearly, the liquid-like magnetic structure factor is organized in vector **Q** into a checkerboard pattern, with the strongest scattering around (002) [33]

magnetic moments obey the Ising-like anisotropy in which they would be restricted to point directly into or out of the tetrahedra. The expectation of this anisotropy was a consequence of the eigenfunctions associated with the ground state doublet. However, later theoretical analysis [34] which includes mixing of the lowest energy excited state doublet ~ 20 K above the ground state in $Tb_2Ti_2O_7$, can restore magnetic diffuse scattering at the (002) position. Nonetheless, this result indicates that a picture of strict adherance to local [111] anistropy for the $Tb^{3+}$ moments may not provide a satisfactory description of the experiments.

Some evidence for spin freezing has been found in other, independent neutron scattering measurements [35] on single crystal $Tb_2Ti_2O_7$, in the form of hysteretic scattering at temperatures below ~ 1.7 K. This may be expected from weak disorder which is present in all real materials, in which case the spin glassiness would be expected to exhibit variability from sample to sample. However, it is clear that most of the magnetic scattering signal remains both diffuse and dynamic even down to very low temperatures, as reported in neutron measurements down to $T = 50$ mK [36].

Problems associated with the local-[111] Ising-like nature to the Tb spins notwithstanding, a reasonable starting point Hamiltonian to understand $Tb_2Ti_2O_7$ would be:

$$\mathcal{H} = -\sum_{\langle i,j\rangle} J_{ij}\mathbf{S}_i^{z_i}\cdot\mathbf{S}_j^{z_j} + Dr_{\text{nn}}^3 \sum_{i>j} \frac{\mathbf{S}_i^{z_i}\cdot\mathbf{S}_j^{z_j}}{|\mathbf{R}_{ij}|^3} - \frac{3(\mathbf{S}_i^{z_i}\cdot\mathbf{R}_{ij})(\mathbf{S}_j^{z_j}\cdot\mathbf{R}_{ij})}{|\mathbf{R}_{ij}|^5} \tag{8.1}$$

The exchange interactions between the $Tb^{3+}$ moments are expected to be relatively weak, given the localized nature of the 4f electrons and relatively small overlap which arises. In addition, the ground state moments are rather large, $\sim 5\,\mu_B$, consequently dipolar interactions should be relatively large. In fact, a system described by such a classical Hamiltonian on the pyrochlore lattice has been considered and solved by den Hertog and Gingras [37], and their resulting phase diagram is shown in Fig. 8.5. The $y$-axis shows temperature in units of the near-neighbour dipolar interaction strength, while the $x$-axis shows the ratio of near-neighbour exchange to near-neighbour dipolar interactions. The phase diagram consists of two low temperature phases: a $Q = 0$ four-sublattice antiferromagnetically-ordered phase, provided the antiferromagnetic exchange interactions are adequately strong (on the left hand side of the phase diagram), and a disordered spin ice ground state – provided the near neighbour exchange is either ferromagnetic or antiferromagnet but not sufficiently strong. On the pyrochlore lattice with a hard [111] anisotropy for the moments, the near neighbour part of the dipolar interaction (which falls off as $r^{-3}$) favours ferromagnetic correlations, so the spin ice ground state encroaches onto the antiferromagnetic exchange (x < 0) side of this phase diagram.

Putting in all the information we have on $Tb_2Ti_2O_7$, we can place $Tb_2Ti_2O_7$ as shown in Fig. 8.5, which implies a finite temperature phase transition to four sublattice antiferromagnetic order near $T \sim 1$ K. Consequently we are left with another quandry, which is why does $Tb_2Ti_2O_7$ maintain a disordered spin liquid state down

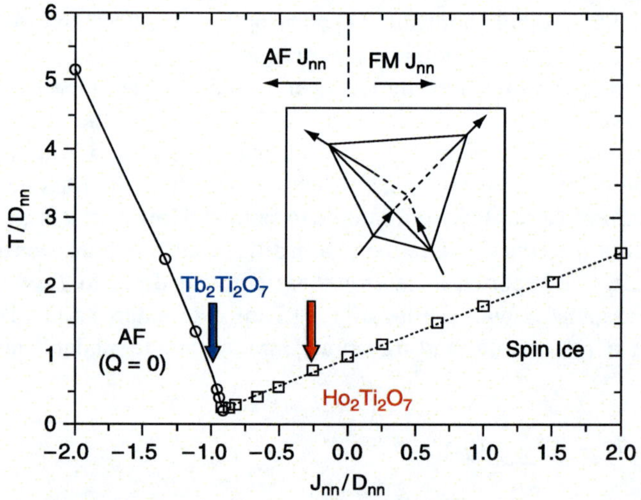


**Fig. 8.5** The general phase diagram [37] for classical Ising spins with local [111] anisotropy, such that they can point either directly into or out of the tetrahedra whose corner they share, coupled with nearest-neighbour exchange and dipolar interactions. The $y$-axis shows the temperature in units of the near-neighbour dipolar interaction strength, while the $x$-axis shows the ratio of near-neighbour exchange to near-neighbour dipolar interaction strength. The spin liquid ground state magnet $Tb_2Ti_2O_7$ and spin ice ground state magnet $Ho_2Ti_2O_7$ are approximately located on this phase diagram

to temperatures as low as 0.03 K [12, 36]? There are deficiencies in this treatment – most notably that the Hamiltonian is not quantum mechanical. In addition, the zero temperature transition from the Q = 0 antiferromagnetically long range ordered state to the spin ice ground state in Fig. 8.5 is not far removed from where we have placed $Tb_2Ti_2O_7$ on the phase diagram. Nonetheless, the low temperature spin liquid state in $Tb_2Ti_2O_7$ remains enigmatic.

## 8.4 Ordered Ground States in $Tb_2Ti_2O_7$

The spin liquid ground state in $Tb_2Ti_2O_7$ can be brought to order with relatively modest perturbations, applied with pressure [38–41] and a magnetic field [30]. Perhaps this underscores the relative fragility of the spin liquid state, and the fact that it is not far removed from ordered states.

This was first achieved with the application of hydrostatic pressure [38, 39], followed by combinations of hydrostatic and uniaxial pressure [40, 41]. Figure 8.6 shows powder neutron diffraction data taken in a pressure cell, which clearly shows the evolution of the powder diffraction pattern from one characterized by diffuse magnetic scattering, to one in which a large number of relatively weak magnetic Bragg peaks are evident as external pressure is applied.

Later work, which we will focus on here, was performed on single crystals under the application of a $[1\bar{1}0]$ magnetic field [30]. As mentioned previously, the selection of a particular [110] direction within the cubic pyrochlore structure, naturally decomposes the system into orthogonal $\alpha$ and $\beta$ chains of spins running along [110] directions. This field geometry is straightforward for neutron scattering experiments on single crystals, as such experiments typically use split-coil vertical field magnet cryostats, and scatter neutrons in a horizontal plane. Such constraints, with the magnetic field aligned along a vertical $[1\bar{1}0]$ direction, results in the horizontal scattering plane being the [HHL] plane, which contains three high symmetry directions.

Such single crystal neutron scattering data is shown in Fig. 8.7. This time-of-flight neutron scattering data was taken with the Disk Chopper Spectrometer at the NIST Center for Neutron Research. This type of measurement has the advantage that extensive maps of the scattering can be straightforwardly made, and that one can simultaneously study elastic and inelastic scattering – that is, it allows for very efficient measurements. Panels (a) and (d) of Fig. 8.7 illustrate this nicely. Panel (a) shows an elastic scattering map, actually integrating in energy from −0.5 meV to 0.5 meV, in the [HHL] plane of reciprocal space. The strongest "checker", centred at (002), of the checkerboard pattern shown in Fig. 8.4 is clearly evident. Panel (d), however, shows the energy dependence of the diffuse scattering around (002). This clearly identifies the same diffuse scattering as inelastic in nature with a characteristic energy of $\hbar\omega \sim 0.3$ meV. This is not so surprising. The disordered short range ordered $Tb^{3+}$ moments in the spin liquid state are also dynamic on a time scale that the neutron scattering measurement is easily sensitive to.

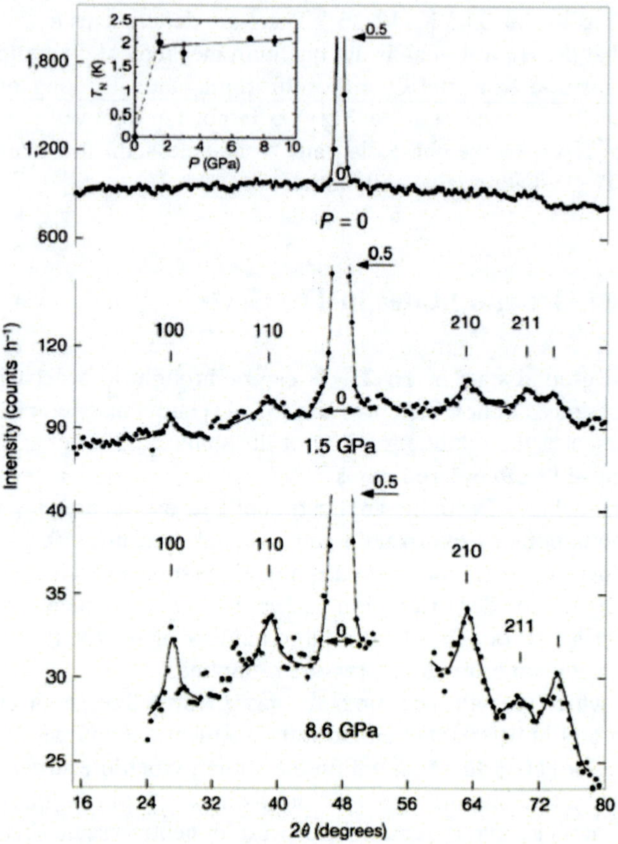


**Fig. 8.6** The evolution of pressure-induced magnetic Bragg peaks in $Tb_2Ti_2O_7$ is shown [39–41]. The inset shows the dependence of the induced ordering temperature, $T_N$ on pressure

More remarkable, however, is the effect of relatively weak magnetic fields applied along the $[1\bar{1}0]$ direction, shown in Panels (b) and (c) of Fig. 8.7. Panel (b) shows the elastic scattering map in reciprocal space taken at a magnetic field of $\mu_o H = 2$ T. The diffuse magnetic scattering around (002) has been cleaned up and a Bragg peak at (002) has appeared. At higher applied fields, as shown for $\mu_o H = 8$ T in Fig. 8.7c, a further sequence of magnetic Bragg peaks has appeared, with the strongest of these at the (112) Bragg position. These results point to at least one, and possibly two phase transitions to a magnetically ordered state under application of a $[1\bar{1}0]$ magnetic field. The energy dependence of the scattering around (002) shown in panel (d) shows the magnetic diffuse scattering to decrease strongly as a function of field, with the scattering evolving from quasi-elastic to inelastic as the field strength increases. A gap of $\sim$ 0.1–0.2 meV develops in the energy dependence of the diffuse scattering as a function of applied $[1\bar{1}0]$ magnetic field.

Figures 8.8 and 8.9 show the results of parametric studies of both the field-induced Bragg peaks and the magnetic diffuse scattering in $Tb_2Ti_2O_7$. Figure 8.8

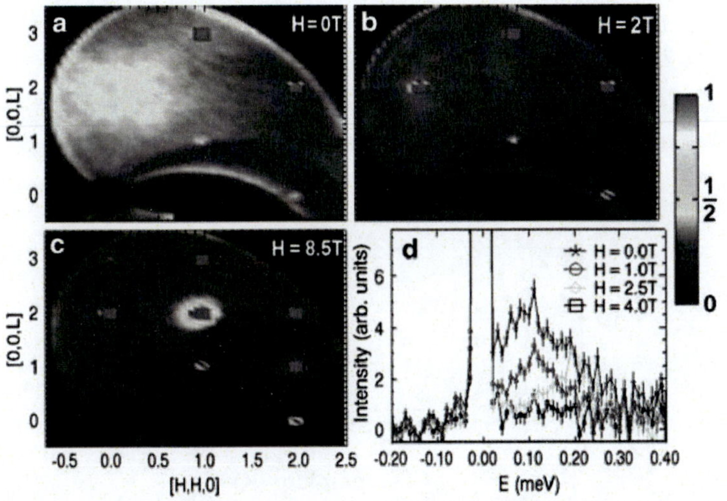


**Fig. 8.7** Reciprocal space maps of elastic scattering integrating from −0.5 meV to +0.5 meV in energy transfer are shown for $Tb_2Ti_2O_7$ at $T = 0.4$ K [30]. (**a**) shows the spin liquid state at $\mu_o H = 0$, while (**b**) and (**c**) show data taken from the polarized paramagnetic state at $\mu_o H = 2$ T, and the magnetic field-induced ordered state at $\mu_o H = 8$ T. All magnetic fields are applied along the $[1\bar{1}0]$ direction. (**d**) shows cuts in energy going through the magnetic diffuse scattering near (002) for different field strength. This high resolution time-of-flight neutron scattering data shows the magnetic diffuse scattering to be quasi-elastic with an extent in energy of ∼0.3 meV in the spin liquid state

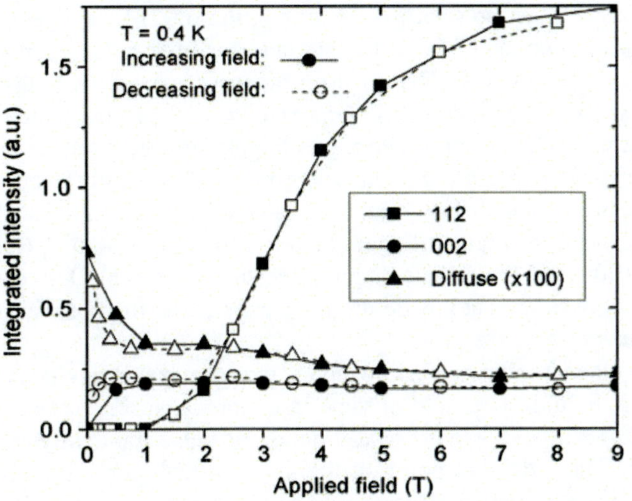


**Fig. 8.8** The field dependence of the magnetic field-induced Bragg scattering at (112) and (002) are shown for $Tb_2Ti_2O_7$ at $T = 0.4$ K, along with the field dependence of the magnetic diffuse scattering characterizing the spin liquid state [30]. All magnetic fields are applied along the $[1\bar{1}0]$ direction

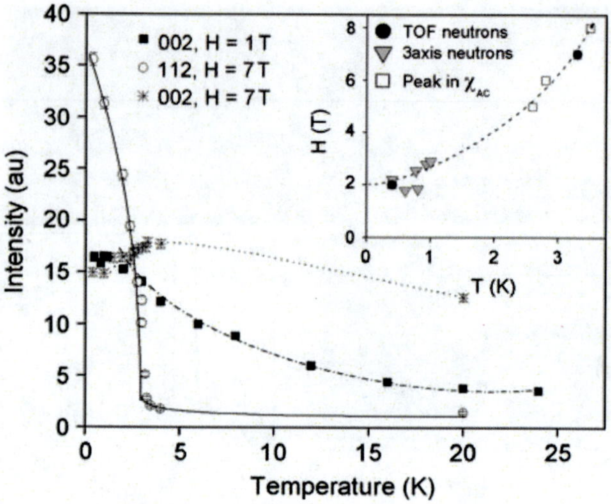


**Fig. 8.9** The temperature dependence of the magnetic field-induced Bragg scattering, at (112) and (002) is shown for $Tb_2Ti_2O_7$ for magnetic fields applied along the [$1\bar{1}0$] direction [30]. A single phase transition is identified associated with the (112) Bragg scattering. The phase diagram locating this ordered magnetic state on an H-T diagram, resulting from these and related measurements, is shown

shows the field dependence at low temperatures, $T = 0.4$ K, while Fig. 8.9 shows the temperature dependence of the field-induced Bragg peaks at selected applied magnetic fields. As can be seen from Fig. 8.8, the (002) Bragg peak rises almost immediately as a function of field strength, and has little field dependence to it for fields beyond $\mu_o$H $\sim$ 0.5 T. The diffuse scattering around (002) displays a field dependence which is the precise complement to the (002) Bragg scattering, strongly implying that the (002) Bragg scattering is largely derived from a sweeping up of the diffuse scattering into a coherent Bragg peak. This clearly occurs with rather low magnetic fields. The diffuse scattering falls off quickly on the scale of $\mu_o$H $\sim$ 0.5 T, plateaus, and then falls off again to zero near $\mu_o$H $\sim$ 4–5 T. Both the Bragg scattering at (002) and the diffuse magnetic scattering around (002) show the same complementary hysteresis at low fields, again consistent with both sets of scattering having the same origin.

The (112) magnetic Bragg peak, which is the strongest Bragg peak seen at high fields (see Fig. 8.7c) in this field of view in reciprocal space, turns on above $\mu_o$H $\sim$ 2 T and grows in a "S" shape as a function of field, apparently saturating near $\mu_o H = 8$ T. The inflection point on the (112) growth curve near $\mu_o H = 4$ T coincides with the second drop in the diffuse scattering intensity. The high field ordered state is therefore conventional in the sense that it does not coexist with a disordered component to the magnetic structure – as would be the case for fields below ~4 T.

The temperature dependence to the field-induced Bragg peaks, shown in Fig. 8.9 is very revealing. The field induced (112) magnetic Bragg peak falls off in intensity

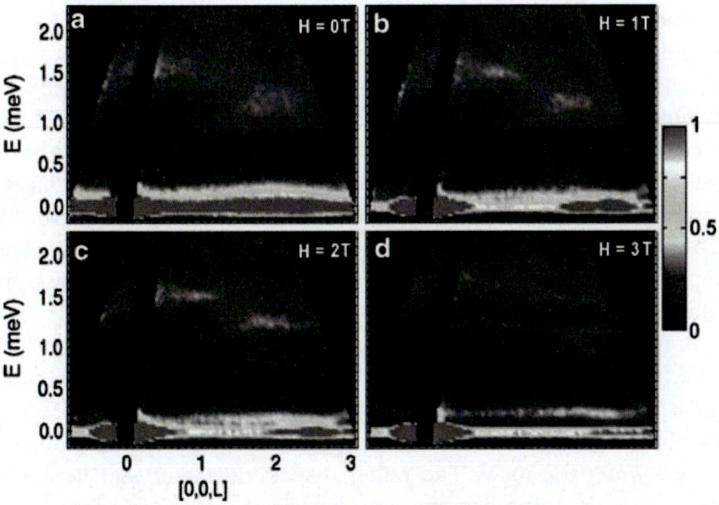


**Fig. 8.10** Dispersion surfaces determined from time-of-flight neutron scattering measurements on $Tb_2Ti_2O_7$ at $T = 0.4$ K are shown [30]. Energy-transfer is plotted on the $y$-axis and [00L] on the $x$-axis. Data from the spin liquid state at $\mu_o H = 0$ T and polarized paramagnetic state at $\mu_o H = 1$ T, (**a**) and (**b**), respectively, are qualitatively similar. Data in the field-induced ordered state at $\mu_o H = 2$ T and 3 T, (**c**) and (**d**), respectively, are qualitatively similar show clear, dispersive spin wave excitations appear within the gap, a hallmark of magnetic long range order

with strong downwards curvature as a function of temperature, displaying a well-defined phase transition near $T = 3.1$ K at $\mu_o H = 7$ T. This contrasts strongly with the temperature dependence of the (002) field-induced Bragg peak. At $\mu_o H = 1$ T, where the (002) Bragg peak is the only field-induced Bragg peak in our field of view in reciprocal space, the Bragg scattering evolves with upwards curvature as a function of temperature, and still has easily measureable intensity for temperatures as high at $T = 24$ K, the highest temperature we measured. This is not the characteristic of the order parameter associated with an ordered state, but is much closer to the signature expected for a polarized paramagnet.

As mentioned previously, the time-of-flight neutron scattering technique allows the simultaneous measurement of both maps of the elastic scattering in reciprocal space, or measurements across other surfaces in the four dimensional $\mathbf{Q}$-$\hbar\omega$ space. So we can also perform measurements which map out the dispersion of the low lying excitations as a function of energy and a particular direction in reciprocal space. This is what is shown for $Tb_2Ti_2O_7$ in Fig. 8.10. This data is taken at $T = 0.4$ K, and at various $[1\bar{1}0]$ fields up to 3 T. The top two panels of Fig. 8.10a, b show the low lying magnetic inelastic scattering along the (00L) direction in $\mu_o H = 0$ and 1 T, respectively. These are within the spin liquid state and the polarized paramagnetic state respectively – which we expect to differ little, if at all, from the spin liquid state. Indeed we see largely the same inelastic spectra at both fields. We can describe this inelastic scattering as a function of increasing energy, starting with the quasielastic scattering near $\hbar\omega = 0$ meV. We see quasi-elastic scattering, peaking near (002) and extending out in energy to $\hbar\omega\sim$ 0.3 meV, followed by an energy gap. At higher energies we observe rather diffuse inelastic scattering with weak dispersion near $\hbar\omega\sim$ 1–1.5 meV. The scattering in the $\hbar\omega\sim$ 1–1.5 meV range we previously identified as originating from the lowest energy crystal field excitation, the doublet excited state appropriate to $Tb^{3+}$, in $Tb_2Ti_2O_7$. Here, we note that with the relatively high energy resolution afforded by the time-of-flight technique, these excitations are poorly defined in energy, corresponding to relatively short lifetimes. We will return to this issue later, but for now we note that the spin liquid state in zero and low applied fields is characterized by rather "blobby", that is poorly defined, low lying magnetic excitations.

The excitation spectra in the field induced ordered state for $\mu_o H = 2$ T and 3 T in Fig. 8.10c, d, respectively, are very different from those within the spin liquid state, as one might expect. We identify three features. The quasi-elastic scattering breaks off from the elastic scattering and becomes a distinct, dispersionless excitation near $\hbar\omega\sim$ 0.2 meV. The rather dispersionless crystal field excitations in the $\hbar\omega\sim$ 1–2 meV range become very noticibly sharper in energy – implying longer lifetimes. Finally, and most spectacularly, we see dispersive spin wave excitations appear within the energy gap which characterized the spin liquid state. These all point to the development of a field-induced long range ordered state for fields applied along the $[1\bar{1}0]$ direction and in excess of ~2 T. That is, not only does the field-induced phase beyond $\mu_o H = 2$ T display a set of new magnetic Bragg peaks, the most prominent of which is the (112), but this state also displays collective spin wave excitations – the hallmark of an ordered phase.

In principle, the dispersion of the spin wave excitations can be theoretically modelled to inform on the low energy spin Hamiltonian appropriate to $Tb_2Ti_2O_7$. This has yet to be done; however simply counting the number of observed excitations tells us that our picture of a doublet ground state and doublet low lying excited state for $Tb^{3+}$ in $Tb_2Ti_2O_7$ is correct. This is shown in Fig. 8.11, which shows cuts in energy through the energy – (00L) maps shown in Fig. 8.10a, d. Five excitations are clearly observed in finite field, as would be expected provided appropriate matrix elements connect four non-degenerate eigenfunctions within the long range ordered state. This is also good evidence for a small splitting of the doublet ground state, as is allowed by the non-Kramers nature of the $Tb^{3+}$ ground state.

## 8.5 Structural Fluctuations in the Spin Liquid State of $Tb_2Ti_2O_7$

The main focus of our studies has been the spin degrees of freedom in geometrically frustrated magnets, and this is true of the research field as a whole. However, the geometry of the lattice, and therefore the lattice itself, is a key ingredient to the physics of geometrically frustrated magnets. Consequently, although little effort has been devoted to it, it should not be surprising that the lattice degrees of freedom are not merely bystanders to the spin liquid physics, but that they participate in some

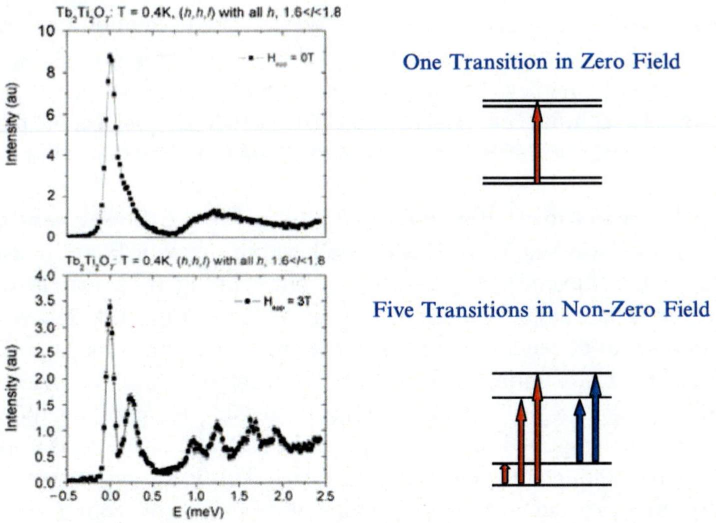


**Fig. 8.11** Cuts through the dispersion surfaces shown in Fig. 8.10 at $\mu_o H = 0$ and $\mu_o H = 3$ T of the form (HH1.7) are shown. These clearly show the appearance of sharp, spin wave peaks on entering the field-induced ordered state. The right side of the figure makes the qualitative argument as to how one broad inelastic feature can evolve to five sharp features on the basis of a doublet ground state and doublet excited state which split on entering the field-induced ordered state of $Tb_2Ti_2O_7$

way. In fact, this realization has now been made in several different contexts within geometrically-frustrated magnetism – in particular with regard to a spin-Peierls like first order phase transition [42,43], as well as magnetization plateaus [44,45], which occur in certain spinels. Also, general theoretical arguments have been made with respect to the manner in which lattice degrees of freedom can relax the constraints imposed by competing magnetic interactions on frustrated lattices [46, 47].

Motivation for studying the lattice degrees of freedom in $Tb_2Ti_2O_7$ is straightforward to come by. As discussed previously, the nature of the $Tb^{3+}$ ground state in $Tb_2Ti_2O_7$ is very sensitive to the local geometry at the $Tb^{3+}$ site, as $Tb^{3+}$ is not a Kramers ion. For example, $Tb^{3+}$ sitting within a perfect cube of eight $O^{2-}$ neighbouring ions results in a singlet, non magnetic ground state. Further, much earlier measurements show strong magnetostriction and related effects at low temperatures in $Tb_2Ti_2O_7$ [48, 49]. Finally, as just discussed, the low lying crystal field excitations in the spin liquid state of $Tb_2Ti_2O_7$ are not sharp in energy, which suggests something unusual in the nature of the local crystal structure which the $Tb^{3+}$ crystal field states are sensitive to.

We carried out high resolution X-ray scattering diffraction measurements [50] on single crystal $Tb_2Ti_2O_7$ down to temperatures as low as $T = 0.3$ K, using a rotating anode X-ray source and a perfect single crystal Ge(111) monochromator to achieve the high resolution. The crystal was mounted in a $^3$He refridgerator in a four circle goiniometer at McMaster University, and the scattered X-ray intensity was measured with a HiStar area detector. We focussed attention on a relatively simple measurement – that of the Bragg intensity and lineshape of the (12 0 0) and (8 8 0) allowed chemical Bragg peaks of the Fd$\bar{3}$m spacegroup. We performed measurements on both single crystals of $Tb_2Ti_2O_7$ and $Y_2Ti_2O_7$, a cubic pyrochlore where neither the A nor the B site is magnetic – that is a magnetic vegetable, for comparison.

Maps in reciprocal space around the (12 0 0) and (8 8 0) Bragg peak positions in $Tb_2Ti_2O_7$ are shown at $T = 0.3$ K, well within the spin liquid phase and at $T = 20$ K – in the conventional paramagnetic phase in Fig. 8.12. Panels (a) and (c) of Fig. 8.12 show data at 20 K, while that in panels (b) and (d) show data at $T = 0.3$ K. The temperature dependence seen in these reciprocal space maps presented in Fig. 8.12 is somewhat subtle. The most obvious feature is that the peak intensity at both (12 0 0) and (8 8 0) decreases with decreasing temperature. What is not obvious is that the integrated scattering around both Bragg positions is temperature independent, even though the peak intensity varies. This is the case as the scattering actually becomes broader in **Q** as the temperature decreases from $T = 20$ K to $T = 0.3$ K. In the case of (12 0 0) the scattering broadens in the longitudinal direction (along the arrow shown in Fig. 8.12a, b). In the case of (8 8 0) it broadens in the transverse direction (along the line drawn normal to the arrow in Fig. 8.12c, d)

We can quantify this behaviour. Figure 8.13 shows cuts through the reciprocal space contour maps of Fig. 8.12, in the longitudinal direction for the (12 0 0) and in the transverse direction for the (8 8 0). These plots are normalized to each other

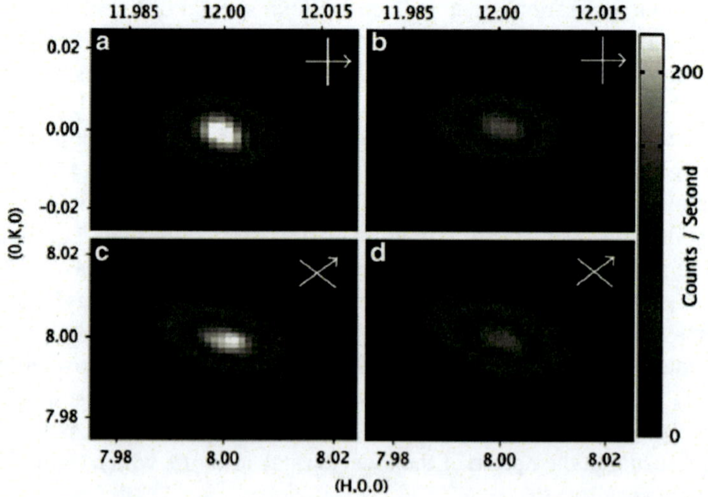


**Fig. 8.12** High resolution X-ray diffraction maps of the (12 0 0) (**a**) and (**b**) and (8 8 0) (**c**) and (**d**) chemically-allowed Bragg peaks in $Tb_2Ti_2O_7$ [50]. The data sets were taken at $T = 20$ K (left side panels, (**a**) and (**c**)) and $T = 0.3$ K (right side panels, (**b**) and (**d**)). The *arrows* show the longitudinal direction in reciprocal space, while the *orthogonal bar, without an arrowhead,* indicate the transverse direction. The decrease in the peak intensities of both Bragg reflections is clear

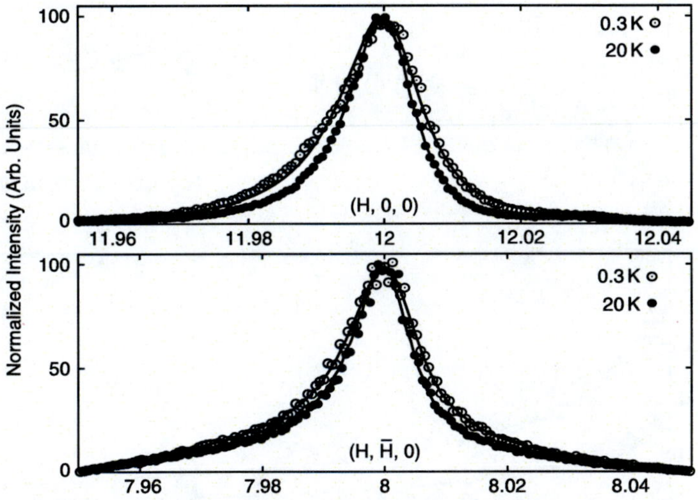


**Fig. 8.13** Cuts taken in the longitudinal direction of the (12 0 0) Bragg peaks (*top*) and the transverse direction of the (8 8 0) Bragg peaks (*bottom*) in $Tb_2Ti_2O_7$ are shown [50]. The peak intensities of the data sets at $T = 20$ K and $T = 0.3$ K have been normalized so as to allow a comparison of the widths of the scattering. The scattering at both wavevectors clearly broadens on lowering the temperature into the spin liquid phase

so that their widths can be compared, and this is what Fig. 8.13 shows. It is clear from this data that the scattering at both (12 0 0) and (8 8 0) is substantially broader at $T = 0.3$ K as compared with $T = 20$ K. In itself this is very unconventional behaviour. Most materials show little temperature dependence to their Bragg scattering at low temperatures. Lattice parameters are largely temperature independent below 20 K, and any temperature dependence to the Bragg intensities comes from Debye-Waller effects, which are weak at low temperature and which act so that the intensity increases with decreasing temperature.

Cuts of the form shown in Fig. 8.13 were made at all temperatures below 25 K. This data was fit to a resolution convolution of an Ornstein-Zernike form to the scattering, for the purpose of extracting correlation lengths appropiate to this Bragg scattering. A data set at $T = 30$ K was used as the resolution function, which assumes that the Bragg scattering for temperatures of 30 K and above is resolution limited. The results of this analysis is shown in Fig. 8.14. The top panel of Fig. 8.14 shows the peak intensity of the Bragg scattering for each of (12 0 0) and (8 8 0) in $Tb_2Ti_2O_7$, and the peak intensity of (12 0 0) in $Y_2Ti_2O_7$. The temperature dependence of the Bragg scattering in $Y_2Ti_2O_7$, the magnetic vegetable, is as expected – there is none at low temperatures. In contrast, the peak intensity at both (12 0 0) and (8 8 0) decrease by about 20% from 25 K to 0.3 K.

The bottom two panels of Fig. 8.14 show the fascinating temperature dependence of the inverse correlation lengths, along both longitudinal and transverse directions for the (12 0 0) (middle panel) and (880) (lower panel). The scattering around both Bragg peaks takes on finite inverse correlation lengths as the temperature decreases

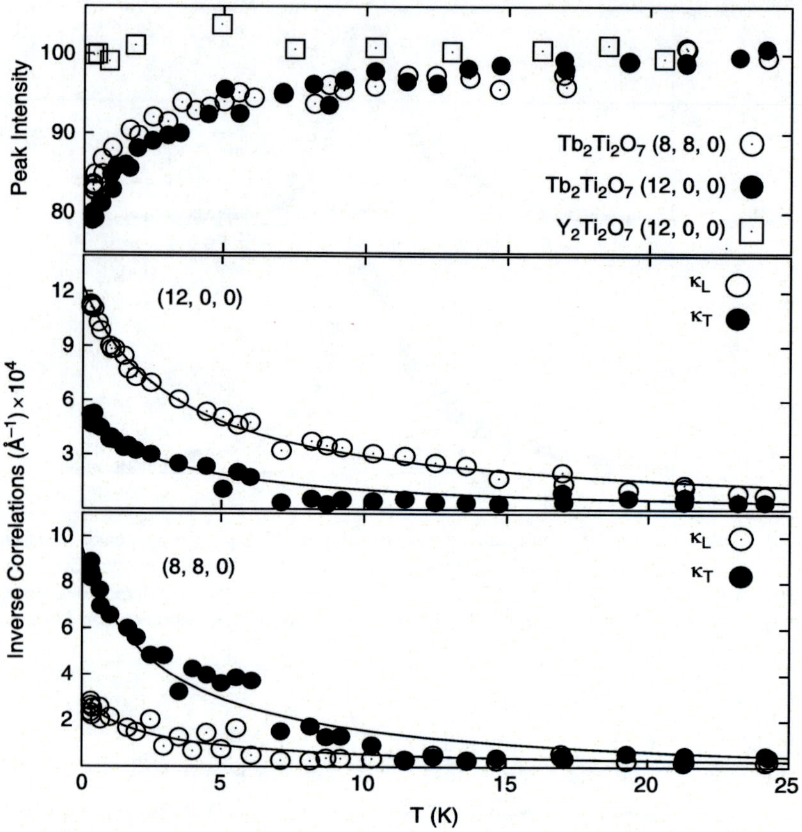


**Fig. 8.14** The top panel shows the peak intensity of the (12 0 0) and (8 8 0) Bragg peaks in $Tb_2Ti_2O_7$ as a function of temperature, and compares it to the temperature dependence of the (12 0 0) Bragg intensity of non-magnetic $Y_2Ti_2O_7$ [50]. The bottom two panels show the inverse correlation lengths extracted from a resolution convolution of Bragg peaks of the form seen in Fig. 8.13. The broadening of the Bragg peaks is clearly anisotropic, with the two Bragg peaks displaying different anisotropy

below 25 K. The temperature dependence shows upwards curvature (and that of the peak intensity shows downwards curvature only) indicating that no phase transition takes place. Yet it appears that the system is trying to approach a phase transition, albeit at unattainably low temperatures. Remarkably the broadening of the scattering is clearly anisotropic in **Q**, and this anisotropy varies from Bragg position to Bragg position. The broadening is preferentially along the longitudinal direction for the (12 0 0) and preferentially along the transverse direction for the (8 8 0) Bragg peak.

This behaviour is very reminisient of the broadening of Bragg peaks above Jahn–Teller phase transitions in other Tb-based transition metal oxides, such as $TbVO_4$ and $TbAsO_4$, where well characterized Jahn–Teller phase transitions occur at $T_C =$ 33.1 and 28 K, respectively [51]. The Jahn–Teller transitions in these materials take them from tetagonal to orthorhombic structures, so the analogy with $Tb_2Ti_2O_7$ is

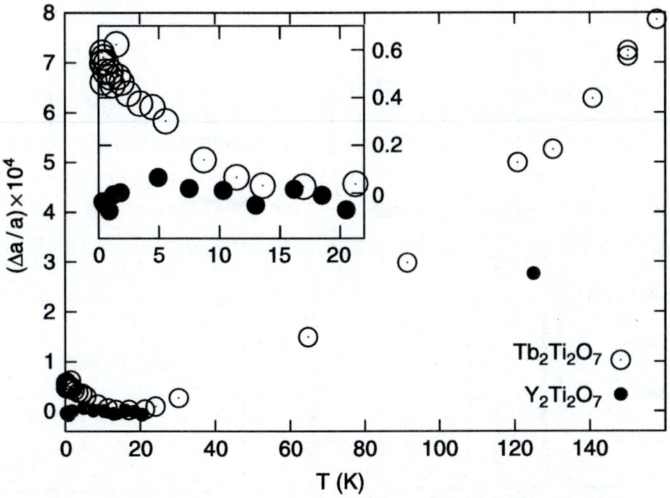


**Fig. 8.15** The temperature dependence of the cubic lattice parameter in $Tb_2Ti_2O_7$ is shown and compared to that measured in non-magnetic $Y_2Ti_2O_7$. $Tb_2Ti_2O_7$ exhibits anomolous thermal expansion on cooling into the spin liquid state below $\sim$ 20 K [50]

incomplete. Nevertheless, Bragg reflections exist for $TbVO_4$ and $TbAsO_4$ which split in either the longitudinal or transverse directions. We can imagine a similar scenario for $Tb_2Ti_2O_7$, although no phase transition occurs. Thus, the anisotropic broadening of the Bragg scattering can be interpreted as being due to fluctuations above a Jahn–Teller-like phase transition.

## 8.6 Magnetic Order and Fluctuations in $Tb_2Sn_2O_7$

The pyrochlore material $Tb_2Sn_2O_7$ possesses the same crystal structure as $Tb_2Ti_2O_7$. Once again, the ion at the B-site, $Sn^{4+}$ in this case, is nonmagnetic, leaving the magnetic $Tb^{3+}$ ions decorating the pyrochlore lattice. However, in contrast to the spin liquid state which $Tb_2Ti_2O_7$ displays at low temperatures, $Tb_2Sn_2O_7$ has been shown to exhibit a form of magnetic order, albeit a complicated one at sufficiently low temperatures. Bulk measurements revealed a single phase transition at 0.87 K [52], which was confirmed by neutron diffraction and heat capacity measurements. These neutron measurements further determined that large ferromagnetic spin clusters, with moments canted off from the local <111> axes, described the Bragg diffraction data well, see Fig. 8.16b [53, 54]. This "ordered spin-ice" phase has similarities to that seen in $Ho_2Ru_2O_7$ [55, 56]. Mirebeau et al. [53, 57] noted a difference in the magnitude of the ordered moment deduced from the magnetic neutron refinement and heat capacity measurements, and that consistent with the intense diffuse magnetic scattering which they observed. This led them to conclude

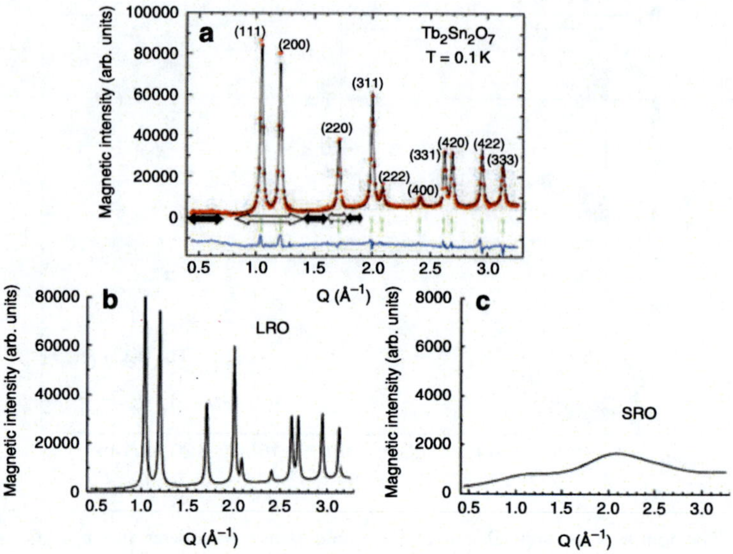


**Fig. 8.16** Magnetic diffraction from $Tb_2Sn_2O_7$ at 100 mK. (**a**) The full spectrum with a 1.2 K data set subtracted to account for the nuclear scattering. The *solid line* is a fit to the data accounting for the short and long range order. The long and short range contributions to the pattern are plotted separately in (**b**) and (**c**) respectively [53]

that the ordered state in $Tb_2Sn_2O_7$ is complicated and coexists with slow, collective fluctuations on the time scale of $10^{-4}$ to $10^{-8}$ s down to mK temperatures. Such slow collective spin fluctuations would be responsible for the broad, diffuse magnetic scattering seen in Fig. 8.16c.

Muon spin relaxation experiments [54, 58] on $Tb_2Sn_2O_7$ in this ground state failed to observe the hallmark of static magnetic order, oscillations in the muon asymmetry in zero longitudinal field conditions, ruling out static magnetic order on the appropriate muon time scale, microseconds. Instead characteristic relaxation due to persistent spin dynamics on the time scale of between $8 \times 10^{-11}$ s and $5 \times 10^{-9}$ s, depending on detailed assumptions in both measurements, was observed. Later neutron scattering measurements by Rule et al. [59] showed the presence of co-existing static and dynamic spins in $Tb_2Sn_2O_7$ at the lowest temperatures, while very high energy resolution neutron spin echo measurements [60] at low momentum transfer (0.08 Å$^{-1}$) confirmed the existence of a dynamical component in this ground state. In this study, the signal below 660 mK, is static out to 10 ns but data above this temperature clearly relaxes in the spin echo window. This is the clearest evidence to-date of the slowing down of spin dynamics and their eventual freezing in this system. Recently, Giblin et al. [61] measured the the muon spectra with the muons stopped outside the sample, and performed ultralow field bulk magnetization measurements. The field cooled state shows conventional static order, while the zero field cooled state may be interpreted in terms of conventional magnetic

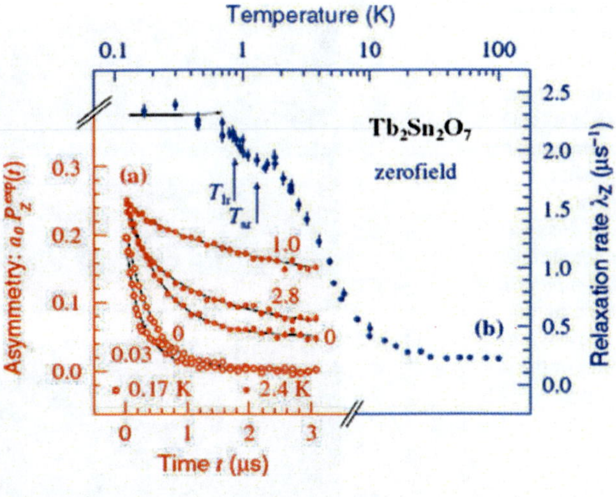


**Fig. 8.17** (a) Muon spectra at 0.17 and 2.4 K in $Tb_2Sn_2O_7$ are shown, with the latter ones vertically shifted by 0.05 for clarity. The numbers written next to each spectrum correspond to the longitudinal magnetic field, expressed in Tesla. (b) The temperature dependence of the spin–lattice relaxation rate. Evidence of short and long range order is indicated at 1.3 K and 0.87 K [54]

domains. These results rule out a purely dynamical ground state suggested by the earlier muon experiments.

The question of why the ground state in $Tb_2Sn_2O_7$ is different from that observed in $Tb_2Ti_2O_7$ remains outstanding. In order to shed light on this, new inelastic neutron scattering measurements were carried out on polycrystalline $Tb_2Ti_2O_7$and $Tb_2Sn_2O_7$ [59, 62, 63]. The results of these studies are consistent with the much earlier inelastic neutron data data collected on $Tb_2Ti_2O_7$ the spin liquid [30, 36, 62, 64, 65]. The low energy inelastic magnetic scattering from both compounds is dominated by crystalline electric field-like spin excitations. The lowest of these in energy lies approximately 18 K (1.5 meV) above the ground state in $Tb_2Sn_2O_7$ and, like $Tb_2Ti_2O_7$, it partially softens below 20 K. The next set of crystal field excitations are above 10 meV.

On closer inspection, important differences between these crystal field excitations in $Tb_2Sn_2O_7$ and $Tb_2Ti_2O_7$ become clear. Detailed modelling of these spin excitations shows that the wavefunctions associated with the ground state doublet and the first excited state doublet are inverted in $Tb_2Sn_2O_7$ compared with $Tb_2Ti_2O_7$. The temperature dependence of these spin excitations show related differences. While the size of the gap between these two doublets is similar, as the temperature is lowered towards the ordered state of $Tb_2Sn_2O_7$, a low lying collective spin excitation builds out of the quasielastic scattering, presumably due to the buildup of static, ferromagnetically coupled spins within $Tb_2Sn_2O_7$'s ground state (Figs. 8.17 and 8.18) [59, 62].

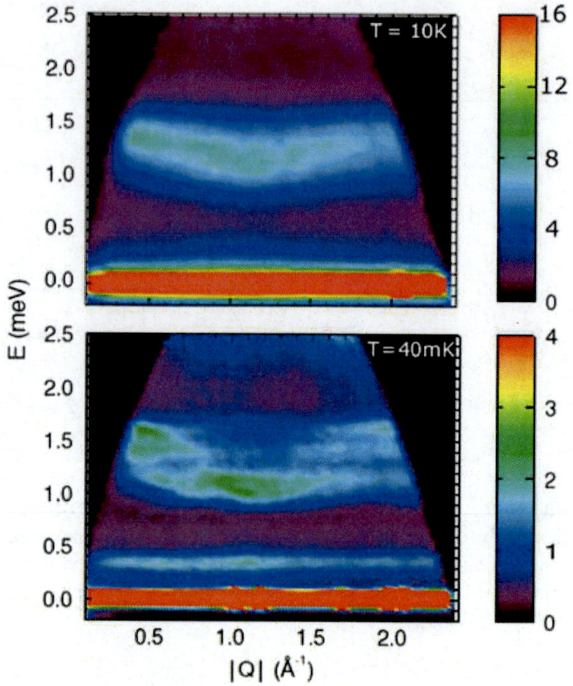


**Fig. 8.18** Inelastic neutron spectra taken at 10 K and 0.04 K in $Tb_2Sn_2O_7$ is shown. At both temperatures the ≈1.4 meV mode softens at the first antiferromagnetic correlation wave vector. In the high temperature paramagnetic state, this is a Q-independent mode. In the ordered state (*bottom panel*), structure can be seen in this broad mode and a low lying excitation grows out of the quasielastic scattering

### *8.6.1 Phase Transitions and Fluctuations in $Gd_2Ti_2O_7$ and $Gd_2Sn_2O_7$*

Antiferromagnetically coupled Heisenberg spins on a pyrochlore lattice are known to give rise to a highly frustrated ground state which is macroscopically degenerate. Calculations for classical [66] and quantum spins [67] show that such a system should remain in a spin liquid state with short-range spin correlations at any non-zero temperature [68]. The addition of dipolar interactions in the presence of antiferromagnetic Heisenberg exchange on the pyrochlore lattice was shown by Palmer and Chalker [69] to stabilize a four-sublattice ordered state with an ordering vector $\mathbf{k} = 0$. $Gd_2Ti_2O_7$ and $Gd_2Sn_2O_7$, with the spin only, $S = 7/2$ $Gd^{3+}$ ion decoraring the pyrochlore lattice, should provide excellent natural venues for such a Heisenberg-coupled spin system on the pyrochlore lattice with significant dipolar interactions. The large moments at the $Gd^{3+}$ sites guarantees that the dipolar interactions will be relatively strong, while the exchange interactions in these insulating 4f systems are relatively weak.

The bulk magnetic susceptibility for both $Gd_2Ti_2O_7$ and $Gd_2Sn_2O_7$ follows a Curie–Weiss law, with an effective moment close to the theoretical value of 7.94 $\mu_B$ appropriate to the free ion [68, 70] and a Curie–Weiss temperature of −10 K. Both compounds, however, enter magnetically ordered states at ≈ 1 K.

Specific heat and ac susceptibility measurements showed that $Gd_2Ti_2O_7$ entered an ordered phase at 0.97 K [68]. Later specific heat studies taken to lower temperatures showed there are two transitions in zero applied field, at 0.97 and 0.7 K [71] (see Fig. 8.19). Ramirez et al. also observed additional phase transitions, induced in applied magnetic fields, which were later confirmed by single crystal specific heat studies by Petrenko et al. [72]. Figure 8.20 shows the temperature-field phase diagram obtained from a series of low temperature specific heat measurements on single crystals of $Gd_2Ti_2O_7$ for three different directions of an applied magnetic field. The measurements reveal an unexpected anisotropy for $Gd_2Ti_2O_7$ in a magnetic field, in light the anticipated absence of a significant single-ion anisotropy for $Gd^{3+}$.

The nature of these ordered states was difficult to determine using neutron diffraction due to the high absorption cross section of gadolinium. However, a

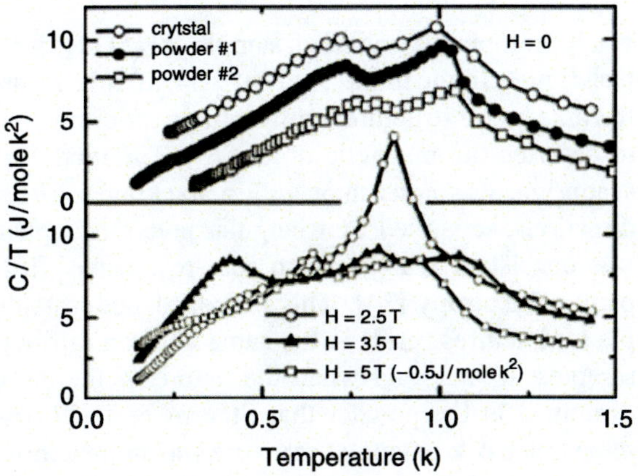


**Fig. 8.19** The temperature dependance of the specific heat divided by temperature of $Gd_2Ti_2O_7$ under various applied magnetic fields. In zero field, three different samples revealed two phase transitions that are magnetic in origin. The *bottom panel* shows how these transitions behave in moderate applied magnetic fields [71]

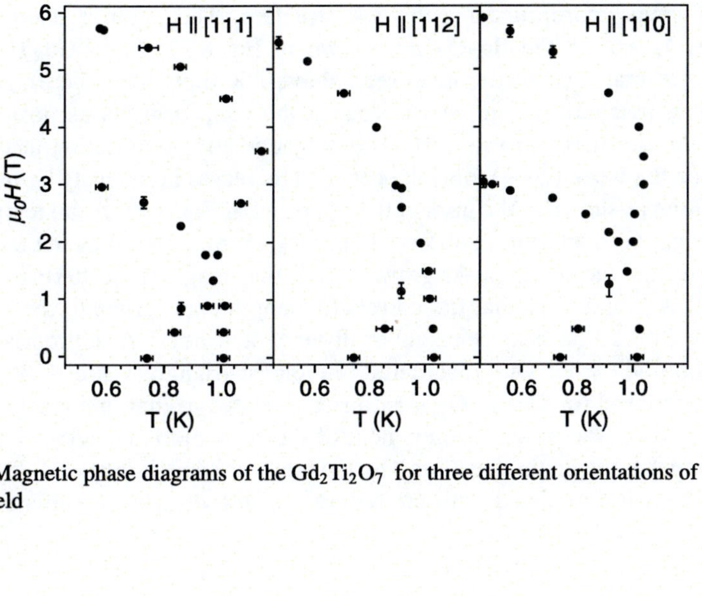


**Fig. 8.20** Magnetic phase diagrams of the $Gd_2Ti_2O_7$ for three different orientations of an applied magnetic field

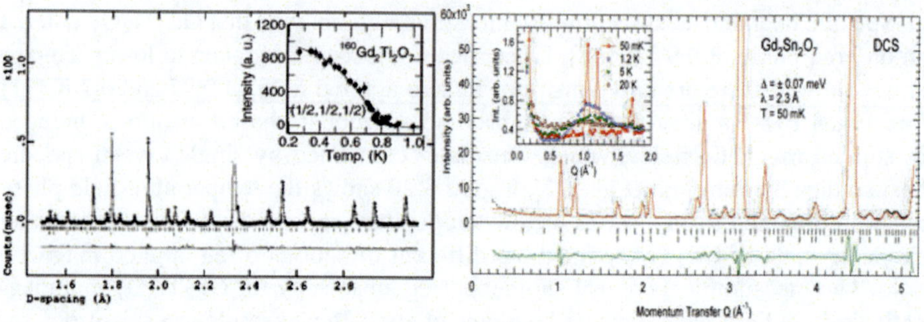


**Fig. 8.21** *Left*: Diffraction from $Gd_2Ti_2O_7$ by Champion et al. [73] with the inset from Stewart et al. [75]. *Right*: Neutron scattering from $Gd_2Sn_2O_7$ from Stewart et al. [81]. Inset shows the broad diffuse scattering centred around $Q = 1.05$ Å$^{-1}$ that goes away as the temperature is lowered through the magnetic ordering temperature [81]

small, isotopically enriched, sample of $^{160}Gd_2Ti_2O_7$ has been studied using neutron techniques. In the initial study at $T = 50$ mK, a partially ordered, noncollinear antiferromagnetic structure, with a single ordering wavevector $\mathbf{k} = [\frac{1}{2}\frac{1}{2}\frac{1}{2}]$ was thought to describe the magnetic order [73]. The model used to understand this magnetic scattering, was consistent with a stacking of ordered kagomé planes along [111] directions, separated by triangular layers occupied by disordered $Gd^{3+}$ spins and was consistent with electron spin resonance (ESR) results that saw large global planar anisotropy [74]. This proposed 1-**k** structure and more complicated multiple **k** structures result in the same neutron diffraction pattern from polycrystalline samples. It was only when the nature of the the diffuse magnetic scattering was examined in the powder that Stewart et al. [75] were able to conclusively determine that a 4-**k** magnetic structure was more appropriate. Later, Sosin et al. [76,77] repeated the ESR measurements and found only a local anisotropy and no ordered planes of spins appropriate for the 1-**k** structure. The neutron scattering from $^{160}Gd_2Ti_2O_7$ and $^{160}Gd_2Sn_2O_7$ are shown in Fig. 8.21. In $^{160}Gd_2Ti_2O_7$, very broad diffuse magnetic scattering is seen above 1 K (not shown here). Below this temperature, magnetic Bragg peaks occur at the reciprocal lattice positions indicated by the tic marks in Fig. 8.21. The magnetic Bragg reflection at the lowest wavevector, the weak $(\frac{1}{2}\frac{1}{2}\frac{1}{2})$, only appears and increases in intensity below the second transition as shown in the inset of the left panel of Fig. 8.21. At all temperatures, diffuse magnetic scattering is observed, although it diminishes as the temperature is decreased and the Bragg peaks grow. This diffuse magnetic scattering, centred at $|\mathbf{Q}| \approx 1.2$ Å$^{-1}$ indicates that the correlation length appropriate to the disordered spins is $\approx 3.5$ Å, the nearest neighbor distance and not 7 Å appropriate for the model proposed earlier with disordered interplanar triangular sites [73]. The new structure proposed for $Gd_2Ti_2O_7$ is intimately related to that previously proposed by [73]; however the ordered component of the structure is described by a more-complicated 4-**k** magnetic structure. The lower phase transition at $\sim 0.75$ K in zero magnetic field is associated with the ordering of the disordered sublattice within the 4-**k** magnetic structure. Neither ordered state is that proposed by Palmer and Chalker [69]. The presence of two distinct, coexisting subsets of Gd moments on the pyrochlore lattice is consistent with $^{155}$Gd Mössbauer experiments [78] but the physical origin for such a magnetic sublattice remains elusives.

The isostructural stannate, $Gd_2Sn_2O_7$, has received comparatively less attention. As mentioned above the paramagnetic properties of $Gd_2Sn_2O_7$ are similar to $Gd_2Ti_2O_7$ [79]. However, only a single phase transition, at 1.0 K, is observed on cooling and it involves the entire spin system. Recent neutron diffraction measurements by two groups [80, 81] confirmed that $Gd_2Sn_2O_7$ possesses only one phase transition down to the lowest temperatures and that the ground state is indeed that predicted theoretically by Palmer and Chalker [69], see Fig. 8.21. Wills [80] has proposed that a specific type of third-neighbour superexchange interaction was responsible for the differing ground states in the these two Gd-based pyrochlore materials. These studies highlight the extreme sensitivity of the ground states to small changes in the microscopic properties of these highly frustrated materials. It is also worth noting that, the broad diffuse magnetic scattering seen in $^{160}Gd_2Ti_2O_7$ at all temperatures below 10 K, consistent with a complicated coexistence of ordered and disordered $Gd^{3+}$ moments, is absent below the phase transition in $Gd_2Sn_2O_7$ [81].

Once again, to further understand the magnetic ground states and perhaps definitively identify the cause underlying the differences between $Gd_2Ti_2O_7$ and $Gd_2Sn_2O_7$ one can profitably investigate the spin excitation spectrum of these compounds. Specific heat is a bulk probe that often provides reliable insight into the low lying excitations. The specific heat, $C_v$, of these two compounds just below the transition temperature is described by an anomalous power law, $C_v \propto T^2$, associated with an unusual energy-dependence of the density of magnetic states [68, 71, 78]. Electron spin resonance [74, 76, 77], Mössbauer [82], muon spin relaxation [83, 84] and neutron spin echo [85] have been performed to probe this region of frequency space. In many of these measurements, spin fluctuations were interpreted to exist significantly below the ordering temperature of both compounds, consistent with the unusual energy-dependence of the density of states seen in the specific heat measurements but inconsistent with the recent neutron work by Stewart et al. [81] on $Gd_2Sn_2O_7$. Sosin et al.[76, 77] found a small gap $\approx 1$ K in the spin excitation spectrum which coexists with the paramagnetic signal in the $Gd_2Ti_2O_7$, but not in the $Gd_2Sn_2O_7$ sample, again consistent with the recent neutron scattering work mentioned above. Very recently the specific heat of $Gd_2Sn_2O_7$ was measured by Quillium et al. [86] to much lower temperatures than previously attained, and it was found that the proportionality of the heat capacity to $T^2$ did *not* hold below 400 mK, see Fig. 8.22. Rather $C_v$ was observed to decrease exponentially below 350 mK. This is very strong evidence for a gapped spin-wave spectrum in $Gd_2Sn_2O_7$. Stewart et al. [81] neutron scattering data on $Gd_2Sn_2O_7$ is consistent with such a model (see Fig. 8.23). Indeed, this neutron scattering data is well described by linear spin wave calculations performed on the Palmer-Chalker model of spins [87].

Our current understanding of these Heisenberg-like antiferromagnets on the pyrochlore lattice are at a tantalizing stage. The $Gd_2Sn_2O_7$ ground state appears

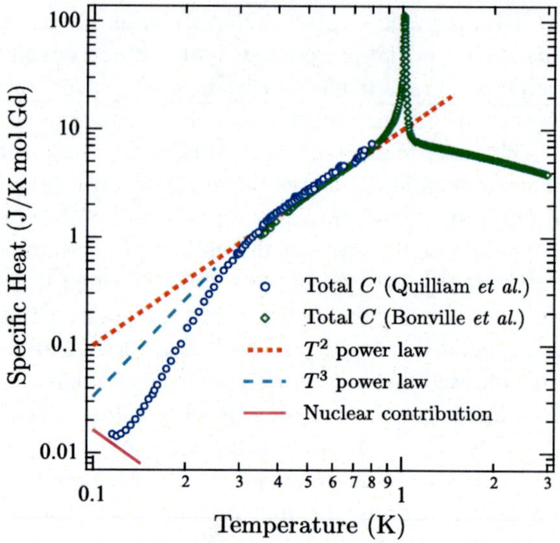


**Fig. 8.22** Specific heat of $Gd_2Sn_2O_7$ as a function of temperature from Quilliam et al. [86] (*blue circles*) between 115 and 800 mK. Previously measured data at higher temperatures (*green diamonds*) [78], the $T^2$ power law previously proposed (*dotted red line*) and a $T^3$ power law (*dashed blue line*) are also plotted. The upturn seen below 150 mK results from the nuclear electric quadrupole interaction

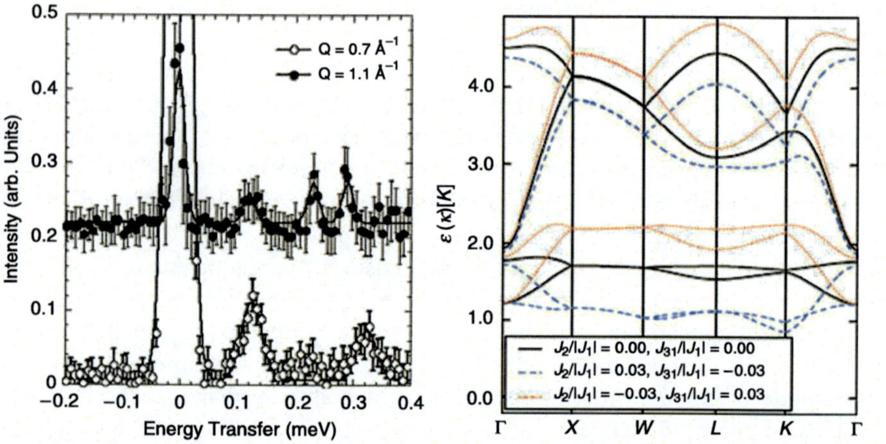


**Fig. 8.23** High energy resolution inelastic neutron scattering from $Gd_2Sn_2O_7$ at 50 mK. For clarity the spectra have been shifted vertically. A gap of 0.13 meV is seen in both data sets along with higher energy transfer modes. Right, the calculated spin wave excitation spectrum in K for $J_2/|J_1|$, $J_{31}/|J_1|$ equal to (0.0,0.0), (0.03, −0.03), and (−0.03, 0.03) plotted along a high-symmetry path in the first Brillouin zone of the fcc lattice. For the three parameter sets shown there exists a finite gap to spin wave excitations throughout the zone

to be well described by the Palmer and Chalker model [69] while that of $Gd_2Ti_2O_7$ is somehow perturbed away from this model. Understanding the ubiquitous low temperature spin dynamics in a relatively simple model magnet, such as those represented by Gd-pyrochlores will be very helpful in understanding the nature of the low temperature spin dynamics in a variety of other frustrated magnets [87].

## 8.7 Conclusions

The low temperature properties of several antiferromagnetic pyrochlores were reviewed, with a particular emphasis on how scattering techniques have informed on the enigmatic spin liquid and ordered states, and their sensitivity to relatively modest external perturbations. While much has been discovered in the recent past, many new questions have been opened up. The role of the lattice, for example, on which the frustrated spins reside has arisen as a relatively important one in the field of highly frustrated magnetism in general, and one can fairly safely predict that it will continue to play an important role in the near future.

Advances in experimental infrastructure are expected to impact geometrically frustrated magnetism in a very significant manner in the near term. Advances in spallation-based neutron sources, particularly at the Spallation Neutron Source at Oak Ridge, J-PARC in Japan, and the Second Target Station at ISIS in the UK will usher in a new generation of neutron instrumentation with vastly superior figures-of-merit for performance. Highly frustrated magnetism, with magnetic scattering distributed over large fractions of the Brillouin zone are very well positioned to benefit from this new instrumentation. Technical advances are also coming into place with regard to our ability to expose samples to extreme magnetic fields while maintaining compatibility with scattering experiments. These should allow more sophisticated scattering experiments, both neutron and synchrotoron X-ray, to be carried out with magnetic fields much larger than those presently available. All of these suggest that our ability to carry out sophisticated new scattering experiments which can provide definitive insights into the structure and dynamics of new materials, and in particular into new magnetic materials with exotic ground states, will be much increased in the near future. For these reasons, there is every reason to be optimistic for significant advances in the experimental study of highly frustrated magnets in the near term.

**Acknowledgements**

This work was supported at McMaster University by NSERC of Canada and by the Canadian Institute for Advanced Research. All of our work in frustrated magnetism has been performed in collaboration with a large number of talented scientists and it is a pleasure to acknowledge their contributions. These include C.P. Adams, A.J. Berlinsky, S.T. Bramwell, C.L. Broholm, W.J.L. Buyers, J.P.Castellan, J.P. Clancy, M.F. Collins, J.R.D. Copley, H.A. Dabkowska, P. Deen, S.R. Dunsiger, G. Ehlers, T. Fennell, M.J.P. Gingras, J.E. Greedan, H. Kawamura, R.F. Keifl, S.-H. Lee, M.J. Lewis, G.M. Luke, T.E. Mason, M. Mao, I.Mirebeau, R. Moessner, Y. Qiu, N.Raju, M. Ramazanoglu, J.N. Reimers, K.A. Ross, J.P.C. Ruff, K.C. Rule, J.R. Stewart and J. van Duijn.

## References


1. For reviews see: Magnetic Systems with Competing Interactions, ed. by H.T. Diep (World Scientific, Singapore, 1994)
2. A.P. Ramirez, Ann. Rev. Math. Sci. 24, 453 (1994)
3. P. Shiffer, A.P. Ramirez, Commun. Cond. Math. Phys. 18, 21 (1996); Can. J. Phys. 79 (2001)
4. G.H. Wannier, Phys. Rev. 79, 357 (1950).
5. J.A. Mydosh, see for example: Spin Glasses (Taylor and Francis, London, 1995)
6. S.T. Bramwell, M.J.P. Gingras, see for example: Science, 294, 1495 (2001)
7. G. Lawes, M. Kenzelmann, N. Rogado, K.H. Kim, G.A. Jorge, R.J. Cava, A. Aharony, O. Entin-Wohlman, A.B. Harris, T. Yildirim, Q.Z. Huang, S. Park, C. Broholm, A.P. Ramirez, Phys. Rev. Lett. 93, 247201 (2004)
8. Y. Okamoto, M. Nohara, H. Aruga-Katori, H. Takagi, Phys. Rev. Lett. 99, 137207 (2007)
9. J.M. Hopkinson, S.V. Isakov, H.Y. Kee, Y.B. Kim, Phys. Rev. Lett. 99, 037201 (2007)
10. P. Schiffer, A.P. Ramirez, D.A. Huse, A.J. Valentino, Phys. Rev. Lett. 73, 2500 (1994)
11. S.R. Dunsiger, J.S. Gardner, J.A. Chakhalian, A.L. Cornelius, M. Jaime, R.F. Kiefl, R. Movshovich, W.A. MacFarlane, R.I. Miller, J.E. Sonier, B.D. Gaulin, Phys. Rev. Lett. 85, 3504 (2000)
12. J.S. Gardner, S.R. Dunsiger, B.D. Gaulin, M.J.P. Gingras, J.E. Greedan, R.F. Kiefl, M.D. Lumsden, W.A. MacFarlane, N.P. Raju, J.E. Sonier, I. Swainson, Z. Tun, Phys. Rev. Lett. 82, 1012 (1999)
13. I. Mirebeau, A. Apetrei, J. Rodriguez-Carvajal, P. Bonville, A. Forget, D. Colson, V. Glazkov, J.P. Sanchez, O. Isnard, E. Suard, Phys. Rev. Lett. 94, 246402 (2005)
14. P. Dalmas de Reotier, A. Yaouanc, L. Keller, A. Cervellino, B. Roessli, C. Baines, A. Forget, C. Vaju, P.C.M. Gubbens, A. Amato, P.J.C. King, Phys. Rev. Lett. 96, 127202 (2006)
15. M.J. Harris, S.T. Bramwell, D.F. McMorrow, T. Zeiske, K.W. Godfrey, Phys. Rev. Lett. 79, 2554 (1997)
16. J.S. Gardner, B.D. Gaulin, S.-H. Lee, C. Broholm, N.P. Raju, J.E. Greedan, Phys. Rev. Lett. 83, 211 (1999)
17. B.D. Gaulin, J.N. Reimers, T.E. Mason, J.E. Greedan, Z. Tun, Phys. Rev. Lett. 69, 3244 (1992)
18. D.K. Singh, J.S. Helton, S. Chu, T.H. Han, C.J. Bonnoit, S. Chang, H.J. Kang, J.W. Lynn, Y.S. Lee Phys. Rev. B 78, 0220405(R) (2008)
19. B.Z. Malkin, A.R. Zakirov, M.N. Popova, S.A. Klimin, E.P. Chukalina, E. Antic-Fidancev, P. Goldner, P. Aschehoug, G. Dhalenne, Phys. Rev. B 70, 075112 (2004)
20. J.S. Gardner, B.D. Gaulin, D.M. Paul, J. Crystal Growth 191, 740 (1998)
21. G. Balakrishnan, O.A. Petrenko, M.R. Lees, D.M. Paul, J. Phys. Condens. Matter 10, L723 (1998)
22. J.E. Greedan, M. Sato, N. Ali, W.R. Datars, J. Solid State Chem. 68, 300 (1987)
23. T. Katsufuji, H.Y. Hwang, S.W. Cheong, Phys. Rev. Lett. 85, 1998 (2000)
24. J.P. Castellan, B.D. Gaulin, J. van Duijn, M.J. Lewis, M.D. Lumsden, R. Jin, J. He, S.E. Nagler, D. Mandrus, Phys. Rev. B 66, 134528 (2002)
25. S. Yonezawa, Y. Muraoka, Y. Matsushita, Z. Hiroi, J. Phys.: Condens. Matter 16, L9 (2004)
26. S. Yonezawa, Y. Muraoka, Y. Matsushita, Z. Hiroi, J. Phys. Soc. Jpn. 73, 819 (2004)
27. K.A. Ross, J.P.C. Ruff, C.P. Adams, J.S. Gardner, H.A. Dabkowska, Y. Qiu, J.R.D. Copley, B.D. Gaulin, Phys. Rev. Lett. 103, 227202 (2009)
28. C.R. Wiebe, P.L. Russo, A.T. Savici, Y.J. Uemura, G.J. MacDougall, G.M. Luke, S. Kuchta, J.E. Greedan, J. Phys.: Condens. Matter 17, 6469 (2005)
29. J.P. Clancy, J.P.C. Ruff, S.R. Dunsiger, Y. Zhao, H.A. Dabkowska, J.S. Gardner, Y. Qiu, J.R.D. Copley, T. Jenkins, B.D. Gaulin, Phys. Rev. B 79, 014408 (2009)
30. K.C. Rule, J.P.C. Ruff, B.D. Gaulin, S.R. Dunsiger, J.S. Gardner, M.J. Lewis, H.A. Dabkowska, I. Mirebeau, P. Manual, Y. Qiu, J.R.D. Copley, Phys. Rev. Lett. 96, 177201 (2006)
31. M.J.P. Gingras, B.C. den Hertog, M. Faucher, J.S. Gardner, S.R. Dunsiger, L.J. Chang, B.D. Gaulin, N.P. Raju, J.E. Greedan, Phys. Rev. B 62, 6496 (2000)
32. I. Mirebeau, P. Bonville, M. Hennion, Phys. Rev. B 76, 184436 (2007)

33. J.S. Gardner, B.D. Gaulin, A.J. Berlinsky, P. Waldron, S.R. Dunsiger, N.P. Raju, J.E. Greedan, Phys. Rev. B. 64, 224416 (2001)
34. Y.-J. Kao, M. Enjalran, A.D. Maestro, H.R. Molavian, M.J.P. Gingras, Phys. Rev. B 68, 172407 (2003)
35. Y. Yasui, M. Kanada, M. Ito, H. Harashina, M. Sato, H. Okumura, K. Kakurai, H. Kadowaki, J. Jap. Phys. Soc. 71, 599 (2002)
36. J.S. Gardner, A. Keren, G. Ehlers, C. Stock, E. Segal, J.M. Roper, B. Fåk, P.R. Hammar, M.B. Stone, D.H. Reich, B.D. Gaulin, Phys. Rev. B 68, 134410 (2003)
37. B.C. den Hertog, M.J.P. Gingras, Phys. Rev. Lett. 84, 3430 (2000)
38. I. Mirebeau, I.N. Goncharenko, P. Cadavez-Peres, S.T. Bramwell, M.J.P. Gingras, J.S. Gardner, Nature 420, 54 (2002)
39. I. Mirebeau, I.N. Goncharenko, J. Phys.: Condens. Matter 16, S653 (2004)
40. I. Mirebeau, I.N. Gonchararenko, G. Dhalenne, A. Revcolevschi, Phys. Rev. Lett. 93, 187204 (2004)
41. I. Mirebeau, I.N. Goncharenko, J. Phys.: Condens. Matter 17, S771 (2005)
42. S.H. Lee, C. Broholm, T.H. Kim, W. Ratcliff II, S.W. Cheong, Phys. Rev. Lett. 84, 3718 (2000)
43. S.H. Lee, C. Broholm, W. Ratcliff II, G. Gasparovic, Q. Huang, T.H. Kim, S.W. Cheong, Nature, 418, 856 (2002)
44. H. Ueda, H.A. Katori, H. Mitamura, T. Goto, H. Takagi, Phys. Rev. Lett. 94, 047202 (2005)
45. M. Matsuda et al., Nat. Phys. 3, 397 (2007)
46. O. Tchernyshyov, R. Moessner, S.L. Sondhi, Phys. Rev. B, 66, 064403 (2002)
47. K. Penc, N. Shannon, Y. Motome, H. Shiba, J. Phys.: Condens. Matter 19, 145267 (2007)
48. I.V. Aleksandrov et al., JETP 62, 1287 (1985)
49. L.G. Mamsurova, K.G. Pigal'skii, K.K. Pukhov, JEPT Lett. 43, 755 (1986)
50. J.P.C. Ruff, B.D. Gaulin, J.P. Castellan, K.C. Rule, J.P. Clancy, J. Rodriguez, H.A. Dabkowska, Phys. Rev. Lett. 99, 237202 (2007)
51. K.C. Rule, M.J. Lewis, H.A. Dabkowska, D.R. Taylor, B.D. Gaulin, Phys. Rev. B 77, 134116 (2008)
52. K. Matsuhira, Y. Hinatsu, K. Tenya, H. Amitsuka, T. Sakakibara, J. Phys. Soc. Jpn. 71, 1576 (2002)
53. I. Mirebeau, A. Apetrei, J. Rodriguez-Carvajal, P. Bonville, A. Forget, D. Colson, V. Glazkov, J.P. Sanchez, O. Isnard, E. Suard, Phys. Rev. Lett. 94, 246402 (2005)
54. P.D. de Reotier, A. Yaouanc, L. Keller, A. Cervellino, B. Roessli, C. Baines, A. Forget, C. Vaju, P.C.M. Gubbens, A. Amato, P.J.C. King, Phys. Rev. Lett. 96, 127202 (2006)
55. C.R. Wiebe, J.S. Gardner, S.J. Kim, G.M. Luke, A.S. Wills, B.D. Gaulin, J.E. Greedan, I. Swainson, Y. Qiu, C.Y. Jones, Phys. Rev. Lett. 93, 076403 (2004)
56. J.S. Gardner, A.L. Cornelius, L.J. Chang, M. Prager, Th. Brckel, G. Ehlers, J. Phys.: Condens. Matter 17, 7089 (2005)
57. I. Mirebeau, A. Apetrei, I.N. Goncharenko, R. Moessner, Physica B 385–386, 307 (2006)
58. F. Bert, P. Mendels, A. Olariu, N. Blanchard, G. Collin, A. Amato, C. Baines, A.D. Hillier, Phys. Rev. Lett. 97, 117203 (2006)
59. K.C. Rule, G. Ehlers, J.R. Stewart, A.L. Cornelius, P.P. Deen, Y. Qiu, C.R. Wiebe, J.A. Janik, H.D. Zhou, D. Antonio, B.W. Woytko, J.P. Ruff, H.A. Dabkowska, B.D. Gaulin, J.S. Gardner, Phys. Rev. B. 76, 212405 (2007)
60. Y. Chapuis, A. Yaouanc, P. Dalmas de Reotier, S. Pouget, P. Fouquet, A. Cervellino, A. Forget, J. Phys.: Condens. Matter 19, 446206 (2007)
61. S.R. Giblin, J.D. Champion, H.D. Zhou, C.R. Wiebe, J.S. Gardner, I. Terry, S. Calder, T. Fennell, S.T. Bramwell, Phys. Rev. Lett. 101, 237201 (2008)
62. I. Mirebeau, P. Bonville, M. Hennion, Phys. Rev. B 76, 184436 (2007)
63. I. Mirebeau, H. Mutka, P. Bonville, A. Apetrei, A. Forget, Phys. Rev. B 78, 174416 (2008)
64. J.S. Gardner, S.R. Dunsiger, B.D. Gaulin, M.J.P. Gingras, J.E. Greedan, R.F. Kiefl, M.D. Lumsden, W.A. MacFarlane, N.P. Raju, J.E. Sonier, I. Swainson, Z. Tun, Phys. Rev. Lett. 82, 1012 (1999)
65. J.S. Gardner, B.D. Gaulin, A.J. Berlinsky, P. Waldron, S.R. Dunsiger, N.P. Raju, J.E. Greedan, Phys. Rev. B 64, 224416 (2001)

66. R. Moessner, J.T. Chalker, Phys. Rev. B 58, 12049 (1998)
67. B. Canals, C. Lacroix, Phys. Rev. B 61, 1149 (2000)
68. N.P. Raju, M. Dion, M.J.P. Gingras, T.E. Mason, J.E. Greedan, Phys. Rev. B 59, 14489 (1999)
69. S.E. Palmer, J.T. Chalker, Phys. Rev. B 62, 488 (2000)
70. V. Bondah-Jagalu, S.T. Bramwell, Can. J. Phys. 79, 1381 (2001)
71. A.P. Ramirez, B.S. Shastry, A. Hayashi, J.J. Krajewski, D.A. Huse, R.J. Cava, Phys. Rev. Lett. 89, 067202 (2002)
72. O.A. Petrenko, M.R. Lees, G. Balakrishnan, D. McK Paul, Phys. Rev. B 70, 012402 (2004)
73. J.D.M. Champion, A.S. Wills, T. Fennell, S.T. Bramwell, J.S. Gardner, M.A. Green, Phys. Rev. B 64, 140407 (2001)
74. A.K. Hassan, L.P. Levy, C. Darie, P. Strobel, Phys. Rev. B 67, 214432 (2003)
75. J.R. Stewart, G. Ehlers, A.S. Wills, S.T. Bramwell, J.S. Gardner, J. Phys.: Condens. Matter 16, L321 (2004)
76. S.S. Sossin, A.I. Smirnov, L.A. Prozorova, O.A. Petrenko, M.E. Zhitomirsky, J.P. Sanchez, J. Mag. Mag. Mater. 310, 1590 (2006)
77. S.S. Sossin, A.I. Smirnov, L.A. Prozorova, G. Balakrishnan, M.E. Zhitomirsky, Phys. Rev. B 73, 212402 (2006)
78. P. Bonville, J.A. Hodges, M. Ocio, J.P. Sanchez, P. Vulliet, S. Sosin, D. Braithwaite, J. Phys.: Condens. Matter 15, 7777 (2003)
79. K. Matsuhira, Y. Hinatsu, K. Tenya, H. Amitsuka, T. Sakakibara, J. Phys. Soc. Jpn. 71, 1576 (2002)
80. A.S. Wills, M.E. Zhitomirsky, B. Canals, J.P. Sanchez, P. Bonville, P. Dalmas de Réotier A. Yaouanc, J. Phys.: Condens. Matter 18, L37 (2006)
81. J.R. Stewart, J.S. Gardner, Y. Qiu, G. Ehlers, Phys. Rev. B 78, 132410 (2008)
82. E. Bertin, P. Bonville, J.P. Bouchaud, J.A. Hodges, J.P. Sanchez, P. Vulliet, Eur. Phys. J. B 27, 347 (2002)
83. P. Bonville, J.A. Hodges, E. Bertin, J.P. Bouchaud, P. Dalmas de Rotier, L.P. Regnault, H.M. Ronnow, J.P. Sanchez, S. Sosin, A. Yaouanc, Hyperfine Inter. 156, 102 (2004)
84. A. Yaouanc, P. Dalmas de Rotier, V. Glazkov, C. Marin, P. Bonville, J.A. Hodges, P.C.M. Gubbens, S. Sakarya, C. Baines, Phys. Rev. Lett. 95, 047203 (2005)
85. G. Ehlers, J. Phys.: Condens. Matter 18, R231 (2006)
86. J.A. Quilliam, K.A. Ross, A.G. Del Maestro, M.J.P. Gingras, L.R. Corruccini, J.B. Kycia, Phys. Rev. Lett. 99, 097201 (2007)
87. A.D. Maestro, M.J.P. Gingras, Phys. Rev. B 76, 064418 (2007)

# Chapter 9
# Kagomé Antiferromagnets: Materials Vs. Spin Liquid Behaviors

**Philippe Mendels and Andrew S. Wills**

**Abstract** We review the most studied families of kagomé antiferromagnets, including the jarosites, SCGO(x), volborthite and herbertsmithite. Emphasis is given to their detailed structures and the principal experiments which singled out the originality of their kagomé-based physics; deviations from ideality and relevance to the idealized kagomé cases are also discussed.

After the first proposal by Fazekas and Anderson in 1973 of a resonating valence bond ground state (RVB) in triangular systems [1], and its revival in the context of High Temperature superconductors [2], kagomé antiferromagnets have played a central role in the development of new concepts and the discovery of new materials which display a spin-liquid character. These magnets combine the effects of low dimensionality with a simple corner-sharing geometry and have served as a playground for theoretical approaches, especially in the quantum $S = 1/2$ limit.

The discovery and study of quantum frustrated materials has represented a real challenge for the experimentalists, and has even led to a "Grail quest" following the seminal studies by Ramirez et al. in 1990, of $SrCr_8Ga_4O_{19}$, a "pyrochlore slab" compound that bridges the kagomé and pyrochlore geometries [3]. This compound, the first to display persistent fluctuations at low $T$, has been considered as the archetype of the Heisenberg model on a corner sharing antiferromagnet for long and has been the subject of intensive studies. Six years later, the Jarosites joined the clan when they were shown to display the possibility of unconventional spin glass properties and interesting fluctuating behavior through $\mu$SR experiments which have played a crucial role in the detection of fluctuating states – see the Chapter by Keren and Carretta, this book. Only recently did the gap between theory and experiment start to be filled with $S = 1/2$ magnets, Volborthite ($Cu_3V_2O_7(OH)_2.2H_2O$) [6, 90], in 2001 and an apparently simple "perfect" compound, Herbertsmithite ($ZnCu_3(OH)_6Cl_2$) [4, 5], in 2005, which discovery gave a strong impulse to research on quantum kagomé lattices.

In this chapter, we present a survey of a selection of most studied model kagomé antiferromagnets. We focus on their detailed structures and their deviations from ideality, reviewing the main experiments which revealed the originality of the kagomé physics. The materials are categorized from the most magnetically ordered to the most fluctuating cases, but it should be remembered that fluctuations characteristic of the frustration remain a common feature of many kagomé antiferromagnets that show long-range magnetic order. We also emphasize the resulting impact on the low $T$ physics of the deviations away from the idealized kagomé antiferromagnet. Finally, some less studied systems are shortly presented, including emergent magnets based on the Langasite structure and organic frameworks.

## 9.1 A Short Theoretical Survey: What would be the Ideal Kagomé Antiferromagnet?

Contrary with the initial proposal by Anderson, corner sharing is a prominent ingredient for stabilizing fluctuating ground states such as, but not limited to, the RVB state [8, 9]. It should also be noted that some more connected edge-sharing lattices might also display the dynamical modes associated with liquid-type physics over a broad $T$-range [10, 11], making the search for new model frustrated magnets all the more interesting.

From a pure classical approach, the corner sharing geometry of the kagomé antiferromagnet generates a macroscopic ground state degeneracy and branches of zero energy excitations (Fig. 9.1) that prevent long range order, even at $T = 0$, and limit the magnetic correlation length to the very short values typical of a "spin liquid state". Exact diagonalization calculations of finite clusters in the $S = 1/2$ quantum limit create a picture of a fluctuating ground state with an unusual continuum of

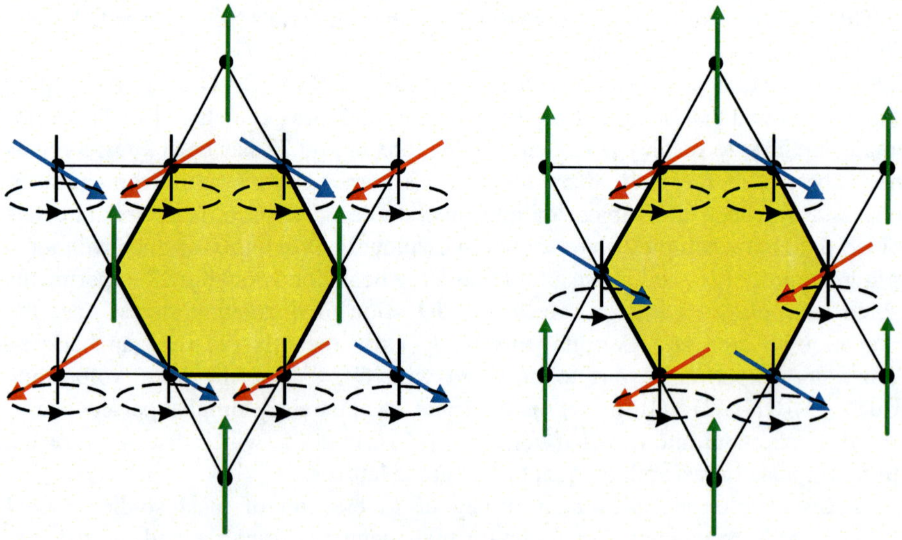

**Fig. 9.1** *Left*: $\mathbf{q} = 0$ and $\mathbf{q} = \sqrt{3} \times \sqrt{3}$ local modes on a classical kagomé lattice with neighbouring spins at 120°: simultaneous *local* rotation of spins belonging to two sublattices, around the direction of the third has no energy cost

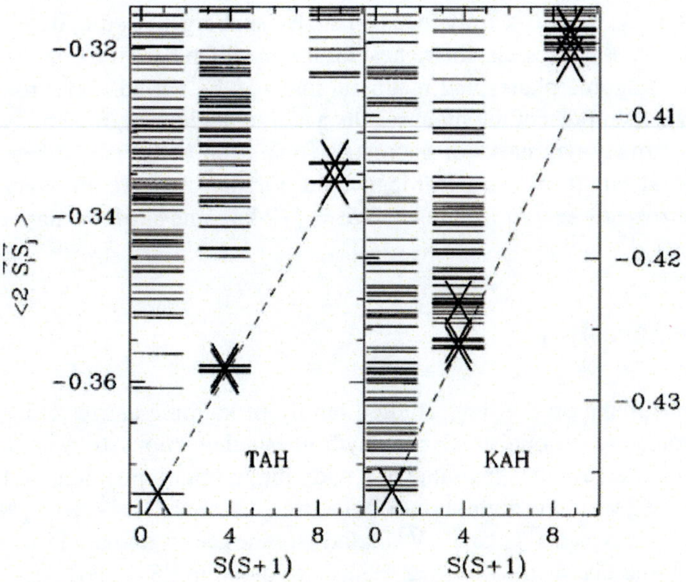


**Fig. 9.2** Energy spectrum using exact diagonalizations techniques. The striking difference between the triangular and kagomé antiferromagnets, labeled TAH and KHAF, is the continuum of excitations in the singlet channel. The gap between the first singlet and triplet states is estimated at $\Delta < J/20$, if non-zero [17]. From [8]

singlet excitations between the ground state and the first excited triplet, at variance with the triangular HAF (Fig. 9.2) [8, 9]. A remarkable result is that an effective dimer-model yields a similar energy landscape with the same entropy [12]. These are also quite strong indications that the system breaks into singlets which resonate to lower the total energy, possibly pointing at a RVB state. Quite recently, more proposals have been made for the ground state of the KHAF, which can be divided into two broad classes: valence bond crystal or resonating valence bond spin liquid. These are reviewed in the lecture notes by Lhuillier [13], to which the reader should refer for an introduction to the related theories.

From an experimental point of view, the ideal kagomé antiferromagnet should obey the following criteria:

- *Decoupled kagomé planes*, i.e. interplane coupling should be weak or, best, negligible
- *No perturbation* such as anisotropy of the interactions, long-range dipolar interactions, second neighbour interactions or dilution
- *Quantum* spins ($S = 1/2$) to increase the weight of fluctuations

In terms of anisotropy, the pure Ising kagomé case has been treated theoretically. Again the absence of order is predicted, but there are at present no experimental realizations [14, 15], except perhaps in the low-T regime of the recently studied Nd-Langasite ($Nd_3Ga_5SiO_{14}$) [16].

In this survey, we will mainly focus on the Heisenberg-like compounds. We begin our review with the jarosites, the richest family in which both the diamagnetic spacers between kagomé planes and magnetic ions can be varied. After reviewing the jarosites, we then present the main results accumulated on kagomé bilayers, which demonstrate many hallmarks of geometric frustration and serve as key references in the field of Highly Frustrated Magnetism. Finally, we give an overview of the rapidly progressing search for the ideal $S = 1/2$ kagomé antiferromagnet.

## 9.2 The Jarosites

The jarosites make up the most studied family of kagomé antiferromagnets. Synthesised from conventional and Redox hydrothermal reactions, they crystallise with a highly flexible structure that allows a wide range of compositions to be formed. They have the general formula $AB_3(SO_4)_2(OH)_6$ ($A = Na^+, K^+, Rb^+, NH_4^+, Ag^+, H_3O^+, \frac{1}{2}Pb^{2+}$; $B = Fe^{3+}, Cr^{3+}, V^{3+}$) and provide access to model frustrated magnets in both the classical, $S = 5/2$, and more quantum, $S = 3/2$, 1 limits. They have been found to display a range of both conventional long-ranged magnetic order and more exotic unconventional orderings that are summarized in Table 9.1 and will be discussed in this section.

### 9.2.1 *Synthesis and the Jarosite Crystal Structure: Idealized and Disordered*

The jarosites are a subgroup of the alunite mineral family [18]. Their crystal structure is typically described in the space group $R\bar{3}m$ [19] and is shown in Fig. 9.3. The moment bearing transition metals are coordinated by a distorted octahedron made up of equatorial hydroxide groups and axial capping by the oxygens of the $SO_4^{2-}$ units. It is the hydroxide groups that link the transition metals together to form the kagomé network and mediate exchange interactions between them. Two layers of the $SO_4^{2-}$ units are then locked together by the A-cation before the sequence is repeated to form the hexagonal ...ABC... stacking sequence.

The jarosites are prone to unstoichiometry with synthesis by traditional hydrothermal techniques [23] from acidic $Fe^{3+}$ sulfate bearing aqueous solution leading to hydronium incorporation at the A-site of the non-hydronium jarosites and loss of $Fe^{3+}$. This substitution creates a mechanism for disorder as the hydronium ions are not necessarily well defined rigid $C_{3v}$ units, but can distort depending on the local environment, perhaps even to lose a proton to some other region of the crystal structure to leave a $H_2O$ molecule on the A-site. Loss of $Fe^{3+}$ and charge-balancing through the taking up of additional protons occurs readily under normal acidic synthesis conditions. Unfortunately, the extent and the effects of the resultant site disorder and the exchange disorder are difficult to estimate and caution should be

**Table 9.1** Summary tables of the different types of orderings found in the jarosites, adapted from [22], values of the intercept of the inverse susceptibility with the temperature axis $\theta$, the critical temperature $T_C$

| Formula | $\theta$ /K | $T_C$ /K | Ordering type | Reference |
|---|---|---|---|---|
| $KFe_3(SO_4)_2(OH)_6$ | −800 | 65 | $\mathbf{k} = 00\frac{3}{2}, q = 0$ | [24] |
| $KFe_3(SO_4)_2(OH)_6$ | −663 | 64.5, 57.0 | $\mathbf{k} = 00\frac{3}{2}, q = 0$ | [25] |
| $KFe_3(SO_4)_2(OH)_6$ | −828 | 65 | $\mathbf{k} = 00\frac{3}{2}, q = 0$ | [26] |
| $KFe_3(SO_4)_2(OH)_6$ | −828 | 65 | $\mathbf{k} = 00\frac{3}{2}, q = 0$ | [27] |
| $NaFe_3(SO_4)_2(OH)_6$ | −667 | 62, 42 | $\mathbf{k} = 00\frac{3}{2}, q = 0$ | [28] |
| $NaFe_3(SO_4)_2(OH)_6$ | −825 | 61 | $\mathbf{k} = 00\frac{3}{2}, q = 0$ | [26] |
| $NaFe_3(SO_4)_2(OH)_6$ | −825 | 62 | $\mathbf{k} = 00\frac{3}{2}, q = 0$ | [27] |
| $(NH_4)Fe_3(SO_4)_2(OH)_6$ | −670 | 61.7, 57.1 | $\mathbf{k} = 00\frac{3}{2}, q = 0$ | [31] |
| $(ND_4)Fe_3(SO_4)_2(OD)_6$ | −640 | 62, 46 | $\mathbf{k} = 00\frac{3}{2}, q = 0$ | [28] |
| $(NH_4)Fe_3(SO_4)_2(OH)_6$ | −812 | 62 | $\mathbf{k} = 00\frac{3}{2}, q = 0$ | [27] |
| $TlFe_3(SO_4)_2(OH)_6$ | – | – | $\mathbf{k} = 00\frac{3}{2}, q = 0$ | [32] |
| $TlFe_3(SO_4)_2(OH)_6$ | −813 | 63 | $\mathbf{k} = 00\frac{3}{2}, q = 0$ | [26] |
| $AgFe_3(SO_4)_2(OD)_6$ | −677 | 51 | $\mathbf{k} = 00\frac{3}{2}, q = 0$ | [28] |
| $AgFe_3(SO_4)_2(OH)_6$ | −803 | 60 | $\mathbf{k} = 00\frac{3}{2}, q = 0$ | [26] |
| $RbFe_3(SO_4)_2(OD)_6$ | −688 | 47 | $\mathbf{k} = 00\frac{3}{2}, q = 0$ | [28] |
| $RbFe_3(SO_4)_2(OH)_6$ | −829 | 64 | $\mathbf{k} = 00\frac{3}{2}, q = 0$ | [27] |
| $RbFe_3(SO_4)_2(OH)_6$ | −829 | 64 | $\mathbf{k} = 00\frac{3}{2}, q = 0$ | [26] |
| $Pb_{0.5}Fe_3(SO_4)_2(OH)_6$ | – | – | unknown | [32] |
| $Pb_{0.5}Fe_3(SO_4)_2(OH)_6$ | −700 to −900 | 28, 42 | unknown | [20] |
| $Pb_{0.5}Fe_3(SO_4)_2(OH)_6$ | −832 | 56 | unknown | [26] |
| $(D_3O)Fe_3(SO_4)_2(OD)_6$ | −700 | 13.8 | Unconventional spin glass | [33–35] |
| $(D_3O)Fe_{3-x}Al_y(SO_4)_2(OD)_6$ | −720 | 41.1 | $\mathbf{k} = 00\frac{3}{2}, q = 0$ | [28] |
| $KCr_3(SO_4)_2(OH)_6$ | – | 4 | $\mathbf{k} = 000, q = 0$ | [37] |
| $KCr_3(SO_4)_2(OH)_6$ | −70 | 1.8 | $\mathbf{k} = 000, q = 0$ | [38] |
| $KCr_3(SO_4)_2(OH)_6$ | −54 | 1.5 | $\mathbf{k} = 000, q = 0$ | [38] |
| $(H_3O)Cr_3(SO_4)_2(OD)_6$ | −78 | 1.2 | Unknown | [22,36] |
| $(H_3O)V_3(SO_4)_2(OD)_6$ | +45 | 21 | $\mathbf{k} = 00\frac{3}{2}, q = 0$ | [39] |
| $NaV_3(SO_4)_2(OD)_6$ | +45 | 21 | $\mathbf{k} = 00\frac{3}{2}, q = 0$ | [22] |
| $NaV_3(SO_4)_2(OH)_6$ | +52 | 33 | $\mathbf{k} = 00\frac{3}{2}, q = 0$ | [40] |
| $KV_3(SO_4)_2(OD)_6$ | +45 | 21 | $\mathbf{k} = 00\frac{3}{2}, q = 0$ | [22] |
| $KV_3(SO_4)_2(OH)_6$ | +52 | 50 | $\mathbf{k} = 00\frac{3}{2}, q = 0$ | [41] |

The values of the Weiss temperature for the Fe-jarosites are difficult to determine accurately as the strong magnetic interactions lead to deviation from the Curie–Weiss law.

exercised over the analysis and discussion of specific magnetic properties. Part of this difficulty arises from the question over whether the rogue protons within the crystal structure localise on the oxygens of the $SO_4^{2-}$ bilayer or on the bridging hydroxide groups [43] that connect the magnetic $B^{3+}$ ions. Evidence is difficult

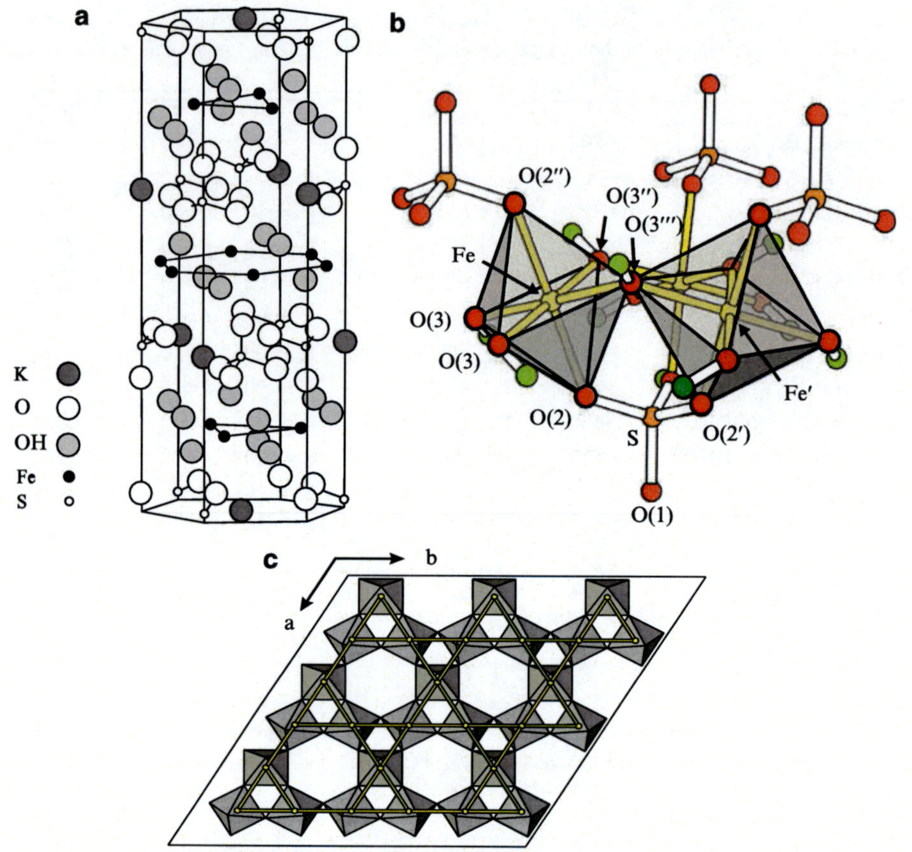


**Fig. 9.3** (**a**) The crystal structure of the majority of the jarosites is described in the space group $R\bar{3}m$ [19] exemplified here for $KFe_3(SO_4)_2(OH)_6$. An exception to this is $Pb_{0.5}Fe_3(SO_4)_2(OH)_6$, where segregation of the $Pb^{2+}$ ions may occur depending on the synthesis conditions leading to a doubling of the unit cell along the *c*-direction [20, 21]. (**b**) and (**c**) The local coordination of the Fe octahedra and their coupling to make up the kagomé network. From [22]

to interpret and frequently inconclusive, but it is clear that detailed aspects of the crystal structure such as these will play an important role in the magnetism of these frustrated magnets.

A synthetic breakthrough was the development of an alternate route based on the oxidation of bulk iron, rather than aqueous solutions of $Fe^{3+}$. This allowed the synthesis of several Fe-jarosite samples with very little reported substitution of $H_3O^+$ for $K^+$ and high coverages of the Fe-site [26, 44]. The reaction proceeds via an $Fe^{2+}$ intermediate, Szomolnokite $FeSO_4 \cdot H_2O$ [45]:

$$Fe + H_2SO_4 + \frac{1}{2}O_2 + H_2O \rightarrow Fe(SO_4) \cdot H_2O + H_2O \tag{9.1}$$

$$6Fe(SO_4) \cdot H_2O + K_2SO_4 + 6H_2O \rightarrow 2KFe_3(SO_4)_2(OH)_6 + 3H_2 + 3H_2SO_4 \tag{9.2}$$

Subsequent oxidation of the $Fe^{2+}$ to $Fe^{3+}$ in the Szomolnokite precipitate causes hydrolysis of the $H_2O$ molecules, drawing the OH units closer to the $Fe^{3+}$ centres, and to allowing bridging of the $Fe^{3+}$ octahedrons to make up the kagomé plane.

This Redox reaction mechanism is not followed by hydronium jarosite, as the high hydration number of the $H_3O^+$ cation lowers its mobility too much for it to occur. The traditional hydrothermal reaction solution of $Fe^{3+}$ and $SO_4^{2-}$ is therefore required:

$$3Fe_2(SO_4)_3 + 2[H_3O(H_2O)_6]^+ \rightarrow 2[H_3OFe_3(SO_4)_2(OH)_6] + 5H_2SO_4 + 2H^+ \tag{9.3}$$

where $[H_3O(H_2O)_6]^+$ corresponds to solvated hydronium ion.

#### 9.2.1.1 Néel States and Short-ranged Order

Interest in the jarosites began with initial studies on jarosite itself, [46] which has the idealised formula $KFe_3(SO_4)_2(OH)_6$. While frustrated magnetism had not at that time been developed as a context within which to understand its magnetic properties, this early work revealed a canted magnetic ordering that can now be understood in terms of competing exchange interactions and magnetic frustration. Errors in the original magnetic structure were corrected by Inami and later groups [24,28] who showed that $KFe_3(SO_4)_2(OH)_6$ orders at low temperature into the 120° '$q = 0$' magnetic structure made from the propagation of triangular motifs with the chirality $\kappa = +1$. The observation of this in-plane structure, as opposed to the $\sqrt{3} \times \sqrt{3}$ structure predicted to be stabilised by quantum fluctuations and linear spin wave theory [29, 30], suggests the presence of additional magnetic interactions that are beyond the simple n.n. kagomé model. One additional interaction that is certainly present is inter-plane coupling, as evidenced by the antiferromagnetic stacking of these magnetic layers along the $c$-axis, i.e. the propagation vector is $\mathbf{k} = (00\frac{3}{2})$ with respect to the hexagonal setting of the $R\bar{3}m$ space group. Further work using powder neutron diffraction showed that there are two successive transitions [47] in the range $50 < T < 65$ K in which the moments order firstly as an umbrella structure that is canted slightly out of the kagomé plane before dropping into the plane at the lower temperature transition [25, 39], a complex behaviour that points to a subtle interplay between the crystal structure and the magnetic ordering – similar low temperature structures are observed in the $ND_4^+$, $Na^+$, $Ag^+$, and $Rb^+$ jarosites (Fig. 9.4). Spin wave measurements on a single crystal indicate that the Dzyaloshinski–Moriya (D-M) interactions [51,52] may be the dominant anisotropic interaction in $KFe_3(SO_4)_2(OH)_6$ [53,54]. The D-M interaction could therefore provide a mechanism whereby changes to the bonding geometry upon cooling rapidly affect the effective spin anisotropy of the Fe-jarosites.

As well as canting the atomic spins when Néel order occurs, frustration in the jarosites commonly leads to a significant amount of diffuse scattering in the neutron

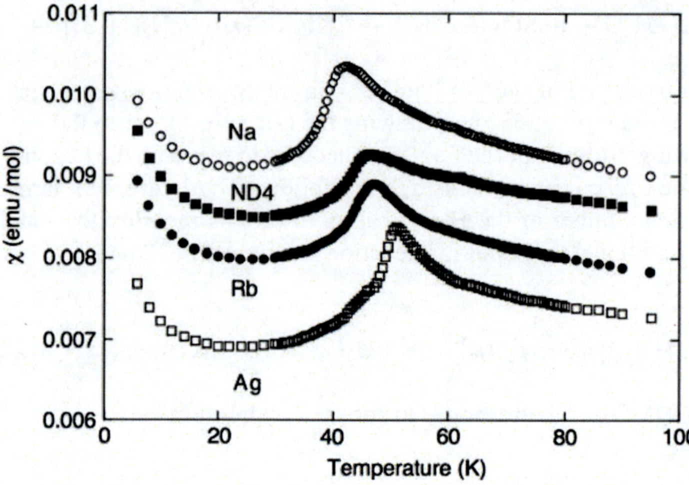


**Fig. 9.4** Susceptibility of the Fe-jarosites $AFe_3(OH)_6(SO_4)_2$ (A = Na, $ND_4$, Rb, and Ag) measured in an applied field of 10,000 Oe. Neighbouring datasets have been offset by successive increments of 0.0005 emu.mol$^{-1}$. From [28]

scattering spectra that is characteristic of short-ranged ordered spins. In these classical magnets it is tempting to think of this as being demonstrative of a mixture of both $\kappa = +1$ and $\kappa = -1$ correlations, and so of emergent $\sqrt{3} \times \sqrt{3}$ ordering from within the dominant $\kappa = +1, q = 0$ structures, but details are difficult to deduce as synthetic problems have so far prevented the manufacture of large samples of deuterated materials that are required for diffuse single crystal inelastic neutron scattering. However, recent improvements to the synthetic methods have allowed single crystal studies of spin waves in the ordered phases of $KFe_3(SO_4)_2(OD)_6$, and will be discussed in Sect. 9.2.2.

Long and short ranged order is also observed in the Cr analogue, $KCr_3(SO_4)_2(OH)_6$ by neutron scattering, which revealed that below an antiferromagnetic transition at $T_N \sim 2$ K into a Néel state, though the ordering is only partial with the sublattice magnetisation being far below that expected for $Cr^{3+}$ [38]. The remaining fraction of spins forms a gapless spin liquid state. It is noted that the propagation vector observed in $KCr_3(SO_4)_2(OD)_6$ is $\mathbf{k} = (000)$, in contrast with the Fe-analogue, which corresponds to a ferromagnetic stacking of the kagomé layers. This change in the propagation of the magnetic structure demonstrates that the interlayer coupling in these materials is modified by substitution of the B-site metal.

The final magnetic structure to be considered is that of the vanadium jarosites $KV_3(SO_4)_2(OH)_6$ which features a transition to long range magnetic order at $T_N \sim 50$ K [22]. First reports indicated that the structure was a canted ferromagnet with antiferromagnetic stacking of the kagomé layers, $\mathbf{k} = (00\frac{3}{2})$, and so may have revealed features characteristic of frustrated ferromagnetism [39]. Later work indicated that the moments were simply collinear and no effects of frustration are evidenced [41,42].

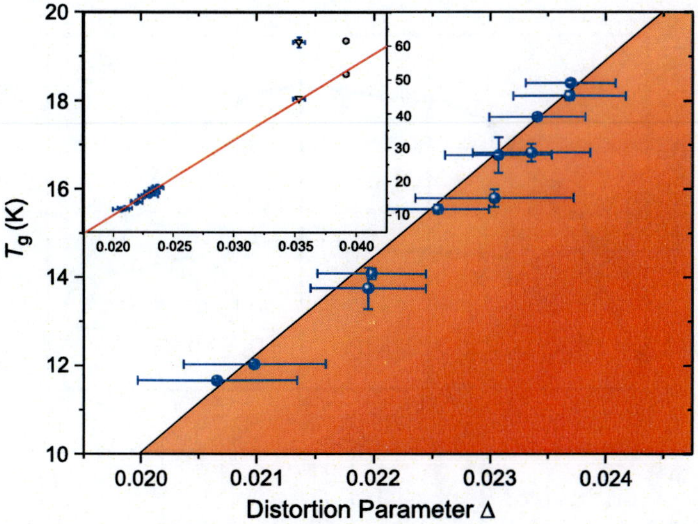


**Fig. 9.5** The spin glass freezing temperature of $(H_3O)Fe_3(SO_4)_2(OD)_6$ as a function of the ratio of bond lengths $\Delta = 1 - R(\mathrm{Fe-O})_{\mathrm{equitorial}}/R(\mathrm{Fe-O})_{\mathrm{apical}}$. The inset shows that the same linear function fitted for hydronium jarosite can be extended to the antiferromagnetic Néel ordering temperatures of $KFe_3(SO_4)_2(OD)_6$ (*grey circles*) and $(ND_4)Fe_3(SO_4)_2(OD)_6$ (*grey triangles*). From [49]

## 9.2.2 *Fe jarosites: $S = \frac{5}{2}$ Kagomé Antiferromagnets*

A key feature of the Fe-jarosites is the difference in behaviour between hydronium jarosite, which displays a spin glass-like state at low temperature, and all the other jarosites that show long range Néel order. While this was initially explained in terms of the naturally high Fe-coverage of hydronium jarosite, there have been more recent suggestions that protonation of the hydroxide units that bridge the Fe centres, [48] or changes to the Fe-coordination are important (Fig. 9.5) [49]. The latter has the remarkable feature of showing that the distortion around the Fe-site maps well to the spin glass freezing temperature in $(H_3O)Fe_3(SO_4)_2(OH)_6$ and the Néel ordering of $KFe_3(SO_4)_2(OH)_6$, and explains the ordering in hydronium jarosite that is induced by substitution of $Al^{3+}$ onto the $Fe^{3+}$ site [28, 50].

Understanding the nature of any magnetic anisotropy is a key challenge for differentiating the drives for the types of magnetic orderings. Unfortunately, synthetic difficulties have made such determinations difficult and unclear, with spin wave measurements of small single crystals of potassium jarosite being compatible with both single ion anisotropy and D-M components (Fig. 9.6).

### 9.2.2.1 Is Hydronium Jarosite a Topological Spin Glass?

The spin glass-like state of hydronium jarosite brings to the fore questions over whether glassy spin dynamics can exist in a magnet without randomness. Whilst

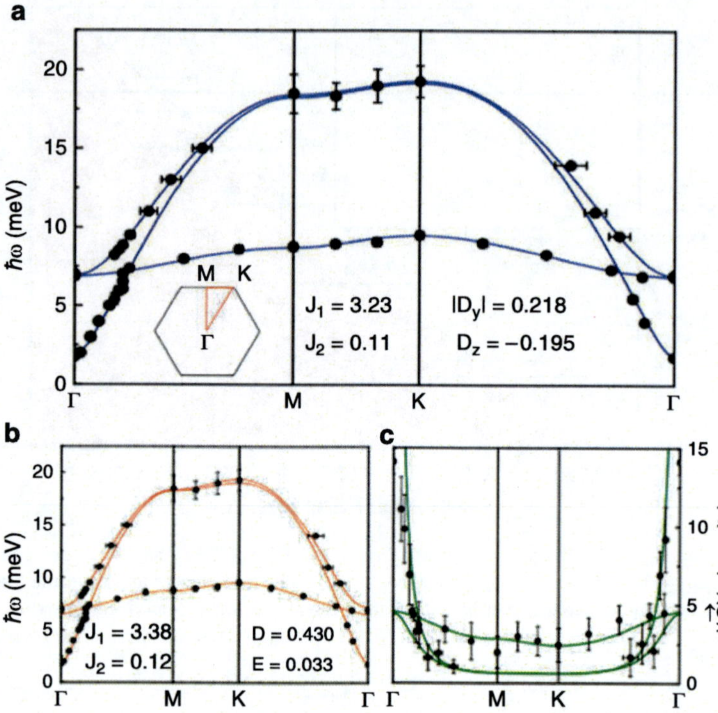


**Fig. 9.6** Spin wave dispersion along the high symmetry directions for $KFe_3(SO_4)_2(OD)_6$ with the *solid lines* showing a fit to a model including a DM component. From [53]

disorder is likely to be always present due to the flexibility of the crystallographic structure, the thermodyanmics and kinetics of this spin glass state appear quite different to those in site- or bond-disordered spin glasses.

As all of the Fe-jarosites feature Weiss temperatures falling in the range $-900 < \theta < -1{,}200$ K, the strong antiferromagnetic exchange cannot explain why hydronium jarosite does not order magnetically. Magnetic susceptibility measurements indicate a transition to a spin glass-like state in the range $12 \leq T_g \leq 17$ K [33], as a bifurcation between zero-field cooled and field cooled magnetization data. Specific heat shows that the underlying thermodynamics of the glassy state are unconventional with a quadratic $T^2$ dependence reminiscent of the Goldstone modes in a 2-dimensional antiferromagnet being seen below $T_g$ rather than the linear relation characteristic of site-disorder spin glasses [55]. This observation is compatible with Chandra, Coleman and Ritchey's model of an evolution via a sequence of zero-energy modes that is retarded by spin or exchange anisotropy [56, 57]. As the progression is through ground states, the energy of the individual triangular plaquettes is uniform and Goldstone modes are expected. Of course it is difficult to determine which energy, whether D-M interaction, single ion *XY* anisotropy, or anisotropic superexchange are responsible for this highly frustrated magnet ordering into an unconventional spin glass state as only powder samples of hydronium jarosite are known and spin waves cannot be measured in such a system. However, $^{57}$Fe Mossbauer spectra do suggest the opening of a gap below the freezing transition at which the moments lie in the kagomé planes, supporting the hypothesis of an anisotropy driven freezing transition [58]. Relaxation, tentatively attributed to spin cluster tunneling was also related to a low temperature anomaly in the specific heat, at $T \sim 0.7$ K [59]. The length scale of the spin-spin correlations as determined by powder neutron diffraction is $\xi \sim 19 \pm 2$ Å at 1.9 K, approximately the distance across a kagomé star [34,60]. $\mu$SR gives an upper limit of the static component of the moment in the spin glass-like phase to be estimated as 3.4 $\mu_B$ per $Fe^{3+}$ with a time window of $\sim 10^{-6}$ s, which indicates that a significant fraction of the spin only moment expected for high-spin $Fe^{3+}$, 5.92 $\mu_B$, remains fluctuating at low temperature [61].

Dynamic scaling of the freezing temperature shows the spin glass transition is critical, contrary to the dynamic transitions seen in 2-dimensional site disordered spin glasses. Ageing at fixed temperature below $T_g$ obeys the same scaling law as in spin glasses, but at the same time is remarkably insensitive to temperature changes [35], an unusual behaviour that is perhaps related to the uniform energy landscape that disordered chiral states would still be based on [35]. Comparison of the temperature dependences of hydronium jarosite with a site disordered spin glass, $SrCr_{8.6}Ga_{3.4}O_{19}$ and the pyrochlore $Y_2Mo_2O_7$, further reveal that temperature selective ageing characteristic of conventional spin glasses is abnormally weak in the jarosite, again pointing to a different underlying mechanism [62,63].

### 9.2.3 Cr Jarosites- $S = \frac{3}{2}$ Kagomé Antiferromagnets

Much less is known about the chromium jarosites which provide examples of $S = \frac{3}{2}$ kagomé antiferromagnets. The most studied is, again, the potassium member, $KCr_3(OH)_6(SO_4)_2$ which features a magnetic transition in the range $1.5 < T_N < 4$ K, depending on synthesis conditions [37, 38, 64]. NMR measurements indicate a deviation from the Curie–Weiss law due to the short-range spin correlation below about 150 K (with a corresponding $T_N$ value of 4.2 K). Below this temperature $\mathbf{k} = (000)$ Néel order occurs [37, 38]. The long range ordering is weak with only 1.1(3) $\mu_B$ of the expected spin-only moment of $g\mu_B S = 3\mu_B$ [38]. The remainder of the spins are observed to be in gapless quantum spin fluctuations, $\Delta/k_B < 0.25$ K, with a bandwidth of 60 K. $\mu$SR studies below $T_N$ show that spin fluctuations persist without any clear signature of spin freezing, even at $T = 25$ mK [64].

The $Rb^+$, $NH_4^+$, $Na^+$ analogues show similar behaviour, though a small hysteresis in the magnetisation at 2 K is indicative of a small ferromagnetic component, proposed to result from a canting of the 120° spin structure [65].

Little is at present known about the hydronium member, although early reports indicate an absence of long range order, suggesting that again the hydronium member is intrinsically different to the other A-site materials [22, 36].

### *9.2.4 Conclusion*

The jarosites represent a very rich family of classical and semi-classical kagomé magnets where the spin value can be tuned to make a range of magnetic ground states for study that are based on ordered and/or fluctuating spins. Efforts to understand their magnetic behaviour led to the first discussion over the effects of the DM interaction and suggestions that its perturbation was very important and could in fact drive the ordering to a Néel ground state, a scenario which could be further tested owing to the recent synthesis of high quality crystals. There remain several open questions about these magnets, such as the roles played by disorder and another perturbing parameter, namely the inter-kagomé layer magnetic coupling, in the stabilization of the ground state.

## 9.3 Pyrochlore Slabs

Among all Highly Frustrated Magnets, the chromium-based $S = 3/2$ pyrochlore slab compounds $SrCr_{9p}Ga_{12-9p}O_{19}$ (SCGO($p$)) [66] and $Ba_2Sn_2ZnCr_{7p}Ga_{10-7p}O_{22}$ (BSZCGO($p$)) [67] of the magnetoplumbite family appear as good realizations of a Heisenberg Hamiltonian on a corner sharing lattice and have been found to retain the essence of a spin liquid ground state despite a spin glass-like freezing. Indeed, the observation of the unusual properties of SCGO, and models developed to explain them, played an important role in opening the field of Highly Frustrated Magnetism.

### *9.3.1 Synthesis*

Only a few attempts to produce single crystals of SCGO have been successful, and these unfortunately led to a poor magnetic coverage of the kagomé planes [3, 68]. For this reason, most of the work has been performed on better quality powder samples. Initially limited to a Cr content of up to $p \simeq 0.9$, powders could be then successfully synthesized up to a content $p = 0.96$ [69, 74]. These were obtained by mixing $SrCO_3$, $Cr_2O_3$, $Ga_2O_3$ in stoichiometric quantities and firing at 1350 °C for 4–5 days with several intermediate grindings. This process was continued until laboratory x-ray powder diffraction confirmed that all the $Cr_2O_3$ had reacted. Samples with a good mosaicity could be obtained by this method and their high quality was assessed through magnetization and NMR local measurements (see below). Attempts to synthesise a fully Cr-occupied pyrochlore slab have to date been unsuccessful, leading above $p = 0.96$ to increasing amounts of $Cr_2O_3$ impurity scaling with $p - 0.96$ [74].

### 9.3.2 Magnetic Network

In both SCGO and BSZCGO, octahedrally $Cr^{3+}$, $S = 3/2$, ions are located at corners either of tetrahedra or triangles (Fig. 9.7). The $3d^3$ electronic configuration has a negligible single-ion anisotropy. As is the case in spinels, Cr–Cr exchange is direct in nature which limits any coupling with second neighbours [72]. This therefore yields a Heisenberg Hamiltonian with n.n. exchange only, on a corner sharing lattice. The various Cr–Cr distances within bilayers only differ by 3.7% and the average coupling, $J \sim 40$ K, calculated from a high temperature series expansion fit of the high $T$ susceptibility, is almost two orders of magnitude larger than the perturbation terms, like single-ion anisotropy $\sim 0.08$ K estimated from ESR measurements [70] and dipolar interactions $\sim 0.1$ K, or next nearest neighbor interactions $< 1$ K [71].

There are two minor differences between these compounds; in SCGO, in between two consecutive bilayers there are layers of Cr pairs which were demonstrated to turn into a singlet state at quite high $T$, hence do not impact on the low-$T$ physics of the bilayers [72]. In BSZCGO, which looks like a simpler case from a magnetic point of view, the natural random 1:1 Ga-Zn substitution present inside the pyrochlore slab and required for charge balance, leads to local random structural distortions which might induce some (minor) modulation in the exchange constants [71, 73]. Overall, the structure of both compounds is essentially 2D.

The main drawback is that both compounds have Ga/Cr substitutional disorder, and a Cr-coverage of the frustrated bilayer higher than $p = 0.96$ and $p = 0.97$ cannot be reached in SCGO($p$) and in BSZCGO($p$). This results respectively in 4 and 3% spin vacancies ($Ga^{3+}$ is diamagnetic) in the magnetic sublattice. The

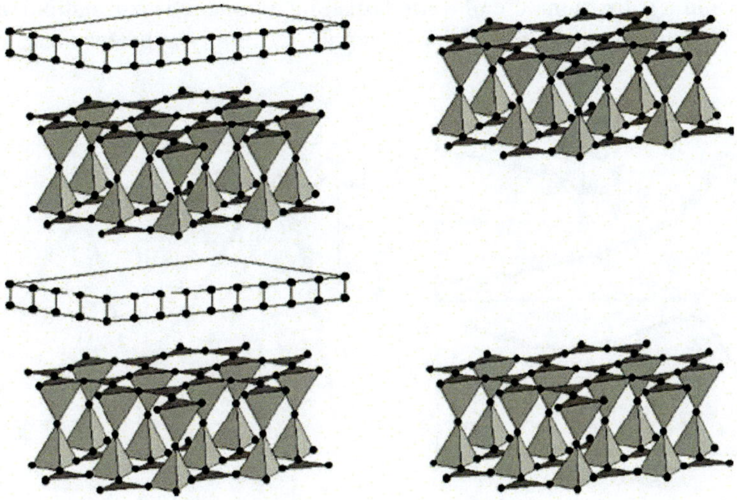

**Fig. 9.7** Pyrochlore slabs: Only Cr ions have been represented. *Left*: SCGO, pyrochlore slabs are separated by Cr–Cr pairs which dimerize below 180 K. *Right*: BSZCGO which features a simpler and purer structure of well-separated pyrochlore slabs. From [71]

related magnetic defects that they produce, turn out to be insufficient to destroy the fluctuating character of the ground state. One way to get around the dilution problems and to extrapolate to the pure case is to perform studies versus $p$ which is quite well controlled in this family and to distinguish between intrinsic properties and those related to dilution of the frustrated network. Yet, one has to bear in mind that diamagnetic defects are always present.

### 9.3.3 Generic Physics

#### High $T$

The linear dependence of the inverse susceptibility of SCGO measured using bulk magnetometry, $\chi^{-1}_{\mathrm{macro}}$, was found to extend much further below the Weiss temperature than would be expected from mean-field models of unfrustrated systems (Fig. 9.8), and led to estimates for the antiferromagnetic Weiss-temperature of 500-600 K. Only at very low-$T$ can the upturn of the susceptibility be seen that is associated with the defects (see next section) and a transition to a spin-glass state with a remarkably high ratio $|\Theta|/T_g \sim 150-200$.

One of the greatest advantages of local techniques, such as NMR, is that they enable the intrinsic physics of the frustrated magnet to be distinguished from the effects of defects that can easily dominate the bulk susceptibility. In SCGO the intrinsic susceptibility measured at the Ga nucleus lying inside the pyrochlore slab is marked by a maximum $\sim J/2$ (Fig. 9.9) which is largely independent of Cr-content, indicating that it is an intrinsic response that is robust in character [73–75]. This maximum is the signature of a strengthening of magnetic correlations whereas the finite $T \to 0$ extrapolation of the susceptibility simply indicates that the corre-

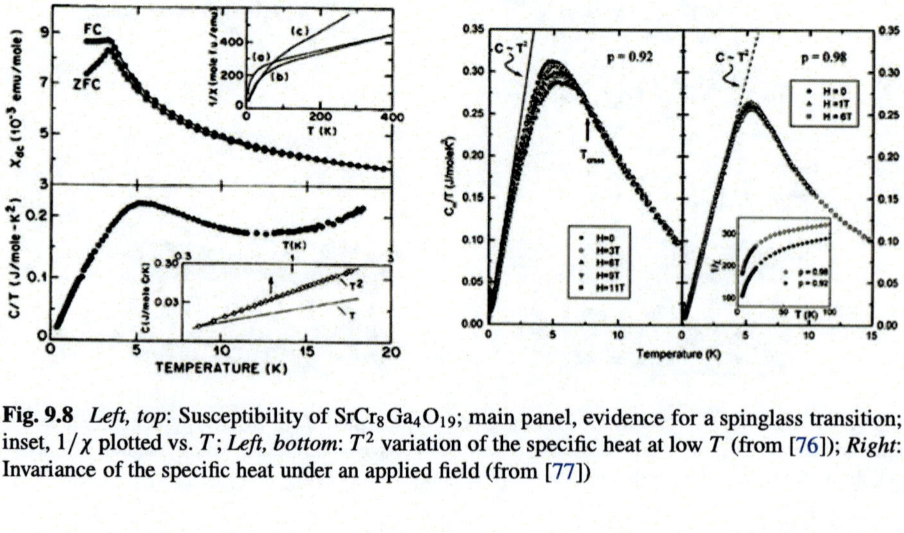


**Fig. 9.8** *Left, top*: Susceptibility of $SrCr_8Ga_4O_{19}$; main panel, evidence for a spinglass transition; inset, $1/\chi$ plotted vs. $T$; *Left, bottom*: $T^2$ variation of the specific heat at low $T$ (from [76]); *Right*: Invariance of the specific heat under an applied field (from [77])

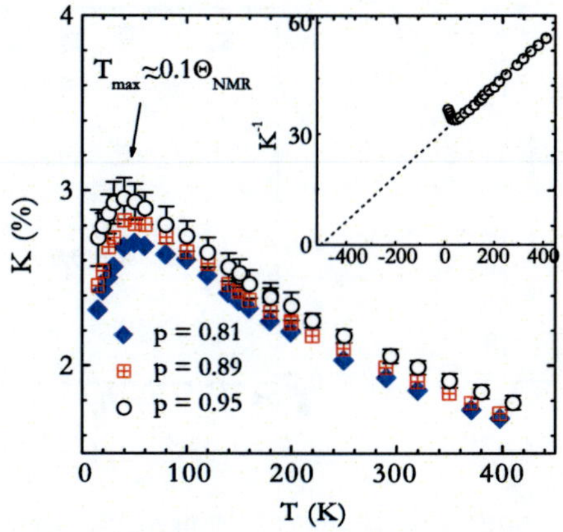


**Fig. 9.9** $T$-variation of NMR line shift for various lattice covering in SCGO. Inset: the inverse shift plotted vs. $T$ yields a Weiss temperature of $-450$ K. From [74]

lation length remains finite at $T = 0$ and that no gap opens. The former finding is further confirmed by neutron scattering results [69].

### A Spin Liquid State?

The spin-glass state was studied in detail using both macroscopic and local measurements. Whereas there is a divergence of the non-linear susceptibility with conventional critical exponents [76], the specific heat displays the $T^2$ dependence typical of classical spin waves in a 2D antiferromagnet [3] rather than the linear T-dependence observed in canonical spin glasses. In addition, the specific heat was found to be field independent, providing evidence for low energy singlet excitations [77].

The most striking evidence for a liquid-like ground state in the SG phase comes from the persistence of dynamics in muon relaxation measurements down to $T = 20$ mK [78]. A slowing down of the fluctuations around $T_g$ is clearly observed together with a very uncommon plateau of the muon relaxation rate (Fig. 9.10) with a Gaussian early time dependence of the relaxing asymmetry. This represents a hallmark of particular fluctuations that have now been observed in many highly frustrated magnets. The dynamical character was deduced from the absence of decoupling under an applied longitudinal field – see chapter by Carretta and Keren. Neutron scattering studies show the correlation length to be short and the static moment to be only 25% of what would be expected for a completely frozen picture (Fig. 9.10) [79]. Although these studies were performed on a sample with only 89% of the magnetic sites occupied, they were further confirmed by careful studies on the least diluted samples [69, 81]. Finally, one should note that Ga NMR cannot be followed at low temperature because of a wipe-out of the signal related to the slowing down of the dynamics which yields too fast a relaxation to be observed in the NMR time-window. The fact that the signal is not recovered even well below $T_g$ is further evidence for the persistence of fluctuations far into the spin

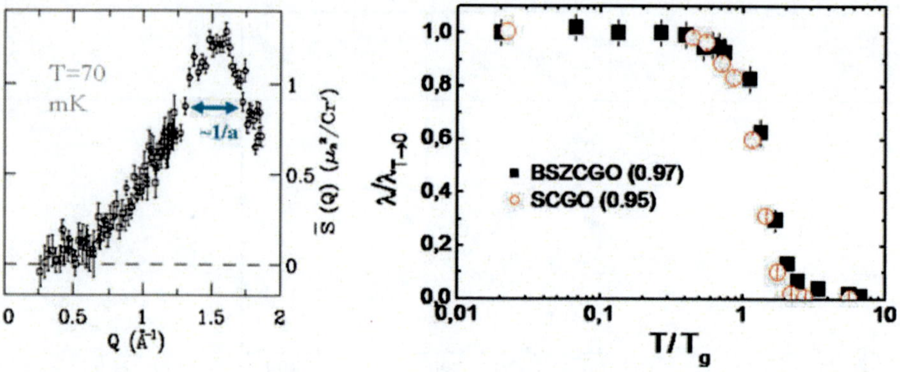


**Fig. 9.10** *Left panel*: In the frozen state, neutron scattering yields a correlation length of the order of lattice spacing (from [79]). *Right panel*: $\mu$SR relaxation rate for SCGO and BSZCGO. The freezing temperatures, respectively 3.9 K and 1.5 K differ by a factor 2. A clear slowing down of the electronic fluctuations occur at $T_g$ but the plateau observed at low $T$ contrasts with the peak expected in a conventional spinglass transition (from [80])

glass phase [74, 75]. Such wipe-out effects have been reported in various contexts and are indeed commonly linked to fast relaxation processes [82, 83].

## 9.3.4 Non-magnetic Defects

### Staggered Static Response

As emphasized in Sect. 9.3.2, non-magnetic defects originate mainly from spin vacancies in the magnetic lattice. In the well established case of unfrustrated antiferromagnetic correlated systems, such as high-$T_C$ cuprates or $S = 1/2$ spin chains, a staggered susceptibility increases quickly at low $T$ around the vacancy, and appears to be peaked at the n.n. sites [84]. The spatial response decreases quite steeply on a length characteristic of short ranged correlations typically expanding beyond a few lattice spacings (Fig. 9.11). NMR, which probes locally the susceptibility, was used to establish a map of these susceptibilities by singling out typical spectra for n.n., and through a line broadening found symmetric because of the staggered characteristic of the response of defects. Its width is directly linked to the low-$T$ increase of the macroscopic susceptibility due to defects. It therefore played a central role in unveiling the physics induced by non-magnetic defects well explained by theories modeling these responses. This approach which could be coined "perturb to reveal" proved to be quite fruitful in this field and might prove to be promising as well in the context of HFM.

In SCGO, the Ga NMR linewidth scales with the low-$T$ susceptibility and the proportion of defects, $(1 - p)$ when their concentrations are low, typically $(1 - p) < 0.2$ [74]. Moreover the symmetric lineshape was the first clear indication

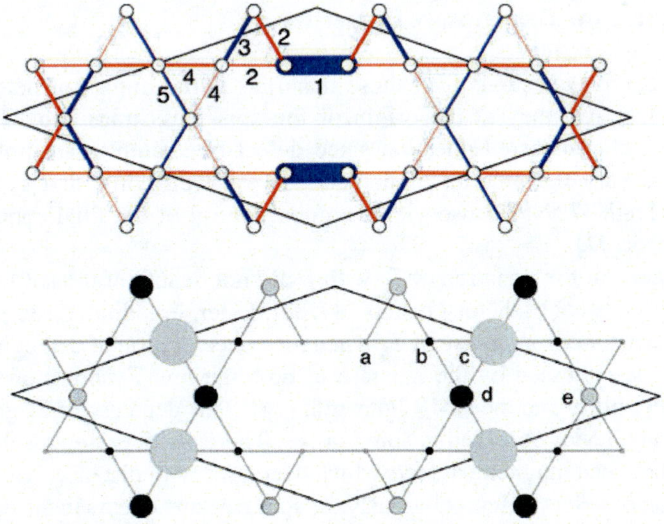


**Fig. 9.11** Exact diagonalization study of the magnetic texture induced by a single impurity in a 15-site cluster. The *black lines* denote the boundary of the cluster. *Top*: Bond spin-spin correlations. Bar width represents the strength of correlation functions on each bond, measured as a deviation from the pure-system values on a linear scale where the strongest correlation function is $\langle S_i \cdot S_j \rangle = -2.994$, approximately twice that in the pure-system. *Blue (red)* lines denote bonds on which the deviation is negative (positive). *Bottom*: Site magnetization profile of the lowest triplet state. *Circle radius* represents the magnitude of the local moment on a linear scale, where the largest circle corresponds to a moment of 0.722 $\mu_B$ 0.5 $\mu_B$ is the moment of a spin 1/2, and the *dark gray circles* represent sites with induced moments opposite to the effective field direction. From [86]

of a staggered response. There is an interesting difference with the unfrustrated case which came first out of exact diagonalizations for $S = 1/2$ [85]: on a given triangle around a spin vacancy, AF bonds will be locally satisfied, leading to a release of frustration and the creation of a spin singlet. The staggered response comes out further than the 1st n.n where a singlet is localized and is also found for $S = 3/2$ [86]. This could explain well the observed drastic low-$T$ broadening. The depressed response in the immediate vicinity of the defect was not observed through NMR experiments and should have been visible as a reduced shift. The more complicated structure of pyrochlore slab, where Ga probes 12 Cr sites, or the dynamical wipe-out of the NMR signal below 20 K does not allow probing of the ground state. It is worth noting that in SCGO, the linewidth extrapolates to zero when $p \to 1$, whereas in BSZCGO it extrapolates to a finite value. This is the signature that in the latter compound, an additional kind of defect plays a role, which is likely to be associated with the existence of bond defects induced by the 50% random occupation by $Zn^{2+}$ of the $Ga^{3+}$ site [73].

**Impact on the Low Temperature Properties**

One of the central questions over these materials is the impact of defects on the low-$T$ physics. Are they at the origin of the spin-glass transition? This is still not a completely answered question since defect-free samples are not available. Yet, one can vary the amount of defects in very well controlled syntheses for $p < p_{\max}$. High -$T$ NMR also yields a good control of the final content for the purest samples [71].

Early studies of the variation of $T_g$ with $p$ did not benefit of the least substituted samples and gave contradictory results [87, 88]. Later, less diluted samples clearly pointed to a very weak *decrease* of $T_g$ when the concentration of defect is increased (Fig. 9.15), as witnessed by the increase of both the low-$T$ upturn of the macroscopic susceptibility and the NMR linewidth [74]. This decrease of $T_g$ goes against the common behavior observed in spin glasses. There, the freezing results from the interaction between magnetic defects which decreases with distance, hence with the concentration of defects. Here, obviously the spin freezing occurs in the defect channel (Curie-like susceptibility) but the effective interaction between defects is either not set by their mutual distance or does not play any role in the temperature scale of the spin glass transition.

The $T \to 0$ relaxation rate has a smooth variation with $p$ through the percolation threshold of the pyrochlore slab [81]. This points at local excitation modes as expected in the KHAF.

The interpretation of the variation of the muon relaxation rate with $T$, $p$ and the applied field still represents a real challenge. It could be reasonably fitted assuming a phenomenological model where both sporadic dynamics due to unconfined spinon -like excitations $\propto p$ and more conventional spin-glass dynamics $\propto (1 - p)$ contribute to the relaxation [89]. The observation of a relaxation requires a gapless state which would therefore give an upper bound of a hypothetical singlet-triplet gap of less than 30 mK $\sim J/100$. Sticking to this simple picture it seems that spinon-like excitations would be destroyed above a characteristic temperature $\sim T_g$. This raises questions about what property sets the energy scale that causes the system to turn into such a state, and why the transition temperature is only weakly dependent on the amount of defects. No theoretical approach has been further developed to interpret this set of data.

### *9.3.5 Concluding Remarks*

While SCGO gave a very solid foundation to the field of geometrically frustrated magnetism, its complicated corner sharing lattice does not allow a refined comparison with theoretical models and does not tackle the important case of fluctuations generated but the quantum character. A new route to investigate such effects has been opened by Volborthite and Herbertsmithite which are described in the following.

## 9.4 Towards S = 1/2 Ideal Compounds

Copper (II) materials are natural candidates for realising $S = 1/2$ quantum antiferromagnets, yet most of them form square-based network, built from edge-sharing or corner sharing $CuO_4$ square plaquettes. The situation is far from hopeless since several compounds have been isolated recently which feature nearly apparently good realizations of a kagomé lattice, Volborthite, Herbertsmithite and the recently discovered Kapellasite. These have opened new experimental avenues with which to probe our understanding of kagomé physics and inspired new theoretical work.

### *9.4.1 Volborthite*

#### Structure

Volborthite is a natural copper vanadate mineral $Cu_3V_2O_7(OH)_2.2H_2O$ [6], that features a 2D structure with Cu-O(OH) layers separated by 7Å pillars built from pyrovanadate groups $V_2O_7$, which are surrounded by water molecules (Fig. 9.12). $V^{5+}$ ions are diamagnetic and are good NMR probes of the physics of the Cu planes through equivalent hyperfine coupling to 6 sites located at the vertices of one Cu hexagon of the kagomé lattice. A monoclinic distortion leads to two slightly different Cu-Cu bonds which yield two sets of magnetic interactions the relative

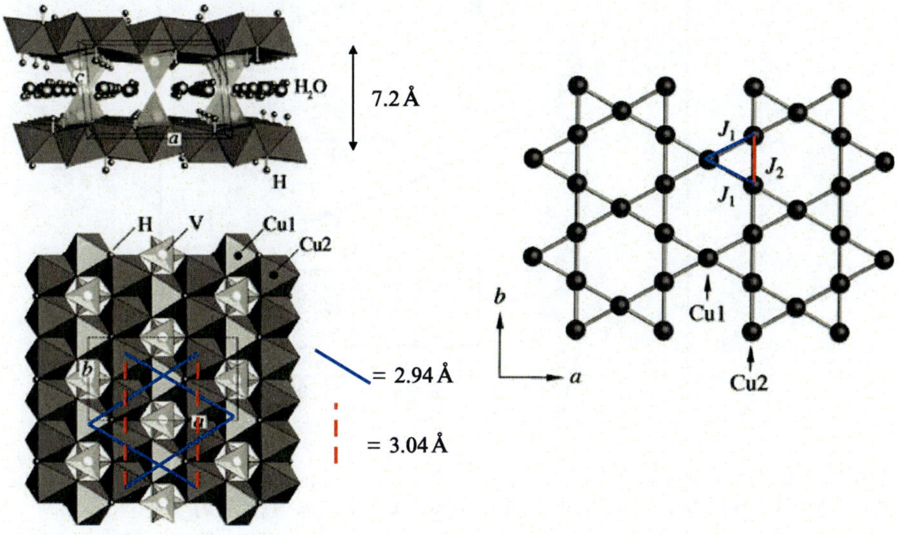


**Fig. 9.12** *Left*: structure of volborthite, side view and top view with two slightly different Cu–Cu bonds. From a magnetic point of view the kagomé lattice features isosceles rather than equilateral triangles. From [90]

magnitude of which is unknown. The triangles of interactions are therefore isosceles rather than equilateral and display an average value of the exchange constant ~85 K. The discovery of this compound and the experimental results triggered some theoretical interest in this type of distorted kagomé lattice and it has been shown that a great deal of frustration survives the lowering of symmetry, though the degeneracy is subextensive within a classical approach [91,92].

### $T > 2$ K

As with many frustrated networks, the linear behavior of inverse susceptibility extends well below the Weiss-temperature. A small upturn of $\chi$ observed in the purest samples indicates less than 1% of impurities treated as quasi-free $S = 1/2$ spins with as low as a 0.07% record content [93,94].

These paramagnetic-like impurities are responsible for the NMR line broadening observed at low $T$, typically below 10 K. Both V NMR shift and noticeably, because of the small impurity content in comparison with SCGO, macroscopic susceptibility, display a maximum around 20 K ~ $J/4$ (Fig. 9.13). Specific heat data on this compound is not accurate enough to allow a definite statement on the ultimate $T \rightarrow 0$ variation [90]. Overall these behaviors are quite similar to those observed in SCGO.

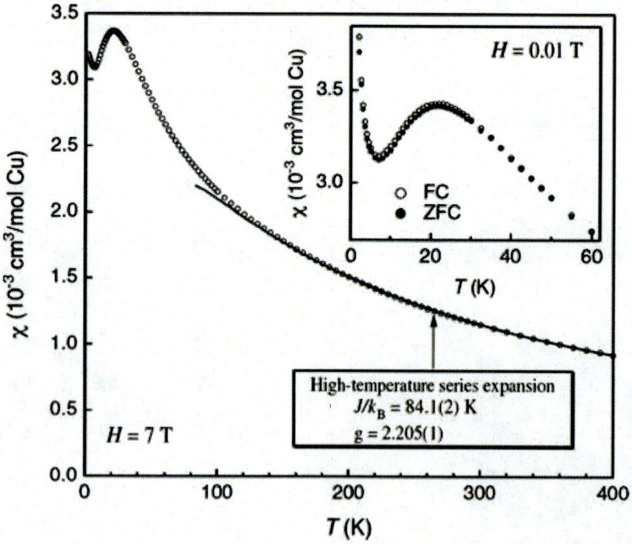


**Fig. 9.13** Macroscopic susceptibility of Volborthite. Note the maximum of susceptibility and a Curie upturn at low-$T$ likely related to 0.5% defects in the sample. From [90]

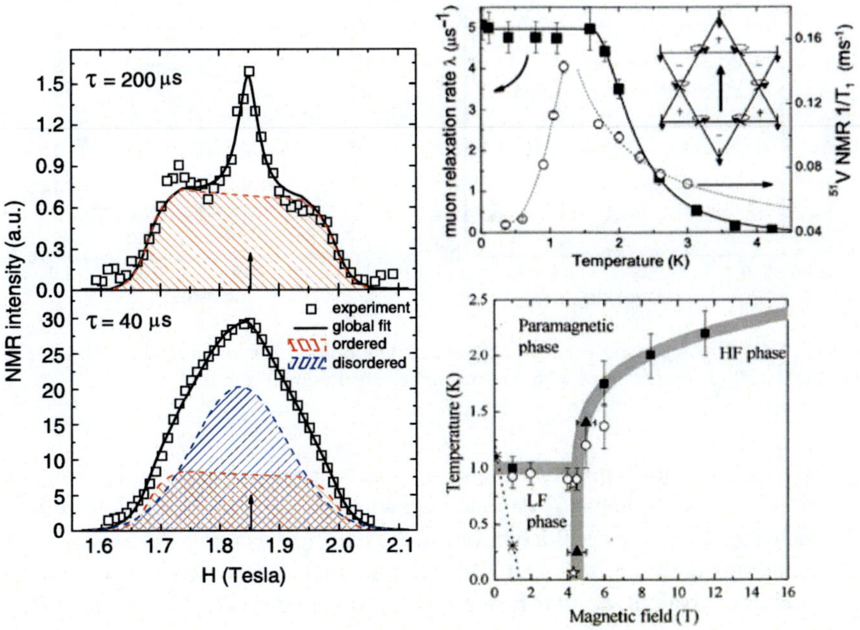


**Fig. 9.14** *Left panel*: Low $T$ V NMR spectra (from [96]). *Right panel*: Relaxation rates measured through V NMR and $\mu$SR have a different $T$-dependence (from [95]); $H - T$ phase diagram (from [94])

## Frozen State

A transition to a frozen state was first observed between 1 and 1.5 K both through a separation between the field cooled and zero-field cooled branches of the susceptibility and a drastic broadening of the NMR spectra [96], the temperatures of which only weakly depend on impurity contents less than 1%, which reveals an intrinsic character. A complete $H - T$ phase diagram was sketched through NMR [94] which displays a transition from a low field frozen phase to a high field more ordered phase (Fig. 9.14). Low $T$ magnetization steps have been also observed [93], among which the first corresponds to the 4 Teslas transition and await some theoretical interpretation.

Through NMR one can track the local structure in the frozen state. Two different average internal fields at the V sites were evidenced for a 1.8 T field Fig. 9.14. This was attributed to the spatially anisotropic character of this compound which yields two non-connected ground state spin configurations, one of which (75%) gives a cancellation of the internal field at the V site [96]. Much alike the data from SCGO, a plateau of the muon relaxation rate is found below $T_g$ [97]. Interestingly, one can vary the rate of non-magnetic defects in the Cu planes by substituting Zn in a controlled manner on the Cu sites up to a content $\sim 5\%$. Again the transition temperature is found to decrease when disorder increases but, unlike SCGO, the decrease is quite sharp, as well as the value of the $T = 0$ relaxation rate [98] (Fig. 9.15). The origin for this is unclear. Finally, the relaxation plateau is consistent with a picture

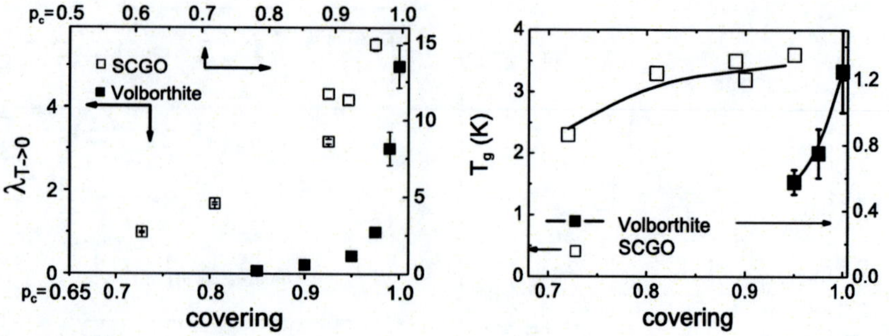


**Fig. 9.15** Low $T$ comparison between SCGO and Volborthite: Evolution of the relaxation rate and of the transition temperatures with non-magnetic substitution. From [98]

of persistent dynamics while $1/T_{1-\mathrm{NMR}}$ displays a peak at the transition temperature [94, 96] and has a linear $T$-dependence when $T \to 0$ which as well indicates the persistence of low energy fluctuations in the frozen phase (Fig. 9.15). These contrasting findings between $\mu$SR and NMR can be qualitatively reconciled by noting that V probes equivalently 6 sites hence zero energy modes in a $\sqrt{3}\times\sqrt{3}$ are filtered out and do not contribute to the NMR relaxation, while muons which usually bind to an oxygen 1Å away, implant at a non symmetric site and could sense such excitations. Finally, some neutron work in zero field has been performed on deuterated sample which corroborates these findings. Diffuse scattering reveals the presence of short range order involving approximately 40% of the spins, which fails to develop into long range order even at millikelvin temperatures. Correlations evolve from $\mathbf{q} = 0$ type around 5 K to a $\mathbf{q} = \sqrt{3} \times \sqrt{3}$ at the freezing temperature observed in NMR.

Inelastic neutron scattering shows a non-dispersive magnetic excitation centered around 4.5 meV (or $\sim J/2$) which might signal the existence of a Dzyaloshinskii–Moriya anisotropy much as in the Jarosites [99].

### *9.4.2 Herbertsmithite: "An end to the Drought of Quantum Spin Liquids [100]"*

Herbertsmithite $ZnCu_3(OH)_6Cl_2$ is an end member of the paratacamite family $Zn_xCu_{4-x}(OH)_6Cl_2$ [4], and is perhaps the most promising candidate to-date for an idealized quantum kagomé AFM. Despite the unprecedented absence of any observed order for a kagomé compound down to $J/9000$, three years after its discovery this status is still not confirmed. While $Cu^{2+}$ ions form a kagomé plane, additional Dzialoshinsky–Moriya interactions have to be taken into account and the presence of diamagnetic defects is a very real possibility.

The discovery of this compound triggered a remarkable revival of both theoretical and experimental studies on the kagomé lattice [101–104]. In the following paragraph we give an overview of our present understanding and attempt to stress the open debates which are currently going on.

**Structure and Interactions**

The $x = 1$ compound of the Zn-paratacamite family $Zn_xCu_{4-x}(OH)_6Cl_2$ can be viewed as a double variant of the parent clinoatacamite compound ($x = 0$); first, the symmetry relaxes from monoclinic to rhombohedral ($R\overline{3}m$) around $x = 0.33$, leading to a kagomé lattice in the $a$-$b$ plane; then, in the $c$- elongated $x > 0.33$ pyrochlore structure, the magnetic bridge along $c$-axis between $a$-$b$ kagomé planes is progressively suppressed by replacing the apical $Cu^{2+}$ by a diamagnetic $Zn^{2+}$. Due to a more favorable electrostatic environment, $Cu^{2+}$ is expected to preferentially occupy the distorted octahedral kagomé sites. When $x = 1$ the $S = 1/2$ ions should therefore form well defined kagomé layers that are themselves well separated by diamagnetic $Zn^{2+}$ (Fig. 9.16).

**Phase Diagram of the paratacamite Family**

In clinoatacamite ($x = 0$), two transitions are observed in $\mu$SR, at 18 K and 6 K [105]. The lower, leading to long-ranged order with a gap in the excitation spectrum of $\sim$1.2 meV [106]. At $x = 0.33$, a partial ordering is observed and finally, the ordering disappears above $x = 0.5 - 0.6$. No order has ever been observed in the $x = 1$ compound down to 20 mK and the upper bound value of a hypothetical frozen moment would be less than $10^{-3}\mu_B$ (Fig. 9.17) [107]. These results are completely consistent with ac susceptibility measurements which behaves monotonously for $x = 1$ down to the lowest measured $T$ and neutron scattering where no magnetic diffraction Bragg peak was detected [108].

From macroscopic magnetization measurements, a clear ferromagnetic component is observed when the system gets in the $T < 6$ K ordered phase ($x < 0.5$). Combined with the moderate decrease of the Weiss temperature when $x$ decreases, this indicates that the coupling between Cu within the kagomé planes is antiferromagnetic and dominant, $J \sim 180$ K, whereas that between a Cu on the Zn sites and a Cu inside the kagomé layer is weakly ferromagnetic ($\sim$ few tens K). This family therefore offers an interesting possibility to explore weakly ferromagnetically coupled kagomé layers [4] for $0.33 < x < 1$.

**Susceptibility**

Probing the existence of a gap through susceptibility has been a major challenge since the discovery of the compound. Various results are now reported through macroscopic susceptibility, and local techniques, Cl [109], $^{17}$O NMR [110]. The $\mu$SR shift results are still under debate [111]. $^{17}$O NMR is certainly the most sensitive probe of the susceptibility of kagomé planes since oxygen is responsible for the exchange coupling bridge between two adjacent Cu. The hyperfine coupling of Cl to Cu is at least an order of magnitude smaller. In macroscopic susceptibility

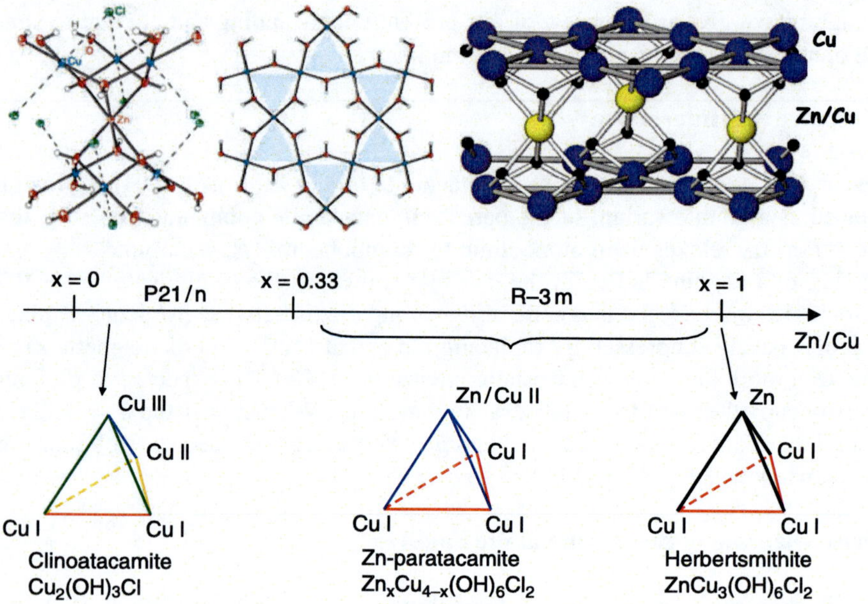


**Fig. 9.16** *Top left*: Structure of Herbertsmithite from [4]. *Top right*: Simple sketch of the structure where Zn and Cu sites only have been represented. Bottom: Evolution of the structure of paratacamites when Zn content is increased. A structural transition accurs around $x = 0.33$ which yields well defined perfect kagomé planes, assumed to be filled with Cu only

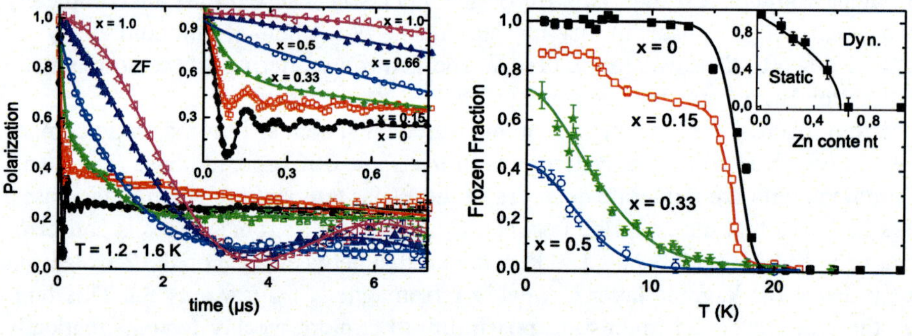


**Fig. 9.17** *Left*: Evolution of the low-$T$ $\mu$SR asymmetry when Zn content is progressively increased. The oscillations, as detailed in the inset progressively disappear and vanish for $x > 0.5$. *Right*: Phase diagram of the paratacamite family. From [107]

experiments as well as for $^{17}$O NMR, a Curie–Weiss behavior is observed at high $T$, extending well below the Weiss temperature. Below 150 K, the two quantities start to markedly differ. A clear upturn is observed in $\chi_{macro}$ whereas the variation of the oxygen shift $^{17}K$ goes the opposite direction. This is certainly the best evidence that the compound is not free from defects. If it were, the only contribution to both measurements would come from fully occupied $Cu^{2+}$ kagomé planes and would be identical provided that no staggered component dominates the O signal. Through O shift, a maximum in the susceptibility at $\sim J/2$ is evident which is certainly related to the kagomé physics and also observed in other corner-sharing kagomé based lattices [73, 75, 96], yet with slightly different shapes of the $T$-variation. A rapid decrease of the shift follows but the $T = 0$ shift remains *finite* (Fig. 9.18).

One can invoke two possible theoretical scenarios:

- A *non-singlet* ground state, such as appears to be the experimental case, with a decreasing susceptibility at low $T$ associated with the strengthening of the AF correlations below $T \simeq J/2$ which then level off as $T \to 0$, and surprisingly, do not yield a gapped ground state.
- A *singlet-triplet gap*: The observation that the $T = 0$ susceptibility is finite in herbertsmithite which could arise from
  - A filling of the hypothetical gap – expected to be at most $J/20$ – under the modest applied field 6.5 T of the NMR experiment. It is however not clear whether the magnetic energy brought to the system $\sim J/30$ should be compared to the gap value or to the larger one at which a magnetization plateau should occur. In addition, the value of the maximum of the oxygen shift $^{17}K$ around 50 K is not in line with the expected small value of the gap.
  - The existence of an additional DM interaction, $\vec{D} \cdot (\vec{S}_i \times \vec{S}_j)$, between $Cu^{2+}$ spins which results from the absence of an inversion center between Cu ions, as suggested first by [112]. This would modify the $T = 0$ susceptibility by *mixing singlet and triplet* states [118].

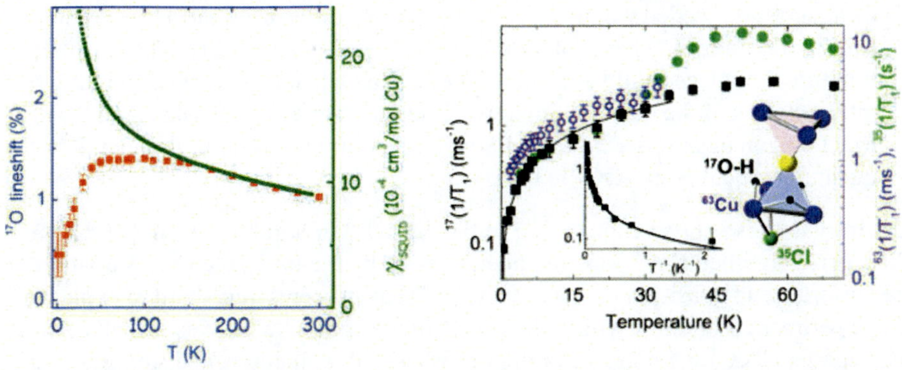


**Fig. 9.18** *Left*: $^{17}O$ NMR shift versus temperature (red) compared to macroscopic susceptibility (green). *Right*: Comparison of the relaxation rates for various nuclear probes. The semi-log (vertical axis) plot allows to rescale by the hyperfine constants which are quite different for the three nuclei. The low $T$ evolution is found remarkably similar although the symmetry of hyperfine couplings are very different, consistent with an expected flat $q$ dependence of the excitation modes. Inset: $^{17}T_1^{-1}$ plotted versus $1/T$. The non-linear variation is evidence for non-gapped excitations

**Are there Defects? What is their Nature?**

Beyond the previously reported discrepancy between $^{17}$O shift and macroscopic susceptibility, there are several arguments which fit with a scenario of Zn"diluted" kagomé planes for compounds even with the ideal stoechiometry. In turn, this means that some $Cu^{2+}$ reside on the $Zn^{2+}$ sites, which were previously expected to be weakly coupled. Both contributions would yield a Curie-like upturn at low $T$ and one can expect that the latter one is a classical case of weakly paramagnetic centers:

- A specific line is observed in $^{17}$O spectra which fits at high $T$ with an environment of one non-magnetic Zn and one Cu [110]
- The steep low-$T$ upturn of $\chi_{\text{macro}}$ can easily be interpreted in terms of weakly interacting paramagnetic centers on the Zn site which fits well with the idea of Zn-Cu inter-site exchange between the two sites [113]
- A Schottky-like anomaly which characteristic energy scales with the field seems observed in specific heat, again in agreement with an inter-site exchange scenario [114]
- The magnetization of paramagnetic-like defects is easily saturated in moderate fields [115]
- Low-$T$ $\mu$SR experiments yield a plateau of relaxation below 1 K, a temperature which is found to be typical of the interaction between the paramagnetic-like sites [107]
- Explaining the low-$T$ Curie-like tail by invoking Dzyaloshinskii–Moriya (DM) interactions only as calculated for $T > 0.25J$ through exact diagonalizations [112], requires values for the DM interaction as high as $0.2J$, much too large to fit the ESR lineshape dominated by DM interaction at high $T$ [116]. Moreover the magnitude needed would as well induce some order, a scenario first pointed at for classical spins [51] and later derived for $S = 1/2$ spins [118]. These interactions are though certainly present and need to be taken in consideration for a full interpretation of the data. An interesting phase diagram can be sketched through exact diagonalizations where Herbertsmithite is found to lie in the vicinity of a quantum critical point (QCP) [118].

Contrary to these observations is the initial Jahn-Teller argument from [4] that predicts perfectly filled $Cu^{2+}$ kagomé planes driven by the favorable Cu occupation of the distorted and quasi-planar kagomé sites. Today, it seems quite hard to explain all the experimental findings within the initial framework of what was initially hailed as a "perfect" kagomé magnet. Further synthesis of well controlled samples might yield a way to clarify the responses of this material.

Experimental data suggest that as much as 6–10% of the kagomé sites could be occupied by $Zn^{2+}$. For $D = 0$, such non-magnetic impurities are found to induce a dimer freezing in the two adjacent triangles it belongs to, as expected from the relief of frustration [85]. Further from the impurity position a staggered response is generated, for all antiferromagnets [84]. The major difference with the unfrustrated case is that the response is not peaked on the sites next to the impurity. This scenario is strongly supported by $^{17}$O NMR spectral analysis [110, 117]. Both the tendency to dimer freezing and the staggered response hold for as large values of $D/J$ as 1, while the critical point at $D/J = 0.1$ does not seem much affected by the presence of these spinless defects [119].

### Dynamical Behavior

At variance with the susceptibility, relaxation measurements yield similar behavior of $1/T_1$ at low-$T$. Given the filtering form factors which are quite different for all probes, this indicates that excitations are consistent with a flat mode in $q$-space. The analysis of the $T$-dependence clearly demonstrates the absence of a gap in the excitations and rather a subinverse linear law which is quite puzzling (Fig. 9.18). A very different behavior is observed in $\mu$SR experiments for $T < 1$ K but which might be rather related to Cu on the Zn site.

## 9.5 Other Compounds

### *9.5.1 Organic Materials*

Organic and organometallic chemistry provide many building blocks which can be used to construct novel kagomé materials that are capable of being tuned by chemical methods. In these the spin may reside on radical groups or metal centres, if present.

### TTF Salts

In the $[Re_6Se_8(CN)_6]^{4-}$ TetraThioFulvaleme (TTF) salt partly oxidised dimers form a kagomé lattice [120]. This compound is a 2D metallic conductor at high temperature (space group $R\bar{3}$), but undergoes a structural transition in which charge ordering occurs and the 3 fold symmetry is broken. At low-$T$, susceptibility confirms the organisation of these mixed valent dimers into $S = 1/2$ chains.

### Nitronylnitroxide Salt

In one series of compounds with $S = 1/2$ nitronylnitroxide radicals,m-N-methylpyridinium - nitronyl nitroxide (m-MPYNN).X. $\frac{1}{3}$(acetone), where X = $BF_4$, $ClO_4$ and $I_3$, the unpaired electrons on the MPYNN molecules are at the origin of magnetism [121]. Structurally, this family has a very unique packing structure of MPYNN cation radicals in a 2D hexagonal layer. The double contact between the NO group and the pyridine ring between two molecules yields a ferromagnetic coupling, 23.3 K, for the $BF_4^-$ salt. Below this temperature, the dimers become effective $S = 1$ units arranged in a kagomé lattice with an inter-dimer antiferromagnetic interaction ($\sim$ 3 K for the $BF_4^-$ salt). The susceptibility clearly shows a non-magnetic ground state, with a gapped behavior [122] but this compound also undergoes a structural transition at high $T$ which destroys the three-fold symmetry [123] and casts doubt over its relevance to the kagomé physics.

### 9.5.2 $Y_{0.5}Ca_{0.5}BaCo_4O_7$

In $Y_{0.5}Ca_{0.5}BaCo_4O_7$, 75% of all Co are believed to occupy a kagomé site with a $Co^{2+}$, $S = 3/2$ state whereas the remaining 25% sites are non magnetic ($Co^{3+}$) and located between the kagomé layers [124]. Only a diffuse scattering neutron study with polarization analysis has been performed on this compound up to now. No three-dimensional long-range antiferromagnetic is observed at 1.2 K, well below the apparent Weiss temperature $\sim$ 2,200 K. The measured scattering intensities are consistent with the valence assignment for the different sites and the diffuse magnetic scattering indicate the presence of short range 2D chiral spin correlations for a spin hexagon, at low $T$.

### 9.5.3 Langasites

If restricted to n.n. interactions, the magnetic lattice of Nd-Langasite $Nd_3Ga_5SiO_{14}$ is topologically equivalent to a kagomé lattice, the first one to be decorated with rare earth ions [16, 125]. This has the potential to open a new field of research of the type seen in the pyrochlores where anisotropy effects tuned by selection of the rare-earth ion play a leading role. Indeed, neutron scattering, magnetization, and specific heat measurements performed on very dilute samples indicate that crystal electric field effects on the $Nd^{3+}$ ion play a definitive role but might also complicate the interpretation of macroscopic magnetic and specific heat measurements [126] in terms of frustration. Indeed, with a CEF $\sim$ 100 K, the few K exchange energy is not expected to play the leading role. Local $\mu$SR, NQR andNMR measurements indicate the persistence of rather slow fluctuations down to 60 mK [127] and neutron measurements do not indicate any long-range order down to 46 mK order, hence there is the possibility of a spin-liquid ground state [128]. Due to the smallness of the exchange, probing the nature of the true ground state represents a real challenge and, as a prerequisite, more work needs to be performed in order to have a reliable estimate of $J$.

It has been theoretically predicted that for classical Ising spins $S > 3/2$ a crossover is to be expected from a cooperative paramagnet state suggested for long [14, 15] to a semiclassical spin-liquid state [129]. Indeed, transverse quantum fluctuations to the easy-axis should lift the extensive classical degeneracy. Further, magnetization, neutron diffraction and specific heat measurements indicate that the fluctuations are considerably suppressed under an applied field. A crossover to a more ordered ground state is observed for fields of $\sim$ 1 T [128]. Whether this relates to a nematic-like order as suggested theoretically [130] needs further studies.

Quite recently Pr-Langasite $Pr_3Ga_5SiO_{14}$ has been investigated, which shows all the potential of this new family [131]. Again, no long-range order is evident down to 35 mK but, as proven from a $\mu$SR study, $Pr^{3+}$ ion is in a singlet state, well explained by its Kramers character and the symmetry [132]. This calls for specific care about the rare earth character of the magnetic ions in the Langasite family. Certainly, pyrochlore studies performed for the last 10 years will be a good source of inspiration. The subtle balance between dipolar interactions, single-ion anisotropy and exchange might be crucial in the definition of the ground state.

## 9.6 Conclusion

We have provided a brief overview of most deeply studied kagomé-based magnets in the past 20 years. Clearly the kagomé geometry favors what are arguably the most original low-energy excitations in condensed matter and yields singular responses in all the measured quantities, as well as marginal orders (if any), such as unconventional spin glass states or fluctuating ordered magnets which are still unsatisfactorily modeled.

While each system has its specific deviations from the ideal case, three of them are commonly identified:

(1) Moderate ($\sim$ a few %) non-magnetic dilution of the frustrated network which does not seem to impact much the nature of the ground state such as in SCGO, Herbertsmithite and some of the jarosites. These can mask much of the intrinsic physics in macroscopic measurements through an effective paramagnetic response of the perturbed magnetic lattice,
(2) Non-negligible Dzyaloshinskii–Moriya anisotropy such as found in the Jarosites or Herbertsmithite, which in contrast with site disorder, clearly affects the nature of the ground state by mixing a hypothetic singlet ground state and excited triplet states,
(3) Lattice defects such as non-perfect n.n. kagomé magnetic structure like in volborthite or 3-D coupling of the kagomé layers.

The two former ingredients are starting to be incorporated into models and will certainly provide new tests of theories. This "perturb to reveal strategy" might turn out to be the right one to cope with the non-ideality and a rewarding way to experimentally unveil the hidden nature of the ideal kagomé lattice.

The recent discovery of quite simple kagomé systems such as Herbertsmithite, where no freezing has been observed up to now down to $J/9000$, has given a new impetus to the field on both the theoretical and the experimental sides, which are presently flourishing. For the first time, the possibility of probing the ultimate ground state of the quantum KHAF without any marginal order exists, but deviations from the simple model caused by the presence of 5–10% dilution and

$\sim 0.1J$ DM anisotropy, continues to power a tremendous experimental search for yet more ideal two dimensional, spin 1/2, kagomé compound, which has no inter-plane interactions and no impurities on the kagomé plane. In this spirit, it is natural to welcome the promising recently discovered Kapellasite, a polymorph of Herbertsmithite [135–137] and of which study is underway. Also, novel exciting tracks are explored, with mixed organic-mineral compounds which seem to approach the ideal case [133, 134].

Finally, the synthesis of single crystals of such new $S = 1/2$ promising materials such as Herbertsmithite, Volborthite and subsequent compounds as well as charge doping are routes for future synthesis work which will certainly open major avenues to unveil the many facets of the kagomé physics.

## References


1. P.W. Anderson, Math. Res. Bull. **8**, 153 (1973)
2. P.W. Anderson, Science **235**, 1196 (1987)
3. A.P. Ramirez, G.P. Espinosa, A.S. Cooper, Phys. Rev. Lett. **64**, 2070 (1990)
4. M.P. Shores, E.A. Nytko, B.M. Barlett, D.G. Nocera, J. Am. Chem. Soc. **127**, 13462 (2005)
5. For an introduction see B.G. Levy, Phys. Today 16 (2007)
6. M.A. Lafontaine, A.L. Bail, G. Férey, J. Solid State Chem. **85**, 2020 (1990)
7. Z. Hiroi et al., J. Phys. Soc. Jpn. **70**, 3377 (2001)
8. P. Lecheminant, B. Bernu, C. Lhuillier, L. Pierre, P. Sindzingre, Phys. Rev. B **56**, 2521 (1997)
9. G. Misguich, C. Lhuillier, in Two Dimensional Quantum Antiferromagnets, ed. by H.T. Diep. Frustrated Spin Systems (World Scientific Publishing, Singapore, 2004)
10. A. Olariu et al., Phys. Rev. Lett. **97**, 167203 (2006)
11. S. Nakatsuji et al., Science **309**, 1697 (2005)
12. F. Mila, Phys. Rev. Lett. **81**, 2356 (1998)
13. C. Lhuillier, Lecture notes from the Ecole de troisieme cycle de Suisse Romande, arxiv.org/abs/cond-mat/0502464v1 (2005)
14. J.H. Barry, M. Khatum, Int. J. Mod. Phys. B **11**, 93 (1997) and references therein
15. R. Liebmann, Statistical Mechanics of Periodic Frustrated Ising Systems (Springer, Berlin, 1986)
16. P. Bordet et al., J. Phys. Condens. Matter **18**, 5147 (2006)
17. P. Sindzingre, C. Lhuillier, to be published in Europhys. Lett. (2009), arXiv:0907.4164
18. C. Palache, H. Berman, C. Frondel, **Dana's System of Mineralogy** (Wiley, London, 1951)
19. S.B. Hendricks, Am. Miner. **22**, 773 (1937)
20. A.S. Wills, A.M.L. Smith et al., J. Magn. Magn. Mater. **272–276**, 1300 (2004)
21. J.T. Szymanski, Can. Miner. **23**, 659 (1985)
22. A.S. Wills, Can. J. Phys. **79**, 1501 (2001)
23. J.E. Dutrizac, NATO Conf. Ser. (Hydrometall Process Fund, 1984)
24. T. Inami, M. Nishiyama et al., Phys. Rev. B **61**, 12181 (2000)
25. J. Frunzke, T. Hansen et al., J. Mater. Chem. **11** 179 (2001)
26. B.M. Bartlett, D.G. Nocera, J. Am. Chem. Soc. **127**, 8985 (2005)
27. D. Grohol, D.G. Nocera, D. Papoutsakis, Phys. Rev. B **67**, 064401 (2003)
28. A.S. Wills, A. Harrison et al., Phys. Rev. B **61**, 6156 (2000)
29. A. Chubukov, Phys. Rev. Lett. **69**, 832 (1992)
30. S. Sachdev, Phys. Rev. B **45**, 12377 (1992)
31. S. Maegawa et al., J. Phys. Soc. Jpn. **65**, 2776 (1996)
32. M.F. Collins, unpublished work.

33. A.S. Wills, A. Harrison, J. Chem. Soc. Faraday Trans. **92**, 2161 (1996)
34. A.S. Wills, A. Harrison et al., Europhys. Lett. **42**, 325 (1998)
35. A.S. Wills, V. Dupuis et al., Phys. Rev. B **62**, R9264 (2000)
36. A.S. Wills, Ph.D. Thesis, The University of Edinburgh (1997)
37. T. Inami et al., Phys. Rev. B **64**, 054421 (2001)
38. S.-H. Lee et al., Phys. Rev. B **56**, 8091 (1997)
39. A.S. Wills, Phys. Rev. B **63**, 064430 (2001)
40. D. Grohol et al., Phys. Rev. B, **68**, 094404 (2003)
41. M. Kato, T. Hori et al., Phys. B Condens. Matter **329**, 1042 (2003)
42. D. Grohol, Q.Z. Huang et al., Phys. Rev. B **68**, 094404 (2003)
43. M. Nishiyama, S. Maegawa, Physica B **329–333**, 1065 (2003)
44. D. Grohol, D.G. Nocera, J. Am. Chem. Soc. **124**, 2640 (2002)
45. W. Bisson, A.S. Wills, Z. Kristallogr. Suppl. **26**, 511 (2007)
46. M.G. Townsend, G. Longworth, E. Roudaut, Phys. Rev. B **33**, 4919 (1986)
47. T. Inami, S. Maegawa et al., J. Magn. Magn. Mater. **177**, 752 (1998)
48. D. Grohol, D.G. Nocera, Chem. Mater. **19**, 3061 (2007)
49. W.G. Bisson, A.S. Wills, J. Phys.: Condens. Matter **20**, 452204 (2008)
50. A. Harrison, A.S. Wills et al., Physica B **241**, 722 (1997) "Long-range order induced by diamagnetic dilution of jarosites, model kagomé antiferromagnets."
51. M. Elhajal, B. Canals et al., Phys. Rev. B **66**, 014422 (2002)
52. M. Elhajal, B. Canals et al., Physica B- Condens. Matter **312**, 716 (2002)
53. K. Matan, et al., Phys. Rev. Lett. **96**, 247201 (2006)
54. T. Yildirim, A.B. Harris, Phys. Rev. B **73**, 214446 (2006)
55. J.A. Mydosh, Spin Glasses: An Experimental Introduction (Taylor & Francis Ltd, London, 1993)
56. I. Ritchey, P. Chandra, P. Coleman, Phys. Rev. B **47**, 15342 (1993)
57. P. Chandra, P. Coleman, I. Ritchey, J.Physique I **3**, 591 (1993)
58. P. Bonville, V. Dupuis et al., Hyperfine Inter. **168**, 1085 (2006)
59. J. Majzlan, R. Stevens et al., Phys. Chem. Miner. **31**, 518 (2004)
60. G.S. Oakley, D. Visser et al., Physica B **268**, 142 (1999)
61. A. Harrison, K.M. Kojima et al., Physica B- Condens. Matter **289**, 217 (2000)
62. V. Dupuis, E. Vincent et al., J. Appl. Phys. **91**, 8384 (2002)
63. F. Ladieu, F. Bert et al., J. Phys. Condens. Matter **16**, S735 (2004)
64. A. Keren, K. Kojima et al., Phys. Rev. B **53**, 6451 (1996)
65. T. Morimoto, M. Nishiyama et al., J. Phys. Soc. Jpn. **72**, 2085 (2003)
66. X. Obradors et al., Solid State Commun. **65**, 189 (1988)
67. I.S. Hagemann et al., Phys. Rev. Lett. **86**, 894 (2001)
68. A.P. Ramirez, Annu. Rev. Mater. Sci. **24**, 453 (1994)
69. C. Mondelli, H. Mutka, C. Payen et al., J. Can. Phys. **79**, 1401 (2001)
70. H. Ohta et al., J. Phys. Soc. Jpn. **65**, 848 (1996)
71. For a detailed discussion and a review, see D. Bono et al., Low Temp. Phys. **8–9**, 704 (2005) and references therein.
72. S.-H. Lee et al., Phys. Rev. Lett. **76**, 4424 (1996)
73. D. Bono et al., Phys. Rev. Lett. **92**, 217202 (2004)
74. L. Limot et al., Phys. Rev. B **65**, 144447 (2002)
75. P. Mendels et al., Phys. Rev. Lett. **85**, 3496 (2000)
76. A.P. Ramirez, G.P. Espinosa, A.S. Cooper, Phys. Rev. B **45**, 2505 (1992)
77. A.P. Ramirez, B. Hessen, M. Winklemann, Phys. Rev. Lett. **84**, 2957 (2000)
78. Y.J. Uemura et al., Phys. Rev. Lett. **73**, 3306 (1994)
79. S.-H. Lee et al., Europhys. Lett. **35**, 127 (1996)
80. D. Bono, P. Mendels, G. Collin, N. Blanchard, C. Baines, A. Amato, J. Phys.: Condens. Matter **16**, S817 (2004)
81. A. Keren et al., Phys. Rev. Lett. **84**, 3450 (2000)
82. D.E. MacLaughlin, H. Alloul, Phys. Rev. Lett. **36**, 1158 (1976)
83. P. M. Singer, T. Imai, Phys. Rev. Lett. **88**, 187601 (2002)

84. H. Alloul, J. Bobroff, M. Gabay, P.J. Hirschfeld, Rev. Mod. Phys. **81**, 45 (2009)
85. S. Dommange et al., Phys. Rev. B **68**, 224416 (2003)
86. A. Lauchli et al., Phys. Rev. B **76**, 144413 (2007)
87. B. Martinez et al., Phys. Rev. B **46**, 10786 (1992)
88. For a discussion see, A.P. Ramirez Handbook on Magnetic Materials, ed. by K.J.H. Busch (Elsevier Science, Amsterdam, 2001), vol. 13, p. 423
89. D. Bono et al., Phys. Rev. Lett. **93**, 187201 (2004)
90. Z. Hiroi, M. Hanawa, N. Kobayashi, M. Nohara, H. Takagi, Y. Kato, M. Takigawa, J. Phys. Soc. Jpn. **70**, 3377 (2001)
91. F. Wang, A. Vishwanath, Y.B. Kim, Phys. Rev. B **76**, 094421 (2007)
92. T. Yavors'kii, W. Apel, H.-U. Everts, Phys. Rev. B **76**, 064430 (2007)
93. M. Yoshida et al., J. Phys. Soc. Jpn. **78**, 043704 (2009)
94. M. Yoshida et al., Phys. Rev. Lett. **103**, 077207 (2009)
95. P. Mendels, A. Olariu, F. Bert, D. Bono, L. Limot, G. Collin, B. Ueland, P. Schiffer, R.J. Cava, N. Blanchard, F. Duc, J.C. Trombe, J. Phys.: Condens. Matter **19**, 145224 (2007)
96. F. Bert et al., Phys. Rev. Lett. **95**, 087203 (2005)
97. A. Fukaya et al., Phys. Rev. Lett. **91**, 207603 (2003)
98. F. Bert, P. Mendels, D. Bono, A. Olariu, F. Ladieu, J.-C. Trombe, F. Duc, C. Baines, A. Amato, A. Hillier, Physica B **374–375**, 134 (2006)
99. G. Nilsen et al., ArXiv 1001.2462 (2010)
100. P.A. Lee, Science, Perspectives **321**, 1306 (2008)
101. Y. Ran, M. Hermele, P.A. Lee, X.G. Wen, Phys. Rev. Lett. **98**, 117205 (2007)
102. R.R.P. Singh, D.A. Huse, Phys. Rev. B **76**, 180407 (2007)
103. R.R.P. Singh, D.A. Huse, Phys. Rev. B **77**, 144415 (2008)
104. S. Ryu, O.I. Motrunich, J. Alicea, M.P.A. Fisher, Phys. Rev. B **75**, 184406 (2007)
105. X.G. Zheng et al., Phys. Rev. Lett. **95**, 057201 (2005)
106. A.S. Wills et al., J. Magn. Magn. Matter **272–276**, 850 (2004)
107. P. Mendels et al., Phys. Rev. Lett. **98**, 077204 (2007)
108. J.S. Helton et al., Phys. Rev. Lett. **98**, 107204 (2007)
109. T. Imai et al., Phys. Rev. Lett. **100**, 077203 (2008)
110. A. Olariu et al., Phys. Rev. Lett. **100**, 087202 (2008)
111. O. Ofer, A. Keren, Phys. Rev. B **79**, 134424 (2009)
112. M. Rigol, R.R.P. Singh, Phys. Rev. Lett. 98, 207204 (2007)
113. G. Misguich, P. Sindzingre, Eur. Phys. J. B **59**, 305 (2007)
114. M.A. de Vries et al., Phys. Rev. Lett. **100**, 157205 (2008)
115. F. Bert et al., Phys. Rev. B **76**, 132411 (2007)
116. A. Zorko et al., Phys. Rev. Lett. **101**, 026405 (2008)
117. P. Mendels, F. Bert, to be published in J. Phys. Soc. Japan, **79**, 011001 (2010)
118. O. Cépas, C.M. Fong, P.W. Leung, C. Lhuillier, Phys. Rev. B **78**, 140405 (2008)
119. I. Rousochatzakis et al., Phys. Rev. B **79**, 214415 (2009)
120. S.A. Baudron et al., J. Am. Chem. Soc. **127**, 11785 (2005)
121. K. Awaga et al., Phys. Rev. B **49**, 3975 (1994)
122. I. Watanabe et al., Phys. Rev. B **58**, 2438 (1998)
123. T. Kambe et al., J. Phys. Soc. Jpn. **73**, 796 (2004)
124. W. Schweika, M. Valldor, P. Lemmens, Phys. Rev. Lett. **98**, 067201 (2007)
125. J. Robert et al., Phys. Rev. Lett. **96**, 197205 (2006)
126. V. Simonet et al., Phys. Rev. Lett. **100**, 237204 (2008)
127. A. Zorko et al., Phys. Rev. Lett. **100**, 147201 (2008)
128. H.D. Zhou et al., Phys. Rev. Lett. **99**, 236401 (2007)
129. A. Sen, F. Wang, K. Damle, R. Moessner, Phys. Rev. Lett. **102**, 227001 (2009)
130. A. Sen, K. Damle, A. Vishwanath, Phys. Rev. Lett. **100**, 097202 (2008)
131. H.D. Zhou et al., Phys. Rev. Lett. **102**, 067203 (2009)
132. A. Zorko et al., Phys. Rev. Lett. **104**, 057202 (2010)
133. L. Marcipar et al., Phys. Rev. B **58**, 2438 (1998)
134. E.A. Nytko, J.S. Helton, P. Müller, D.G. Nocera, J. Am. Chem. Soc. **130**, 2922 (2008)
135. W. Krause, H. Bernhardt, R.S.W. Braithwaite, U. Kolitsch, R. Pritchard, Mineral. Mag. **70**, 329 (2006)
136. R.H Colman, C. Ritter, A.S. Wills, Chem. Mater. **20**, 6897 (2008)
137. O. Janson, J. Richter, H. Rosner, Phys. Rev. Lett. **101**, 106403 (2008)

# Part IV
# Specific Effects in Frustrated Magnets

# Chapter 10
# Magnetization Plateaus

**Masashi Takigawa and Frédéric Mila**

**Abstract** This chapter is devoted to a review of the phenomenon of magnetization plateaus in frustrated quantum magnets. We explain why frustration can introduce 'accidents' in the magnetization process of quantum antiferromagnets, in the form of plateaus occurring at rational values of the magnetization. These are driven by two mechanisms: the stabilization of a commensurate, classical state by quantum fluctuations or the occurrence of a superfluid-insulator transition in effective hard-core-boson models. The first mechanism always leads to a 'classical' plateau, i.e. a plateau with a simple classical analog, while the second often drives the system into a 'quantum' plateau which has no classical analog. Experimental examples of such plateaus are discussed in detail, with special emphasis on the compound $SrCu_2(BO_3)_2$, in which a remarkable series of plateaus has been discovered and is currently under intensive investigation. Related topics including magnetization jumps, supersolid phases, and lattice distortions are discussed briefly.

## 10.1 Introduction

The magnetization of a classical Néel antiferromagnet at zero temperature is a very simple process. To minimize the total energy, the spins lie primarily in the plane perpendicular to the field, but their field-driven polarization causes a tilt into the field direction, resulting in a magnetization which increases linearly with the field up to saturation (Fig. 10.1). In a real antiferromagnet, this simple picture is modified in several respects. In particular, the magnetization usually depends on the orientation of the field, as a consequence of anisotropies arising from various sources, and thus can jump when the spins undergo a spin-flop transition. However, even without anisotropy, the magnetization of a real antiferromagnet at zero temperature will be different from the classical paradigm because of the quantum nature of spin.

In unfrustrated geometries, such as the square lattice with nearest-neighbor interactions, quantum fluctuations have a relatively modest impact on the magnetization. They reduce the magnetization somewhat below its classical value and typically induce a curvature in $M(H)$ [1], as indicated in the left panel of Fig. 10.1. This

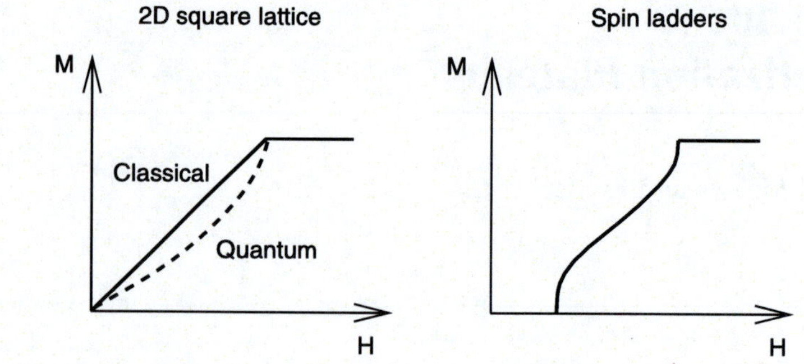


**Fig. 10.1** Schematic representation of the magnetization curve of typical unfrustrated antiferromagnets

is the case provided that the dimensionality is greater than one. In one dimension (1D), quantum fluctuations can open a gap, as shown by Haldane for spin-1 chains [2], and this Haldane gap between the singlet ground state and the first triplet excitation forces the magnetization to remain exactly equal to zero up to a critical field proportional to the gap. This result applies to all gapped antiferromagnets, such as spin-1/2 ladders [3] (Fig. 10.1) or weakly coupled dimers (below).

In frustrated geometries, quantum fluctuations are enhanced significantly. The consequences for the magnetization curve can be rather dramatic. In particular, under appropriate circumstances, quantum fluctuations can induce magnetization plateaus at rational values of the magnetization, as first predicted in some pioneering papers [4–7, 9, 10]. With the development of high-field facilities, such plateaus have been detected in a significant number of compounds. One example is shown in Fig. 10.2 [11]. As a consequence, understanding the mechanisms leading to these plateaus, and the nature and magnetic structure of the states realized in the plateau regions, has emerged as a very active field of research in the context of highly frustrated magnets.

The goal of this chapter is to present an overview of the field, with an emphasis on the basic mechanisms behind plateau formation and on the experimental consequences beyond the plateaus themselves.

## 10.2 Mechanisms for Formation of Magnetization Plateaus

Although parts of this discussion could be applied to other cases, we concentrate in this section on antiferromagnets with pure Heisenberg interactions in a magnetic field. These are described by the Hamiltonian

$$\mathcal{H} = \sum_{i,j} J_{ij}\, \boldsymbol{S}_i \cdot \boldsymbol{S}_j - g\mu_{\mathrm{B}} H \sum_i S_i^z, \tag{10.1}$$

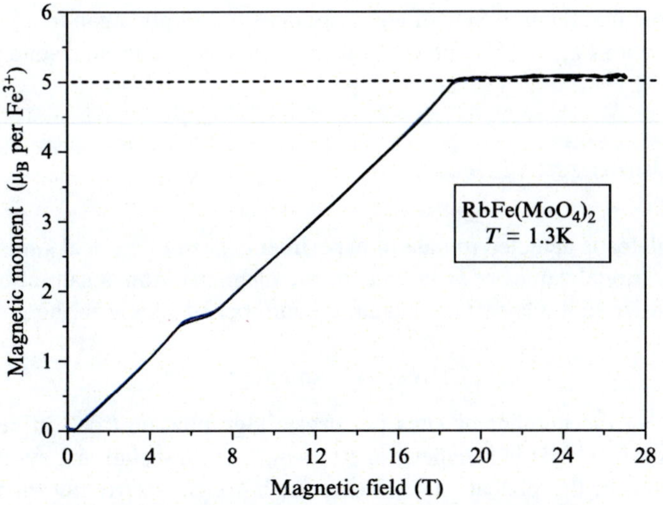


**Fig. 10.2** The magnetization plateau at 1/3 of saturation observed in $RbFe(MoO_4)_2$. The magnetic field was applied perpendicular to the $c$-axis of the crystal (Fig. 10.5). (After Smirnov et al. [11])

where $\boldsymbol{S}_i$ is a spin-$S$ operator. We are interested in those situations where a plateau appears at a certain value $m = g\mu_B s$ of the magnetization per site.

## 10.2.1 Spin Gap

The presence of a plateau implies some kind of incompressibility, in the sense that it is impossible to increase the total spin $S^z_{\text{total}} = \sum_i S^z_i$ by increasing the magnetic field. This in turn implies the presence of a gap to magnetic excitations. Indeed, if one considers a system of $N$ sites and supposes that there is a magnetization plateau for $H_{c1} < H < H_{c2}$, inside which the magnetization corresponds to $S^z_{\text{total}} = Ns$, then at $H_{c1}$ ($H_{c2}$), there is a level-crossing with a state of lower (higher) magnetization, for example $Ns - n_1$ ($Ns + n_2$). Because the energy of a state is a linear function of the field with slope $-g\mu_B S^z$, the state with magnetization $Ns - n_1$ ($Ns + n_2$) is separated from the ground state inside the plateau by an energy $g\mu_B n_1(H - H_{c1})$ ($g\mu_B n_2(H_{c2} - H)$). Thus the first magnetic excitation (i.e. the first state with $S^z_{\text{total}} \neq Ns$) is separated from the ground state throughout the plateau by a gap $\Delta(H)$ given by

$$\Delta(H) = g\mu_B S \min\left(n_1(H - H_{c1}), n_2(H_{c2} - H)\right). \tag{10.2}$$

This gap can be detected by nuclear magnetic resonance (NMR) or inelastic neutron scattering experiments, and indeed it has been detected in a number of systems (below). Note that the presence of a plateau does not imply a gap to all excitations: other states with $S^z_{\text{total}} = Ns$ can be arbitrarily close to the ground state [12].

### 10.2.2 Quantized Plateaus

All of the plateaus detected to date in experiments or in numerical simulations correspond to rational values of $m$ in units of the magnetization at saturation. In 1D, it has been shown that, inside the plateau, the number of sites $n$ in the unit cell must be such that

$$n(S - s) \text{ is an integer.} \tag{10.3}$$

Let us call $n_0$ the number of sites per crystallographic unit cell in zero field. If the condition of (10.3) is satisfied for $n = n_0$, then translational symmetry need not be broken in the plateau state. However, if $n_0(S - s)$ is not an integer, the number of sites per unit cell will be larger in the plateau phase than in zero field. In that case, translational symmetry is broken inside the plateau, and the plateau state is accordingly a phase in the thermodynamic sense, separated from the high-temperature paramagnetic phase by a phase transition.

The core of the argument is an extension of the Lieb–Schultz–Mattis theorem [13]. Following Oshikawa, Yamanaka, and Affleck [7], let us consider the case with one site per crystallographic unit cell in zero field and suppose that $S - s = p/q$, with $p$ and $q$ coprime. Then one can construct $q$ states which satisfy the following conditions: (1) they are degenerate with the ground state in the thermodynamic limit; (2) they are eigenvectors of the translation operators with eigenvalues $\exp[2\pi i(S - s)q]$; (3) they are mutually orthogonal. Point (3) is a simple consequence of point (2), which itself follows easily from the form of these states. Then, if these states are linear combinations of states with broken translational symmetry, these latter must have $q$ sites per unit cell. Thus $n = q$, and hence $n(S - s) = p$ is indeed an integer.

In 2D, similar arguments have been articulated [8], and if the translational symmetry is broken, (10.3) also applies. In 2D, the degeneracy can be of topological origin however, as in the quantum dimer models discussed in the Chap. 17 by Moessner and Raman, and it is in principle possible to have a rational plateau without breaking the translational symmetry. Such a possibility has not been realized yet in frustrated magnets.

According to this formula, the (minimal) number of sites in the unit cell depends not only on the fraction of the magnetization with respect to saturation, but also on the spin. As an example, for a 1/8-plateau, meaning a plateau occurring at a magnetization which is 1/8 of $M$ at saturation, this result implies a unit cell with a number of sites which is a multiple of 16 for $S = 1/2$, 8 for $S = 1$, 16 for $S = 3/2$, 4 for $S = 2$, and so on.

### 10.2.3 Order by Disorder

We turn next to a discussion of the physical mechanisms which can produce a plateau. The first one is related conceptually to the notion of order by disorder, discussed in the Chap. 1 by Chalker in this volume. In a semiclassical ($1/S$) approach, the ground-state energy of a quantum antiferromagnet can be written as

$$E_0 = E_{\text{classical}} + \frac{1}{2}\sum_q \omega_q + O(1), \tag{10.4}$$

where the first term (which has contributions of order $S^2$ and $S$) is the same for all classical ground states if there are several of these, while the zero-point energy (the second term) is of order $S$ and usually depends on the classical configuration around which the expansion is made. Thus if the classical ground state is degenerate, this degeneracy can be lifted by the zero-point energy. The general rule is that coplanar (and even collinear if there are any) ground states are favored because their fluctuations are softer.

This mechanism appears to control entirely the magnetization process of the triangular-lattice Heisenberg antiferromagnet in a magnetic field. This model is described by the Hamiltonian

$$\mathcal{H} = J\sum_{\langle i,j\rangle} \boldsymbol{S}_i \cdot \boldsymbol{S}_j - H\sum_i S_i^z, \tag{10.5}$$

where the first sum is over nearest-neighbor pairs on the triangular lattice and the factor $g\mu_{\mathrm{B}}$ has been included for simplicity in the field $H$. In zero field, the classical ground state is the planar, 3-sublattice structure in which the spin orientations are at 120 degrees to each other on each triangle. Up to a global rotation, this ground state is unique. In a field, the energy can be minimized as usual by the 'umbrella' structure, where the spins remain primarily in the plane perpendicular to the field, but have a component tilted towards the field. However, unlike on the square lattice, there are many other ground states in the case of the triangular lattice, a result which may be traced to the fact that the Hamiltonian can be written in terms of the total spin of elementary triangles (as explained in the Chap. 1 by Chalker). An exhaustive discussion of the classical ground-state manifold can be found in [14]. In particular, there are several coplanar ground states for all fields. In agreement with the general expectation, among all of the classical ground states for a given field, thermal or quantum fluctuations act to select coplanar ground states; among these they select the state which has one spin antiparallel to the field for $H < H_{\mathrm{c}}$ and the state with two parallel spins for $H_{\mathrm{c}} < H < H_{\mathrm{sat}}$, where $H_{\mathrm{sat}} = 18JS$ is the saturation field and $H_{\mathrm{c}} = H_{\mathrm{sat}}/3$ is the critical field at which the classical magnetization is exactly 1/3 of its saturation value (Fig. 10.3).

Because, for a given field, thermal or quantum fluctuations select a unique state, one might be tempted to believe that this is the complete story, in which case the

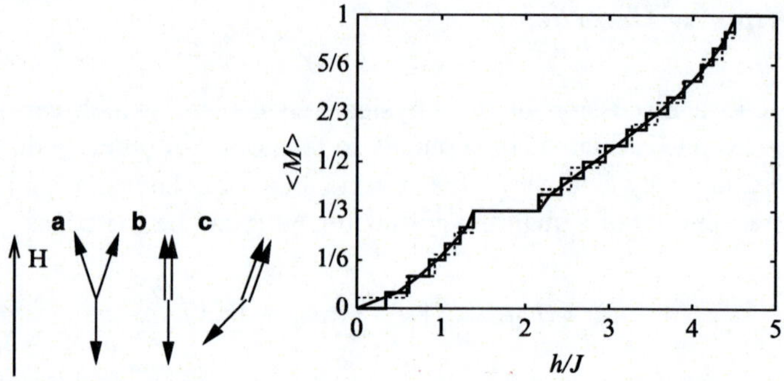


**Fig. 10.3** *Left*: classical ground-state configurations on the triangular lattice. **(a)** $0 < H < H_c$, **(b)** $H = H_c$, **(c)** $H_c < H < H_{sat}$. *Right*: magnetization curve of the $S = 1/2$ Heisenberg antiferromagnet on the triangular lattice. The *thin dashed* and *solid curves* are respectively for $N = 27$ and 36 sites. The *bold curve* is an extrapolation to the thermodynamic limit. (After Honecker et al. [15])

magnetization of the Heisenberg model on the triangular lattice would be a smooth curve similar to that on the square lattice (Fig. 10.1). However, comparing the energies of various candidate ground states for a given field is not sufficient: if in a given configuration the fluctuations are suitably soft, the zero-point energy gain can be large enough to overcome the classical energy difference and stabilize the state with soft fluctuations in a parameter range where it is not the classical ground state. On the triangular lattice, the configuration stabilized at $H_c$ is a collinear one with two spins oriented along the field direction and one spin opposite to it. According to the general rule, this state is expected to have very soft fluctuations. The $1/S$ expansion performed in [4] confirms this expectation, leading to the conclusion that this configuration is indeed stable in a parameter range around $H_c$ limited by the critical fields $H_{c1} = H_c(1 - 0.084/S)$ and $H_{c2} = H_c(1 + 0.215/S)$. Thus one has a clear case where quantum fluctuations lead to a magnetization plateau at 1/3. This conclusion has been confirmed by exact-diagonalization calculations performed for finite clusters with $S = 1/2$, which provide very strong evidence in favor of such a plateau (Fig. 10.3). We note here that thermal fluctuations also lead to a stabilization of the collinear structure, as first discovered in [14]. The 1/3-plateau in triangular-lattice antiferromagnets has been observed experimentally (Fig. 10.2), as we will discuss in Sect. 10.3.1. The same mechanism has been investigated also in other models, with similar conclusions [16].

### 10.2.4 Superfluid-Insulator Transition

The incompressible nature of the plateau phase suggests that it should be possible to interpret it as some kind of insulating phase. In principle, this can be achieved by converting the spin operators into bosons or fermions. In view of the symmetries of the problem, the bosonic approach is more natural. In the presence of a magnetic field, the SU(2) rotational symmetry of the Heisenberg model is reduced to a U(1) symmetry corresponding to the rotational symmetry around the field direction, and magnetic ordering is in the same universality class as the Bose–Einstein condensation of bosons into a superfluid phase. For $S = 1/2$, this description is easily implemented using the Matsubara–Matsuda transformation into hard-core-bosons, which allows one to express the spin operators in terms of bosonic operators acting in a reduced Hilbert space with at most one boson per site,

$$S_i^+ = b_i^\dagger, \quad S_i^- = b_i, \quad S_i^z = n_i - 1/2. \tag{10.6}$$

This mapping transforms the Heisenberg model with nearest-neighbor spin exchange $J$ into a hard-core-boson model with nearest-neighbor hopping and inter-site repulsion on the same lattice,

$$\mathcal{H} = t \sum_{\langle i,j \rangle} (b_i^\dagger b_j + H.c.) + V \sum_{\langle i,j \rangle} n_i n_j - \mu \sum_i n_i, \tag{10.7}$$

where $t = J/2$, $V = J$, and $\mu = H - zJ/2$, $z$ being the coordination number of the lattice. Note that zero magnetic field, where the total spin along the field vanishes in the ground state, corresponds to half-filling of the lattice by bosons. This model has explicit particle-hole symmetry, densities larger (smaller) than 1/2 corresponding to positive (negative) field.

In this language, a magnetization plateau is a commensurate charge-density wave of bosons. For specificity, consider again the triangular lattice of the previous section. The state inside the 1/3-plateau, with two spins per unit cell oriented along the field and one antiparallel to it, corresponds to a charge-density wave at 2/3 filling with two sites occupied and one empty. This is indeed a natural instability of hard-core-bosons or of fermions on the triangular lattice. The symmetrical state at 1/3 filling, with one occupied site and two empty sites per unit cell, corresponds to the 1/3-plateau in negative field.

In such a plateau, and more generally in all plateaus that can be described by a semiclassical approach starting from a classical configuration, the order parameter is purely local in both descriptions: it is the spin component along the field, $S_i^z$, in the spin language, and the boson density, $n_i$, in the bosonic language. Following Hida and Affleck [17], we refer to such plateaus as 'classical' plateaus, even if their stabilization at zero temperature is due to quantum fluctuations.

### 10.2.5 'Quantum' Plateaus

That all plateaux are not simply connected to a classical state can be seen easily by considering the frustrated spin-1/2 ladder described by the Hamiltonian

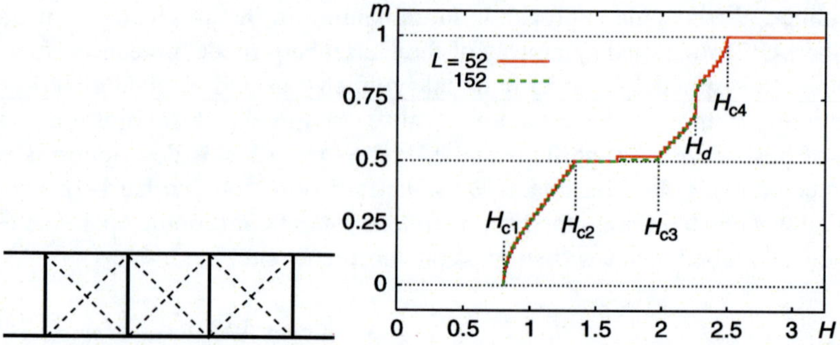


**Fig. 10.4** *Left*: representation of the frustrated ladder of (20.2). The couplings entering (20.2) are denoted by the different line types, *thick solid* ($J_\perp$), *thin solid* ($J_\parallel$), and *dashed* ($J_\times$). *Right*: magnetization curve for the spin ladder of (20.2) obtained by density-matrix renormalization-group (DMRG) calculations. (After Fouet et al. [18])

$$\mathcal{H} = \sum_n J_\perp \boldsymbol{S}_{1n} \cdot \boldsymbol{S}_{2n} - H \sum_n (S^z_{1n} + S^z_{2n}) + \sum_n \sum_{i=1,2} J_\parallel \boldsymbol{S}_{in} \cdot \boldsymbol{S}_{in+1}$$
$$+ \sum_n J_\times (\boldsymbol{S}_{1n} \cdot \boldsymbol{S}_{2n+1} + \boldsymbol{S}_{2n} \cdot \boldsymbol{S}_{1n+1}). \quad (10.8)$$

In the spin operators $\boldsymbol{S}_{in}$, the index $i$ refers to the ladder leg while the index $n$ refers to the rung (Fig. 10.4). If the two inter-rung couplings $J_\parallel$ and $J_\times$ are equal, the simple product of $S_z = +1$ triplets and singlets on alternating rungs is an eigenstate with magnetization 1/2. This state can be proven to be the ground state in a finite field range, and hence to correspond to a 1/2-plateau. However, this wavefunction has no classical counterpart, because every second rung is in a singlet state. Again following [17], we refer to such plateaus as 'quantum' plateaus.

The difference between quantum and classical plateaus lies in the symmetry of the system. A classical plateau in this situation would have 3 spins up and 1 spin down per unit cell, thus breaking both the translational symmetry along the ladder leg and the mirror symmetry perpendicular to the rungs, leading to a four-fold degeneracy. By contrast, the 1/2-plateau wave function realized in the frustrated ladder breaks only the translational symmetry, and leads only to a two-fold degeneracy.

Beyond this ladder system, the inadequacy of the semiclassical approach to describe some magnetization plateaus has been identified in a trimerized chain [17], as well as in coupled-dimer systems such as the two-dimensional orthogonal dimer model realized in the compound $SrCu_2(BO_3)_2$ (below). It is also clearly inadequate to account for the 2/3-plateau discovered recently in the antiferroquadrupolar phase of the bilinear–biquadratic spin-1 model on the triangular lattice [19] (discussed in the Chap. 13 by K. Penc and A. Läuchli in this volume). In this latter case, the difficulty comes from the fact that, like the singlet state of two $S = 1/2$ entities, quadrupolar states have no classical analog.

To understand the physical mechanism behind the 1/2-plateau in the frustrated ladder, and more generally in frustrated arrangements of coupled spin-1/2 dimers, it is very useful to begin from the limit of isolated dimers to derive an effective hard-core-boson Hamiltonian [9, 10]. The starting point of this expansion is the Hamiltonian $H_0$ describing dimers in the critical field $H_c = J_\perp$, where the singlet is degenerate with the triplet which is polarized along the field direction. The ground state of this problem is $2^N$-fold degenerate, where $N$ is the number of dimers, and, to first-order in $J_\|$, $J_\times$, and $H - H_c$, the effective Hamiltonian takes the form of a 1D hard-core-boson model,

$$\mathcal{H}_{\text{eff}} = -t \sum_n (b_n^\dagger b_{n+1} + b_{n+1}^\dagger b_n) + v \sum_n n_n n_{n+1} - \mu \sum_n n_n, \tag{10.9}$$

with $t = (J_\| - J_\times)/2$, $v = (J_\| + J_\times)/2$, and $\mu = H - H_c - (J_\| + J_\times)/4$ (as discussed in the Chap. 20 by Mila and Schmidt). This effective model has been investigated thoroughly over the years. For the purposes of this review, its most important property is the superfluid-insulator transition which takes place upon increasing the nearest-neighbor repulsion at $v = 2t$. Because the hopping amplitude $t$ is proportional to the difference between the mutually frustrating intra-leg and inter-leg couplings, while the repulsion is proportional to their sum, frustration decreases the kinetic energy and hence increases the effective relative repulsion, leading to the appearance of a plateau if $1/3 < J_\|/J_\times < 3$. This result has been confirmed numerically.

To summarize, frustration suppresses the kinetic energy in coupled-dimer systems, and the appearance of a ('quantum') plateau can be thought of as a superfluid-insulator transition of hard-core-bosons, a boson corresponding to a triplet polarized along the field, and an empty site to a singlet. This is to be contrasted with 'classical' plateaus, in which spins are polarized parallel or antiparallel to the field. Accordingly, all spins are expected to have a significant polarization in classical plateaus, while in quantum plateaus there should be a strong contrast between the spins composing a singlet and those in a triplet. This difference has indeed been detected by NMR in $SrCu_2(BO_3)_2$ (below).

### *10.2.6 High-Order Plateaus*

High-order plateaus, by which is meant plateaus with a large denominator in the fractional saturation, correspond to charge-density waves of hard-core-bosons with a large unit cell. This situation is in fact to be expected in the presence of long-range correlations which keep the particles away from each other at large distances. For classical plateaus, by contrast, such correlations can only come from long-range exchange processes, which are a rather unlikely possibility in typical antiferromagnets where the exchange is dominated by short-range kinetic processes.

In this respect, dimer systems are significantly more promising as candidates for exotic plateaus. In fact the hard-core-boson effective Hamiltonian is not simply the translation of a spin Hamiltonian, but the result of a strong-coupling expansion in which high-order terms lead to long-range interactions. These terms are expected to decrease exponentially with distance, but because the kinetic energy can be reduced dramatically by frustration effects, their effective magnitude can be large enough to induce charge-density waves with long periods. This is believed to be the mechanism for the 1/8-plateau of $SrCu_2(BO_3)_2$. Here it should, however, be noted that, beyond first order, the hard-core-boson effective Hamiltonian contains not only hopping terms and two-body interactions, but also all possible types of $n$-site terms, including correlated hopping (processes for which the hopping amplitude for a particle depends on the occupancy of sites not in the direct process) and $n$-body interactions. The simplest investigation of these bosonic models is to convert them back into a spin model by the Matsubara–Matsuda transformation. This new spin model is in general completely different from the original one, and much insight may be gained by a simple classical approximation: because the local quantum physics was included when deriving the effective model, this 'classical' approximation is able to reproduce quantum effects such as quantum plateaus.

When performing high-order expansions, it is necessary also to transform the operators. As a consequence, spins inside a singlet can acquire some magnetization components once again, a result in qualitative agreement with NMR observations of the 1/8-plateau of $SrCu_2(BO_3)_2$.

### 10.2.7 Transition into Plateaus

Spatial symmetries are usually broken in fractional plateau phases, while the spin-rotation symmetry around the field direction is broken outside them. According to the Landau theory of phase transitions for such a case, the transition between the two phases is generically either of first order, or it involves an intermediate phase where both types of order coexist.

In the first case, a magnetization jump is expected to be present at the transition, as indeed clearly observed at the low-field transition into the 1/8-plateau of $SrCu_2(BO_3)_2$. Magnetization jumps have been predicted to occur in other situations as well, notably a large jump to full magnetic saturation expected in several geometries due to the rigorously localized nature of magnons in their fully polarized state. Jumps not connected at all to a plateau are also possible. As one example, such a jump has been discovered between the 1/2-plateau and saturation for the frustrated ladder discussed above (Fig. 10.4). At this transition, non-magnetic rungs undergo a transition from the singlet to the $S^z = 0$ triplet.

If there is no magnetization jump upon entering or leaving the plateau, the expected intermediate phase should break both a spatial symmetry and the rotational spin symmetry. In terms of bosons, this is the equivalent of a supersolid phase, where bosons are condensed but nevertheless break a spatial symmetry. The possibility of stabilizing such phases in frustrated quantum magnets has been discussed recently in the context of both spin-1/2 [21–23] and spin-1 models [20].

## 10.3 Experimental Observation of Magnetization Plateaus

We turn now to a discussion of the experimental observation of magnetization plateaus and of associated phenomena in real materials. A representative (but not exhaustive) list of examples of magnetization plateaus can be found in Table 10.1. As mentioned in Sect. 10.2.2, there are two types of plateau. In the first, the translational symmetry of the crystal is *not* broken, corresponding to the case where the periodicity of the ground state is equal to that of the crystal structure, i.e. $n$ in (10.3) is equal to the number of spins in the crystal unit cell $n_0$ (see above). A typical example is the half-magnetization plateau in $S = 1$ bond-alternating chains such as $Ni_2(Medpt)_2(\mu\text{-}N_3)(\mu\text{-ox})\cdot ClO_4\cdot 0.5H_2O$ [24]. Because the exchange couplings alternate along the chain, a unit cell contains at least two spins, and it is clear that $n = 2$ satisfies (10.3) for the half-magnetization plateau ($s = 1/2$) for $S = 1$. The physical mechanism for the half-magnetization plateau is obvious in the limit of strong alternation, i.e. the case of isolated $S = 1$ dimers. The ground state changes from the singlet at low fields to the $S^z = 1$ triplet at intermediate fields, where the 1/2-plateau state appears, before saturation to the $S^z = 2$ state takes over. As no change of symmetry is involved, this type of plateau can be reached from the high-temperature paramagnetic state without a phase transition.

By contrast, in the second type of plateau the magnetic unit cell of the ground state is larger than the crystalline unit cell, i.e. $n$ is larger than $n_0$. Such a plateau

**Table 10.1** Examples of magnetization plateaus

| Material | Description | Fraction | Symmetry breaking? | Reference |
|---|---|---|---|---|
| $Ni_2(Medpt)_2(\mu\text{-}N_3)(\mu\text{-ox})\cdot ClO_4\cdot 0.5H_2O$ | $S = 1$ bond-alternating chain | 1/2 | No | [24] |
| $F_2PNNNO$ | $S = 1$ coupled dimers | 1/2 | No | [25] |
| $NH_4CuCl_3$ | $S = 1/2$ coupled dimers | 1/4, 3/4 | No | [26] |
| $Cu_3(CO_3)_2(OH)_2$ | $S = 1/2$ diamond chain | 1/3 | No | [28] |
| $RbFe(MoO_4)_2$ | $S = 5/2$ triangular lattice | 1/3 | Yes | [30, 31] |
| $Cs_2CuBr_4$ | $S = 1/2$ distorted triangular lattice | 1/3 | Yes | [33] |
| $MCr_2O_4$ ($M$ = Cd, Hg) | $S = 3/2$ pyrochlore lattice | 1/2 | | [41, 42] |
| $SrCu_2(BO_3)_2$ | $S = 1/2$ Shastry–Sutherland lattice | 1/8, 1/4, 1/3 | Yes | [47, 61] |
| $RB_4$ ($R$ = Tb, Tm) | large $\mathcal{J}$ Shastry–Sutherland lattice | Many | Yes | ([86–88], Siemensmeyer et al., unpublished) |

is usually associated with a magnetic superstructure, which breaks the translational symmetry. The plateau is therefore separated from the paramagnetic state by a well defined phase transition.

There are now numerous examples of magnetization plateau. Most of them, however, are of the first type, and only a relatively small number of symmetry-breaking plateaus is known to date. The plateaus at 1/4 and 3/4 of the saturation magnetization in $NH_4CuCl_3$ [26], a system composed of coupled spin 1/2-dimers, were once considered to be an example of the second type because the monoclinic unit cell of the crystal contains only four $Cu^{2+}$ ions, whereas (10.3) requires $n$ to be a multiple of eight. However, neutron scattering experiments later revealed the presence of a structural transition which doubles the periodicity of the crystal along the $b$-direction [27]. Indeed no phase transition has been detected as a function of temperature in the field range where the plateaus are formed at low temperature. For the remainder of this chapter, we will focus mostly on the second type of plateau.

Although we have considered only isotropic Heisenberg interactions in the previous sections, real materials have various types of anisotropic interaction. Of these, the most important is the antisymmetric Dzyaloshinskii–Moriya (DM) interaction, which is expressed as

$$\mathcal{H}_{\mathrm{DM}} = \boldsymbol{D}_{ij} \cdot \left(\boldsymbol{S}_i \times \boldsymbol{S}_j\right) \qquad (10.10)$$

and is allowed if the center of the bond connecting $\boldsymbol{S}_i$ and $\boldsymbol{S}_j$ does not have inversion symmetry. Because anisotropic interactions generally do not commute with the total spin, the magnetization inside the plateaus may not be strictly constant in real materials. Further, the anisotropic interactions lower the spin rotation symmetry and often change dramatically the nature of quantum phase transitions between the plateau and its neighboring phases [29]. This behavior has been observed in $SrCu_2(BO_3)_2$.

### 10.3.1 'Classical' Plateaus in Triangular and Pyrochlore Lattices

As mentioned in Sect. 10.2.3, Heisenberg spin systems on a triangular lattice are expected to show the 120-degree spin structure at zero field and to develop 1/3-plateaus over a finite range of magnetic field. Such behavior has indeed been observed in the semi-classical $S = 5/2$ spin system $RbFe(MoO_4)_2$ [30, 31]. This material has a layered structure (Fig. 10.5), in which the $Fe^{3+}$ ions form an equilateral triangular lattice. As shown in Fig. 10.2, the magnetization curve exhibits a 1/3-plateau for $\mathbf{H} \perp c$. Because there is a weak anisotropy of easy-plane type, the magnetization for $\mathbf{H} \parallel c$ does not show a plateau. Figure 10.5 shows the magnetic phase diagram for $\mathbf{H} \perp c$ determined from a variety of experiments, including $^{87}$Rb NMR measurements, which support an up-up-down structure spin structure (Fig. 10.3) in the plateau phase [32]. The experimental phase diagram shows good agreement with numerical simulations of the Heisenberg model on a triangular lattice [14]. In particular, the width in field of the plateau phase becomes significantly

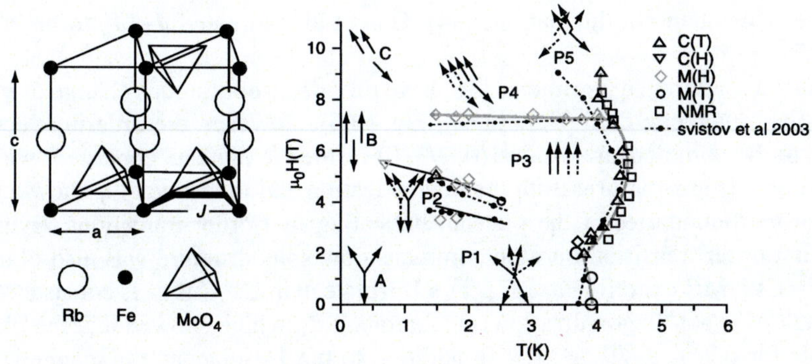


**Fig. 10.5** *Left*: crystal structure of $RbFe(MoO_4)_2$. *Right*: magnetic phase diagram for $\mathbf{H} \perp c$ (after Svistov et al. [32])

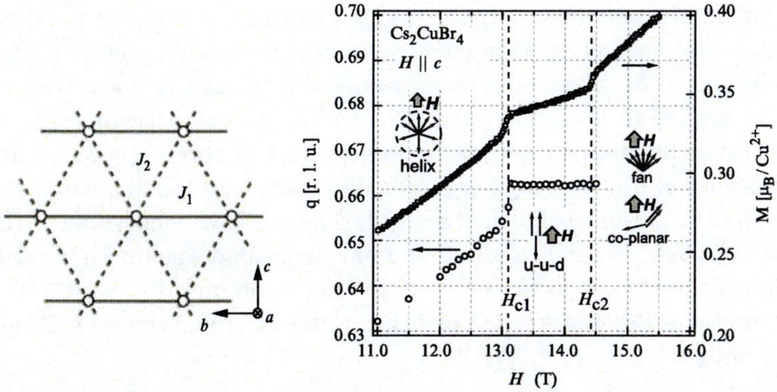


**Fig. 10.6** *Left*: distorted isosceles triangular lattice of $Cs_2CuBr_4$. *Right*: magnetization curve of $Cs_2CuBr_4$ with the field oriented parallel to the $c$-direction, measured at 400 mK (*right-hand scale*), shown together with the field-dependence of the magnetic ordering vector $\mathbf{Q} = (0, q, 0)$ measured at 60 mK (*left-hand scale*). (After Ono et al. [36])

greater at higher temperatures, indicating that thermal fluctuations play an important role in stabilizing the plateau, a result consistent with the order-by-disorder mechanism.

Because both thermal and quantum fluctuations stabilize the 1/3-plateau, it is expected to be most robust in spin-1/2 systems. In fact, a 1/3-plateau has been observed in $Cs_2CuBr_4$, where the $Cu^{2+}$ ions form a less frustrated isosceles triangular lattice in the $bc$-plane (left panel, Fig. 10.6) [33]: the coupling $J_1$ in the $b$-direction is stronger than the coupling $J_2$ in the other directions. The magnetization data in the vicinity of the 1/3-plateau are shown in the right panel of Fig. 10.6. Because of the lattice distortion, $Cs_2CuBr_4$ shows an incommensurate helical spin order at zero field with wave vector $\mathbf{Q} = (0, q, 0)$, where $q = 0.575$. By comparing this value with the theoretical calculation of $q$ as a function of $J_2/J_1$, including the effects of quantum fluctuations [34], Ono et al. estimated $J_2/J_1$ to be 0.74 in $Cs_2CuBr_4$ [35].

Neutron scattering experiments [36] have revealed a continuous change of $q$ with the applied magnetic field (right panel, Fig. 10.6). However, the ordering vector is locked to the commensurate value (0, 2/3, 0) inside the plateau, consistent with the up-up-down spin structure. Both the magnetization and the value of $q$ show a finite jump at the boundaries of the plateau, indicating first-order transitions. Evidence for a first-order transition into the commensurate spin structure was also obtained from $^{133}Cs$ NMR experiments [37]. The 1/3-plateau in $Cs_2CuBr_4$ is consistent with the result of exact-diagonalization calculations [38], which showed that the plateau is stable for $0.7 \lesssim J_2/J_1 \lesssim 1.3$. In addition to the 1/3-plateau, the magnetization data also show a small feature indicative of another plateau at 2/3 of the saturation field [36], although this is much narrower than the 1/3-plateau. There appears to be no satisfactory explanation as yet for this 2/3-plateau. From results obtained for the isostructural compound $Cs_2CuCl_4$ [39], a weak DM interaction is expected on the $J_2$ bond, with the vector $\boldsymbol{D}$ oriented parallel to the $a$-direction. This forces the spins to lie in the $bc$-plane, and therefore the 1/3-plateau is not observed when the field is applied in the $a$-direction [33, 36]. In this case, the ordering remains incommensurate up to saturation, with a non-coplanar, umbrella-type spin structure.

The 1/3-plateau in the triangular lattice illustrates how the fluctuations select a unique ground state from a choice of many degenerate, low-energy states. There are in fact several routes for the selection of a unique ground state in highly frustrated antiferromagnets. One of these is the spin-lattice coupling [40], which is responsible for the remarkably stable 1/2-plateaus in the Cr spinel oxides $CdCr_2O_4$ and $HgCr_2O_4$, shown in Fig. 10.7a [41,42].

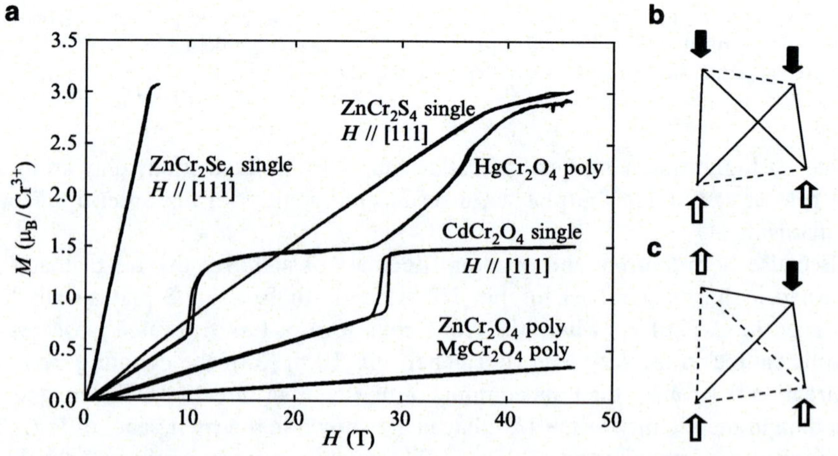


**Fig. 10.7** *Left*: (**a**) magnetization curves for various Cr spinel oxides measured at 1.8 K (for Hg and Cd compounds) and at 4.2 K (for the other materials). (After Ueda et al. [42].) *Right*: the spin order and lattice distortion are illustrated in (**b**) for zero field and in (**c**) for the 1/2-plateau state, where the solid (*dashed*) lines represent shortened (elongated) bonds

In these materials, the $Cr^{3+}$ ions, which have spin 3/2, form a pyrochlore lattice, a corner-sharing network of tetrahedra known as one of the most frustrated 3D geometries (further details are presented in Chap. 7 on spinel compounds Takagi and Niitaka and Chap. 8 on pyrochlore compounds Gaulin and Gardner). Both the classical ($S = \infty$) and the quantum ($S = 1/2$) spin system on the pyrochlore lattice are believed to remain disordered down to $T = 0$. In reality, however, many Cr spinel oxides show antiferromagnetic order at a finite temperature in zero field. The magnetic order is often driven by the spin-lattice coupling through the "order-by-distortion" mechanism [43] (discussed in the Chap. 11 by Tchernyshyov and Chern), which is similar to the Jahn–Teller effect. The exchange coupling depends linearly on the distance between spins, while the elastic energy has a quadratic dependence. The system may therefore always gain exchange energy by distorting the lattice to stabilize a collinear Néel order, as illustrated in Fig. 10.7b. Here the four antiferromagnetic bonds (solid lines) are shortened, while the two ferromagnetic bonds (dashed lines) become elongated. Simultaneous magnetic and structural phase transitions are indeed observed quite frequently in spinel oxides.

When the spin configuration changes under a strong magnetic field, so does the lattice structure. The half-magnetization plateaus correspond to the three-up-one-down structure on every tetrahedron shown in Fig. 10.7c. This spin structure can be stabilized by a trigonal distortion of the tetrahedra, resulting in three short bonds and three long bonds. Such a simultaneous magnetic and structural phase transition in magnetic field has actually been observed by combined neutron and synchrotron X-ray scattering experiments in $HgCr_2O_4$ [44]. The field-temperature diagram containing the highly robust 1/2-plateau can be reproduced very well by a theoretical analysis of the coupled spin-lattice model on a pyrochlore lattice [45, 46].

### 10.3.2 $SrCu_2(BO_3)_2$ and the Shastry–Sutherland Model

Among various materials showing magnetization plateaus, $SrCu_2(BO_3)_2$ exhibits by far the most spectacular phenomena, and thus has attracted enduring experimental and theoretical attention ([47]; for a review see [48]). The crystal structure of $SrCu_2(BO_3)_2$ consists of alternating stacks of magnetic $CuBO_3$ layers and non-magnetic Sr layers. The magnetic layers depicted in Fig. 10.8a, b contain a network of orthogonal dimers formed by pairs of $Cu^{2+}$ ions, each carrying a spin 1/2. The simplest spin model appropriate for this material includes isotropic exchange interactions for the nearest-neighbor (intra-dimer) bonds $J$ and the next-nearest-neighbor (inter-dimer) bonds $J'$ (Fig. 10.8c),

$$\mathcal{H}_0 = J \sum_{n.n.} \boldsymbol{S}_i \cdot \boldsymbol{S}_j + J' \sum_{n.n.n.} \boldsymbol{S}_i \cdot \boldsymbol{S}_j. \tag{10.11}$$

This model had been considered by Shastry and Sutherland [49] long before $SrCu_2(BO_3)_2$ was discovered. The ground state of the model is obvious in the two

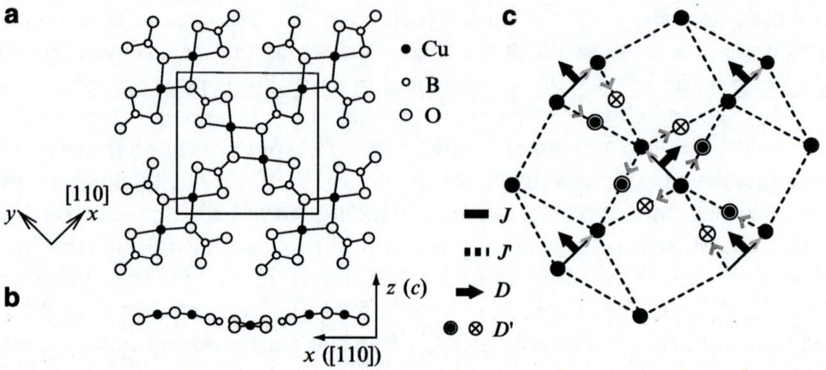


**Fig. 10.8** *Left*: magnetic layer of $SrCu_2(BO_3)_2$ viewed along (**a**) the *c*-direction and (**b**) the $[1\bar{1}0]$ direction. *Right*: spin model (**c**) appropriate for $SrCu_2(BO_3)_2$. In addition to the Shastry–Sutherland model, which contains only Heisenberg exchange terms $J$ and $J'$, intra- and inter-dimer DM interactions ($D$ and $D'$) are also included. The directions of the DM vectors are defined from (10.10) with the bond direction $i \rightarrow j$ shown by the *arrows*

limiting cases. When $J'/J \ll 1$, the system reduces to a collection of dimer singlets, and in fact this product of dimer singlets remains the exact ground state for $J'/J < 0.68$ [50] (as discussed in the Chap. 19 by Miyahara). On the other hand, when $J'/J \gg 1$, the model is equivalent to an antiferromagnet on a square lattice, and hence the ground state has Néel order. Between these two limiting phases, there is theoretical evidence for the presence of an intermediate phase, first predicted to be an ordered helical phase [51] but later suggested to be more probably another type of singlet phase based on four-site plaquettes [52, 53].

A wide range of experiments, namely susceptibility [54], specific heat [55], nuclear quadrupole resonance (NQR) [47], NMR [56], electron spin resonance (ESR) [57], and inelastic neutron scattering [58] have established that $SrCu_2(BO_3)_2$ has a dimer-singlet ground state with an excitation gap of 35 K. The values of the exchange parameters were estimated as $J = 85$ K and $J'/J = 0.64$ by fitting the susceptibility data to numerical calculations performed for the Shastry–Sutherland model [59]. Slightly different values of $J = 71$ K and $J'/J = 0.60$ were obtained from an analysis of the excitation spectra [60].

The most striking property of $SrCu_2(BO_3)_2$ is the sequence of magnetization plateaus at 1/8, 1/4, and 1/3 of the saturation magnetization. Magnetization curves measured in a pulsed magnetic field are shown in Fig. 10.9 (left panel) [61]. Because the crystal unit cell contains two dimers (four Cu sites) per layer, (10.3) states that the translational symmetry should be broken in all the plateaus. The physical mechanism stabilizing these plateaus is believed to be the strong suppression of the triplet kinetic energy due to the frustration. As in the case of the frustrated-ladder model discussed in Sect. 10.2.5, frustration in the Shastry–Sutherland model inhibits inter-dimer hopping of the triplets. In fact, one-body hopping of triplets is allowed only from 6th-order processes in the perturbative expansion with respect to $J'/J$ [50]. This leads to the stability of unusually high-order plateaus (Sect. 10.2.6).

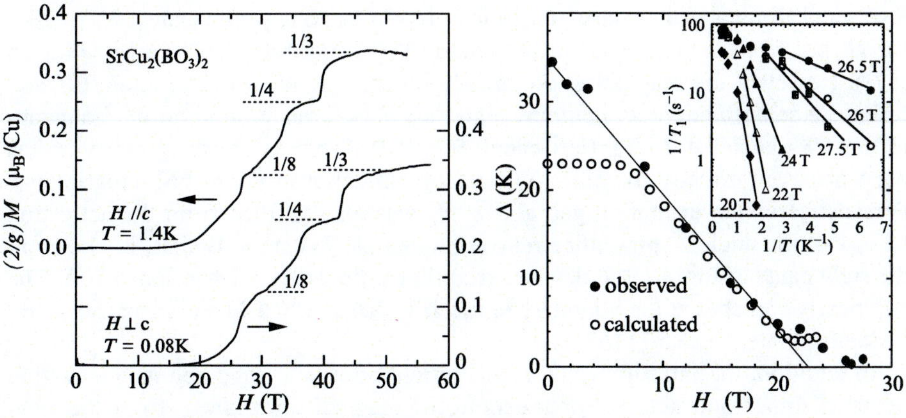
SrCu2(BO3)2
1/3
1/4
1/8
1/3
1/4
1/8
H //c
T = 1.4K
H ⊥ c
T = 0.08K
(2/g) M (μB/Cu)
H (T)
Δ (K)
H (T)
• observed
○ calculated
1/T1 (s^-1)
1/T (K^-1)
26.5T
26T
27.5T
24T
22T
20T

levels and eliminates the phase transition at $H_c$. In fact one would expect $H_c \sim 23\,\mathrm{T}$ for $H \parallel c$ from the gap value at zero field (35 K) and the value of $g$ ($g_c = 2.28$ [57]). However, the magnetization already begins a gradual rise around 16 T and follows a smooth curve without any sign of a phase transition up to the boundary to the 1/8-plateau at 26.5 T (left panel, Fig. 10.9). This anticrossing behavior has been observed directly by ESR [65] and by optical absorption [66] experiments. Persistence of the excitation gap above $H_c$ was also verified from the activation energy of the nuclear spin-lattice relaxation rate ($1/T_1$) at the B sites, as shown in the right panel of Fig. 10.9. It deviates from linear field dependence above 20 T. The gap nearly vanishes at the boundary to the 1/8-plateau (26.5 T) and increases again inside the plateau (27.5 T) [67].

In addition, the intra-dimer DM interaction induces a staggered magnetization on each dimer and acts to tilt the uniform magnetization away from the field direction [68, 69]. For isolated dimers, the staggered and uniform magnetizations are given in the low-field limit and for $D/J \ll 1$ by the compact expressions $\boldsymbol{m}_s = \left(g\mu_B/2J^2\right)\boldsymbol{D} \times \boldsymbol{H}$ and $\boldsymbol{m}_u = \left(g\mu_B/4J^3\right)(\boldsymbol{D} \times \boldsymbol{H}) \times \boldsymbol{D}$ [69]. The former can be detected by NMR and the latter contributes to the torque signal. The field-induced staggered magnetization has been observed in $SrCu_2(BO_3)_2$ by $^{11}B$-NMR experiments [67]. The magnitude of $D$ determined from various experiments [66, 67, 70, 71] is consistently in the range 2–4 K. An important consequence of the intra-dimer DM interaction is the absence of a Bose-condensation transition near $H_c$. This is because the U(1) spin-rotation symmetry is already broken at lower fields. The buckling of the $CuBO_3$ planes also induces anisotropic components of the $g$-tensor, with a sign alternation, which has the same effect as the intra-dimer DM interaction. Given the modest magnitude estimated from structural data, the effects of the staggered $g$-tensor are considered to be rather minor compared with $\boldsymbol{D}$ [66]. Some spectroscopic results [66, 71], most notably the detection of the direct singlet–triplet transition by ESR for fields oriented parallel to the $c$-axis [57, 65], seem to require a finite $c$-component of $\boldsymbol{D}$. However, there is no indication of a structural change with temperature that would allow such a component. An alternative explanation has been proposed based on the idea of dynamical modulation of $\boldsymbol{D}$ due to spin-phonon coupling [72]. Finally, recent ab initio estimates of the DM interactions are in remarkably good agreement with experiment regarding both the relative orientation and the magnitude of the DM vectors [73].

### 10.3.3 'Quantum' Plateaux and Spin Superstructure in $SrCu_2(BO_3)_2$

The existence of a spin superstructure at the 1/8-plateau in $SrCu_2(BO_3)_2$ has been confirmed directly by NMR experiments on both the Cu and the B sites [74, 75]. Because the magnetic hyperfine field acting on a nuclear spin is determined by the local magnetization in the neighborhood of the sites being probed, the NMR spectrum represents an effective histogram of the distribution of the magnetic moments

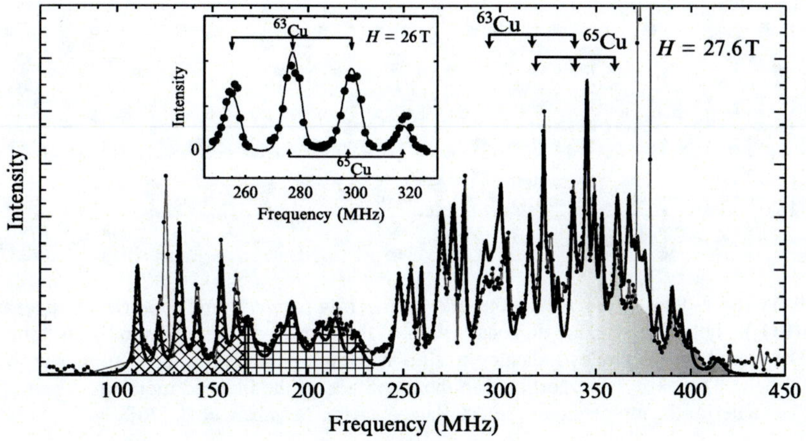


**Fig. 10.10** Cu NMR spectra (*solid dots*) in $SrCu_2(BO_3)_2$, measured at $T = 35\,$mK in fields of 26 T (*inset*) and 27.6 T (*main panel*) oriented parallel to the $c$-axis. In the *inset*, the spectrum is decomposed into two sets of three quadrupole-split lines from $^{63}$Cu and $^{65}$Cu, as indicated by the *lines* and *arrows*. In the main panel, the *arrows* show the resonance positions in the absence of the magnetic hyperfine field. The *thin line* connects the data points while the *thick line* represents the result of a fit to the spectrum using 11 sites with distinct hyperfine fields. (After Kodama et al. [74])

(as explained in the Chap. 4 by Carretta and Keren). Figure 10.10 shows the Cu NMR spectra measured in static magnetic fields of 26 T (inset: below the onset of the 1/8-plateau) and at 27.6 T (main panel: inside the 1/8-plateau) at $T = 35\,$mK. The spectrum at 26 T can be reproduced as a superposition of six lines, all of which arise from a single Cu site, because each of the two isotopes $^{63}$Cu and $^{65}$Cu generates three lines split by the quadrupolar interaction (indicated by the arrows). Thus the magnetization is uniform at 26 T. By contrast, the spectrum at 27.6 T shows many sharp peaks distributed over an extremely wide frequency range, indicating many distinct Cu sites with different magnetizations. This constitutes direct evidence of a magnetic superstructure with high-order commensurability.

In spite of this apparent complexity, some important features can be recognized immediately in this spectrum [74]. The six sharp lines in the frequency range 110–165 MHz are assigned to a single Cu site with a large, negative hyperfine field of $-16\,$T, which corresponds to a magnetization $\langle S_z \rangle \sim 0.3$. These sites are identified as the most strongly polarized "triplet dimer" sites. Their intensity is indeed approximately 1/8 of the whole spectrum. However, other sites also have finite hyperfine fields, indicating that one triplet is not confined to a single dimer: the broad lines in the range 165–230 MHz are assigned to another site with magnetization $\langle S_z \rangle \sim 0.2$. In addition, some lines show positive hyperfine fields, which correspond to sites whose moments lie antiparallel to the magnetic field. The Cu-NMR spectrum in the 1/8-plateau was reproduced by assuming 11 distinct Cu sites with different magnetization [74]. Such a large number of sites is compatible with the unit cell shown in the left panel of Fig. 10.11.

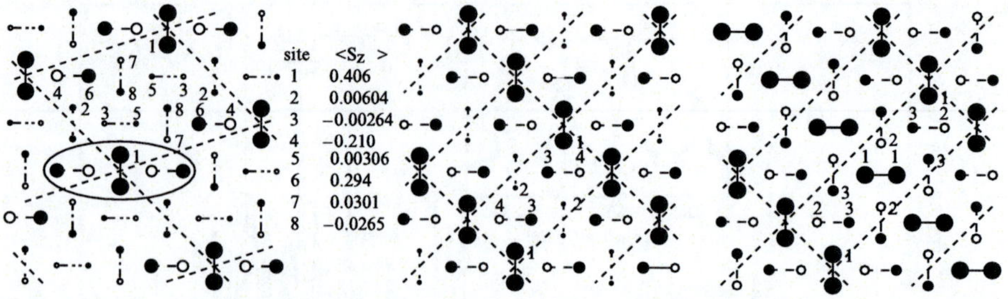


**Fig. 10.11** Spin-density profile for the 1/8- (*left*), 1/4- (*center*), and 1/3- (*right*) plateaus in $SrCu_2(BO_3)_2$, obtained by exact diagonalization of the Shastry–Sutherland model on finite clusters. The closed (*open*) circles indicate the sites with spin polarization along (opposite to ) the external field. Their size is proportional to the amplitude of the induced moments. The numbers show the values of the moments for the 1/8-plateau. (After Miyahara et al. [76])

The wide distribution in the magnitude of the ordered moment over different sites revealed by the NMR experiment is a distinctive signature of quantum plateaus (Sect. 10.2.5). This behavior has been reproduced by exact-diagonalization calculations on finite clusters (up to 24 sites) using the Shastry–Sutherland model with parameter values appropriate for $SrCu_2(BO_3)_2$ [76]. The calculated spin-density distribution is shown in Fig. 10.11 for various plateaus. To select a unique ground state with broken symmetry on a finite cluster, a spin-phonon coupling was included with periodic boundary conditions appropriate for the predetermined geometry of the supercell. The rectangular supercells for the 1/4- and the 1/3-plateaus were obtained by earlier hard-core-boson calculations [77]. The inclusion of spin-phonon coupling can be justified based on the large anomalies exhibited by the elastic constants at the boundaries of the plateaus [78]. However, there is to date no direct evidence for a static lattice distortion at the phase transitions. Whether or not the spin-phonon coupling plays an essential role in stabilizing the superstructure within the plateau regions remains an open question.

A prominent feature of the calculated spin-density profile is that one "triplet" unit is spread over three dimers, as shown by the ellipse in the left panel of Fig. 10.11; the most strongly polarized triplet dimer has neighboring dimers with oscillating spin density on both sides. The negative polarization is due to the (unfrustrated) antiferromagnetic exchange field from the central dimer. While the other dimers also have finite magnetizations, their magnitudes are relatively small. The hard-core-boson calculations [77] show "stripe" order of the triplets in the 1/4- and 1/3-plateaus. Note that the rhombic ordering pattern in the 1/8-plateau is obtained by removing every other triplet from the stripes of the 1/4-plateau. The distribution of hyperfine fields obtained from the calculated results for the red 1/8-plateau is in good agreement with the NMR experiment [74, 76]. We stress that, while the local magnetization in the plateaus is always collinear with the external field for the Shastry–Sutherland model, it should have a finite transverse component in $SrCu_2(BO_3)_2$ due to the intra-dimer DM interaction. The distribution of the transverse magnetization is not yet known. Similar spin superstructures have been detected by very recent NMR experiments for the 1/4- (Takigawa et al., unpublished) and the 1/3-plateaus [79].

### 10.3.4 Phase Diagram of $SrCu_2(BO_3)_2$

The phase transition into the 1/8-plateau has been investigated by NMR at the B sites [74, 75]. Similar to the Cu NMR results, the B NMR spectrum consists of one set of quadrupole-split lines below 26.5 T but shows a large number of peaks originating from the spin superstructure above 27 T. The two distinct types of spectrum coexist in a narrow field range 26.5–27 T at the transition. A similar coexistence has also been observed for the transition as a function of temperature: at 27.5 T, it occurs over the temperature range 0.36–0.52 K. Such behavior indicates that the transitions in both field and temperature are of first order, and are affected by some form of disorder. It should be noted that observation of the 1/8-plateau at 1.4 K (left panel, Fig. 10.9) is apparently incompatible with the disappearance of the spin superstructure in the NMR experiments above 0.52 K. A natural explanation is the adiabatic cooling caused by the application of a pulsed magnetic field [80]. For spin systems with a singlet ground state, the spin gap to the triplet excitations is reduced by a magnetic field. If the magnetic field is applied rapidly enough, the process is adiabatic, i.e. the population distribution among different levels remains unchanged. This causes significant cooling.

More detailed information on the phase diagram has been obtained from recent magnetization and torque measurements in steady magnetic fields up to 31 T [80] (Fig. 10.12a). A tiny cantilever was used to detect the force acting on the sample, which has contributions from both the magnetic force $-\boldsymbol{M} \cdot \nabla \boldsymbol{H}$, and the torque $\tau \propto \boldsymbol{M} \times \boldsymbol{H}$; in homogeneous fields, $\nabla \boldsymbol{H} = 0$. When the field is tilted slightly from the $c$-axis, the force is dominated by the torque effect due to the transverse magnetization, $M_\perp \propto \tau / H$, which is generated both by the intra-dimer DM interaction and by the anisotropy of the $g$-tensor. Alternatively, the longitudinal magnetization $\langle S_z \rangle$ can be detected if the sample is moved off the center of the magnet to make the field sufficiently inhomogeneous. As shown in Fig. 10.12a, $M_\perp$ and $\langle S_z \rangle$ exhibit quite different field-dependences even at low fields. In fact, the field-dependence of $M_\perp$ is similar to the staggered magnetization measured by NMR.

The magnetization curve exhibits a large and nearly discontinuous step on entering the 1/8 phase. However, this is actually a very steep increase of the magnetization over a narrow field range, pointing to a coexistence of the two phases which is consistent with the NMR results. No hysteresis was observed between up and down field-sweeps. In addition, the torque and the magnetization data exhibit two clear steps above the 1/8-plateau, indicating successive phase transitions (the arrows in the inset of Fig. 10.12a). The first transition at 28.4 T marks the upper boundary of the 1/8-plateau, which is followed by another transition near 29.6 T. The temperature-field phase diagram may then be constructed by plotting the extrema of the second derivative of $M_\perp$, and is shown in Fig. 10.12b.

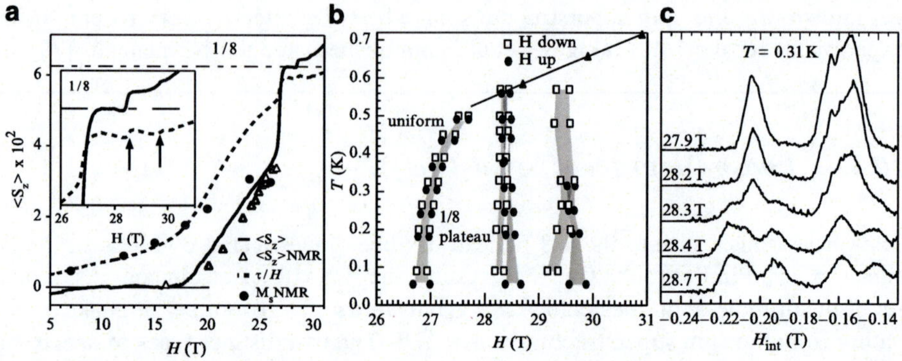


**Fig. 10.12** (**a**) Longitudinal magnetization (*solid line*) and torque divided by field (*dashed line, arbitrary scale*), which is proportional to the uniform transverse magnetization, $M_{\perp}$, measured in $SrCu_2(BO_3)_2$ under steady magnetic fields. (After Levy et al. [80].) The open triangles (*solid circles*) show the uniform longitudinal (*staggered transverse*) magnetization obtained from the NMR shift at the Cu (B) sites [67, 74]. (**b**) Phase diagram in the vicinity of the 1/8-plateau. (After Levy et al. [80].) The *solid circles* (*open squares*) show the positions of the extrema of $d^2M_{\perp}/dH^2$ measured with up (*down*) field-sweeps. The separation between the extrema represents the width of the transition. The *solid triangles* indicate transitions to the paramagnetic phase determined from the disappearance of the large internal field in the B NMR spectrum [81]. (**c**) Change of the B NMR spectrum for $H \parallel c$ across the upper boundary of the 1/8-plateau. Only that part of the spectrum corresponding to the most negative hyperfine field ($H_{\text{int}}$) is displayed. (After Takigawa et al. [81])

The upper boundary of the 1/8-plateau is also marked by a sudden change of the B NMR spectrum [81]. A part of the B NMR spectra corresponding to the largest negative hyperfine fields is shown for various fields in Fig. 10.12c. The lowest field (27.8 T) is in the 1/8-plateau. From earlier experiments, the peaks shown in the figure are assigned to the B sites next to the strongly polarized triplet dimer (site 1 in the left panel of Fig. 10.11) [75]. These peaks exhibit a sudden splitting near 28.3 T, coinciding with the upper boundary of the plateau. An important observation is that, in spite of the splitting, the value of the largest hyperfine field is virtually unchanged across the phase boundary. This means that there are still highly polarized triplet dimers above the plateau, providing direct evidence for the persistence of a spin superstructure up to at least 31 T. This is in strong contrast to the behavior at the lower boundary of the plateau. The transition to the paramagnetic phase with increasing temperature was determined by the disappearance of the large hyperfine field, as shown in Fig. 10.12b. The transition temperature extrapolates smoothly to the plateau region and also to the specific-heat anomaly observed at higher fields (up to 33 T) (Tsujii et al., unpublished).

At the time of writing, the microscopic nature of the phases intermediate between the 1/8- and the 1/4-plateaus, where a continuous change of the magnetization coexists with the symmetry-breaking spin superstructure, has not been fully understood. If the anisotropic interactions were absent, the increase of magnetization would translate directly into an increase of the boson density. The persistence of the spin superstructure then implies a coexistence of doped mobile bosons on top of the charge-density-wave order of the plateaus. The doped bosons would undergo Bose condensation at a sufficiently low temperature, resulting in a supersolid phase. A supersolid phase has indeed been proposed for the Shastry–Sutherland model above the 1/3-plateau [82]. Although one-body hopping of triplets is inhibited in the Shastry–Sutherland model, a correlated hopping is allowed whereby a triplet can hop from one site to another if a second triplet is located next to both sites. The gain of kinetic energy due to correlated hopping is presumably important in the proposed supersolid phases. The role of correlated hopping in stabilizing the supersolid phase has been demonstrated theoretically in a simpler frustrated dimer model [23]. As already mentioned, the situation changes dramatically in the presence of DM interactions. The broken spin-rotation symmetry in the $ab$-plane eliminates the possibility of Bose condensation, and therefore a true supersolid phase cannot occur. A complete understanding of the intriguing phase diagram of $SrCu_2(BO_3)_2$ may reveal new physics resulting from the interplay between the spin frustration and anisotropic interactions.

To conclude this section, we mention that phases at even higher fields have been explored recently. Sebastian et al. have extended the magnetization measurements in pulsed magnetic fields up to 85 T [83] and have found several new anomalies suggesting additional plateaus. These authors also found that the 1/3-plateau ends near 70 T, above which the magnetization increases continuously and appears to reach the 1/2-plateau above 80 T. Recent NMR experiments up to 34 T (Takigawa et al., unpublished) confirmed several distinct phases between the 1/8- and 1/4-plateaus. The possibility of still further plateaus with high commensurability, such as 1/6 [84, 85] and 2/15 [84] has been proposed theoretically on the basis of accurate effective models derived by high-order perturbation theory [84] and by the CORE (Contractor Renormalization) technique [85]. Despite the difficulty of performing experiments in very high magnetic fields, remarkable progress in magnet technology has enabled the community to make steady progress in understanding the fascinating properties of the quantum phases of $SrCu_2(BO_3)_2$.

### *10.3.5 $RB_4$: A New Family of Shastry–Sutherland System*

Very recently, the new family of compounds $RB_4$, where $R$ represents the rare-earth elements, has attracted considerable attention because the trivalent rare-earth ions in these materials form layers with the geometry of the Shastry–Sutherland lattice. In particular, $TmB_4$ ([86, 87], Siemensmeyer et al., unpublished) and $TbB_4$ [88, 89] exhibit series of magnetization plateaus, suggesting the formation of symmetry-breaking superstructures. Both $Tm^{3+}$ and $Tb^{3+}$ have large total angular momenta of $\mathcal{J} = 6$, and thus these may be regarded as classical spin systems. However, they also have complications, such as a strong anisotropy due to the crystalline electric-field effect and quadrupolar interactions, in addition to the conventional spin-exchange interactions. $TmB_4$ has Ising anisotropy with an easy $c$-axis. In fields oriented along the $c$-axis, it shows narrow magnetization plateaus at 1/9, 1/8, 1/7, and possibly at other fractions in addition to the stable 1/2-plateau, before saturation takes place above 4 T ([87], Siemensmeyer et al., unpublished). By contrast, $TbB_4$ has easy-plane anisotropy with a hard $c$-axis, and for fields in this direction it shows a spectacular series of plateaus at 1/9, 1/3, 4/9, 1/2, and other fractions over the field range 15–30 T [88]. It is certainly an important future problem to understand the mechanism leading to the formation of these plateaus, and to clarify whether they have features in common with those observed in $SrCu_2(BO_3)_2$.

## 10.4 Conclusion

We have provided a brief overview of the mechanisms and experimental observations of magnetization plateaus, and more generally of their relation to the physics of frustrated spin systems. We have discussed two mechanisms for the formation of plateaus. The order-by-disorder scenario gives a good description for plateaus with classical spin configurations, of which the spin-1/2 Heisenberg model on a triangular lattice provides a typical example. The bosonic description is more suitable for 'quantum' plateaus, whose formation is associated with the localization of bosons into a symmetry-breaking charge-density-wave order. On the theoretical side, such quantum plateaus have been established in various types of spin model, particularly in coupled-dimer models, and in some cases supersolid phases are predicted in the neigborhood of the plateaus. On the experimental side, however, $SrCu_2(BO_3)_2$ is to date the only material where symmetry-breaking quantum plateaus have been investigated in depth. Studies of this material are complicated by its anisotropic interactions, which are presumed to have siginificant effects on the high-field properties and the phase diagram. An important future direction is to understand the role of anisotropic interactions more precisely in simple spin models, as well as in the Shastry–Sutherland model. An important experimental direction is the effort to find symmetry-breaking quantum plateaus in other materials with the ideal Shastry–Sutherland lattice, and also in other lattices with simpler geometry.

We would like to acknowledge all of the colleagues with whom we have had the pleasure to discuss and often to collaborate on the issue of magnetization plateaus, in particular F. Becca, C. Berthier, J. Dorier, M. Horvatic, K. Kodama, A. Läuchli, N. Laflorencie, S. Miyahara, K. Penc, and K.-P. Schmidt. We thank A. Honecker and M. Oshikawa for their critical reading of the manuscript.

## References

1. M.E. Zhitomirsky, T. Nikuni, Phys. Rev. B **57**, 5013 (1998)
2. F.D.M. Haldane, Phys. Lett. **93A**, 464 (1983)
3. E. Dagotto, T.M. Rice, Science **271** 618 (1996)
4. A.V. Chubukov, D.I. Golosov, J. Phys. Condens. Matter **3**, 69 (1991)
5. K. Hida, J. Phys. Soc. Jpn. **63**, 2359 (1994)

6. D.C. Cabra, A. Honecker, P. Pujol, Phys. Rev. Lett. **79**, 5126 (1997)
7. M. Oshikawa, M. Yamanaka, I. Affleck, Phys. Rev. Lett. **78**, 1984 (1997)
8. M. Oshikawa, Phys. Rev. Lett. **90**, 236401 (2003)
9. K. Totsuka, Phys. Rev. B **57**, 3454 (1998)
10. F. Mila, Eur. Phys. J. B **6**, 201 (1998)
11. A.I. Smirnov, H. Yashiro, S. Kimura, Y. Narumi, K. Kindo, A. Kikkawa, K. Katsumata, A.Ya. Shapiro, L.N. Demianets, Phys. Rev. B **75**, 134412 (2007)
12. D.C. Cabra, M.D. Grynberg, P.C. Holdsworth, A. Honecker, P. Pujol, J. Richter, D. Schmalfuss, J. Schulenburg, Phys. Rev. B **71**, 144420 (2005)
13. E.H. Lieb, T. Schultz, D.J. Mattis, Ann. Phys. (N.Y.) **16**, 407 (1961)
14. H. Kawamura, S. Miyashita, J. Phys. Soc. Jpn. **54**, 4530 (1985)
15. A. Honecker, J. Phys. Condens. Matter **11**, 4697 (1999)
16. M.E. Zhitomirsky, A. Honecker, O.A. Petrenko, Phys. Rev. Lett. **85**, 3269 (2000)
17. K. Hida, I. Affleck, J. Phys. Soc. Jpn. **74**, 1849 (2005)
18. J.-B. Fouet, F. Mila, D. Clarke, H. Youk, O. Tchernyshyov, P. Fendley, R.M. Noack, Phys. Rev. B **73**, 214405 (2006)
19. A. Läuchli, F. Mila, K. Penc, Phys. Rev. Lett. **97**, 087205 (2006)
20. P. Sengupta, C.D. Batista, Phys. Rev. Lett. **98**, 227201 (2007)
21. K.-K. Ng, T.-K. Lee, Phys. Rev. Lett. **97**, 127204 (2006)
22. N. Laflorencie, F. Mila, Phys. Rev. Lett. **99**, 027202 (2007)
23. K.P. Schmidt, J. Dorier, A.M. Läuchli, F. Mila, Phys. Rev. Lett. **100**, 090401 (2008)
24. Y. Narumi, M. Hagiwara, R. Sato, K. Kindo, H. Nakano, M. Takahashi, Phys. B **246–247**, 509 (1998)
25. Y. Hosokoshi, Y. Nakazawa, K. Inoue, K. Takizawa, H. Nakano, M. Takahashi, T. Goto, Phys. Rev. B **60**, 12924 (1999)
26. W. Shiramura, K. Takatsu, B. Kurniawan, H. Tanaka, H. Uekusa, Y. Ohashi, K. Takizawa, H. Mitamura, T. Goto, J. Phys. Soc. Jpn. **67**, 1548 (1998)
27. Ch. Rüegg, M. Oettli, J. Shefer, O. Zaharko, A. Furrer, H. Tanaka, K.W. Krämer, H.-U. Güdel, P. Vorderwisch, K. Habicht, T. Polinski, M. Meissner, Phys. Rev. Lett. **93**, 037207 (2004)
28. H. Kikuchi, Y. Fujii, M. Chiba, S. Mitsudo, T. Idehara, T. Tonegawa, K. Okamoto, T. Sakai, T. Kuwai, H. Ohta, Phys. Rev. Lett. **94**, 227201 (2005)
29. K. Penc, J.-B. Fouet, S. Miyahara, O. Tchernyshyov, F. Mila, Phys. Rev. Lett. **99**, 117201 (2007)
30. T. Inami, Y. Ajiro, T. Goto, J. Phys. Soc. Jpn. **65**, 2374 (1996)
31. L.E. Svistov, A.I. Smirnov, L.A. Prozorova, O.A. Petrenko, L.N. Demianets, A.Ya. Shapiro, Phys. Rev. B **67**, 094434 (2003)
32. L.E. Svistov, A.I. Smirnov, L.A. Prozorova, O.A. Petrenko, A. Micheler, N. Büttgen, A.Ya. Shapiro, L.N. Demianets, Phys. Rev. B **74**, 024412 (2006)
33. T. Ono, H. Tanaka, H. Aruga-Katori, F. Ishikawa, H. Mitamura, T. Goto, Phys. Rev. B **67**, 104431 (2003)
34. Z. Weihong, R.H. McKenzie, R.R.P. Singh, Phys. Rev. B **59**, 14367 (1999)
35. T. Ono, H. Tanaka, T. Nakagomi, O. Kolomiyets, H. Mitamura, F. Ishikawa, T. Goto, K. Nakajima, A. Oosawa, Y. Koike, K. Kakurai, J. Klenke, P. Smeibidle, M. Meissner, H. Aruga-Katori, J. Phys. Soc. Jpn. **74**, Suppl. 135 (2005)
36. T. Ono, H. Tanaka, O. Kolomiyets, H. Mitamura, T. Goto, K. Nakajima, A. Oosawa, Y. Koike, K. Kakurai, J. Klenke, P. Smeibidle, M. Meissner, J. Phys. Condens. Matter **16**, S773 (2004)
37. Y. Fujii, T. Nakamura, H. Kikuchi, M. Chiba, T. Goto, S. Matsubara, K. Kodama, M. Takigawa, Phys. B **346–347**, 45 (2004)
38. S. Miyahara, K. Orino, N. Furukawa, Phys. B **378–380**, 587 (2006)
39. R. Coldea, D.A. Tennant, K. Habicht, P. Smeibidl, C. Wolters, Z. Tylcznski, Phys. Rev. Lett. **88**, 137203 (2002)
40. Y. Yamashita, K. Ueda, Phys. Rev. Lett. **85** 4960 (2000)
41. H. Ueda, H. Aruga-Katori, H. Mitamura, T. Goto, H. Takagi, Phys. Rev. Lett. **94**, 047202 (2005)

42. H. Ueda, H. Mitamura, Y. Ueda, Phys. Rev. B **73**, 094415 (2006)
43. O. Tchernyshyov, R. Moessner, S.L. Sondhi, Phys. Rev. B **66**, 064403 (2002)
44. M. Matsuda, H. Ueda, A. Kikkawa, K. Katsumata, Y. Narumi, T. Inami, Y. Ueda, S.-H. Lee, Nat. Phys. **3**, 397 (2007)
45. K. Penc, N. Shannon, H. Shiba, Phys. Rev. Lett. **93**, 197203 (2004)
46. Y. Moyome, K. Penc, N. Shannon, J. Mag. Mag. Mat. **300**, 57 (2006)
47. H. Kageyama, K. Yoshimura, R. Stern, N. Mushnikov, K. Onizuka, M. Kato, K. Kosuge, C. Slichter, T. Goto, Y. Ueda, Phys. Rev. Lett. **82**, 3168 (1999)
48. S. Miyahara, K. Ueda, J. Phys. Condens. Matter **15**, R327 (2003)
49. B.S. Shastry, B. Sutherland, Phys. B **108**, 1069 (1981)
50. S. Miyahara, K. Ueda, Phys. Rev. Lett. **82**, 3172 (1999)
51. M. Albrecht, F. Mila, Europhys. Lett. **34**, 145 (1996)
52. A. Koga, N. Kawakami, Phys. Rev. Lett. **84**, 4461 (2000)
53. A. Läuchli, S. Wessel, M. Sigrist, Phys. Rev. B **66**, 014401 (2000)
54. H. Kageyama, K. Onizuka, T. Yamauchi, Y. Ueda, S. Hane, H. Mitamura, T. Goto, K. Yoshimura, K. Kosuge, J. Phys. Soc. Jpn. **68**, 1821 (1999)
55. H. Kageyama, H. Suzuki, M. Nohara, K. Onizuka, H. Takagi, and Y. Ueda, Physica B **281–282**, 667 (2000)
56. K. Kodama, J. Yamazaki, M. Takigawa, H. Kageyama, K. Onizuka, Y. Ueda, J. Phys. Condens. Matter **14**, L319 (2002)
57. H. Nojiri, H. Kageyama, K. Onizuka, Y. Ueda, M. Motokawa, J. Phys. Soc. Jpn. **68**, 2906 (1999)
58. H. Kageyama, M. Nishi, N. Aso, K. Onizuka, T. Yosihama, K. Nukui, K. Kodama, K. Kakurai, Y. Ueda, Phys. Rev. Lett. **84**, 5876 (2000)
59. S. Miyahara, K. Ueda, J. Phys. Soc. Jpn. **69**, Suppl. B 72 (2000)
60. C. Knetter, A. Bühler, E. Müller-Hartmann, G.S. Uhrig, Phys. Rev. Lett. **85**, 3958 (2000)
61. K. Onizuka, H. Kageyama, Y. Narumi, K. Kindo, Y. Ueda, T. Goto, J. Phys. Soc. Jpn. **69**, 1016 (2000)
62. B.D. Gaulin, S.H. Lee, S. Haravifard, J.P. Castellan, A.J. Berlinsky, H.A. Dabkowska, Y. Qiu, J.R.D. Copley, Phys. Rev. Lett. **93**, 267202 (2004)
63. K. Kakurai, K. Nukui, N. Aso, M. Nishi, H. Kadowaki, H. Kageyama, Y. Ueda, L.-P. Regnault, O. Cépas, Prog. Theor. Phys. Suppl. **159**, 22 (2005)
64. O. Cépas, K. Kakurai, L. P. Regnault, T. Ziman, J. P. Boucher, N. Aso, M. Nishi, H. Kageyama, Y. Ueda, Phys. Rev. Lett. **87**, 167205 (2001)
65. H. Nojiri, Kageyama, Y. Ueda, M. Motokawa, J. Phys. Soc. Jpn. **72**, 3243 (2003)
66. T. Rõõm, D. Hüvonen, U. Nagel, J. Hwang, T. Timusk, H. Kageyama, Phys. Rev. B **70**, 144417 (2004)
67. K. Kodama, S. Miyahara, M. Takigawa, M. Horvatić, C. Berthier, F. Mila, H. Kageyama, Y. Ueda, J. Phys. Condens. Matter **17**, L61 (2005)
68. S. Miyahara, F. Mila, K. Kodama, M. Takigawa, M. Horvatić, C. Berthier, H. Kageyama, Y. Ueda, J. Phys. Condens. Matter **16**, S911 (2004)
69. S. Miyahara, J.-B. Fouet, S.R. Manmana, R.M. Noack, H. Mayaffre, I. Sheikin, C. Berthier, F. Mila, Phys. Rev. B **75**, 184402 (2007)
70. A. Zorko, D. Arčon, H. van Tol, L.C. Brunel, H. Kageyama, Phys. Rev. B **69**, 174420 (2004)
71. G.A. Jorge, R. Stern, M. Jaime, N. Harrison, J. Bonča, S. El Shawish, C.D. Batista, H.A. Dabkowska, B.D. Gaulin, Phys. Rev. B **71**, 092403 (2005)
72. O. Cépas, T. Ziman, Phys. Rev. B **70**, 024404 (2004)
73. V.V. Mazurenko, S.L. Skornyakov, V.I. Anisimov, F. Mila, Phys. Rev. B **78**, 195110 (2008)
74. K. Kodama, M. Takigawa, M. Horvatić, C. Berthier, H. Kageyama, Y. Ueda, S. Miyahara, F. Becca, F. Mila, Science **298**, 395 (2002)
75. M. Takigawa, K. Kodama, M. Horvatić, C. Berthier, H. Kageyama, Y. Ueda, S. Miyahara, F. Becca, F. Mila, Phys. B **346–347**, 27 (2004)
76. S. Miyahara, F. Becca, F. Mila, Phys. Rev. B **68**, 024401 (2003)
77. S. Miyahara, K. Ueda, Phys. Rev. B **61**, 3417 (2000)

78. B. Wolf, S. Zherlitsyn, S. Schmidt, B. Lüthi, H. Kageyama, Y. Ueda, Phys. Rev. Lett. **86**, 4847 (2001)
79. R. Stern, I. Heinmaa, P.L. Kuhns, A.P. Reyes, W.G. Moulton, H.A. Dabkowska, B.D. Gaulin, National High Magnetic Field Laboratory, 2004 Research Report 278
80. F. Levy, I. Sheikin, C. Berthier, M. Horvatić, M. Takigawa, H. Kageyama, T. Waki, and Y. Ueda, Europhys. Lett. **81**, 67004 (2008)
81. M. Takigawa, S. Matsubara, M. Horvatić, C. Berthier, H. Kageyama, Y. Ueda, Phys. Rev. Lett. **101**, 037202 (2008)
82. T. Momoi, K. Totsuka, Phys. Rev. B **62**, 15067 (2000)
83. S.E. Sebastian, N. Harrison, P. Sengupta, C.D. Batista, S. Francoual, E. Palm, T. Murphy, H.A. Dabkowska, B.D. Gaulin, Proc. Nat. Acad. Sci. **105**, 20157 (2008)
84. J. Dorier, K.P. Schmidt, F. Mila, Phys. Rev. Lett. **101**, 250402 (2008)
85. A. Abendschein, S. Capponi, Phys. Rev. Lett. **101**, 227201 (2008)
86. F. Iga, A. Shigekawa, Y. Hasegawa, S. Michimura, T. Takabatake, S. Yoshii, T. Yamamoto, M. Hasegawa, K. Kindo, J. Mag. Mag. Mat. **310**, e443 (2007)
87. S. Gabáni, S. Mat'aš, P. Priputen, K. Flachbart, K. Siemensmeyer, E. Wulf, A. Evdokimova, N. Shitsevalova, Acta Phys. Pol. A **113**, 227 (2008)
88. S. Yoshii, T. Yamamoto, M. Hagiwara, S. Michimura, A. Shigekawa, F. Iga, T. Takabatake, K. Kindo, Phys. Rev. Lett. **101**, 087202 (2008)
89. S. Yoshii, K. Ohoyama, K. Kurosawa, H. Nojiri, M. Matsuda, P. Frings, F. Duc, B. Vignolle, G.L.J.A. Rikken, L.-P. Regnault, S. Michimura, F. Iga, Phys. Rev. Lett. **103**, 077203 (2009)

# Chapter 11
# Spin-Lattice Coupling in Frustrated Antiferromagnets

**Oleg Tchernyshyov and Gia-Wei Chern**

**Abstract** We review the mechanism of spin-lattice coupling in relieving the geometrical frustration of pyrochlore antiferromagnets, in particular spinel oxides. The tetrahedral unit, which is the building block of the pyrochlore lattice, undergoes a spin-driven Jahn–Teller instability when lattice degrees of freedom are coupled to the antiferromagnetism. By restricting our considerations to distortions which preserve the translational symmetries of the lattice, we present a general theory of the collective spin–Jahn–Teller effect in the pyrochlore lattice. One of the predicted lattice distortions breaks the inversion symmetry and gives rise to a chiral pyrochlore lattice, in which frustrated bonds form helices with a definite handedness. The chirality is transferred to the spin system through spin-orbit coupling, resulting in a long-period spiral state, as observed in spinel $CdCr_2O_4$. We discuss explicit models of spin-lattice coupling using local phonon modes, and their applications in other frustrated magnets.

## 11.1 Introduction

As explained in Chap. 1 by Chalker, sufficiently strong frustration in a magnet results in a large degeneracy of its ground-state manifold. Prime examples of this behavior are Heisenberg antiferromagnets on the kagomé [1,2] and pyrochlore [3,4] lattices with interactions restricted to nearest-neighbor sites. In the classical limit of a large spin $S$, the ground states of these magnets exhibit very high, continuous degeneracies and possess numerous zero modes, which correspond to moving the system through its manifold of ground states [5]. The pyrochlore antiferromagnet represents a particularly striking example of high ground-state degeneracy: at least half of its spin-wave modes have zero frequencies in any collinear ground state [4].

A large degeneracy means enhanced sensitivity to perturbations, even when these are nominally weak. In this chapter we will consider a coupling between spins and the underlying lattice, which has its origin in the dependence of the exchange integrals on the atomic positions, $J(\mathbf{r}_1, \mathbf{r}_2)\, \mathbf{S}_1 \cdot \mathbf{S}_2$, and is known as magnetoelastic exchange [6]. In the pyrochlore antiferromagnet, this coupling lifts the degeneracy of the classical ground states and induces a symmetry-lowering distortion of the lattice, in analogy with the spin-Peierls effect in antiferromagnetic spin chains [7]. A spin-Peierls-like phase transition has been observed in several antiferromagnetic spinels where the magnetic ions form the pyrochlore lattice [8–10].

The problem of coupled spins and lattice degrees of freedom in a pyrochlore antiferromagnet is reminiscent of the collective Jahn–Teller effect [11] in crystalline solids. We therefore begin the discussion by studying the Jahn–Teller distortion in a tetrahedral "molecule" with four spins, which is the structural unit of the pyrochlore lattice. A symmetry-based analysis will be supplemented by models with specific spin-phonon coupling mechanisms. We will then extend the analysis to an infinite lattice, to examine some of the possible ground states of the classical spin system. In concluding this chapter, we will test the theory of spin–phonon coupling on the example of $CdCr_2O_4$, a frustrated Heisenberg antiferromagnet with $S = 3/2$ spins residing on the pyrochlore lattice.

## 11.2 Spin-Driven Jahn–Teller Effect in a Tetrahedron

Consider the basic structural unit of the pyrochlore lattice, a cluster in the shape of a regular tetrahedron with four spins of length $S$ at the corners (Fig. 11.1). The Heisenberg exchange energy depends on the total spin of the cluster, $\mathbf{S}_{\text{tot}} = \sum_{i=1}^{4} \mathbf{S}_i$, according to

$$H_0 = J \sum_{i<j} \mathbf{S}_i \cdot \mathbf{S}_j = \frac{J\mathbf{S}_{\text{tot}}^2}{2} - \text{const.} \tag{11.1}$$

For antiferromagnetic exchange ($J > 0$), this energy is minimized when the total spin of the cluster is 0. The degeneracy of the ground state is thus equal to the number of distinct spin-singlet states in a system of four spins.

For spins of length $S$, the number of linearly independent singlet ground states is $2S + 1$. Indeed, the pair of spins 1 and 2 can have a total combined spin $\mathbf{S}_{12}$ ranging from 0 to $2S$. The same is true of the spin pair 3 and 4. A state with a total spin $S_{\text{tot}} = 0$ can be obtained by combining states with $S_{12} = S_{34} = 0$, $S_{12} = S_{34} = 1$,

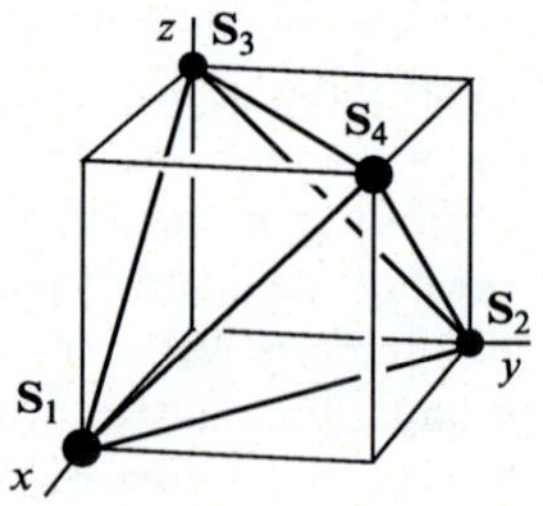


**Fig. 11.1** Representation of four spins in the corners of a regular tetrahedron

and so on, up to $S_{12} = S_{34} = 2S$. Thus one observes that, for any spin length $S$, the ground state of the four spins is degenerate. The high symmetry of the cluster means that, in accordance with the Jahn–Teller theorem,[1] the ground-state energy can be lowered through a distortion.

Assume for simplicity (a more general case will be considered below) that the exchange coupling $J$ between spins $i$ and $j$ has a dependence on their separation, $r_{ij} = |\mathbf{r}_i - \mathbf{r}_j|$, given by

$$J(r_{ij}) = J(\bar{r}_{ij}) + J'(\bar{r}_{ij})\delta r_{ij} + J''(\bar{r}_{ij})\delta r_{ij}^2/2 + \ldots, \tag{11.2}$$

where $\bar{r}_{ij}$ is a reference distance. At zeroth order in the displacements $\delta\mathbf{r}_i$, we recover the unperturbed Heisenberg Hamiltonian (11.1) with $2S + 1$ degenerate singlet ground states.

The first-order term,

$$H_1 = J' \sum_{i<j} (\mathbf{S}_i \cdot \mathbf{S}_j)\delta r_{ij}, \tag{11.3}$$

lifts the degeneracy of the ground-state manifold. As long as the displacements involved remain small enough to satisfy $J'\delta r_{ij} \ll J$, excited states with $S_{\mathrm{tot}} > 0$ at energies $J$ and higher may be neglected. Thus it is necessary to determine the energy levels of $H_1$ in the Hilbert space of the singlet ground states.

### *11.2.1 Generalized Coordinates and Forces*

The perturbation Hamiltonian (11.3) depends on the atomic displacements $\delta\mathbf{r}_i$, which comprise $4 \times 3 = 12$ degrees of freedom. However, not all of these will influence the exchange energy of the spins. As one example, it does not change under rigid translations of the tetrahedron (3 modes) or under global rotations (a further 3 modes). The remaining 6 modes represent various deformations of the four-site cluster. To facilitate a further analysis, we classify these modes in terms of the irreducible representations (irreps) of the tetrahedral point group $T_d$ [13].

The 6 modes belong to three irreps of $T_d$. The breathing mode (irrep $A$) leaves the symmetry of the tetrahedron fully intact. A doublet of tetragonal and orthorhombic distortions, $\mathbf{Q}^E = (Q_1^E, Q_2^E)$, transforms under irrep $E$. Finally, a triplet $\mathbf{Q}^{T_2} = (Q_1^{T_2}, Q_2^{T_2}, Q_3^{T_2})$, transforming as irrep $T_2$, elongates and compresses opposing bonds; equal-amplitude superpositions of the triplet components yield trigonal distortions. The coordinates of these modes can be expressed in terms of Cartesian displacements of the spins with the coefficients listed in Table 11.1.

By expressing the changes in bond lengths, $\delta r_{ij}$, in terms of the generalized coordinates, we reduce the perturbation Hamiltonian (11.3) to the form

[1] It is important to note that the degeneracy is not caused by the symmetry of time reversal, so the conditions of the theorem are fulfilled [12].

**Table 11.1** Coefficients relating the 6 distortions of a tetrahedron to the displacements $\delta\mathbf{r}_i$ of its vertices. The reference frame is shown in Fig. 11.1

| | | $\delta x_1$ | $\delta y_1$ | $\delta z_1$ | $\delta x_2$ | $\delta y_2$ | $\delta z_2$ | $\delta x_3$ | $\delta y_3$ | $\delta z_3$ | $\delta x_4$ | $\delta y_4$ | $\delta z_4$ |
|---|---|---|---|---|---|---|---|---|---|---|---|---|---|
| $A$ | $Q^A$ | $+\frac{1}{\sqrt{12}}$ | $-\frac{1}{\sqrt{12}}$ | $-\frac{1}{\sqrt{12}}$ | $-\frac{1}{\sqrt{12}}$ | $+\frac{1}{\sqrt{12}}$ | $-\frac{1}{\sqrt{12}}$ | $-\frac{1}{\sqrt{12}}$ | $-\frac{1}{\sqrt{12}}$ | $+\frac{1}{\sqrt{12}}$ | $+\frac{1}{\sqrt{12}}$ | $+\frac{1}{\sqrt{12}}$ | $+\frac{1}{\sqrt{12}}$ |
| $E$ | $Q_1^E$ | $-\frac{1}{\sqrt{24}}$ | $+\frac{1}{\sqrt{24}}$ | $-\frac{1}{\sqrt{6}}$ | $+\frac{1}{\sqrt{24}}$ | $-\frac{1}{\sqrt{24}}$ | $-\frac{1}{\sqrt{6}}$ | $+\frac{1}{\sqrt{24}}$ | $+\frac{1}{\sqrt{24}}$ | $+\frac{1}{\sqrt{6}}$ | $-\frac{1}{\sqrt{24}}$ | $-\frac{1}{\sqrt{24}}$ | $+\frac{1}{\sqrt{6}}$ |
| | $Q_2^E$ | $+\frac{1}{\sqrt{8}}$ | $+\frac{1}{\sqrt{8}}$ | 0 | $-\frac{1}{\sqrt{8}}$ | $-\frac{1}{\sqrt{8}}$ | 0 | $-\frac{1}{\sqrt{8}}$ | $+\frac{1}{\sqrt{8}}$ | 0 | $+\frac{1}{\sqrt{8}}$ | $-\frac{1}{\sqrt{8}}$ | 0 |
| $T_2$ | $Q_1^{T_2}$ | 0 | $+\frac{1}{\sqrt{8}}$ | $+\frac{1}{\sqrt{8}}$ | 0 | $+\frac{1}{\sqrt{8}}$ | $-\frac{1}{\sqrt{8}}$ | 0 | $-\frac{1}{\sqrt{8}}$ | $+\frac{1}{\sqrt{8}}$ | 0 | $-\frac{1}{\sqrt{8}}$ | $-\frac{1}{\sqrt{8}}$ |
| | $Q_2^{T_2}$ | $+\frac{1}{\sqrt{8}}$ | 0 | $-\frac{1}{\sqrt{8}}$ | $+\frac{1}{\sqrt{8}}$ | 0 | $+\frac{1}{\sqrt{8}}$ | $-\frac{1}{\sqrt{8}}$ | 0 | $+\frac{1}{\sqrt{8}}$ | $-\frac{1}{\sqrt{8}}$ | 0 | $-\frac{1}{\sqrt{8}}$ |
| | $Q_3^{T_2}$ | $+\frac{1}{\sqrt{8}}$ | $-\frac{1}{\sqrt{8}}$ | 0 | $-\frac{1}{\sqrt{8}}$ | $+\frac{1}{\sqrt{8}}$ | 0 | $+\frac{1}{\sqrt{8}}$ | $+\frac{1}{\sqrt{8}}$ | 0 | $-\frac{1}{\sqrt{8}}$ | $-\frac{1}{\sqrt{8}}$ | 0 |

**Table 11.2** Coefficients relating the bond elongations $\delta r_{ij}$ to the distortion coordinates $Q^A$, $\mathbf{Q}^E = (Q_1^E, Q_2^E)$, and $\mathbf{Q}^{T_2} = (Q_1^{T_2}, Q_2^{T_2}, Q_3^{T_2})$

| | $Q^A$ | $Q_1^E$ | $Q_2^E$ | $Q_1^{T_2}$ | $Q_2^{T_2}$ | $Q_3^{T_2}$ |
|---|---|---|---|---|---|---|
| $\delta r_{14}$ | $+\sqrt{\frac{2}{3}}$ | $+\frac{1}{\sqrt{12}}$ | $-\frac{1}{2}$ | $-1$ | 0 | 0 |
| $\delta r_{23}$ | $+\sqrt{\frac{2}{3}}$ | $+\frac{1}{\sqrt{12}}$ | $-\frac{1}{2}$ | $+1$ | 0 | 0 |
| $\delta r_{24}$ | $+\sqrt{\frac{2}{3}}$ | $+\frac{1}{\sqrt{12}}$ | $+\frac{1}{2}$ | 0 | $-1$ | 0 |
| $\delta r_{13}$ | $+\sqrt{\frac{2}{3}}$ | $+\frac{1}{\sqrt{12}}$ | $+\frac{1}{2}$ | 0 | $+1$ | 0 |
| $\delta r_{34}$ | $+\sqrt{\frac{2}{3}}$ | $-\frac{1}{\sqrt{3}}$ | 0 | 0 | 0 | $-1$ |
| $\delta r_{12}$ | $+\sqrt{\frac{2}{3}}$ | $-\frac{1}{\sqrt{3}}$ | 0 | 0 | 0 | $+1$ |

$$H_1 = \sum_\alpha J^{\alpha\prime} \sum_n Q_n^\alpha f_n^\alpha, \tag{11.4}$$

where the index $\alpha$ enumerates the irreps and $n$ its components. The variable $f_n^\alpha$ is the generalized force which is conjugate to the coordinate $Q_n^\alpha$ and has the same symmetry properties. $J^{\alpha\prime}$ is a coupling constant appropriate for the irrep $\alpha$. (Table 11.2)

For illustration, the breathing mode $Q_A$ couples to the spin operator

$$f^A = \frac{1}{\sqrt{6}} \sum_{i<j} \mathbf{S}_i \cdot \mathbf{S}_j, \tag{11.5}$$

which is invariant under all symmetry operations of $T_d$. Furthermore, up to a trivial factor, this is the unperturbed Hamiltonian (11.1) and so has the same value in any of the degenerate ground states. Consequently, the term $-J^{A\prime} Q^A f^A$ in the perturbation Hamiltonian (11.3) produces a trivial energy shift of the degenerate ground states, but does not split them.

The triplet mode and the associated triplet force also do not induce a splitting. To demonstrate this, we note that $Q_1^{T_2}$ couples to the operator

$$f_1^{T_2} = (\mathbf{S}_2 \cdot \mathbf{S}_3 - \mathbf{S}_1 \cdot \mathbf{S}_4)/\sqrt{2}, \tag{11.6}$$

which vanishes in any singlet state.[2] In the presence of an applied magnetic field, the triplet forces cannot be neglected because of the nonzero total spin $S_{\mathrm{tot}}$. The triplet forces and the corresponding trigonal distortions play an important role in the stabilization of the half-magnetization plateaus observed in some spinel chromites (Sect. 11.3.1).

The only two modes involved in the splitting of the ground state are the components of the doublet $(Q_1^E, Q_2^E)$ of tetragonal and orthorhombic distortions. These couple, respectively, to the spin operators

$$f_1^E = \frac{\mathbf{S}_1 \cdot \mathbf{S}_4 + \mathbf{S}_2 \cdot \mathbf{S}_3 + \mathbf{S}_2 \cdot \mathbf{S}_4 + \mathbf{S}_1 \cdot \mathbf{S}_3 - 2\mathbf{S}_1 \cdot \mathbf{S}_2 - 2\mathbf{S}_3 \cdot \mathbf{S}_4}{\sqrt{12}},$$
$$f_2^E = \frac{\mathbf{S}_2 \cdot \mathbf{S}_4 + \mathbf{S}_1 \cdot \mathbf{S}_3 - \mathbf{S}_1 \cdot \mathbf{S}_4 - \mathbf{S}_2 \cdot \mathbf{S}_3}{2}. \tag{11.7}$$

In what follows, we omit the irrep superscript because only the doublet $E$ participates in lifting the degeneracy of the ground-state manifold.

In addition to the magnetoelastic exchange (11.3), which is linear in the distortion amplitude, it is necessary to consider also the elastic energy of the deformation. We thus obtain the spin-distortion Hamiltonian

$$H = J'(\mathbf{Q} \cdot \mathbf{f}) + k|\mathbf{Q}|^2/2 \equiv J'(Q_1 f_1 + Q_2 f_2) + k(Q_1^2 + Q_2^2)/2, \tag{11.8}$$

where $k$ is the elastic constant of the doublet distortion. Having established this, the next task is to minimize the energy (11.8) with respect to both the coordinates and the spins.

### 11.2.2 Four $S = 1/2$ Spins on a Tetrahedron

The problem of four $S = 1/2$ spins on a deformable tetrahedron was first analyzed by Yamashita and Ueda [14]. The ground state of the unperturbed exchange Hamiltonian (11.1) is two-fold degenerate. As a basis in this Hilbert space one may use singlet states with a well-defined total spin on bonds 12 and 34, $S_{12} = S_{34} = \sigma$, where $\sigma = 0$ or 1 [15]. In this basis, the force operators $f_1$ and $f_2$ are proportional to the Pauli matrices $\sigma_1$ and $\sigma_3$, respectively, so that the Hamiltonian (11.8) reduces to

$$H = J'(Q_1\sigma_1 + Q_2\sigma_3)\sqrt{3}/2 + k(Q_1^2 + Q_2^2)/2. \tag{11.9}$$

For a given distortion $\mathbf{Q}$, the ground-state manifold is split into the two energy levels

[2] Indeed, from $\mathbf{S}_1 + \mathbf{S}_2 + \mathbf{S}_3 + \mathbf{S}_4 = 0$ it follows that $(\mathbf{S}_1 + \mathbf{S}_4)^2 = (\mathbf{S}_2 + \mathbf{S}_3)^2$ and thus $\mathbf{S}_1 \cdot \mathbf{S}_4 = \mathbf{S}_2 \cdot \mathbf{S}_3$.

$$E_{1,2} = \pm|J'|Q\sqrt{3}/2 + kQ^2/2, \tag{11.10}$$

and the energy of the system is minimized when $Q = |J'|\sqrt{3}/(2k)$. Note that this quantity depends on the magnitude of the distortion, $Q = \sqrt{Q_1^2 + Q_2^2}$, but not on its "direction:" it can be tetragonal, purely orthorhombic, or any combination of these. This degeneracy is associated with a continuous symmetry of the Hamiltonian (11.9) that involves simultaneous "rotations" of both the distortion coordinates and the Pauli matrices,

$$\begin{pmatrix} Q_1 \\ Q_2 \end{pmatrix} \mapsto \begin{pmatrix} \cos\theta & \sin\theta \\ -\sin\theta & \cos\theta \end{pmatrix} \begin{pmatrix} Q_1 \\ Q_2 \end{pmatrix}, \quad \begin{pmatrix} \sigma_1 \\ \sigma_3 \end{pmatrix} \mapsto \begin{pmatrix} \cos\theta & \sin\theta \\ -\sin\theta & \cos\theta \end{pmatrix} \begin{pmatrix} \sigma_1 \\ \sigma_3 \end{pmatrix}. \tag{11.11}$$

The invariance of the energy under this transformation does not reflect an underlying symmetry, and applies at the level of approximation used here. Terms of higher order in $\mathbf{Q}$ break this symmetry to leave only a three-fold degeneracy, as may be expected on symmetry grounds [14]. The lowest-order anharmonic term allowed by the symmetry is proportional to

$$Q_x Q_y Q_z \equiv \left(-\frac{1}{2}Q_1 + \frac{\sqrt{3}}{2}Q_2\right)\left(-\frac{1}{2}Q_1 - \frac{\sqrt{3}}{2}Q_2\right)Q_1 = \frac{1}{4}Q^3 \cos 3\alpha, \tag{11.12}$$

where $Q_x$, $Q_y$, and $Q_z$ measure tetragonal distortions along the respective axes and $\alpha$ is the polar angle in the $(Q_1, Q_2)$ plane. Depending on the sign of the cubic term, it favors three distinct ground states at $\alpha = 0, \pm 2\pi/3$ or at $\alpha = \pi, \pm\pi/3$. The former "vacua" have spin singlets on two opposing bonds (e.g. $S_{12} = S_{34} = 0$ for $\alpha = 0$) while the latter have spin triplets on two opposing bonds ($S_{12} = S_{34} = 1$ for $\alpha = \pi$). These ground states exhibit valence-bond order, which violates the point-group symmetry of the cluster but not the SO(3) symmetry of the exchange interaction. The two-component order parameter (11.7), introduced by Harris, Berlinsky and Bruder [16], measures the differences in spin correlations on the different bonds.

The ground states of the cluster exhibit a tetragonal lattice distortion along one of the three major axes with $\mathbf{Q} = -J'\langle\mathbf{f}\rangle\sqrt{3}/(2k)$. If $J' < 0$, as would be expected for direct antiferromagnetic exchange, the tetrahedron is flattened (elongated) in a ground state with triplets (singlets) on opposing bonds.

For spins of length $S > 1/2$, the analysis proceeds by a similar route [15]. The lowest-order perturbation (11.3) yields three degenerate singlet ground states with the largest possible spins on two opposing bonds, such as $S_{12} = S_{34} = 2S$, and a tetragonal distortion (a flattening of the tetrahedron for $J' < 0$). This result is most easily understood in the classical limit $S \to \infty$, to which we turn next.

### 11.2.3 Four Classical Spins on a Tetrahedron

For classical spins, the total energy $E(\mathbf{f}, \mathbf{Q})$ (11.8) can be minimized in two steps. We minimize it first with respect to the distortion $\mathbf{Q}$ at a fixed $\mathbf{f}$.[3] A minimum is achieved when $\mathbf{Q} = -J'\mathbf{f}/k$, yielding the energy

$$E(\mathbf{f}) = -J'^2 f^2/(2k), \tag{11.13}$$

whence the total energy is minimized by states with the largest magnitude of the force doublet $\mathbf{f}$. Thus it is necessary to quantify the magnetoelastic forces in those states of the ground-state manifold with $S_{\text{tot}} = 0$.

The domain of attainable $\mathbf{f}$ values forms a regular triangle in the $(f_1, f_2)$ plane (Fig. 11.2). Its three corners correspond to the three distinct collinear states with four satisfied bonds ($\mathbf{S}_i \cdot \mathbf{S}_j = -S^2$) and two frustrated ones ($\mathbf{S}_i \cdot \mathbf{S}_j = +S^2$). States elsewhere on the perimeter of the triangle have coplanar spins.

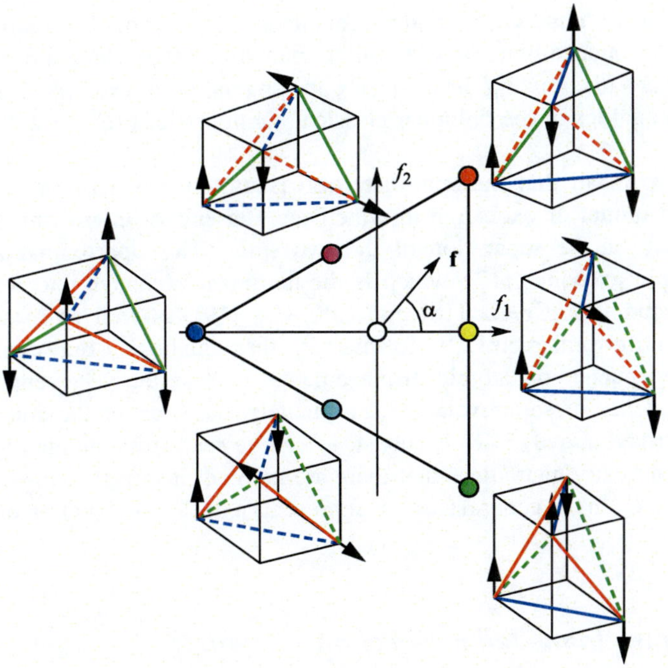


**Fig. 11.2** The domain of attainable values of the force doublet $\mathbf{f} = (f_1, f_2)$ (11.7) for classical spins. *Dashed lines* indicate frustrated bonds ($\mathbf{S}_i \cdot \mathbf{S}_j \geq 0$). Reprinted with permission from [15]

[3] This method cannot be applied to quantum spins because the operators $f_1$ and $f_2$ do not commute [15], and so their values cannot be measured simultaneously.

Not unexpectedly, the doublet force is maximized (and the total energy minimized) in the collinear states. Indeed, in such states antiparallel spins attract and parallel spins repel each other with forces $F = -J' \mathbf{S}_i \cdot \mathbf{S}_j$ of the largest possible magnitude, $|J'|S^2$. The large forces result in large distortions, yielding a large decrease in the total energy. Thus in the classical limit one expects collinear ground states in which the tetrahedron is flattened along one of the principal axes for $J' < 0$.[4]

### 11.2.4 Color Notation and Other Useful Analogies

We find it convenient here to introduce an analogy with the color triangle, where the vertices correspond to the primary colors red, green, and blue, the midpoints of the edges to the secondary colors cyan, magenta, and yellow, and the center to the absence of color. If we color bonds perpendicular to the major axes $a$, $b$, and $c$ respectively red, green, and blue, then the color of the state in Fig. 11.2 reflects the color of the frustrated bonds.

The collinear ground states obtained for classical spins provide a simple rationalization for the valence-bond states found for quantum spins of length $S > 1/2$ [15]. Quantum states maximizing the spins of opposing pairs (for example $S_{12} = S_{34} = 2S$) are the analogs of the collinear classical configurations ($S_1 = S_2 = -S_3 = -S_4$).

Lastly, we recall that the spin-distortion Hamiltonian (11.8) was derived for a simplified model of exchange in which the Heisenberg interaction strength $J$ depends only on the separation of the two spins. This approximation is good for direct spin exchange [17], which is the dominant exchange interaction in the chromium spinels $ZnCr_2O_4$ [18] and $CdCr_2O_4$ [19], as well as in some other chromium antiferromagnets [20]. However, in other situations $J$ may exhibit a more complex dependence on atomic displacements, such as the very sensitive bond-angle-dependence of superexchange. Fortunately, the form of the spin-distortion coupling derived above (11.8) is robust, as can be seen from symmetry considerations: group theory guarantees that there are no other invariant terms of the same order in $\mathbf{f}$ and $\mathbf{Q}$. In the general case, $J'$ represents a linear combination of exchange derivatives.

### 11.2.5 Spin–Jahn–Teller Effect on a Triangle

Another well-known lattice producing strong frustration in an antiferromagnet is the kagomé geometry [1,2], a network of corner-sharing triangles in a plane, and its three-dimensional variant, the hyperkagomé lattice [21]. It is natural to ask whether

[4] If $J' > 0$, the spins are still collinear but the tetrahedron is elongated along the same axis.

spin-lattice coupling is also an effective mechanism for relieving frustration in this case. We answer this question by considering the building block of the kagomé lattice, a triangle with three spins.

Classical Heisenberg spins interacting through antiferromagnetic Heisenberg exchange minimize their energy by making angles of 120° with one another. An analysis along the lines of Sect. 11.2 shows that the correction to the exchange energy from the magnetoelastic term is quadratic in the spin displacements: the linear term cancels. Without this linear term, a spontaneous distortion does not occur. The absence of the linear term can be understood simply from the magnetoelastic forces between the spins: the three forces, being proportional to the scalar products of the spins, are equal in a state where the spins make equal angles with each other. These symmetrical forces only shrink the triangle without distorting it.

The argument against the Jahn–Teller distortion fails if the quadratic term in the magnetoelastic correction is negative and large enough to overcome the purely elastic cost of the distortion, a scenario proposed recently by Wang and Vishwanath [22]. In our view, however, empirical evidence indicates that a Jahn–Teller instability of this sort would be a rare exception. The strength of the quadratic magnetoelastic term can be estimated from the splitting of degenerate phonons in antiferromagnets with a spin-induced Jahn–Teller distortion [18]. Such splittings, observed in $ZnCr_2O_4$ [18], $CdCr_2O_4$ [19], and MnO [23], do not exceed 12% of the phonon frequencies, which indicates the dominance of the purely elastic contribution.

While the spin-Jahn–Teller distortion of a triangle appears unlikely in the classical limit, the opposite extreme – quantum spins of length $S = 1/2$ – are a completely different situation. The ground state of three such spins interacting through antiferromagnetic Heisenberg exchange is any state with a total spin $S_\Delta = 1/2$. Such a state is fourfold degenerate: part of this degeneracy is of the Kramers type, as the projection of the total spin on an arbitrary axis can be $S^z_\Delta = \pm 1/2$, and there is an additional twofold degeneracy related to the symmetry of the triangle. This degeneracy can be understood in terms of valence-bond states, in which two of the spins on the triangle form a singlet bond (a quantum dimer) while the third remains free. Figure 11.3 shows three such states, although they are not mutually orthogonal; in fact only two of these states are linearly independent.

The presence of a non-Kramers degeneracy leads to the spin–Jahn–Teller effect on a triangle with spins $S = 1/2$. The analysis is similar to that for four $S = 1/2$ spins on a tetrahedron (Sect. 11.2.2), with three distinct ground states. Depending on the sign of the cubic term, the distorted triangle will have either two long bonds and one short bond with a singlet on it, or two short bonds and one long bond with a triplet.

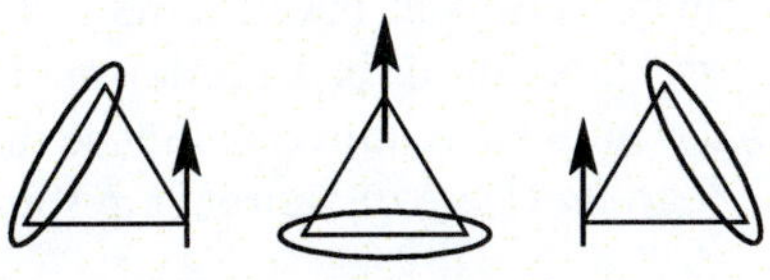

**Fig. 11.3** Valence-bond ground states of three spins $S = 1/2$ with antiferromagnetic Heisenberg exchange interactions. The ellipses designate singlet bonds

## 11.3 Models with Local Phonon Modes

The symmetry-based analysis of the previous section is basically exact regardless of the underlying microscopic model for the phonons. In this section, we review some specific models of spin-lattice coupling based on local phonon modes and also discuss their applications.

Probably the simplest situation is the "bond-phonon model," in which the exchange integral (11.3) and also the elastic energy depend only on the bond length $r_{ij}$. The elastic energy is approximated by the sum of individual terms $k\delta r_{ij}^2/2$. After integrating out the bond variables $\delta r_{ij}$, the model generates a biquadratic spin Hamiltonian

$$-(J'^2/2k)\sum_{\langle ij\rangle}(\mathbf{S}_i\cdot\mathbf{S}_j)^2, \tag{11.14}$$

which clearly favors collinear spin configurations. Because of its simplicity, this model has been applied in numerous studies of frustrated magnets. As one example, Becca et al. found using the bond-phonon model that magnetoelastic coupling leads to a spin-Peierls transition in the frustrated $J_1$-$J_2$ antiferromagnet on the square lattice [24, 25]. This study may be relevant to the nature of the transition to a phase of collinear order observed in the quasi-2D antiferromagnet $Li_2VOSiO_4$.

On the pyrochlore lattice, the bond-phonon model gives a ground state with $3^N$-fold degeneracy, where $N$ is the total number of tetrahedra: each tetrahedron can be flattened along one of the 3 major axes, independently of the other tetrahedra. A more realistic phonon model can be formulated in terms of the independent displacements of each atom, with the bond lengths determined from these atomic displacements by $\delta r_{ij} = (\mathbf{u}_i - \mathbf{u}_j)\cdot\hat{\mathbf{r}}_{ij}$. This is known as the site-phonon (or Einstein phonon) model of spin-lattice coupling [26,27], in the simplest version of which the elastic energy is approximated by a sum of individual terms $k|\mathbf{u}_i|^2/2$, an assumption which leads to a constant dispersion similar to the long-wavelength limit of optical phonons. In addition to (11.14), after integrating out the displacements the model introduces couplings between bond variables,

$$-(J'^2/2k)\sum_{i,j\neq k}(\hat{\mathbf{r}}_{ij}\cdot\hat{\mathbf{r}}_{ik})(\mathbf{S}_i\cdot\mathbf{S}_j)(\mathbf{S}_i\cdot\mathbf{S}_k). \tag{11.15}$$

Because of these coupling terms, coherent long-range distortions are possible in this model. By using the site-phonon model, Wang and Vishwanath [22] showed that a zigzag collinear order of the triangular antiferromagnet $CuFeO_2$ is a ground state of the spin-lattice Hamiltonian. However, because the $Fe^{3+}$ ion in this compound has spin $S = 5/2$ and $L = 0$, resulting in a rather small magnetic anisotropy, a relatively large spin-lattice coupling is required to induce the zigzag collinear order from the non-collinear 120° ground state of Heisenberg spins on this lattice.

In the context of the pyrochlore lattice, the interaction term of (11.15) corresponds to an antiferromagnetic coupling between the force doublets of nearest-neighbor tetrahedra, $K\sum_{\langle\alpha\beta\rangle}\mathbf{f}_\alpha\cdot\mathbf{f}_\beta$, where the coupling constant $K > 0$. As a result, neighboring tetrahedra tend to be flattened along different directions (e.g. $\langle 100\rangle$ and $\langle 010\rangle$). While this reduces the total number of ground states, it still leaves a very high accidental degeneracy which grows exponentially with $N$.

### 11.3.1 Half-Magnetization Plateau in $ACr_2O_4$ Spinels

Local phonon models provide an explicit description with which one may study the spin-lattice stabilization of the half-magnetization plateau observed in some spinel chromites. The low-temperature magnetization curves of the spinels $CdCr_2O_4$ and $HgCr_2O_4$ exhibit a sharp transition into a wide plateau region where the magnetization is equal to half its saturation value [28, 29]. Each tetrahedron in the half-plateau state is in one of the four 3-up-1-down collinear spin configurations. While thermal and quantum fluctuations act in general to favor collinear spins, and indeed in some cases help to stabilize the magnetization plateaus in frustrated magnets, the observed half-magnetization plateau in spinels arises most likely due to a coupling between the spins and the lattice, a scenario also supported by the observation of a spontaneous structural distortion accompanying the half-plateau transition.

Because the total spin $S_{\text{tot}}$ is nonzero in the presence of a magnetic field, coupling of the moments to the singlet ($A$) and triplet ($T_2$) phonon modes can no longer be neglected. Still, when the applied magnetic field is weak, the distortion is such that the crystal retains a tetragonal symmetry. The spins develop a canted antiferromagnetic order with two frustrated bonds and four antiferromagnetic bonds. As the field strength increases, the doublet force $|\mathbf{f}^E| = (4/\sqrt{3})\,(S^2 - S_{\text{tot}}^2/16)$ decreases as a result of the increasing total spin. At a critical field, the trigonal distortion, accompanying a 3-up-1-down collinear spin configuration maximizing the triplet forces, $f_i^{T_2} = \sqrt{2}S^2$, becomes more favorable energetically. The tetragonal distortion thus gives way to the trigonal one through a discontinuous transition.

Using the bond-phonon model, Penc and co-workers [30] obtained a classical phase diagram for the spin–Jahn–Teller effect in a single tetrahedron. The transition to the collinear 3:1 states takes place at a field strength $H \approx 3J$ and $J'^2/k \gtrsim 0.05J$. A general symmetry analysis also showed that the collinear 3:1 states are always stabilized over a finite range of magnetic field, provided that $J^{E'} < 2J^{T_2'}$ [30].

Similar to the ground-state manifold at zero field, the model with independent bonds retains an extensively degenerate manifold of half-magnetization states, because the trigonal distortion of individual tetrahedra can be along any of the four $\langle 111\rangle$ axes. This accidental degeneracy is lifted by the additional coupling term (11.15) introduced by the site-phonon model, which favors an antidistortive coupling between neighboring tetrahedra. Using this rule, Bergman et al. [27] showed that a unique spin configuration (up to discrete lattice symmetries) with a quadrupled, cubic unit cell is the ground state of the system in the half-magnetization plateau. The resulting space group, $P4_332$, is consistent with the X-ray diffraction pattern of the spinel $HgCr_2O_4$ [31].

## 11.4 Collective Spin–Jahn–Teller Effect on the Pyrochlore Lattice

Attempts to extend the analysis of the spin–Jahn–Teller effect on a single tetrahedron to an infinite pyrochlore lattice encounter a conceptual problem: there are *infinitely many* phonon modes coupled to the spins (one may expect two for every tetrahedron). There are also technical difficulties: detailed knowledge of the crystal's elastic properties is required. As a result of these difficulties, the problem lacks a general solution.

Some progress may, however, be made through the use of local phonon models, as described in the previous section. Still, a massive accidental degeneracy remains in the ground states of these simplified models. Further, the lattice modes in real crystals are plane waves, and thus a lattice distortion involves only a small number of phonons with specific lattice momenta. For example, the distortion in $ZnCr_2O_4$ shows superlattice peaks in a diffraction experiment with wavenumbers $[\frac{1}{2}\frac{1}{2}\frac{1}{2}]$ [10]. This does make it possible to take an alternative, phenomenological approach in which only a small number of lattice modes is considered. Such an approach was taken by Tchernyshyov et al. [15], who considered the simplest case where spin displacements preserve the translational symmetries of the lattice and break only point-group symmetries.

The pyrochlore lattice is made up of tetrahedra of two different orientations. Because all tetrahedra of the same orientation are related by lattice translation (which is assumed to remain a good symmetry), it is necessary to consider only two tetrahedra of opposite orientations $A$ and $B$ (Fig. 11.5). The symmetry group must be expanded from $T_d$ by including the inversion $I$ exchanging the two sublattices of tetrahedra, $T_d \otimes I = O_h$ [13]. The irreps remain largely unaltered, with the exception of a newly added parity index, which enters because these are either even ($g$) or odd ($u$) under the inversion.

At linear order in the displacements, the only modes which couple to the spins are the doublets $E_g$ and $E_u$. The former represents an overall tetragonal or orthorhombic distortion of the lattice, while the latter is an optical phonon with wavenumber $\mathbf{q} = 0$ that distorts tetrahedra of opposite orientations in exactly opposite ways (e.g. by flattening tetrahedra $A$ and elongating tetrahedra $B$ by the same amount and in the same direction). These modes can be expressed in terms of linear combinations of distortions on tetrahedra of types $A$ and $B$,

$$\mathbf{Q}^g = \frac{\mathbf{Q}^A + \mathbf{Q}^B}{\sqrt{2}}, \quad \mathbf{Q}^u = \frac{\mathbf{Q}^A - \mathbf{Q}^B}{\sqrt{2}}. \tag{11.16}$$

The spin-lattice energy (11.8) generalizes to

$$E(\mathbf{f}^A, \mathbf{f}^B, \mathbf{Q}^A, \mathbf{Q}^B) = J'(\mathbf{Q}^A \cdot \mathbf{f}^A + \mathbf{Q}^B \cdot \mathbf{f}^B) + k_g |\mathbf{Q}^g|^2/2 + k_u |\mathbf{Q}^u|^2/2, \quad (11.17)$$

where $k_g$ and $k_u$ are the elastic constants of the even and odd $E$ doublets. Minimization with respect to the lattice modes $\mathbf{Q}^g$ and $\mathbf{Q}^u$ yields the energy as a function of the spin variables in the form

$$\begin{aligned} E(\mathbf{f}^A, \mathbf{f}^B) &= -\frac{K_g |\mathbf{f}^A + \mathbf{f}^B|^2}{4} - \frac{K_u |\mathbf{f}^A - \mathbf{f}^B|^2}{4} \\ &= -\frac{(K_g + K_u)\left(|\mathbf{f}^A|^2 + |\mathbf{f}^B|^2\right)}{4} \\ &\quad -\frac{(K_g - K_u)\left(\mathbf{f}^A \cdot \mathbf{f}^B\right)}{2}, \end{aligned} \quad (11.18)$$

where we have introduced the effective magnetoelastic exchange couplings $K_{g,u} = J'^2/k_{g,u}$.

The second line in (11.18) is the result familiar from (11.13): the magnitude of the doublet force $\mathbf{f}$ is maximized on tetrahedra of both sublattices. Thus one expects to find states with collinear spins and (for $J' < 0$) tetrahedra flattened along one of the three $\langle 100 \rangle$ directions. The third line in (11.18) represents a coupling between the $\mathbf{f}$ variables of the two sublattices, whose consequences depend on which of the two lattice modes is softer.

If $K_g > K_u$ ($E_g$ mode softer), the energy is minimized when $\mathbf{f}^A$ and $\mathbf{f}^B$ are in the same corner of the force triangle (Fig. 11.2). Tetrahedra of both sublattices are flattened along the same $\langle 100 \rangle$ direction, so that only an $E_g$ distortion is present. The spin configuration is shown in Fig. 11.4. The magnetic unit cell coincides with the structural one. Because we are considering an O(3)-symmetric Heisenberg model, the global orientation of the spins can be arbitrary.

If, on the other hand, $K_g < K_u$ (softer $E_u$ mode), the two $\mathbf{f}$ vectors are located in different corners of the triangle, so that tetrahedra of types $A$ and $B$ are flattened along two different $\langle 100 \rangle$ directions, giving six possible ground states. The distortion is a superposition of the $E_u$ and $E_g$ modes. The presence of the even mode is understood in a straightforward manner: if tetrahedra of type $A$ are flattened along $\langle 100 \rangle$ and tetrahedra of type $B$ along $\langle 010 \rangle$, the lattice is on average *elongated* along $\langle 001 \rangle$. The presence of the $E_u$ component of the distortion means that the inversion symmetry is broken spontaneously. This has important consequences for the magnetic order, shown in Fig. 11.5, as we discuss in detail below. Here we note only that frustrated bonds form left-handed spirals in this particular state.

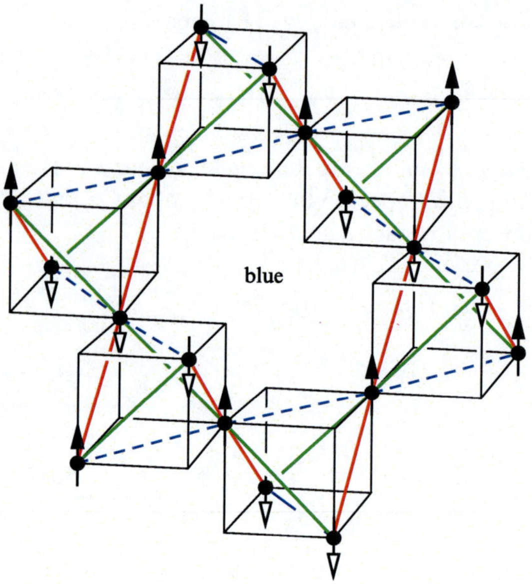


**Fig. 11.4** Magnetic order in a state with a softer $E_g$ mode. Frustrated bonds are shown as dashed lines. The lattice is flattened uniformly (for $J' < 0$). Reprinted with permission from [15]

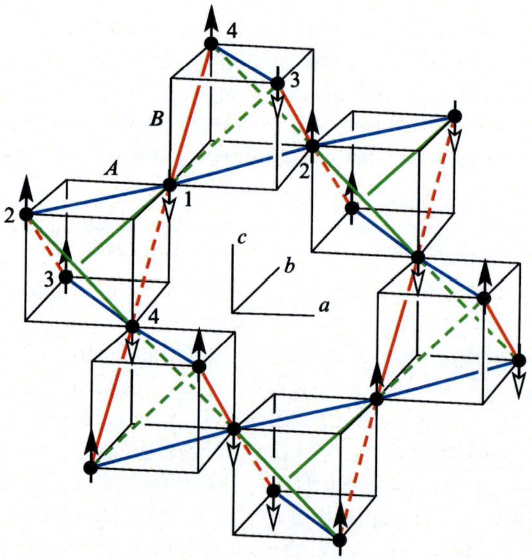


**Fig. 11.5** Magnetic order in a state with a softer $E_u$ mode. Frustrated bonds are shown as *dashed lines*. Tetrahedra of the two different orientations (labeled $A$ and $B$) are flattened along axes $a$ and $b$, so that the net distortion of the lattice is (for $J' < 0$) an elongation along axis $c$. Reprinted with permission from [32]

## 11.5 Collective Jahn–Teller Effect in $CdCr_2O_4$

The normal spinels $ACr_2O_4$ (where $A$ is a nonmagnetic Zn, Mg, Cd, or Hg ion) are strongly frustrated antiferromagnets exhibiting the spin-induced Jahn–Teller distortion. The magnetic $Cr^{3+}$ ions forming the pyrochlore lattice have electronic configuration $3d^3$. The oxygen octahedron surrounding the chromium ion splits the $3d$ levels into a high-energy $e_g$ doublet and a low-energy $t_{2g}$ triplet; the former are unoccupied while the latter are singly occupied, with the 3 electrons having parallel spins. Thus the orbital degree of freedom is quenched and the spins form a nearly isotropic magnetic moment with total spin $S = 3/2$ and a gyromagnetic ratio of $g = 2$ [8]. Interactions between the spins are mediated mostly by direct antiferromagnetic exchange between neighboring Cr sites [18] (with the exception of $HgCr_2O_4$, where this contribution is comparable to the ferromagnetic superexchange term mediated by the oxygen ions).

All of these compounds have a strongly correlated paramagnetic state well below their Curie–Weiss temperatures, $\Theta$, and order magnetically at a temperature $T_N \ll \Theta$. The phase transition is discontinuous and is accompanied by a lattice distortion [8–10]. The lack of orbital degrees of freedom points to a magnetoelastic origin for the lattice distortion.

$CdCr_2O_4$ is the best-understood system to date. In the distorted state below $T_N$ it shows no superlattice peaks [9], indicating that the translational symmetry of the high-temperature cubic phase (space group $Fd\bar{3}m$) remains unchanged. The point-group symmetry is lowered: the lattice exhibits a tetragonal distortion with lattice constants $a = b < c$ (an overall elongation). The low-temperature structure was identified by Chung et al. as the pyrochlore lattice with a uniform tetragonal elongation (space group $I4_1/amd$) [9], i.e. a pure $E_g$ distortion. However, as explained below, there are good reasons to believe that this distortion also involves a staggered component $E_u$, which breaks the inversion symmetry and lowers the space group down to $I4_122$.

1. The dominance of direct antiferromagnetic exchange between adjacent chromium spins means that the exchange constant decreases with increasing ionic separation, i.e. $J' < 0$. If the distortion were of pure $E_g$ type, the crystal would flatten along one axis, yielding $a = b > c$ in contradiction to the experimental data [9]. As explained in the previous section, an $E_u$-driven distortion would lead to an overall elongation of the lattice, $a = b < c$, in agreement with the experiment.
2. An $E_u$ distortion breaks the inversion symmetry of the crystal, making the crystalline lattice *chiral*. Indeed, the elongated (frustrated) bonds in Fig. 11.5 form spirals of one helicity. Spin-orbit coupling in the form of the Dzyaloshinskii–Moriya interaction would then spread the chirality indexchirality from the lattice to the spins, generating a spiral magnetic order. The observed magnetic order in $CdCr_2O_4$ is in fact a weakly incommensurate spiral [9].

### *11.5.1 Spiral Magnetic Order in $CdCr_2O_4$*

Chung et al. reported an incommensurate magnetic order with magnetic Bragg peaks at $\mathbf{q} = 2\pi(0, \delta, 1)$ in a crystal with an elongated $c$ axis ($a = b = 0.995c$) and $\delta = 0.09$. The magnetization lies in the $ac$-plane. Because the incommensurability $\delta$ is small, the magnetic order can be understood as a commensurate state with $\mathbf{q} = 2\pi(0, 0, 1)$ twisting slowly along the $b$-direction.

The same authors found two such structures which would be consistent with the data they obtained from elastic neutron scattering. One of the proposed ordering patterns is derived from the commensurate state shown in Fig. 11.5, and is precisely what one expects when the magnetoelastic effect is driven by the $E_u$ phonon. The other candidate is derived from an "orthogonal" state where the spins on every tetrahedron are oriented, for example, in directions $+\hat{\mathbf{x}}$, $-\hat{\mathbf{x}}$, $+\hat{\mathbf{y}}$, and $-\hat{\mathbf{y}}$. Such a state is very hard to obtain through the spin-driven Jahn–Teller effect [15], and no other justification for this state is currently known.

The small value of $\delta$ makes it possible to treat this incommensurability as the result of a weak perturbation on top of the Heisenberg exchange and magnetoelastic coupling. Chern et al. [32] derived possible magnetic spiral states that arise when the Dzyaloshinskii–Moriya interaction is added to these two energy terms. These authors found two candidate solutions, one of which is entirely consistent with the data of Chung et al. This analysis and its results are described in the next section.

### 11.5.2 Theory of Spiral Magnetic Order

The Dzyaloshinskii–Moriya (DM) interaction gives a contribution

$$H_{\mathrm{DM}} = \sum_{\langle ij \rangle} \mathbf{D}_{ij} \cdot [\mathbf{S}_i \times \mathbf{S}_j], \tag{11.19}$$

to the Hamiltonian, where the sum extends over pairs of nearest neighbors. This term is allowed on the ideal pyrochlore lattice, where the bonds are not centrosymmetric, a necessary condition for a non-vanishing coupling constant $\mathbf{D}_{ij}$. At the same time, the high symmetry of the lattice constrains the direction of this vector [33]: for a bond oriented along the [110] lattice direction, the vector must point along $[1\bar{1}0]$ ($\mathbf{D}_{ij} = (\pm D, \mp D, 0)/\sqrt{2}$). The value of $\mathbf{D}_{ij}$ on any other bond is then found through the symmetry transformations of the system. In a collinear magnetic state, the expectation value of $H_{\mathrm{DM}}$ is zero, but its contribution can be lowered by twisting the spins into a spiral.

The pitch of the spiral is determined by the competition between the DM coupling strength $D$ and a spin stiffness. In most antiferromagnets, the spin stiffness is set by the exchange interaction, but the pyrochlore antiferromagnet with nearest-neighbor exchange is an exception: the large degeneracy of its ground state leads to a vanishing stiffness. As a result, the ground state in the presence of the DM interaction is not a long-range spiral but is rather a commensurate state with noncollinear spins [33].

The presence of magnetoelastic interactions changes the situation. Recall that the spin-induced Jahn–Teller effect selects a state with collinear spins. This tendency towards collinear states results in a finite spin stiffness. As a result, the pitch of the spiral is a quantity of order $D/K$, where $K = J'^2/k$ is the effective magnetoelastic interaction.

A problem where the nearest-neighbor exchange $J$, the magnetoelastic energy scale $K$, and the DM coupling $D$ are each arbitrary is hard to solve analytically. However, it simplifies if these energy scales are well separated, the conventional hierarchy being

$$JS^2 \gg KS^4 \gg DS^2. \tag{11.20}$$

A quantitative analysis indicates that in $CdCr_2O_4$ these scales are of similar magnitude (as discussed at the end of this section), but still have the expected order, $JS^2 > KS^4 > DS^2$. Thus while a theory based on the assumption (11.20) may not provide a quantitative account of magnetic order in $CdCr_2O_4$, it presents, at a minimum, a good point of departure for understanding it.

Minimization of the dominant term, the exchange energy (11.1), yields a constraint that the total spin be zero on every tetrahedron. The remaining degrees of freedom of a single tetrahedron can be parametrized using the staggered magnetizations

$$\mathbf{L}_1 = \frac{\mathbf{S}_1 - \mathbf{S}_2 - \mathbf{S}_3 + \mathbf{S}_4}{4S}, \mathbf{L}_2 = \frac{-\mathbf{S}_1 + \mathbf{S}_2 - \mathbf{S}_3 + \mathbf{S}_4}{4S}, \mathbf{L}_3 = \frac{-\mathbf{S}_1 - \mathbf{S}_2 + \mathbf{S}_3 + \mathbf{S}_4}{4S}.$$

Because each spin belongs to two tetrahedra, the staggered magnetizations on one sublattice of tetrahedra determine completely those of the other sublattice. We use the staggered magnetizations of sublattice $A$, $\{\mathbf{L}_i^A\}$, to express the magnetizations of sublattice $B$, and, except in cases of possible confusion, suppress the sublattice index to simplify the notation.

Even for a single tetrahedron, the three staggered magnetizations are not completely independent. Vanishing of the total spin in the ground state makes them mutually orthogonal and imposes on their lengths the constraint [34]

$$\mathbf{L}_i = L_i \hat{\mathbf{l}}_i, \quad \hat{\mathbf{l}}_i \cdot \hat{\mathbf{l}}_j = \delta_{ij}, \quad \sum_{i=1}^{3} L_i^2 = 1. \tag{11.21}$$

The lengths $L_i$ parametrize the angles between the spins and are related to the bond doublet $\mathbf{f}$ by

$$f_1 = 2S^2 (L_1^2 + L_2^2 - 2L_3^2)/\sqrt{3}, \quad f_2 = 2S^2 (L_1^2 - L_2^2). \tag{11.22}$$

Thus five parameters are required to describe the magnetic state of a tetrahedron in the ground state of (11.1): three Euler angles for the triad $\{\hat{\mathbf{l}}_i\}$ and two further parameters for the bond doublet, e.g. $\mathbf{f} = (f_1, f_2)$.

The magnetic order of $CdCr_2O_4$ in the commensurate limit $\delta \to 0$ (Fig. 11.5) has staggered site magnetizations: $\mathbf{L}_2 = \mathbf{L}_3 = 0$ and $\mathbf{L}_1 = e^{i\mathbf{q}\cdot\mathbf{r}}\hat{\mathbf{n}}_1$, where $\mathbf{q} = 2\pi(0, 0, 1)$ and $\hat{\mathbf{n}}_1$ is an arbitrary unit vector. In terms of the three staggered magnetizations, the DM term for a single tetrahedron is

$$E_{\mathrm{DM}} = -DS^2 (\hat{\mathbf{a}} \cdot \mathbf{L}_2 \times \mathbf{L}_3 + \hat{\mathbf{b}} \cdot \mathbf{L}_3 \times \mathbf{L}_1 + \hat{\mathbf{c}} \cdot \mathbf{L}_1 \times \mathbf{L}_2). \tag{11.23}$$

It is easy to see that the DM energy is exactly zero for the collinear state. For either sign of $D$, this term can be made negative by allowing a small component of $\mathbf{L}_2$ or $\mathbf{L}_3$, which describes a twisting spin configuration. The lowering of the DM energy (11.23) is accompanied by an increase of the magnetoelastic energy. However, further analysis shows that the former is linear in $\mathbf{L}_2$ and $\mathbf{L}_3$ while the latter is quadratic, so that such a twisting distortion always occurs.

To pass to a continuum description, we express the rapidly oscillating unit vectors $\hat{\mathbf{l}}_i = \hat{\mathbf{n}}_i \, \mathrm{e}^{\mathrm{i}\mathbf{q}\cdot\mathbf{r}}$ in terms of a slowly varying orthonormal triad $\{\hat{\mathbf{n}}_i\}$, and use the length constraint (11.21) to eliminate $L_1$ by

$$\mathbf{L}_1 = \left(1 - \frac{L_2^2 + L_3^2}{2}\right) \hat{\mathbf{n}}_1 \, \mathrm{e}^{\mathrm{i}\mathbf{q}\cdot\mathbf{r}}, \quad \mathbf{L}_2 = L_2 \, \hat{\mathbf{n}}_2 \, \mathrm{e}^{\mathrm{i}\mathbf{q}\cdot\mathbf{r}}, \quad \mathbf{L}_3 = L_3 \, \hat{\mathbf{n}}_3 \, \mathrm{e}^{\mathrm{i}\mathbf{q}\cdot\mathbf{r}}. \quad (11.24)$$

At this point, the magnetic structure is described in terms of five slowly varying fields, $L_2$, $L_3$, and the triad $\{\hat{\mathbf{n}}_i\}$.

The number of independent fields can be further reduced by examining tetrahedra of sublattice $B$: the vector of total magnetization on each $B$ tetrahedron must vanish, giving three more constraints. The total spin of a $B$ tetrahedron centered at $\mathbf{r}^B = \mathbf{r}^A + (1/4, 1/4, 1/4)$ is given by

$$\mathbf{M}^B(\mathbf{r}^B) = \mathbf{S}_1(\mathbf{r}^A + \mathbf{a}_1) + \mathbf{S}_2(\mathbf{r}^A + \mathbf{a}_2) + \mathbf{S}_3(\mathbf{r}^A + \mathbf{a}_3) + \mathbf{S}_4(\mathbf{r}^A),$$

where $\mathbf{a}_1 = (0, 1/2, 1/2)$, $\mathbf{a}_2 = (1/2, 0, 1/2)$, and $\mathbf{a}_3 = (1/2, 1/2, 0)$ are primitive lattice vectors (the centers of the tetrahedra form a diamond lattice; the Bravais lattice is fcc). By expressing the spins $\{\mathbf{S}_i\}$ in terms of the staggered magnetizations $\{\mathbf{L}_i\}$ and by using the gradient expansion, one obtains, to the lowest orders in $L_2$, $L_3$, and the gradients, the constraint

$$\mathbf{M}^B = L_3 \, \hat{\mathbf{n}}_3 - \frac{1}{4}\frac{\partial \hat{\mathbf{n}}_1}{\partial y} = 0. \quad (11.25)$$

Thus it is clear that $L_3$ and $\hat{\mathbf{n}}_3$ are determined by the gradient $\nabla \hat{\mathbf{n}}_1$.

In a similar way, expressing the staggered magnetizations on sublattice $B$ to lowest order in $\nabla \hat{\mathbf{n}}_1$ leads to

$$\mathbf{L}_1^B = L_2 \hat{\mathbf{n}}_2 - \frac{1}{4}\frac{\partial \hat{\mathbf{n}}_1}{\partial z}, \quad \mathbf{L}_2^B = \hat{\mathbf{n}}_1, \quad \mathbf{L}_3^B = -\frac{1}{4}\frac{\partial \hat{\mathbf{n}}_1}{\partial x}. \quad (11.26)$$

Substituting these expressions into (11.23) and adding contributions from tetrahedra of both types yields

$$E_{\mathrm{DM}} = -\frac{1}{4} D S^2 \, \hat{\mathbf{n}}_1 \cdot \left(\hat{\mathbf{a}} \times \frac{\partial \hat{\mathbf{n}}_1}{\partial x} + \hat{\mathbf{b}} \times \frac{\partial \hat{\mathbf{n}}_1}{\partial y} - \hat{\mathbf{c}} \times \frac{\partial \hat{\mathbf{n}}_1}{\partial z}\right). \quad (11.27)$$

The DM energy contains terms linear in gradients of $\hat{\mathbf{n}}_1$, indicating that the spins are unstable against the formation of spiral configurations. As an example, the first term $\hat{\mathbf{n}}_1 \cdot \hat{\mathbf{a}} \times \partial \hat{\mathbf{n}}_1/\partial x$ favors a magnetic order with the unit vector $\hat{\mathbf{n}}_1$ perpendicular to the $a$-axis and spiraling about it.

As discussed previously, the spin stiffness has its origin in the magnetoelastic energy. For simplicity, here we consider only distortions due to $E_u$ phonons and neglect the effect of $E_g$ distortions (a procedure definitely appropriate in the limit $K_g \ll K_u$). The linear decrease of the DM energy due to gradients of $\hat{\mathbf{n}}_1$ must be balanced by the increase in magnetoelastic energy, which on symmetry grounds must be quadratic in $\nabla \hat{\mathbf{n}}_1$. From (11.18), the increase of magnetoelastic energy is $E_{\mathrm{me}} = -K_u \mathbf{u}_0 \cdot \delta \mathbf{u}/2$, where $K_u = J'^2/k_u$ and $\mathbf{u} = \mathbf{f}^A - \mathbf{f}^B$. The unperturbed odd doublet is $\mathbf{u}_0 = 4S^2(0, 1)$. Because we are describing the spiral magnetic order in terms of the staggered magnetizations, it is convenient to use (11.22) for the calculation of $\delta \mathbf{f}^A$ and $\delta \mathbf{f}^B$. Retaining terms to second order in $\nabla \hat{\mathbf{n}}_1$ leads to

$$E_{\mathrm{me}} = \frac{K_u S^4}{4}\left[\left(\frac{\partial \hat{\mathbf{n}}_1}{\partial x}\right)^2 + \left(\frac{\partial \hat{\mathbf{n}}_1}{\partial y}\right)^2 + 2\left(\frac{\partial \hat{\mathbf{n}}_1}{\partial z}\right)^2 - L_2 \hat{\mathbf{n}}_2 \cdot \frac{\partial \hat{\mathbf{n}}_1}{\partial z} + 4L_2^2\right]. \quad (11.28)$$

Because the DM energy (11.27) does not depend on $L_2$, minimization of the total energy with respect to $L_2$ affects only the magnetoelastic term (11.28) and yields

$$L_2 = \frac{1}{8}\hat{\mathbf{n}}_2 \cdot \frac{\partial \hat{\mathbf{n}}_1}{\partial z}; \quad (11.29)$$

thus $L_2$ is also eliminated. The minimized magnetoelastic energy is

$$E_{\mathrm{me}} = \frac{K_u S^4}{4}\left[\left(\frac{\partial \hat{\mathbf{n}}_1}{\partial x}\right)^2 + \left(\frac{\partial \hat{\mathbf{n}}_1}{\partial y}\right)^2 + 2\left(\frac{\partial \hat{\mathbf{n}}_1}{\partial z}\right)^2 - \left(\hat{\mathbf{n}}_2 \cdot \frac{\partial \hat{\mathbf{n}}_1}{\partial z}\right)^2\right]. \quad (11.30)$$

The total energy of the spiral state, now expressed as a functional of the vector fields $\hat{\mathbf{n}}_1(\mathbf{r})$ and $\hat{\mathbf{n}}_2(\mathbf{r})$, is the sum of (11.27) and (11.30). Its minimization yields a second-order partial differential equation. While we are unable to find the most general solution to this equation, we can find three highly symmetrical spiral solutions in which the spins remain perpendicular to, while twisting about, one of the $\langle 100 \rangle$ axes. As one example, a spiral state along the $b$ axis is described by $\hat{\mathbf{n}}_1 = (\cos\theta(y), 0, \sin\theta(y))$ and has total energy

$$E = -(DS^2\theta' + K_u S^4 \theta'^2)/4, \quad (11.31)$$

where $\theta' = \mathrm{d}\theta/\mathrm{d}y$. Minimization of this quantity gives the pitch of the spiral,

$$\theta' = 2\pi\delta = \frac{D}{2K_u S^2}. \quad (11.32)$$

Equation (11.29) implies that $\mathbf{L}_2^A = 0$; the $A$ tetrahedra therefore have coplanar spins spanned by two orthogonal Néel vectors, $\mathbf{L}_1^A$ and $\mathbf{L}_3^A$. The angles between spin pairs are given by $\theta_{14} = \theta_{23} = 2L_3$. From (11.25), this angle is related to the pitch by $2L_3 = \pi\delta$. The spiral magnetic state has the structure

$$\begin{aligned}\mathbf{L}_1^A &= \cos(\pi\delta/2)\,(\cos(2\pi\delta y),\ 0,\ \sin(2\pi\delta y))\ \mathrm{e}^{2\pi i z},\\ \mathbf{L}_2^A &= 0,\\ \mathbf{L}_3^A &= \sin(\pi\delta/2)\,(-\sin(2\pi\delta y),\ 0,\ \cos(2\pi\delta y))\ \mathrm{e}^{2\pi i z},\end{aligned} \tag{11.33}$$

producing Bragg scattering at wavevector $\mathbf{q} = 2\pi(0, \pm\delta, 1)$, while the ordered magnetic moments lie in the $ac$-plane (Fig. 11.6b). All of this is consistent with the experimental data of [9].

It is worth noting that the distorted crystal structure preserves certain symmetries on interchanging the $A$ and $B$ sublattices, such as inversion at a Cr site followed by a $\pi/2$ rotation in the $ab$-plane. However, the magnetic order described here breaks these symmetries. From (11.26) one has $\mathbf{L}_1^B = \mathbf{L}_3^B = 0$, which means that every tetrahedron on sublattice $B$ has collinear spins, whereas the spins of the $A$ tetrahedra are twisted into a (weakly) non-collinear state. This disparity between the two sublattices should result in different distortions of the two types of tetrahedra, thus further lowering the symmetry of the crystal. However, the magnitude of the additional distortion is expected to be small because the degree of non-collinearity is small, $\delta \ll 1$.

Spiral states in which the spins rotate about the $a$-axis (Fig. 11.6a) can be obtained similarly, by using the ansatz $\hat{\mathbf{n}}_1 = (0, \cos\theta(x), \sin\theta(x))$. The resulting solution can also be obtained from (11.33) through symmetry operations which exchange the two sublattices of tetrahedra, such as inversion on a Cr site. This spiral produces a magnetic Bragg peak at $\mathbf{q} = (\mp\delta, 0, 1)$.

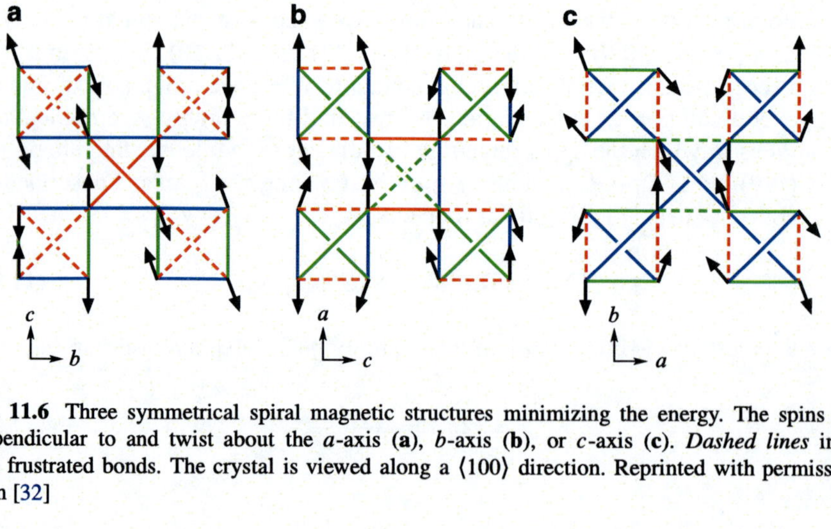


**Fig. 11.6** Three symmetrical spiral magnetic structures minimizing the energy. The spins are perpendicular to and twist about the $a$-axis (**a**), $b$-axis (**b**), or $c$-axis (**c**). *Dashed lines* indicate frustrated bonds. The crystal is viewed along a $\langle 100 \rangle$ direction. Reprinted with permission from [32]

Finally, there is a third spiral solution, shown in Fig. 11.6c, where the spins twist about the $c$-axis. In this state, which is not related to the other two solutions by any symmetry, the magnetic Bragg peak occurs at $\mathbf{q} = 2\pi(0, 0, 1 + \delta)$ and both sublattices have tetrahedra with coplanar, rather than collinear, spins. That this spiral state has the same energy as the previous two is a coincidence: its total energy is also given by (11.31) with $\theta' = \mathrm{d}\theta/\mathrm{d}z$. This degeneracy is lifted when other perturbations, such as further-neighbor interactions, are taken into account. $CdCr_2O_4$ has a significant *third*-neighbor antiferromagnetic exchange interaction, which acts to favor strongly the states with spirals twisting along the $a$- or $b$-axis [32].

In closing this section, we comment on the assumption (11.20) of well-separated energy scales in the problem. *Ab initio* calculations [32] yield a nearest-neighbor exchange energy $JS^2 = 1.1\,\mathrm{meV}$ and a magnetoelastic energy $K_u S^4 = 0.76\,\mathrm{meV}$. The strength of the DM interaction can be estimated from the measured pitch of the spiral, $\delta = 0.09$, using (11.32), which gives $DS^2 = 0.21\,\mathrm{meV}$. While the three energy scales are not vastly different, they do appear in the correct order of decreasing magnitude, $JS^2 > K_u S^4 > DS^2$.

## 11.6 Summary and Open Questions

The spinel compound $CdCr_2O_4$ provides an opportunity to test our understanding of the ground state of the Heisenberg antiferromagnet on the pyrochlore lattice. When the lattice degrees of freedom are included, both the selected magnetic order and the lattice distortion are in agreement with a theoretical model [15] based on two vibrational doublets of the crystal, the $\mathbf{q} = 0$ optical phonon $E_u$ and the uniform lattice distortion $E_g$. The model ties the incommensurate nature of the spiral magnetic order to a spontaneous breaking of the inversion symmetry in the crystal, which has not yet been observed directly in $CdCr_2O_4$.

The magnetoelastic phase transition between the high-$T$ correlated paramagnet [35] and the low-$T$ ordered phase remains poorly understood. It is strongly discontinuous in both $ZnCr_2O_4$ [8] and $CdCr_2O_4$ [9], with both the lattice distortion and the ordered moment reaching their $T = 0$ values immediately below the ordering temperature. A Landau free-energy approach based on the spin-Peierls order parameter (11.7) appears to be the only candidate approach available for modeling the underlying physics [15]. However, it is not evident that this phenomenology can provide a full description. In $CdCr_2O_4$, the order parameter has the $E_u$ symmetry and does not allow for a cubic invariant in the free energy, thus excluding the most obvious cause for a discontinuous phase transition. The phase transition may yet turn out to be first-order if the even ($E_g$-symmetric) order parameter is nearly soft [15], but this question has not yet been clarified. In fact it remains unclear whether the valence-bond variables (11.7) represent a good choice of the order parameter for these systems: the low-$T$ phase in both $ZnCr_2O_4$ and $CdCr_2O_4$ is magnetically ordered, suggesting that a type of staggered magnetization may be a more appropriate choice.

Finally, a realistic model of this phase transition must take into account the entropy of the correlated paramagnetic state [4]. The very high entropy of the disordered phase may be responsible for the discontinuous nature of the phase transition, as has been demonstrated for the case of lattice models with large numbers of flavors [36].

This material is based upon work supported by the US National Science Foundation under Grant No. DMR-0348679.

## References

1. A.P. Ramirez, J. Appl. Phys. **70**, 5952 (1991)
2. D.A. Huse, A.D. Rutenberg, Phys. Rev. B **45**, 7536 (1992)
3. M.J. Harris, M.P. Zinkin, Z. Tun, B.M. Wanklyn, I.P. Swainson, Phys. Rev. Lett. **73**, 189 (1994)
4. R. Moessner, J.T. Chalker, Phys. Rev. B **58**, 12049 (1998)
5. J.T. Chalker, P.C.W. Holdsworth, E.F. Shender, Phys. Rev. Lett. **68**, 855 (1992)
6. C. Kittel, Phys. Rev. **120**, 335 (1960)
7. I.S. Jacobs, J.W. Bray, H.R. Hart, L.V. Interrante, J.S. Kasper, G.D. Watkins, D.E. Prober, J.C. Bonner, Phys. Rev. B **14**, 3036 (1976)
8. S.H. Lee, C. Broholm, T.H. Kim, W. Ratcliff, S.W. Cheong, Phys. Rev. Lett. **84**, 3718 (2000)
9. J.H. Chung, M. Matsuda, S.H. Lee, K. Kakurai, H. Ueda, T.J. Sato, H. Takagi, K.P. Hong, S. Park, Phys. Rev. Lett. **95**, 247204 (2005)
10. H. Ueda, H. Mitamura, T. Goto, Y. Ueda, Phys. Rev. B **73**, 094415 (2006)
11. I. Bersuker, *The Jahn-Teller effect* (Cambridge University Press, Cambridge, 2006)
12. H.A. Jahn, E. Teller, Proc. Roy. Soc. London Ser. A **161**, 220 (1937)
13. L.D. Landau, E.M. Lifshitz, *Quantum Mechanics: Non-Relativistic Theory* (Butterworth-Heinemann, 1981)
14. Y. Yamashita, K. Ueda, Phys. Rev. Lett. **85**, 4960 (2000)
15. O. Tchernyshyov, R. Moessner, S.L. Sondhi, Phys. Rev. B **66**, 064403 (2002)
16. A.B. Harris, A.J. Berlinsky, C. Bruder, J. Appl. Phys. **69**, 5200 (1991)
17. K. Motida, S. Miyahara, J. Phys. Soc. Jpn. **28**, 1188 (1970)
18. A.B. Sushkov, O. Tchernyshyov, W. Ratcliff II, S.W. Cheong, H.D. Drew, Phys. Rev. Lett. **94**, 137202 (2005)
19. R.V. Aguilar, A.B. Sushkov, Y.J. Choi, S.W. Cheong, H.D. Drew, Phys. Rev. B **77**, 092412 (2008)
20. A. Olariu, P. Mendels, F. Bert, B.G. Ueland, P. Schiffer, R.F. Berger, R.J. Cava, Phys. Rev. Lett. **97**(16), 167203 (2006)
21. Y. Okamoto, M. Nohara, H. Aruga-Katori, H. Takagi, Phys. Rev. Lett. **99**(13), 137207 (2007)
22. F. Wang, A. Vishwanath, Phys. Rev. Lett. **100**, 077201 (2008)
23. T. Rudolf, C. Kant, F. Mayr, A. Loidl, Phys. Rev. B **77**, 024421 (2008)
24. F. Becca, F. Mila, Phys. Rev. Lett. **89**, 037204 (2002)
25. C. Weber, F. Becca, F. Mila, Phys. Rev. B **72**, 024449 (2005)
26. C. Jia, J.H. Nam, J.S. Kim, J.H. Han, Phys. Rev. B **71**, 212406 (2005)
27. D.L. Bergman, R. Shindou, G.A. Fiete, L. Balents, Phys. Rev. B **74**, 134409 (2006)
28. H. Ueda, H. Katori, H. Mitamura, T. Goto, H. Takagi, Phys. Rev. Lett. **94**, 047202 (2005)
29. H. Ueda, H. Mitamura, T. Goto, Y. Ueda, Phys. Rev. B **73**, 094415 (2006)
30. K. Penc, N. Shannon, H. Shiba, Phys. Rev. Lett. **93**, 197203 (2004)
31. M. Matsuda, H. Ueda, A. Kikkawa, Y. Tanaka, K. Katsumata, Y. Narumi, T. Inami, Y. Ueda, S.H. Lee, Nature Physics **3**, 397 (2007)
32. G.W. Chern, C.J. Fennie, O. Tchernyshyov, Phys. Rev. B **74**, 060405 (2006)
33. M. Elhajal, B. Canals, R. Sunyer, C. Lacroix, Phys. Rev. B **71**, 094420 (2005)

34. C.L. Henley, J. Appl. Phys. **61**, 3962 (1987)
35. S.H. Lee, C. Broholm, W. Ratcliff, G. Gasparovic, Q. Huang, T.H. Kim, S.W. Cheong, Nature **418**, 856 (2002)
36. S.B. Shlosman, R. Kotecký, Comm. Math. Phys. **83**, 493 (1981)

# Chapter 12
# Spin Ice

**Michel J.P. Gingras**

**Abstract** Geometrical frustration usually arises in systems composed of magnetic moments (spins) which reside on the sites of a lattice whose antiferromagnetic exchange interactions form triangular or tetrahedral units. A less well-known form of geometrical frustration appears in systems with *ferromagnetically* coupled spins in the presence of strong, non-collinear, single-ion, easy-axis (Ising-like) anisotropies. This is the situation in a number of pyrochlore oxide materials, where Ising-like magnetic rare-earth moments ($Ho^{3+}$, $Dy^{3+}$) occupy a lattice of corner-shared tetrahedra and are coupled by effectively ferromagnetic (dipolar) interactions. These systems possess a macroscopic number of quasi-degenerate classical ground states and display an extensive low-temperature entropy closely related to the extensive proton-disorder entropy in common water ice. For this reason, these magnetic systems are known as *spin ice*. This chapter reviews the essential ingredients of spin-ice phenomenology in magnetic pyrochlore oxides.

## 12.1 Introduction

In some geometrically frustrated magnetic systems, there exists an exponentially large number, $\Omega_0$, of degenerate classical ground states. This results in an extensive residual, or zero-point, ground-state entropy, $S_0 = k_B \ln(\Omega_0)$. The most celebrated example of such a system is the triangular lattice of Ising spins which have nearest-neighbor antiferromagnetic exchange interactions: as shown by Wannier in 1950, this system remains disordered at all temperatures. Any state that has two spins up and one spin down (or vice versa) per triangle is a ground state and the zero-point entropy per site is $S_0 \approx 0.323\ k_B$ [1]. However, the triangular Ising antiferromagnet was not the first condensed matter system to be identified as having a residual entropy. In fact, the first such system was not even a magnetic one. Some fifteen years before Wannier's work, William Giauque (Nobel Prize for Chemistry, 1949) and co-workers had performed thermodynamic measurements and determined that the solid phase of common water ice possesses a residual entropy with no conventional origin [2,3]. This result was soon explained by Linus Pauling (Nobel Prize for

Chemistry, 1954) in terms of a macroscopic number of proton ($H^+$) configurations in water ice arising from the mismatch between the crystalline symmetry of ice and the local hydrogen-bonding requirement of the water molecule [4].

Over the past decade, a class of insulating magnetic materials where the configurational disorder in the orientations of the magnetic moments is precisely the same as that of water ice has been the subject of extensive experimental and theoretical studies. Because of the analogy with water ice, the term "spin ice" [5–9] has been coined for these systems. Most of the chapters in this volume focus on geometrically frustrated antiferromagnets, the interest in which lies primarily in the pursuit of novel quantum ground states whose exotic properties, most notably the absence of conventional, semi-classical, long-range Néel order, arise from the increased quantum zero-point motion caused by the frustration. This chapter differs in that the spin ices are frustrated Ising ferromagnets, and are systems where quantum fluctuations do not play a significant role. However, the experimental and theoretical studies discussed here reveal a rich variety of equilibrium and non-equilibrium thermodynamic behavior in spin-ice systems [9].

This chapter reviews some of the key elements of the spin-ice phenomenology. It draws particular attention to the problem of water ice and the semi-formal origin of the Ising nature of the magnetic moments in spin-ice materials, two topics not usually considered in detail in standard solid-state textbooks. Also reviewed in some depth is the mean-field theory of spin ices, as this simple approach played an essential role in uncovering the microscopic origin behind the emergence of the "ice rules" in real dipolar spin-ice materials. A brief discussion of research topics in the field of spin-ice systems that are of current interest concludes the chapter.

## 12.2 Water Ice, Pauling Entropy, and Anderson Model

### *12.2.1 Water Ice and Pauling Model*

Water ice is a fascinating, strongly correlated condensed matter system, not least because it exhibits an apparent violation of the third law of thermodynamics. In the early 1930s, a series of remarkable specific-heat experiments by William Giauque and co-workers found that the limiting low-temperature state of the common (hexagonal) structure of water ice, referred to as $I_h$, is characterized by a residual entropy, $S_0 = 0.82 \pm 0.05$ Cal/deg·mol, that differed from zero by a value far in excess of experimental errors [2, 3]. In a famous 1935 paper, Linus Pauling showed that, because of its configurational proton disorder, $I_h$ possesses a finite entropy at zero temperature, which he estimated as 0.806 Cal/deg·mol [4], i.e. very close to the experimental value.

The ice problem is a classic example of how the separation of energy scales in an interacting system can leave some effective low-energy degrees of freedom inoperative, and hence ultimately frustrated in the overall scheme by which the system uses local dynamical processes (taking a finite, as opposed to infinite, amount of time) to minimize its energy at low temperature. In ice, the chemical binding energy of the water molecule is so strong, 221 kCal/mol, that its chemical structure is left essentially unaltered in the solid phase. Consequently, the ground state of water ice does not, or more precisely cannot dynamically, involve the full minimization of the electrostatic energy of a globally neutral ensemble of $O^{2-}$ and $H^+$ ions. Rather, in the hexagonal ("wurtzite") and cubic ("sphalerite") phases of ice [11], the $O^{2-}$ ions form an open tetrahedral structure, whose 109-degree angles accommodate almost perfectly the preferred $H-O-H$ bonding configuration of an isolated $H_2O$ molecule.

In the wurtzite phase, the bond length between two distinct $O^{2-}$ ions is 2.76 Å, while the covalent $O-H$ bond of a $H_2O$ molecule is a much smaller 0.96 Å. Because the integrity of the $H_2O$ molecular structure is maintained in the solid phase, the minimum-energy position of a proton is not at the midpoint between two $O^{2-}$ ions, and instead there are two equivalent positions along the $O-O$ bond for each proton (with the four-fold oxygen coordination in the hexagonal wurtzite structure, there is in effect one proton per $O-O$ bond on average). The constraint imposed by the energetically robust $H_2O$ structure therefore results in the two "ice rules", due to Bernal and Fowler, which govern what are acceptable low-energy proton configurations in the hexagonal wurtzite structure [10]. The first ice rule states that there should on average be only one proton per $O-O$ bond. The second rule states that, in order for ice to consist of a hydrogen-bonded solid of water molecules, for each $O^{2-}$ ion, two protons must be in a "near position" and two in a "far position" (Fig. 12.1). Although on a purely electrostatic basis the protons should maximize their separation, the ice rules, which are based on the dominance

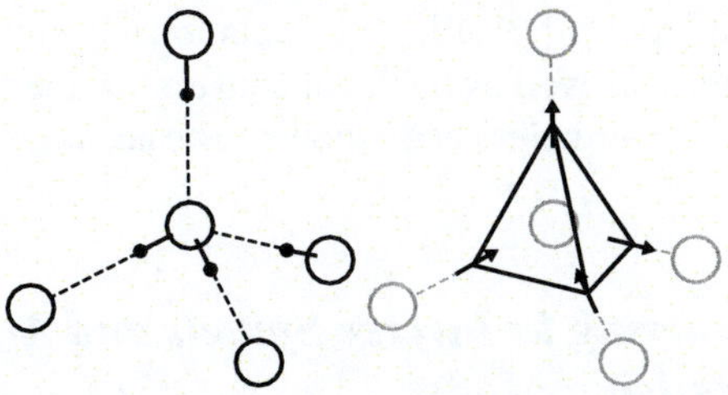

**Fig. 12.1** *Left*: local proton arrangement in water ice, showing $O^{2-}$ ions (*large white circles*) and protons (hydrogen ions, $H^+$, *small black circles*). Each $O^{2-}$ ion is coordinated tetrahedrally by four other $O^{2-}$ ions, but nearby, covalently bonded protons are present on only two of these links while on the other two the protons (covalently bonded to different $O^{2-}$ ions) have only a hydrogen bond with the central $O^{2-}$ ion. In the hexagonal phase of ice, $I_h$, the low-energy configurations are those obeying these "Bernal–Fowler ice rules" [10], whereby the protons around each $O^{2-}$ ion have a "two-near/two-far" configuration. *Right*: as left, but with the position of a proton represented by a displacement vector (*arrow*) located at the mid-point of the $O^{2-}$–$O^{2-}$ (oxide–oxide) bond. The "two-near/two-far" Bernal–Fowler ice rule corresponds to a "two-in/two-out" configuration of the displacement vectors. The displacement vectors become the Ising magnetic moments in spin ice (Fig. 12.2)

of the unaltered integrity of the $H_2O$ molecule, render the proton-proton interaction an effectively low-energy part of the problem, and frustrates it.

The ice rules were proposed by Bernal and Fowler in 1933 [10]. At that time, X-ray diffraction could determine only the lattice structure of the oxygen ions. From the first ice rule, Bernal and Fowler argued that the protons must lie along the direct $O-O$ line of contact (bond). They proposed a regular, crystalline proton structure, expecting that this would be the case. However, at this time Giauque and co-workers had already obtained compelling evidence for a residual zero-point entropy in water ice [2], and this led Pauling to his proposal, published in 1935, that the open tetrahedral structure of ice leads to many equivalent ways of satisfying the Bernal-Fowler ice rules, and hence to an extensive entropy [4].

Pauling put forward an elegant argument to estimate the configurational proton entropy, which proceeds as follows. First consider one mole of ice containing $N_0$ $O^{2-}$ ions and, therefore, $2N_0$ $O-O$ bonds for the hexagonal structure of ice in which no two protons lie on any given $O-O$ bond. That is, all bonds are taken to obey the first Bernal–Fowler ice rule. Each $O-O$ bond can be taken as having two possible positions for a proton, giving $2^{(2N_0)}$ possible proton configurations for the whole system. However, out of the 16 possible configurations associated with each $O^{2-}$ ion, ten are energetically unfavorable: the $OH_4^{2+}$ configuration, the four $OH_3^+$ configurations, the four $OH^-$ configurations, and the $O^{2-}$ configuration. This leaves six configurations that satisfy the second Bernal–Fowler rule as the allowed local proton configurations around each $O^{2-}$ ion. An upper bound on the number of ground-state configurations, $\Omega_0$, can therefore be estimated by reducing the $2^{2N_0}$ configurations by a simple 6/16 "deflation" weight factor for each oxygen ion, which gives $\Omega_0 \leq 2^{2N_0}(6/16)^{N_0} = (3/2)^{N_0}$. The corresponding configurational entropy, $S_0 = k_B \ln(\Omega_0) = N_0 k_B \ln(3/2) = 0.806$ Cal/deg·mol, is in excellent agreement with the residual entropy of $0.82 \pm 0.05$ Cal/deg·mol determined by Giauque and Stout [3]. Pauling's calculation neglects the global constraint on the number of protons, as well as the local constraints due to closed loops on the wurtzite lattice, but has nevertheless been shown to be accurate to within 1–2% [12].

### 12.2.2 *Cation Ordering in Inverse Spinels and Antiferromagnetic Pyrochlore Ising Model*

In a 1956 paper [13], Philip Anderson (Nobel Prize for Physics, 1977) investigated the problems of cation ordering in the class of materials known as inverse spinels and of antiferromagnetic ordering in normal spinels. To a first approximation, both of these problems map onto that of an Ising model with antiferromagnetic nearest-neighbor exchange interactions on the B-site of the spinel lattice. This lattice is structurally identical to the pyrochlore lattice shown in Fig. 12.2. A discussion of magnetic ordering in spinels may be found in the Chap. 7 by H. Takagi and S. Niitaka in this volume.

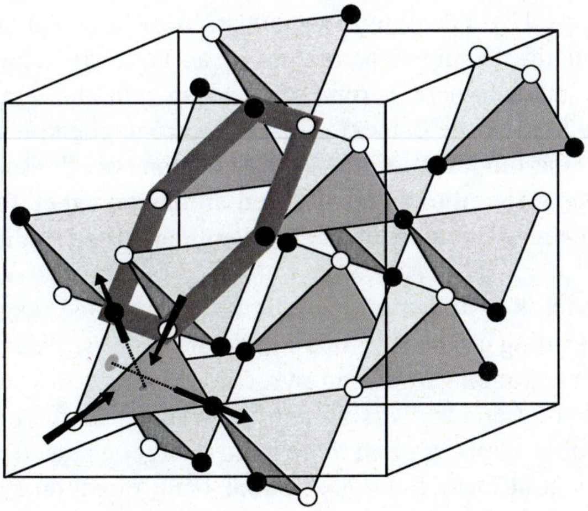

**Fig. 12.2** Pyrochlore lattice of corner-sharing tetrahedra, as occupied by $Ho^{3+}$ and $Dy^{3+}$ ions in the spin-ice materials $Ho_2Ti_2O_7$ and $Dy_2Ti_2O_7$. The rare-earth magnetic moments, which have Ising symmetry, occupy the corners of the tetrahedra, as shown on the lower left "downward" tetrahedron of the lattice (*arrows*). The spins shown here are the equivalents of the proton displacement vectors in Fig. 12.1. Each spin axis is oriented along the local $\langle 111 \rangle$ quantization axis, which is a line joining the site to the middle of the opposing triangular face (a point represented by the disks) and meets with the three other $\langle 111 \rangle$ axes in the center of the tetrahedron. In the spin-ice materials, the "two-in/two-out" condition arises from the combined effect of magnetic (exchange and dipole–dipole) interactions and the strong Ising anisotropy. For clarity, other spins on the lattice are denoted by black and white circles, where white represents a spin pointing into a downward tetrahedron while black is the opposite. The entire lattice can be seen to be in an "ice-rules" state (two black and two white sites for every tetrahedron). The hexagon (*thick gray line*) pertains to the loop excitations used in the "loop Monte Carlo" simulations discussed in Sect. 12.4.3. The hexagon illustrated is the smallest possible loop involving multiple spins, and a "loop move" corresponds to reversing all spins on the loop to produce a new ice-rules state. The reversal of all spins on closed loops represents the lowest-energy excitations which allow the system to explore the quasi-degenerate ice-rule manifold of states relevant for dipolar spin ice at low temperatures. (Figure reprinted with permission from Melko et al. [46]. Copyright 2001 by the American Physical Society.)

Consider first the problem of cation ordering in the inverse spinel. The associated minimum-energy problem consists in placing + and − signs on the pyrochlore lattice (Fig. 12.2) in a ratio 1:1 such that the number of +/− pairs is maximized. This condition is satisfied with two + and two − signs for each tetrahedron. The centers of the tetrahedra in the spinel lattice are arranged with respect to each other on the same diamond lattice as the $O^{2-}$ ions in the cubic phase of water ice ($I_c$). As shown in Fig. 12.2, the spinel lattice and, equivalently, the pyrochlore lattice, consist of alternating "upward" and "downward"-oriented tetrahedra. Consequently, the problem of + and − charge organization is mapped directly onto an ice-rule problem where a + sign corresponds to a "proton-near" configuration on an upward tetrahedron and a "proton far" one on a downward tetrahedron, and vice-versa for a − sign. Neglecting the same constraints as Pauling for the water-ice problem, the problem of cation ordering in inverse spinels therefore maps onto an ice-like problem, and is therefore characterized by a Pauling zero-point entropy [13]. The problem of cation ordering is also manifestly the same as that of an antiferromagnetic Ising model on the pyrochlore lattice, where + represents an up spin and − a down spin. It is interesting to note that, in the context of an antiferromagnetic Ising model on the pyrochlore lattice, the model considered by Anderson was the second example of what would now be referred to as a frustrated antiferromagnet, the first being the problem of the Ising antiferromagnet on the triangular lattice studied by Wannier in 1950 [1].

Anderson's model is an important paradigm for frustrated magnetic systems of interacting spins residing on the sites of a lattice of corner-shared triangles or tetrahedra, as occurs in kagomé, garnet, and pyrochlore lattices. However, there are no known realizations of Anderson's Ising antiferromagnet on the pyrochlore lattice, because this has cubic symmetry and there is no energetic reason permitted by this symmetry which would favor a unique, global Ising direction $\hat{z}$ to the detriment of any other global direction. However, a system of antiferromagnetically coupled, isotropic Heisenberg spins on the pyrochlore lattice is realistic. This problem was the topic of an important 1979 paper [14] in which Villain anticipated the failure of the classical Heisenberg pyrochlore antiferromagnet to develop long-range order down to zero temperature [15, 16] and introduced the term "collective paramagnet" for such a system – a terminology one may wish to employ for the classical variant of the more modern term "spin liquid" used in quantum spin systems.

Because they lack the intrinsic propensity to develop classical, long-ranged magnetic order, antiferromagnetic materials with Heisenberg spins on a pyrochlore lattice are expected to be excellent candidates for exotic quantum mechanical ground states. It was in anticipation of observing unconventional magnetic and thermodynamic behavior in the broad family of magnetic pyrochlore oxide materials, of generic formula $A_2B_2O_7$, that rapid growth occurred some two decades ago in experimental and theoretical efforts devoted to the study of these systems [17, 18]. It was this general scientific endeavor that led to the discovery of "spin ice" [5–7], a novel class of frustrated ferromagnetic Ising systems which are close magnetic analogs of both water ice and the cation ordering in spinels. This discovery also led to the rebirth of a disguised variant of Anderson's antiferromagnetic Ising model [13] on the pyrochlore lattice.

## 12.3 Discovery of Spin Ice

### *12.3.1 Rare-Earth Pyrochlore Oxides: Generalities*

Before reviewing the discovery of spin-ice materials and the spin-ice problem, a brief discussion is in order of some general aspects of the pyrochlore oxides, the broader class of materials to which spin ices belong. It is also worthwhile to review at an elementary level the background to the origin of the magnetism in the rare-earth ions present in the spin ices, how they acquire their strong Ising nature, and what are the predominant interactions between the magnetic ions.

The magnetic $A_2B_2O_7$ pyrochlore oxides form a broad family of materials that exhibit a wide range of thermodynamic and magnetic phenomena, a number of which remain at best partially understood [17, 18]. In this crystal structure, which gave its name to the geometry, both the trivalent $A^{3+}$ ion (A is a rare-earth element such as Gd, Tb, Dy, Ho, or Y) and the tetravalent $B^{4+}$ ion (B = Ti, Sn, Mo, Mn) reside on two independent and interpenetrating pyrochlore sublattices. Figure 12.2 shows only one of those two sublattices, for example the A sublattice. In $A_2B_2O_7$, one may have either the A or the B sublattice occupied by a magnetic ion, as in $Tb_2Ti_2O_7$ [19] and $Y_2Mo_2O_7$ [20], respectively, or both types of site may be occupied by magnetic ions, as is the case in $Tb_2Mo_2O_7$ [21].

As discussed in the seminal work of Villain [14], the pyrochlore lattice is highly frustrated when the spin of the magnetic ions is of the isotropic, Heisenberg type, and the interaction with nearest-neighbor sites is an antiferromagnetic exchange coupling. The pyrochlore oxides can be metallic (such as $Nd_2Mo_2O_7$, which displays an anomalous Hall effect [22]) or they can be insulating, as is the case for the $A_2Ti_2O_7$ and $A_2Sn_2O_7$ series [17, 18]. In this chapter, we restrict our considerations to the insulating $A_2Ti_2O_7$ and $A_2Sn_2O_7$ series with Dy and Ho as the magnetic ion A. Because $Ti^{4+}$ and $Sn^{4+}$ are non-magnetic, we deal here with only one magnetic pyrochlore sublattice, the A sublattice in $A_2B_2O_7$, to which we refer henceforth simply as "the pyrochlore lattice." If spin–spin interactions are neglected [23], the $Tb^{3+}$ ion in $Tb_2Ti_2O_7$ and $Tb_2Sn_2O_7$ would also be described as an Ising spin at sufficiently low temperatures [24, 25], but it is found that spin interactions render a purely Ising-model description of these two materials invalid [23].

The pyrochlore lattice can be described conveniently as a face-centered cubic (FCC) lattice with a primitive ("upward" or "downward") tetrahedral basis cell of four sites (Fig. 12.2). The pyrochlore lattice possesses a trigonal (threefold rotational) symmetry with respect to any of the four equivalent $\langle 111 \rangle$ cubic lattice directions (the diagonals of the cubic cell in Fig. 12.2). For each of the four sites in the tetrahedral unit cell, it will prove convenient to use as local axes ($\hat{z}_i$) of spin quantization the specific $\langle 111 \rangle$ cube diagonal which passes through a given site $i$ and the middle of the opposite triangular face (Fig. 12.2).

It is worth noting that common water ice at atmospheric pressure, ice $I_h$, has a hexagonal structure while the magnetic pyrochlore lattice has cubic symmetry. Strictly, the Ising pyrochlore problem is equivalent to cubic ice, $I_c$, and not the hexagonal $I_h$ phase. However, this does not modify the "ice-rule" analogy (or mapping), or the close connection between the orientational configurations of the magnetic moments in spin ice and the local proton coordination in water ice.

### *12.3.2 Microscopic Hamiltonian: Towards an Effective Ising Model*

In rare-earth ions, the total angular momentum, $\mathbf{J} = \mathbf{L} + \mathbf{S}$, can be taken as a good quantum number. For a given ion, the Hund rules can be applied to determine the isolated (vacuum) electronic ground state. As an example, Tb has electronic configuration [Xe]$4f^9 6s^2$ and $Tb^{3+}$ has ground-state configuration [Xe]$4f^8$. Following the Hund rules, one finds that $L = 3$ and $S = 3$, hence $J = 6$ for $Tb^{3+}$ and the spectroscopic notation for the ground multiplet is $^7f_6$. Similarly, $J = 8$ for $Ho_2Ti_2O_7$ and $J = 15/2$ for $Dy_2Ti_2O_7$.

Electrostatic and covalent bonding effects, which originate from the crystalline environment, lead to a lifting of the $(2J + 1)$-fold electronic degeneracy of the otherwise free-ion ground state. This is known as a crystal-field effect. The theoretical description of crystal-field effects going beyond the simple, Coulombic point-charge description of the ionic charges surrounding a rare-earth ion, which includes covalency effects and mixing between electronic multiplets, is a rather involved technical matter into which we do not enter here. In the discussion to follow, we will assume simply that the single-ion, crystal-field energy levels (of the non-interacting ions) have been determined appropriately, for example by experimental spectroscopic techniques (such as optical spectroscopy or inelastic neutron scattering). Such a spectroscopic approach allows the determination of transition frequencies between crystal-field levels, and the associated intensities, which can then be described by an effective crystal-field Hamiltonian, $\mathcal{H}_{\rm cf}$, as described for example in [24–26]. We return to this point below.

Ideally, one would wish to ensure that the energy levels determined in this way are not strongly "dressed" (by which is meant altered, or renormalized) by inter-ion interactions, which are described by a further Hamiltonian term $\mathcal{H}_{\rm int}$. One way of effecting this is to consider a highly magnetically diluted variant of the system of interest [25]. This leads one to introduce the minimal Hamiltonian, $\mathcal{H}$, required to describe the essential physics,

$$\mathcal{H} = \mathcal{H}_{\rm cf} + \mathcal{H}_{\rm Z} + \mathcal{H}_{\rm int}. \qquad (12.1)$$

Here, $\mathcal{H}_{\rm cf}$ is the crystal-field Hamiltonian responsible for lifting the degeneracy of the otherwise free, single-ion electronic ground state. Its energy eigenstates are the crystal-field energy levels of the previous paragraph. As a first approximation, one may express $\mathcal{H}_{\rm cf}$ in terms of polynomial functions of the $J_{i,z}$ and $J_{i,\pm} = J_{i,x} \pm \imath J_{i,y}$ components of the angular-momentum operator $\mathbf{J}$. For the local symmetry of the rare-earth ions at the A sites of the $A_2B_2O_7$ structure, $\mathcal{H}_{\rm cf}$ is written as [24–26]

$$\mathcal{H}_{\rm cf} = \sum_i \sum_{l,m} B_l^m O_l^m(\mathbf{J}_i), \qquad (12.2)$$

where $B_l^m$ are the crystal-field coefficients and $O_l^m(\mathbf{J}_i)$ the equivalent crystal-field operators. Spectroscopic measurements allow the coefficients $B_l^m$ to be determined by a process of fitting to the energy levels augmented by further constraints from the observed transition intensities [24, 26]. From the Wigner–Eckart theorem, $l \leq 6$ for $L = 3$ 4f elements. As an illustration, $O_2^0 = 3J_z^2 - J(J+1)$, $O_4^0 = 35J_z^4 - (30J(J+1) - 25)J_z^2 + 3J^2(J+1) - 6J(J+1))$, and $O_4^{\pm 3} = c_\pm [J_z, J_+^3 \pm J_-^3]_+$ with $[A, B]_+ = (AB + BA)/2$, $c_+ = 1/2$ and $c_- = -\imath/2$ [27–29].

In (12.1), $\mathcal{H}_Z = -g_L \mu_B \sum_i \mathbf{J}_i \cdot \boldsymbol{B}$ is the Zeeman energy describing the interactions of the rare-earth magnetic ions with the magnetic field $\boldsymbol{B}$; $g_L$ is the Landé factor and $\mu_B$ is the Bohr magneton.

The final term,, $\mathcal{H}_{\text{int}}$, describes the interactions between the ions. We consider $\mathcal{H}_{\text{int}}$ to have the form

$$\mathcal{H}_{\text{int}} = -\frac{1}{2} \sum_{(i,j)} \mathcal{J}_{ij}\, \mathbf{J}_i \cdot \mathbf{J}_j + \left(\frac{\mu_0}{4\pi}\right) \frac{(g_L \mu_B)^2}{2 r_{\text{nn}}{}^3} \sum_{(i,j)} \frac{(\mathbf{J}_i \cdot \mathbf{J}_j - 3 \mathbf{J}_i \cdot \hat{\boldsymbol{r}}_{ij} \hat{\boldsymbol{r}}_{ij} \cdot \mathbf{J}_j)}{(r_{ij}/r_{\text{nn}})^3}, \tag{12.3}$$

where $r_{ij}\hat{\boldsymbol{r}}_{ij} = \boldsymbol{r}_j - \boldsymbol{r}_i$, with $\boldsymbol{r}_i$ the position of ion $i$. $\mathcal{J}_{ij}$ denotes the microscopic quantum mechanical exchange constant between ions $i$ and $j$, and is defined with the convention that $\mathcal{J}_{ij} < 0$ is antiferromagnetic and $\mathcal{J}_{ij} > 0$ is ferromagnetic. Note that in (12.3), the two sums are performed over all ions $i$ and $j$, resulting in a double-counting of bonds canceled by the prefactor $1/2$ of each sum.

While the "origin" of the first term lies in the interactions between the "real" electronic spins, $\boldsymbol{S}_i$ and $\boldsymbol{S}_j$, here we consider for simplicity only an effective isotropic "exchange" between the total angular momenta $\mathbf{J}_i$ and $\mathbf{J}_j$, with interaction coefficients $\mathcal{J}_{ij}$. This type of description has proven adequate to describe the physics of spin-ice materials [30–33]. The four distinct types of symmetry-allowed anisotropic nearest-neighbor exchange interactions on the pyrochlore lattice are described in [34, 35]. The second term is the long-ranged magnetostatic dipole-dipole interaction, with $r_{\text{nn}}$ the distance between nearest neighbors. The pyrochlore oxides have a conventional, cubic unit cell of size $a \sim 10$Å with 16 ions (i.e. 4 primitive upward or downward tetrahedral basis cells of the type shown in Fig. 12.2) per cell and $r_{\text{nn}} = (a\sqrt{2})/4$.

We discuss next the relative energy scales set by $\mathcal{H}_{\text{cf}}$, $\mathcal{H}_Z$, and $\mathcal{H}_{\text{int}}$. For the $(Ho,Dy)_2(Ti,Sn)_2O_7$ materials, which we will show below to be spin ices, $\mathcal{H}_{\text{cf}}$ is by far the largest energy scale in the problem. In these four materials, the spectrum of $\mathcal{H}_{\text{cf}}$ consists of a ground-state doublet separated by an energy gap, of typical size $\Delta \sim 300$ K, to the first excited state. The Landé factors $g_L$ for $Ho^{3+}$ and $Dy^{3+}$ are respectively 5/4 and 4/3, while $\mu_B = 9.27 \times 10^{-24}$ J/T $\approx 0.671$ K/T in $\mathcal{H}_Z$. Thus even in a field of 20 T, which is the approximate size of the largest fields accessible from commercial laboratory magnets, the Zeeman energy scale is of the order of 10 K, and hence is much smaller than $\Delta$. Consequently, when calculating the properties of $(Ho,Dy)_2(Ti,Sn)_2O_7$ at temperatures below approximately 10 K and in magnetic fields below 20 T, it is safe to consider only the magnetic crystal-field ground-state doublet of $Ho^{3+}$ and $Dy^{3+}$, and to neglect both the susceptibility contributions from the excited states and the van Vleck susceptibility.

The interaction part of the Hamiltonian, $\mathcal{H}_{\text{int}}$, nevertheless deserves particular attention. In insulating magnetic rare-earth materials, the unfilled 4f orbitals of the rare-earth ion, which carry the electronic spin, are shielded by the 5s and 5p orbitals, as a result of which 4f-orbital overlap, both between the rare-earth ions and between the rare-earth and $O^{2-}$ ions, is small. The effective exchange coupling between rare-earth ions is thus much smaller for insulating, rare-earth oxides than for transition-metal oxides: a typical value of $\mathcal{J}_{ij}$ for nearest-neighbor spins is $\mathcal{J} \sim 66$ mK in $Dy_2Ti_2O_7$ [30, 33]. One may also estimate the dipolar energy scale, $\mathcal{D} \equiv \mu_0(g_L\mu_B)^2/(4\pi r_{nn}{}^3)$, using values $r_{nn} \sim 3.5$Å and $g_L = 4/3$ to obtain $\mathcal{D} \sim 25$ mK [30, 33]. Comparing $\mathcal{J}$ and $\mathcal{D}$ with $\Delta \sim 300$ K shows again that the interaction part of the full Hamiltonian $\mathcal{H}$ is very small compared to the crystal-field part.[1] In practice, this means that the independent (non-interacting) single-ion, crystal-field eigenstates of $\mathcal{H}_{cf}$ form a very convenient basis to describe the Hilbert space of interacting $Dy^{3+}$ and $Ho^{3+}$ ions in $(Ho,Dy)_2(Ti,Sn)_2O_7$. In particular, because the scale of $\mathcal{H}_{int}$ is so small compared to $\Delta$, for any practical purposes one may neglect all the excited states of $\mathcal{H}_{cf}$ and work with a reduced Hilbert space spanned solely by the two degenerate states of the ground doublet for each magnetic site.

In such a problem, where the high-energy sector of the theory is so well separated from the low-energy sector, it is convenient to introduce an effective Hamiltonian, $\mathcal{H}_{eff}$, that operates only within the low-energy sector. Chapter 20 of Mila and Schmidt in this volume describes the general methodology for deriving an appropriate $\mathcal{H}_{eff}$ for any given interacting system where there are well separated "reference" energy sectors. For the spin ices, it is amply sufficient to use the lowest order of the perturbation theory which defines $\mathcal{H}_{eff}$,

$$\mathcal{H}_{eff} \approx P\mathcal{H}_{int}P, \tag{12.4}$$

where $P$, the projector in the low-energy sector, is defined as

$$P = \sum_{\{k\}} |\Phi_{0,\{k\}}\rangle\langle\Phi_{0,\{k\}}| \tag{12.5}$$

with

$$|\Phi_{0,\{k\}}\rangle = \prod_{i=1}^{N} |\phi_{i,0}^{(k_i)}\rangle \tag{12.6}$$

and $|\phi_{i,0}^{(k_i)}\rangle$ the $k_i^{th}$ state of the single-ion, crystal-field ground doublet of rare-earth ion $i$. Because $\mathcal{H}_{int}$ is a pairwise Hamiltonian, $\mathcal{H}_{eff}$ is, to lowest order as given by (12.4), also a pairwise effective Hamiltonian. $\mathcal{H}_{eff}$ can then be determined by considering two ions, each in one of the two states, $|\phi_{i,0}^{(k_i=+)}\rangle$ or $|\phi_{i,0}^{(k_i=-)}\rangle$, of their respective crystal-field ground doublets. The effective Hamiltonian, $\mathcal{H}_{eff}$, is therefore a $4 \times 4$ matrix. At this point, further progress requires a return to the crystal-field problem to discuss the nature of the non-interacting states $|\phi_{i,0}^{(k_i=\pm)}\rangle$ which make up each of the single-ion ground doublets.

[1] To obtain realistic figures for the comparison between the spin-spin interactions and $\mathcal{H}_{cf}$, the coupling constants $\mathcal{J}_{ij}$ and $\mathcal{D}$ should be multiplied by the relevant products of matrix elements. Taking as a crude estimate for such products the magnitude squared of $\mathbf{J}$, $|\mathbf{J}|^2 \sim 60$ for both $Dy^{3+}$, with J = 15/2, and $Ho^{3+}$, with J = 8, the results $\mathcal{J}_{ij}|\mathbf{J}|^2 \ll \Delta$ and $\mathcal{D}|\mathbf{J}|^2 \ll \Delta$ remain.

From inelastic neutron scattering measurements, it is found that for $Ho_2Ti_2O_7$, $|\phi_{i,0}^{(k_i=\pm)}\rangle \approx |J = 8, m_J = \pm 8\rangle$, with negligible contributions from other components $|J = 8, m_J\rangle$. The crystal-field parameters for $Dy_2Ti_2O_7$ can be obtained by rescaling those determined for $Ho_2Ti_2O_7$ [24]. One finds for the former material $|\phi_{i,0}^{(k_i=\pm)}\rangle \approx |J = 15/2, m_J = \pm 15/2\rangle$, again with negligible weight from other components $|J = 15/2, m_J\rangle$ [24]. It is important to recall the previous discussion concerning the choice of the $\hat{z}_i$ quantization direction: the eigenstate of $J_z$ with magnetic quantum number $m_J$ refers here to the component of $\mathbf{J}_i$ along the local $\hat{z}_i$ direction oriented along one of the four cubic $\langle 111\rangle$ directions. Calculating the matrix elements of $P\mathcal{H}_{\rm int}P$ for both $Ho_2Ti_2O_7$ and $Dy_2Ti_2O_7$ shows that the only significant matrix elements are $\langle\phi_{i,0}^{(k_i=\pm)}|J_z|\phi_{i,0}^{(k_i=\pm)}\rangle \approx \pm J$. The physical meaning of this result is simple: in both $Ho_2Ti_2O_7$ and $Dy_2Ti_2O_7$, the magnetic moment can, for all practical purposes, only be aligned parallel or antiparallel to the local $\langle 111\rangle$ direction. Thus one has truly a local $\langle 111\rangle$ Ising model.

It is instructive to write out this set of local quantization axes ($\hat{z}_i \parallel \langle 111\rangle$). We take $\hat{z}_1 = \frac{1}{\sqrt{3}}(\hat{x} + \hat{y} + \hat{z})$, $\hat{z}_2 = \frac{1}{\sqrt{3}}(-\hat{x} - \hat{y} + \hat{z})$, $\hat{z}_3 = \frac{1}{\sqrt{3}}(-\hat{x} + \hat{y} - \hat{z})$, and $\hat{z}_4 = \frac{1}{\sqrt{3}}(\hat{x} - \hat{y} - \hat{z})$, which have the property $\hat{z}_\mu \cdot \hat{z}_\nu = -1/3$ for $\mu \neq \nu$. To proceed formally, one could reexpress the $4 \times 4$ matrix representing $\mathcal{H}_{\rm eff}$ in terms of tensor products of Pauli matrices, $\sigma_i^{\alpha_i} \otimes \sigma_j^{\beta_j}$, where $\alpha_i = x_i, y_i, z_i$ in a local orthogonal frame $\{x_i, y_i, z_i\}$ [23]. Had we retained the full $|J, m_J\rangle$ decomposition of $|\phi_{i,0}^{(k_i=\pm)}\rangle$ in the treatment above, we would have found that the coefficient of the $\sigma_i^{z_i}\sigma_j^{z_j}$ term has by far the largest coefficient in $\mathcal{H}_{\rm eff}$, and hence would have obtained again a classical model with Ising spins oriented parallel or antiparallel to their local $\langle 111\rangle$ direction.

Alternatively, one could have proceeded much more straightforwardly by simply making the replacement

$$\mathbf{J}_i \to |\langle J_z\rangle|\sigma_i^{z}\hat{z}_i \tag{12.7}$$

at the outset and using $|\langle J_z\rangle| \approx 8$ for $Ho_2Ti_2O_7$ and $|\langle J_z\rangle| \approx 15/2$ for $Dy_2Ti_2O_7$. Indeed, this is the approximation made implicitly in most previous numerical studies of dipolar spin ice [7,30–32,36].[2] Because there are no other components $\sigma_i^{\alpha_i}$ with $\alpha_i \neq z_i$ remaining in $\mathcal{H}_{\rm eff}$ which do not commute among themselves, we obtain in this way a strictly classical Ising model for $H_{\rm eff}$. One may therefore consider only the basis of eigenstates of $\sigma_i^{z_i}$, and henceforth we will indeed treat $\sigma_i^{z_i} = \pm 1$ merely as a classical variable.

While the discussion leading to $\mathcal{H}_{\rm eff}$ as a classical Ising model for the Dy-based and Ho-based oxide pyrochlores, as presented here, is rather academic, this is not (as mentioned in Sect. 12.3.1) the case for $Tb_2Ti_2O_7$ [19] and $Tb_2Sn_2O_7$ [37]. Indeed, for these two materials, the energy gap separating the ground doublet from the excited doublet is not very large when compared to the exchange and dipole-dipole interactions. In this case, higher-order terms in the perturbation expansion leading

[2] However, estimating $D$ using the crystal-field ground-state doublet wave functions gives a moment $\mu = g_L\mu_B\langle J_z\rangle$ slightly less than $10\mu_B$ [33].

to $\mathcal{H}_{\rm eff}$ must be retained (as described in Chap. 20 by Mila and Schmidt) in order to derive an appropriate quantum theory of these two materials; this exercise was performed for a simple model of $Tb_2Ti_2O_7$ in [23].

Returning to $\mathcal{H}_{\rm int}$ above, and making the replacement $\mathrm{J}_i \rightarrow |\langle \mathrm{J}_z \rangle| \sigma_i^{z} \hat{z}_i$, one obtains the effective, classical, $\langle 111 \rangle$ pyrochlore Ising model

$$\mathcal{H}_{\rm DSM} = -\frac{1}{2}\sum_{(i,j)} J_{ij}\,(\hat{z}_i\cdot\hat{z}_j)\,\sigma_i^{z_i}\sigma_j^{z_j} + \frac{D}{2}\sum_{(i,j)} \frac{(\hat{z}_i\cdot\hat{z}_j - 3\hat{z}_i\cdot\hat{r}_{ij}\hat{r}_{ij}\cdot\hat{z}_j)}{(r_{ij}/r_{\rm nn})^3}\,\sigma_i^{z_i}\sigma_j^{z_j}\,, \tag{12.8}$$

to which we refer henceforth as the "dipolar spin-ice model" (DSM). In this expression, $J_{ij} = \mathcal{J}_{ij}\langle \mathrm{J}_z\rangle^2$ and $D = \mu_0(g_{\rm L}\mu_{\rm B}\langle \mathrm{J}_z\rangle)^2/(4\pi r_{\rm nn}{}^3)$. It should be emphasized that the variables $\sigma_i^z = \pm 1$ have become simple labels indicating whether the $z_i$-component of $\mathbf{J}_i$ is oriented "in" or "out" of a primitive tetrahedral unit cell characterized by the set $\{\hat{z}_1, \hat{z}_2, \hat{z}_3, \hat{z}_4\}$ (Fig. 12.2). Having established that the magnetic $Dy^{3+}$ and $Ho^{3+}$ ions in $Dy_2Ti_2O_7$ and $Ho_2Ti_2O_7$ should be described at temperatures much lower than the lowest crystal-field gap $\Delta$ by effective classical Ising spins, we may now proceed to a discussion of the experimental behavior of these two materials, and of how the spin-ice phenomenology arises.

### 12.3.3 Discovery of Spin Ice in $Ho_2Ti_2O_7$

In a 1997 paper, Harris, Bramwell, and collaborators reported the results of a neutron scattering study of $Ho_2Ti_2O_7$ [5]. They found that, in zero applied magnetic field, no evidence of a transition to long-range order could be detected down to 0.35 K, while muon spin relaxation results could exclude a transition down to 0.05 K [6]. The most surprising part of these results was that the Curie–Weiss temperature, $\theta_{\rm CW}$, was found to be positive, $\theta_{\rm CW} \approx +1.9$ K, indicating overall ferromagnetic interactions. Naively, one would have expected such a three-dimensional cubic system with ferromagnetic interactions to develop long-range order at a critical temperature of the same order as $\theta_{\rm CW}$. It was also found that the magnetic field-dependence of the neutron scattering intensity depends on the protocol followed to magnetize the sample [38]: the system displayed a history-dependence reminiscent of random spin glasses [39], although no significant random disorder is present in $Ho_2Ti_2O_7$.

The authors of [5, 6] proposed that the strong local $\langle 111 \rangle$ Ising anisotropy of $Ho^{3+}$ in $Ho_2Ti_2O_7$ frustrates the development of ferromagnetic order. They considered a simple model of ferromagnetically coupled $\langle 111 \rangle$ Ising spins on the pyrochlore lattice [40], establishing the connection between their model and that of Pauling for the problem of proton disorder in water ice, and hence coining the term "spin-ice model" for this system (Fig. 12.1). For the sake of clarity, and because we discuss below the important role of long-range dipole-dipole interactions in spin-ice materials, here we relabel the model of Harris et al. as the "nearest-neighbor spin-ice model" to distinguish it from the "dipolar spin-ice model" of (12.8). In the next section, we review this nearest-neighbor spin-ice model and show that it has the same residual entropy as water ice.

### 12.3.4 *Nearest-Neighbor Ferromagnetic* $\langle 111 \rangle$ *Ising Model and Pauling's Entropy*

#### 12.3.4.1 Nearest-Neighbor Spin-Ice Model

Consider a simplified version of the Ising Hamiltonian in (12.8) where the dipolar interaction coefficient is set first to $D = 0$ and where the exchange interactions, $J_{ij}$, are restricted solely to nearest neighbors, whence

$$\mathcal{H}_{\text{nn}} = -J \sum_{\langle i,j \rangle} (\hat{z}_i \cdot \hat{z}_j)\, \sigma_i^{z_i} \sigma_j^{z_j} . \tag{12.9}$$

Taking $\hat{z}_i \cdot \hat{z}_j = -1/3$ for nearest neighbors on the pyrochlore lattice,

$$\begin{aligned} \mathcal{H}_{\text{nn}} &= -J \sum_{\langle i,j \rangle} (\hat{z}_i \cdot \hat{z}_j)\, \sigma_i^{z_i} \sigma_j^{z_j} \\ &= \frac{J}{3} \sum_{\langle i,j \rangle} \sigma_i^{z_i} \sigma_j^{z_j} . \end{aligned} \tag{12.10}$$

This can be rewritten as [16]

$$\mathcal{H}_{\text{nn}} = \frac{J}{6} \sum_{\Delta} (L_\Delta)^2 - \frac{2N_\Delta J}{3}, \tag{12.11}$$

where the sum $\sum_\Delta$ is performed over each upward and downward tetrahedral unit (labeled by $\Delta$), with a total of $N_\Delta$ tetrahedra of each type, and where $L_\Delta$ is the total spin on unit $\Delta$, defined as

$$L_\Delta \equiv \sigma_\Delta^{z_1} + \sigma_\Delta^{z_2} + \sigma_\Delta^{z_3} + \sigma_\Delta^{z_4} . \tag{12.12}$$

From (12.10), one can appreciate the "first magic" of spin ice. (we will discuss the "second magic" in Sect. 12.4.1): starting with a ferromagnetic nearest-neighbor exchange, $J > 0$ in (12.9), one finishes with Ising variables $\sigma_i^{z_i}$ which have an effective antiferromagnetic coupling constant $J/3$ in (12.10). This is the same model as the frustrated antiferromagnetic pyrochlore Ising model used by Anderson to describe the problem of cation ordering in inverse spinels [13]. The ferromagnetic $\langle 111 \rangle$ pyrochlore Ising model is therefore frustrated and, just as in Anderson's antiferromagnet Ising model, it must possess a residual Pauling entropy.

In terms of $\mathcal{H}_{\text{nn}}$ (12.11), the ground-state configuration of a single tetrahedron for antiferromagnetic interactions consists of all $\sigma_i^{z_i} = +1$ or all $\sigma_i^{z_i} = -1$. In terms of the "real" spins $\mathbf{J}_i$ in (12.3), this corresponds to all $\mathbf{J}_i$ oriented "in" or all "out" of a reference tetrahedral unit cell. Globally, there are only two such states and, upon cooling the system, a second-order transition occurs to a four-sublattice, antiferromagnetic, Néel ordered state in the three-dimensional Ising universality class. The four-sublattice Néel ordered phase described in terms of the variables $\mathbf{J}_i$ maps onto a ferromagnetic ground state in terms of the variables $\sigma_i^{z_i}$. The pyrochlore lattice with antiferromagnetically coupled $\langle 111 \rangle$ spins is, therefore, not frustrated. Conversely, for ferromagnetic $J$, the Ising variables $\sigma_i^{z_i}$ are effectively coupled antiferromagnetically and the ground state on a single tetrahedron is six-fold degenerate, a result arising from the condition $\sigma_\Delta^{z_1} + \sigma_\Delta^{z_2} + \sigma_\Delta^{z_3} + \sigma_\Delta^{z_4} = 0$ on each tetrahedron $\Delta$ (12.12) i.e. two positive and two negative values of the relevant $\sigma_i^{z_i}$.

This is the incipient spin configuration leading to the extensive Pauling entropy of the nearest-neighbor spin-ice model in the thermodynamic limit [7], the subject we discuss next. The reader is invited to refer also to Fig. 12.1, where a more direct physical, or geometrical, connection between the $\sigma_i^{z_i}$ configurations in the spin-ice problem and that of the "real" proton positions in water ice is illustrated.

#### 12.3.4.2 Pauling Entropy of the Nearest-Neighbor Spin Ice Model

The macroscopically degenerate ground states which are often characteristic of frustrated systems can be understood in terms of an underconstraint argument [16]: the extensive degeneracy arises from the difference between the number of constraints necessary to determine a ground state and the number of degrees of freedom that the system possesses. By following the heuristic argument of Pauling for water ice, one may calculate the residual entropy of spin ice. Consider Anderson's Ising pyrochlore antiferromagnet, to which the local $\langle 111 \rangle$ Ising pyrochlore model can be mapped, as discussed in Sect. 12.2.2. The ground-state condition is said to be "underconstrained", demanding only that the total magnetization, $L_\Delta$, of the four Ising spins on each tetrahedron should satisfy the condition $L_\Delta = 0$. This is achieved by six of the $2^4 = 16$ possible spin configurations. Counting $2^4$ configurations for each tetrahedron gives, for a system of $N$ spins and $N/2$ upward and downward tetrahedra, $(2^4)^{N/2} = 4^N$ microstates. This number overestimates very seriously the exact total of $2^N$ microstates for $N$ Ising spins, the reason being that each spin is shared between two tetrahedra, and hence the 16 configurations on each tetrahedron are not independent.

Adopting the reasoning of Pauling, one may allocate instead $2^2$ states per tetrahedron and, assuming that $6/16$ of them satisfy the constraint, obtain a ground-state degeneracy $\Omega_0 = \{(2^2)^{N/2}(6/16)\}^{N/2} = (3/2)^{N/2}$. The corresponding entropy, $S_0 = k_\mathrm{B} \ln(\Omega_0) = (N k_\mathrm{B})/2 \ln(3/2)$, is of course simply Pauling's original result retrieved for the pyrochlore spin-ice problem. Another way to obtain the residual entropy is as follows: for a pyrochlore lattice with $N$ spins, there are $N/4$ tetrahedral unit cells (for example all the upward tetrahedra in Fig. 12.2). If all of these tetrahedra are in a spin-ice state, there can be $6^{N/4}$ independent spin-ice configurations in the system. However, not all of these states are valid ground states, because the $N/4$ downward tetrahedra, which are formed by the corners of the upward tetrahedra must also satisfy the ice rules. The probability that a random tetrahedron satisfies the ice rule is 6/16, and so the number $6^{N/4}$ should be deflated by this factor for each downward tetrahedron; the result is a total of $6^{N/4} \times (6/16)^{N/4} = (3/2)^{N/2}$ states of the whole system obeying the ice rules, as obtained above.

Finally, one could also have simply borrowed Pauling's result directly: in the mapping from water ice to Anderson's antiferromagnet, or alternatively to spin ice, $N\mathrm{O}^{2-}$ ions correspond to $2N$ spins. We will demonstrate next that the estimate of Pauling, $S_0 = (Nk_\mathrm{B}/2)\ln(3/2)$, is in good agreement with experimental results for spin-ice materials.

### 12.3.5 Residual Entropy of $Dy_2Ti_2O_7$ and $Ho_2Ti_2O_7$

The first compelling thermodynamic evidence for the existence of a spin-ice state in Ising pyrochlore systems was obtained from measurements of the magnetic specific heat, $C(T)$, in $Dy_2Ti_2O_7$ [7]. Like $Ho_2Ti_2O_7$, this material, which has magnetic $Dy^{3+}$ ions, is characterized by a strongly Ising-like ground-state doublet that is separated from the excited doublet by a large energy of approximately 300 K. The temperature-dependence of $C(T)$ is shown in the upper panel of Fig. 12.3. To determine the residual magnetic entropy, Ramirez and co-workers followed an approach similar to the one used by Giauque and colleagues to determine the entropy of water ice [2,3].

In general, it is possible only to measure the change of entropy between two temperatures. Giauque and collaborators calculated the entropy change of water between 10 K and the gas phase by integrating the specific heat divided by temperature, $C(T)/T$, and adding to it the latent heat at the melting and vaporization transitions. They then compared this value with the absolute value expected for the calculated entropy of an ideal (non-interacting) gas phase, using as input the results of spectroscopic measurements which allowed them to determine the rotational and vibrational energy spectrum, and hence the roto-vibrational entropy, and adding to this the translational entropy of an ideal gas given by the Sackur–Tetrode equation. The difference between the measured value and the value expected theoretically gave the residual entropy of water ice, whose origin was then explained by Pauling [4].

Ramirez et al. [7] measured the magnetic specific heat of a powder sample of $Dy_2Ti_2O_7$ between $T_1 = 300$ mK, "deep" inside the frozen (spin-)ice regime, and $T_2 = 10$ K, in the paramagnetic regime, where the expected entropy per mole should be $R\ln(2)$ for a two-state system ($R = N_0 k_\mathrm{B}$ is the molar gas constant and $N_0$ is the Avogadro number). The entropy change between $T_1$ and $T_2$, $\Delta S_{1,2}$, was found by integrating $C(T)/T$ between these two temperatures,

$$\Delta S_{1,2} = S(T_2) - S(T_1) = \int_{T_1}^{T_2} \frac{C(T)}{T} \mathrm{d}T. \qquad (12.13)$$

The lower panel of Fig. 12.3 shows that the magnetic entropy recovered is approximately 3.9 $\mathrm{J\,mol^{-1}\,K^{-1}}$, a value considerably smaller than $R\ln(2) \approx 5.76$ $\mathrm{J\,mol^{-1}\,K^{-1}}$. The difference, 1.86 $\mathrm{J\,mol^{-1}\,K^{-1}}$, is quite close to the Pauling estimate

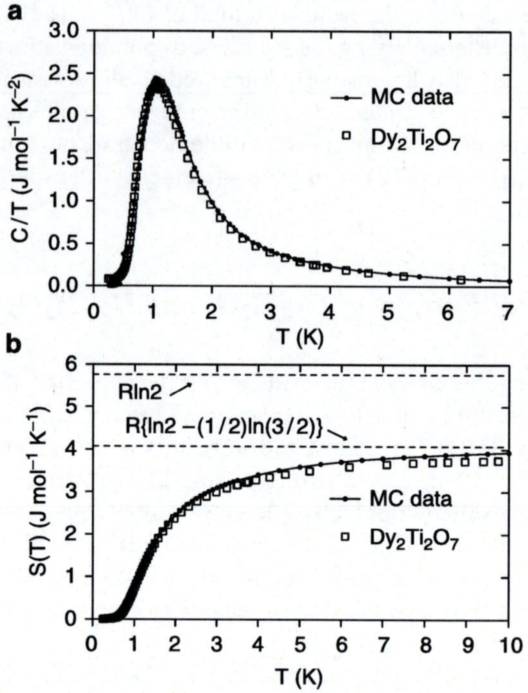


**Fig. 12.3** (**a**) Specific heat and (**b**) entropy data for $Dy_2Ti_2O_7$ from [7], compared with the results of Monte Carlo simulations [30] for the dipolar spin-ice model, with $J/3 = -1.24$ K and $5D/3 = 2.35$ K. Two regimes of temperature can be identified. (**i**) At a temperature $T$ much higher than the peak temperature, $T_{\text{peak}} \sim 1.24$ K, the system is in the paramagnetic regime, is weakly correlated, and individual tetrahedra do not obey the "two-in/two-out" ice rules. As the temperature approaches $T_{\text{peak}}$, the system evolves progressively to satisfy the ice rules. The Schottky-like peak in $C$ arises when the temperature drops below the energy gap between the ice-rule-obeying states and the excited "three-in/one-out" and "all-in/all-out" states. (**ii**) As $T$ drops below $T_{\text{peak}}$, the spin-flip rate drops exponentially rapidly [45] as the system settles into an ice-rule-obeying state with two spins "in" and two "out" on each tetrahedron. There is no phase transition between the high-temperature paramagnetic state ($T > T_{\text{peak}}$) and the spin-ice regime at $T < T_{\text{peak}}$. The spin-ice regime can therefore be described as a collective paramagnet [14]. (Figure reprinted with permission from B. C. den Hertog and M. J. P. Gingras [30]. Copyright 2000 by the American Physical Society.)

for the entropy associated with the extensive degeneracy of ice, $(R/2)\ln(3/2) = 1.68\,\text{Jmol}^{-1}\,\text{K}^{-1}$, consistent with the existence of a spin-ice state in $Dy_2Ti_2O_7$. More recent measurements [41] on $Dy_2Ti_2O_7$ have found a specific heat below 1.5 K that has quite significant quantitative differences from that of [7], which lead ultimately to a considerably better agreement between the experimentally determined residual entropy of the material and the estimate of Pauling.

As discussed above, $Ho_2Ti_2O_7$ was the first material to be proposed as a spin-ice system. It transpires that it is rather less convenient to perform low-temperature specific heat measurements in $Ho_2Ti_2O_7$ than in $Dy_2Ti_2O_7$. The origin of this difficulty lies in the unusually large hyperfine interaction between the nuclear and electronic spins of $Ho^{3+}$. This interaction leads to a Schottky anomaly in the specific heat at a temperature $T \sim 0.3$ K which largely obscures the broad and otherwise purely electronic specific-heat feature arising from the formation of the ice-rule-obeying low-energy manifold. Once the nuclear contribution to the specific heat has been subtracted, the electronic contribution and the residual Pauling entropy of the spin-ice state in $Ho_2Ti_2O_7$ can be revealed [31, 42].

In summary, $Ho_2Ti_2O_7$ and $Dy_2Ti_2O_7$ possess a residual low-temperature entropy compatible with that estimated on the basis of a Pauling argument applied to the ferromagnetic nearest-neighbor $\langle 111 \rangle$ Ising model on the pyrochlore lattice. The related materials $Ho_2Sn_2O_7$ [43] and $Dy_2Sn_2O_7$ [44], which are not discussed here, are also spin-ice systems.

## 12.4 Dipolar Spin-Ice Model

### *12.4.1 Competing Interactions in the Dipolar Spin-Ice Model*

In Sect. 12.3.4, we invoked a straightforward, nearest-neighbor ferromagnetic $\langle 111 \rangle$ Ising model to rationalize the presence of spin-ice phenomenology in real materials. However, in Sect. 12.3.2 it was argued that magnetic dipole–dipole interactions are often sizeable in rare-earth magnetic systems. In the cases of $Dy_2Ti_2O_7$ and $Ho_2Ti_2O_7$, the value of $D = \mu_0(g_L\mu_B\langle J_z \rangle)^2/(4\pi {r_{nn}}^3)$ is estimated as $D \sim 1.4$ K [30, 36](see footnote 2), which is comparable to the Curie–Weiss temperature $\theta_{CW}$ measured experimentally in these materials. Thus if one were to assume that the nearest-neighbor exchange interaction $J \sim \theta_{CW} \sim 1$ K, one would deduce a regime where the dipolar interactions are comparable in magnitude [7, 30, 36].

This observation raises a paradox: the existence of a ground state with extensive degeneracy should in principle result from the underconstrained physics, i.e. the lack of constraints imposed by the Hamiltonian on the spin configurations that minimize the classical ground-state energy [16]. Considering the dipolar interactions of (12.8) in this context, first they are of a rather complicated nature because they couple the spin and spatial directions through the $(\hat{z}_i \cdot \hat{r}_{ij})(\hat{r}_{ij} \cdot \hat{z}_j)$ term, and secondly they are very long-ranged, decaying as $1/{r_{ij}}^3$ with the separation $r_{ij}$ of ions $i$ and $j$. This poses the question of how to understand the emergence of an extensively degenerate spin-ice state at a temperature $T$ of order $D$ in the presence of spin interactions as "complicated" (by which is meant anisotropic and long-ranged) as a dipolar coupling. How this occurs is what we refer to as the "second magic" of dipolar spin ice.

An initial approach to this question is to go beyond the nearest-neighbor ferromagnetic $\langle 111 \rangle$ Ising model and perform Monte Carlo simulations where the dipolar interactions of (12.8) are included [7, 30–33, 36, 45, 46]. The first studies of this problem considered dipolar interactions truncated beyond a certain cut-off distance [7, 36]. These were followed by another analysis [30] which incorporated the true long-range part of the dipolar interactions using the Ewald summation method [45] which is commonly used to treat long-range Coulomb and dipolar interactions [47].

Consider $\mathcal{H}_{\rm DSM}$ in (12.8), with long-range dipole–dipole interactions ($D \neq 0$) and only nearest-neighbor exchange $J_{ij} = J$, as a starting model to describe $Dy_2Ti_2O_7$. One may take the dipolar interaction coupling $D$ to be *a priori* a quantity known to reasonably high accuracy from the (good) approximation that the crystal-field ground-state doublet is composed essentially only of the $|{\rm J} = 15/2, m_{\rm J} = \pm 15/2\rangle$ components. This gives $D \sim 1.4$ K [30] and the exchange coupling, $J$, as the only unknown parameter in the model. It was found in Monte Carlo simulations [30] that fitting either the height of the specific-heat peak or the temperature at which this peak occurs allows for a unique and consistent determination of $J$. By these methods, the values $J/3 \approx -1.24$ K [30][3] and $J/3 \approx -0.55$ K [31] were estimated respectively for $Dy_2Ti_2O_7$ [30] and $Ho_2Ti_2O_7$ [31]. It is noteworthy that $J$ is antiferromagnetic in both cases, which if taken alone should, from the discussion of Sect. 12.3.4, give rise to a four-sublattice long-range Néel-ordered phase [30]. This observation indicates already that it must be the dipolar interactions which are responsible for the spin-ice phenomenology. Figure 12.3 shows that a Monte Carlo simulation of the DSM (12.8), with $J/3 \approx -1.24$ K and $5D/3 \approx 2.35$ K, gives an excellent description of the magnetic specific-heat data $C(T)$ of $Dy_2Ti_2O_7$. As a consequence, integration of $C(T)/T$ reproduces the Pauling-like experimental residual entropy (lower panel, Fig. 12.3). Similar agreement between experiment and Monte Carlo data for the specific heat was found for $Ho_2Ti_2O_7$ using $J/3 \approx -0.55$ K and $5D/3 \approx 2.35$ K [31]. Further, it was possible in $Ho_2Ti_2O_7$ to perform neutron scattering studies on a single crystal [31], where, with $J$ fixed from the $C(T)$ data, it was found that the dependence of the measured intensity on the scattering wave vector $\boldsymbol{q}$ is well reproduced by the results from Monte Carlo simulations [31], as illustrated in Fig. 12.4.

Hence, despite their complex structure, not only are dipolar interactions compatible with the existence of a degenerate state, but in fact they appear to be responsible for it, given that the nearest-neighbor exchange interaction $J$, which in $Ho_2Ti_2O_7$ and $Dy_2Ti_2O_7$ is antiferromagnetic, would by itself lead to long-range order. One must therefore ask where the spin-ice phenomenology comes from in the DSM (12.8) with long-range dipole–dipole interactions. One way to address this question is to take the dipolar interactions and truncate them beyond the nearest-neighbor separation: on the pyrochlore lattice, $\hat{z}_i \cdot \hat{z}_j = -1/3$ and $(\hat{z}_i \cdot \hat{r}_{ij})(\hat{r}_{ij} \cdot z_j) = -2/3$ for nearest-neighbor ions $i$ and $j$, whence

[3] In [30,31], the quantities $J_{\rm nn} \equiv J/3$ and $D_{\rm nn} = 5D/3$ were employed to describe the interactions between the Ising variables $\sigma_i^{z_i}$. The reason for introducing these two rescaled couplings becomes clear when considering (12.14)

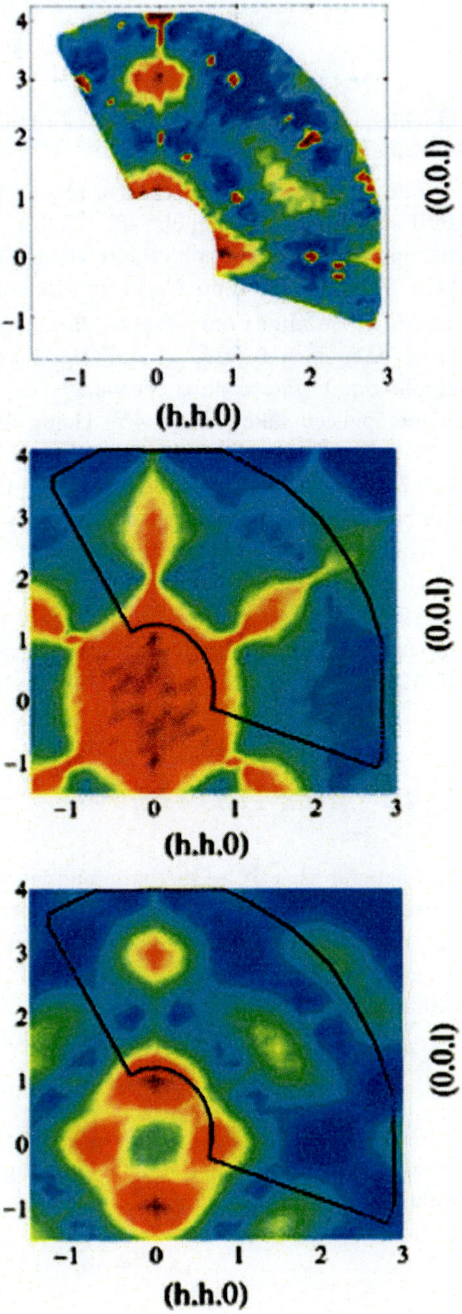


**Fig. 12.4** $Ho_2Ti_2O_7$: neutron scattering in the $(hhl)$ plane showing experimental data (*upper panel*; the sharp spots are nuclear Bragg scattering with no magnetic component) compared with Monte Carlo simulations of the nearest-neighbor spin-ice model (*middle panel*) and of the dipolar model (*lower panel*) [31]. Blue (*light*) indicates the weakest and red-brown (*dark*) the strongest intensity. (Figure reprinted with permission from S.T. Bramwell et al. [31]. Copyright 2001 by the American Physical Society.)

$$H = \sum_{\langle i,j \rangle} \left( \frac{J}{3} + \frac{5D}{3} \right) \sigma_i^{z_i} \sigma_j^{z_j} + H^{\text{dip}}_{>r_{\text{nn}}}, \tag{12.14}$$

in which all dipolar interactions beyond the nearest-neighbor distance $r_{\text{nn}}$ have been incorporated into $H^{\text{dip}}_{>r_{\text{nn}}}$.

The first term in (12.14) is that of an effective nearest-neighbor pyrochlore Ising antiferromagnet [13], which for $(J + 5D)/3 > 0$ is frustrated and possess a Pauling entropy, while a four-sublattice "all-in/all-out" long-range Néel order occurs for $(J + 5D)/3 < 0$. Monte Carlo simulations [45, 46] and mean-field theory [48] find that $H^{\text{dip}}_{>r_{\text{nn}}}$ modifies only slightly the location of the Néel/spin-ice boundary from $(J + 5D)/3 = 0$ to $(J + 4.53D)/3 \approx 0$. Alternatively stated, the long-range dipole–dipole interactions act weakly to stabilize the Néel order to the detriment of the spin-ice state [30, 45, 46]. Using the values of $J/3$ and $5D/3$ given above for $Dy_2Ti_2O_7$ and $Ho_2Ti_2O_7$, one finds that both materials fulfil the criterion $(J + 4.53D)/3 > 0$, as well as the less accurate $(J + 5D)/3 > 0$ criterion, and as such can be characterized as dipolar spin-ice systems.

Having established a zeroth-order criterion, $(J + 5D)/3 > 0$, to determine whether a dipolar pyrochlore Ising system exhibits spin-ice behavior, one now has a definitive question to address, namely how the long-range $1/{r_{ij}}^3$ tail of $H^{\text{dip}}_{>r_{\text{nn}}}$ appears not to lift the extensive degeneracy created by the nearest-neighbor part of (12.14). We use a mean-field theory to begin addressing this question.

### 12.4.2 Mean-Field Theory

The general idea of a Ginzburg-Landau theory is to determine when the paramagnetic phase, where all the components $m_i^{a,u}$ of the local (on-site) magnetization vanish, becomes spontaneously unstable (critical) against the development of nonzero $m_i^{a,u}$. We follow the approach of [48–50]. Consider a general, bilinear spin Hamiltonian,

$$H = \frac{1}{2} \sum_{(i,j)} S_i^{a,u} \mathcal{K}_{uv}^{ab}(i,j) S_j^{b,v}, \tag{12.15}$$

where $S_i^{a,u}$ is the $u = x, y, z$ component of spin $\boldsymbol{S}_i^a$ on the $a^{\text{th}}$ sublattice of the $i^{\text{th}}$ primitive basis vector and $\mathcal{K}_{uv}^{ab}(i,j)$ is a generalized spin-spin interaction. Making the mean-field ansatz

$$(S_i^{a,u} - m_i^{a,u})(S_i^{b,v} - m_j^{b,v}) = 0, \tag{12.16}$$

where $m_i^{a,u}$ is the thermal average $\langle S_i^{a,u} \rangle$, allows $H$ to be decoupled and written as an effective one-particle problem from which the free-energy, $\mathcal{F}(\{m_i^{a,u}\}) \equiv -\beta^{-1} \ln(Z)$, where $Z = \text{Tr}[\exp(-\beta H)]$ is the partition function, can be expanded as a Taylor series of the order parameters $m_i^{a,u}$. Here $\beta = 1/T$, where $T$ is the temperature, and the choice of units is such that the Boltzmann constant $k_{\rm B} = 1$. The leading term in $\mathcal{F}$ is quadratic in $m_i^{a,u}$,

$$\mathcal{F} \approx \frac{1}{2}\sum_{i,j}\sum_{a,b}\sum_{u,v} m_i^{a,u}\left\{nT\delta_{i,j}\delta^{a,b}\delta^{u,v} - \mathcal{K}_{uv}^{ab}(i,j)\right\} m_j^{b,v} + \mathcal{F}_0(T), \quad (12.17)$$

where $\mathcal{F}_0(T)$ is an $m_i^{a,u}$-independent, temperature-dependent function and $n$ is the number of spin components of $\boldsymbol{S}_i^a$. We introduce the Fourier-transform representation of $m_i^{a,u}$,

$$m_i^{a,u} = \sum_{\boldsymbol{q}} m_{\boldsymbol{q}}^{a,u} \mathrm{e}^{-\imath \boldsymbol{q}\cdot\boldsymbol{R}_i^a}, \quad (12.18)$$

$$\mathcal{K}_{uv}^{ab}(i,j) = \frac{1}{N_{\rm cell}}\sum_{\boldsymbol{q}} \mathcal{K}_{uv}^{ab}(\boldsymbol{q})\mathrm{e}^{\imath \boldsymbol{q}\cdot\boldsymbol{R}_{ij}^{ab}}, \quad (12.19)$$

where $N_{\rm cell}$ is the number of FCC Bravais-lattice points and $\boldsymbol{R}_i^a$ denotes the position of a magnetic ion on sublattice $a$ of basis cell $i$. Equations (12.18) and (12.19) applied to $\mathcal{F}$ give

$$\frac{(\mathcal{F}-\mathcal{F}_0)}{N_{\rm cell}} = \frac{1}{2}\sum_{\boldsymbol{q}}\sum_{a,b}\sum_{u,v} m_{\boldsymbol{q}}^{a,u}\left(nT\delta^{a,b}\delta^{u,v} - \mathcal{K}_{uv}^{ab}(\boldsymbol{q})\right) m_{-\boldsymbol{q}}^{b,v}. \quad (12.20)$$

To diagonalize $\mathcal{K}_{uv}^{ab}(\boldsymbol{q})$ requires a transformation to normal modes

$$m_{\boldsymbol{q}}^{a,u} = \sum_{\alpha=1}^{4}\sum_{\mu=1}^{3} U_{u,\mu}^{a,\alpha}(\boldsymbol{q})\phi_{\boldsymbol{q}}^{\alpha,\mu}, \quad (12.21)$$

where the indices $(\alpha,\mu)$ label the normal modes, $\phi_{\boldsymbol{q}}^{\alpha,\mu}$. $U(\boldsymbol{q})$ is the unitary matrix which diagonalizes $\mathcal{K}_{uv}^{ab}(\boldsymbol{q})$ in the spin and sublattice spaces with eigenvector $\lambda(\boldsymbol{q})$,

$$U^{\dagger}(\boldsymbol{q})\mathcal{K}(\boldsymbol{q})U(\boldsymbol{q}) = \lambda(\boldsymbol{q}), \quad (12.22)$$

where, in component form, $U_{u,\mu}^{a,\alpha}(\boldsymbol{q})$ represents the $(a,u)$ component of the $(\alpha,\mu)$ eigenvector at $\boldsymbol{q}$ with eigenvalue $\lambda_{\mu}^{\alpha}(\boldsymbol{q})$. The mean-field free energy, to quadratic order in the normal-mode variables, is then

$$\mathcal{F}(T) \sim \frac{1}{2}\sum_{\boldsymbol{q}}\sum_{\alpha,\mu} \phi_{\boldsymbol{q}}^{\alpha,\mu}\left(nT - \lambda_{\mu}^{\alpha}(\boldsymbol{q})\right)\phi_{-\boldsymbol{q}}^{\alpha,\mu}. \quad (12.23)$$

For the Ising case of interest here, $n = 1$ and the indices representing the spin components $(u, v)$, in addition to the label $\mu$, can be dropped from (12.23). Our primary goal is to diagonalize $\mathcal{K}^{ab}$, which, by taking a tetrahedral primitive basis with local ⟨111⟩ Ising spins, becomes simply a $4 \times 4$ matrix, and to determine the spectrum $\lambda^{\alpha}(\boldsymbol{q})$. We define as $\lambda^{\max}(\boldsymbol{q})$ the largest of the four eigenvalues $\lambda^{\alpha}(\boldsymbol{q})$ at each $\boldsymbol{q}$. Let us assume that the three-dimensional surface [as a function of $(q_x, q_y, q_z)$] $\lambda^{\alpha}(\boldsymbol{q})$ reaches an absolute maximum value within the full Brillouin zone for some wave vector $\boldsymbol{q}_{\text{ord}}$, with $\boldsymbol{q}_{\text{ord}}$ called the ordering wave vector, which we shall label $\lambda^{\max}(\boldsymbol{q}_{\text{ord}})$. $\lambda^{\max}(\boldsymbol{q}_{\text{ord}})$ is the critical temperature, $T_c$, because, when the temperature $T$ is reduced below this value, $(T - \lambda^{\max}(\boldsymbol{q}_{\text{ord}}))$ in (12.23) changes sign, causing the mode $\phi^{\max}_{\boldsymbol{q}_{\text{ord}}}$ to go soft (critical) and to develop a nonzero thermal expectation value, at least at the level of mean-field theory. A key property of highly frustrated systems, such as the classical Heisenberg antiferromagnet on pyrochlore, kagomé, and FCC lattices, is that an infinitely large number of modes become critical simultaneously at $T_c$ [49].

For the nearest-neighbor ⟨111⟩ Ising pyrochlore ferromagnet [first term on the right-hand side of (12.14)], or Anderson's pyrochlore Ising antiferromagnet, one finds that there are two exactly degenerate branches of soft modes that have an identical eigenvalue $\lambda^{\max}(\boldsymbol{q})$ for every wave vector $\boldsymbol{q}$ in the Brillouin zone (Fig. 12.5). This result is obtained after constructing the $4 \times 4$ matrix $\mathcal{K}^{ab}(\boldsymbol{q})$ obtained from (12.19) and finding its four eigenvalues. The mean-field theory of the nearest-neighbor model therefore predicts, in the reciprocal-space description, that there is no unique ordering wavevector $\boldsymbol{q}_{\text{ord}}$ which develops at $T_c$. In fact, there are $2N_{\text{cell}}$ modes (i.e. an extensive number of modes) that go soft simultaneously at $T_c = 2(J/3 + 5D/3)$, where $N_{\text{cell}} = N/4$ is the number of primitive basis cells and $N$ is the number of magnetic moments. This result agrees with the discussion presented above that the nearest-neighbor spin-ice model should have a degenerate ground state.

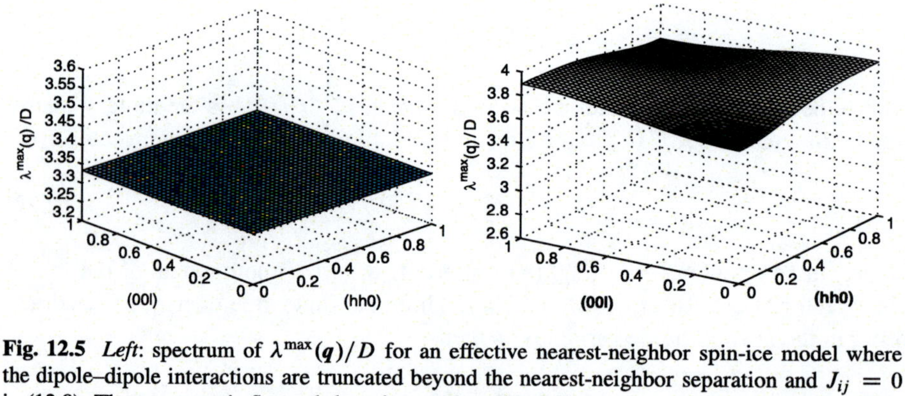


**Fig. 12.5** *Left*: spectrum of $\lambda^{\max}(\boldsymbol{q})/D$ for an effective nearest-neighbor spin-ice model where the dipole–dipole interactions are truncated beyond the nearest-neighbor separation and $J_{ij} = 0$ in (12.8). The spectrum is flat and there is no selected ordering wave vector $\boldsymbol{q}_{\text{ord}}$. Because there is on average one uncompensated spin out of the three neighboring a reference spin on a given tetrahedron, the net "mean field" acting at a site is twice the effective nearest-neighbor interaction for the Ising variable, which is $5D/3$. Hence, the mean-field transition temperature, $T_c$, is $10D/3 \sim 3.33$ in units of $D$, corresponding here to a $\boldsymbol{q}$-independent $\lambda^{\max}(\boldsymbol{q})$. *Right*: $\lambda^{\max}(\boldsymbol{q})/D$ for the full dipolar spin-ice model with $J_{ij} = 0$ in (12.8). Note the slightly dispersive nature of $\lambda^{\max}(\boldsymbol{q})/D$ which displays a maximum at $\boldsymbol{q}_{\text{ord}} = [001]$

Turning to the role of the long-range dipole–dipole interactions beyond nearest neighbors, contained in the term $H^{\rm dip}_{>\rm nn}$ in (12.14), because we are still considering only an Ising model, the $4 \times 4$ matrix $\mathcal{K}^{ab}(\boldsymbol{q})$ is constructed simply by adding the Fourier transform of $H^{\rm dip}_{>\rm nn}$. The long-range $(1/r_{ij}{}^3)$ nature of the dipolar interactions renders somewhat technical the calculation of the Fourier transform in (12.19) [50, 51]. The simplest, although not particularly well controlled, way to proceed is to truncate $H^{\rm dip}_{>\rm nn}$ at some cut-off distance $R_c$ and to determine $\lambda^{\rm max}(\boldsymbol{q}_{\rm ord})$ for that $R_c$ [48]. One may then monitor the evolution of the $\lambda^{\rm max}(\boldsymbol{q})$ "surface" and the corresponding $\boldsymbol{q}_{\rm ord}$ with progressive increases of $R_c$. For "small" $R_c$, for example less than ten times the nearest-neighbor distance, one finds that $\lambda^{\rm max}(\boldsymbol{q})$ acquires sizeable dispersion and a specific value of $\boldsymbol{q}_{\rm ord}$ is selected. However, this value depends on the chosen $R_c$, a physically unacceptable result given that ultimately one must proceed to $R_c = \infty$ [48]. As the value of $R_c$ is increased, $\lambda^{\rm max}(\boldsymbol{q})$ becomes progressively less dispersive and the associated $\boldsymbol{q}_{\rm ord}$ moves towards [001]. This result is confirmed by calculations of $\mathcal{K}^{ab}(\boldsymbol{q})$ performed directly in $\boldsymbol{q}$ space using the Ewald-summation method. Figure 12.5 shows the $\boldsymbol{q}$-dependence of $\lambda_{\rm max}(\boldsymbol{q})$ in the $(hhl)$ reciprocal plane.

One interpretation of the mean-field-theory calculation is that there exists a type of self-screening, such that the selection of an ordering wave vector $\boldsymbol{q}_{\rm ord}$ "promoted" by the shells of spins of smaller radii is canceled by the contributions obtained as the cut-off radius $R_c$ is increased progressively. However, this is an incomplete description of the true microscopic reasons why dipolar interactions cause the ice rules to be satisfied. We return in Sect. 12.4.4 to a full discussion of why dipolar spin ice obeys the ice rules [51]. Here, we discuss only the results of the mean-field calculations, which show that the final selection of an ordering wave vector $\boldsymbol{q}_{\rm ord}$ is "fragile" in the DSM (12.8). By this we refer to the flatness of the spectrum of $\lambda^{\rm max}(\boldsymbol{q})$, which shows that dipolar spin ice has very little propensity for the selection of an ordered state. Indeed, a slight modification of the exchange interactions beyond nearest-neighbors in (12.8) can alter quite dramatically the spin correlations in the spin-ice regime [33]. However, as noted above, careful consideration of $\lambda^{\rm max}(\boldsymbol{q})$ reveals that it does have an absolute maximum, which occurs at $\boldsymbol{q}_{\rm ord} = [001]$. Thus dipolar spin ice is characterized by an ordered phase with a unique, commensurate propagation wave vector and, as the system is cooled from the paramagnetic phase, it should undergo a phase transition to a long-range-ordered ground state with no extensive degeneracy. Because this is a classical system, quantum fluctuations cannot inhibit the development of long-range order, as occurs in a number of the systems described in other chapters of this volume. Even at the level of mean-field theory, the third law of thermodynamics is obeyed in the DSM without having to invoke the effects of quantum mechanics at low temperature.

The fate of this predicted, long-range ordered phase both in Monte Carlo simulations of the DSM and in real materials, is easier to address by turning first to the simulations.

### 12.4.3 Loop Monte Carlo Simulations and Phase Diagram of Dipolar Spin Ice

In Monte Carlo simulations that employ conventional, Metropolis, single-spin-flip dynamics [30, 31], the rate of accepted spin-flips falls exponentially fast with decreasing temperature once the system has entered the spin-ice regime where the "two-in/two-out" ice rules are obeyed rigorously [45]. As an example, in simulations of $Dy_2Ti_2O_7$ one finds that it becomes for all practical purposes impossible to equilibrate the system below a "freezing temperature" $T_f$ which is around 0.4 K [45]. It is worth noting that this temperature corresponds rather closely to the freezing temperature found in AC susceptibility measurements on $Dy_2Ti_2O_7$ [52]. The results from the mean-field theory may then be reinterpreted in the light of the Monte Carlo simulations. The flatness of $\lambda^{\max}(\boldsymbol{q})$ illustrates the competition among all the quasi-degenerate, ice-rule-obeying, quasi-critical modes that exist in the DSM. In real space, once the system is cooled into the ice regime, i.e. below the temperature at which the specific heat peaks (top panel, Fig. 12.3) and at a value much lower than the energy barrier of approximately $2(J/3 + 5D/3)$, it becomes extremely difficult to flip spins in a way that would violate the ice rules. Thus although the low-temperature spin dynamics within the spin-ice state is extremely slow, the system has not yet reached the critical temperature for the transition to long-range order. The spin-ice regime must, therefore, be seen as smoothly connected to the high-temperature paramagnetic phase: alternatively stated, the spin-ice state is a paramagnetic state, or to be more precise, in the terminology of Villain [14] it is a collective paramagnetic state. Hence, spin ice is a classical spin liquid, albeit one with extraordinarily slow spin dynamics. One is therefore led to ask the questions of how the out-of-equilibrium freezing of spin ice can be overcome, at least within a computational approach to the DSM, what is the long-range ordered phase, what is the nature of the phase transition, and at what temperature it ultimately occurs.

In order to explore the low-temperature ordering properties of dipolar spin ice, a Monte Carlo algorithm is required which allows non-local updates that by-pass effectively the energy barriers which separate nearly degenerate states and allows the simulation to explore the restricted, ice-rules phase space which prevents ordering in the model [45, 46]. To construct such update processes, one must first identify the true, zero-energy modes which can take the nearest-neighbor spin-ice model from one ice state to another, exactly energetically degenerate, ice state. An example of these zero modes, or loops, is shown in Fig. 12.2. With interactions beyond nearest-neighbor terms, these "loop moves", where spins are flipped on closed loops without violating the ice rules, become only quasi-zero modes, but still act to take the DSM from one ice-rules state to another without introducing ice-rule defects into the tetrahedra. This type of loop move allows all of the quasi-degenerate spin-ice states to be sampled ergodically, and facilitates the development of a long-range-ordered state at low temperatures. The algorithm used to identify loops of "potentially flipable spins" is described in [45, 46, 53, 54]. Once a loop has been identified, one calculates the energy difference between the original spin configuration for that loop and

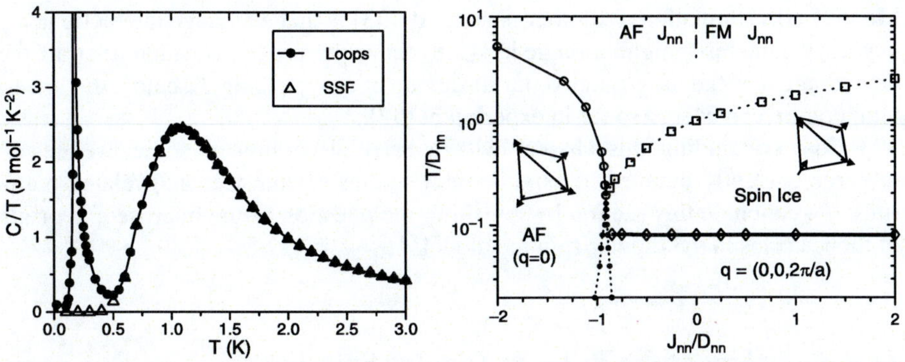


**Fig. 12.6** Results from loop Monte Carlo simulations. *Left*: low-temperature magnetic specific heat for systems of $L \times L \times L$ cubic unit cells with $L = 4$, using the values $J/3 = -1.24$ K and $5D/3 = 2.35$ K suitable for $Dy_2Ti_2O_7$. Filled circles are data obtained from loop Monte Carlo simulations [45, 46]. Open triangles are data obtained using a standard, single-spin-flip Metropolis algorithm [30]. *Right*: phase diagram of the DSM with $J_{nn} \equiv J/3$ and $D_{nn} = 5D/3$. The antiferromagnetic, four-sublattice Néel ground state is the "all-in/all-out" configuration on each tetrahedron. The spin-ice configuration, which includes the $\mathbf{q} = (0, 0, 2\pi/a)$ ground state, is a "two-in/two-out" configuration on each tetrahedron [45, 46]. The region contained between the quasi-vertical, dotted lines displays hysteresis in the selected long-range-ordered state [$\mathbf{q} = 0$ competes with $\mathbf{q} = (0, 0, 2\pi/a)$] as $J_{nn}/D_{nn}$ is varied at fixed temperature $T$

the configuration with flipped spins, which arises due to the long-range part of the dipolar interactions. Whether or not to flip the spins on the loop is then decided according to a standard Metropolis test. One finds from loop Monte Carlo simulations [45, 46] with $J/3 = -1.24$ K and $5D/3 = 2.35$ K a sharp, first-order transition at $T_c \approx 180$ mK, or $T_c/D \approx 0.128$ (Fig. 12.6). A detailed analysis reveals a number of essential properties of the system at the transition [45, 46]:

1. The transition is an extremely strong, first-order process.
2. The entropy removed on the high- and low-temperature "wings" of the transition plus the latent heat associated with the transition equals Pauling's residual entropy within a numerical uncertainty of 1%–2%.
3. The state below $T_c$ is a long-range-ordered ice-rules-obeying state and is identical to the one predicted by mean-field theory [48].
4. $T_c$ is independent of $D/J$, with $T_c/D \approx 0.128$.

One may now return to the question posed above, of what happens to the numerically predicted phase transition to long-range order in real systems. The numerical evidence for a first-order transition to long-range order in the DSM is compelling [45, 46]. However, there is to date no reported experimental evidence for a phase transition in either $Dy_2Ti_2O_7$ [52] or $Ho_2Ti_2O_7$ [6, 9] down to temperatures of approximately 60 mK. A possible, if not likely, reason for this failure of real spin-ice materials to develop long-range order at low temperature is that the spins in real materials are frozen and cannot equilibrate thermally because real systems are unable to benefit from the non-local type of spin dynamics employed in loop

Monte Carlo simulations. Another possibility [33] is that exchange interactions $J_{ij}$ beyond the nearest-neighbor range in (12.8) act as perturbations which frustrate the development of the long-range order discussed above, pushing $T_c$ below the lowest temperature considered so far in experiment [52].

Before concluding this chapter with a brief discussion of some avenues of research currently pursued in experimental studies of spin ices and related materials, we conclude this section by revisiting the question of the microscopic origin of the ice rules in the dipolar spin ice model, $\mathcal{H}_{\mathrm{DSM}}$, in (12.8).

### 12.4.4 Origin of Ice Rules in Dipolar Spin Ice

The Monte Carlo simulation results of Sect. 12.4.1 show that the spin-ice phenomenology is the consequence of long-range dipolar interactions. The mean-field-theory calculations of Sect. 12.4.2 provide an indication of the self screening mechanism enforcing the relevance of the ice rules. However, neither method really explains why dipolar spin ice obeys the ice rules. An approach which was successful in answering this question was reported in [51], and is summarized here.

The first important ingredient [55] is to note that the ice-rules condition on the pyrochlore lattice which favors "two-in/two-out" states can be formulated as

$$\sum_{\Delta} \sigma_{\Delta}^{z1} + \sigma_{\Delta}^{z2} + \sigma_{\Delta}^{z3} + \sigma_{\Delta}^{z4} = 0, \tag{12.24}$$

where the sum is performed over upward and downward tetrahedra, which is equivalent to a divergence-free "spin field" $\boldsymbol{\Omega}$ obeying $\nabla \cdot \boldsymbol{\Omega} = 0$. These "lattice fluxes" are link variables on the diamond lattice (dual to the pyrochlore lattice) and $\nabla\cdot$ is the lattice divergence. Such a phase with local constraint that can be mapped to a divergence-free flux is called a "Coulomb phase" [56]. We return to this in Sect. 12.5.8. By introducing a weight $\rho[\boldsymbol{B}(\boldsymbol{r})]$ of the form

$$\rho[\boldsymbol{\Omega}(\boldsymbol{r})] \propto \exp\left(-\frac{K}{2}\int [\boldsymbol{\Omega}(\boldsymbol{r})]^2 \mathrm{d}^3x\right) \tag{12.25}$$

for nonzero flux, which corresponds to local violation of the ice rules, one may solve for the correlations

$$\langle \Omega_u(0)\Omega_v(\boldsymbol{x})\rangle \sim \frac{3x_u x_v - r^2\delta_{uv}}{r^5}, \tag{12.26}$$

where $u$ and $v$ are Cartesian components. Equation (12.26) reveals that the local constraint, namely the ice rules, leads to spin correlations that are dipolar in nature at large distances.

The second important observation [51] is that a simplified model of the spin-ice type may be constructed whose eigenmodes $\phi_q^\alpha$ can be used to form a projector onto the ice-rule-obeying ground states. Because of the asymptotic dipolar correlations possessed by such states (c.f. 12.26), the real-space matrix elements of this projector turn out to be the same as the dipolar part (the second term) of the interactions in $\mathcal{H}_{\mathrm{DSM}}$ in (12.8), plus a small and rapidly converging $(1/{r_{ij}}^5)$ correction term, $(V(r_{ij}))$. The dipolar interactions between the real magnetic moments are therefore, up to small and rapidly decaying short-range corrections, a projector onto the manifold of all possible ice-rule ground states. Thus, when a dipolar $\langle 111 \rangle$ Ising system is cooled below a temperature on the order of $D$, the spin configurations are forced energetically into ice-rules states. When $V(r_{ij}) < T < D$, this forcing is almost indiscriminate with regard to the precise state, and it is only when the temperature reaches the scale of $V(r_{ij})$ that the transition to a long-range-ordered spin-ice state (that discussed in Sect. 12.4.3) occurs, provided that ergodicity can be maintained using non-local spin-flip dynamics [45, 46].

A somewhat simpler explanation for the origin of the ice rules in dipolar spin ice has been proposed recently [57]. "Exploding" each point dipole into its constitutive magnetic (monopole) charges and imposing that each tetrahedron be neutral (in magnetic charge) leads automatically to the conclusion that all ice-rule-obeying are, again, up to a small and rapidly decaying short-range correction, ground states of the microscopic magnetostatic dipolar interactions of (12.8) [57]. These magnetic charges have interactions which decay as $1/r$ for large inter-charge separations $r$. Magnetic-field-driven [57] and thermally driven [58] nucleation of these objects has been invoked to explain both the thermodynamic [57] and dynamical [58] properties of spin-ice materials, and their possible experimental manifestations are discussed in Sect. 12.5.8.

## 12.5 Current Research Topics in Spin Ices and Related Materials

Since the discovery of spin-ice behavior in $Ho_2Ti_2O_7$ [5] and $Dy_2Ti_2O_7$ [7], there has been a sustained research drive aimed at exploring the interesting thermodynamic and magnetic phenomena offered by spin ices and by a number of closely related systems. This section serves as an update regarding some of the topics that are of current interest in the study of spin ice.

### *12.5.1 Magnetic-Field Effects*

Because of the large magnetic moment of $Dy^{3+}$ and $Ho^{3+}$ in spin-ice materials, even rather small magnetic fields can induce dramatic effects. Of particular interest is the "pinning" of the spins by a sufficiently strong field acting, as a function of field orientation, on one [57, 59–61], two [32], three [32, 62], or all four [63] sites of a primitive tetrahedral cell. This gives rise to collective behavior including phase transitions to long-range order [32,62], topological Kasteleyn transitions [59,61,63], magnetization plateau states [57, 59], and a sort of magnetic-monopole gas-solid transition [57]. Some of the salient features characterizing the behavior of the system according to the direction of the magnetic field, $\boldsymbol{B}$, are discussed in brief.

#### 12.5.1.1 Field Parallel to $[\alpha, \beta, (\bar{\alpha} + \bar{\beta})]$, $[11\bar{2}]$

The case of very strong $\boldsymbol{B}$ in the direction $\alpha\hat{x} + \beta\hat{y} - (\alpha + \beta)\hat{z}$ (e.g. parallel to $[11\bar{2}]$), with $\alpha + \beta \neq 0$ (i.e. unlike $[1\bar{1}0]$), is perhaps the simplest one from the point of view of field-induced collective behavior in dipolar spin-ice systems [32]. For $\boldsymbol{B}$ in this direction, the magnetic moments on sublattice #1, with easy-axis Ising direction $\hat{z}_1 = \frac{1}{\sqrt{3}}(\hat{x} + \hat{y} + \hat{z})$ (Sect. 12.3.2), are perpendicular to $\boldsymbol{B}$ and therefore not coupled to it. The moments on the three other sublattices are frozen (pinned) in a "one-in/two-out" configuration by sufficiently strong field. In the limit $|\boldsymbol{B}| \to \infty$, the field-decoupled moments on sublattice #1, which form a regular FCC lattice, interact only among themselves through dipolar and exchange interactions at and beyond third-nearest-neighbor distances. Numerical simulations [32, 33] and experiments [62] find that, because of these interactions, the field-decoupled moments undergo a phase transition to a ferromagnetic state in $Dy_2Ti_2O_7$ at approximately 0.3 K.

#### 12.5.1.2 Field Parallel to $[1\bar{1}0]$

The situation with $\boldsymbol{B}$ parallel to $[1\bar{1}0]$ is a special case of $[\alpha, \beta, (\bar{\alpha} + \bar{\beta})]$ discussed just above where moments on sublattices #1 and #2, with $\hat{z}_1 = \frac{1}{\sqrt{3}}(\hat{x} + \hat{y} + \hat{z})$ and $\hat{z}_2 = \frac{1}{\sqrt{3}}(-\hat{x} - \hat{y} + \hat{z})$, are decoupled from $\boldsymbol{B}$. At the same time, the moments on sublattices #3 and #4, where $\hat{z}_3 = \frac{1}{\sqrt{3}}(-\hat{x} + \hat{y} - \hat{z})$ and $\hat{z}_4 = \frac{1}{\sqrt{3}}(\hat{x} - \hat{y} - \hat{z})$, are pinned by a strong $[1\bar{1}0]$ field. The pyrochlore lattice can be viewed as two sets of perpendicular chains, each set composed respectively of the (#1,#2) sublattices, which form $\beta$ chains, and the (#3,#4) sublattices, which form $\alpha$ chains [64]. In this situation, one expects theoretically that the field-decoupled spins on the $\beta$ chains undergo a collective phase transition, again driven by dipolar and exchange interactions at and beyond the third-neighbor separation [32, 65]. Results from specific-heat [64] and neutron scattering measurements [5, 38, 66] provide evidence that strong correlations develop among the spins on the $\beta$ chains even at moderate fields $|\boldsymbol{B}| \sim 0.5$ T. However, in contrast to theoretical predictions, no experiment has yet found true long-range order on the $\beta$ chains. Numerical [45] and experimental [66] evidence suggests that failure to observe such order may be the result of small field misalignments away from the perfect $[1\bar{1}0]$ orientation. Such an offset in the field direction may be sufficient to frustrate the three-dimensional correlations among $\beta$ chains, causing the system to remain in a quasi-one-dimensional, short-range-correlated state and failing to develop long-range order.

#### 12.5.1.3 Field Parallel to [111]

The pyrochlore lattice can also be viewed as an assembly of stacked kagomé-lattice planes (planes of corner-shared triangles) separated by alternate triangular planes aligned such that each kagomé triangle is decorated with an additional spin alternating between positions above and below the kagomé plane to give the upward and downward tetrahedra. At sufficiently low temperatures, for $\boldsymbol{B} \parallel [111]$, the apical moments on each tetrahedron are aligned with the field direction: if $\boldsymbol{B}$ is "up", then the spins are oriented "out" on the upward tetrahedra and "in" on the downward tetrahedra. For $|\boldsymbol{B}|$ below a critical field value, $B_c$, of order of 1 T for $Dy_2Ti_2O_7$ [67] and 1.7 T for $Ho_2Ti_2O_7$ [42, 61], the moments on the kagomé plane can still maintain ice-rules configurations, respectively "in-in-out" or "in-out-out" on the upward and downward tetrahedra. This leaves an intra-tetrahedron degree of freedom associated with the kagomé planes (the apical moments are polarized by the field), which is analogous to that in Pauling's model and therefore possesses a residual, zero-point entropy. When $B < B_c$, this state, referred to as "kagomé ice," is characterized by a field-independent [111] magnetization plateau [42, 59, 67]. The magnetic entropy has been predicted [59], and also found in experiment [68], to be non-monotonic as a function of $|\boldsymbol{B}|$ and to exhibit a peak at $B_c$. A very recent analysis interprets the transition from the plateau regime to the high-field saturated-magnetization regime above $B_c$ as a condensation of magnetic monopole-like defects [57]. For $B < B_c$, the available degrees of freedom on the kagomé planes can be mapped to hard-core dimers on a honeycomb lattice. Using this description, the kagomé ice state is found to be critical. Theory predicts that a tilting of the magnetic field away from perfect [111] alignment allows one to control the entropy of this critical state, which is ultimately terminated above a critical tilting angle by a topological phase transition known as a "Kasteleyn transition" [59]. Intensity data from neutron scattering on $Ho_2Ti_2O_7$ in a magnetic field tilted away from [111] yield to a rather compelling interpretation in terms of such a Kasteleyn transition from a critical kagomé-ice state [61].

#### 12.5.1.4 Field Parallel to [100]

A field $\boldsymbol{B} \parallel [100]$ has an equal projection along each of the Ising axes $\hat{z}_i$. A field $|\boldsymbol{B}|$ larger than 0.04 T overwhelms the correlations caused by the long-range magnetostatic dipolar interactions and gives rise to a [100]-polarized state that satisfies the ice rules [45]. On the basis of Monte Carlo simulations, it was thought initially that for fields below a critical value, the system would undergo a first-order liquid-gas transition between a state of low [100] magnetization and a state with higher [100] magnetization upon cooling [69]. However, a recent study suggests that this problem is more subtle [63]: at low temperatures, the approach to the state of saturated magnetization is another example of a Kasteleyn transition, which is topological in nature because the magnetization can be changed only by correlated spin excitations that span the whole system. Comparison of simulation data with the magnetic field-dependence of the [100] magnetization for $Dy_2Ti_2O_7$ provides good evidence that the physics of the Kasteleyn transition is indeed relevant in this problem [63].

### *12.5.2 Dynamical Properties and Role of Disorder*

Experiments have been conducted to investigate the nature of the persistent spin dynamics observed down to the lowest temperatures [70–72]. Substitution of the magnetic $Ho^{3+}$ and $Dy^{3+}$ ions by non-magnetic ions affects the spin dynamics and leads to a partial and non-monotonic lifting of the ground-state degeneracy [41]. However, the problem of the final transition from spin ice to dipolar spin glass as the magnetic rare-earth ion is substituted by $Y^{3+}$ has not yet been explored. A full understanding of the origin of the spin dynamics in spin ices is still awaited [72]. A recent study of $Ho_2Ti_2O_7$ proposes that the excitation of nuclear states can perturb the electronic Ising spin states to give rise to persistent spin dynamics [73]. One challenging aspect of the spin dynamics in spin-ice materials such as $Dy_2Ti_2O_7$ is the fact that the relaxation rate $\tau(T)$ remains quite independent of temperature between 4 K and 13 K [74, 75]. This $T$-independent regime had been interpreted previously in terms of a quantum mechanical tunneling of the spins between their "in" and "out" directions. However, recent analysis suggests that the magnetic relaxation can be interpreted entirely in terms of the diffusion of thermally-nucleated topological defects, again akin to magnetic monopoles, on trajectories constrained to lie on a network of "Dirac strings" [58].

### *12.5.3 Beyond the Dipolar Spin-Ice Model*

The large number of high-quality experiments on spin ices now permits a refinement of the spin Hamiltonian and the extraction of exchange interactions beyond nearest neighbors, allowing an excellent, quantitative description of many bulk measurements and neutron scattering experiments [33, 60]. In particular, exchange interactions beyond nearest neighbors are argued to induce weak spin-spin correlations that are responsible for the fact that the neutron scattering intensity spans the Brillouin zone boundaries [33, 76].

### *12.5.4 Metallic Spin Ice*

The metallic pyrochlore material $Pr_2Ir_2O_7$ exhibits Kondo-type effects in the resistivity and magnetic susceptibility, with logarithmic increases of both quantities at low temperatures [77,78]. However, one expects the magnetic moments on the $Pr^{3+}$ ions in $Pr_2Ir_2O_7$ to be well described as Ising spins, for which Kondo physics is not readily explicable. Further, the predominant $Pr^{3+}$–$Pr^{3+}$ interactions in metallic $Pr_2Ir_2O_7$ would normally be expected to be of RKKY type. In contrast to experimental findings, Monte Carlo simulations on RKKY-coupled $\langle 111 \rangle$ spins find a transition to long-range order at temperatures close to the Curie–Weiss temperature [79]. Thus it is not currently clear what, if any, is the role of magnetic frustration in determining the exotic thermodynamic and transport properties of $Pr_2Ir_2O_7$. Very recent work reports an anomalous Hall effect (AHE) in this material [80].

### 12.5.5 Artificial Spin Ice

Lithographically fabricated, single-domain ferromagnetic islands can be arranged such that the magnetic dipolar interactions create a two-dimensional analog of spin ice. The magnetic moments can be imaged directly, allowing the study of topics including the local accommodation of frustration and the emergence of the ice rules [81,82]. It has also been proposed that a colloidal version of artificial ice can be realized using optical lattices in traps [83]. The availability of more such artificial systems will be of prime interest in the field. The out-of-equilibrium properties of lithographically prepared systems in magnetic fields promise to be one of the first areas in which these systems can offer new insight.

### 12.5.6 Stuffed Spin Ice

It is possible to alter $Ho_2Ti_2O_7$ spin ice chemically by "stuffing" additional $Ho^{3+}$ magnetic moments into the normally non-magnetic Ti sites [84, 85]. The resulting series of samples, $Ho_2(Ti_{2-x}Ho_x)O_{7-x/2}$, displays an increased connectivity compared to the standard pyrochlore and raises the question of how the Pauling entropy of "normal" spin ice evolves with stuffing [84, 86]. At the time of writing, the question of homogeneity of stuffed spin-ice materials remains to be ascertained, and further studies of these systems are required.

### 12.5.7 Quantum Mechanics, Dynamics, and Order in Spin Ices

The compounds $Tb_2Ti_2O_7$ [25, 26] and $Tb_2Sn_2O_7$ [26] should both possess an Ising ground-state doublet, similar to their Ho and Dy counterparts [24]. However, $Tb_2Ti_2O_7$ remains in a collective paramagnetic (spin-liquid) state down to 50 mK despite an antiferromagnetic Curie–Weiss temperature of $-14$ K [19, 25] and is seemingly not a spin-ice state [23]. By contrast, neutron scattering measurements [37] find a long-range-ordered spin-ice configuration in $Tb_2Sn_2O_7$, although muon spin relaxation studies find significant spin dynamics in this compound [87–89]. Determining the role of the deviation from a strictly Ising-like description of the magnetic moments in these two materials, and the effects of the resulting quantum fluctuations, is a task which remains to be completed [23]. In this context, it is worth noting that the pyrochlore material $Pr_2Sn_2O_7$ has been proposed to display a "dynamical spin-ice state" [90]. Little is known about the microscopic physics operating in this system.

### 12.5.8 Coulomb Phase, Monopoles and Dirac Strings in Spin Ices

The exploration of the "Coulomb phase" nature of the spin-ice state and of its monopole-like excitations is a very recent and exciting development in the theoretical [56–58] and experimental [91–93] study of spin-ice physics.

The "Coulomb phase" is an emergent state of lattice models of highly frustrated magnetic systems which have local constraints that can be mapped to a divergence-free "flux". The perfectly ice-rule obeying state of spin ices is thus a Coulomb phase where the local "two-in/two-out" constraints are represented by a divergence-free $\boldsymbol{\Omega}(\boldsymbol{r})$ field, with the coarse-grained version of $\boldsymbol{\Omega}(\boldsymbol{r})$ behaving as an electric field. In particular, defects in the otherwise perfect "two-in/two-out" spin configuration, behave as effective charges with Coulomb interactions between them. It has rapidly become common practice in the field to refer to these defects in the spin-ice states as "monopoles" – however, these are *not* Dirac's elementary monopoles [94, 95]. A thermal fluctuation which flips a spin amounts to the creation of a pair of positive and negative monopoles on adjoining tetrahedra. Once created, these two monopoles can diffuse away from each other, leaving in their wake a chain of flipped spins with respect to the configuration that existed prior to the fluctuation, with this chain popularly referred to as "Dirac string" [58,91,92]. While the spin-ice state is really a paramagnetic phase, albeit a collective paramagnetic one [14], the underlying spin-spin correlations do not decay exponentially with distance, $r$. Rather, this Coulomb phase is characterized by spin-spin correlations that are dipolar at large distance, both in terms of their $1/r^3$ decay and their spatially anisotropic structure. This real space power-law decay of the correlations is reflected in "pinch-points" singularities in reciprocal space which are observable in neutron scattering experiments [90]. At low-temperature, the density of monopoles is low and their RMS separation, $\xi_{\text{ice}}$, provides a cut-off to the $1/r^3$ spin-spin correlations and hence a finite-width to the reciprocal space pinch-points.

Following the original work of Castelnovo et al. [57], Jaubert and Holdsworth used Monte Carlo simulations to show that the temperature-independent relaxation rate extracted from an analysis of the a.c. magnetic susceptibility of the $Dy_2Ti_2O_7$ spin-ice material [74, 75] can be understood in terms of the freezing-out of monopoles [58]. Recent experiments provide an even more direct evidence for the underlying Coulomb phase nature of the spin-ice state and of its monopole-like excitations. Fennell et al. performed a polarized neutron-scattering experiment on a single crystal of the $Ho_2Ti_2O_7$ spin-ice compound where the scattering signal is decomposed in two (spin flip and non-spin-flip) components [91]. The pinch-points predicted by theory are revealed in the component where the neutron spin is flipped. The pinch points are obscured in the more intense non-spin-flip signal and this provides an explanation as to why their observation in previous unpolarized neutron scattering experiments was inconclusive [31, 76]. In the work of Morris et al. [92], an applied magnetic field $\boldsymbol{B}$ along the [100] direction of a single crystal of the $Dy_2Ti_2O_7$ spin-ice material induces a magnetically polarized state that simultaneously satisfies the ice-rules and minimize the Zeeman energy [63,69]. The strength of $\boldsymbol{B}$ can be tuned near the Kastelyn transition for this field direction [63] where thermally excited monopole-antimonopole pairs start to appear. The flipped spins forming the "Dirac string" are then oriented antiparallel to $\boldsymbol{B}$, with the strings producing cone-like features in the neutron scattering intensity pattern. These conic features progressively transform in inclined sheets of scattering when $\boldsymbol{B}$ is tilted away from perfect [100] alignment, in close concordance with the calculations [92]. Moreover, the magnetic specific heat in zero magnetic field can also be well described in terms of a Debye-Hückel theory for a dilute gas of thermally excited monopoles [92].

In a weak electrolyte, the so-called second Wien effect describes the nonlinear increase in the dissociation constant (or equivalently the conductance) in an applied electric field. In a seminar paper, Onsager derived a general equation for the Wien effect that provides an excellent description of experimental conductivity measurements. Bramwell et al. [93] borrowed Onsager's theory to calculate the thermally-induced monopole-antimonopole pair dissociation rate in spin ice and to estimate the increase in the magnetic moment fluctuation rate, $\nu(B)$, induced by a small magnetic field, $B$, of the order of 1 mT. Bramwell and co-workers argued that the field-dependence of the transverse relaxation rate, $\lambda$, of the muon spin polarization in a muon spin relaxation (muSR) is proportional to $\nu(B)$. By measuring the field and temperature dependence of $\lambda$, the authors of [93] reported an estimate of the charge $Q \sim 5\ \mu_B$/Å for the monopoles, in close agreement with the theoretical estimate [57]. At the time where this book is going into print, it is noted that discussions pertaining to the definitive interpretation of the experimental results reported in [93] have very recently arisen in the research community.

One may expect more exciting experiments probing the Coulomb phase nature and the monopole-like excitations of the low-temperature regime of spin-ice materials to be performed in the near future. These are sure to deepen our understanding of the fascinating properties of these systems.

## 12.6 Conclusion

Since the discovery of spin-ice materials a little over ten years ago, their experimental and theoretical investigation has raised a range of profound questions concerning several aspects of the physics of frustrated magnetic systems. While these studies have led to an enhanced global understanding of many fundamental issues pertaining to frustration in condensed matter systems, some unanswered questions remain, particularly in the context of low-temperature spin dynamics, random disorder, and the properties of metallic Ising pyrochlore systems. These questions can be expected to form the focus of spin-ice research for several years to come.

**Acknowledgements**

I thank Steve Bramwell, Benjamin Canals, Adrian Del Maestro, Byron den Hertog, Sarah Dunsiger, Matt Enjalran, Tom Fennell, Jason Gardner, Bruce Gaulin, John Greedan, Ying-Jer Kao, Roger Melko, Hamid Molavian, Jacob Ruff, and Taras Yavors'kii for their collaboration on projects related to spin ice. I am grateful to Tom Devereaux, Chris Henley, Rob Hill, Peter Holdsworth, Rob Kiefl, Jan Kycia, Graeme Luke, Paul McClarty, Roderich Moessner, Jeffrey Quilliam, Peter Schiffer, Andreas Sorge, Ka-Ming Tam, Oleg Tchernyshyov, and Jordan Thompson for many valuable and stimulating conversations over the past few years. For funding support I thank the NSERC of Canada, the Canada Research Chair Program, the Research Corporation, the Canadian Institute for Advanced Research, the Ontario Innovation Trust, the Ontario Challenge Fund, and Materials and Manufacturing Ontario. I acknowledge the hospitality of the Kavli Institute of Theoretical Physics (KITP) at the University of California in Santa Barbara, where most of this review was written. Work at the KITP was supported in part by the National Science Foundation under Grant No. NSF PHY05-51164.

## References

1. G.H. Wannier, Phys. Rev. **79**, 357 (1950); erratum Phys. Rev. B **7**, 5017 (1973)
2. W.F. Giauque, M.F. Ashley, Phys. Rev. **43**, 81 (1933)
3. W.F. Giauque, J.W. Stout, J. Am. Chem. Soc. **58**, 1144 (1936)
4. L. Pauling, J. Am. Chem. Soc. **57**, 2680 (1935)
5. M.J. Harris, S.T. Bramwell, D.F. McMorrow, T. Zeiske, K.W. Godfrey, Phys. Rev. Lett. **79**, 2554 (1997)
6. M.J. Harris, S.T. Bramwell, T. Zeiske, D.F. McMorrow, P.J.C. King, J. Magn. Magn. Mater. **177**, 757 (1998)
7. A.P. Ramirez, A. Hayashi, R.J. Cava, R. Siddharthan, B.S. Shastry, Nature **399**, 333 (1999)
8. S.T. Bramwell, M.J.P. Gingras, Science **294**, 1495 (2001)
9. S.T. Bramwell, M.J.P. Gingras, P.C.W. Holdsworth, in *Frustrated Spin Systems*, ed. by H.T. Diep (World Scientific Publishing, Singapore, 2004), p. 367
10. J.D. Bernal, R.H. Fowler, J. Chem. Phys. **1**, 515 (1933)
11. J.M. Ziman, *Models of Disorder:The Theoretical Physics of Homogeneously Disordered Systems* (Cambridge University Press, Cambridge, 1976)
12. J.F. Nagle, J. Math. Phys. **7**, 1484 (1966)
13. P.W. Anderson, Phys. Rev. **102**, 1008 (1956)
14. J. Villain, Z. Phys. B **33**, 31 (1979)
15. J.N. Reimers, Phys. Rev. B **45**, 7287 (1992)
16. R. Moessner, J.T. Chalker, Phys. Rev. B **58**, 12049 (1998)

17. J.E. Greedan, J. of Materials Chemistry **11**, 37 (2001)
18. J.S. Gardner, M.J.P. Gingras, J.E. Greedan, Rev. Mod. Phys. **82**, 53 (2010)
19. J.S. Gardner, S.R. Dunsiger, B.D. Gaulin, M.J.P. Gingras, J.E. Greedan, R.F. Kiefl, M.D. Lumsden, W.A. MacFarlane, N.P. Raju, J.E. Sonier, I. Swainson, Z. Tun, Phys. Rev. Lett. **82**, 1012 (1999)
20. J.E. Greedan, M. Sato, Y. Xu, F.S. Razavi, Solid State Commun. **59**, 895 (1986)
21. B.D. Gaulin, J.N. Reimers, T.E. Mason, J.E. Greedan, Z. Tun, Phys. Rev. Lett. **69**, 3244 (1992)
22. Y. Taguchi, Y. Oohara, H. Yoshizawa, N. Nagaosa, Y. Tokura, Science **291**, 2573 (2001)
23. H.R. Molavian, M.J.P. Gingras, B. Canals, Phys. Rev. Lett. **98**, 157204 (2007)
24. S. Rosenkranz, A.P. Ramirez, A. Hayashi, R.J. Cava, R. Siddharthan, B.S. Shastry, J. Appl. Phys. **84**, 5914 (2000)
25. M.J.P. Gingras, B.C. den Hertog, M. Faucher, J.S. Gardner, S.R. Dunsiger, L.J. Chang, B.D. Gaulin, N.P. Raju, J.E. Greedan, Phys. Rev. B **62**, 6496 (2000)
26. I. Mirebeau, P. Bonville, M. Hennion Phys. Rev. B **76**, 184436 (2007)
27. A. Abragam, B. Bleaney, *Electron Paramagnetic Resonance of Transition Ions* (Dover, 1986)
28. I.D. Ryabov, J. Magn. Reson. **140**, 141 (1999)
29. C. Rudowicz, C.Y. Chung, J. Phys. Condens. Matter **16**, 1 (2004)
30. B.C. den Hertog, M.J.P. Gingras, Phys. Rev. Lett. **84**, 3430 (2000)
31. S.T. Bramwell, M.J. Harris, B.C. den Hertog, M.J.P. Gingras, J.S. Gardner, D.F. McMorrow, A.R. Wildes, A.L. Cornelius, J.D.M. Champion, R.G. Melko, T. Fennell, Phys. Rev. Lett. **87**, 047205 (2001)
32. J.P.C. Ruff, R.G. Melko, M.J.P. Gingras, Phys. Rev. Lett. **95**, 097202 (2005)
33. T. Yavors'kii, T. Fennell, M.J.P. Gingras, S.T. Bramwell, Phys. Rev. Lett. **101**, 037204 (2008)
34. S.H. Curnoe, Phys. Rev. B **78**, 094418 (2008)
35. P.A. McClary, S.H. Curnoe, M.J.P. Gingras, J. Phys. Conf. Ser. **145**, 012032 (2009)
36. R. Siddharthan, B.S. Shastry, A.P. Ramirez, A. Hayashi, R.J. Cava, S. Rosenkranz, Phys. Rev. Lett. **83**, 1854 (1999)
37. I. Mirebeau, A. Apetrei, J. Rodríguez-Carvajal, P. Bonville, A. Forget, D. Colson, V. Glazkov, J.P. Sanchez, O. Isnard, E. Suard, Phys. Rev. Lett. **94**, 246402 (2005)
38. T. Fennell, O.A. Petrenko, B. Fåk, J.S. Gardner, S.T. Bramwell, B. Ouladdiaf, Phys. Rev. B **72**, 224411 (2005)
39. K. Binder, A.P. Young, Rev. Mod. Phys. **58**, 801 (1986)
40. S.T. Bramwell, M.J. Harris, J. Phys. Condens. Matter **10**, L215 ((1998)
41. X. Ke, R.S. Freitas, B.G. Ueland, G.C. Lau, M.L. Dahlberg, R.J. Cava, R. Moessner, P. Schiffer, Phys. Rev. Lett. **99**, 137203 (2007)
42. A.L. Cornelius, J.S. Gardner, Phys. Rev. B, **64**, 060406 (2001)
43. H. Kadowaki, Y. Ishii, K. Matsuhira, Y. Hinatsu, Phys. Rev. B **65**, 144421 (2002)
44. X. Ke, B.G. Ueland, D.V. West, M.L. Dahlberg, R.J. Cava, P. Schiffer, Phys. Rev. B **76**, 214413 (2007)
45. R.G. Melko, M.J.P. Gingras, J. Phys. Condens. Matter **16**, R1277 (2004)
46. R.G. Melko, B.C. den Hertog, M.J.P. Gingras, Phys. Rev. Lett. **87**, 067203 (2001)
47. C. Kittel, *Introduction to Solid State Physics*, 8th edn. (Wiley, 2005), Appendix B
48. M.J.P. Gingras, B.C. den Hertog, Can. J. Phys. **79**, 1339 (2001)
49. J.N. Reimers, A.J. Berlinsky, A.C. Shi, Phys. Rev. B **43**, 865 (1991)
50. M. Enjalran, M.J.P. Gingras, Phys. Rev. B **70**, 174426 (2004)
51. S.V. Isakov, R. Moessner, S.L. Sondhi, Phys. Rev. Lett. **95**, 217201 (2005)
52. H. Fukazawa, R.G. Melko, R. Higashinaka, Y. Maeno, M.J.P. Gingras, Phys. Rev. B **65**, 054410 (2002)
53. G.T. Barkema, M.E.J. Newman, Phys. Rev. E **57**, 1155 (1998)
54. M.E.J. Newmann, G.T. Barkema, *Monte Carlo Methods in Statistical Physics* (Clarendon, Oxford, 1999)
55. S.V. Isakov, K. Gregor, R. Moessner, S.L. Sondhi, Phys. Rev. Lett. **93**, 167204 (2004)
56. C.L. Henley, Annu. Rev. Condens. Matter Phys. 1:179–210 (2010)
57. C. Castelnovo, R. Moessner, S.L. Sondhi, Nature **451**, 42 (2008)
58. L.D.C. Jaubert, P.C.W. Holdsworth, Nature Phys. **5**, 258 (2009)

59. R. Moessner, S.L. Sondhi, Phys. Rev. B **68**, 064411 (2003)
60. Y. Tabata, H. Kadowaki, K. Matsuhira, Z. Hiroi, N. Aso, E. Ressouche, B. Fåk, Phys. Rev. Lett. **97**, 257205 (2006)
61. T. Fennell, S.T. Bramwell, D.F. McMorrow, P. Manuel P.A.R. Wildes, Nature Phys. **8**, 566 (2007)
62. R. Higashinaka, Y. Maeno Phys. Rev. Lett. **95**, 237208 (2005)
63. L. Jaubert, J.T. Chalker, P.C.W. Holdsworth, R. Moessner, Phys. Rev. Lett. **100**, 067207 (2008)
64. Z. Hiroi, K. Matsuhira, M. Ogata, J. Phys. Soc. Jpn. **72**, 3045 (2003)
65. S. Yoshida, K. Nemoto, K. Wada, J. Phys. Soc. Jpn. **73**, 1619 (2004)
66. J.P. Clancy, J.P.C. Ruff, S.R. Dunsiger, Y. Zhao, H.A. Dabkowska, J.S. Gardner, Y. Qiu, J.R.D. Copley, T. Jenkins, B.D. Gaulin, Phys. Rev. B **79**, 014408 (2009)
67. T. Sakakibara, T. Tayama, Z. Hiroi, K. Matsuhira, S. Takagi, Phys. Rev. Lett. **90**, 207205 (2003)
68. H. Aoki, T. Sakakibara, K. Matsuhira, Z. Hiroi, J. Phys. Soc. Jpn. **73**, 2851 (2004)
69. M.J. Harris, S.T. Bramwell, P.C.W. Holdsworth, J.D. Champion, Phys. Rev. Lett. **81**, 4496 (1998)
70. G. Ehlers, J.S. Gardner, C.H. Booth, M. Daniel, K.C. Kam, A.K. Cheetham, D. Antonio, H.E. Brooks, A.L. Cornelius, S.T. Bramwell, J. Lago, W. Häussler, N. Rosov, Phys. Rev. B **73**, 174429 (2006)
71. J.P. Sutter, S. Tsutsui, R. Higashinaka, Y. Maeno, O. Leupold, A.Q. Baron, Phys. Rev. B **75**, 140402 (2007)
72. J. Lago J, S.J. Blundell, C. Baines, J. Phys. Condens. Matter **19**, 326210 (2007)
73. G. Ehlers, E. Mamontov, M. Zamponi, K.C. Kam, J.S. Gardner, Phys. Rev. Lett. **102**, 016405 (2009)
74. J. Snyder, B.G. Ueland, J.S. Slusky, H. Karunadasa, R.J. Cava, A. Mizel, P. Schiffer, Phys. Rev. Lett. **91**, 107201 (2003)
75. J. Snyder, B.G. Ueland, J.S. Slusky, H. Karunadasa, R.J. Cava, P. Schiffer, Phys. Rev. B **69**, 064414 (2004)
76. T. Fennell, O.A. Petrenko, B. Fåk, S.T. Bramwell, M. Enjalran, T. Yavors'kii, M.J.P. Gingras, R.G. Melko, G. Balakrishnan, Phys. Rev. B **70**, 134408 (2004)
77. S. Nakatsuji, Y. Machida, Y. Maeno, T. Tayama, T. Sakakibara, J. van Duijn, L. Balicas, J.N. Millican, R.T. Macaluso, J.Y. Chan, Phys. Rev. Lett. **96**, 087204 (2006)
78. Y. Machida, S. Nakatsuji, Y. Maeno, T. Tayama, T. Sakakibara, S. Onoda, Phys. Rev. Lett. **98**, 057203 (2007)
79. A. Ikeda, H. Kawamura, J. Phys. Soc. Jpn. **77**, 073707 (2008)
80. Y. Machida, S. Nakatsuji, S. Onoda, T. Tayama, T. Sakakibara, Nature **463**, 210 (2009)
81. R.F. Wang, C. Nisoli, R.S. Freitas, J. Li, W. McConville, B.J. Cooley, M.S. Lund, N. Samarth, C. Leighton, V.H. Crespi, P. Schiffer, Nature **439**, 303 (2006)
82. C. Nisoli, R.F. Wang, J. Li, W.F. McConville, P.E. Lammert, P. Schiffer, V.H. Crespi, Phys. Rev. Lett. **98**, 217203 (2007)
83. A. Libál, C. Reichhardt, C.J. Reichhardt Phys. Rev. Lett. **97**, 228302 (2006)
84. G.C. Lau, R.S. Freitas, B.G. Ueland, B.D. Muegge, E.L. Duncan, P. Schiffer, R.J. Cava, Nature Phys. **2**, 249 (2006)
85. G.C. Lau, B.D. Muegge, T.M. McQueen, E.L. Duncan, R.J. Cava, J. Sol. State Chem. **179**, 3136 (2006)
86. G.C. Lau, R.S. Freitas, B.G. Ueland, M.L. Dahlberg, Q. Huang, H.W. Zandbergen, P. Schiffer, R.J. Cava, Phys. Rev. B **76**, 054430 (2007)
87. P. Dalmas de Réotier, A. Yaouanc, L. Keller, A. Cervellino, B. Roessli, C. Baines, A. Forget, C. Vaju, P.C. Gubbens, A. Amato, P.J. King, Phys. Rev. Lett. **96**, 127202 (2006)
88. F. Bert, P. Mendels, A. Olariu, N. Blanchard, G. Collin, A. Amato, C. Baines, A.D. Hillier, Phys. Rev. Lett. **97**, 117203 (2006)
89. S.R. Giblin, J.D. Champion, H.D. Zhou, C.R. Wiebe, J.S. Gardner, I. Terry, S. Calder, T. Fennell, S.T. Bramwell, Phys. Rev. Lett. **101**, 237201 (2008)
90. H.D. Zhou, C.R. Wiebe, J.A. Janik, L. Balicas, Y.J. Yo, Y. Qiu, J.R. Copley, J.S. Gardner, Phys. Rev. Lett. **101**, 227204 (2008)

91. T. Fennell, P.P. Deen, A.R. Wildes, K. Schmalzl, D. Prabhakaran, A.T. Boothroyd, R.J. Aldus, D.F. McMorrow, S.T. Bramwell, Science **326**, 415 (2009)
92. D.J.P. Morris, D.A. Tennant, S.A. Grigera, B. Klemke, C. Castelnovo, R. Moessner, C. Czternasty, M. Meissner, K.C. Rule, J.-U. Hoffmann, K. Kiefer, S. Gerischer, D. Slobinsky, R.S. Perry, Science **326**, 411 (2009)
93. S.T. Bramwell, S.R. Giblin, S. Calder, R. Aldus, D. Prabhakaran, T. Fennell, Nature **461**, 956 (2009)
94. P.A.M. Dirac, Proc. R. Soc. London Ser. A **133**, 60 (1931)
95. R.J. Jackson, *Classical Electrodynamics* (Wiley, New York, 1998), pp. 273–280

# Chapter 13
# Spin Nematic Phases in Quantum Spin Systems

**Karlo Penc and Andreas M. Läuchli**

**Abstract** In this chapter, we explore the possibility for spin systems to develop a type of order that breaks the O(3) spin symmetry but does not have a magnetic moment. Such ordering is usually referred to as multipolar or nematic, with quadrupolar being the simplest example. These phases have been found in $S = 1$ Heisenberg models extended with biquadratic exchange, in certain $S = 1/2$ Heisenberg models with both ferromagnetic and antiferromagnetic exchange couplings, and in models with cyclic ring-exchange terms. We present theoretical and numerical methods which can be used to understand and characterize quadrupolar and nematic phases. While quadrupolar/nematic ordering is well documented in model systems, it has not yet been identified unambiguously in real materials, although there exist some promising candidates which we also review.

## 13.1 Introduction and Materials

Magnetism and spins are usually thought to be inseparable. It is the quantum mechanical spin of individual electrons which, through Hund's rules, forms the moment of a magnetic ion. When these magnetic ions interact, the spins usually order in the ground state, and the individual spins are oriented in a given direction in spin space. Such a state breaks spontaneously both spin-rotation and time-reversal symmetries. An essential property of quantum mechanical systems is the possibility of discovering phases which do not carry a magnetic moment and do not break time-reversal symmetry. One paradigm is that spin-1/2 entities may pair into singlets, realizing valence-bond phases of different kinds [1] – in such a quantum paramagnetic phase, neither the spin-rotation nor the time-reversal symmetry is broken, as discussed in Chap. 2 by Lhuillier and Misguich. One might ask whether yet more exotic possibilities exist, for example a phase without magnetic order, but which nevertheless breaks spin-rotation symmetry. This chapter explores such a situation.

In the following, we call a *spin nematic* any state that has no magnetic order, i.e. $\langle \mathbf{S}_i \rangle = 0$, but still breaks the spin rotational symmetry, by virtue of a more complicated order parameter. The simplest such example is onsite quadrupolar order, where the spin fluctuations along different axes are anisotropic: e.g. $\langle (S^x)^2\rangle \neq \langle (S^y)^2\rangle = \langle (S^z)^2\rangle$. More complicated possibilities involving pseudovectors $(\mathbf{S}_i \times \mathbf{S}_j)$ and higher rank tensors (octupoles, hexadecapoles) are possible and will be discussed in this chapter.

Historically the possibility of quadrupolar ordering in $S = 1$ Heisenberg models with additional biquadratic $[(\mathbf{S}_i \cdot \mathbf{S}_j)^2]$ spin-exchange interactions was raised by Blume and Hsieh in 1969 [2]. In 1984, Andreev and Grishchuk noted that the notion of quadrupoles can also be extended to $S = 1/2$ spins, albeit not with on-site, but rather with bond-based order parameters [3].

The possible observation of quadrupolar and nematic phases is challenging: in thermodynamic measurements they behave in the same way as antiferromagnets [3]. Careful analysis by a technique such as neutron scattering study would be able to discern a difference [4]. While to date there is no well-established case for a spin-nematic ordered phase in a real material on which many of the relevant ideas could be tested, there are several prospective cases that we introduce below.

Interest in spin-quadrupolar ordering has been raised recently by experimental findings in the layered compound $NiGa_2S_4$, where the $Ni^{2+}$ ions are in the $S = 1$ state, and form a (two-dimensional) triangular lattice [5]. Specific-heat measurements in this system show a $T^2$ behavior at low temperatures – an indication for a **k**-linear Goldstone mode. However, neutron scattering shows only short-ranged magnetic order. This led Tsunetsugu and Arikawa to suggest that a hidden order is present, which is actually antiferroquadrupolar [6, 7]. Later, an alternative scenario for the system based on ferroquadrupolar order was suggested [8, 9].

Another candidate for nonmagnetic nematic order is a recently synthesized family of vanadates which realize a frustrated square-lattice $J_1$-$J_2$ model with ferromagnetic $J_1$ interactions [10, 11]. In these compounds, which are also structurally layered and have chemical formula AA'VO$(PO_4)_2$ (where (A,A') = Pb, Zn, Sr, or Ba), the effective interactions between the magnetically active $S = 1/2$ $V^{4+}$ ions depend on the A and A' elements. In some materials the exchange constants are actually rather close to the theoretical phase boundary to the $n$-type nematic phase, raising the possibility that with a suitable (A,A') combination this boundary may actually be reached.

A further square lattice $J_1$–$J_2$ compound of interest in this regard is (CuCl) $LaNb_2O_7$ [12], in which the $Cu^{2+}$ ions carry the $S = 1/2$ spins. The ground state of this material has a spin gap, a situation which may result from the fact that a type of quadrumerization is observed. Theoretically, a perturbation expansion [13] indicates that under favorable conditions the gap from the singlet to the quintuplet state of the quadrumer units can be closed in an applied magnetic field, leading to a field-induced quadrupolar phase.

More general multipolar ordering is present in rare-earth and actinide compounds, and has been studied for some time. We briefly introduce these systems for completeness, without discussing them further in the remainder of the chapter. In these systems, the crystal field and the strong spin-orbit coupling act to mix the orbital and spin quantum numbers of the $f$ electrons into multiplets which can be classified according to the irreducible representations of the relevant point group. The possible multipolar degrees of freedom are then determined by the nature of the ground-state multiplet. While thermodynamic measurements can be used to map out the boundaries of different ordered phases of $f$-electron systems, the identification of the order parameter is often very difficult. Unlike magnetic-dipole and electric-quadrupole ordering, both of which are relatively easy to observe, evidence for multipolar ordering is mostly indirect, and is often based on induced secondary order parameters.

A well-known example of this situation is the hidden order in $NpO_2$ which occurs within a phase discovered by specific-heat measurements: despite the fact that the lowest multiplet of the $Np^{4+}$ ion is a Kramers doublet, no magnetic order has been observed. Recent resonant X-ray experiments [14, 15], supplemented by theoretical arguments, suggested a triple-**q** octupolar ordering with wave vectors $Q_x = (2\pi/a)(1,0,0)$, $Q_y = (2\pi/a)(0,1,0)$, $Q_z = (2\pi/a)(0,0,1)$. In addition to this primary order parameter, a triple-**q** quadrupolar ordering with the same wave vector is induced. This scenario has been confirmed in $^{17}O$ NMR studies [16], supplemented by appropriate theoretical interpretation [17].

Another example worth highlighting is $CeB_6$, where the Ce ions have a simple cubic structure. The suggested primary order parameter of this system is an antiferroquadrupolar order with $(2\pi/a)(1/2,1/2,1/2)$. The uniform magnetic field induces antiferro-aligned octupoles with $(2\pi/a)(1/2,1/2,1/2)$ [18]. Finally, rare-earth skutterudites have also provided very rich and complex physics, the materials $PrRu_4P_{12}$ and $PrFe_4P_{12}$ even supporting a hexadecapole order [19, 20].

To conclude the experimentally motivated part of this introduction, ultracold atomic gases, in which the trapped atom species often have high internal (pseudo) spin states, offer another area with great promise of producing quadrupolar and nematic ordering phenomena [21–23].

Returning to its theoretical foundations, nematic (directional) ordering has also been found in the compass model, which is motivated by systems with orbital order [24]. In this model, the coupling between orbitals along a given bond direction is Ising-like, but with different Ising axes in the different bond directions. In the classical compass model, the competition between bonds in different directions leads to frustration and to a macroscopic degeneracy of states, from which thermal fluctuations may stabilize a "directional ordering of fluctuations," actually a nematic order [25]. In the quantum case, it is the zero-point fluctuations which stabilize the nematic order [26–28].

In this chapter, we will focus on the first class of systems introduced above, illustrating the emergence and properties of nematic phases by considering systems of small, quantum spins ($S = 1$ and $S = 1/2$).

## 13.2 Multipolar States of a Single Spin

We begin by addressing the question of whether there exist single-spin states which have no magnetic moment. To answer this, we investigate in more detail the spins as quantum mechanical objects, starting with the case of spin 1/2. The wave function of an $S = 1/2$ object has the general form $(a_\uparrow + ib_\uparrow)\,|\uparrow\rangle + (a_\downarrow + ib_\downarrow)\,|\downarrow\rangle$, with four real parameters. By fixing the overall phase and normalizing the wave function, the number of parameters is reduced to two, so that an arbitrary spin-1/2 wave function can be written conveniently as

$$|\hat{\Omega}\rangle = \cos\frac{\theta}{2}\mathrm{e}^{-\mathrm{i}\varphi/2}\,|\uparrow\rangle + \sin\frac{\theta}{2}\mathrm{e}^{\mathrm{i}\varphi/2}\,|\downarrow\rangle.$$

This wave function has particular properties: it satisfies the equation $(\hat{\mathbf{\Omega}}\cdot\mathbf{S})|\hat{\Omega}\rangle = S|\hat{\Omega}\rangle$ and describes a spin oriented in the direction of the unit vector

$$\hat{\mathbf{\Omega}} = \begin{pmatrix} \sin\vartheta\cos\varphi \\ \sin\vartheta\sin\varphi \\ \cos\vartheta \end{pmatrix}.$$

This is called a spin coherent state. Any $S = 1/2$ state can be obtained by rotating the state $|\uparrow\rangle$, and will always break time-reversal invariance.[1]

For an $S = 1$ spin, the number of free parameters is $6-2 = 4$: while the number of constraints remains two, the complex amplitudes of each of the $S^z = -1, 0$ and 1 states (which we denote by $|\bar{1}\rangle$, $|0\rangle$ and $|1\rangle$, respectively) give six parameters. The four free parameters allow a greater flexibility than in the case of $S = 1/2$ spins: in this case, the spin coherent states

$$|\hat{\Omega}\rangle = \frac{1+\cos\vartheta}{2}\mathrm{e}^{-\mathrm{i}\varphi}|1\rangle + \frac{\sin\vartheta}{\sqrt{2}}|0\rangle + \frac{1-\cos\vartheta}{2}\mathrm{e}^{\mathrm{i}\varphi}|\bar{1}\rangle \tag{13.1}$$

do not exhaust all the possibilities. For example, the state $|0\rangle$ cannot be obtained by a rotation of the state $|1\rangle$. In fact, in the state $|0\rangle$ the spin has no orientation at all: $\langle 0|S^\alpha|0\rangle = 0$. Nevertheless, the $O(3)$ rotational symmetry of the spin space is broken, because $\langle 0|(S^z)^2|0\rangle = 0$ and $\langle 0|(S^x)^2|0\rangle = \langle 0|(S^y)^2|0\rangle = 1$, which can be understood as describing a spin that "fluctuates" in the plane perpendicular to the $z$-axis. Furthermore, the spin coherent state is (up to a phase factor) invariant under the time-reversal operation.[2] As a consequence, the nonmagnetic state $|0\rangle$ is known as a quadrupolar state with a director[3] parallel to the $z$-axis.

By using the completeness relation $\hat{1} = 3\int \mathrm{d}\hat{\mathbf{\Omega}}|\hat{\Omega}\rangle\langle\hat{\Omega}|$, where $\mathrm{d}\hat{\mathbf{\Omega}} = \mathrm{d}\cos\vartheta\mathrm{d}\varphi/4\pi)$, an arbitrary spin-1 state $|\Psi\rangle$ can be expressed as a superposition of spin coherent states with amplitude $\langle\hat{\Omega}|\Psi\rangle$, because $|\Psi\rangle = \hat{1}|\Psi\rangle = 3\int \mathrm{d}\hat{\mathbf{\Omega}}|\hat{\Omega}\rangle\langle\hat{\Omega}|\Psi\rangle$. As examples,

---

[1] The action of the antiunitary time-reversal operator is $\mathcal{T}|\uparrow\rangle = |\downarrow\rangle$, $\mathcal{T}|\downarrow\rangle = -|\uparrow\rangle$. In addition, $\mathcal{T}$ changes the value of the wave function amplitude to its complex conjugate.

[2] $\mathcal{T}$ changes the sign of $|0\rangle$ and interchanges $|1\rangle$ and $|\bar{1}\rangle$: $\mathcal{T}|1\rangle = |\bar{1}\rangle$, $\mathcal{T}|0\rangle = -|0\rangle$, and $\mathcal{T}|\bar{1}\rangle = |1\rangle$, with complex conjugation of the associated amplitudes.

[3] The director is also used in the description of a nematic liquid crystal.

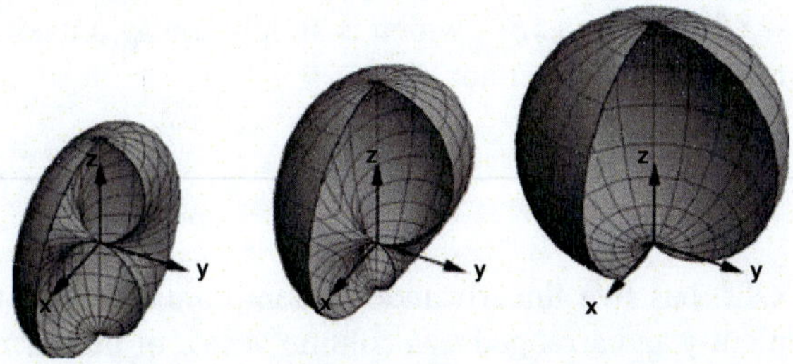


**Fig. 13.1** Representation of spin-component distributions for the states (from left to right) $|\langle\hat{\boldsymbol{\Omega}}|y\rangle|^2$ (quadrupole), $|\langle\hat{\boldsymbol{\Omega}}|(\cos\frac{\pi}{12}|y\rangle - \mathrm{i}\sin\frac{\pi}{12}|x\rangle)|^2$ (*half-polarized state*, $m = 1/2$), and $|\langle\hat{\boldsymbol{\Omega}}|1\rangle|^2$ (*fully polarized spin*)

$$|0\rangle = 3\int \mathrm{d}\hat{\boldsymbol{\Omega}}\frac{\sin\vartheta}{\sqrt{2}}|\hat{\Omega}\rangle \quad \text{and} \quad |1\rangle = 3\int \mathrm{d}\hat{\boldsymbol{\Omega}}\mathrm{e}^{\mathrm{i}\varphi}\frac{1+\cos\vartheta}{2}|\hat{\Omega}\rangle\,. \tag{13.2}$$

expressions which are very useful to visualize the nature of the spin state [8, 29]. Illustrations are shown in Fig. 13.1.

To description of quadrupolar states for $S = 1$ spins is facilitated by choosing the time-reversal-invariant basis

$$|x\rangle = \mathrm{i}\frac{|1\rangle - |\bar{1}\rangle}{\sqrt{2}}, \quad |y\rangle = \frac{|1\rangle + |\bar{1}\rangle}{\sqrt{2}}, \quad |z\rangle = -\mathrm{i}|0\rangle, \tag{13.3}$$

in which the director of $|x\rangle$ is parallel to the $x$-axis, $|y\rangle$ to $y$, and $|z\rangle$ to $z$. A linear combination $|\mathbf{d}\rangle = \sum_\alpha d_\alpha|\alpha\rangle$ of these basis states with real coefficients $d_\alpha$ is also a quadrupolar state, with a director parallel to $(d_x, d_y, d_z)$ in the three-dimensional space. Such states also satisfy $(\mathbf{d}\cdot\mathbf{S})^2\,|\mathbf{d}\rangle = 0$, which is the mathematical version of the statement that there are no spin fluctuations in the direction $\mathbf{d}$. In fact an arbitrary spin state may be expressed in this basis by allowing for complex vectors $\mathbf{d} = \mathbf{u} + \mathrm{i}\mathbf{v}$, with $u^2 + v^2 = 1$ (normalization, where $u = |\mathbf{u}|$ and $v = |\mathbf{v}|$) and $\mathbf{u}\cdot\mathbf{v} = 0$ (overall phase-fixing) [3, 30]. The spin operator in the basis (13.3) is

$$S^\alpha = -\mathrm{i}\sum_{\beta,\gamma}\varepsilon_{\alpha\beta\gamma}|\beta\rangle\langle\gamma|, \tag{13.4}$$

and the expectation value is $\langle\mathbf{d}|\mathbf{S}|\mathbf{d}\rangle = 2\mathbf{u}\times\mathbf{v}$, which has squared amplitude $|\langle\mathbf{d}|\mathbf{S}|\mathbf{d}\rangle|^2 = 4u^2v^2$. The complex amplitudes encode the possibility of time-reversal-breaking (magnetic) states, for example $|1\rangle = (|y\rangle - \mathrm{i}|x\rangle)/\sqrt{2}$.

One must then ask what type of operator can be used to detect such a state. Because a quadrupole does not break time-reversal symmetry, it must be a product of an even number of spin operators. In fact, the $3\times 3 = 9$-component quadratic form $S^\alpha S^\beta$ can, for any $S \geq 1$,[4] be decomposed into a scalar (the term $\mathrm{Tr}\, S^\alpha S^\beta = S[S+1]$), a three-component vector operator (the antisymmetric part $S^\alpha S^\beta$

[4] For $S = 1/2$ spins, $Q^{\alpha\beta} = 0$.

$-S^\beta S^\alpha \propto \mathrm{i}\varepsilon_{\alpha\beta\gamma} S^\gamma$, which is simply the spin itself), and a symmetric and traceless rank-2 tensor operator

$$Q^{\alpha\beta} = S^\alpha S^\beta + S^\beta S^\alpha - \frac{2}{3} S(S+1)\delta_{\alpha\beta}. \tag{13.5}$$

$Q^{\alpha\beta}$ has five linearly independent components,[5] which can be combined conveniently and arranged as a column vector of the form

$$\hat{\mathbf{Q}} = \begin{pmatrix} \hat{Q}^{x^2-y^2} \\ \hat{Q}^{3z^2-r^2} \\ \hat{Q}^{xy} \\ \hat{Q}^{yz} \\ \hat{Q}^{xz} \end{pmatrix} = \begin{pmatrix} (S^x)^2 - (S^y)^2 \\ \frac{1}{\sqrt{3}}\left[3(S^z)^2 - S(S+1)\right] \\ S^x S^y + S^y S^x \\ S^y S^z + S^z S^y \\ S^x S^z + S^z S^x \end{pmatrix}. \tag{13.6}$$

The quantity $\sum_{\alpha,\beta} Q^{\alpha\beta} Q^{\alpha\beta} = 2\hat{\mathbf{Q}} \cdot \hat{\mathbf{Q}}$ is a scalar. In the case of $S = 1$ spins,

$$Q^{\alpha\beta} = \frac{2}{3}\delta_{\alpha\beta} - |\beta\rangle\langle\alpha| - |\alpha\rangle\langle\beta| \tag{13.7}$$

in the basis of (13.3). The expectation values $\langle \mathbf{d}|Q^{\alpha\beta}|\mathbf{d}\rangle$ are given by $\frac{2}{3}\delta_{\alpha\beta} - 2u_\alpha u_\beta - 2v_\alpha v_\beta$ and the sum of the amplitudes of the spin and of the quadrupole is $|\langle \mathbf{d}|\mathbf{S}|\mathbf{d}\rangle|^2 + |\langle \mathbf{d}|\mathbf{Q}|\mathbf{d}\rangle|^2 = 4/3$.

For higher spins, $S > 1$, the space of possible on-site order parameters becomes still larger. Local order parameters of this type include multipolar states of degrees $k$ up to $2S$ ($k = 0$ is a scalar, 1 a dipole (vector), 2 a quadrupole, 3 an octupole, and so on), i.e. these order parameters are rank-$k$ tensor operators. Even-$k$ cases are time-reversal invariant, while odd-$k$ cases break time-reversal symmetry.

## 13.3 Competition Between Dipoles and Quadrupoles

### *13.3.1 The Bilinear–Biquadratic Model*

The minimal model to describe the competition between magnetic and quadrupolar degrees of freedom is the bilinear–biquadratic Hamiltonian

---

[5] The $2k+1$ components of rank-$k$ tensor operator $\mathbf{T}^{(k)}$ satisfy

$$\left[J^z, T_q^{(k)}\right] = q T_q^{(k)} \quad \text{and} \quad \left[J^\pm, T_q^{(k)}\right] = \sqrt{k(k+1) - q(q\pm 1)}\, T_{q\pm 1}^{(k)}.$$

For quadrupoles, $T_q^{(2)} = Q^q$, whence $Q^{++} = S^+S^+$, $Q^+ = -(S^+S^z + S^zS^+)$, $Q^0 = \sqrt{2/3}\left[3(S^z)^2 - S(S+1)\right]$ and similarly.

$$\mathcal{H} = \sum_{\langle i,j \rangle} \left[ J_1 \, \mathbf{S}_i \cdot \mathbf{S}_j + J_2 \left( \mathbf{S}_i \cdot \mathbf{S}_j \right)^2 \right]. \tag{13.8}$$

Occasionally we adopt the single parameter notation $J_1 = J \cos \vartheta$ and $J_2 = J \sin \vartheta$, which is common in the literature. Because the simplest isotropic interaction between quadrupoles takes the form

$$\hat{\mathbf{Q}}_i \cdot \hat{\mathbf{Q}}_j = 2 \left( \hat{\mathbf{S}}_i \cdot \hat{\mathbf{S}}_j \right)^2 + \hat{\mathbf{S}}_i \cdot \hat{\mathbf{S}}_j - \frac{2}{3} S^2 (S+1)^2, \tag{13.9}$$

an expression valid for arbitrary $S$, the Hamiltonian of (13.8) is in fact

$$\mathcal{H} = \sum_{\langle i,j \rangle} \left[ \left( J_1 - \frac{J_2}{2} \right) \mathbf{S}_i \cdot \mathbf{S}_j + \frac{J_2}{2} \mathbf{Q}_i \cdot \mathbf{Q}_j + \frac{J_2}{3} S^2 (S+1)^2 \right]. \tag{13.10}$$

The conventional antiferromagnetic Heisenberg model is recovered at $\vartheta = 0$, while for $\vartheta = \pi$ one obtains its ferromagnetic analog. In both of these limits the spins order: in two dimensions, antiferromagnetic long-range order of the two- (square lattice) or three-sublattice (triangular lattice) type is realized for simple lattices when $\vartheta = 0$, and ferromagnetic order occurs for $\vartheta = \pi$. Between these magnetic phases one may expect on the basis of the previous section to find quadrupolar ordering once the $\mathbf{Q}_i \cdot \mathbf{Q}_j$ interactions become dominant, and this is indeed the case.

For $S = 1$ spins, the case we discuss in detail here, one has a remarkable (although not unexpected) identity

$$\hat{\mathbf{Q}}_i \cdot \hat{\mathbf{Q}}_j + \hat{\mathbf{S}}_i \cdot \hat{\mathbf{S}}_j = 2\mathcal{P}_{i,j} - \frac{2}{3}, \tag{13.11}$$

where $\mathcal{P}_{i,j}$ is a permutation operator interchanging the spin configurations at sites $i$ and $j$. Because $\mathcal{P}_{i,j}$ does not distinguish between the states of a spin-1 entity, the symmetry of the term $\hat{\mathbf{Q}}_i \cdot \hat{\mathbf{Q}}_j + \hat{\mathbf{S}}_i \cdot \hat{\mathbf{S}}_j$ is increased from SU(2) to SU(3), and the three $S = 1$ states form the three-dimensional fundamental irreducible representation of the SU(3) algebra. For the Hamiltonian (13.8), this corresponds to $\vartheta = \pi/4$ or $-3\pi/4$. The SU(3) points, as we will show, are of particular significance in the phase diagram. The SU(3) symmetry can also be exploited in order to extend spin-wave theory to describe the quadrupolar degrees of freedom. To demonstrate the SU(3) symmetry most clearly, the Hamiltonian can be written as

$$\mathcal{H} = \sum_{\langle i,j \rangle} \left[ (J_1 - J_2) \mathbf{S}_i \cdot \mathbf{S}_j + J_2 \left( \mathbf{1} + \mathcal{P}_{i,j} \right) \right], \tag{13.12}$$

which is a type of "anisotropic" SU(3) Heisenberg model.

Here, we remark also that on bipartite lattices there are additional SU(3)-symmetric points at $\vartheta = \pm\pi/2$. These correspond to choosing the fundamental representation **3** (the quark) on one of the sublattices and the antifundamental representation

$\bar{\mathbf{3}}$ (antiquark) on the other sublattice [31,32]. In the language of Young tableaus, the fundamental representation **3** is a single box[6] and is denoted as (1,0). The antifundamental representation $\bar{\mathbf{3}}$ corresponds to a column of 2 boxes and is denoted by (0,1). On two sites, the product $(1,0) \otimes (0,1) = (0,0) \oplus (1,1)$ contains the $(0,0)$ singlet, which is the statement that one may have a singlet valence bond composed of two spins. By contrast, to make an SU(3) singlet one must add up three spins belonging to the fundamental representation.

### 13.3.2 Energy Spectra of Small Clusters

Here, we discuss the spectra of a two-site bond and a three-site triangle, in order to gain some insight into the model and to understand certain features, such as the crossing of the energy levels. For these purposes it is convenient to use the Hamiltonian (13.12). A straightforward numerical diagonalization produces the spectra shown in Fig. 13.2.

Both the bond and the triangle are called complete graphs, where all sites are connected to all others. For such graphs there is an algebraic identity

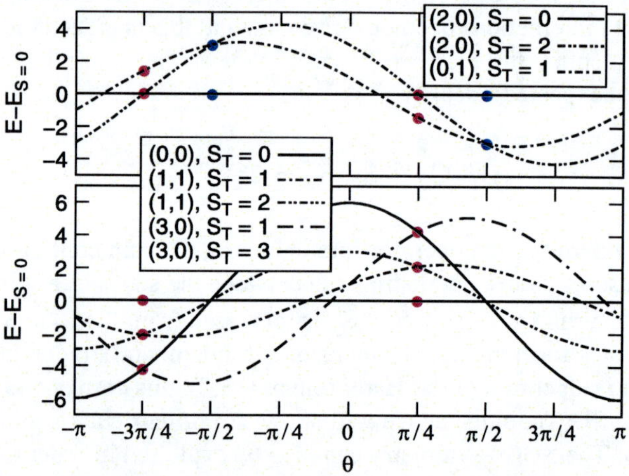


**Fig. 13.2** Energy spectra of a bond (*upper panel*) and a triangle (*lower panel*). Energy levels are labeled by the Young tableaus $(n_1, n_2)$ and by the total SU(2) spin $S_T$. Solid circles denote the degeneracies arising at the SU(3)-symmetric points

[6] A Young tableau – an array of boxes arranged in left-justified rows – is used to label the irreducible representations (or multiplets) of SU(n) algebras. The notation $(n_1, n_2)$ refers to the number of boxes in each row which extend beyond the row below. Alternatively stated, if a Young tableau has $\{m_1, m_2, m_3\}$ boxes in the first, second, and third rows, $(n_1, n_2) = (m_1 - m_2, m_2 - m_3)$.

$$\sum_{\langle i,j\rangle} \mathbf{S}_i \cdot \mathbf{S}_j = \frac{1}{2}\Big(\sum_i \mathbf{S}_i\Big)^2 - \frac{L}{2} S(S+1) = \frac{1}{2} S_T(S_T+1) - L, \tag{13.13}$$

where $L$ is the number of sites and $S_T$ is the total spin. Similarly, for the SU(3) part of the Hamiltonian,

$$\sum_{\langle i,j\rangle} \boldsymbol{\lambda}_i \cdot \boldsymbol{\lambda}_j = \frac{1}{2}\Big(\sum_i \boldsymbol{\lambda}_i\Big)^2 - \frac{1}{2}\sum_i \boldsymbol{\lambda}_i^2 = 2C_{(n_1,n_2)} - 2LC_{(1,0)}. \tag{13.14}$$

The components of $\boldsymbol{\lambda}$ are the 8 generators of the SU(3) algebra (the $3 \times 3$ Gell–Mann matrices), so that $\boldsymbol{\lambda}_i \cdot \boldsymbol{\lambda}_j = 2\mathcal{P}_{i,j} - \frac{2}{3}$. $C_{(n_1,n_2)}$, which plays a role similar to $S_T(S_T+1)$, is the eigenvalue of the quadratic Casimir operator of the SU(3) algebra for an $(n_1, n_2)$ Young tableau,

$$C_{(n_1,n_2)} = n_1 + n_2 + \frac{1}{3}\left(n_1^2 + n_2 n_1 + n_2^2\right). \tag{13.15}$$

Because for complete graphs the SU(2) and SU(3) parts of the Hamiltonian commute with each other, both $(n_1, n_2)$ and $S_T$ are good quantum numbers and can be used to label the states. The energy of such an $|(n_1,n_2), S_T\rangle$ state is the sum of the SU(2) and SU(3) terms, and is given by

$$E = \left[C_{(n_1,n_2)} - \frac{4}{3}L\right] J_2 + \left[\frac{S_T(S_T+1)}{2} - L\right](J_1 - J_2) + \frac{4}{3}\frac{L(L-1)}{2} J_2 \tag{13.16}$$

for a complete $L$-site graph (this includes the bond, the triangle, the tetrahedron, and higher clusters). We refer to Table 13.1 for more details of the spectrum.

The SU(3)-multiplet content of a bond can be obtained by applying the branching rules of the SU(3) algebra: $(1,0) \otimes (1,0) = (0,1) \oplus (2,0)$. These multiplets are in fact also closely related to the geometrical symmetry of the clusters. Specifically, the point group of the bond is $D_2$, which is homomorphic with the symmetric group $S_2$, and has a symmetric ($\mathsf{A}_1$) and an antisymmetric ($\mathsf{A}_2$) irreducible representation. Out of $3 \times 3 = 9$ spin states, three transform according to $\mathsf{A}_2$, $|10\rangle - |01\rangle$, $|\bar{1}0\rangle - |0\bar{1}\rangle$, and $|1\bar{1}\rangle - |\bar{1}1\rangle$, and these are denoted by the $(0,1)$ Young tableaus. These states also have $S_T = 1$, so are denoted as $|(n_1,n_2), S_T\rangle = |(0,1), 1\rangle$. The remaining six states, $|11\rangle$, $|00\rangle$, $|\bar{1}\bar{1}\rangle$, $|10\rangle + |01\rangle$, $|\bar{1}0\rangle + |0\bar{1}\rangle$, and $|1\bar{1}\rangle + |\bar{1}1\rangle$, form a symmetric combination of the $\mathsf{A}_1$ irreducible representation and are denoted by the $(2,0)$ Young tableau (two boxes making a row). This SU(3) multiplet contains the SU(2) singlet $|1\bar{1}\rangle + |\bar{1}1\rangle - |00\rangle$, so in the notation introduced above is the state $|(2,0), 0\rangle$. The remaining five states make up an SU(2) quintuplet, $|(2,0), 2\rangle$. The energetic contribution from the SU(3)-symmetric part of the Hamiltonian depends only on the Young tableau, while the SU(2) "anisotropy" lifts the degeneracy within the SU(3) multiplet. This can be seen in Fig. 13.2 close to the points $\vartheta = \pi/4$ and $-3\pi/4$.

**Table 13.1** Energy spectra of the two-site bond using the **3** ⊗ **3** (top table) and the **3** ⊗ $\bar{\mathbf{3}}$ decompositions (middle table), and of the triangle (bottom table)

| Site irreps | Young tableau | $(n_1, n_2)$ | $C_{(n_1,n_2)}$ | $G$ | $S_T$ | $E$ | Degeneracy |
|---|---|---|---|---|---|---|---|
| | | Two sites, **3** ⊗ **3** decomposition, point group $G = D_2$ | | | | | |
| □⊗□ | □<br>□ | (0, 1) | 4/3 | A₂ | 1 | $-J_1 + J_2$ | 3 |
| | □□ | (2, 0) | 10/3 | A₁ | 0 | $-2J_1 + 4J_2$ | 1 |
| | | | | | 2 | $J_1 + J_2$ | 5 |
| | | Two sites, **3** ⊗ $\bar{\mathbf{3}}$ decomposition | | | | | |
| □⊗□<br>□ | □<br>□<br>□ | (0, 0) | 0 | – | 0 | $-2J_1 + 4J_2$ | 1 |
| | □□<br>□ | (1, 1) | 3 | – | 1 | $-J_1 + J_2$ | 3 |
| | | | | | 2 | $J_1 + J_2$ | 5 |
| | | Three sites, **3** ⊗ **3** ⊗ **3** decomposition, point group $G = D_3$ | | | | | |
| □⊗□⊗□ | □<br>□<br>□ | (0, 0) | 0 | A₂ | 0 | $-3J_1 + 3J_2$ | 1 |
| | □□<br>□ | (1, 1) | 3 | E | 1 | $-2J_1 + 5J_2$ | 6 |
| | | | | | 2 | $3J_2$ | 10 |
| | □□□ | (3, 0) | 6 | A₁ | 1 | $-2J_1 + 8J_2$ | 3 |
| | | | | | 3 | $3J_1 + 3J_2$ | 7 |

Similar considerations can be applied to the triangle: $(1,0) \otimes (1,0) \otimes (1,0) = (0,0) \oplus (1,1) \oplus (1,1) \oplus (3,0)$ (Table 1.1). The $D_3$ point group of the triangle is homomorphic with $S_3$, and one may also deduce the SU(2)-multiplet content of a given Young tableau. Here, we note only that the state $(0,0)$ (three boxes in a column) is the SU(3) singlet mentioned above.

## 13.4 Quadrupolar Ordering in $S = 1$ Systems

In this section, we review the known results for the square and triangular lattices. A detailed study of the square lattice was undertaken by Papanicolaou [33]. The triangular lattice has been discussed more recently, in [6] and [8]. Because these are two-dimensional systems, continuous symmetry-breaking occurs only at zero temperature. Here, we use these systems also as test cases on which to discuss the different analytical and numerical methods which can be applied.

### *13.4.1 Variational Phase Diagram*

To obtain some initial insight into the phase diagram, we assume an $n_\Lambda$-sublattice, long-range-ordered state ($n_\Lambda = 2$ for the square and $n_\Lambda = 3$ for the triangular lattice), which we describe using a simple, site-factorized variational wave function of the form

$$|\Psi\rangle = \prod_{j=1}^{n_\Lambda} \prod_{i \in \Lambda_j} |\mathbf{d}_j\rangle_i, \tag{13.17}$$

where $i$ is a site index, $j$ is a sublattice index, and $\Lambda_j$ is the set of sites belonging to the $j$th sublattice. In (13.17) we assume that the on-site wave function is identical for all spins on the same sublattice. The problem reduces to a minimization of $\langle\Psi|\mathcal{H}|\Psi\rangle$ with respect to $4n_\Lambda$ real parameters (4 for each sublattice). The expectation values used in the evaluation of $\langle\Psi|\mathcal{H}|\Psi\rangle$ are

$$\langle\mathbf{d}_i|\mathbf{S}|\mathbf{d}_i\rangle \cdot \langle\mathbf{d}_j|\mathbf{S}|\mathbf{d}_j\rangle = \left|\langle\mathbf{d}_i|\mathbf{d}_j\rangle\right|^2 - \left|\langle\mathbf{d}_i|\bar{\mathbf{d}}_j\rangle\right|^2, \tag{13.18}$$

$$\langle\mathbf{d}_i|\mathbf{Q}|\mathbf{d}_i\rangle \cdot \langle\mathbf{d}_j|\mathbf{Q}|\mathbf{d}_j\rangle = \left|\langle\mathbf{d}_i|\mathbf{d}_j\rangle\right|^2 + \left|\langle\mathbf{d}_i|\bar{\mathbf{d}}_j\rangle\right|^2 - \frac{2}{3}. \tag{13.19}$$

The variational phase diagrams obtained for the square and triangular lattices are shown in Fig. 13.3, and below we discuss in detail the different phases. First, however, let us contrast the Heisenberg and the quadrupolar exchange interactions. The term $\mathbf{S}_i \cdot \mathbf{S}_j$ is maximal when the two spins are parallel and minimal when they are antiparallel. The situation is different for quadrupoles: the expectation value of $\hat{\mathbf{Q}}_i \cdot \hat{\mathbf{Q}}_j$ in the site-factorized wave-function subspace of the pure (both $\mathbf{d}_i$ and $\mathbf{d}_j$ real) quadrupolar states is ($i \neq j$)

$$\langle\hat{\mathbf{Q}}_i \cdot \hat{\mathbf{Q}}_j\rangle = 2(\mathbf{d}_i \cdot \mathbf{d}_j)^2 - \frac{2}{3}, \tag{13.20}$$

which is thus maximal for collinear directors ($\mathbf{d}_i \| \mathbf{d}_j$) and minimal for perpendicular ones ($\mathbf{d}_i \perp \mathbf{d}_j$).

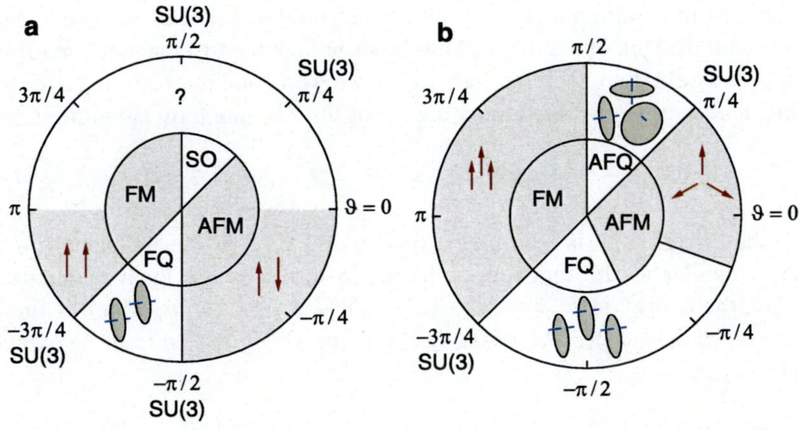


**Fig. 13.3** Phase diagram of the $S = 1$ bilinear–biquadratic model on (**a**) the square and (**b**) the triangular lattice. The inner circle shows the variational, the outer the numerical results. FM and AFM: ferro- and antiferromagnetic; FQ and AFQ: ferro- and antiferroquadrupolar; SO stands for "semi-ordered" and "?" denotes the region for which no numerical results are available. Further details are presented in the text

#### 13.4.1.1 Ferromagnetic and Ferroquadrupolar Phases

We begin by considering the case when the wave functions on each of the sublattices are equal, $\mathbf{d}_j = \mathbf{d} = \mathbf{u}+i\mathbf{v}$, i.e. the ordering wave vector is $\mathbf{q} = 0$. Then $\langle \mathbf{Q}_i \cdot \mathbf{Q}_j \rangle = 4/3 - 4u^2v^2$ and $\langle \mathbf{S}_i \cdot \mathbf{S}_j \rangle = 4u^2v^2$, and the energy per site is

$$E = zJ_2 + 2z(J_1 - J_2)u^2v^2. \tag{13.21}$$

The minimum depends on the sign of $J_1 - J_2$: for $J_1 - J_2 < 0$ (more precisely, $\pi/2 < \vartheta < 5\pi/4$), one obtains $u = v = 1/\sqrt{2}$, which is simply the ferromagnetic state with time-reversal-breaking order parameter $\mathbf{S} = 2\mathbf{u} \times \mathbf{v}$. When $J_1 - J_2 > 0$, the minimum is achieved with $u = 1$ and $v = 0$, i.e. $\mathbf{d}$ is real. This describes a nonmagnetic, ferroquadrupolar phase, whose director is oriented in the arbitrary direction $\mathbf{u}$ (an alternative solution is $u = 0$ and $v = 1$, with director $\mathbf{v}$). The actual extent of this phase is $-3\pi/4 < \vartheta < -\pi/2$ for the square lattice and $-3\pi/4 < \vartheta < \Theta$ for the triangular lattice, where $\Theta = \arctan(-2) \approx -0.352\pi$.

It is interesting to note the nature of the quantum phase transition between ferromagnetic and ferroquadrupolar order at the SU(3)-symmetric point ($\vartheta = -3\pi/4$). At this single point, the high symmetry enlarges the degeneracy of the ground state, so that both the ferromagnetic and the ferroquadrupolar state are among the ground states [34]. Moving $\vartheta$ away from this point selects one of these ordered states, with a jump in order parameter, giving the appearance of a first-order phase transition, albeit with no hysteresis (which accompanies a true first-order phase transition). Because of the increased symmetry group at the SU(3) point, this phase transition lies beyond the framework of the Landau theory of phase transitions. The same situation arises also at the other SU(3) point, $\vartheta = \pi/4$.

Let us now consider the effect of the magnetic field on the quadrupolar state. We assume that the magnetic field is parallel to the $z$-axis, and add the Zeeman term, $-h\sum_i S_i^z$, to the Hamiltonian. The pure quadrupolar state has vanishing matrix elements with the spin operators, and thus does not couple to a magnetic field. However, once the field is applied, the state $|\mathbf{d}\rangle$ may deform and may develop a magnetic moment. The energy (13.21) in the presence of the Zeeman term is modified to

$$E = zJ_2 + 2z(J_1 - J_2)u^2v^2 - 2h(u_x v_y - u_y v_x), \tag{13.22}$$

whence the energy gain is maximal if both $\mathbf{u}$ and $\mathbf{v}$ lie in the $xy$-plane (i.e. they are perpendicular to the magnetic field $\mathbf{h}$). In terms of the local magnetization $m = 2uv$, the energy is $E = zJ_2 + (z/2)(J_1 - J_2)m^2 - hm$, which is minimal for $m = h/(z(J_1 - J_2))$. The resulting state (see also Fig. 13.1) can be expressed conveniently as

$$|\mathbf{d}\rangle = \cos\frac{\mu}{2}\left(\cos\phi|x\rangle + \sin\phi|y\rangle\right) + i\sin\frac{\mu}{2}\left(-\sin\phi|x\rangle + \cos\phi|y\rangle\right), \tag{13.23}$$

with $u = \cos(\mu/2)$ and $v = \sin(\mu/2)$, so that $m = \sin\mu$, while $u_x = u\cos\phi$ and $u_y = u\sin\phi$. Thus, the director turns perpendicular to the applied magnetic field [30], and the quadrupolar state develops a magnetic moment both linearly proportional to and parallel to the field by shifting the center point of its fluctuations. Time-reversal symmetry is broken by a linear combination of quadrupolar states with complex amplitudes. The broken continuous O(2) symmetry associated with the angle $\phi$ gives rise to a Goldstone mode (Sect. 13.4.3). With increasing magnetic field, the magnetic moment of the ferroquadrupolar phase saturates and then goes over continuously (second-order phase transition) to the ferromagnetic phase ($\mu = \pi/2$) at $h_{c2}^{\mathrm{var}} = z(J_1 - J_2)$.

#### 13.4.1.2 Antiferroquadrupolar Phase on the Square Lattice

The variational solution for the square lattice was discussed by Papanicolaou in [33]. For $-\pi/2 \le \vartheta \le \pi/4$, a two-sublattice Néel antiferromagnet is realized, with fully developed spins. More complex is the region $\pi/4 \le \vartheta \le \pi/2$, where the small coordination number of the lattice leads to a highly degenerate "semi-ordered" phase. Assuming first that the ground state contains only pure quadrupoles, the requirement that the directors be perpendicular is not sufficient to select an ordered state, but leaves a highly degenerate ground-state manifold. Further, the term $\hat{\mathbf{Q}}_i \cdot \hat{\mathbf{Q}}_j$ is not minimized only when the directors are perpendicular, but also by having a pure quadrupole on one of the sites and a partly or fully magnetized spin on the other, so that the director and the magnetization are parallel: one example is $|\mathbf{d}_i\rangle = |z\rangle$ and $|\mathbf{d}_j\rangle = \cos(\mu/2)|x\rangle + \mathrm{i}\sin(\mu/2)|y\rangle$, a combination with a finite magnetic moment along $z$ (13.23). At the SU(3) point $\pi/4 = \vartheta$, these states can again be rotated into the Néel state within the manifold $\langle\mathbf{d}_i|\mathbf{d}_j\rangle = \delta_{i,j}$. Clearly, this large degeneracy at the variational level is lifted by quantum effects, the most likely scenario being that zero-point fluctuations select some specific types of ordered state out of this manifold.

We also stress here that the variational wave function does allow us to capture some of the quantum mechanical nature of the ordered phases. The quadrupolar states as presented above do not exist for classical spins, where instead the competition of the bilinear and biquadratic interactions leads to an angle $\alpha$ different from $\pi$ between the spins on neighboring sites for $J_2 > J_1/2$; $\alpha$ is obtained by minimizing the bond energy $J_1\cos\alpha + J_2\cos^2\alpha$. With $\alpha \ne \pi$, the state of classical spins is degenerate, and, unless a spiral order with wave vector $\mathbf{q} \propto (1,1)$ sets in, again one is faced with some type of disordered state

#### 13.4.1.3 Three-Sublattice Phases on the Triangular Lattice

In a triangular geometry, when $\Theta < \vartheta < \pi/4$, a parameter range which includes the pure Heisenberg model, the result of the minimization procedure is a wave function describing a system of partially developed magnetic moments on three different sublattices, whose orientations all differ by the angle $2\pi/3$,

$$\begin{aligned}
|\mathbf{d}_1\rangle &= \cos\frac{\eta}{2}|z\rangle - \mathrm{i}\sin\frac{\eta}{2}|y\rangle, \\
|\mathbf{d}_2\rangle &= \cos\frac{\eta}{2}|z\rangle - \mathrm{i}\sin\frac{\eta}{2}\left(\frac{\sqrt{3}}{2}|x\rangle - \frac{1}{2}|y\rangle\right), \\
|\mathbf{d}_3\rangle &= \cos\frac{\eta}{2}|z\rangle - \mathrm{i}\sin\frac{\eta}{2}\left(-\frac{\sqrt{3}}{2}|x\rangle - \frac{1}{2}|y\rangle\right).
\end{aligned} \tag{13.24}$$

Here, we have taken the moments to lie in the $xy$-plane, while the parameter $\eta$ depends on the ratio $J_2/J_1$, $\cos\eta = -3J_2/(8J_1 + J_2)$. In this state, the nearest-neighbor spin-spin correlation functions are equal on all bonds, as also are the quadrupole-quadrupole correlations,

$$\langle \mathbf{S}_i \cdot \mathbf{S}_j\rangle = -\frac{1}{2}\sin^2\eta \quad \text{and} \quad \langle \mathbf{Q}_i \cdot \mathbf{Q}_j\rangle = -\frac{1}{24} + \frac{3}{4}\cos\eta + \frac{5}{8}\cos^2\eta; \tag{13.25}$$

the amplitude of the ordered spin moment is $|\langle\mathbf{S}\rangle| = \sin\eta$ (here $i$ and $j$ belong to different sublattices). For the pure Heisenberg model, $\eta = \pi/2$ and $|\langle\mathbf{S}\rangle| = 1$. On moving away from $J_2 = 0$ point in either direction in the parameter range, the moment decreases.

Approaching the boundary to the ferroquadrupolar phase, $\vartheta = \Theta$, $\eta \propto \sqrt{2J_1 - J_2}$ as $J_2 \to -2J_1$, while the spin moment decreases accordingly, $|\langle S\rangle| \propto \sqrt{2J_1 - J_2}$. The wave function evolves smoothly into the ferroquadrupolar wave function, $|\mathbf{d}_i\rangle = |0\rangle$, whose director is perpendicular to the plane spanned by the spins. Thus, the phase boundary between the ferroquadrupolar and the antiferromagnetic phase is of second order, with the spin order parameter vanishing at the phase boundary, while the quadrupolar order parameter persists across the boundary (but becomes coupled to the spin order parameter in the antiferromagnetic phase).

At $\vartheta = \pi/4$, we expect to find the three-sublattice SU(3) antiferromagnetic phase. The variational wave function (13.24) reproduces such a state: at the SU(3)-symmetric point, $\cos\eta = -1/3$ and $\langle\mathbf{d}_i|\mathbf{d}_j\rangle = \delta_{i,j}$. This state can be rotated continuously within the SU(3) antiferromagnetic manifold into a state where all three vectors $\mathbf{d}$ are real and mutually perpendicular. This three-sublattice, antiferroquadrupolar phase gives the minimum energy for $\pi/4 \le \vartheta \le \pi/2$, as may be deduced from (13.20). An example of such a state is $|\mathbf{d}_1\rangle = |x\rangle$, $|\mathbf{d}_2\rangle = |y\rangle$, and $|\mathbf{d}_3\rangle = |z\rangle$. Quadrupole moments are perfectly adapted to the geometry of the triangular lattice: while antiferromagnetic Heisenberg exchange frustrates spins on a triangle, quadrupoles can minimize $\hat{\mathbf{Q}}_i \cdot \hat{\mathbf{Q}}_j$ on each bond. Further, there is no degeneracy of the type encountered for the square lattice.

The ordering of the spins and the quadrupole moments is summarized in Fig. 13.4, where we show the quantities $S(\mathbf{k}) = \langle\mathbf{S}_\mathbf{k} \cdot \mathbf{S}_{-\mathbf{k}}\rangle$ and $Q(\mathbf{k}) = \langle\mathbf{Q}_\mathbf{k} \cdot \mathbf{Q}_{-\mathbf{k}}\rangle$,

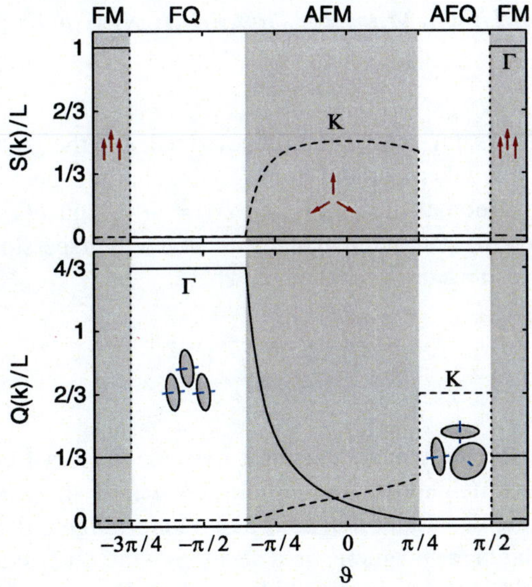


**Fig. 13.4** Spin ($S(\mathbf{k})$) and quadrupole ($Q(\mathbf{k})$) structure factors on the triangular lattice as a function of $\vartheta$, obtained from the variational calculation. The shaded regions represent magnetic ordering, either ferromagnetic (FM) or antiferromagnetic (AFM), while in the unshaded regions only quadrupolar order is realized (cf. Fig. 13.3b). $\Gamma$ is the center of the Brillouin zone [$\mathbf{k} = (0, 0)$] and the $K$ points are situated at the corners of the hexagonal Brillouin zone associated with the three-sublattice ordering [$\mathbf{k} = (4\pi/3, 0)$ and $\mathbf{k} = \left(2\pi/3, 2\pi/\sqrt{3}\right)$]. While the spin order parameters vanish in the quadrupolar phases, the quadrupolar ones may be finite in the magnetic phases

respectively the spin and quadrupole structure factors.[7] It is evident that, in the ferromagnetic phase, both the ferromagnetic and the ferroquadrupolar order parameters are finite, because the ferromagnetically aligned spins select an axis that breaks the O(3) symmetry. In the antiferromagnetic phase, both the ferro- and antiferroquadrupolar order parameters take finite values [using the variational wave function of (13.24), $Q(\mathbf{k})/L = (1 + 3\cos\eta)^2/12$ at the $\Gamma$ point and $(1 - \cos\eta)^2/8$ at the $K$ point, while $S(\mathbf{k}) = \sin^2(\eta)/2$ at the $K$-point]. The ferroquadrupolar order is associated with the plane spanned by the spins in the antiferromagnetic phase, while the antiferroquadrupolar order is associated with the breaking of $O(3)$ symmetry on the sublattices.

The phase diagram of the triangular-lattice model in a finite magnetic field is rather complicated [8], and will not be discussed here.

[7] Here, we have defined the Fourier transform as $A_{\mathbf{k}} = L^{-1/2}\sum_{j=1}^{L} A_j e^{i\mathbf{k}\cdot\mathbf{r}_j}$, where $\mathbf{r}_j$ is the coordinate of the $j$th lattice site and $L$ is the number of sites.

### 13.4.2 One- and Two-Magnon Instability of the Fully Polarized State

At high magnetic fields, the spins are fully polarized, and the ground state is simply $|1111\ldots\rangle$ (the field is taken parallel to $z$). The excitation spectrum is gapped, the gap of the single-magnon excitations being $\Delta_1 = h + \min \omega(\mathbf{k})$: the $S_k^-$ operator creates a delocalized $|0\rangle$ state with an energy cost $h$ and dispersion relation $\omega(\mathbf{k}) = zJ_1[\gamma(\mathbf{k}) - 1]$ due to Heisenberg exchange, where

$$\gamma(\mathbf{k}) = \frac{1}{z}\sum_{\boldsymbol{\delta}} e^{i\mathbf{k}\cdot\boldsymbol{\delta}} \tag{13.26}$$

is a geometrical factor,[8] in which the sum is over the vectors $\boldsymbol{\delta}$ connecting a given site with its $z$ nearest neighbors. As the field is lowered, the gap is closed, below which there is a finite density of magnons and a long-range order of the spin component perpendicular to the magnetic field develops – the $O(2)$ symmetry remaining after the introduction of the magnetic field is broken. For $J_1 > 0$, in the square lattice the dispersion is minimal at $\mathbf{k} = (\pi, \pi)$, and the critical field is $h_{c1} = 8J_1$. In the triangular lattice, the critical field is $h_{c1} = 9J_1$, with an instability towards the three-sublattice state. These results are exact in the case that the phase transition is continuous.

When the biquadratic term is present, bound states of magnons are formed if $J_2 < 0$. The simplest two-magnon state is $|\bar{1}\rangle$, and if one neglects the dissociation of the state $|\bar{1}\rangle$ into two dispersive $|0\rangle$ states, the energy gap is

$$\Delta_2(\mathbf{k}) = 2h - 2zJ_1 + zJ_2[\gamma(k) + 1]. \tag{13.27}$$

The gap closes at the $\Gamma$ point at the field $h_{c2}^{\text{var}} = z(J_1 - J_2)$, which is the same phase boundary as that found in the variational calculation in Sect. 13.4.1. The presence of magnon bound states implies either a second-order boundary to the ferroquadrupolar phase [35] or a first-order transition into a spin-canted state – a situation which can be clarified by examining whether the bound magnon pairs attract or repel each other.

However, the situation is more subtle than this: the state $|\bar{1}\rangle$ can dissociate into $|00\rangle$ with amplitude $J_1 - J_2$, so that even when the biquadratic term is dominant, the states $|\bar{1}\rangle$ are strongly coupled to and inseparable from the nearest-neighbor $|00\rangle$ states. Strictly speaking, the simple reasoning leading to (13.27) is valid only when

---

8 For the square lattice, $\gamma(\mathbf{k}) = \left(\cos k_x + \cos k_y\right)/2$, and for the triangular lattice,

$$\gamma(\mathbf{k}) = \frac{1}{3}\left(\cos k_x + \cos\frac{k_x + k_y\sqrt{3}}{2} + \cos\frac{k_x - k_y\sqrt{3}}{2}\right).$$

In the limit $k \to 0$, $1 - \gamma(\mathrm{k}) = \frac{1}{2z}\sum_{\boldsymbol{\delta}}(\mathbf{k}\cdot\boldsymbol{\delta})^2 = k^2/4$ for both lattices.

$J_1 = J_2$. Away from this point, the $|00\rangle$ states are mixed with $|\bar{1}\rangle$. Clearly, the two-body scattering problem is exactly solvable by analytical means, but its solution is tedious and is omitted here. We note only that for the triangular lattice, a bound state appears below the two-magnon continuum for $-5 < J_1/J_2 < 1$ (as long as $J_2 < 0$), and the ferroquadrupolar phase is slightly extended in comparison with the results of the variational calculation. A particularly simple result is obtained for $J_1 = 0$, where the magnons cannot propagate and the exact ground state is the $\mathbf{k} = 0$ singlet superposition $|1\bar{1}\rangle + |\bar{1}1\rangle - |00\rangle$ on one bond in a background of the $|1\rangle$ spins. The energy of this state is $2h + (2z+1)J_2$, and the critical field is $h_{c2} = -(z+1/2)J_2$. Compared to the variational result, $(h_{c2} - h_{c2}^{\text{var}})/h_{c2} = 1/(2z+1)$, the discrepancy is of order $1/z$. Thus one may demonstrate explicitly the validity of the variational result in the limit of large connectivity.

This concept, of Bose-condensing bound states of magnons or triplons by the application of a magnetic field as a means of stabilizing spin nematic phases, is also very valuable in a number of other systems. Chubukov discussed two-magnon bound-state formation in higher-spin chains [35], as well as in the frustrated, ferromagnetic $S = 1/2$ Heisenberg chain [36]. It was found more recently that even larger bound states involving up to four magnons are thermodynamically stable, and thus give rise to field-induced quadrupolar, octupolar, and even hexadecupolar critical correlations [37–42]. Two-magnon bound states with a $d$-wave form factor have been found for the square lattice with diagonal frustration (possibly also including ring-exchange terms) [43]. These bound states form the basis for a field-induced bond quadrupolar phase. Similarly, on the triangular lattice with nearest-neighbor interactions and ring-exchange terms, three-magnon bound states have been observed which condense into an octupolar phase at high magnetic fields.

A related bound-state scenario occurring at low fields has been put forward for the dimer phase of the Heisenberg model on the Shastry–Sutherland lattice [44, 45]. In this phase, the elementary triplon excitations are quite immobile. Bound states of two triplons, however, are much more mobile due to effective correlated hopping terms. This leads to the formation of $S = 2$ bound states which will condense before the triplons when the magnetic field is applied, making possible a field-induced quadrupolar phase. Such a scenario has been confirmed both at the mean-field level and numerically in a simplified effective model involving correlated hopping terms [46, 47].

### 13.4.3 *Spin-Wave Theory for the Ferroquadrupolar Phase*

The excitation spectrum of a quadrupolar phase may be treated by an appropriate extension of spin-wave theory. In the standard theory, the spins in the ordered magnetic state are assumed to be spin coherent states of amplitude $S$. To describe fluctuations about the ordered state, one introduces the Holstein–Primakoff bosons, which with $z$-axis quantization can be used to express the spin operators as $S^z = S - a^\dagger a$, $S^+ = \sqrt{2S - a^\dagger a}\,a$, and $S^- = a^\dagger \sqrt{2S - a^\dagger a}$. This representation preserves the commutation relations of the spin operators. Expanding the square root in $1/S$ provides a controlled means of including fluctuations.

In order to extend the theory to include quadrupoles, we introduce the representation of the spin operators using Schwinger bosons, $S^z = (b_\uparrow^\dagger b_\uparrow - b_\downarrow^\dagger b_\downarrow)/2$, $S^+ = b_\uparrow^\dagger b_\downarrow$, and $S^- = b_\downarrow^\dagger b_\uparrow$. The algebra of the spin operators is again satisfied, and the number of Schwinger bosons determines the value of the spin, $b_\uparrow^\dagger b_\uparrow + b_\downarrow^\dagger b_\downarrow = 2S$. Schwinger bosons are very convenient for the representation of spin coherent states, because the classical spins can be obtained by replacing the bosons with complex numbers (i.e. by "condensing" the boson). The replacement of $b_\uparrow^\dagger$ and $b_\uparrow$ by $\sqrt{2S - a^\dagger a}$, and identifying the spin-$\downarrow$ Schwinger bosons with the Holstein–Primakoff bosons, $b_\downarrow^\dagger = a^\dagger$ and $b_\downarrow = a$, leads to the large-$S$ spin-wave theory mentioned above. Unfortunately, this expansion is not useful to describe quadrupoles, because these are entangled states of the spin-$\uparrow$ and spin-$\downarrow$ Schwinger bosons. This problem can be remedied by the introduction of one additional boson [33,48]. A convenient choice is to select the three bosons $a_\nu$ according to the basis of (13.3), with $\nu = x, y, z$ and a fixed on-site occupation number $M = \sum_\nu a_\nu^\dagger(j) a_\nu(j)$. Although $S = 1$ spins have $M = 1$, here we treat $M$ as a parameter.[9] Following (13.4) and (13.7) then leads to the expression of the spin and quadrupole operators as

$$S_j^\alpha = -\mathrm{i}\varepsilon_{\alpha\beta\gamma} a_\beta^\dagger(j) a_\gamma(j), \quad Q_j^{\alpha\beta} = \frac{2M}{3}\delta_{\alpha\beta} - a_\alpha^\dagger(j) a_\beta(j) - a_\beta^\dagger(j) a_\alpha(j). \tag{13.28}$$

An ordered state of SU(3) spin coherent states corresponds to condensing the appropriate boson, while the remaining, uncondensed ones play the role of the Holstein–Primakoff bosons. The information about which boson to condense is obtained from the variational wave function considered in Sect. 13.4.1, which actually represents the mean-field result in the limit $M \to \infty$, where each boson is replaced by its average value.[10] As an example, if the $\nu = y$ bosons condense on site $j$, the appropriate replacement in (13.28) is

$$a_y^\dagger(j), a_y(j) \to \sqrt{M - a_x^\dagger(j) a_x(j) - a_z^\dagger(j) a_z(j)}, \tag{13.29}$$

where $a_x$ and $a_z$ have the role of Holstein–Primakoff bosons. Care should be taken with the fact that, while $a_y$ commutes with $a_z$ and $a_x$, the expression (13.29) does not, and the replacement should be made without changing the order of operators written in (13.28). After this replacement, the spin operators satisfy the usual spin

[9] On each site, the fundamental representation of the SU(3) symmetry is enlarged to a single-row Young tableau $(M, 0)$ of length $M$ boxes.

[10] In contrast, the SU(3) Schwinger-boson mean-field theory [49] corresponds to averaging over quadratic forms of the bosons.

commutation relations. Expanding in powers of $1/M$ gives a 'spin-wave' theory in which the excitations consist both of spin and quadrupolar degrees of freedom.

The details of such a spin-wave analysis are presented below for the case of ferroquadrupolar order. The formalism can also be used to calculate excitation spectra in the more complicated antiferroquadrupolar state. Tsunetsugu and Arikawa [6] have used a similar technique in calculations for the triangular lattice. This type of calculation has been extended to octupoles by Chubukov [50].

#### 13.4.3.1 Technical Details

The Hamiltonian connecting sites $i$ and $j$ is given by

$$\mathcal{H}(i,j) = J_2 + (J_1 - J_2)\mathbf{S}_i \cdot \mathbf{S}_j + J_2 \mathbf{A}_i \cdot \mathbf{A}_j, \tag{13.30}$$

where

$$\mathbf{A}_i \cdot \mathbf{A}_j = \sum_{\mu,\nu} a^\dagger_\mu(i) a_\nu(i) a^\dagger_\nu(j) a_\mu(j) \tag{13.31}$$

is the SU(3)-symmetric part of the Hamiltonian, which reduces to the permutation operator in the case of the $M = 1$ fundamental representation. Substituting (13.29) into the terms of the Hamiltonian, summing over all nearest-neighbor pairs of sites $\langle i,j\rangle$, and performing a Fourier transformation leads in the harmonic approximation to

$$\sum_{\langle i,j\rangle} \mathbf{A}_i \cdot \mathbf{A}_j = \frac{z}{2} L M^2 - zM \sum_{\nu=x,z} \sum_{\mathbf{k}\in\mathrm{BZ}} [1-\gamma(\mathbf{k})]\, a^\dagger_\nu(\mathbf{k}) a_\nu(\mathbf{k}), \tag{13.32}$$

$$\begin{aligned}\sum_{\langle i,j\rangle} \mathbf{S}_i \cdot \mathbf{S}_j = \frac{z}{2} M \sum_{\nu=x,z} \sum_{\mathbf{k}\in\mathrm{BZ}} \gamma(\mathbf{k}) \Big[ & a^\dagger_\nu(\mathbf{k}) a_\nu(\mathbf{k}) + a^\dagger_\nu(-\mathbf{k}) a_\nu(-\mathbf{k}) \\ & - a_\nu(\mathbf{k}) a_\nu(-\mathbf{k}) - a^\dagger_\nu(\mathbf{k}) a^\dagger_\nu(-\mathbf{k}) \Big].\end{aligned} \tag{13.33}$$

$L$ is the number of lattice sites, $z$ is the coordination number ($z = 4$ for the square and 6 for the triangular lattice), and $\gamma(k)$ is the geometrical factor introduced in (13.26). Combining these terms and neglecting a constant energy shift yields the Hamiltonian in the form

$$\begin{aligned}\mathcal{H} = \frac{zM}{2} \sum_{\nu=x,z} \sum_{\mathbf{k}\in\mathrm{BZ}} & 2A_\nu(\mathbf{k}) a^\dagger_\nu(\mathbf{k}) a_\nu(\mathbf{k}) \\ & + B_\nu(\mathbf{k}) \Big[ a_\nu(\mathbf{k}) a_\nu(-\mathbf{k}) + a^\dagger_\nu(\mathbf{k}) a^\dagger_\nu(-\mathbf{k}) \Big],\end{aligned} \tag{13.34}$$

with $A_\nu(\mathbf{k}) = J_1\gamma(\mathbf{k}) - J_2$ and $B_\nu(\mathbf{k}) = (J_2 - J_1)\gamma(\mathbf{k})$. The conventional Bogoliubov transformation

$$a^\dagger_\nu(\mathbf{k}) = \cosh\xi_\nu(\mathbf{k}) \alpha^\dagger_\nu(\mathbf{k}) + \sinh\xi_\nu(\mathbf{k}) \alpha_\nu(-\mathbf{k}) \tag{13.35}$$

$$a_\nu(\mathbf{k}) = \cosh\xi_\nu(\mathbf{k}) \alpha_\nu(\mathbf{k}) + \sinh\xi_\nu(\mathbf{k}) \alpha^\dagger_\nu(-\mathbf{k}) \tag{13.36}$$

with $\tanh 2\vartheta_\nu(\mathbf{k}) = -B_\nu(\mathbf{k})/A_\nu(\mathbf{k})$ ensures bosonic commutation relations for the operators $\alpha_\nu$ and diagonalizes the Hamiltonian (13.34), giving

$$\mathcal{H} = \sum_{\nu=x,z} \sum_{\mathbf{k}} \left\{ \omega_\nu(\mathbf{k})\alpha^\dagger_\nu(\mathbf{k})\alpha_\nu(\mathbf{k}) - \frac{zM}{4}\left[A_\nu(\mathbf{k}) - \sqrt{A_\nu(\mathbf{k})^2 - B_\nu(\mathbf{k})^2}\right]\right\}, \tag{13.37}$$

where the dispersion of the generalized spin waves is

$$\omega_\nu(\mathbf{k}) = \frac{zM}{2}\sqrt{A_\nu(\mathbf{k})^2 - B_\nu(\mathbf{k})^2}. \tag{13.38}$$

This dispersion is shown in Fig. 13.5b for a doubly degenerate mode in the Brillouin zone, the degeneracy corresponding to the two $a_\nu$ Holstein–Primakoff bosons.

In order to access the signatures of the excitations in inelastic neutron scattering experiments we determine the dynamical spin correlation functions. The calculation of the imaginary part proceeds through the steps

$$\begin{aligned} S^{xx}(\mathbf{k},\omega) &= \sum_f |\langle f|S^x(\mathbf{k})|\mathrm{GS}\rangle|^2 \delta(\omega - E_f + E_{\mathrm{GS}}) \\ &= M\sum_f |\langle f|a^\dagger_z(\mathbf{k}) - a_z(-\mathbf{k})|\mathrm{GS}\rangle|^2 \delta(\omega - E_f + E_{\mathrm{GS}}) \\ &= M\sqrt{\frac{A_z(\mathbf{k}) + B_z(\mathbf{k})}{A_z(\mathbf{k}) - B_z(\mathbf{k})}}\,\delta(\omega - \omega_z(\mathbf{k})), \end{aligned} \tag{13.39}$$

where $|\mathrm{GS}\rangle$ denotes the ground state. We draw attention to the fact that the $\nu = z$ branch contributes to $S^{xx}(\mathbf{k},\omega)$ – the spin fluctuations are perpendicular to both the direction of the director ($y$) and the direction of the quadrupolar excitation ($z$). Correspondingly, the $\alpha^\dagger_x$ bosons contribute to $S^{zz}(\mathbf{k},\omega)$. Because of the remaining $O(2)$ symmetry, $S^{xx}(\mathbf{k},\omega) = S^{zz}(\mathbf{k},\omega)$, whereas $S^{yy}(\mathbf{k},\omega)$ is higher-order in $1/M$. Thus the spin fluctuates in the $xz$-plane perpendicular to the director (which we have chosen to be parallel to $y$), similar in a sense to the fluctuations in a spin system with easy-plane anisotropy. The structure factors $S^{\nu\nu}(\mathbf{k}) = \int \mathrm{d}\omega S^{\nu\nu}(\mathbf{k},\omega)$ are shown in Fig. 13.5a.

We now apply a magnetic field along the $z$-axis. The director of the ferroquadrupolar state is perpendicular to the magnetic field, oriented for example along the $y$-axis again. The single-site wave function is then given by (13.23), and is created by applying the $a^\dagger_\parallel$ boson to the vacuum, $a^\dagger_\parallel|0\rangle = \cos\frac{\mu}{2}|y\rangle - \mathrm{i}\sin\frac{\mu}{2}|x\rangle$. The different bosons are related by an SU(3) rotation in the boson space,

$$\begin{pmatrix} a^\dagger_x \\ a^\dagger_y \\ a^\dagger_z \end{pmatrix} = \begin{pmatrix} \cos\frac{\mu}{2} & \mathrm{i}\sin\frac{\mu}{2} & 0 \\ \mathrm{i}\sin\frac{\mu}{2} & \cos\frac{\mu}{2} & 0 \\ 0 & 0 & 1 \end{pmatrix} \begin{pmatrix} a^\dagger_\perp \\ a^\dagger_\parallel \\ a^\dagger_z \end{pmatrix}. \tag{13.40}$$

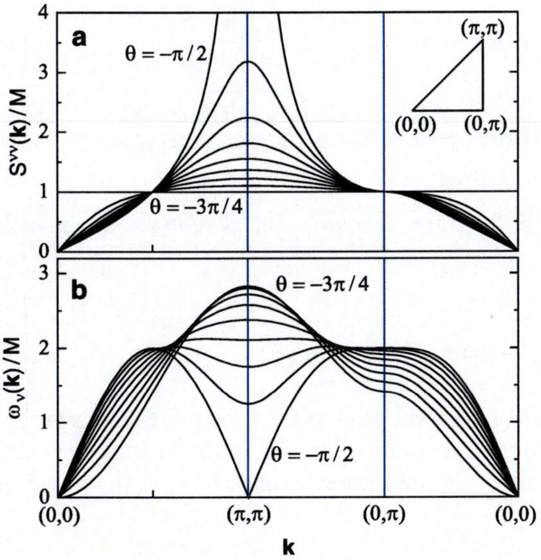


**Fig. 13.5** (a) Spin structure factor $S(\mathbf{k})$ and (b) dispersion relation $\omega(\mathbf{k})$ of spin-wave excitations in the ferroquadrupolar phase for $\vartheta$ values between $-3\pi/4$ and $-\pi/2$, in steps of $\pi/32$. One observes clearly the linear $k$-dependence of both the dispersion and the spin structure factor for $k \to 0$, except at the boundary with the ferromagnetic phase ($\vartheta = -3\pi/4$). Softening of the dispersion and divergence of the structure factor are observed at $\mathbf{k} = (\pi, \pi)$ as $\vartheta$ approaches the boundary to the antiferromagnetic phase, $\vartheta = -\pi/2$

The ordering effect of the field is described by condensing the $a_{\parallel}^{\dagger}$ bosons: $(a_{\parallel}^{\dagger})^M|0\rangle$ has a finite expectation value, $m = M \sin\mu$, with the operator $S^z$, and gives 0 with $S^x$ and $S^y$. While an SU(3) rotation leaves $\mathbf{A}_i \cdot \mathbf{A}_j$ invariant, the form of $\mathbf{S}_i \cdot \mathbf{S}_j$ is changed. In the Hamiltonian, a new term of order $M^{3/2}$ appears, which contains single bosons, and which originates from the terms $\mathbf{S}_i \cdot \mathbf{S}_j$ and $-h[S_i^z + S_j^z]/z$. The $M^{3/2}$ contributions of these two terms cancel, if $\mu$ is fixed by the constraint $h = z(J_1 - J_2)m$ obtained from the mean-field result. The net result of these considerations is an analog of (13.34) with $A_z = J_1\gamma(\mathbf{k}) - J_2$, $B_z = (J_2 - J_1)\cos\mu\gamma(\mathbf{k})$, and

$$A_\perp = \left(J_1 \cos^2\mu + J_2 \sin^2\mu\right)\gamma(\mathbf{k}) - J_2,$$
$$B_\perp = (J_2 - J_1)\cos^2\mu\gamma(\mathbf{k}). \tag{13.41}$$

The spectrum is obtained from (13.38) by substituting the appropriate $A$ and $B$ components. Similarly, the spin-spin correlation functions are

$$S^{xx}(\mathbf{k},\omega) = M\cos^2\frac{\mu}{2}\sqrt{\frac{A_z(\mathbf{k}) + B_z(\mathbf{k})}{A_z(\mathbf{k}) - B_z(\mathbf{k})}}\delta(\omega - \omega_z(\mathbf{k})), \tag{13.42}$$

$$S^{yy}(\mathbf{k},\omega) = M\sin^2\frac{\mu}{2}\sqrt{\frac{A_z(\mathbf{k})-B_z(\mathbf{k})}{A_z(\mathbf{k})+B_z(\mathbf{k})}}\delta(\omega-\omega_z(\mathbf{k})), \tag{13.43}$$

$$\tilde{S}^{zz}(\mathbf{k},\omega) = M\cos^2\mu\sqrt{\frac{A_\perp(\mathbf{k})+B_\perp(\mathbf{k})}{A_\perp(\mathbf{k})-B_\perp(\mathbf{k})}}\delta(\omega-\omega_\perp(\mathbf{k})). \tag{13.44}$$

In the function $S^{zz}$, the finite average value is subtracted to give $\tilde{S}^z(j) = S^z(j) - M\sin\mu$.

#### 13.4.3.2 Long-Wavelength Limit

Some insight into the physical properties of the finite-field state is obtained by considering the situation close to the $\Gamma$ point, i.e. in the long-wavelength limit $k \to 0$. Here, the dispersion of the spin wave is linear in $k$, $\omega_\nu(\mathbf{k}) \approx v_0 k$, with velocity

$$v_0 = \frac{zM}{2\sqrt{2}}\sqrt{J_2(J_2-J_1)}. \tag{13.45}$$

This linear dispersion, which is otherwise typical for an antiferromagnet, was first calculated by Matveev [51]. The spin structure factor $S^{xx}(\mathbf{k}) = S^{zz}(\mathbf{k})$ also vanishes linearly with $k$,

$$S^{xx}(\mathbf{k}) = S^{zz}(\mathbf{k}) = \frac{M}{2}\sqrt{\frac{-J_2}{2(J_1-J_2)}}k = \chi v_0 k, \tag{13.46}$$

where $\chi = 1/[z(J_1 - J_2)]$ is the mean-field susceptibility. At the quadrupolar ordering wave vector the spin structure factor vanishes – this is yet another possible definition of the spin-liquid state (see Chap. 2 by Lhuillier and Misguich). On approaching the SU(3)-symmetric point, $\vartheta = -3\pi/4$, which separates the ferroquadrupolar and ferromagnetic phases, the $k$-linear part of the spectrum decreases. At the SU(3) point the effective spin-wave velocity vanishes, and only the quadratic dispersion, $\omega_\nu(\mathbf{k}) \approx zMJ_1k^2/4$, which is typical of a ferromagnet, remains.

In a magnetic field, the degeneracy of the two modes is lifted. The $O(3)$ symmetry is reduced to $O(2)$ (the direction of the director may be chosen freely in the plane perpendicular to the magnetic field), leaving a single Goldstone mode, $\omega_\perp$, which has dispersion $\omega_\perp(\mathbf{k}) = v_0\cos\mu\, k$. A gap proportional to the magnetic field, $\omega_z(\mathbf{k}=0) = h/2$, opens in the other mode, which as a consequence is raised into the multimagnon continuum arising from the mode $\omega_\perp$, and therefore is no longer well defined.

#### 13.4.3.3 Instability to an Antiferromagnetic State at $\vartheta = -\pi/2$

For the square lattice, the dispersion becomes soft at the point $\mathbf{k} = (\pi, \pi)$ in the Brillouin zone as $J_1 \to 0^-$, a process accompanied by a diverging spin structure factor. The forms

$$\omega_\nu(\mathbf{k}) \approx 4M\sqrt{J_1 J_2}\sqrt{1 + \lambda^2|\mathbf{k} - (\pi, \pi)|^2}, \tag{13.47}$$

$$S(\mathbf{k}) \approx 2M\sqrt{\frac{J_2}{J_1}}\frac{1}{\sqrt{1 + \lambda^2|\mathbf{k} - (\pi, \pi)|^2}}, \tag{13.48}$$

announce the instability of the ferroquadrupolar state towards the conventional, two-sublattice, antiferromagnetic Néel order. Within the ferroquadrupolar phase, the spins show Néel-type short-range order with a strong easy-plane anisotropy, where the easy plane is perpendicular to the ordered directors of the quadrupoles. The typical length scale associated with this short-range order is $\lambda = \sqrt{\frac{J_2}{8J_1}}$, which diverges as $J_1 \to 0^-$. In a finite magnetic field, $\omega_z(\mathbf{k})$ becomes gapless for $h = 8\sqrt{-J_1 J_2}$ and $S^{xx}(\mathbf{k}, \omega)$ diverges: as expected from above, the spin fluctuations are perpendicular to both the director (set in the $y$-direction) and to the direction of the applied magnetic field. The point $J_1 = 0$ is also unstable for the bipartite cubic and body-centered-cubic lattices.

In case of frustrated (non-bipartite) lattices, this instability occurs later, in the sense that a more extended ferroquadrupolar phase is allowed. On the triangular lattice, the critical value is $J_1 = -J_2/2$, where the instability is towards a three-sublattice antiferromagnetic order. In a magnetic field, the instability occurs along the line $h = 6\sqrt{-3(J_1 + 2J_2)J_2}$ in parameter space, a result which is confirmed by the variational calculation only for large fields; at low fields the phase transition occurs earlier, and as a first-order transition to a 2:1-type state [8]. Finally, in the fcc lattice, the dispersion softens along a whole line in the Brillouin zone when $J_1 = -J_2$.

### *13.4.4 Numerical Approach*

The bilinear–biquadratic $S = 1$ model (13.8) can be simulated in detail, on bipartite lattices and for $\vartheta \in [-\pi, 0]$, i.e. $J_2 \leq 0$, by advanced Quantum Monte Carlo methods [52] (Harada, unpublished). These algorithms have been used to map the phase diagram of the model on the square and cubic lattice for negative $J_2$, where the infamous sign problem is absent. The results of these calculations show unambiguously the presence of a ferroquadrupolar phase between the SU(3)-symmetric points **33** and **3$\bar{3}$**, meaning when $-3\pi/4 < \vartheta < -\pi/2$ [53], at $T = 0$ for the square lattice (as shown in Fig. 13.3a, lower, outer semicircle) and below a critical temperature $T_c(\vartheta)$ for the cubic lattice. In this context we mention also QMC calculations performed as a function of N at $\vartheta = -\pi/2$ for the SU(N) model with

**Table 13.2** The lowest states in the Anderson towers of the antiferromagnetic (AFM), antiferroquadrupolar (AFQ), and ferroquadrupolar (FQ) ordered states (for even site numbers $N_s$). $\Gamma_1 \equiv (0,0)$, $s-$wave; $\Gamma_2 \equiv (0,0)$, $f-$wave; $\Gamma_3 \equiv (4\pi/3, 0)$, $s-$wave

| | AFM | | | AFQ | | | FQ |
|---|---|---|---|---|---|---|---|
| $S$ | $\Gamma_1$ | $\Gamma_2$ | $\Gamma_3$ | $\Gamma_1$ | $\Gamma_2$ | $\Gamma_3$ | $\Gamma_1$ |
| 0 | 1 | | | 1 | | | 1 |
| 1 | | 1 | 1 | | | | |
| 2 | 1 | | 2 | | | 1 | 1 |
| 3 | 1 | 2 | 2 | | 1 | | |
| 4 | 2 | 1 | 3 | 1 | | 1 | 1 |
| 5 | 1 | 2 | 4 | | | 1 | |
| 6 | 3 | 2 | 4 | 1 | 1 | 1 | 1 |
| 7 | 2 | 3 | 5 | | 1 | 1 | |
| 8 | 3 | 2 | 6 | 1 | | 2 | 1 |
| 9 | 3 | 4 | 6 | 1 | 1 | 1 | |

the fundamental-antifundamental representation, where two-sublattice ordering was confirmed for $N \leq 4$ [54].

On the triangular lattice, QMC algorithms face a sign problem, and exact diagonalization of small clusters is the method of choice. The different phases can be identified by computing correlation functions of different operators (although due to the small cluster sizes this approach is of somewhat limited utility), and by examining the low-energy spectrum to search for the features known as Anderson towers[11]. These towers are the energies of a particular set of eigenstates, with spin and spatial symmetries defined by the long-range order they describe, which become degenerate in the thermodynamic limit [55] and are the fingerprints of an SU(2)-symmetry-breaking state on a finite-size cluster. These sets of states were used in a pioneering study by Bernu et al. [56, 57] to confirm the presence of three-sublattice antiferromagnetic order in the $S = 1/2$ triangular-lattice Heisenberg model. In the case of nematics, one may also search for the corresponding states, and Table 13.2 shows the expected low-energy spectrum of the different ordered states. One possible means of obtaining the quantum numbers and the multiplicities of the states in the tower is from the variational wave function of (13.17), by a simple decomposition of the $S$ content of such a state when it describes a certain type of order.

Numerical exact diagonalization results for a 12-site cluster are shown in Fig. 13.6 for three selected values of $\vartheta$. In the ferroquadrupolar phase ($\vartheta = -\pi/2$), we observe a tower of states consisting of a single multiplet per even spin sector, and these levels are completely symmetric under the action of the space group. Note that, at the transition point $\vartheta = -3\pi/4$ between the ferroquadrupolar and the ferromagnetic phase, the model is an SU(3) ferromagnet, whose ground state on a finite system belongs to the $(N_s, 0)$ multiplet of SU(3) [58]. The decomposition of this

[11] Please refer to Chap. 18 on "Numerical Simulations of Frustrated Systems" for a more detailed discussion of Anderson towers.

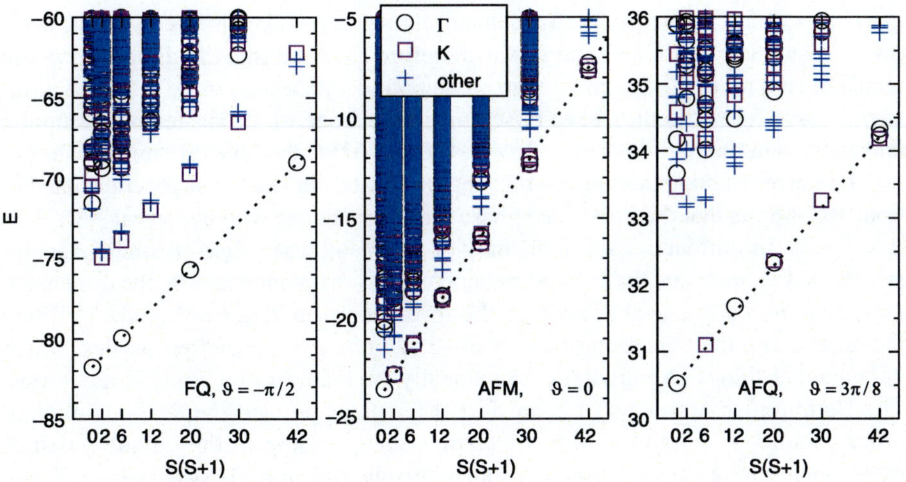


**Fig. 13.6** Energy spectra as a function of total spin for a 12-site triangular-lattice cluster with periodic boundary conditions. The values of $\vartheta$ are chosen to yield representative FQ ($\vartheta = -\pi/2$), AFM ($\vartheta = 0$), and AFQ ($\vartheta = 3\pi/8$) phases

large SU(3) representation into SU(2) representations yields a completely flat tower of states with one level for each of $S = N_s, N_s - 2, N_s - 4, \ldots$. The tower of states in the semiclassical 120° Néel state at $\vartheta = 0$ is expected to have the same structure as that derived for the $S = 1/2$ problem in [56, 57]. Indeed, we recover the required structure up to $S = 4$ ($m/m_{\text{sat}} = 1/3$) on the 12-site cluster. The tower of states in the antiferroquadrupolar phase ($\vartheta = 3\pi/8$) has a rich structure, but one which currently lacks a straightforward analytical explanation. However, the $S$ content of the site-factorized wave functions matches perfectly the tower of states observed in exact diagonalization calculations performed for $N_s = 12$ and 21 spins, as shown in Fig. 13.6.

## 13.5 From Chains to the Square Lattice

The bilinear–biquadratic $S = 1$ model on the chain is a prototypical interacting spin model in one dimension. In addition to the celebrated Haldane-gap phase, it has many points ($\vartheta = -3\pi/4, -\pi/2, -\pi/4$, arctan[1/3], $+\pi/4$) which are exactly solvable either by Bethe ansatz or as matrix-product states. The model has also served as a test-bed for many numerical approaches. It is not our goal here to review the literature of the chain model, but merely to address the relevance of spin-nematic correlations in the phase diagram. In 1991, Chubukov [59] drew attention to this issue by remarking that the ferromagnet becomes unstable towards the formation of bound states of two ferromagnetic spin waves as $\vartheta \to -3\pi/4^{-}$. He argued that beyond the ferromagnet there would be a disordered spin-nematic phase with a finite gap, but with no dimerization. Subsequent numerical studies [58] could not, however, substantiate this claim and it was therefore assumed that the dimerized phase would extend up to the ferromagnetic phase boundary. Despite these results, a number of more recent studies have raised again the issue of strong ferroquadrupolar correlations in the regime $\vartheta \to -3\pi/4^+$ [30, 60–62] (Läuchli et al., unpublished).

The current understanding of this region, based on the most recent contributions [63–66], is that the dimerized phase indeed spans the whole interval $-3\pi/4 < \vartheta < -\pi/4$. Important in the establishing of this result was a field-theoretical analysis which demonstrated that a hypothetical ferroquadrupolar phase in one dimension (1D) is generically unstable towards the formation of a dimerized phase [66]. On the other hand, the transition point $\vartheta = -3\pi/4$ has very special properties, which among other things obscure quite dramatically the features of a numerical analysis. The Hamiltonian at this point is an $SU(3)$ ferromagnet, and the ferroquadrupolar order parameter commutes with the Hamiltonian, leading to the unique possibility of having long-range ferroquadrupolar correlations in a 1D system [34]. When moving away from the SU(3) point by increasing $\vartheta$, the large SU(3) representation is split into separate even-spin sectors, and the singlet becomes the nondegenerate ground state on a finite-size system. Because this singlet wavefunction is connected adiabatically to the exact ferroquadrupolar ground state at $\vartheta = -3\pi/4$, the correlations in this state in the region $\vartheta \gtrsim -3\pi/4$ appear essentially ferroquadrupolar in small systems. Only on very large length scales does the exponential decrease of the ferroquadrupolar correlations concomitant with the appearance of a very small dimerization become observable.

While much activity has focused on the transition from the dimerized to the ferromagnetic region (above), the gapless regime $\vartheta \in [+\pi/4, \pi/2)$ has received considerably less attention. Despite the early consensus on the presence of an extended gapless phase [67], the nature of the dominant correlations in this phase has not been established definitively. The initial conjecture concerning this gapless phase favored an almost "trimerized" ground state [68, 69], featuring dominant singlet correlations involving three consecutive spins. Although models can be constructed which have exactly trimerized ground states [70, 71], it was shown in [72] that these are not the dominant fluctuations in the standard bilinear–biquadratic model. It was realized only recently, by combining numerical simulations with earlier field-theoretical results [73], that the extended critical phase actually has dominant antiferroquadrupolar correlations [64]. It is quite significant that the model on a simple, bipartite system, such as the chain, develops dominant antiferroquadrupolar correlations with a period of three lattice units. For two chains with zig-zag coupling, a geometry which can be considered as a finite strip of the triangular lattice, the critical phase persists [74], and one could therefore argue that the critical phase of the single chain constitutes the fluctuating remnants of the fully developed antiferroquadrupolar order on the triangular lattice.

The dimensional crossover from the one-dimensional chain to the square lattice in a spatially anisotropic version of the bilinear–biquadratic $S = 1$ model was the subject of a recent Quantum Monte Carlo study, and, strikingly, a direct quantum phase transition from the dimerized to the quadrupolar phase was observed [75].

Such a direct, continuous transition is a challenge to the conventional Landau–Ginzburg theory, which would predict in general either a first-order transition or two distinct second-order transitions. A subsequent analytical study has confirmed the theoretical possibility of a direct, continuous transition between the two phases [66].

## 13.6 Nematic Ordering in $S = 1/2$ Systems

Andreev and Grishchuk noted in 1984 that the physics of quadrupoles can also be relevant in $S = 1/2$, albeit not as on-site entities, but rather by considering bond order parameters [3]. In Sect. 13.2, we considered the product of operators $S^\alpha S^\beta$ on the same site, and the same quantity, $S_1^\alpha S_2^\beta$, can be formed from spin operators on different sites (labeled 1 and 2). Its decomposition into irreducible tensor operators proceeds by noting that the trace of $S_1^\alpha S_2^\beta$ is a scalar, $\mathbf{S}_1 \cdot \mathbf{S}_2$, the antisymmetric part of $S_1^\alpha S_2^\beta$ is a vector, $\mathbf{P}_{12} = \mathbf{S}_1 \times \mathbf{S}_2$, and the traceless, symmetric part is the tensor

$$Q_{12}^{\alpha\beta} = S_1^\alpha S_2^\beta + S_1^\beta S_2^\alpha - \frac{2}{3}(\mathbf{S}_1 \cdot \mathbf{S}_2)\delta_{\alpha\beta}. \tag{13.49}$$

Because both $\mathbf{P}_{12}$ and $\mathbf{Q}_{12}$ are bilinear in the spin operators, they describe a type of order where time-reversal symmetry is not broken. When $\langle \mathbf{P} \rangle$ takes a finite value, it is called $p$-nematic, and a state with $\langle Q \rangle \neq 0$ is an $n$-nematic, which is the analog of the quadrupoles discussed in the previous sections.

In the time-reversal-invariant basis for two $S = 1/2$ spins,

$$\begin{aligned} |s\rangle &= \frac{1}{\sqrt{2}}(|\uparrow\downarrow\rangle - |\downarrow\uparrow\rangle), \quad |x\rangle = \frac{i}{\sqrt{2}}(|\uparrow\uparrow\rangle - |\downarrow\downarrow\rangle), \\ |y\rangle &= \frac{1}{\sqrt{2}}(|\uparrow\uparrow\rangle + |\downarrow\downarrow\rangle), \quad |z\rangle = -\frac{i}{\sqrt{2}}(|\uparrow\downarrow\rangle + |\downarrow\uparrow\rangle), \end{aligned} \tag{13.50}$$

the time-reversal-breaking (magnetic) order parameters are

$$S_1^\alpha + S_2^\alpha = -i \sum_{\beta,\gamma \in \{x,y,z\}} \varepsilon_{\alpha\beta\gamma} |\beta\rangle\langle\gamma| \quad \text{and} \quad S_1^\alpha - S_2^\alpha = i|\alpha\rangle\langle s| - i|s\rangle\langle\alpha|, \tag{13.51}$$

and the nonmagnetic order parameters are

$$\begin{aligned} \mathbf{S}_1 \cdot \mathbf{S}_2 &= \frac{1}{4} \sum_{\alpha \in \{x,y,z\}} |\alpha\rangle\langle\alpha| - \frac{3}{4}|s\rangle\langle s|, \\ Q_{12}^{\alpha\beta} &= -\frac{1}{2}(|\alpha\rangle\langle\beta| + |\beta\rangle\langle\alpha|) + \frac{1}{3}\delta_{\alpha\beta}|\gamma\rangle\langle\gamma|, \\ P_{12}^\alpha &= \frac{1}{2}(|\alpha\rangle\langle s| + |s\rangle\langle\alpha|). \end{aligned} \tag{13.52}$$

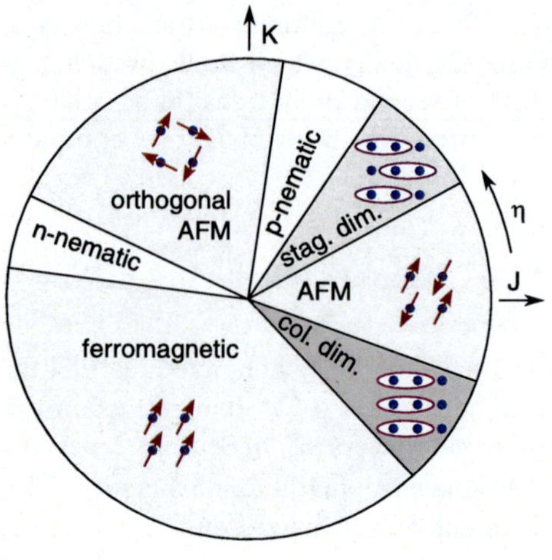


**Fig. 13.7** Schematic phase diagram of the cyclic-ring-exchange model on the square lattice, based on [43, 77]. The angle $\eta$ parameterizes the ratio of the Heisenberg coupling $J = \cos\eta$ to the ring-exchange coupling $K = \sin\eta$

A dimer wave function can be constructed as $c|s\rangle + \sum_\alpha d_\alpha|\alpha\rangle$ ($\alpha = x, y, z$). A state with $c$ and $d_\alpha$ real does not break time-reversal symmetry; $c = 1$ and $\mathbf{d} = 0$ is a singlet bond, while $c = 0$ and $|\mathbf{d}| > 0$ is an $n$-type nematic composed of triplet bonds with a director set by $\mathbf{d}$. When both $c > 0$ and $|\mathbf{d}| > 0$, the state is a $p$-type nematic with $\langle P^\alpha_{12}\rangle = cd^\alpha$. A detailed discussion of these order parameters for the case of the spin ladder can be found in [76].

However, because the order is defined on a bond, a mean-field (or variational) theory of the kind we have presented in the previous section for $S = 1$ quadrupolar states is not feasible. For $S = 1$, the order parameter is a site variable, and can be optimized by the construction of a site wave function, a process which neglects the entanglement between the sites. For $S = 1/2$ nematics, this entanglement is an essential feature. With the exception of trivial cases, such as weakly coupled dimers, it is not at present known how to construct a variational wave function to describe a spin-1/2 nematic phase. By contrast, significant progress has been made on the numerical side in recent years, and some nontrivial microscopic models are now known which show strong evidence for the presence of spin-nematic phases, of both $p$- and $n$-type.

Numerical evidence for a $p$-type nematic is found in an exact-diagonalization study [77] of a square-lattice model where, in addition to the Heisenberg exchange, a four-site ring exchange is present. The phase diagram of this model is quite rich, featuring several dimerized and magnetic phases (Fig. 13.7). In addition, it offers a clean realization of a two-step SU(2) symmetry-restoration scenario put forward by Chandra and Coleman in [78, 79] (see also [80, 81]). The starting point is a semi-classical Néel order with four coplanar sublattices, such that the spin orientations on nearest-neighbor bonds differ by 90° in the Néel phase, and thus secondary order parameters of the type $\langle \mathbf{S}_{(x,y)} \times \mathbf{S}_{(x+1,y)}\rangle$ are also finite. Upon frustrating this Néel state with antiferromagnetic nearest-neighbor Heisenberg exchange, the magnetic order is lost, $\langle \mathbf{S}_{(x,y)}\rangle = \mathbf{0}$, while the vector chirality remains finite $\langle \mathbf{S}_{(x,y)} \times \mathbf{S}_{(x+1,y)}\rangle \neq 0$ in a small region of parameter space. The symmetry group of this $p$–type spin nematic (two-sublattice, "collinear") is larger than that of the original Néel order (four-sublattice, coplanar), and therefore this process can be considered as a partial restoration of SU(2) symmetry. The full restoration is accomplished on continuing into the paramagnetic, staggered-dimer phase adjacent to the spin-nematic phase.

A complementary $n$-type spin-nematic phase has also been found in the frustrated $J_1$–$J_2$ square-lattice model with ferromagnetic nearest-neighbor ($J_1$) and antiferromagnetic next-nearest-neighbor ($J_2$) interactions, as well as close to the ferromagnetic phase in the ring-exchange model mentioned above [43]. The appearance of this phase in proximity to a ferromagnetic state is currently best understood by considering the formation of a $d$-wave bound state of two flipped spins within the completely polarized state. The spin-nematic phase can then be described as a field-induced Bose-Einstein condensation of bound states, a sense in which it is closely related to the paired state of hard-core bosons stabilized by correlated hopping processes [46, 47]; these two states differ only in the spatial symmetry of the bound-state wave function. We remark also that nematic order is not unique to quantum spins, but appears also for classical spins in the form of coplanar or collinear spin states with no long-range magnetic order [82–84].

Finally, we mention also that $S = 1/2$ spin systems on the triangular lattice may support even more complicated forms of multipole-related ordering: recent numerical calculations including ring-exchange terms have provided evidence for triatic (octupolar) ordering in a magnetic field [85]. The classical analog of this octupolar order was studied for the kagomé lattice in [86].

Another recent development in this direction concerns the phase diagram of the frustrated ferromagnetic Heisenberg chain in a magnetic field. This problem has been addressed in a number of recent studies [37–42], and a rich phase diagram containing at least three different, 1D analogs of spin-nematic phases (quadrupolar, octupolar, hexadecupolar) has emerged. Because many experimental systems are available which can in a reasonable approximation be modeled by such a frustrated ferromagnetic chain (such as $LiCuVO_4$ to name one example [87]), these might be among the best candidate systems for the definitive observation of field-induced spin-nematic behavior in experiments.

## 13.7 Conclusions

The possibility of having SO(3) symmetry-breaking in the absence of a magnetic moment is a fascinating example of the richness of quantum mechanical ordering phenomena. On the theoretical side, there is now a sound understanding of the phases themselves, and of a number of microscopic interactions and models which can stabilize spin-nematic phases are known. While unambiguous experimental evidence for spin-nematic phases in real materials is still lacking, a number of promising systems are currently under very active investigation, which hopefully will lead to such evidence emerging in the near future.

## Acknowledgements

We are grateful to M. Arikawa, Ph. Corboz, J.-C. Domenge, P. Fazekas, S. Korshunov, C. Lhuillier, F. Mila, T. Momoi, S. Nakatsuji, J. Romhányi, G. Schmid, N. Shannon, H. Shiba, P. Sindzingre, T.A. Tóth, K. Totsuka, S. Trebst, M. Troyer, and H. Tsunetsugu for discussions. We thank A. Chubukov and B. Normand for a critical reading of the manuscript. Our research on spin nematics has been supported by the Hungarian OTKA Grant No.K62280, by the Swiss National Fund, and by the Swiss Center for Scientific Computing (CSCS Manno).


## References


1. G. Misguich, C. Lhuillier, in *Frustrated Spin Systems*, ed. by H.T. Diep (World Scientific, Singapore, 2005)
2. M. Blume, Y.Y. Hsieh, J. Appl. Phys. **40**, 1249 (1969)
3. A.F. Andreev, I.A. Grishchuk, Zh. Eksp. Teor. Fiz. **87**, 467 (1984) [Sov. Phys. JETP **60**, 267 (1984)]
4. V. Barzykin, L.P. Gor'kov, Phys. Rev. Lett. **70**, 2479 (1993)
5. S. Nakatsuji et al., Science **309**, 1697 (2005)
6. H. Tsunetsugu, M. Arikawa, J. Phys. Soc. Jpn. **75**, 083701 (2006)
7. H. Tsunetsugu, M. Arikawa, J. Phys. Cond. Mat. **19**, 145248 (2007)
8. A. Läuchli, F. Mila, K. Penc, Phys. Rev. Lett. **97**, 087205 (2006)
9. S. Bhattacharjee, V.B. Shenoy, T. Senthil, Phys. Rev. B **74**, 092406 (2006)
10. E.E. Kaul, H. Rosner, N. Shannon, R.V. Shpanchenko, C. Geibel, J. Magn. Magn. Mater. **272–276**, 922 (2004)
11. N. Kini, E.E. Kaul, C. Geibel, J. Phys. Condens. Matter **18**, 1303 (2006)
12. H. Kageyama, T. Kitano, N. Oba, M. Nishi, S. Nagai, K. Hirota, L. Viciu, J.B. Wiley, J. Yasuda, Y. Baba, Y. Ajiro, K. Yoshimura, J. Phys. Soc. Jpn. **74**, 1702 (2005)
13. H.T. Ueda, K. Totsuka, Phys. Rev. B **76**, 214428 (2007)
14. P. Santini, S. Carretta, G. Amoretti, R. Caciuffo, N. Magnani, G.H. Lander, Rev. Mod. Phys. **81**, 807 (2009)
15. J.A. Paixao, C. Detlefs, M.J. Longfield, R. Caciuffo, P. Santini, N. Bernhoeft, J. Rebizant, G.H. Lander, Phys. Rev. Lett. **89**, 187202 (2002)
16. Y. Tokunaga, D. Aoki, Y. Homma, S. Kambe, H. Sakai, S. Ikeda, T. Fujimoto, R.E. Walstedt, H. Yasuoka, E. Yamamoto, A. Nakamura, Y. Shiokawa, Phys. Rev. Lett. **97**, 257601 (2006)
17. O. Sakai, R. Shiina, H. Shiba, J. Phys. Soc. Jpn. **74**, 457 (2005)
18. R. Shiina, H. Shiba, P. Thalmeier, J. Phys. Soc. Jpn. **66**, 1741 (1997)
19. T. Takimoto, J. Phys. Soc. Jpn. **75**, 034714 (2006)
20. Y. Kuramoto, H. Kusunose, A. Kiss, J. Phys. Soc. Jpn. **78**, 072001 (2009)
21. C.V. Ciobanu, S.-K. Yip, T.L. Ho, Phys. Rev. A **61**, 33607 (2000)
22. H.H. Tu, G.M. Zhang, L. Yu, Phys. Rev. B **74**, 174404 (2006)
23. A.M. Turner, R. Barnett, E. Demler, A. Vishwanath, Phys. Rev. Lett. **98**, 190404 (2007)
24. K.I. Kugel, D.I. Khomskii, Sov. Phys. Usp. **25** 231 (1982)
25. A. Mishra, M. Ma, F.-C. Zhang, S. Guertler, L.-H. Tang, S. Wan, Phys. Rev. Lett. **93**, 207201 (2004); Phys. Rev. B **72**, 024448 (2005)
26. B. Douçot, M.V. Feigel'man, L.B. Ioffe, A.S. Ioselevich, Phys. Rev. B **71**, 024505 (2005)
27. Z. Nussinov, M. Biskup, L. Chayes, J. van den Brink, Europhys. Lett. **67**, 990 (2004)
28. J. Dorier, F. Becca, F. Mila, Phys. Rev. B **72**, 024448 (2005)
29. R. Barnett, A. Turner, E. Demler, Phys. Rev. Lett. **97**, 180412 (2006)
30. B.A. Ivanov, A.K. Kolezhuk, Phys. Rev. B **68**, 052401 (2003)

31. B. Sutherland, Phys. Rev. B **12**, 3795 (1975)
32. I. Affleck, Phys. Rev. Lett. **54**, 966 (1985)
33. N. Papanicolaou, Nucl. Phys. B **305**, 367 (1988)
34. C.D. Batista, G. Ortiz, Phys. Rev. B **65**, 180402(R) (2002)
35. A.V. Chubukov, D.V. Khveschenko, J. Phys. C Solid State Phys. **20**, L505 (1987)
36. A.V. Chubukov, Phys. Rev. B **44**, 4693 (1991)
37. F. Heidrich-Meisner, A. Honecker, T. Vekua, Phys. Rev. B **74**, 020403(R) (2006)
38. T. Vekua, A. Honecker, H.-J. Mikeska, F. Heidrich-Meisner, Phys. Rev. B **76**, 174420 (2007)
39. L. Kecke, T. Momoi, A. Furusaki, Phys. Rev. B **76**, 060407(R) (2007)
40. T. Hikihara, L. Kecke, T. Momoi, A. Furusaki, Phys. Rev. B **78**, 144404 (2008)
41. J. Sudan, A. Lüscher, A.M. Läuchli, Phys. Rev. B **80**, 140402(R) (2009)
42. A.M. Läuchli, J. Sudan, A. Lüscher, J. Phys. Conf. Ser. **145**, 012057 (2009)
43. N. Shannon, T. Momoi, P. Sindzingre, Phys. Rev. Lett. **96**, 027213 (2006)
44. C. Knetter, A. Bühler, E. Müller-Hartmann, G.S. Uhrig, Phys. Rev. Lett. **85**, 3958 (2000);
45. K. Totsuka, S. Miyahara, K. Ueda, Phys. Rev. Lett. **86**, 520 (2001)
46. R. Bendjama, B. Kumar, F. Mila, Phys. Rev. Lett. **95**, 110406 (2005);
47. K.P. Schmid, J. Dorier, A. Läuchli, F. Mila, Phys. Rev. B **74**, 174508 (2006)
48. A. Joshi, M. Ma, F. Mila, D.N. Shi, F.C. Zhang, Phys. Rev. B **60**, 6584 (1999)
49. P. Li, G.M. Zhang, S.Q. Shen, Phys. Rev. B **75**, 104420 (2007)
50. A. Chubukov, J. Phys. Condens. Matter **2** 1593 (1990)
51. V.M. Matveev, Zh. Eksp. Teor. Fiz. **65**, 1626 (1973) [Sov. Phys. JETP **38**, 813 (1974)]
52. K. Harada, N. Kawashima, J. Phys. Soc. Jpn. **70**, 13 (2001)
53. K. Harada, N. Kawashima, Phys. Rev. B **65**, 052403 (2002)
54. K. Harada, N. Kawashima, M. Troyer, Phys. Rev. Lett. **90**, 117203 (2003)
55. P.W. Anderson, *Basic Notions of Condensed Matter Physics*, Frontiers in Physics, vol. 55. (Benjamin Cummings, London, 1984)
56. B. Bernu, C. Lhuillier, L. Pierre, Phys. Rev. Lett. **69**, 2590 (1992)
57. B. Bernu, P. Lecheminant, C. Lhuillier, L. Pierre, Phys. Rev. B **50**, 10048 (1994)
58. G. Fáth, J. Sólyom, Phys. Rev. B **51**, 3620 (1995)
59. A.V. Chubukov, Phys. Rev. B **43**, 3337 (1991)
60. N. Kawashima, Prog. Theor. Phys. Suppl. **145**, 138 (2002)
61. D. Porras, F. Verstraete, J.I. Cirac, Phys. Rev. B **73**, 014410 (2006)
62. M. Rizzi, D. Rossini, G. De Chiara, S. Montangero, R. Fazio, Phys. Rev. Lett. **95**, 240404 (2005)
63. K. Buchta, G. Fáth, Ö. Legeza, J. Sólyom, Phys. Rev. B **72**, 054433 (2005)
64. A. Läuchli, G. Schmid, S. Trebst, Phys. Rev. B **74**, 144426 (2006)
65. S. Bergkvist, I. McCulloch, A. Rosengren, Phys. Rev. A **74**, 053419 (2006)
66. T. Grover, T. Senthil, Phys. Rev. Lett. **98**, 247202 (2007)
67. G. Fáth, J. Sólyom, Phys. Rev. B **44**, 11836 (1991)
68. K. Nomura, S. Takada, J. Phys. Soc. Jpn. **60**, 389 (1991);
69. Y. Xian, J. Phys. Condens. Matter **5** 7489, (1993)
70. J. Sólyom, J. Zittartz, Europhys. Lett. **50** 389 (2000);
71. M. Greiter, S. Rachel, D. Schuricht, Phys. Rev. B **75**, 060401(R) (2007)
72. A. Schmitt, K.-H. Mütter, M. Karbach, Y. Yu, G. Müller, Phys. Rev. B **58**, 5498 (1998)
73. C. Itoi, M. Kato, Phys. Rev. B **55**, 8295 (1997)
74. Ph. Corboz, A. Läuchli, K. Totsuka, H. Tsunetsugu, Phys. Rev. B **76**, 220404(R) (2007)
75. K. Harada, N. Kawashima, M. Troyer, J. Phys. Soc. Jpn. **76**, 013703 (2007)
76. P. Lecheminant, K. Totsuka, Phys. Rev. B **74**, 224426 (2006)
77. A. Läuchli, J.-C. Domenge, C. Lhuillier, P. Sindzingre, M. Troyer, Phys. Rev. Lett. **95**, 137206 (2005)
78. P. Chandra, P. Coleman, Phys. Rev. Lett. **66**, 100 (1991)
79. P. Chandra, P. Coleman, A.I. Larkin, J. Phys. Condens. Matter **2**, 7933 (1990)
80. A.V. Chubukov, Phys. Rev. B **44**, 5362 (1991)
81. P. Azaria, B. Delamotte, D. Mouhanna, Phys. Rev. Lett. **68**, 1762 (1992)
82. M.E. Zhitomirsky, Phys. Rev. Lett. **88**, 057204 (2002)

83. B. Canals, M.E. Zhitomirsky, J. Phys. Condens. Matter **16**, S759 (2004)
84. N. Shannon, K. Penc, Y. Motome, Phys. Rev. B **81**, 184409 (2010)
85. T. Momoi, P. Sindzingre, N. Shannon, Phys. Rev. Lett. **97**, 257204 (2006)
86. M.E. Zhitomirsky, Phys. Rev. B **78**, 094423 (2008)
87. M. Enderle, C. Mukherjee, B. Fåk, R.K. Kremer, J.-M. Broto, H. Rosner, S.-L. Drechsler, J. Richter, J. Malek, A. Prokofiev, W. Assmus, S. Pujol, J.-L. Raggazzoni, H. Rakoto, M. Rheinstädter, H.M. Rønnow, Europhys. Lett. **70**, 237 (2005)

# Part V
# Advanced Theoretical Methods and Concepts in Frustrated Magnetism

# Chapter 14
# Schwinger Bosons Approaches to Quantum Antiferromagnetism

**Assa Auerbach and Daniel P. Arovas**

**Abstract** This chapter reviews the Schwinger Boson Mean Field Theory (SBMFT) for the quantum Heisenberg model. For bipartite lattices, the SBMFT is set up as the large $N$ limit of the Heisenberg model, where the SU(N) spin operators are represented by $N$ flavors of Schwinger bosons. For frustrated lattices, an Sp(N) representation is used. For the bipartite lattices, the mean field equations are solved, and the ground state wave function, order parameter and excitations dispersions are calculated. The SBMFT is used to calculate the temperature dependent staggered magnetization of the weakly coupled layered antiferromagnet.

## 14.1 SU($N$) Heisenberg Models

The use of large $N$ approximations to treat strongly interacting quantum systems been very extensive in the last decade. The approach originated in elementary particles theory, but has found many applications in condensed matter physics. Initially, the large $N$ expansion was developed for the Kondo and Anderson models of magnetic impurities in metals. Soon thereafter it was extended to the Kondo and Anderson lattice models for mixed valence and heavy fermions phenomena in rare earth compounds [1, 2].

In these notes, we shall formulate and apply the large $N$ approach to the quantum Heisenberg model [3–6]. This method provides an additional avenue to the static and dynamical correlations of quantum magnets. The mean field theories derived below can describe both ordered and disordered phases, at zero and at finite temperatures, and they complement the semiclassical approaches.

Generally speaking, the parameter $N$ labels an internal SU (N) symmetry at each lattice site (i.e., the number of "flavors" a Schwinger boson or a constrained fermion can have). In most cases, the large $N$ approximation has been applied to treat spin Hamiltonians, where the symmetry is SU(2), and $N$ is therefore not a truly large parameter. Nevertheless, the $1/N$ expansion provides an easy method for obtaining simple mean field theories. These have been found to be surprisingly successful as well.

The large $N$ approach handles strong local interactions in terms of constraints. It is not a perturbative expansion in the size of the interactions but rather a saddle point expansion which usually preserves the spin symmetry of the Hamiltonian. The Hamiltonians are written as a sum of terms $\mathcal{O}^\dagger_{ij}\mathcal{O}_{ij}$, which are biquadratic in the Schwinger boson creation and annihilation operators, on each bond on the lattice. This sets up a natural mean field decoupling scheme using one complex Hubbard Stratonovich field per bond.

At the mean field level, the constraints are enforced only on average. Their effects are systematically reintroduced by the higher-order corrections in $1/N$.

It turns out that different large $N$ generalizations are suitable for different Heisenberg models, depending on the sign of couplings, spin size, and lattice. Below, we describe two large $N$ generalizations of the Heisenberg antiferromagnet.

## 14.2 Schwinger Representation of SU($N$) Antiferromagnets

The SU(2) algebra is defined by the familiar relations $[S^\alpha, S^\beta] = i\epsilon_{\alpha\beta\gamma}S^\gamma$. The spin operators commute on different sites, and admit a bosonic representation. Since the spectrum of a bosonic oscillator includes an infinite tower of states, a constraint is required in order to limit the local Hilbert space dimension to $2S+1$. In the Holstein–Primakoff representation, one utilizes a single boson $h$, writing $S^+ = h^\dagger\sqrt{2S - h^\dagger h}$, $S^- = (S^+)^\dagger$, and $S^z = h^\dagger h - S$, together with the non-holonomic constraint $0 \le h^\dagger h \le 2S$. The square roots prove inconvenient, and practically one must expand them as a power series in $h^\dagger h/2S$. This generates the so-called spin-wave expansion.

Another representation, due to Schwinger, makes use of two bosons, $a$ and $b$. We write

$$S^+ = a^\dagger b, \quad S^- = b^\dagger a, \quad S^z = \tfrac{1}{2}(a^\dagger a - b^\dagger b), \tag{14.1}$$

along with the holonomic constraint,

$$a^\dagger a + b^\dagger b = 2S, \tag{14.2}$$

where the boson occupation, $2S$, is an integer which determines the representation of SU(2). This scheme is depicted graphically in Fig. 14.1.

There are three significant virtues of the Schwinger representation. The first is that there are no square roots to expand. The second is that the holonomic constraint (14.2) can be elegantly treated using a Lagrange multiplier. The third is that it admits a straightforward and simple generalization to SU($N$). That generalization involves adding additional boson oscillators – $N$ in all for SU($N$) – which we write as $b_\mu$ with $\mu = 1, \ldots, N$. The generators of SU($N$) are then

$$S_{\mu\nu} = b^\dagger_\mu b_\nu. \tag{14.3}$$

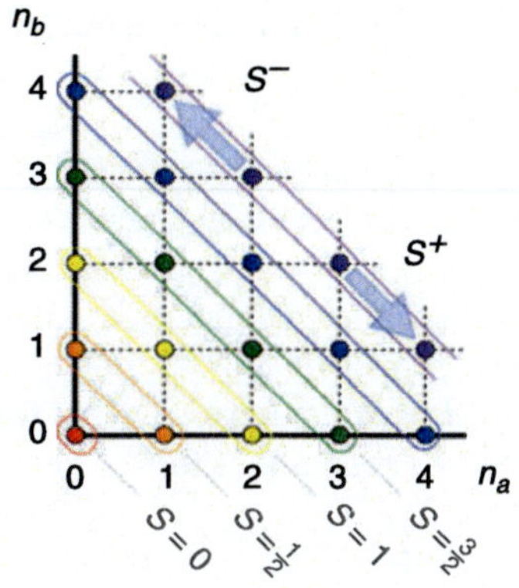


**Fig. 14.1** Schwinger representation of SU(2)

These satisfy the SU($N$) commutation relations

$$\left[S_{\mu\nu}\,,\,S_{\mu'\nu'}\right] = S_{\mu\nu'}\,\delta_{\mu'\nu} - S_{\mu'\nu}\,\delta_{\mu\nu'}. \tag{14.4}$$

The constraint is then

$$\sum_{\mu=1}^{N} b_\mu^\dagger b_\mu = n_b, \tag{14.5}$$

which specifies the representation of SU($N$). The corresponding Young tableau is one with $n_b$ boxes in a single row.

### 14.2.1 Bipartite Antiferromagnet

We consider the case of nearest neighbor SU(2) antiferromagnet, with interaction strength $J > 0$, on a bipartite lattice with sublattices $A$ and $B$. A bond $\langle ij \rangle$ is defined such that $i \in A$ and $j \in B$. The antiferromagnetic bond operator is defined as

$$\mathcal{A}_{ij} = a_i b_j - b_i a_j. \tag{14.6}$$

This is antisymmetric under interchange of the site indices $i$ and $j$, and transforms as a singlet under a global SU(2) rotation.

Consider now a rotation by $\pi$ about the $y$ axis on sublattice $B$ only, which sends

$$a_j \to -b_j, \qquad b_j \to a_j. \tag{14.7}$$

This is a canonical transformation which preserves the constraint (14.5). The antiferromagnetic bond operator takes the form

$$\mathcal{A}_{ij} \longrightarrow a_i a_j + b_i b_j. \tag{14.8}$$

The $\mathsf{SU}(2)$ Heisenberg model is written in the form

$$\begin{aligned} \mathcal{H} &= J \sum_{\langle ij \rangle} \mathbf{S}_i \cdot \mathbf{S}_j \\ &= -\frac{J}{2} \sum_{\langle ij \rangle} \left( \mathcal{A}_{ij}^\dagger \mathcal{A}_{ij} - 2S^2 \right). \end{aligned} \tag{14.9}$$

The extension to $\mathsf{SU}(N)$ for $N > 2$ is straightforward. With $N$ species of bosons, (14.8) generalizes to

$$\mathcal{A}_{ij} = \sum_{\mu=1}^{N} b_{i\mu} b_{j\mu}. \tag{14.10}$$

The nearest-neighbor $\mathsf{SU}(N)$ antiferromagnetic Heisenberg model is then

$$\begin{aligned} \mathcal{H} &= -\frac{J}{N} \sum_{\langle ij \rangle} \left( \mathcal{A}_{ij}^\dagger \mathcal{A}_{ij} - NS^2 \right) \\ &= \frac{J}{N} \sum_{\langle ij \rangle} \left( \sum_{\mu,\nu} S_i^{\mu\nu} \widetilde{S}_j^{\nu\mu} - NS^2 \right), \end{aligned} \tag{14.11}$$

where

$$\widetilde{S}_j^{\mu\nu} = -b_{j\nu}^\dagger b_{j\mu} \tag{14.12}$$

are the generators of the *conjugate representation* on sublattice $B$. One should note that $\mathcal{H}$ of (14.11) is not invariant under uniform $\mathsf{SU}(N)$ transformations $U$ but only under staggered conjugate rotations $U$ and $U^\dagger$ on sublattices $A$ and $B$, respectively.

### *14.2.2 Non-bipartite (Frustrated) Antiferromagnets*

For the group $\mathsf{SU}(2)$, one can always form a singlet from two sites in an identical spin-$S$ representation. That is, the tensor product of two spin-$S$ states always contains a singlet:

$$S \otimes S = 0 \oplus 1 \oplus \cdots \oplus 2S. \tag{14.13}$$

For $\mathsf{SU}(N)$ this is no longer the case. For example, for two $\mathsf{SU}(3)$ sites in the fundamental representation, one has $\mathbf{3} \otimes \mathbf{3} = \overline{\mathbf{3}} \oplus \mathbf{6}$. One needs three constituents to make an $\mathsf{SU}(3)$ singlet, as with color singlets in QCD, and $N$ constituents in the case of $\mathsf{SU}(N)$. This is why, in the case of the antiferromagnet, one chooses the conjugate representation on the B sublattice – the product of a representation and its conjugate always contains a singlet.

But what does one do if the lattice is not bipartite? This situation was addressed by Read and Sachdev [7], who extended the Schwinger boson theory to the group $\mathsf{Sp}(N)$. This amounts to generalizing the link operator $\mathcal{A}_{ij}$ in (14.6) to include a flavor index $m$:

$$\mathcal{A}_{ij} = \sum_{m=1}^{N} \left(a_{im}b_{jm} - b_{im}a_{jm}\right)$$
$$\equiv \sum_{\alpha,\alpha'=1}^{2N} \Lambda_{\alpha\alpha'}\, b_{i\alpha}\, b_{j\alpha'}. \tag{14.14}$$

Here, the indices $\alpha$ and $\alpha'$ run from 1 to $2N$. They may be written in composite form as $\alpha \to (m, \mu)$, where $m$ runs from 1 to $N$ and $\mu$ from 1 to 2 (or $\uparrow$ and $\downarrow$). In this case, on each site one has $b_{m\uparrow} = a_m$ and $b_{m\downarrow} = b_m$. The matrix $\Lambda_{\alpha\alpha'}$ is then $\Lambda_{m\mu,n\nu} = \delta_{mn}\,\epsilon_{\mu\nu}$, where $\epsilon_{\mu\nu} = i\sigma^y_{\mu\nu}$ is the rank two antisymmetric tensor.

If we make a global transformation on the Schwinger bosons, with $b_{i\alpha} \to U_{\alpha\alpha'}\, b_{i\alpha'}$, then we find

$$\mathcal{A}_{ij} \to (U^{\mathrm{t}}\Lambda U)_{\alpha\alpha'}\, b_{i\alpha}\, b_{j\alpha'}. \tag{14.15}$$

Thus, the link operators remain invariant under the class of complex transformations which satisfy $U^{\mathrm{t}}\Lambda U = \Lambda$. This is the definition of the group $\mathsf{Sp}(2N,\mathbb{C})$. If we further demand that $U \in \mathsf{U}(2N)$, which is necessary if the group operations are to commute with the total occupancy constraint, we arrive at the group

$$\mathsf{Sp}(N) = \mathsf{Sp}(2N,\mathbb{C}) \cap \mathsf{U}(2N). \tag{14.16}$$

The particular representation is again specified by the local boson occupation, $n_b = \sum_\alpha b^\dagger_{i\alpha} b_{i\alpha}$. The Hamiltonian is

$$\mathcal{H} = -\frac{1}{2N} \sum_{i<j} J_{ij}\, \mathcal{A}^\dagger_{ij} \mathcal{A}_{ij}. \tag{14.17}$$

Here, we have allowed for further neighbor couplings, which can be used to introduce frustration in the square lattice antiferromagnet, e.g. the $J_1 - J_2 - J_3$ model [7]. For each distinct coupling $J_{ij}$ (assumed translationally invariant), a new Hubbard–Stratonovich decomposition is required.

## 14.3 Mean Field Hamiltonian

Within a functional integral approach, one introduces a single real field $\lambda_i(\tau)$ on each site to enforce the occupancy constraint, and a complex Hubbard-Stratonovich field $Q_{ij}(\tau)$ on each link to decouple the interaction. At the mean field level it is assumed that these fields are static. This results in the mean field Hamiltonian

$$\mathcal{H}^{\mathrm{MF}} = \frac{pN}{J}\sum_{i<j}|Q_{ij}|^2 + \sum_{i<j}\left(Q_{ij}\,\mathcal{A}^{\dagger}_{ij} + Q^{*}_{ij}\,\mathcal{A}_{ij}\right) \\ + \sum_i \lambda_i\left(b^{\dagger}_{i\alpha}b_{i\alpha} - n_b\right) + (\mathcal{V}N)^{-1/2}\sum_{i\alpha}\left(\phi^{*}_{i\alpha}b_{i\alpha} + \phi_{i\alpha}b^{\dagger}_{i\alpha}\right), \tag{14.18}$$

where $\mathcal{V}$, the volume, is the number of Bravais lattice sites, and where $a$ runs from 1 to $N$ for the $\mathsf{SU}(N)$ or $\mathrm{O}(N,\mathbf{C})$ models (for which $p = 1$), and from 1 to $2N$ for the $\mathsf{Sp}(N)$ models (for which $p = 2$). The field $\phi_{i\alpha}$, which couples linearly to the Schwinger bosons, is conjugate to the condensate parameter $\langle b^{\dagger}_{i\alpha}\rangle$, which means

$$\frac{\partial F}{\partial \phi^{*}_{i\alpha}} = \frac{\langle b_{i\alpha}\rangle}{\sqrt{\mathcal{V}N}}. \tag{14.19}$$

Let us further assume that the mean field solution has the symmetry of the underlying lattice, and that the interactions are only between nearest neighbor sites on a Bravais lattice. Then, after Fourier transforming, we have

$$\mathcal{H}^{\mathrm{MF}} = \mathcal{V}N\left(\frac{pz}{2J}|Q|^2 - \frac{n_b}{N}\lambda\right) \\ + \frac{z}{2}\sum_{\mathbf{k},\alpha,\alpha'}\left[Q\,\xi_{\mathbf{k}}\,K_{\alpha\alpha'}\,b^{\dagger}_{\mathbf{k},\alpha}b^{\dagger}_{-\mathbf{k},\alpha'} + Q^{*}\,\xi^{*}_{\mathbf{k}}\,K_{\alpha\alpha'}\,b_{\mathbf{k},\alpha}b_{-\mathbf{k},\alpha'}\right] \\ + \lambda\sum_{\mathbf{k},\alpha} b^{\dagger}_{\mathbf{k},\alpha}b_{\mathbf{k},\alpha} + (\mathcal{V}N)^{-1/2}\sum_{\mathbf{k},\alpha}\left(\phi^{*}_{\mathbf{k},\alpha}\,b_{\mathbf{k},\alpha} + \phi_{\mathbf{k},\alpha}\,b^{\dagger}_{\mathbf{k},\alpha}\right), \tag{14.20}$$

where $z$ is the lattice coordination number, and where $K_{\alpha\alpha'} = \delta_{\alpha\alpha'}$ for $\mathsf{SU}(N)$ and $K_{\alpha\alpha'} = \Lambda_{\alpha\alpha'}$ for $\mathsf{Sp}(N)$. We define

$$\xi_{\mathbf{k}} \equiv \frac{2}{z}\sum_{\delta}{}' \eta_{\delta}\,\mathrm{e}^{\mathrm{i}\mathbf{k}\cdot\delta}, \tag{14.21}$$

where the sum is over all *distinct* nearest neighbor vectors in a unit cell. That is, $-\delta$ is not included in the sum. The quantity $\eta_{\delta} = \pm 1$ is a sign about which we shall have more to say presently. On the square lattice, for example, $\xi_{\mathbf{k}} = \eta_x\,\mathrm{e}^{\mathrm{i}k_x} + \eta_y\,\mathrm{e}^{\mathrm{i}k_y}$. For symmetric $K_{ab}$, owing to the sum on $\mathbf{k}$, we can replace $\xi_{\mathbf{k}}$ with its real part, while for antisymmetric $K_{ab}$ we must replace it with $i$ times its imaginary part. We, therefore, define

$$\gamma_{\mathbf{k}} = \frac{2}{z}\sum_{\delta}{}' \eta_{\delta}\,\cos(\mathbf{k}\cdot\delta) \quad \text{if } K = K^{\mathrm{t}} \tag{14.22}$$

$$= \frac{2i}{z}\sum_{\delta}{}' \eta_{\delta}\,\sin(\mathbf{k}\cdot\delta) \quad \text{if } K = -K^{\mathrm{t}} \tag{14.23}$$

The sign $\eta_\delta$ is irrelevant on bipartite lattices, since it can be set to unity for all $\delta$ simply by choosing an appropriate center for the Brillouin zone. But on frustrated lattices, the signs matter.

It is now quite simple to integrate out the Schwinger bosons. After we do so, we make a Legendre transformation to replace the field $\phi_{i\alpha}$ with the order parameter $\beta_{i\alpha} = \langle b_{i\alpha}\rangle/\sqrt{\mathcal{V}N}$, by writing

$$G = F - \sum_{i\alpha} \left(\phi_{i\alpha}\,\beta^*_{i\alpha} + \phi^*_{i\alpha}\,\beta_{i\alpha}\right). \tag{14.24}$$

The final form of the free energy per site, per flavor, is

$$g \equiv \frac{G}{\mathcal{V}N} = \frac{pz}{2J}|Q|^2 - \left(\kappa + \tfrac{p}{2}\right)\lambda + p\int_{\mathrm{BZ}} \frac{\mathrm{d}^d k}{(2\pi)^d}\left[\tfrac{1}{2}\omega_{\mathbf{k}} + T\,\ln\left(1 - \mathrm{e}^{-\omega_{\mathbf{k}}/T}\right)\right] + E_{\mathrm{con}}, \tag{14.25}$$

where $\kappa = n_b/N$, and $E_{\mathrm{con}}$ is the condensation energy,

$$E_{\mathrm{con}} = \lambda \sum_{\mathbf{k},\alpha} |\beta_{\mathbf{k},\alpha}|^2 + \tfrac{1}{2}z \sum_{\mathbf{k},\alpha,\alpha'} K_{\alpha\alpha'}\left(Q\,\gamma_{\mathbf{k}}\,\beta^*_{\mathbf{k},\alpha}\beta^*_{-\mathbf{k},\alpha'} + Q^*\,\gamma^*_{\mathbf{k}}\,\beta_{\mathbf{k},\alpha}\beta_{-\mathbf{k},\alpha'}\right). \tag{14.26}$$

The dispersion is given by

$$\omega_{\mathbf{k}} = \sqrt{\lambda^2 - \left|zQ\gamma_{\mathbf{k}}\right|^2}. \tag{14.27}$$

The fact that $g$ is formally of order $N^0$ (assuming $\kappa$ is as well) allows one to generate a systematic expansion of the free energy in powers of $1/N$.

### 14.3.1 Mean Field Equations

The mean field equations are obtained by extremizing the free energy $G$ with respect to the parameters $\lambda$, $Q$, and $\beta_{\mathbf{k},a}$. Thus,

$$\kappa + \tfrac{p}{2} = p\int_{\mathrm{BZ}} \frac{\mathrm{d}^d k}{(2\pi)^d}\,\frac{\lambda}{\omega_{\mathbf{k}}}\left(n_{\mathbf{k}}(T) + \tfrac{1}{2}\right) + \sum_{\mathbf{k},\alpha} |\beta_{\mathbf{k},\alpha}|^2 \tag{14.28}$$

$$\frac{pz}{J}|Q|^2 = p\int_{\mathrm{BZ}} \frac{\mathrm{d}^d k}{(2\pi)^d}\,\frac{|zQ\gamma_{\mathbf{k}}|^2}{\omega_{\mathbf{k}}}\left(n_{\mathbf{k}}(T) + \tfrac{1}{2}\right) + \lambda\sum_{\mathbf{k},\alpha} |\beta_{\mathbf{k},\alpha}|^2 \tag{14.29}$$

$$0 = \lambda\beta_{\mathbf{k},\alpha} + zQ\gamma_{\mathbf{k}} \sum_{\alpha'} K_{\alpha\alpha'}\,\beta^*_{-\mathbf{k},\alpha'}. \tag{14.30}$$

Here, $n_{\mathbf{k}}(T) = \left(e^{\omega_{\mathbf{k}}/T} - 1\right)^{-1}$ is the thermal Bose occupancy function. In deriving the second of the above mean field equations, we have also invoked the third. Assuming that the condensate occurs at a single wavevector **k**, the last equation requires that $\omega_{\mathbf{k}} = 0$ at the ordering wavevector, ensuring gaplessness of the excitation spectrum. When there is no condensate, $\beta_{\mathbf{k},\alpha} = 0$ for all **k** and $\alpha$.

It is instructive to compute $\langle S_{\mathbf{R}}^{\alpha\alpha'} \rangle = \langle b_{\mathbf{R},\alpha}^{\dagger} \, b_{\mathbf{R},\alpha'} - \kappa\, \delta_{\alpha\alpha'} \rangle$, which serves as the local order parameter. After invoking the mean field equations, one finds

$$\langle S_{\mathbf{R}}^{\alpha\alpha'} \rangle = N \sum_{\mathbf{k},\mathbf{k}'} e^{i(\mathbf{k}'-\mathbf{k})\cdot\mathbf{R}}\, \beta_{\mathbf{k}\alpha}^{*}\, \beta_{\mathbf{k}'\alpha'} - \sum_{\mathbf{k},\alpha''} \left|\beta_{\mathbf{k}\alpha''}\right|^{2} \delta_{\alpha\alpha'} . \tag{14.31}$$

Note that the trace of the above expression vanishes on average (*i.e.* upon summing over **R**), and vanishes locally provided that the condensate satisfies the orthogonality condition

$$\sum_{\alpha} \beta_{\mathbf{k}\alpha}^{*}\, \beta_{\mathbf{k}'\alpha} = \delta_{\mathbf{k}\mathbf{k}'} \sum_{\alpha} \left|\beta_{\mathbf{k}\alpha}\right|^{2} . \tag{14.32}$$

In the case of an $\mathsf{SU}(N)$ antiferromagnet on a (bipartite) hypercubic lattice, the condensate occurs only at the zone center $\mathbf{k} = 0$ and the zone corner $\mathbf{k} = \pi$. One then has

$$\langle S_{\mathbf{R}}^{\alpha\alpha'} \rangle = N\left(\beta_{0\alpha}^{*}\, \beta_{\pi\alpha'} + \beta_{\pi\alpha}^{*}\, \beta_{0\alpha'}\right) e^{i\pi\cdot\mathbf{R}} . \tag{14.33}$$

Thus, Bose condensation of the Schwinger bosons is equivalent to long-ranged magnetic order [11–14].

At $T = 0$, there is a critical value of $\kappa$ above which condensation occurs. To find this value, we invoke all three equations, but set the condensate fraction to zero. For the $\mathsf{SU}(N)$ models, the minimum of the dispersion occurs at the zone center, $\mathbf{k} = 0$. Setting $\omega_{\mathbf{k}=0} = 0$, we obtain the relation $\lambda = z|Q|$. The first equation then yields

$$\kappa_{c} = \tfrac{1}{2} \int_{\mathrm{BZ}} \frac{d^{d}k}{(2\pi)^{d}} \left(1 - |\gamma_{\mathbf{k}}|^{2}\right)^{-1/2} - \tfrac{1}{2} . \tag{14.34}$$

For $d = 1$, there is no solution, and there is never a condensate. For $d = 2$, one finds $\kappa_{c} = 0.19$ on the square lattice [3]. Since $\kappa = S$ for the $\mathsf{SU}(2)$ case, this suggests that even the minimal $S = \frac{1}{2}$ model is Néel ordered on the square lattice, a result which is in agreement with quantum Monte Carlo studies.

Consider next the $\mathsf{Sp}(N)$ model on the triangular lattice. We first must adopt a set of signs $\eta_{\delta}$. There are three bonds $\delta_{1,2,3}$ per unit cell, along the directions $\mathbf{a}_1$, $\mathbf{a}_2$, and $\mathbf{a}_1 - \mathbf{a}_2$, where the primitive direct lattice vectors are $\mathbf{a}_1 = a\,\hat{\mathbf{x}}$ and $\mathbf{a}_2 = \frac{1}{2} a\,\hat{\mathbf{x}} + \frac{\sqrt{3}}{2} a\,\hat{\mathbf{y}}$. Lattice symmetry suggests $\eta_1 = +1$, $\eta_2 = -1$, and $\eta_3 = +1$ (as opposed to all $\eta_{\delta} = 1$), resulting in [8]

$$\gamma_{\mathbf{k}} = \tfrac{1}{3} \sin\theta_1 - \tfrac{1}{3} \sin\theta_2 + \tfrac{1}{3} \sin(\theta_2 - \theta_1), \tag{14.35}$$

where the wavevector is written as

$$\mathbf{k} = \frac{\theta_1}{2\pi}\mathbf{G}_1 + \frac{\theta_2}{2\pi}\mathbf{G}_2, \tag{14.36}$$

with $\mathbf{G}_{1,2}$ being the two primitive reciprocal lattice vectors for the triangular lattice. The maximum of $|\gamma_{\mathbf{k}}|^2$, corresponding to the minimum of the dispersion $\omega_{\mathbf{k}}$, occurs when $\gamma_{\mathbf{k}}$ lies at one of the two inequivalent zone corners. In terms of the $\theta_i$, these points lie at $(\theta_1, \theta_2) = (\frac{4\pi}{3}, \frac{2\pi}{3})$, where $\gamma_{\mathbf{k}} = -\frac{\sqrt{3}}{2}$, and at $(\theta_1, \theta_2) = (\frac{2\pi}{3}, \frac{4\pi}{3})$, where $\gamma_{\mathbf{k}} = \frac{\sqrt{3}}{2}$. Sachdev [8] has found $\kappa_c = 0.34$ for the triangular structure. As one would guess, frustration increases the value of $\kappa_c$ relative to that on the square lattice. On the kagomé lattice, which is even more highly frustrated, he finds $\kappa_c = 0.53$.

## 14.4 The Mean Field Antiferromagnetic Ground State

For a finite system (no long range order or Bose condensation) one can explicitly write down the ground state of the SU($N$) Schwinger Boson Mean Field Theory $\Psi^{\mathrm{MF}}$. It is simply the vacuum of all the Bogoliubov operators

$$\beta_{\mathbf{k},\alpha}\Psi^{\mathrm{MF}} = 0 \qquad \forall\, \mathbf{k}, \alpha. \tag{14.37}$$

where,

$$\beta_{\mathbf{k}\alpha} = \cosh\theta_{\mathbf{k}}\, b_{\mathbf{k}\alpha} - \sinh\theta_{\mathbf{k}}\, b^{\dagger}_{-\mathbf{k}\alpha}, \tag{14.38}$$

and

$$\tanh 2\theta_{\mathbf{k}} = -\frac{zQ\gamma_{\mathbf{k}}}{\lambda}. \tag{14.39}$$

The ground state wavefunction $\Psi^{\mathrm{MF}}$ can be explicitly written in terms of the original Schwinger bosons as

$$\Psi^{\mathrm{MF}} = C \exp\left[\frac{1}{2}\sum_{ij} u_{ij} \sum_{m} b^{\dagger}_{ia} b^{\dagger}_{ja}\right] |0\rangle,$$
$$u_{ij} = \frac{1}{\mathcal{V}}\sum_{\mathbf{k}} \mathrm{e}^{\mathrm{i}\mathbf{k}\mathbf{R}_{ij}} \tanh\theta_{\mathbf{k}}. \tag{14.40}$$

For $N = 2$, using the *unrotated* operators $a^{\dagger}$ and $b^{\dagger}$, the mean field Schwinger boson ground state $\Psi^{\mathrm{MF}}$ is

$$\Psi^{\mathrm{MF}}_{N=2} = \exp\left[\sum_{\substack{i\in A\\ j\in B}} u_{ij}\left(a_i^\dagger b_j^\dagger - b_i^\dagger a_j^\dagger\right)\right]|0\rangle. \tag{14.41}$$

$\Psi^{\mathrm{MF}}$ contains many configurations with occupations different from $2S$ and is therefore not a pure spin state. As shown in [9], under Gutzwiller projection it reduces to a valence bond state. Since $\tanh\theta_{\mathbf{k}+\boldsymbol{\pi}} = -\tanh\theta_{\mathbf{k}}$, where $\boldsymbol{\pi} = (\pi, \pi, \ldots)$, the bond parameters $u_{ij}$ only connect sublattice $A$ to $B$. Furthermore, one can verify that for the nearest neighbor model above, $u_{ij} \geq 0$, and therefore the valence bond states obey Marshall's sign.

Although $\Psi^{\mathrm{MF}}$ is manifestly rotationally invariant, it may or may not exhibit long-ranged antiferromagnetic (Néel) order. This depends on the long-distance decay of $u_{ij}$. As was shown [9, 10] the SBMFT ground state for the nearest neighbor model is disordered in one dimension, and can exhibit long-range order in two dimensions for physically relevant values of $S$.

For further calculations, it is convenient to introduce the parametrizations:

$$\begin{aligned}
\omega_{\mathbf{k}} &\equiv c\sqrt{(2\xi)^{-2} + \frac{z}{2}(1-\gamma_{\mathbf{k}}^2)}\\
c &\equiv \sqrt{2z}\,|Q|\\
\xi^{-1} &\equiv \frac{2}{c}\sqrt{\lambda^2 - (z|Q|)^2}\\
t &= \frac{T}{z|Q|}.
\end{aligned} \tag{14.42}$$

Here, $c, \xi$, and $t$ describe the spin wave velocity, correlation length, and the dimensionless temperature, respectively. In Fig. 14.2 the dispersion for the one-dimensional antiferromagnet is drawn.

At the zone center and zone corner the mean field dispersion is that of free massive relativistic bosons,

$$\omega_{\mathbf{k}} \approx c\sqrt{(2\xi)^{-2} + |\mathbf{k}-\mathbf{k}_\gamma|^2}, \quad \mathbf{k}_\gamma = 0, \boldsymbol{\pi}. \tag{14.43}$$

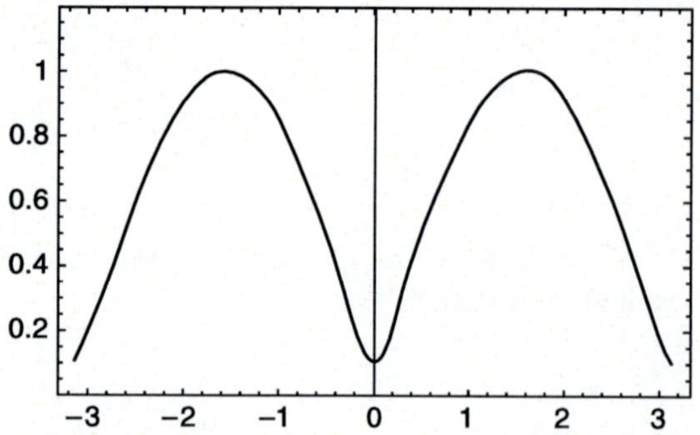


**Fig. 14.2** Mean field dispersion $\omega_k$, in the domain $-\pi < k < \pi$, for the one-dimensional antiferromagnet

When the gap (or "mass" $c/2\xi$) vanishes, $\omega_{\mathbf{k}}$ are Goldstone modes which reduce to dispersions of antiferromagnetic spin waves.

## 14.5 Staggered Magnetization in the Layered Antiferromagnet

Consider now a layered antiferromagnet on a cubic lattice where the in-plane nearest neighbor coupling is $J$ and the interlayer coupling is $\alpha J$, with $\alpha \ll 1$. We expect long range magnetic order at a finite Néel temperature $T_{\mathrm{N}}$. The order parameter, which is the staggered magnetization $\mathbf{M} = \langle(-1)^l \, \mathrm{e}^{\mathrm{i}\pi\cdot\mathbf{R}} \, \mathbf{S}_{\mathbf{R},l}\rangle$ becomes finite when the in-plane correlation length $\xi$, which diverges exponentially at low $T$, produces an effective coupling between neighboring layers $\alpha \, \xi^2(T_{\mathrm{N}})$ which is of order unity. Here, $\mathbf{R}$ locates the site within a plane, and $l$ is the layer index. This means in effect that the coarse grained spins start to interact as if in an isotropic three dimensional cubic lattice which orders at $T_{\mathrm{N}}$. The interlayer mean field theory, introduced by Scalapino, Imry and Pincus (SIP) [15] in the 1970s, can be applied within the SBMFT. Here we follow Keimer et al. [16], and Ofer et al. [17], to compute the temperature dependent staggered magnetization, in the range $T \in [0, T_{\mathrm{N}}]$.

The Hamiltonian is given by

$$\mathcal{H} = \sum_{\mathbf{R},l} \left( \mathbf{S}_{\mathbf{R},l} \cdot \mathbf{S}_{\mathbf{R}+\hat{\mathbf{x}},l} + \mathbf{S}_{\mathbf{R},l} \cdot \mathbf{S}_{\mathbf{R}+\hat{\mathbf{y}},l} + \alpha \, \mathbf{S}_{\mathbf{R},l} \cdot \mathbf{S}_{\mathbf{R},l+1} \right) \tag{14.44}$$

The interplane coupling is decomposed using Hartree-Fock staggered magnetization field:

$$\alpha \, S^z_{\mathbf{R},l} \, S^z_{\mathbf{R},l+1} \longrightarrow (-1)^l \, \mathrm{e}^{\mathrm{i}\pi\cdot\mathbf{R}} \left( S^z_{\mathbf{R},l} - S^z_{\mathbf{R},l+1} \right) h_{\mathbf{R},l} - \frac{h^2_{\mathbf{R},l}}{\alpha}, \tag{14.45}$$

where it is assumed that $\mathbf{M} = M \, \hat{\mathbf{z}}$. Here, $h_{\mathbf{R},l}$ is the local Néel field due to any ordering in the neighboring layers. Assuming a uniform solution, $h_{\mathbf{R},l} = h$, self-consistency is achieved when

$$\frac{h}{\alpha} = 2 \, M(T,h) = \langle \, a^\dagger_{\mathbf{R}} \, a_{\mathbf{R}} \, \rangle - \langle \, b^\dagger_{\mathbf{R}} \, b_{\mathbf{R}} \, \rangle, \tag{14.46}$$

where $M(T,h)$ is the staggered magnetization response to an ordering staggered field $h$ on a single layer.

Extracting $T_{\mathrm{N}}$ is relatively easy, since as $T \to T_{\mathrm{N}}$, $h \to 0$, and the expressions for $\langle \, a^\dagger_{\mathbf{R}} \, a_{\mathbf{R}} \, \rangle$ and $\langle \, b^\dagger_{\mathbf{R}} \, b_{\mathbf{R}} \, \rangle$. In this limit, one finds that the second mean field equation in (14.29) is not affected by the staggered field, which simplifies the calculation. At $T_{\mathrm{N}}$ one finds

$$2\alpha \, \chi^{\mathrm{s}}_{\mathrm{2D}}(T_{\mathrm{N}}) = 1. \tag{14.47}$$

Since we know that $\chi^{s}_{2D} \propto \xi^{2}_{2D}(T_N)$, we recover the ordering temperature of the SIP theory. The more precise calculation yields (restoring the Heisenberg exchange energy scale $J$),

$$T_N = \frac{J}{\log \alpha}\left(\frac{2\pi M_0}{|\log\left(\frac{1}{4}\pi^2 \log(4\alpha/\pi)\, M_0^2\right)|}\right) \tag{14.48}$$

The numerically determined $M(T) = h(T)/2\alpha$ is shown in Fig. 14.3 for various anisotropy parameters.

One can also analyze the layered antiferromagnet using the SBMFT's native decoupling scheme, without proceeding via the interlayer mean field theory of (14.45). Starting from an anisotropic Heisenberg model with in-plane exchange $J_{\parallel}$ and interlayer exchange $J_{\perp}$, one assumes a mean field solution where $Q_{ij} = Q_{\parallel}$ when $\langle ij\rangle$ is an in-plane bond, and $Q_{ij} = Q_{\perp}$ when $\langle ij\rangle$ is an out-of-plane bond. The second mean field equation, (14.29), then becomes two equations. The Néel temperature can be written $T_N = J_{\parallel}\, f(J_{\perp}/J_{\parallel})$, where $f(\alpha)$ is a dimensionless function. To find $f(\alpha)$, we demand that the spectrum be gapless, but the condensate vanishes. This results in two coupled equations,

$$\kappa + \frac{1}{2} = \int_{-1}^{1} \mathrm{d}\gamma_{\parallel}\, \rho_{\parallel}(\gamma_{\parallel}) \int_{-1}^{1} d\gamma_{\perp}\, \rho_{\perp}(\gamma_{\perp})\, \frac{1+\epsilon}{\Omega(\gamma_{\parallel},\gamma_{\perp})} \left(n(\gamma_{\parallel},\gamma_{\perp}) + \tfrac{1}{2}\right) \tag{14.49}$$

$$\frac{J_{\perp}}{2J_{\parallel}} = \frac{\int_{-1}^{1} \mathrm{d}\gamma_{\parallel}\, \rho_{\parallel}(\gamma_{\parallel}) \int_{-1}^{1} d\gamma_{\perp}\, \rho_{\perp}(\gamma_{\perp})\, (\gamma_{\parallel}^2/\Omega)\,(n+\frac{1}{2})}{\int_{-1}^{1} \mathrm{d}\gamma_{\parallel}\, \rho_{\parallel}(\gamma_{\parallel}) \int_{-1}^{1} d\gamma_{\perp}\, \rho_{\perp}(\gamma_{\perp})\, (\gamma_{\perp}^2/\Omega)\,(n+\frac{1}{2})}, \tag{14.50}$$

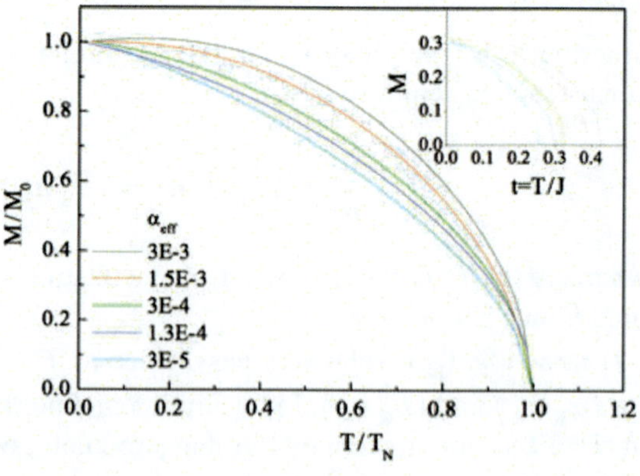


**Fig. 14.3** Numerical solution, from Ofer et al. [17], of the staggered magnetization $M(T)$ of the layered antiferomagnet for various values of anisotropy parameter $\alpha_{\text{eff}}$. $M_0 = S - 0.19660$ is the zero temperature staggered magnetization, and $T_N$ is the Néel temperature

where $\epsilon = |Q_\perp/2Q_\parallel|$ and $\Omega(\gamma_\parallel,\gamma_\perp) = \sqrt{(1+\epsilon)^2-(\gamma_\parallel+\epsilon\,\gamma_\perp)^2}$, and where $n = \left(e^{\Omega/t_c}-1\right)^{-1}$, with $t_c = T_N/4\,|Q_\parallel|$. Once the above two equations are solved for $\epsilon$ and $t_c$, we determine $Q_\parallel$ from

$$Q_\parallel = J_\parallel \int_{-1}^{1} d\gamma_\parallel\, \rho_\parallel(\gamma_\parallel) \int_{-1}^{1} d\gamma_\perp\, \rho_\perp(\gamma_\perp)\, \frac{\gamma_\parallel^2}{\Omega}\left(n+\frac{1}{2}\right). \tag{14.51}$$

The functions $\rho_\parallel$ and $\rho_\perp$ are given by

$$\rho_\parallel(\gamma_\parallel) = \frac{2}{\pi^2}\,\mathrm{K}\left(1-\gamma_\parallel^2\right), \qquad \rho_\perp(\gamma_\perp) = \frac{1}{\pi\sqrt{1-\gamma_\perp^2}}. \tag{14.52}$$

**Acknowledgements**

We are grateful to A. Keren for help in preparing these notes. Support from the Israel Science Foundation, the US Israel Binational Science Foundation, and the fund for Promotion of Research at Technion is acknowledged. We are grateful for the hospitality of Aspen Center for Physics, and Lewiner Institute for Theoretical Physics, Technion, where these notes were written.

## References

1. P. Coleman, Phys. Rev. B 35, 5072 (1987)
2. N. Read, D. Newns, J. Phys. C 16, 3273 (1983)
3. D.P. Arovas, A. Auerbach, Phys. Rev. B 38, 316 (1988)
4. A. Auerbach, D.P. Arovas, Phys. Rev. Lett. 61, 617 (1988)
5. A. Auerbach, Interacting Electrons and Quantum Magnetism (Springer-Verlag, New York, 1994)
6. N. Read, S. Sachdev, Nucl. Phys. B 316, 609 (1989)
7. N. Read, S. Sachdev, Phys. Rev. Lett. 66, 1773 (1991)
8. S. Sachdev, Phys. Rev. B 45, 12377 (1992)
9. M. Raykin, A. Auerbach, Phys. Rev. B 47, 5118 (1993)
10. M. Havilio, A. Auerbach, Phys. Rev. B 62, 324 (2000)
11. J.E. Hirsch, S. Tang, Phys. Rev. B 39, 2850 (1989)
12. M. Raykin, A. Auerbach, Phys. Rev. Lett. 70, 3808 (1993)
13. J.E. Hirsch, S. Tang, Phys. Rev. B 39, 2850 (1989)
14. M. Raykin, A. Auerbach, Phys. Rev. Lett. 70, 3808 (1993)
15. D.J. Scalapino, Y. Imry, P. Pincus, Phys. Rev. B 11 2042 (1975)
16. B. Keimer et al., Phys. Rev. B 45, 7430 (1992)
17. R. Ofer et al., Phys. Rev. 74, 220508 (2006)

# Chapter 15
# Variational Wave Functions for Frustrated Magnetic Models

**Federico Becca, Luca Capriotti, Alberto Parola, and Sandro Sorella**

**Abstract** Variational wave functions containing electronic pairing and suppressed charge fluctuations (i.e., projected BCS states) have been proposed as the paradigm for disordered magnetic systems (including spin liquids). Here, we discuss the general properties of these states in one and two dimensions, and show that different quantum phases may be described with high accuracy by the same class of variational wave functions, including dimerized and magnetically ordered states. In particular, phases with magnetic order may be obtained from a straightforward generalization containing both antiferromagnetic and superconducting order parameters, as well as suitable spin Jastrow correlations. In summary, projected wave functions represent an extremely flexible tool for understanding the physics of low-dimensional magnetic systems.

## 15.1 Introduction

The variational approach is a widely used tool to investigate the low-energy properties of quantum systems with several active degrees of freedom, including electrons and ions. The basic idea is to construct fully quantum many-body states by a physically motivated *ansatz*. The resulting wave function should be simple enough to allow efficient calculations even for large sizes. Most of the variational calculations are traditionally based upon mean-field approximations, where the many-body wave function is constructed by using independent single-particle states. In this respect, even the BCS theory of superconductivity belongs in this category [1]. Although these mean-field approaches have been instrumental in understanding and describing weakly correlated systems, they have proved inadequate whenever the electron–electron interaction dominates the kinetic energy. The generalization of variational states in this regime is not straightforward, and represents an open problem in the modern theory of Condensed Matter. Probably the most celebrated case is the wave function proposed by Laughlin to describe the fractional quantum Hall effect as an incompressible quantum fluid with fractional excitations [2]. One important example in which electron correlations prevent the use of simple, mean-field approaches is provided by the so-called resonating valence-bond (RVB) state. This intriguing phase, which was conjectured many years ago by Fazekas and Anderson [3, 4], has no magnetic order, no broken lattice symmetries, and remains disordered even at zero temperature. It is now commonly accepted that these spin-liquid states may be stabilized in quantum antiferromagnets with competing (frustrating) interactions [5, 6].

Here, we present one possible approach to the definition of accurate variational wave functions which take into account quite readily both strong electron correlations and the frustrated nature of the lattice. The price to pay when considering these effects is that calculations cannot be performed analytically, and more sophisticated numerical methods, such as the quantum Monte Carlo technique, are required.

Let us begin by considering a simple, frustrated spin model, in which the combined effects of a small spin value, reduced dimensionality, and the presence of competing interactions could lead to non-magnetic phases. We consider what is known as the $J_1-J_2$ frustrated Heisenberg model on a chain or a square lattice,

$$\mathcal{H} = J_1 \sum_{\text{n.n.}} \mathbf{S}_R \cdot \mathbf{S}_{R'} + J_2 \sum_{\text{n.n.n.}} \mathbf{S}_R \cdot \mathbf{S}_{R'}, \tag{15.1}$$

where $J_1$ and $J_2$ are the (positive) nearest-neighbor ($n.n.$) and next-nearest-neighbor ($n.n.n.$) couplings, and $\mathbf{S}_R = (S^x_R, S^y_R, S^z_R)$ are $S = 1/2$ operators; periodic boundary conditions are assumed. Besides the purely theoretical interest, this model is also known to describe the relevant antiferromagnetic interactions in a variety of quasi-one-dimensional [7] and quasi-two-dimensional systems [8, 9].

In one dimension, the phase diagram of the $J_1-J_2$ model has been well established by analytical studies and by Density Matrix Renormalization Group (DMRG) calculations [10]. For small values of the ratio $J_2/J_1$, the system is in a Luttinger spin-fluid phase with a gapless spectrum, no broken symmetry, and power-law spin correlations. By increasing the value of the second-neighbor coupling, a gapped phase is stabilized [7, 10]. The value of the critical point has been determined with high accuracy as $(J_2/J_1)_c = 0.241167 \pm 0.000005$ [11]. The gapped ground state is twofold degenerate and spontaneously dimerized, and at $J_2/J_1 = 0.5$ is expressed by the exact Majumdar–Ghosh wave function [12, 13]. Interestingly, for $J_2/J_1 > 0.5$, incommensurate but short-range spin correlations have been found, whereas the dimer–dimer correlations are always commensurate [10].

By contrast, the phase diagram of the two dimensional $J_1-J_2$ model is the subject of much debate. For $J_2/J_1 \ll 0.5$, an antiferromagnetic Néel order with magnetic wave vector $Q = (\pi, \pi)$ is expected. In the opposite limit, $J_2/J_1 \gg 0.5$, the ground state is a collinear antiferromagnetic phase where the spins are aligned ferromagnetically in one direction and antiferromagnetically in the other [$Q = (\pi, 0)$ or $Q = (0, \pi)$]. The nature of the ground state in the regime of strong frustration, i.e., for $J_2/J_1 \sim 0.5$, remains an open problem, and there is no general consensus on its characterization. Since the work of Chandra and Doucot [14], it has been suggested that a non-magnetic phase should be present around $J_2/J_1 = 0.5$. Unfortunately, exact diagonalization calculations are limited to small clusters which cannot provide definitive answers to this very delicate problem [15–17]. By using series-expansion methods [18–21] and field-theoretical approaches [22], it has been argued that a valence-bond solid, with columnar dimer order and spontaneous symmetry-breaking, could be stabilized. More recently, it has been shown that a clear enhancement of plaquette–plaquette correlations is found by introducing a further, third-nearest-neighbor superexchange term $J_3$, thus suggesting a possible plaquette valence-bond crystal [23].

The primary obstacle to the characterization of the phase diagram in two dimensions is that the lack of exact results is accompanied, in the frustrated case, by difficulties in applying standard stochastic numerical techniques. Quantum Monte Carlo methods can be applied straightforwardly only to spin-1/2 Hamiltonians of the form (15.1), with strong restrictions on the couplings (e.g., $J_1 \geq 0$ and $J_2 \leq 0$ or $J_1 \leq 0$ and $J_2 \leq 0$) in order to avoid a numerical instability known as the *sign problem*. This is because, in general, quantum Monte Carlo methods do not suffer from numerical instabilities only when it is possible to work with a basis in the Hilbert space where the off-diagonal matrix elements of the Hamiltonian are all non-positive. As an example, in a quantum antiferromagnet with $J_1 \geq 0$ and $J_2 = 0$ on a bipartite lattice, after the unitary transformation

$$\mathcal{U}^\dagger = \exp\Big[-\mathrm{i}\pi \sum_{R\in \mathrm{B}} S^z_R\Big] \tag{15.2}$$

(B being one of the two sublattices), the transformed Hamiltonian has non-positive off-diagonal matrix elements in the basis $|x\rangle$ whose states are specified by the value of $S^z_R$ on each site, and $\sum_R S^z_R = S$ [24]. This implies that the ground state of $\mathcal{U}\mathcal{H}\mathcal{U}^\dagger$, $|\tilde{\Psi}_0\rangle = \sum_x \tilde{\Psi}_0(x)|x\rangle$, has all-positive amplitudes, $\tilde{\Psi}_0(x) > 0$, meaning that there exists a purely bosonic representation of the ground state. This property leads to the well-known Marshall–Peierls sign rule [24, 25] for the phases of the ground state of $\mathcal{H}$, $\mathrm{sign}\{\Psi_0(x)\} = (-1)^{N_\uparrow(x)}$, where $N_\uparrow(x)$ is the number of up spins on one of the two sublattices. The Marshall–Peierls sign rule holds for the unfrustrated Heisenberg model and even for the $J_1{-}J_2$ chain at the Majumdar–Ghosh point. However, in the regime of strong frustration, the Marshall–Peierls sign rule is violated dramatically [26], and, because no analogous sign rule appears to exist, the ground-state wave function has non-trivial phases. This property turns out to be a crucial ingredient of frustration.

In this respect, a very useful way to investigate the highly frustrated regime is to consider variational wave functions, whose accuracy can be assessed by employing stable (but approximate) Monte Carlo techniques such as the fixed-node approach [27]. Variational wave functions can be very flexible, allowing the description of magnetically ordered, dimerized, and spin-liquid states. In particular, it is possible to construct variational states with non-trivial signs for the investigation of the strongly frustrated regime.

In the following, we will describe in detail the case in which the variational wave function is constructed by projecting *fermionic* mean-field states [28]. Variational calculations can be treated easily by using standard Monte Carlo techniques. This is in contrast to variational states based on a *bosonic* representation, which are very difficult to handle whenever the ground state has non-trivial phases [29]. Indeed, variational Monte Carlo calculations based on bosonic wave functions suffer from the sign problem in the presence of frustration [30], and stable numerical simulations can be performed only in special cases, for example in bipartite lattices when the valence bonds only connect opposite sublattices [29]. Another advantage of the fermionic representation is that the mean-field Hamiltonian allows one to have a simple and straightforward representation also for the low-lying excited states (see the discussion in Sect. 15.6.1, and also [31] for a frustrated model on a three-leg ladder).

## 15.2 Symmetries of the Wave Function: General Properties

We define the class of projected-BCS (pBCS) wave functions on an $N$-site lattice, starting from the ground state of a suitable translationally invariant BCS Hamiltonian

$$\begin{aligned}\mathcal{H}_{\mathrm{BCS}} &= \sum_{R,R'\sigma} (t_{R-R'} - \mu\,\delta_{R-R'})\, c^{\dagger}_{R,\sigma} c_{R',\sigma} - \sum_{R,R'} \Delta_{R-R'}\, c^{\dagger}_{R,\uparrow} c^{\dagger}_{R',\downarrow} + H.c. \\ &= \sum_{k\sigma} (\epsilon_k - \mu)\, c^{\dagger}_{k,\sigma} c_{k,\sigma} - \sum_{k} \Delta_k c^{\dagger}_{k,\uparrow} c^{\dagger}_{-k,\downarrow} + H.c.,\end{aligned} \tag{15.3}$$

where $c^{\dagger}_{R,\sigma}$ ($c_{R,\sigma}$) creates (destroys) an electron at site $R$ with spin $\sigma$, the bare electron band $\epsilon_k$ is a real and even function of $k$, and $\Delta_k$ is also taken to be even to describe singlet electron pairing. In order to obtain a class of non-magnetic, translationally invariant, and singlet wave functions for spin-1/2 models, the ground state $|\mathrm{BCS}\rangle$ of Hamiltonian (15.3) is projected onto the physical Hilbert space of singly-occupied sites by the Gutzwiller operator $P_{\mathrm{G}} = \prod_R (n_{R,\uparrow} - n_{R,\downarrow})^2$, $n_{R,\sigma}$ being the local density. Thus

$$|p\mathrm{BCS}\rangle = P_{\mathrm{G}}\,|\mathrm{BCS}\rangle = P_{\mathrm{G}} \prod_k (u_k + v_k c^{\dagger}_{k,\uparrow} c^{\dagger}_{-k,\downarrow})|0\rangle, \tag{15.4}$$

where the product is over all the $N$ wave vectors in the Brillouin zone. The diagonalization of Hamiltonian (15.3) gives explicitly

$$u_k = \sqrt{\frac{E_k + \epsilon_k}{2E_k}} \qquad v_k = \frac{\Delta_k}{|\Delta_k|}\sqrt{\frac{E_k - \epsilon_k}{2E_k}} \qquad E_k = \sqrt{\epsilon_k^2 + |\Delta_k|^2},$$

while the BCS pairing function $f_k$ is given by

$$f_k = \frac{v_k}{u_k} = \frac{\Delta_k}{\epsilon_k + E_k}. \tag{15.5}$$

The first feature we wish to discuss is the *redundancy* implied by the electronic representation of a spin state, by which is meant the extra symmetries which appear when we write a spin state as the Gutzwiller projection of a fermionic state. This property is reflected in turn in the presence of a *local* gauge symmetry of the fermionic problem [32–34]. Indeed, the original spin Hamiltonian (15.1) is invariant under the local SU(2) gauge transformations

$$\Sigma^z_\phi : \begin{pmatrix} c^\dagger_\uparrow \\ c_\downarrow \end{pmatrix} \rightarrow \mathrm{e}^{\mathrm{i}\phi\sigma_z} \begin{pmatrix} c^\dagger_\uparrow \\ c_\downarrow \end{pmatrix} = \begin{pmatrix} \mathrm{e}^{\mathrm{i}\phi} & 0 \\ 0 & \mathrm{e}^{-\mathrm{i}\phi} \end{pmatrix} \begin{pmatrix} c^\dagger_\uparrow \\ c_\downarrow \end{pmatrix}, \tag{15.6}$$

$$\Sigma^x_\theta : \begin{pmatrix} c^\dagger_\uparrow \\ c_\downarrow \end{pmatrix} \rightarrow \mathrm{e}^{\mathrm{i}\theta\sigma_x} \begin{pmatrix} c^\dagger_\uparrow \\ c_\downarrow \end{pmatrix} = \begin{pmatrix} \cos\theta & \mathrm{i}\sin\theta \\ \mathrm{i}\sin\theta & \cos\theta \end{pmatrix} \begin{pmatrix} c^\dagger_\uparrow \\ c_\downarrow \end{pmatrix}. \tag{15.7}$$

A third transformation can be expressed in terms of the previous ones,

$$\Sigma^y_\psi : \begin{pmatrix} c^\dagger_\uparrow \\ c_\downarrow \end{pmatrix} \rightarrow \mathrm{e}^{\mathrm{i}\psi\sigma_y} \begin{pmatrix} c^\dagger_\uparrow \\ c_\downarrow \end{pmatrix} = \mathrm{e}^{-\mathrm{i}\pi\sigma_z/4}\mathrm{e}^{\mathrm{i}\psi\sigma_x}\mathrm{e}^{\mathrm{i}\pi\sigma_z/4} \begin{pmatrix} c^\dagger_\uparrow \\ c_\downarrow \end{pmatrix}, \tag{15.8}$$

where $\sigma_x$, $\sigma_y$, and $\sigma_z$ are the Pauli matrices. As a consequence, all the different fermionic states connected by a local SU(2) transformation generated by (15.6) and (15.7) with site-dependent parameters give rise to the same spin state after Gutzwiller projection,

$$P_{\mathrm{G}} \prod_R \Sigma^z_{\phi_R} \Sigma^x_{\theta_R} |\mathrm{BCS}\rangle = \mathrm{e}^{\mathrm{i}\Phi} P_{\mathrm{G}} |\mathrm{BCS}\rangle, \tag{15.9}$$

where $\Phi$ is an overall phase. Clearly, the local gauge transformations defined previously change the BCS Hamiltonian, breaking in general the translational invariance. In the following, we will restrict our considerations to the class of transformations which preserve the translational symmetry of the lattice in the BCS Hamiltonian, i.e., the subgroup of global symmetries corresponding to site-independent angles $(\phi, \theta)$. By applying the transformations (15.6) and (15.7), the BCS Hamiltonian retains its form with modified couplings

$$t_{R-R'} \rightarrow t_{R-R'} \tag{15.10}$$

$$\Delta_{R-R'} \rightarrow \Delta_{R-R'}\mathrm{e}^{2\mathrm{i}\phi} \tag{15.11}$$

for $\Sigma^z_\phi$, while the transformation $\Sigma^x_\theta$ gives

$$\begin{aligned} t_{R-R'} &\rightarrow \cos 2\theta\, t_{R-R'} + \mathrm{i}\sin\theta\cos\theta\,(\Delta_{R-R'} - \Delta^*_{R-R'}) \\ &= \cos 2\theta\, t_{R-R'} - \sin 2\theta\, \mathrm{Im}\Delta_{R-R'} \end{aligned} \tag{15.12}$$

$$\begin{aligned} \Delta_{R-R'} &\rightarrow (\cos^2\theta\, \Delta_{R-R'} + \sin^2\theta\, \Delta^*_{R-R'}) + \mathrm{i}\sin 2\theta\, t_{R-R'} \\ &= \mathrm{Re}\Delta_{R-R'} + \mathrm{i}\,(\cos 2\theta\, \mathrm{Im}\Delta_{R-R'} + \sin 2\theta\, t_{R-R'}). \end{aligned} \tag{15.13}$$

These relations are linear in $t_{R-R'}$ and $\Delta_{R-R'}$, and therefore hold equally for the Fourier components $\epsilon_k$ and $\Delta_k$. We note that, because $\Delta_r$ is an even function, the real (imaginary) part of its Fourier transform $\Delta_k$ is equal to the Fourier transform of the real (imaginary) part of $\Delta_r$. It is easy to see that these two transformations generate the full rotation group on the vector whose components are $(\epsilon_k, \mathrm{Re}\Delta_k, \mathrm{Im}\Delta_k)$. As a consequence, the length $E_k$ of this vector is conserved by the full group.

In summary, there is an infinite number of different translationally invariant BCS Hamiltonians that, after projection, give the same spin state. Choosing a specific representation does not affect the physics of the state, but changes the pairing function $f_k$ of (15.5) before projection. Within this class of states, the only scalar under rotations is the BCS energy spectrum $E_k$. Clearly, the projection operator will modify the excitation spectrum associated with the BCS wave function. Nevertheless, its invariance with respect to SU(2) transformations suggests that $E_k$ may reflect the nature of the physical excitation spectrum.

Remarkably, in one dimension it is easy to prove that such a class of wave functions is able to represent faithfully both the physics of Luttinger liquids, appropriate for the nearest-neighbor Heisenberg model, and the gapped spin-Peierls state, which is stabilized for sufficiently strong frustration. In fact, it is known [35] that the simple choice of nearest-neighbor hopping ($\epsilon_k = -2t \cos k$, $\mu = 0$) and vanishing gap function $\Delta_k$ reproduces the exact solution of the Haldane–Shastry model (with a gapless $E_k$), while choosing a next-nearest neighbor hopping ($\epsilon_k = -2t \cos 2k$, $\mu = 0$) and a sizable nearest-neighbor pairing ($\Delta_k = 4\sqrt{2}t \cos k$) recovers the Majumdar–Ghosh state (with a gapped $E_k$) [12, 13].

## 15.3 Symmetries in the Two-dimensional Case

We now specialize to the two-dimensional square lattice and investigate whether it is possible to exploit further the redundancy in the fermionic representation of a spin state in order to define a pairing function which breaks some spatial symmetry of the lattice but which, after projection, still gives a wave function with all of the correct quantum numbers. We will show that, if suitable conditions are satisfied, a fully symmetric projected BCS state is obtained from a BCS Hamiltonian with fewer symmetries than the original spin problem. For this purpose, it is convenient to introduce a set of unitary operators related to the symmetries of the model.

- Spatial symmetries: for example, $\mathcal{R}_x(x, y) = (x, -y)$ and $\mathcal{R}_{xy}(x, y) = (y, x)$. We define the transformation law of creation operators as $\mathcal{R} c^\dagger_{X,\sigma} \mathcal{R}^{-1} = c^\dagger_{\mathcal{R}(X),\sigma}$, and the action of the symmmetry operator on the vacuum is $\mathcal{R}_x|0\rangle = \mathcal{R}_{xy}|0\rangle = |0\rangle$. Note that these operators map each sublattice onto itself.
- Particle–hole symmetry: $P_h c^\dagger_{X,\sigma} P_h^{-1} = i\,(-1)^X c_{X,-\sigma}$, where the action of the $P_h$ operator on the vacuum is $P_h|0\rangle = \prod_X c^\dagger_{X,\uparrow} c^\dagger_{X,\downarrow}|0\rangle$.
- Gauge transformation: $G\, c^\dagger_{X,\sigma} G^{-1} = i\, c^\dagger_{X,\sigma}$ with $G\,|0\rangle = |0\rangle$.

Clearly, $\mathcal{R}_x$ and $\mathcal{R}_{xy}$ are symmetries of the physical problem (e.g., the Heisenberg model). $G$ is a symmetry because the physical Hamiltonian has a definite number of electrons, while $P_h$ leaves invariant every configuration where each site is singly occupied if the total magnetization vanishes ($N_\downarrow = N_\uparrow = n$). With the definition adopted, $P_h$ acts only to multiply every spin state by the phase factor $(-1)^{N_\downarrow}$. Thus all of the operators defined above commute both with the Heisenberg Hamiltonian and, because reflections do not interchange the two sublattices, with each other. The ground state of the Heisenberg model on a finite lattice, if it is unique, must be a simultaneous eigenstate of all the symmetry operators. We will establish the sufficient conditions which guarantee that the projected BCS state is indeed an eigenstate of all of these symmetries.

Let us consider a hopping term which only connects sites in opposite sublattices, whence $\epsilon_{k+Q} = -\epsilon_k$, and a gap function with contributions from different symmetries ($s$, $d_{x^2-y^2}$, and $d_{xy}$), $\Delta = \Delta^s + \Delta^{x^2-y^2} + \Delta^{xy}$. Further, we consider a case in which $\Delta^s$ and $\Delta^{x^2-y^2}$ couple opposite sublattices, while $\Delta^{xy}$ is restricted to the same sublattice. In this case, the BCS Hamiltonian $\mathcal{H}_{\rm BCS} = \mathcal{H}(t, \Delta^s, \Delta^{x^2-y^2}, \Delta^{xy})$ transforms under the different unitary operators according to

$$
\begin{aligned}
\mathcal{R}_x \mathcal{H}(t, \Delta^s, \Delta^{x^2-y^2}, \Delta^{xy}) \mathcal{R}_x^{-1} &= \mathcal{H}(t, \Delta^s, \Delta^{x^2-y^2}, -\Delta^{xy}),\\
\mathcal{R}_{xy} \mathcal{H}(t, \Delta^s, \Delta^{x^2-y^2}, \Delta^{xy}) \mathcal{R}_{xy}^{-1} &= \mathcal{H}(t, \Delta^s, -\Delta^{x^2-y^2}, \Delta^{xy}),\\
P_h \mathcal{H}(t, \Delta^s, \Delta^{x^2-y^2}, \Delta^{xy}) P_h^{-1} &= \mathcal{H}(t, \Delta^{s*}, \Delta^{x^2-y^2*}, -\Delta^{xy*}),\\
G \mathcal{H}(t, \Delta^s, \Delta^{x^2-y^2}, \Delta^{xy}) G^{-1} &= \mathcal{H}(t, -\Delta^s, -\Delta^{x^2-y^2}, -\Delta^{xy}).
\end{aligned}
$$

From these transformations it is straightforward to define suitable composite symmetry operators which leave the BCS Hamiltonian invariant. For illustration, in the case where $\Delta$ is real, one may select $\mathcal{R}_x P_h$ and $\mathcal{R}_{xy}$ if $\Delta^{x^2-y^2} = 0$ or $\mathcal{R}_x P_h$ and $\mathcal{R}_{xy} P_h G$ if $\Delta^s = 0$. It is not possible to set both $\Delta^{x^2-y^2}$ and $\Delta^s$ simultaneously different from zero and still obtain a state with all the symmetries of the original problem. The eigenstates $|\text{BCS}\rangle$ of (15.3) will in general be simultaneous eigenstates of these two composite symmetry operators with given quantum numbers, for example $\alpha_x$ and $\alpha_{xy}$. The effect of projection over these states is

$$
\begin{aligned}
\alpha_x P_{\rm G}|\text{BCS}\rangle &= P_{\rm G}\mathcal{R}_x P_h|\text{BCS}\rangle = \mathcal{R}_x P_h P_{\rm G}|\text{BCS}\rangle\\
&= (-1)^n \mathcal{R}_x P_{\rm G}|\text{BCS}\rangle,
\end{aligned}
\tag{15.14}
$$

where we have used that both $\mathcal{R}_x$ and $P_h$ commute with the projector. Analogously, when a term $\Delta_{x^2-y^2}$ is present,

$$
\begin{aligned}
\alpha_{xy} P_{\rm G}|\text{BCS}\rangle &= P_{\rm G}\mathcal{R}_{xy} P_h G|\text{BCS}\rangle = \mathcal{R}_{xy} P_h G P_{\rm G}|\text{BCS}\rangle\\
&= \mathcal{R}_{xy} P_{\rm G}|\text{BCS}\rangle.
\end{aligned}
\tag{15.15}
$$

These equations show that the projected BCS state with both $s$ and $d_{xy}$, or $d_{x^2-y^2}$ and $d_{xy}$, contributions to the gap has definite symmetry under reflections, in addition to being translationally invariant. The corresponding eigenvalues, for $n = N/2$ even, coincide with the eigenvalues of the modified symmetry operators $\mathcal{R}_x P_h$ and $\mathcal{R}_{xy} P_h G$ on the pure BCS state.

In the previous discussion of quantum numbers, it was assumed that $u_k$ and $v_k$ are well defined for every wave vector $k$. However, this condition is in general violated: singular $k$-points are present whenever both the band structure $\epsilon_k$ and the gap function $\Delta_k$ vanish, as for example with $\mu = 0$, nearest-neighbor hopping and $d_{x^2-y^2}$ pairing at $k = (\pm\frac{\pi}{2}, \pm\frac{\pi}{2})$. However, on finite lattices, this occurrence can be avoided by the choice of suitable boundary conditions. In fact, we are free to impose either periodic or antiperiodic boundary conditions on the fermionic BCS Hamiltonian (15.3), while maintaining all the symmetries of the original lattice. In our studies, we have selected lattices and boundary conditions which do not result in singular $k$-points. We note that the quantum numbers of the projected state do depend in general on the choice of boundary conditions in the fermionic BCS Hamiltonian.

### *15.3.1 The Marshall–Peierls Sign Rule*

Another interesting property of the class of pBCS wave functions is related to the possibility of satisfying the Marshall–Peierls sign rule by means of a suitable choice of the gap function. In particular, we will restrict our considerations to the class of projected wave functions specified in (15.4) when both $t_{R-R'}$ and $\Delta_{R-R'}$ are real and couple sites in opposing sublattices. We begin with the BCS Hamiltonian (15.3) and perform a particle–hole transformation on the down spins alone, $d^{\dagger}_{R,\uparrow} = c^{\dagger}_{R,\uparrow}$ and $d^{\dagger}_{R,\downarrow} = \mathrm{e}^{\mathrm{i}Q\cdot R} c_{R,\downarrow}$, with $Q = (\pi, \pi)$, followed by the canonical transformation (spin rotation) $a_+(k) = (d_{k,\uparrow} + i d_{k,\downarrow})/\sqrt{2}$ and $a_-(k) = -i(d_{k,\uparrow} - i d_{k,\downarrow})/\sqrt{2}$. The BCS Hamiltonian then acquires the form

$$\mathcal{H}_{\mathrm{BCS}} = \sum_k [h_+(k) + h_-(k)] , \tag{15.16}$$

where $h_\pm(k) = \epsilon_k a^{\dagger}_{\pm}(k) a_\pm(k) \pm i \Delta_k a^{\dagger}_{\pm}(k) a_\pm(k+Q)$, and we have used the symmetry $\Delta_k = -\Delta_{k+Q}$. Due to the anticommutation rules of the operators $a_\pm(k)$, the ground state of $\mathcal{H}_{\mathrm{BCS}}$ can be written as a tensor product of free states of $\pm$ fermions. Moreover, if $\sum_R \Psi(R_1 \cdots R_n) a^{\dagger}_{+}(R_1) \cdots a^{\dagger}_{+}(R_n)|0\rangle$ is the ground state of $\sum_k h_+(k)$, then $\sum_R \Psi^*(R_1 \cdots R_n) a^{\dagger}_{-}(R_1) \cdots a^{\dagger}_{-}(R_n)|0\rangle$ is the ground state of $\sum_k h_-(k)$. Here, we have chosen an arbitrary ordering of the lattice sites. The ground state of $\mathcal{H}_{\mathrm{BCS}}$ is therefore

$$\sum_{R,R'} \Psi(R_1 \cdots R_n)\Psi^*(R'_1 \cdots R'_n) a_+^\dagger(R_1) \cdots a_+^\dagger(R_n) a_-^\dagger(R'_1) \cdots a_-^\dagger(R'_n)|0\rangle. \tag{15.17}$$

If this state is expressed in terms of the original electron operators we obtain, up to a factor of proportionality,

$$\sum_{R,R'} \Psi(R_1 \cdots R_n)\Psi^*(R'_1 \cdots R'_n) \left[c^\dagger_{R_1,\uparrow} - \mathrm{i}e^{\mathrm{i}Q\cdot R_1} c_{R_1,\downarrow}\right] \cdots$$
$$\left[c^\dagger_{R'_1,\uparrow} + \mathrm{i}e^{\mathrm{i}Q\cdot R'_1} c_{R'_1,\downarrow}\right] \cdots c^\dagger_{1,\downarrow} \cdots c^\dagger_{N,\downarrow}|0\rangle. \tag{15.18}$$

In projecting over the state of fixed particle number equal to the number of sites, we must take the same number of creation and annihilation operators in the $N$ factors of the product. The suppression of doubly occupied sites mandated by the Gutzwiller projector is effected by creating an up spin on sites where a down spin has already been annihilated. The only terms which survive are then those with $\{R\} = \{R'\}$, namely

$$\sum_{R} |\Psi(R_1 \cdots R_n)|^2 e^{\mathrm{i}\sum_j Q\cdot R_j} c^\dagger_{R_1,\uparrow} c_{R_1,\downarrow} \cdots c^\dagger_{R_n,\uparrow} c_{R_n,\downarrow} \cdots c^\dagger_{1,\downarrow} \cdots c^\dagger_{N,\downarrow}|0\rangle. \tag{15.19}$$

Finally, by moving the down-spin creation operators to the left, one may order the operators according to the specified ordering of the sites in the lattice, independently of the spin, without introducing any further phase factors. On this basis, the resulting wave function has exactly the Marshall–Peierls sign.

### *15.3.2 Spin Correlations*

Finally, we would like to calculate the form of the long-range decay of the spin correlations in a BCS state. Here, we will show only that the pure BCS state before projection is characterized by correlations which maintain the symmetries of the lattice even when the BCS Hamiltonian breaks the reflection symmetries due to the presence of both $\Delta^{x^2-y^2}$ and $\Delta^{xy}$ couplings. Because the BCS state (15.4) is a translationally invariant singlet, it is sufficient to calculate the longitudinal correlations $\langle S^z_r S^z_0\rangle$. A straightforward application of Wick's theorem leads (for $r \neq 0$) to

$$\langle S^z_r S^z_0\rangle \propto -\left[g^2(r) + h^2(r)\right], \tag{15.20}$$

$$g(r) = \int \mathrm{d}^2k \, \frac{\epsilon_k}{E_k} \, e^{\mathrm{i}k\cdot r}, \tag{15.21}$$

$$h(r) = \int \mathrm{d}^2k \, \frac{\Delta_k}{E_k} \, e^{\mathrm{i}k\cdot r}. \tag{15.22}$$

Note that when the gap function $\Delta_k$ has both $d_{x^2-y^2}$ and $d_{xy}$ contributions, the correlation function apparently breaks rotational invariance. Equation (15.20) can be written equivalently in Fourier space as

$$S(q) = \langle S^z_q S^z_{-q} \rangle \propto \int \mathrm{d}^2 k \, \frac{\epsilon_k \epsilon_{k+q} + \Delta_k \Delta_{k+q}}{\sqrt{\left[\epsilon_k^2 + \Delta_k^2\right] \left[\epsilon_{k+q}^2 + \Delta_{k+q}^2\right]}}. \tag{15.23}$$

Now the effect of an $x$-reflection $\mathcal{R}_x$ on the wave vector $q$ can be deduced by setting $\Delta_k = \Delta_k^{x^2-y^2} + \Delta_k^{xy}$ and changing the dummy integration variable $k \rightarrow \mathcal{R}_x\,(k + Q)$, whence $\Delta_{k+Q} = -\Delta_k^{x^2-y^2} + \Delta_k^{xy}$ and $\Delta_{\mathcal{R}_x k} = \Delta_k^{x^2-y^2} - \Delta_k^{xy}$. The net result of these transformations is simply $S(\mathcal{R}_x q) = S(q)$, demonstrating that the spin correlations of a BCS state are isotropic, even if the gap function breaks rotational invariance before Gutzwiller projection.

The explicit evaluation of the long-range decay of $g(r)$ for a $d_{x^2-y^2}$ gap shows that spin correlations in a BCS state (i.e., before projection) display a power-law decay due to the presence of gapless modes: $\langle S^z_r S^z_0 \rangle \sim 1/r^4$ for sites on opposite sublattices, while $\langle S^z_r S^z_0 \rangle$ vanishes for sites on the same sublattice. A similar result is also expected in the presence of a finite $\Delta_k^{xy}$, because gapless modes are present also in this case.

## 15.4 Connection with the Bosonic Representation

We turn now to a detailed discussion of the relation between the fermionic [28] and bosonic [29] representations of the RVB wave function. Recently, bosonic RVB wave functions have been reconsidered by Beach and Sandvik [36–39]. In particular, it has been possible to improve the earlier results of [29], either by assuming some asymptotic form of the bond distribution [39] or by unconstrained numerical methods [36]. This wave function has been demonstrated to be extremely accurate for the unfrustrated model with $J_2 = 0$ [29, 36].

In the fermionic representation, we have

$$|p\mathrm{BCS}\rangle = P_{\mathrm{G}} \exp\left[\sum_{R<R'} f_{R,R'} (c^\dagger_{R,\uparrow} c^\dagger_{R',\downarrow} + c^\dagger_{R',\uparrow} c^\dagger_{R,\downarrow})\right] |0\rangle, \tag{15.24}$$

where $P_{\mathrm{G}}$ projects onto the physical subspace with one electron per site and $f_{R,R'}$ is the pairing function, given by the Fourier transform of (15.5). The constraint $R < R'$ implies the definition of an (arbitrary) ordering of the lattice sites: here and in the following, we will refer to the lexicographical order. For simplicity, let us denote the singlet operator as $\Theta_{R,R'} = (c^\dagger_{R,\uparrow} c^\dagger_{R',\downarrow} + c^\dagger_{R',\uparrow} c^\dagger_{R,\downarrow})$. Once the Gutzwiller projector is taken into account, we have that

$$|pBCS\rangle = \sum_{R_1<\cdots<R_n} \sum_{P(R')} f_{R_1,R'_1} \cdots f_{R_n,R'_n} \Theta_{R_1,R'_1} \cdots \Theta_{R_n,R'_n} |0\rangle, \quad (15.25)$$

where $n = N/2$ and $P(R')$ represents the permutations of the $n$ sites $R'_k$ not belonging to the set $\{R\}$, satisfying $R_k < R'_k$ for every $k$. The sum defines all the $(N-1)!!$ partitions of the $N$ sites into pairs.

On the other hand, the bosonic RVB wave function may be expressed in terms of the spin-lowering operator, $S^-_R$, as

$$|RVB\rangle = \sum_{R_1<\cdots<R_n} \sum_{P(R')} f^{bos}_{R_1,R'_1} \cdots f^{bos}_{R_n,R'_n} (S^-_{R'_1} - S^-_{R_1}) \ldots (S^-_{R'_n} - S^-_{R_n}) |F\rangle, \quad (15.26)$$

where the sum has the same restrictions as before and $|F\rangle$ is the (fully polarized) ferromagnetic state. In the bosonic representation, a valence-bond singlet is antisymmetric on interchanging the two sites and, therefore, a direction must be specified. The condition $R_k < R'_k$ fixes the phase (i.e., the sign) of the RVB wave function. In order to make contact between the two representations, we express $|F\rangle$ and $S^-_R$ in terms of fermionic operators, namely $|F\rangle = c^\dagger_{1,\uparrow} \ldots c^\dagger_{N,\uparrow}|0\rangle$ and $S^-_R = c^\dagger_{R,\downarrow} c_{R,\uparrow}$. Then

$$|RVB\rangle = \sum_{R_1<\cdots<R_n} \sum_{P(R')} \epsilon_{\{R,R'\}} f^{bos}_{R_1,R'_1} \cdots f^{bos}_{R_n,R'_n} \Theta_{R_1,R'_1} \cdots \Theta_{R_n,R'_n} |0\rangle, \quad (15.27)$$

where $\epsilon_{\{R,R'\}} = \pm 1$ is a configuration-dependent sign arising from the reordering of the fermionic operators $(1, \ldots, N) \to (R_1, R'_1, \ldots, R_n, R'_n)$. The two representations are therefore equivalent only if

$$\epsilon_{\{R,R'\}} f_{R_1,R'_1} \cdots f_{R_n,R'_n} = f^{bos}_{R_1,R'_1} \cdots f^{bos}_{R_n,R'_n} \quad (15.28)$$

for all the valence-bond configurations. In general, for a given $f^{bos}_{R,R'}$, this condition cannot be satisfied by any choice of $f_{R,R'}$. Remarkably, this is however possible for the short-range RVB state [40, 41], where only nearest-neighbor sites are coupled by $f^{bos}_{R_k,R'_k} = 1$. Indeed, by using the Kasteleyn theorems [42], it is possible to prove that (15.28) can be fulfilled on all planar graphs (for example in short-range RVB states on lattices with open boundary conditions). In fact, the left-hand side of (15.28) is a generic term in the Pfaffian of the matrix

$$M(R, R') = \begin{cases} f_{R,R'} & \text{for } R < R', \\ -f_{R',R} & \text{for} R > R'. \end{cases} \quad (15.29)$$

As a consequence, following the arguments of Kasteleyn, it is always possible to orient all the bonds in such a way that *in all cycles of the transition graph the number of bonds oriented in either directions is odd* [42]. Notice that the latter way to orient the bonds will in general be different from the one used in (15.26). Thus, we define $f_{R,R'} = 1$ (with $R < R'$) if the bond is oriented from $R$ to $R'$, and $f_{R,R'} = -1$ otherwise. In summary, in order to define the fermionic pairing function $f_{R,R'}$ once we know the oriented planar graph, it is necessary to:

- Label the sites according to their lexicographical order,
- Orient the bonds in order to meet the Kasteleyn prescription, and
- Take $f_{R,R'} = 1$ for the bond oriented from $R$ to $R'$, and $f_{R,R'} = -1$ otherwise.

This construction is strictly valid only for planar graphs, namely for graphs without intersecting singlets, implying that open boundary conditions must be taken. In this case, it is known that a unique short-range RVB state can be constructed. Periodic boundary conditions imply the existence of four degenerate states, which are obtained by inserting a cut (changing the sign of the pairing function on all bonds intersected) that wraps once around the system, in the $x$, $y$ or both directions [41]. These different states have the same bulk properties and, despite the fact that it would be possible to obtain a precise correspondence between bosonic and fermionic states, their physical properties can be obtained by considering a single (bosonic or fermionic) wave function.

## 15.5 Antiferromagnetic Order

In the preceding sections, we have considered the mean-field Hamiltonian (15.3) containing only hopping and pairing terms. In this case, even by considering the local SU(2) symmetries described above, it is not possible to generate a magnetic order parameter. The most natural way to introduce an antiferromagnetic order is by adding to the BCS Hamiltonian of (15.3) a magnetic field

$$\mathcal{H}_{\mathrm{BCS+AF}} = \mathcal{H}_{\mathrm{BCS}} + \mathcal{H}_{\mathrm{AF}}. \tag{15.30}$$

Usually, the antiferromagnetic mean-field order parameter is chosen is chosen to lie along the $z$-direction [43],

$$\mathcal{H}_{\mathrm{AF}} = \Delta_{\mathrm{AF}} \sum_{R} \mathrm{e}^{\mathrm{i}Q\cdot R} (c^{\dagger}_{R,\uparrow} c_{R,\uparrow} - c^{\dagger}_{R,\downarrow} c_{R,\downarrow}), \tag{15.31}$$

where $Q$ is the antiferromagnetic wave vector (e.g., $Q = (\pi, \pi)$ for the Néel state). However, in this case, the Gutzwiller-projected wave function obtained from the ground state of (15.30) overestimates the correct magnetic order parameter (see Sect. 15.6.2), because important quantum fluctuations are neglected. A more appropriate description which serves to mitigate this problem is obtained by the introduction of a spin Jastrow factor $\mathcal{J}$ which generates fluctuations in the direction *orthogonal* to that of the mean-field order parameter [44, 45]. Therefore, we take a staggered magnetic field $\Delta_{\mathrm{AF}}$ along the $x$ axis,

$$\mathcal{H}_{\rm AF} = \Delta_{\rm AF} \sum_R {\rm e}^{{\rm i}\boldsymbol{Q}\cdot\boldsymbol{R}} (c^\dagger_{R,\uparrow} c_{R,\downarrow} + c^\dagger_{R,\downarrow} c_{R,\uparrow}), \tag{15.32}$$

and consider a long-range spin Jastrow factor $\mathcal{J}$

$$\mathcal{J} = \exp\left( \frac{1}{2} \sum_{R,R'} v_{R-R'} S^z_R S^z_{R'} \right), \tag{15.33}$$

$v_{R-R'}$ being variational parameters to be optimized by minimizing the energy. The Jastrow term is very simple to compute by employing a random walk in the configuration space $|x\rangle = c^\dagger_{R_1,\sigma_1} \dots c^\dagger_{R_N,\sigma_N} |0\rangle$ defined by the electron positions and their spin components along the $z$ quantization axis, because it represents only a *classical* weight acting on the configuration. Finally, the variational *ansatz* is given by

$$|p\mathrm{BCS} + \mathrm{AF}\rangle = \mathcal{J} P_{S_z=0} P_{\mathrm{G}} |\mathrm{BCS} + \mathrm{AF}\rangle, \tag{15.34}$$

where $P_{S_z=0}$ is the projector onto the $S_z = 0$ sector and $|\mathrm{BCS} + \mathrm{AF}\rangle$ is the ground state of the Hamiltonian (15.30). It should be emphasized that this wave function breaks the spin symmetry, and thus, like a magnetically ordered state, it is not a singlet. Nevertheless, after projection onto the subspace with $S^z_{tot} = 0$, the wave function has $\langle S^x_R \rangle = \langle S^y_R \rangle = \langle S^z_R \rangle = 0$. Furthermore, the correlation functions $\langle S^x_R S^x_{R'} \rangle$ and $\langle S^y_R S^y_{R'} \rangle$ have the same behavior, and hence the staggered magnetization lies in the $x-y$ plane [44].

The mean-field Hamiltonian (15.30) is quadratic in the fermionic operators and can be diagonalized readily in real space. Its ground state has the general form

$$|\mathrm{BCS} + \mathrm{AF}\rangle = \exp\left( \frac{1}{2} \sum_{R,R',\sigma,\sigma'} f^{\sigma,\sigma'}_{R,R'} c^\dagger_{R,\sigma} c^\dagger_{R',\sigma'} \right) |0\rangle, \tag{15.35}$$

where the pairing function $f^{\sigma,\sigma'}_{R,R'}$ is an antisymmetric $2N \times 2N$ matrix. We note that in the case of the standard BCS Hamiltonian, with $\Delta_{\rm AF} = 0$ or even with $\Delta_{\rm AF}$ along $z$, $f^{\uparrow,\uparrow}_{R,R'} = f^{\downarrow,\downarrow}_{R,R'} = 0$, whereas in the presence of a magnetic field in the $x-y$ plane the pairing function acquires non-zero contributions also in this triplet channel. The technical difficulty when dealing with such a state is that, given a generic configuration with a definite $z$-component of the spin, $|x\rangle = c^\dagger_{R_1,\sigma_1} \dots c^\dagger_{R_N,\sigma_N} |0\rangle$, one has

$$\langle x | \mathrm{BCS} + \mathrm{AF} \rangle = Pf[\mathbf{F}], \tag{15.36}$$

where $Pf[\mathbf{F}]$ is the Pfaffian of the pairing function

$$\mathbf{F} = \begin{pmatrix} \left[ f(\uparrow, R_\alpha; \uparrow, R_\beta) \right] & \left[ f(\uparrow, R_\alpha; \downarrow, R'_\beta) \right] \\ \left[ f(\downarrow, R'_\alpha; \uparrow, R_\beta) \right] & \left[ f(\downarrow, R'_\alpha; \downarrow, R'_\beta) \right] \end{pmatrix}, \tag{15.37}$$

in which the matrix $\mathbf{F}$ has been written in terms of $N \times N$ blocks and $R_\alpha$ and $R'_\alpha$ are respectively the positions of the up and down spins in the configuration $|x\rangle$ [46].

## 15.6 Numerical Results

In this section, we report numerical results obtained by the variational Monte Carlo method for the one- and two-dimensional lattices. The variational parameters contained in the BCS and BCS+AF mean-field Hamiltonians of (15.3) and (15.30), as well as the ones contained in the spin Jastrow factor, (15.33), can be obtained by the optimization technique described in [47].

### *15.6.1 One-dimensional Lattice*

We begin by considering the one-dimensional case, where the high level of accuracy of the pBCS wave function can be verified by comparison with Lanczos and DMRG results. We consider the Hamiltonian (15.1) on a chain with $N$ sites and periodic boundary conditions, and first discuss in some detail the parametrization of the wave function. For $J_2 = 0$, a very good variational state is obtained simply by projecting the free-electron Slater determinant, where $\epsilon_k = -2t \cos k$ [48]. Then, in one dimension, the nearest-neighbor BCS pairing $\Delta_1$ is irrelevant, and, in order to improve the variational energy, a third-neighbor BCS pairing $\Delta_3$ must be considered in addition; a second-neighbor pairing term $\Delta_2$, like the chemical potential, violates the Marshall–Peierls sign rule, which must hold at $J_2 = 0$, and thus is not considered. To give some indication of the accuracy of the wave function, we note that for $N = 30$, the energy per site of the projected Fermi sea is $E/J_1 = -0.443060(5)$, while by optimizing $\Delta_1$ and $\Delta_3$ one obtains $E/J_1 = -0.443934(5)$, the exact result being $E_0/J_1 = -0.444065$.

When both the chemical potential and $\Delta_2$ vanish, the particle–hole transformation $P_h$ (Sect. 15.3) is a symmetry of the BCS Hamiltonian. In finite chains, the BCS ground state is unique only if the appropriate boundary conditions are adopted in $\mathcal{H}_{\mathrm{BCS}}$: if, for example, $N = 4l + 2$ with integer $l$, periodic boundary conditions (PBC) should be used, while the imposition of antiperiodic boundary conditions (APBC) causes four zero-energy modes to appear in the single-particle spectrum. By filling these energy levels, we can form six orthogonal BCS ground states in the $S_z = 0$ subspace, which, in the thermodynamic limit, are degenerate with the ground state of the BCS Hamiltonian with PBC. However, two of these states have the wrong particle–hole quantum number and are therefore annihilated by the Gutzwiller projector. If the remaining four BCS states (three singlets and one triplet) are still orthogonal after projection, one may infer either the presence of a gapless excitation spectrum or of a ground-state degeneracy. We have built these five projected states (one with PBC and four with APBC) for a $N = 30$ chain and variational parameters appropriate for $J_2 = 0$. Two of them belong to the symmetry subspace of the ground state and represent the same physical wave function (their overlap is $|\langle \Psi_1|\Psi_2\rangle| = 0.999$), two of them are singlets with momentum $\pi$ relative to ground state and again show an extremely large overlap ($|\langle \Psi_3|\Psi_4\rangle| = 0.921$), and the remaining state is a triplet with momentum $\pi$ relative to the ground state. Therefore, only three independent states can be obtained by this procedure. It is remarkable that by optimizing the parameters for the ground state, and without any additional adjustable parameters, these three variational states have overlap higher than 98.7% with the exact eigenstates of the Heisenberg Hamiltonian in the lowest levels of the conformal tower of states [49], thereby reproducing with high accuracy the ground state and the lowest singlet and triplet modes.

By increasing the frustrating interaction, the parameters $\Delta_1$ and $\Delta_3$ (both real) grow until a divergence at $J_2/J_1 \sim 0.15$. For larger values of $J_2/J_1$, the band structure changes: here $\epsilon_k = -2t' \cos 2k - \mu$ and a non-vanishing BCS pairing is found, leading to a finite gap in the BCS spectrum, $E_k = \sqrt{\epsilon_k^2 + \Delta_k^2}$. In this regime, although the variational wave function is translationally invariant, it shows a long-range order in the dimer–dimer correlations (see below). Similar behavior has been also discussed in [50] for a complex wave function on ladders with an odd number of legs. The variational parameters appropriate for this regime correspond to a gapped BCS single-particle spectrum for both PBC and APBC, and then only two states can be constructed. However, the symmetry subspace of the variational wave function depends on the choice of boundary conditions, implying a ground-state degeneracy. In a chain of $N = 30$ sites and for $J_2/J_1 = 0.4$, we found that the two singlets which collapse in the thermodynamic limit (due to the broken translational symmetry) have overlaps higher than 99% with the two pBCS wave functions corresponding to the same variational parameters and different boundary conditions. This shows that the pBCS class of wave functions is able to describe valence-bond crystals and broken-symmetry states. By increasing further the ratio $J_2/J_1$, beyond 0.5 we found that, while the bare dispersion is again $\epsilon_k = -2t \cos k$, $\Delta_2$ acquires a finite value (together with $\Delta_1$ and $\Delta_3$), showing both dimerization and short-range incommensurate spin correlations.

The primary drawback of the variational scenario is that the critical point for the transition from the gapless fluid to the dimerized state is predicted around $J_2/J_1 \sim 0.15$ (where the best singlet variational state is the fully projected Fermi sea), considerably smaller than the known critical point $J_2/J_1 \sim 0.241$. This estimate does not change appreciably on considering further parameters in the BCS Hamiltonian (15.3), probably because the variational wave function does not describe adequately the backscattering term which is responsible for the transition [51]. In order to improve this aspect, it is necessary to include the spin Jastrow factor of (15.33) (without the mean-field magnetic parameter $\Delta_{AF}$). In this way, although the variational state is no longer a singlet, the value of the square of the total spin $\langle S^2\rangle$ remains very small (less than 0.002 and 0.02 for 30 and 122 sites, respectively) and no long-range magnetic order is generated. The Jastrow factor is particularly important in the gapless regime: despite the fact that the gain in energy with respect to the singlet state is less than $10^{-4} J_1$ (specifically, for $J_2 = 0$ we obtain $E/J_1 = -0.444010(5)$), this correction is able to shift the transition, always marked by the divergence of the BCS pairings, to $J_2/J_1 \sim 0.21$, a value much closer to the exact result. A finite value of the chemical potential is generated for $0.22 < J_2/J_1 < 0.5$.

Let us now investigate the physical properties of the variational wave function by evaluating some relevant correlation functions. The spin structure factor is defined as

$$S(q) = \frac{1}{N} \sum_{R,R'} e^{iq(R-R')} \langle S^z_R S^z_{R'} \rangle. \qquad (15.38)$$

While true long-range magnetic order cannot be established in one-dimensional systems, for $J_2/J_1 \ll 1$ the ground state is quasi-ordered, by which is meant that it sustains zero-energy excitations and $S(q)$ displays a logarithmic divergence at $q = \pi$. In Fig. 15.1, we show the comparison of the spin structure factor for an exact calculation on $N = 30$ and for the variational wave function. Remarkably, the variational results deliver a very good description of $S(q)$ in all the different regimes: for small $J_2/J_1$, where the spin fluctuations are commensurate and there is a quasi-long-range order, for $0.21 < J_2/J_1 < 0.5$, where the spin fluctuations are still commensurate but short-range, and for $J_2/J_1 > 0.5$, where they are incommensurate and the maximum of $S(q)$ moves from $q = \pi$ at $J_2/J_1 = 0.5$ to $q = \pi/2$ for $J_2/J_1 \to \infty$. Indeed, it is known that the quantum case is rather different from its classical counterpart [10]: while the latter shows a spiral state for $J_2/J_1 > 0.25$, with a pitch angle $\theta$ given by $\cos\theta = -J_1/4J_2$, the former maintains commensurate fluctuations at least up to the Majumdar-Ghosh point. The behavior of $S(q)$

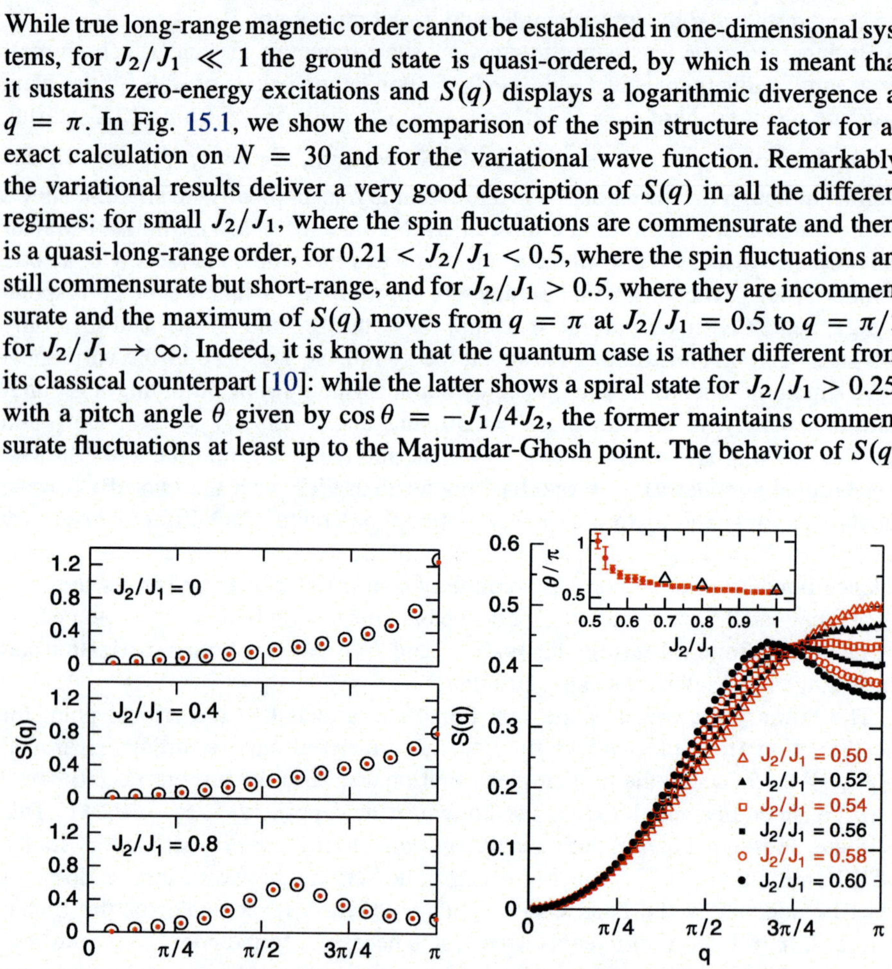


**Fig. 15.1** *Left panels*: comparison between exact (*empty circles*) and variational (*full dots*) results for the spin structure factor $S(q)$ on a chain with 30 sites. *Right panel*: variational results for the spin structure factor $S(q)$ for 122 sites and $J_2/J_1 \geq 0.5$. Inset: position of the maximum of $S(q)$, indicated by $\theta$, as a function of the ratio $J_2/J_1$ (*full dots*). For comparison, the DMRG results of [10] are also shown (*empty triangles*)

for a large lattice with 122 sites and $J_2/J_1 > 0.5$ is shown in Fig. 15.1, where we find good agreement with previous numerical results based upon the DMRG technique [10].

In the one-dimensional $J_1$–$J_2$ model, there is clear evidence for a Berezinskii-Kosterlitz–Thouless transition on increasing the ratio $J_2/J_1$ from a gapless Luttinger liquid to a dimerized state that breaks the translational symmetry. In order to investigate the possible occurrence of a dimerized phase, we analyze the dimer–dimer correlation functions of the ground state,

$$\Theta(R-R') = \langle S^z_R S^z_{R+x} S^z_{R'} S^z_{R'+x}\rangle - \langle S^z_R S^z_{R+x}\rangle\langle S^z_{R'} S^z_{R'+x}\rangle. \tag{15.39}$$

While this definition considers only the $z$ component of the spin operators, in the presence of a broken spatial symmetry the transverse components must also remain finite at large distances, displaying also a characteristic alternation. By contrast, in the gapless regime, the dimer correlations decay to zero at large distances. The differing behavior of these correlations is easy to recognize, with oscillatory power-law decay in the Luttinger regime and constant-amplitude oscillations in the dimerized phase. Figure 15.2 illustrates the comparison of the dimer–dimer correlations (15.39) between the exact and the variational results on a chain with 30 sites. Also for this quantity we obtain very good agreement for all values of the frustrating superexchange $J_2$, both in the gapless and in the dimerized regions. Following [10], it is possible by finite-size scaling to obtain an estimate of the dimer order parameter from the long-distance behavior of the dimer–dimer correlations,

$$d^2 = 9 \lim_{|R|\to\infty} |(\Theta(R-x) - 2\Theta(R) + \Theta(R+x)|, \tag{15.40}$$

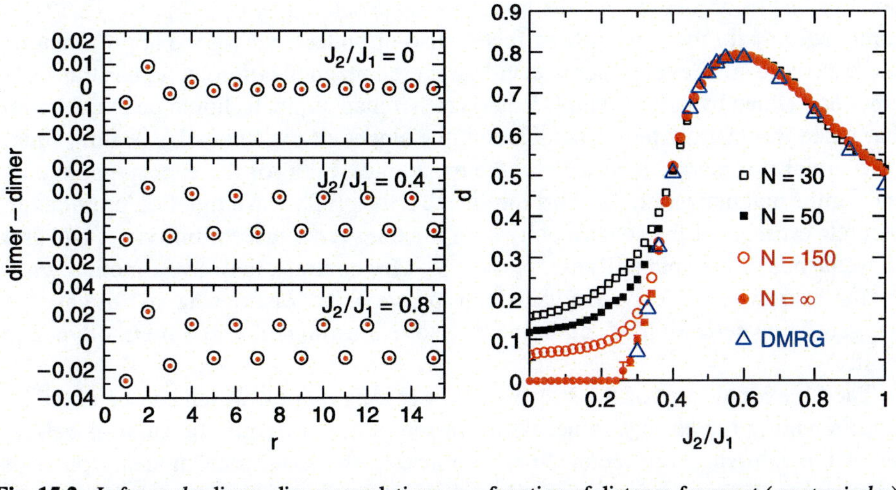


**Fig. 15.2** *Left panels*: dimer–dimer correlations as a function of distance for exact (*empty circles*) and variational (*full dots*) calculations on a chain of 30 sites. *Right panel*: dimer order parameter of (15.40) as a function of the ratio $J_2/J_1$ for $N = 30$, 50, and 150; the extrapolation to the $N \to \infty$ limit is also shown, together with the DMRG results of [10]

where the factor 9 is required to take into account the fact that in (15.39) we considered only the $z$ component of the spin operators. In Fig. 15.2, we present the values of the dimer order parameter as a function of $J_2/J_1$ for three different sizes of the chain, and also the extrapolation in the thermodynamic limit, where the agreement with the DMRG results of [10] is remarkable.

### 15.6.2 Two-dimensional Lattice

We move now to consider the two-dimensional case, starting with the unfrustrated model ($J_2 = 0$), for which exact results can be obtained by Monte Carlo methods [52–54]. In the thermodynamic limit, the ground state is antiferromagnetically ordered with a staggered magnetization reduced to approximately 60% of its classical value, namely $M \simeq 0.307$ [53,54]. This quantity can be obtained both from the spin–spin correlations at the largest distances and from the spin structure factor $S(q)$ at $q = (\pi, \pi)$. In the following, we will consider the former definition and will calculate the isotropic correlations $\langle \mathbf{S}_R \cdot \mathbf{S}_{R'} \rangle$, because this quantity is known to have smaller finite-size effects [52, 53]. For the unfrustrated case, the best wave function has $\epsilon_k = -2t(\cos k_x + \cos k_y)$ and a pairing function with $d_{x^2-y^2}$ symmetry, $\Delta_k^{x^2-y^2} = \Delta_1(\cos k_x - \cos k_y)$ (possibly also with higher harmonics connecting opposite sublattices). The quantity $\Delta_{\mathrm{AF}}$ in (15.32) has a finite value and the spin Jastrow factor (15.33) has an important role.

Figure 15.3 shows the comparison of the variational calculations with the exact results, which are available for rather large system sizes. In the unfrustrated case, the bosonic representation is considerably better than the fermionic one: the accuracy in the energy is around 0.06% and the sublattice magnetization is also very close to the exact value [36,55]. However, the fermionic state defined by (15.30) and (15.32), in combination with the spin Jastrow factor, also provides a very good approximation to the exact results (energy per site and staggered magnetization), whereas the wave function defined by (15.30) and (15.31) is rather inaccurate. It should be emphasized that when the Jastrow factor is included, the slopes of the finite-size scaling functions are also remarkably similar to the exact ones, both for the energy per site $e_0$ and for the magnetization $M$. This implies that the pBCS wave function provides an accurate estimate of the spin velocity $c$, of the transverse susceptibility $\chi_\perp$, and as a consequence of the spin stiffness, $\rho_s = c^2 \chi_\perp$. By contrast, the wave function without the Jastrow factor leads to a vanishing spin velocity. We note that in this case the staggered magnetization $M \simeq 0.365$ is also overestimated in the thermodynamic limit.

The functional form of the Jastrow factor at long ranges, which can be obtained by minimizing the energy, is necessary to reproduce correctly the small-$q$ behavior of the spin-structure factor $S(q)$, mimicking the Goldstone modes typical of a broken continuous symmetry [44]. Indeed, it is clear from Fig. 15.3 that only with a long-range spin Jastrow factor it is possible to obtain $S(q) \sim |q|$ for small momenta, consistent with a gapless spin spectrum. By contrast, with a short-range

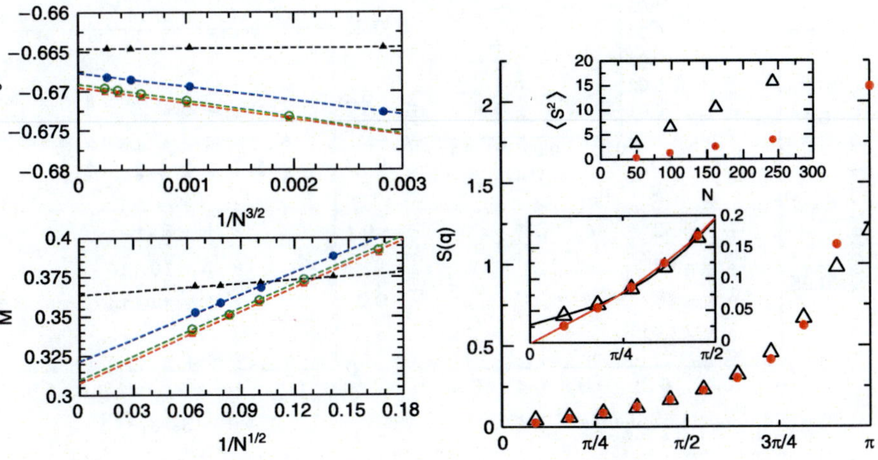


**Fig. 15.3** *Upper left panel*: energy per site as a function of cluster size $N$, showing exact results (*full squares*), variational results obtained by considering (15.30) and (15.32) with the spin Jastrow factor (15.33) (*full circles*), and variational results obtained with (15.30) and (15.31) (*full triangles*). The results obtained by optimizing the pairing function $f^{\text{bos}}_{R,R'}$ of the bosonic representation, described in Sect. 15.4, are also shown (*empty circles*) [55]. *Lower left panel*: staggered magnetization with the same notation as in the upper panel. *Right panel*: static structure factor $S(q)$ for a cluster with $N = 242$ (tilted by 45°): variational results for the state of (15.30) and (15.32) with a long-range Jastrow factor (*full dots*) and for the wave function of (15.30) and (15.31) with a nearest-neighbor Jastrow factor (*empty triangles*). Lower inset: detail at small momenta. Upper inset: square of total spin $\langle S^2\rangle$ as a function of $N$ for the two states, using the same symbols

spin Jastrow factor (for example with a nearest-neighbor term), $S(q) \sim$ const for small $q$, which is clearly not correct [44]. Finally, it should be emphasized that the combined effects of the magnetic order parameter $\Delta_{\text{AF}}$ and the spin Jastrow factor give rise to an almost singlet wave function, strongly reducing the value of $\langle S^2\rangle$ compared to the case without a long-range Jastrow term (see Fig. 15.3).

On increasing the value of the frustrating superexchange $J_2$, the Monte Carlo method is no longer numerically exact because of the sign problem, whereas the variational approach remains easy to apply. In Fig. 15.4, we present the results for the spin–spin correlations at the maximum accessible distances for $J_2/J_1 \leq 0.52$. It is interesting to note that when $J_2/J_1 > 0.4$, a sizable energy gain may be obtained by adding a finite pairing connecting pairs on the same sublattice with $d_{xy}$ symmetry, namely $\Delta_k = \Delta_k^{x^2-y^2} + \Delta_k^{xy}$ [56]. The mean-field order parameter $\Delta_{\text{AF}}$ remains finite up to $J_2/J_1 \sim 0.5$, whereas for $J_2/J_1 > 0.5$ it goes to zero in the thermodynamic limit. Because the Jastrow factor is not expected to destroy the long-range magnetic order, the variational technique predicts that antiferromagnetism survives up to higher frustration ratios than expected [14], similar to the outcome of a Schwinger boson calculation [57]. The magnetization also remains finite, albeit very small, up to $J_2/J_1 = 0.5$ (see Fig. 15.4). We remark here that by using the bosonic RVB state, Beach argued that the Marshall–Peierls sign rule may hold over a rather large range of frustration, namely up to $J_2/J_1 = 0.418$, also

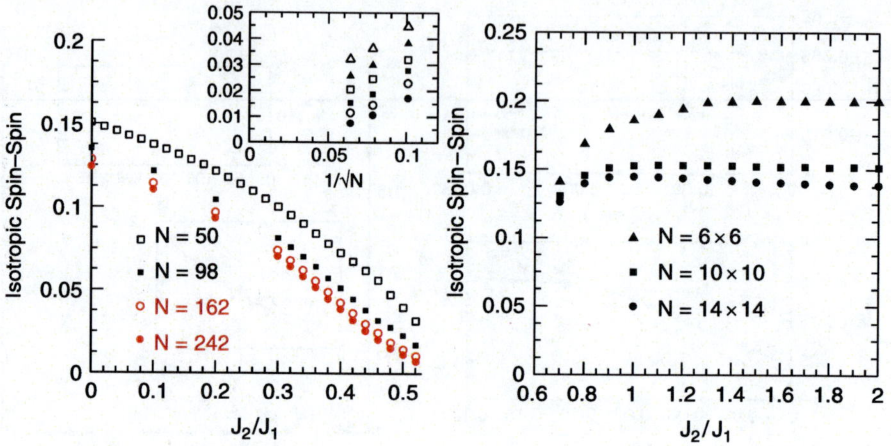


**Fig. 15.4** *Left panel*: spin–spin correlations at the largest distances as a function of the ratio $J_2/J_1$ for different cluster sizes $N$. Inset: finite-size scaling for $J_2/J_1 = 0.42$ (*empty triangles*), 0.44 (*full triangles*), 0.46 (*empty squares*), 0.48 (*full squares*), 0.50 (*empty circles*), and 0.52 (*full circles*). *Right panel*: spin–spin correlations at the largest distances for $J_2/J_1 > 0.7$

implying a finite staggered magnetic moment [39]. In this approach, if one assumes a continuous transition from the ordered to the disordered phase, the critical value is found to be $J_2/J_1 = 0.447$, larger than the value of [14] and much closer to our variational prediction. We note in this context that recent results obtained by coupled cluster methods are also similar, i.e., $J_2/J_1 \sim 0.45$ for a continuous phase transition between a Néel ordered state and a quantum paramagnet [58].

In the regime of large $J_2/J_1$ (i.e., $J_2/J_1 > 0.65$), collinear order with pitch vectors $Q = (0, \pi)$ and $Q = (\pi, 0)$ is expected. The pBCS wave function is also able to describe this phase through a different choice for the bare electron dispersion, namely $\epsilon_k = -2t'[\cos(k_x+k_y)+\cos(k_x-k_y)]$ and $\Delta_k = \Delta_1 \cos k_x + \Delta_2[\cos(k_x+k_y) - \cos(k_x-k_y)]$, with $\Delta_1 \to 0$ for $J_1/J_2 \to 0$. Further, the antiferromagnetic wave vector $Q$ in (15.32) is $Q = (\pi, 0)$. The variational wave function breaks the reflection symmetry of the lattice and, in finite systems, its energy can be lowered by projecting the state onto a subspace of definite symmetry. The results for the spin–spin correlations are shown in Fig. 15.4. By decreasing the value of $J_2/J_1$, we find clear evidence of a first-order phase transition, in agreement with previous calculations using different approaches [19, 21].

For $0.5 < J_2/J_1 < 0.65$, the best variational wave function has no magnetic order ($\Delta_{AF} = 0$ and no Jastrow factor) and the BCS Hamiltonian has $\epsilon_k = -2t(\cos k_x + \cos k_y)$ and $\Delta_k = \Delta_k^{x^2-y^2} + \Delta_k^{xy}$, where $\Delta_k^{x^2-y^2}$ connects pairs on opposite sublattices while $\Delta_k^{xy}$ is for same sublattice. With this specific electron pairing, the signs of the wave function are different from those predicted by the Marshall–Peierls rule and are much more similar to the exact ones. We define

$$\langle s \rangle = \sum_x |\langle x|p\text{BCS}\rangle|^2 \text{sign}\,\{\langle x|p\text{BCS}\rangle\langle x|\Psi_0\rangle\}, \tag{15.41}$$

where $|pBCS\rangle$ and $|\Psi_0\rangle$ are the variational and the exact states, respectively. This quantity is shown in Fig. 15.5, together with the Marshall–Peierls sign, for a $6 \times 6$ lattice. The variational energy, the very large overlap with the exact ground state, and the dimer–dimer correlations shown in Figs. 15.5 and 15.6, all reflect the extremely high accuracy of this state in the strongly frustrated regime. On small clusters, the overlap between the variational wave function and the ground state deteriorates for $J_2/J_1 > 0.55$. This may be a consequence of the proximity to the first order transition, which marks the onset of collinear magnetic order, and implies a mixing of the two finite-size ground states corresponding to the coexisting phases.

In Table 15.1, we report the comparison between the energies of the non-magnetic pBCS wave function and two bosonic RVB states. The first is obtained by a full diagonalization of the $J_1-J_2$ model in the nearest-neighbor valence-bond basis, namely by optimizing all the amplitudes of the independent valence-bond configurations without assuming the particular factorized form of (15.26) [59]. Although this wave function contains a very large number of free parameters, its energy is always higher than that obtained from the pBCS state, showing the importance of having long-range valence bonds. A further drawback of this approach is that it is not possible to perform calculations on large system sizes, the upper limit being $N \sim 40$. The second RVB state is obtained by considering long-range valence bonds, with their amplitudes given by (15.26) and optimized by using the

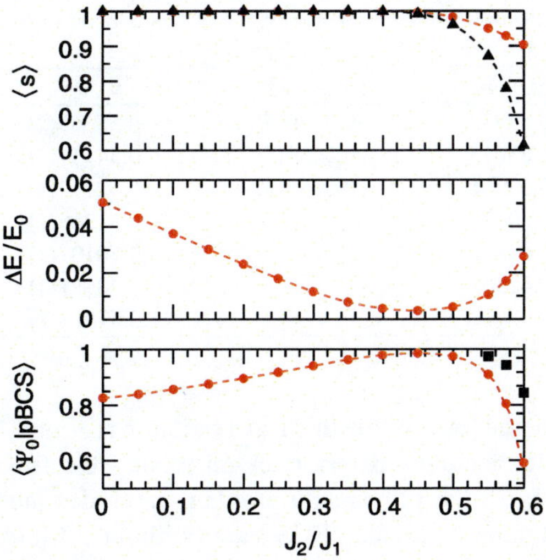


**Fig. 15.5** Comparison between exact and variational results on a $6 \times 6$ lattice. The pBCS wave function has $\Delta_{AF} = 0$ and no Jastrow factor. *Upper panel*: average sign $\langle s \rangle$ of (15.41) (*full circles*); the Marshall–Peierls sign, $\langle s \rangle_{MA} = \sum_x |\langle x|\Psi_0\rangle|^2 \text{sign}\left\{\langle x|\Psi_0\rangle(-1)^{N_\uparrow(x)}\right\}$, is also shown (*full triangles*). *Middle panel*: accuracy of the ground-state energy, $\Delta E/E_0 = (E_0 - E_{pBCS})/E_0$, where $E_0$ and $E_{pBCS}$ are the exact and the variational energies, respectively. *Lower panel*: overlap between the exact $|\Psi_0\rangle$ and variational $|pBCS\rangle$ states (*full circles*). The norm of the projection of the variational state onto the subspace spanned by the two lowest-energy states in the same symmetry sector is also shown (*full squares*) close to the first-order transition to the collinear state

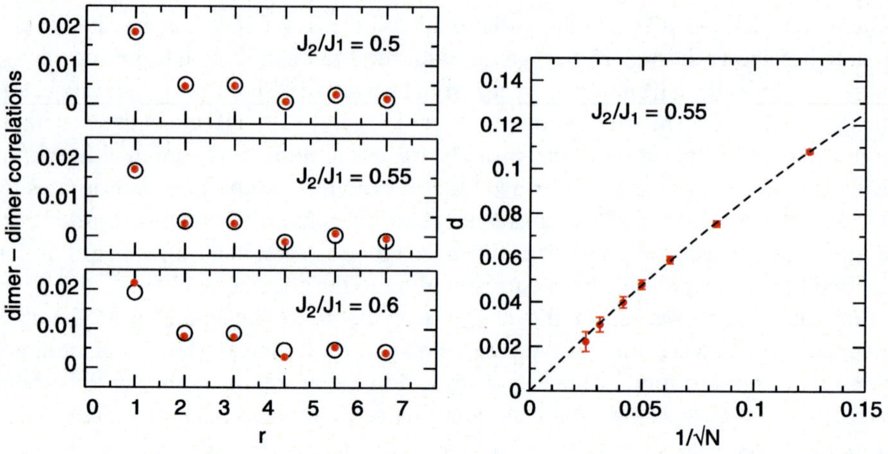


**Fig. 15.6** *Left panels*: dimer–dimer correlations as a function of the Manhattan distance for exact (*empty circles*) and variational (*full dots*) calculations on a 6 × 6 cluster. *Right panel*: finite-size scaling of the dimer order parameter for $J_2/J_1 = 0.55$

**Table 15.1** Energies per site for a $6 \times 6$ lattice. $E_{p\mathrm{BCS}}$ obtained from the pBCS wave function (with $\Delta_{\mathrm{AF}} = 0$ and no Jastrow factor), $E_{\mathrm{LR-RVB}}$ from the long-range bosonic RVB state, optimizing just one parameter using the master-equation method [60], and $E_{\mathrm{SR-RVB}}$ obtained by diagonalizing the $J_1-J_2$ model in the nearest-neighbor valence-bond basis [59]. The exact results $E_0$ are also reported

| $J_2/J_1$ | $E_{\mathrm{SR-RVB}}$ | $E_{\mathrm{LR-RVB}}$ | $E_{p\mathrm{BCS}}$ | $E_0$ |
|---|---|---|---|---|
| 0.30 | −0.54982 | −0.5629(5) | −0.55569(2) | −0.56246 |
| 0.35 | −0.53615 | −0.5454(5) | −0.54134(1) | −0.54548 |
| 0.40 | −0.52261 | −0.5289(5) | −0.52717(1) | −0.52974 |
| 0.45 | −0.50927 | | −0.51365(1) | −0.51566 |
| 0.50 | −0.49622 | | −0.50107(1) | −0.50381 |
| 0.55 | −0.48364 | | −0.48991(1) | −0.49518 |
| 0.60 | −0.47191 | | −0.47983(2) | −0.49324 |

master-equation scheme [60]. While this wave function is almost exact in the weakly frustrated regime, its accuracy deteriorates on raising the frustrating interaction, and for $J_2/J_1 > 0.425$ the minus-sign problem precludes the possibility of reliable results. On the other hand, the pBCS state (without antiferromagnetic order or the Jastrow term) becomes more and more accurate on approaching the disordered region. Remarkably, for $J_2/J_1 = 0.4$, the energy per site in the thermodynamic limit obtained with the long-range bosonic wave function is $E/J_1 = -0.5208(2)$, which is very close to and only slightly higher than that obtained from the fermionic representation, $E/J_1 = -0.5219(1)$.

In the disordered phase, the pBCS wave function does not break any lattice symmetries (Sect. 15.3) and does not show any tendency towards a dimerization. Indeed, the dimer order parameter $d$ (calculated from the correlations at the longest distances) vanishes in the thermodynamic limit, as shown in Fig. 15.6, implying a true spin-liquid phase in this regime of frustration. This fact is in agreement with DMRG calculations on ladders with odd numbers of legs, suggesting a vanishing spin gap for all values of $J_2/J_1$ [61], in sharp contrast to the dimerized phase, which has a finite triplet gap.

Taking together all of the above results, it is possible to draw the (zero-temperature) phase diagram generated by the variational approach, and this is shown in Fig. 15.7.

We conclude by considering the important issue of the low-energy spectrum. In two dimensions, it has been argued that the ground state of a spin-1/2 system is either degenerate or it sustains gapless excitations [62], in analogy to the one-dimensional case [63]. In [64], it has been shown that the wave function with both $d_{x^2-y^2}$ and $d_{xy}$ parameters could have topological order. In fact, by changing the boundary conditions of the BCS Hamiltonian, it should be possible to obtain four different projected states which in the thermodynamic limit are degenerate and orthogonal but, however, not connected by any local spin operator. In this respect, it has been argued more recently that a topological degeneracy may be related to the signs of the wave function and cannot be obtained for states satisfying the Marshall–Peierls rule [65].

In the spin-liquid regime, the simultaneous presence of $\Delta_k^{x^2-y^2}$ and $\Delta_k^{xy}$ could shift the gapless modes of the unprojected BCS spectrum $E_k$ from $(\pm\pi/2, \pm\pi/2)$ to incommensurate $k$-points along the Fermi surface determined by $\epsilon_k = 0$. However, we have demonstrated recently that a particular $\Delta_k^{xy}$ pairing, $\Delta_k^{xy} \propto \sin(2k_x)\sin(2k_y)$, may be imposed, in order to fix the nodes at the commensurate points $(\pm\pi/2, \pm\pi/2)$, without paying an additional energy penalty. Once $E_k$ is connected to the true spin excitations, a gapless spectrum is also expected. At present, within a pure variational technique, it is not possible to assess the possibility of incommensurate, gapless spin excitations being present. An even more challenging problem is to understand if the topological states could survive at all in the presence of a gapless spectrum.

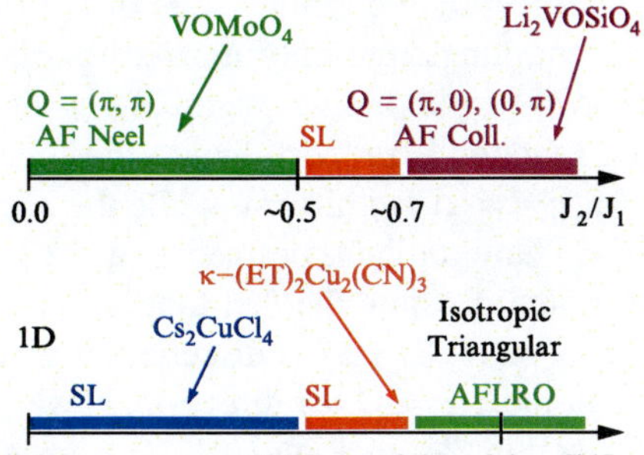


**Fig. 15.7** *Upper panel*: phase diagram of the $J_1-J_2$ model on the square lattice, as deduced from the variational approach. *Lower panel*: phase diagram of the anisotropic triangular lattice from [66]. The approximate locations of some relevant materials are indicated by the arrows

## 15.7 Other Frustrated Lattices

In this last section, we provide a brief overview of related variational studies performed for other lattice structures. In particular, we discuss in some detail the symmetries of the variational wave function on the anisotropic triangular lattice, considered in [66]. In this case, one-dimensional chains with antiferromagnetic interaction $J$ are coupled together by a superexchange $J'$, such that by varying the ratio $J'/J$, the system interpolates between decoupled chains ($J' = 0$) and the isotropic triangular lattice ($J' = J$); the square lattice can also be described in the limit of $J = 0$. The case with $J' < J$ may be relevant for describing the low-temperature behavior of $Cs_2CuCl_4$ [67], whereas $J' \sim J$ may be pertinent to the insulating regime of some organic materials, such as $\kappa - (ET)_2Cu_2(CN)_3$ [68].

In [66], it has been shown that very accurate variational wave functions can be constructed, providing evidence in favor of two different spin-liquid phases, a gapped one close to the isotropic point and a gapless one close to the one-dimensional regime, see Fig. 15.7. We focus our attention on the isotropic point. In this case, a natural variational *ansatz* is the bosonic short-range RVB state of (15.26) [30]. Exact numerical calculations for the $6 \times 6$ isotropic model have shown that the overlap between the short-range RVB wave function and the ground state is very large, $|\langle RVB|\Psi_0\rangle|^2 = 0.891$, and also that the average sign $\langle s\rangle = \sum_x |\langle x|\Psi_0\rangle|^2 \text{sign}\{\langle x|\Psi_0\rangle\langle x|\text{RVB}\rangle\} = 0.971$ [66] is very close to its maximal value, $\langle s\rangle = 1$. We note that both the values of the overlap and of the average sign are much better than those obtained by a wave function that describes a magnetically ordered state, despite the smaller number of variational parameters [69]. Although the short-range RVB state is a very good variational *ansatz*, the bosonic representation of this state is rather difficult to handle in large clusters. Its systematic improvement by the inclusion of long-range valence bonds leads to a very severe sign problem, even at the variational level [30]. In this respect, following the rules discussed in Sect. 15.4, it is possible to obtain a fermionic representation of the short-range RVB state. The signs of the pairing function $f_{R,R'}$ are given in Fig. 15.8 for open boundary conditions. Remarkably, this particular pattern leads to a $2 \times 1$ unit cell, which cannot be eliminated by using local SU(2) transformations of the type discussed in Sect. 15.2. The variational RVB wave function is obtained by projecting the ground state of the BCS Hamiltonian, with a particular choice of the couplings: the only nonzero parameters are the chemical potential $\mu$ and the nearest-neighbor singlet gap $\Delta_{R,R'}$, in the limit $-\mu \gg |\Delta_{R,R'}|$ (so that the pairing function is proportional to the superconducting gap). The amplitude of the gap $|\Delta_{R,R'}| = \Delta$ is uniform, while the appropriate phases are shown in Fig. 15.8. The BCS Hamiltonian is defined on a $2 \times 1$ unit cell and, therefore, is not translationally invariant. Despite the fact that it is invariant under an elementary translation $\mathcal{T}_2$ in the $\tau_2 = (1/2, \sqrt{3}/2)$ direction, it is not invariant under an elementary translation $\mathcal{T}_1$ in the $\tau_1 = (1, 0)$ direction. Nevertheless, this symmetry is recovered after the projection $P_G$, making $|p\text{BCS}\rangle$ translationally invariant. Indeed, one can combine the translation operation $\mathcal{T}_1$ with the SU(2) gauge transformation

$$c^\dagger_{R,\sigma} \to -c^\dagger_{R,\sigma} \tag{15.42}$$

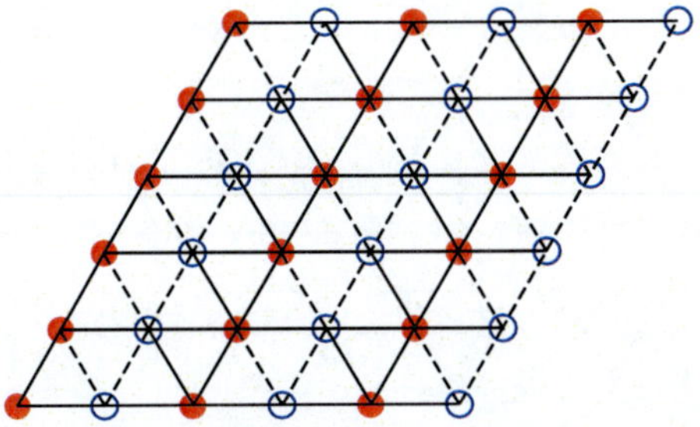

**Fig. 15.8** Nearest-neighbor pairing function consistent with the sign convention of the short-range RVB state in the triangular lattice: *solid (dashed) lines* represent positive (negative) values. Note that the unit cell contains two sites, indicated by *empty* and *full circles*

**Table 15.2** Variational energy estimated in the thermodynamic limit for the antiferromagnetic Heisenberg model on the isotropic triangular lattice ($J' = J$)

| Wave function | $E/J$ |
|---|---|
| Short-range RVB | −0.5123(1) |
| RVB with $\mu = 0$ | −0.5291(1) |
| Best RVB [66] | −0.5357(1) |
| BCS+Néel [70] | −0.532(1) |

for $R = m_1\tau_1 + m_2\tau_2$ with $m_2$ *odd*. Under the composite application of the transformations $\mathcal{T}_1$ and (15.42), the projected BCS wave function does not change. Because the gauge transformation acts as an identity in the physical Hilbert space with singly occupied sites, $|p\text{BCS}\rangle$ is translationally invariant.

Through this more convenient representation of the short-range RVB state by the pBCS wave function, it is possible to calculate various physical quantities using the standard variational Monte Carlo method. One example is the very accurate estimate of the variational energy per site in the thermodynamic limit, $E/J = -0.5123(1)$ [66]. Another important advantage of the fermionic representation is that it is easy to improve the variational *ansatz* in a systematic way. The variational energy can be improved significantly by simply changing the chemical potential $\mu$ from a large negative value to zero, see Table 15.2. We note that in this case $|p\text{BCS}\rangle$ is equivalent to a Gutzwiller-projected free fermion state with nearest-neighbor hoppings defined in a $2 \times 1$ unit cell, because, through the SU(2) transformation of (15.8), the off-diagonal pairing terms are transformed into kinetic terms. Further, the BCS Hamiltonian may be extended readily to include long-range valence bonds by the simple addition of nonzero $\Delta_{R,R'}$ or $t_{R,R'}$ terms. It is interesting to note that, within this approach, it is possible to obtain a variational energy $E/J = -0.5357(1)$ lower than that obtained by starting from a magnetically ordered state and considered in [70], see Table 15.2.

Finally, projected states have been also used to describe the ground state of the Heisenberg Hamiltonian on the kagomé lattice [71, 72]. In this case, different possibilities for the mean-field Hamiltonian have been considered, with no BCS pairing but with non-trivial fluxes through the triangles and the hexagons of which the

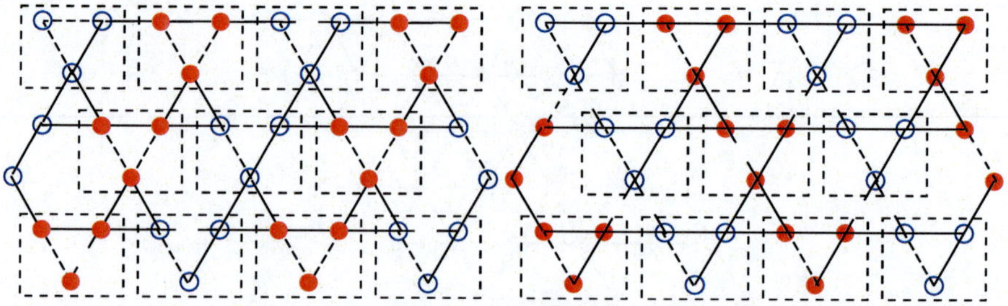

**Fig. 15.9** *Left panel*: signs of the *real* hopping terms $\chi_{R,R'}$ of the U(1) Dirac spin liquid on the kagomé lattice (15.43) [71]. *Solid (dashed) lines* represent positive (negative) values. The unit cell contains six sites (*three empty and three full circles inside the boxes*). *Right panel*: nearest-neighbor pairing function consistent with the sign convention of the short-range RVB state in the kagomé lattice: *solid (dashed) lines* represent positive (negative) values. The unit cell also contains six sites in this case

kagomé structure is composed. In particular, the best variational state in this class can be found by taking

$$\mathcal{H}_{\text{MF}} = -\sum_{\langle R,R'\rangle,\sigma} \chi_{R,R'} c^{\dagger}_{R,\sigma} c_{R',\sigma} + H.c., \tag{15.43}$$

with all the hoppings $\chi_{R,R'}$ having the same magnitude and producing a zero flux through the triangles and $\pi$ flux through the hexagons. One may fix a particular gauge in which all $\chi_{R,R'}$ are real, see Fig. 15.9. In this gauge, the mean-field spectrum has Dirac nodes at $k = (0, \pm\pi/\sqrt{3})$, and the variational state describes a U(1) Dirac spin liquid. Remarkably, this state should be stable against dimerization (i.e., it has a lower energy than simple valence-bond solids), in contrast to mean-field results [73]. Another competing mean-field state [73], which is obtained by giving the fermions chiral masses and is characterized by a broken time-reversal symmetry (with $\theta$ flux through triangles and $\pi-\theta$ flux through hexagons), is also found to have a higher energy than the pure spin-liquid state. In this context, it would be valuable to compare the wave function proposed in [71] with the systematic improvement of the short-range RVB state which has a simple fermionic representation (see Fig. 15.9).

## 15.8 Conclusions

In summary, we have shown that projected wave functions containing both electronic pairing and magnetism provide an extremely powerful tool to study highly frustrated magnetic materials. In particular, these pBCS states may describe all known phases in one-dimensional systems, giving very accurate descriptions when compared to state-of-the-art DMRG calculations. Most importantly, variational wave functions may be easily generalized to treat higher dimensional systems: here, we have presented in detail the case of the two-dimensional $J_1-J_2$ model, as well as some examples of other frustrated lattices which have been considered in the recent past.

The great advantage of this variational approach in comparison with other methods, such as DMRG, is that it can offer a transparent description of the ground-state wave function. Furthermore, the possibility of giving a physical interpretation of the *unprojected* BCS spectrum $E_k$, which is expected to be directly related to the true spin excitations, is very appealing. We demonstrated that this correspondence works very well in one dimension, both for gapless and for dimerized phases. In two dimensions, the situation is more complicated and we close by expressing the hope that future investigations may shed further light one the fascinating world of the low-energy properties of disordered magnetic systems.

**Acknowledgements**

We have had the privilege of discussing with many people over the lifetime of this project, and would like to express our particular thanks to P. Carretta, D. Ivanov, P.A. Lee, C. Lhuillier, F. Mila, G. Misguich, D. Poilblanc, A.W. Sandvik, and X.-G. Wen. We also thank M. Mambrini and K.S.D. Beach for providing us with the energies of the bosonic RVB wave function in Table 15.1, A.W. Sandvik for the bosonic data shown in Fig. 15.3, and S.R. White for the DMRG data shown in Figs. 15.1 and 15.2. We acknowledge partial support from CNR-INFM.

## References

1. J.R. Schrieffer, Theory of Superconductivity (Addison-Wesley, Reading, MA, USA, 1964)
2. R.B. Laughlin, Phys. Rev. Lett. 50, 1395 (1983)
3. P.W. Anderson, Mater. Res. Bull. 8, 153 (1973)
4. P. Fazekas, P.W. Anderson, Philos. Mag. 30, 423 (1974)
5. For a review, see for instance, G. Misguich, C. Lhuillier, in Frustrated Spin Models, ed. by H.T. Diep (World Scientific, New Jersey, 2004)
6. C. Lhuillier, G. Misguich, Introduction to Quantum Spin Liquids. In: *Introduction to Frustrated Magnetism*, ed. by C. Lacroix, P. Mendels, F. Mila (Springer, Heidelberg, 2010)
7. G. Castilla, S. Chakravarty, V.J. Emery, Phys. Rev. Lett. 75, 1823 (1995)
8. R. Melzi, P. Carretta, A. Lascialfari, M. Mambrini, M. Troyer, P. Millet, F. Mila, Phys. Rev. Lett. 85, 1318 (2000)
9. P. Carretta, N. Papinutto, C.B. Azzoni, M.C. Mozzati, E. Pavarini, S. Gonthier, P. Millet, Phys. Rev. B 66, 094420 (2002)
10. S.R. White, I. Affleck, Phys. Rev. B 54, 9862 (1996)
11. S. Eggert, Phys. Rev. B 54, 9612 (1996)
12. C.K. Majumdar, D.K. Ghosh, J. Math. Phys. 10, 1388 (1969)
13. C.K. Majumdar, D.K. Ghosh, J. Math. Phys. 10, 1399 (1969)
14. P. Chandra, B. Doucot, Phys. Rev. B 38, 9335 (1988)
15. E. Dagotto, A. Moreo, Phys. Rev. Lett. 63, 2148 (1989)
16. R.R.P. Singh, R. Narayanan, Phys. Rev. Lett. 65, 1072 (1990)
17. J. Schulz, T.A. Ziman, D. Poilblanc, J. Phys. I 6, 675 (1996)

18. M.P. Gelfand, R.R.P. Singh, D.A. Huse, Phys. Rev. B 40, 10801 (1989)
19. R.R.P. Singh, Z. Weihong, C.J. Hamer, J. Oitmaa, Phys. Rev. B 60, 7278 (1999)
20. V.N. Kotov, J. Oitmaa, O.P. Sushkov, Z. Weihong, Phil. Mag. B 80, 1483 (2000)
21. O.P. Sushkov, J. Oitmaa, Z. Weihong, Phys. Rev. B 63, 104420 (2001)
22. N. Read, S. Sachdev, Phys. Rev. Lett. 62, 1694 (1989)
23. M. Mambrini, A. Lauchli, D. Poilblanc, F. Mila, Phys. Rev. B 74, 144422 (2006)
24. E. Lieb, D. Mattis, J. Math. Phys. 3, 749 (1962)
25. W. Marshall, Proc. R. Soc. Lond. Ser. A 232, 48 (1955)
26. J. Richter, N.B. Ivanov, K. Retzlaff, Europhys. Lett. 25, 545 (1994)
27. D.F.B. ten Haaf, H.J.M. van Bemmel, J.M.J. van Leeuwen, W. van Saarloos, D.M. Ceperley, Phys. Rev. B 51, 13039 (1995)
28. P.W. Anderson, Science 235, 1196 (1987)
29. S. Liang, B. Doucot, P.W. Anderson, Phys. Rev. Lett. 61, 365 (1988)
30. P. Sindzingre, P. Lecheminant, C. Lhuillier, Phys. Rev. B 50, 3108 (1994)
31. F. Becca, L. Capriotti, A. Parola, S. Sorella, Phys. Rev. B 76, 060401 (2007)
32. I. Affleck, Z. Zou, T. Hsu, P.W. Anderson, Phys. Rev. B 38, 745 (1988)
33. F.-C. Zhang, C. Gros, T.M. Rice, H. Shiba, Supercond. Sci. Technol. 36, 1 (1988)
34. X.-G. Wen, Phys. Rev. B 65, 165113 (2002)
35. F.D.M. Haldane, Phys. Rev. Lett. 60, 635 (1988)
36. J. Lou, A.W. Sandvik, Phys. Rev. B 76, 104432 (2007)
37. K.S.D. Beach, A.W. Sandvik, Nucl. Phys. B 750, 142 (2006)
38. A.W. Sandvik, K.S.D. Beach, arXiv:0704.1469
39. K.S.D. Beach, Phys. Rev. B 79, 224431 (2009)
40. B. Sutherland, Phys. Rev. B 37, 3786 (1988)
41. N. Read, B. Chakraborty, Phys. Rev. B 40, 7133 (1989)
42. P.W. Kasteleyn, J. Math. Phys. 4, 287 (1963)
43. C. Gros, Phys. Rev. B 42, 6835 (1990)
44. E. Manousakis, Rev. Mod. Phys. 63, 1 (1991)
45. F. Franjic, S. Sorella, Prog. Theor. Phys. 97, 399 (1997)
46. J.P. Bouchaud, A. Georges, C. Lhuillier, J. Phys. (Paris) 49, 553 (1988)
47. S. Sorella, Phys. Rev. B 71, 241103 (2005)
48. C. Gros, Ann. Phys. 189, 53 (1989)
49. I. Affleck, D. Gepner, H.J. Schulz, T. Ziman, J. Phys. A 22, 511 (1989)
50. S. Sorella, L. Capriotti, F. Becca, A. Parola, Phys. Rev. Lett. 91, 257005 (2003)
51. A. Parola, S. Sorella, F. Becca, L. Capriotti, condmat/0502170.
52. J.D. Reger, A.P. Young, Phys. Rev. B 37, 5978 (1988)
53. A.W. Sandvik, Phys. Rev. B 56, 11678 (1997)
54. M.C. Buonaura, S. Sorella, Phys. Rev. B 57, 11446 (1998)
55. A.W. Sandvik, private communication
56. L. Capriotti, F. Becca, A. Parola, S. Sorella, Phys. Rev. Lett. 87, 097201 (2001)
57. F. Mila, D. Poilblanc, C. Bruder, Phys. Rev. B 43, 7891 (1991)
58. R. Darradi, O. Derzhko, R. Zinke, J. Schulenburg, S.E. Krueger, J. Richter, Phys. Rev. B 78, 214415 (2008)
59. M. Mambrini, private communication
60. K. S. D. Beach, private communication
61. L. Capriotti, unpublished
62. M.B. Hastings, Phys. Rev. B 69, 104431 (2004)
63. E.H. Lieb, T.D. Schultz, D.C. Mattis, Ann. Phys. (N.Y.) 16, 407 (1961)
64. D.A. Ivanov, T. Senthil, Phys. Rev. B 66, 115111 (2002)
65. T. Li, H.-Y. Yang, Phys. Rev. B 75, 172502 (2007)
66. S. Yunoki, S. Sorella, Phys. Rev. B 74, 014408 (2006)
67. R. Coldea, D.A. Tennant, A.M. Tsvelik, Z. Tylczynski, Phys. Rev. Lett. 86, 1335 (2001)
68. Y. Shimizu, K. Miyagawa, K. Kanoda, M. Maesato, G. Saito, Phys. Rev. Lett. 91, 107001 (2003)
69. L. Capriotti, A.E. Trumper, S. Sorella, Phys. Rev. Lett. 82, 3899 (1999)
70. C. Weber, A. Laeuchli, F. Mila, T. Giamarchi, Phys. Rev. B 73, 014519 (2006)
71. Y. Ran, M. Hermele, P.A. Lee, X.-G. Wen, Phys. Rev. Lett. 98, 117205 (2007)
72. M. Hermele, Y. Ran, P.A. Lee, X.-G. Wen, Phys. Rev. B 77, 224413 (2008)
73. M.B. Hastings, Phys. Rev. B 63, 014413 (2000)

# Chapter 16
# Quantum Spin Liquids and Fractionalization

**Grégoire Misguich**

**Abstract** This chapter discusses quantum antiferromagnets which do not break any symmetries at zero temperature – also called "spin liquids" – and focuses on lattice spin models with Heisenberg-like (i.e. SU(2)-symmetric) interactions in dimensions larger than one. We begin by discussing the Lieb–Schultz–Mattis theorem and its recent extension to $D > 1$ by Hastings (2004), which establishes an important distinction between spin liquids with an integer and with a half-integer spin per unit cell. Spin liquids of the first kind, "band insulators", can often be understood by elementary means, whereas the latter, "Mott insulators", are more complex (featuring "topological order") and support spin-1/2 excitations (spinons ). The fermionic formalism (Affleck and Marston, 1988) is described and the effect of fluctuations about mean-field solutions, such as the possible creation of instabilities, is discussed in a qualitative way. In particular, we explain the emergence of gauge modes and their relation to fractionalization. The concept of the projective symmetry group (X.-G. Wen, 2002) is introduced, with the aid of some examples. Finally, we present the phenomenology of (gapped) short-ranged resonating-valence-bond spin liquids, and make contact with the fermionic approach by discussing their description in terms of a fluctuating $Z_2$ gauge field. Some recent references are given to other types of spin liquid, including gapless ones.

## 16.1 Introduction

The concept of "spin liquid" is due to P. W. Anderson, who observed in 1973 [1] that magnetically long-range-ordered (Néel) states were in principle not the only possible ground states for two-dimensional (2D) quantum (and frustrated) antiferromagnets. He explained that such systems could avoid all spontaneous symmetry-breaking, and thus remain "disordered" down to $T = 0$. The picture he provided for such states is the celebrated (short-range) "resonating valence-bond" (RVB) wave function, which is the linear and coherent superposition of a large number of short-range singlet coverings of the lattice. Since then, although our understanding of frustrated quantum spin systems has improved greatly, in general it remains quite incomplete.

First, it is necessary to define precisely what is meant by the term "quantum spin liquid". Depending on the context (experiment, theory, simulation, ...), these words are often applied with rather different meanings. In Sect. 16.2, we will discuss three possible definitions used frequently (and often implicitly) in the literature, and will comment on their implications. The third definition is the most restrictive, having probably no "overlap" with the more "common" states of matter in $D > 1$ magnetic systems. This is the definition we adopt in the remainder of this chapter. It requires the existence of *fractional* excitations, i.e. quasiparticles with quantum numbers (usually the total spin) which are *fractions* of the elementary local degrees of freedom. In spin models, this is essentially equivalent to the existence of spin-$\frac{1}{2}$ excitations (known as spinons). However, such excitations are not easy to realize, because in a system where the local degrees of freedom are spin-flipping proceses which change $S^z_{\rm tot}$ by $\pm 1$, any excitation created in a finite region of the system can only have an *integer* spin. In a one-dimensional system (such as a spin chain), it is well known that domain-wall excitations (or kinks) can carry a half-odd-integer spin. This situation is, however, rather different in $D > 1$, where only some particular states of matter may sustain fractional excitations. We explain, at a qualitative level, in Sect. 16.2.3.2 how these fractional excitations interact with each other through emerging gauge fields, and that such spin liquids sustain a kind of hidden order, called "topological order", a concept due to X.-G. Wen [2–5] which is connected at a profound level to that of fractionalization.

Why is it that spin liquids should be "fractional"? To answer, we will review in Sect. 16.2.4 the Lieb–Schultz–Mattis theorem [6] and its extension by Hastings [7, 8] to $D > 1$. Under certain physically reasonable assumptions, we will argue that this theorem implies essentially (if not mathematically) that a spin-$\frac{1}{2}$ system with an odd number of sites per unit cell (a genuine Mott insulator) and conserved total magnetization $S^z_{\rm tot}$ must either (1) be ordered in a conventional way, meaning with a spontaneously broken symmetry, or (2) have some type of topological order.[1] Because topological order is connected intimately to the existence of fractional excitations, we conclude that a Mott insulator with conserved $S^z_{\rm tot}$ and no spontaneous symmetry-breaking supports topological order and fractional excitations.

In Sect. 16.3, we discuss fractionalized spin liquids in a more rigorous framework, by introducing the basics of the slave-particle formalism and by explaining (Sect. 16.3.4) how gauge fields arise when investigating the fluctuations around mean-field states. Section 16.4 describes phenomenologically the properties of the simplest *gapped* spin liquids, called $\mathbb{Z}_2$ liquids in modern terminology, which correspond essentially to *short-range* RVB states. Their excitations, spinons and visons, are discussed, and a number of realizations in frustrated 2D spin models are reviewed. Section 16.5 is devoted to gapless liquids in $D > 1$, also known

---

[1] A third possibility for $D > 1$ is that the system has been fine-tuned to a critical point, but this does not correspond to a stable *phase*.

as algebraic spin liquids. These states are more complex than the simple $\mathbb{Z}_2$ liquids, and we present an overview of their study using the mean-field approximations discussed in Sect. 16.3. These liquids are closely related to the *long-range* RVB theories of high-temperature superconductors [9–12]). Section 16.6 mentions briefly some of the spin liquids which are not discussed elsewhere in this chapter.

## 16.2 What is a Spin Liquid?

We focus on the zero-temperature properties of lattice quantum spin models with global $U(1)$ (conservation of $S^z_{\text{tot}}$) or $SU(2)$ symmetry (total spin $\mathbf{S}^2_{\text{tot}} = S(S+1)$ conservation).

### 16.2.1 Absence of Magnetic Long-Range Order (Definition 1)

Definition 1: *a quantum spin liquid is a state in which the spin–spin correlations, $\langle S_i^\alpha S_j^\beta \rangle$, decay to zero at large distances $|\mathbf{r}_i - \mathbf{r}_j| \to \infty$.*
This definition is very simple, but it suffers from several limitations. First, any system with continuous spin-rotation symmetry in $D \leq 2$ at finite temperature would satisfy this definition (Mermin–Wagner theorem), even if it is classical and/or ordered at $T = 0$. Second, a spin *nematic* [13] would satisfy this definition, despite the fact that it breaks spin rotation symmetry and has some long-range order in the four-spin correlations (see [14–16] for recent numerical studies of quantum spin nematics). The definition could be made more strict by requiring the global spin-rotation symmetry to be unbroken. In this case, the spin nematics would be excluded. However, a valence-bond crystal[2] (VBC, see [17]) would satisfy this definition, although it possesses certain features of conventional crystalline order.

### 16.2.2 Absence of Spontaneously Broken Symmetry (Definition 2)

Definition 2: *a quantum spin liquid is a state without any spontaneously broken symmetry.*
Such a definition excludes the VBC state discussed above, but still has some unsatisfactory features. Consider a spin-$\frac{1}{2}$ model where the lattice is composed of clusters with an *even* number of spins (for example 2 or 4). Inside each cluster, the exchange

[2] In a VBC, the spins group themselves spontaneously into small clusters (with an even number of sites) which are arranged spatially in a regular pattern. In the crudest approximation, the wave function would be simply a tensor product of singlet states (one for each cluster). Because a VBC wave function is a spin singlet (rotationally invariant) and has short-ranged spin–spin correlations, it is a spin liquid according to definition 1. However, it also possesses some order in the four-spin correlation functions and breaks some of the lattice symmetries.

interactions $J$ are strong and antiferromagnetic. By contrast, the inter-cluster interactions $J'$ are weak.[3] A two-chain spin ladder, in which the rungs form 2-site clusters, realizes this type of geometry. Such models can be understood qualitatively by a perturbation theory around the decoupled limit $J'/J \to 0$. The (unique) ground state is a total-spin singlet (reasonably well approximated by a tensor product of singlets on each cluster) with gapped excitations and no broken symmetry. This state obeys definition 2. However, these systems do not realize new states of matter. These systems undergo a smooth evolution from the $T = 0$ limit (singlet ground state) to the $T = \infty$ limit (free spins), with a characteristic crossover temperature determined by the spin gap. From this point of view, this state would more appropriately be called a "$T = 0$ paramagnet" or a "band insulator" rather than a spin liquid. To our knowledge, almost all gapped Heisenberg spin systems in $D > 1$ which have been observed experimentally belong in this category. These systems are quite similar to *valence-bond solids*, with the spin-1 Haldane chain or AKLT models [18] as standard examples.

### 16.2.3 Fractional Excitations (Definition 3)

Definition 3: *a quantum spin liquid is a state with fractional excitations.*
In the present context, these fractional excitations are usually "spinons ", carrying a half-odd-integer spin (normally $\frac{1}{2}$).

#### 16.2.3.1 What is a Fractional Excitation?

Operations involving any finite number of $S_i^+$ and $S_j^-$ operators may only change the total magnetization $S^z_{\text{tot}}$ by an *integer* (in units where $\hbar = 1$). Thus the creation of a spin-$\frac{1}{2}$ excitation (one "half" of a spin-flip) requires acting in a non-local way on the system. Strictly speaking, such a process is possible only in an infinite system. As a simple example, let us consider the spin-$\frac{1}{2}$ Heisenberg chain with first- and second-neighbor couplings, respectively $J_1$ and $J_2$. For $J_1 = 2J_2 > 0$ (the Majumdar–Ghosh point [19]), the (two-fold degenerate) ground states are given exactly as

$$|a\rangle = \cdots \otimes |[01]\rangle \otimes |[23]\rangle \otimes |[45]\rangle \otimes \cdots , \tag{16.1}$$

$$|b\rangle = \cdots \otimes |[12]\rangle \otimes |[34]\rangle \otimes |[56]\rangle \otimes \cdots , \tag{16.2}$$

where $|[ij]\rangle = |\uparrow_i\downarrow_j\rangle - |\downarrow_i\uparrow_j\rangle$ is a spin singlet state for sites $i$ and $j$. This is an example of a VBC with spontaneous translational symmetry-breaking. Now we

[3] Here, "weak" does not imply necessarily that the coupling is numerically small, but that the system can be understood qualitatively from a weak-coupling limit.

insert in $|a\rangle$ an "up" spin on site 2 by a non-local operation consisting of a shift of the rest of the configuration by one lattice constant to the right,

$$\cdots \otimes |[01]\rangle \otimes |\uparrow_2\rangle \otimes |[34]\rangle \otimes |[56]\rangle \otimes \cdots \tag{16.3}$$

This state has $S^z_{\rm tot} = \frac{1}{2}$ and contains a domain wall between two regions (of types "$a$" and "$b$"). It is not an exact eigenstate of the Hamiltonian, and is clearly a finite-energy excitation. From a variational point of view, it proves the existence of finite-energy spin-$\frac{1}{2}$ excitations in the thermodynamic limit. In a chain of finite length, one may act locally with $S_2^+$ on $|a\rangle$. The resulting state,

$$\cdots \otimes |[01]\rangle \otimes |\uparrow_2\rangle \otimes |\uparrow_3\rangle \otimes |[45]\rangle \otimes \cdots, \tag{16.4}$$

can be viewed as *two* spinons, with parallel spins, on sites 2 and 3. The spin flip has created a pair of spinons which may then propagate to large distances as two independent and elementary excitations.

Another very simple example of fractionalization in a 1D spin chain is the $XY$ chain, which can be mapped exactly onto free, gapless, fermionic spinons by the Jordan-Wigner transformation [20]. Fractionalized (but interacting) spinons are also present in the spin-$\frac{1}{2}$ Heisenberg chain, and spin-charge separation is a general phenomenon in Tomonaga-Luttinger liquids. These examples illustrate that fractionalization is a rather common phenomenon in one dimension. However, ordered states do not in general support fractional excitations in $D > 1$. As an example, any attempt to "separate" two spinons in a $D > 1$ VBC will not lead to a two-spinon state when the entities are far apart. At sufficiently large distances, this results instead in two excitations with *integer* spins (see Fig. 2 in C. Lhuillier's chapter). We will show that the mechanisms leading to fractionalization in $D > 1$ are very different from the "domain wall", "soliton", or "kink" picture valid in 1D.

In 2D, the most famous example of fractionalized systems is provided by fractional quantum Hall fluids. Here the elementary excitations carry an electric charge which is a fraction (for example $\frac{1}{3}$) of that of the electron. As above, local excitations may only have an integer charge. However, if an electron is added to a $\nu = \frac{1}{3}$ quantum Hall fluid, it will decay into three elementary quasiparticles of charge $+\frac{e}{3}$. The property that the system is fractional means that these quasiparticles can be placed far apart from each other with a *finite energy cost*. In the same way, a spin flip (changing $S^z_{\rm tot}$ from 0 to 1, as induced by a neutron scattering process) in the Majumdar-Ghosh chain would decay into two spinons, each carrying half a quantum of magnetization. There may (or may not) be some short-distance bound states between spinons, but the fact that the system is fractionalized means that one can separate two spin-$\frac{1}{2}$ excitations to infinite distances with only a finite energy cost. The spinons are then said to be deconfined. Spinons have the same spin as electrons, but do not carry an electric charge. In this sense, a spinon is a "fraction" of an electron.[4] Removing an electron from a Mott insulator is equivalent to creating a charged hole and removing a spinon (spin). Magnets with deconfined spinons are thus closely related to the problem of spin-charge separation in doped Mott insulators. This said, it is not obvious how to describe effective long-range spinon–spinon interactions in a model where the microscopic interactions are purely local. We will explain – at a rather qualitative level – that gauge theories provide a framework to deal with the question of deconfinement.

#### 16.2.3.2 What is the Connection Between Gauge Theories and Fractional Spin Liquids?

To describe a quantum system with deconfined spinons, it is logical to seek a formalism including single-spinon creation and annihilation operators. On general grounds, such a formalism necessarily involves some *gauge fields*. A spinon creation operator changes the magnetization by $\pm\frac{1}{2}$, and thus cannot be written locally in terms of the spin operators $\boldsymbol{S}_i$. The usual choice is to decompose the spin operators into two spinon operators,

$$S_i^+ = c_{i\uparrow}^\dagger c_{i\downarrow} \ , \ 2S_i^z = c_{i\uparrow}^\dagger c_{i\uparrow} - c_{i\downarrow}^\dagger c_{i\downarrow}, \tag{16.5}$$

and to impose the constraint of one particle per site for all states in the physical Hilbert space,

$$c_{i\uparrow}^\dagger c_{i\uparrow} + c_{i\downarrow}^\dagger c_{i\downarrow} = 1 \ \forall\, i. \tag{16.6}$$

In this review, we focus on the *fermionic*[5] representation, $\{c_{i\sigma}^\dagger, c_{j\sigma'}\} = \delta_{ij}\delta_{\sigma\sigma'}$. Acting with a single $c_{i\uparrow}^\dagger$ operator transforms a physical state into a non-physical one which violates the constraint above. To deal with this, physically it is clear that when inserting a spinon one must "shift" the spin state along some path on the lattice, ending at a point where another spinon is created or destroyed.[6] Thus, $c_{i\uparrow}^\dagger$ must "dressed" with a "string" containing the path information. This will be the role of the gauge field.

Spin operators, and all physical states satisfying (16.6), are invariant under

$$c_{i\sigma}^\dagger \longrightarrow e^{-i\Lambda_i} c_{i\sigma}^\dagger, \ , \ \sigma = \uparrow, \downarrow, \ , \ \Lambda_i \in [0, 2\pi[. \tag{16.7}$$

---

[4] There is, however, no *charge* fractionalization, as in the quantum Hall effect.

[5] To this point, the "bare" spinon operators can be chosen to be fermionic or bosonic. The actual statistics of the *physical* excitations should not depend on this arbitrary choice, which suggests that the fractionalized excitations are not always simply related to the bare creation operators introduced in (16.5).

[6] In a spin chain, there are only two ways to do this, to the right or to the left, but in $D > 1$, many paths are possible.

In fact, invariance under this *gauge transformation* and (16.6) are equivalent.[7] That $c^\dagger_{i\uparrow}$ transforms a physical state into a non-physical one arises because it is not a gauge-invariant operator. The standard solution for this is to introduce[8] a gauge-field operator $A_{ij}$ on each bond of the lattice, which transforms according to $A_{ij} \to A_{ij} + \Lambda_i - \Lambda_j$, so that

$$c^\dagger_{0\uparrow} \exp\big(iA_{01} + iA_{12} + iA_{13} + \cdots + iA_{(n-1)}\big)c_{n\uparrow} \tag{16.8}$$

is gauge-invariant. To understand the physical the meaning of this gauge field, from the discussion of spinons in valence-bond states one may anticipate that $\exp\big(iA_{01} + \cdots + iA_{(n-1)n}\big)$ performs the "shift" operation required to insert or destroy a spinon at each end of the path connecting site 0 to site $n$.

At this stage, the gauge-field operators $A_{ij}$ do not appear explicitly in the spin Hamiltonian [the Heisenberg model can be written using (16.5)]. Thus the transformations above do not yet deliver a gauge theory with a *dynamical* gauge field.[9] However, for many purposes it is important only to derive an effective low-energy theory for the spinons, which is obtained by integrating over some high-energy degrees of freedom (such as gapped fluctuation modes). Such a procedure generally produces all of the local terms which are allowed by symmetry. The simplest terms involving the gauge field, and which are invariant under (16.7), are those of Maxwell type (i.e. analogs of the terms for magnetic and electric energy). The precise nature of the gauge field and its interaction terms depends on details of the spin model, and is (unfortunately) very difficult to predict from microscopic calculations. In some systems, the relevant gauge field will take angular values ($\in [0, 2\pi[$, known as a $U(1)$ gauge field) and in some other cases it is restricted to 0 or $\pi$ (known as $\mathbb{Z}_2$). We refer the reader to Sect. 16.3 for more details, and to the review of Lee, Nagaosa and Wen [21] for a complete discussion.

From the example of electrodynamics, we know that gauge fields can mediate long-range interactions between electric charges (although the Hamiltonian is local). In the present context, the elementary "charges" are the spinons. Generally speaking, a gauge theory can have two kinds of phase: confined phases where excitations with non-zero charge cannot be spatially isolated from each other, and deconfined phases where isolated non-zero charges are finite-energy excitations. Confinement

---

[7] The transformation of (16.7) can be implemented by the operator $\hat{U}(\Lambda) = \exp(\mathrm{i}\sum_{i,\sigma=\uparrow,\downarrow} \Lambda_i c^\dagger_{i\sigma} c_{i\sigma})$. When applied on a state $|\psi\rangle$ satisfying (16.6), this operation gives only a global phase, $\hat{U}(\Lambda)|\psi\rangle = \exp(\mathrm{i}\sum_i \Lambda_i)|\psi\rangle$. It is then convenient to redefine $\hat{U}$ by $\hat{U}(\Lambda) = \exp(\mathrm{i}\sum_i \Lambda_i (c^\dagger_{i\uparrow}c_{i\uparrow} + c^\dagger_{i\downarrow}c_{i\downarrow} - 1))$. Thus, any state $|\psi\rangle$ obeying $\hat{U}(\Lambda)|\psi\rangle = |\psi\rangle$ (gauge invariance, for any $\Lambda$), must satisfy $(c^\dagger_{i\uparrow}c_{i\uparrow} + c^\dagger_{i\downarrow}c_{i\downarrow} - 1)|\psi\rangle = 0$ for any site $i$, which is precisely (16.6).

[8] A more formal construction is presented in Sect. 16.3.2.

[9] By comparison with the Maxwell term in electromagnetism, $\frac{1}{e^2}F^{\mu\nu}F_{\mu\nu}$, the Heisenberg model corresponds formally to infinite coupling, $e = \infty$, which is a non-trivial limit because the gauge-field fluctuations cost no energy and are therefore large.

occurs when the flux $B$ (defined as the circulation of $A_{ij}$) piercing the plaquettes of the lattice fluctuates strongly. In this case, the description of the spin system in terms of spinons interacting with a gauge field is formally correct but not of practical utility, because the gauge field generates an effective, long-range attraction (with a linear potential) between spinons, they are confined in gauge-neutral pairs (with an integer spin, like a magnon), and cannot be elementary excitations of the system.

The situation is qualitatively different if the gauge field is in a deconfined phase, which is realized when the flux fluctuations are small. In this case, the spinons (possibly "dressed" by interactions) are finite-energy states of the model, whose ground state is a fractionalized spin liquid.[10] Thus the existence of fractionalized spin liquids may be formulated as a problem of confinement or deconfinement in certain types of lattice gauge theory coupled to spinons (we refer the reader again to Sect. 16.5 for details).

#### 16.2.3.3 Topological Order

For a conventional type of order associated with a discrete, spontaneous symmetry-breaking (as is the case for a VBC), several ground states $|1\rangle$, $|2\rangle$, ..., $|d\rangle$ are degenerate in the thermodynamic limit. One may look for two (normalized) linear combinations $|a\rangle = \sum_{i=1}^{d} a_i |i\rangle$ and $b\rangle = \sum_i b_i |i\rangle$ of the degenerate ground states, and for a local observable $\hat{O}$ (acting on a finite number of sites) which acts to distinguish them, $\langle a|\hat{O}|a\rangle \neq \langle b|\hat{O}|b\rangle$. If such states $|a\rangle$, $b\rangle$ and such an operator $\hat{O}$ do exist (in the thermodynamic limit), $\hat{O}$ is (by definition [22]) an order parameter for the broken symmetry.

If the topology of the lattice is non-trivial, as for a cylinder or torus, a gapped, fractionalized spin liquid will also exhibit a ground-state degeneracy, *even in the absence of any broken symmetry*. The crucial difference with conventional forms of order is that no *local* observable, $\hat{O}$, can distinguish the ground states in the thermodynamic limit (in $D > 1$). The ground-state degeneracy is suggestive of some form of order, but without an associated local order parameter. This type of non-local order has been named "topological order" in the pioneering works of X.-G. Wen [2–4]. This type of degeneracy is a consequence of fractionalization. The sequence of arguments is not mathematically rigorous, but rather simple and (hopefully) intuitive. We refer to [23] for a more precise discussion.

Consider a spin model with deconfined spinons as elementary excitations and periodic boundary conditions in one direction (taken to be $x$). Starting in a ground state $|1\rangle$, we (1) create locally a pair of spinons, (2) move one of them around the cylinder, (3) annihilate this spinon with its partner, and (4) denote by $|2\rangle$ the resulting

---

[10] *Confinement* should not be confused with the existence of *bound states*. As an example, protons and electrons have bound states, those of the hydrogen atom, but they are *not* confined by the electromagnetic gauge field: because they can be separated to infinite distance by only a finite input energy, they exist as isolated particles. The situation is different for quarks, which are confined by the QCD gauge field and cannot be observed as *isolated* particles at any energy.

final state. Let us further define $\hat{T}$ as the unitary operator describing this process, $|2\rangle = \hat{T}|1\rangle$. If the spectrum remains gapped during the (adiabatically slow) moving process, $\hat{T}$ brings the system back to a ground state, which may however not be the same as the inital ground state. At each intermediate time step, the system contains two spinons. Such an intermediate state may be viewed as being obtained from a spinon-free state by applying some combination of "string" operators which connect the two spinons, as in (16.8). In other words, even when they are far apart, the spinons remain "connected" by a gauge-field string. When the two spinons meet again and annihilate, there remains a non-contractible gauge-field "loop" winding around the cylinder. Intuitively, this is why the new ground state, $|2\rangle$, is generally *different* from $|1\rangle$. To make this schematic argument somewhat more precise, one introduces a second adiabatic process, in which a "twist" $\phi \in [0, 2\pi]$ is applied gradually to the system [23, 24]. This twist amounts to a modification of the boundary conditions for $S_i^+$ (and $S_i^-$): $S^+_{x+L_x,y} \equiv S^+_{x,y}\mathrm{e}^{\mathrm{i}\phi}$. Up to a unitary transformation $U$, the spectrum at $\phi = 0$ is identical to that at $\phi = 2\pi$: $\mathcal{H}_{\phi=2\pi} = U^\dagger \mathcal{H}_{\phi=0} U$. Again we assume that the operation of switching $\phi$ from 0 to $2\pi$ may be performed adiabatically without closing the excitation gap, in which case it defines a unitary operator $\hat{F}$ which transforms ground states (of $\mathcal{H}_{\phi=0}$) into ground states (of $\mathcal{H}_{\phi=2\pi}$). Accordingly, $U\hat{F}$ acts in the ground-state manifold of $\mathcal{H}_{\phi=0}$. Finally, one can show that the two adiabatic processes satisfy $(U\hat{F})\hat{T} = -\hat{T}(U\hat{F})$ [23], because one spinon winding around the cylinder in the presence of a twist $\phi = 2\pi$ experiences an Aharonov–Bohm phase equal to $\mathrm{e}^{\mathrm{i}\phi/2} = -1$ (measured by the gauge-field loop mentioned above). Clearly, this relation cannot be satisfied in the ground-state manifold unless the degeneracy is at least 2.

### 16.2.4 *Half-odd-integer Spins and the Lieb-Schultz-Mattis-Hastings Theorem*

We consider a lattice spin system with periodic boundary conditions, short-range interactions, conserved $S^z_{\mathrm{tot}}$ (global U(1) symmetry) and a half-odd-integer spin (e.g. 1/2) in the unit cell. The lattice dimensions $L_1, L_2, \ldots, L_D$ are taken to be such that each "section" perpendicular to direction 1 has an odd number ($L_2 \times \cdots \times L_D$) of unit cells, and thus has a half-odd-integer spin. The theorem states that, in the thermodynamic limit, the spectrum cannot simultaneously satisfy the two conditions: (1) unique ground state; (2) finite gap to all excitations. Although the proof is quite simple in 1D [6] its generalization to higher dimensions [25], due to Hastings, is quite involved [7, 8]. The argument proposed by Oshikawa [24] is less general,[11] but its simplicity offers deep insight into the LSM theorem for $D > 1$.

[11] It assumes that the gap does not close when twisting the boundary conditions.

What is the relation with the above discussion of the QSL? A conventional reason for a degeneracy of the ground state is spontaneous symmetry-breaking (SBB).[12] However, ordered states do not generally support fractional excitations (consider a VBC), and are thus not QSLs according to the third definition.[13] Thus, in the absence of any SSB, one might conclude that the ground state is unique and should, therefore, sustain *gapless* excitations in order to satisfy the LSM theorem. This is indeed a possibility (the *algebraic* QSL of Sect. 16.5), but the LSM theorem allows another alternative: gapped excitations above a degenerate ground state *without SSB*. In such a case, the states in the degenerate ground-state manifold are locally identical (indistinguishable by any local order parameter) but globally different due to the topological order [4] discussed in the previous paragraph. The LSM theorem is useful because it provides a natural classification for the ground states. A half-odd-integer-spin system is either (1) conventionally ordered (SSB), (2) a gapless QSL, or (3) a topologically ordered, gapped QSL. Only integer-spin systems have the additional possibility of being (4) "quantum paramagnets" (non-degenerate ground state and gapped excitations, as discussed in Sect. 16.2.2).

## 16.3 Mean Fields and Gauge Fields

We review here a formalism for describing deconfined liquids in Heisenberg models, and discuss the possible emergence of gauge fields. The orgin of this approach lies in the slave-particle approaches to the Hubbard and $t$-$J$ models [21].

### *16.3.1 Fermionic Representation of Heisenberg Models*

The group $SU(2)$ can act on the spinon operators of (16.5-16.6) in two different ways, globally, in describing spin rotations, and locally, related to the (gauge) redundancy of the description of spin operators.

**Spin rotations** – A *global spin rotation* is effected by multiplying the doublet $d_1 \equiv \left[c_{i\uparrow}\ c_{i\downarrow}\right]$ to the right by an $SU(2)$ matrix $V$, $d_1 = \left[c_{i\uparrow}\ c_{i\downarrow}\right] \to d_1 V$. By taking the Hermitian transpose of $d_1$ and using $V^\dagger = V^{-1}$, one may show that $d_2 = \left[c^\dagger_{i\uparrow}\ -c^\dagger_{i\downarrow}\right]$ is also transformed by a right-multiplication: $d_2 \to d_2 V$. Thus, $d_1$ and $d_2$ may be grouped into a 2×2 matrix which transforms under $SU(2)$

[12] A magnetically ordered system satisfies the theorem because the spectrum is gapless due to the presence of a spontaneously broken continuous symmetry (Goldstone modes). A VBC has a gapped spectrum, but the ground state is degenerate in the thermodynamic limit, due to translational symmetry-breaking.

[13] In principle, there can be coexistence of some conventional order *and* fractional excitations; this possibility is ignored here.

rotations by right-multiplication,

$$\psi_i = \begin{bmatrix} c_{i\uparrow} & c_{i\downarrow} \\ c^\dagger_{i\downarrow} & -c^\dagger_{i\uparrow} \end{bmatrix} \to \psi_i V. \tag{16.9}$$

From (16.9), $\psi_i \psi_j^\dagger$ is manifestly invariant. This allows one to introduce two rotation-invariant operators, $\chi_{ij}$ and $\eta_{ij}$, for each pair of sites in the system,

$$\psi_i \psi_j^\dagger \equiv \begin{bmatrix} c_{i\uparrow} c^\dagger_{j\uparrow} + c_{i\downarrow} c^\dagger_{j\downarrow} & c_{i\uparrow} c_{j\downarrow} - c_{i\downarrow} c_{j\uparrow} \\ c^\dagger_{i\downarrow} c^\dagger_{j\uparrow} - c^\dagger_{i\uparrow} c^\dagger_{j\downarrow} & c^\dagger_{i\downarrow} c_{j\downarrow} + c^\dagger_{i\uparrow} c_{j\uparrow} \end{bmatrix} = \begin{bmatrix} -\chi^\dagger_{ij} & -\eta^\dagger_{ij} \\ -\eta_{ij} & \chi_{ij} \end{bmatrix}. \tag{16.10}$$

The quantity $\psi_i$ also gives a convenient expression of the spin operators and of the constraint,

$$S_i^a = \frac{1}{2} \mathrm{Tr}\left[\psi_i^\dagger \psi_i (\sigma^a)^T\right], \; a = x, y, z, \tag{16.11}$$

$$\mathrm{Tr}\left[\psi_i^\dagger \sigma^z \psi_i\right] = c^\dagger_{i\uparrow} c_{i\uparrow} - c_{i\downarrow} c^\dagger_{i\downarrow} = c^\dagger_{i\uparrow} c_{i\uparrow} + c^\dagger_{i\downarrow} c_{i\downarrow} - 1 = 0. \tag{16.12}$$

It is useful here to add two other constraints which are consequences of the first, $c_{i\downarrow} c_{i\uparrow} = 0 = \eta^\dagger_{ii}$ (no double site occupancy) and $c^\dagger_{i\downarrow} c^\dagger_{i\uparrow} = 0 = \eta_{ii}$ (no empty sites). Together, the three constraints can be written in the compact form

$$\mathrm{Tr}\left[\psi_i^\dagger \sigma^a \psi_i\right] = 0 \,, \; a = x, y, z. \tag{16.13}$$

**Gauge transformations** – Because of (16.6), $c_{i\uparrow}$ and $c^\dagger_{i\downarrow}$ have the same physical effect, namely of decreasing $S_i^z$ by one unit. They can be placed in a doublet, $p_1$, upon which $SU(2)$ matrices act without changing the physical spin operators. Let $W_i$ be a (site-dependent) $SU(2)$ matrix encoding this *gauge transformation*, $p_1 = \begin{bmatrix} c_{i\uparrow} \\ c^\dagger_{i\downarrow} \end{bmatrix} \to W_i \, p_1$. It is easy to verify that $p_2 = \begin{bmatrix} c_{i\downarrow} \\ -c^\dagger_{i\uparrow} \end{bmatrix}$ transforms by the same left-multiplication. Taken together, these two-column vectors $p_1$ and $p_2$ form once again the matrix $\psi_i$. From (16.11), it is evident that the spin operators are gauge-invariant,

$$\psi_i \to W_i \psi_i, \quad S_i^a \to S_i^a. \tag{16.14}$$

In summary, global spin rotations are described by right-multiplication of $\psi$ and local gauge transformations by left-multiplication. As a specific example of a gauge transformation, we consider the $U(1)$ subgroup of $SU(2)$, which is parameterized by the phase $\Lambda(i)$ as $W_i = \exp(\mathrm{i}\Lambda(i)\sigma^z)$, and which corresponds to (16.7). A spinon ("anti-spinon") carries a charge $+1(-1)$ of this $U(1)$ gauge field. The spin-flip operator $S_i^+ = c_{i\uparrow} c^\dagger_{j\downarrow}$, is *gauge-neutral*.

### 16.3.2 Local $SU(2)$ Gauge Invariance

In a path-integral formulation based on the fermionic representation, the imaginary-time Lagrangian takes the form [11]

$$L = \sum_i \mathrm{Tr}\left[\psi_i^\dagger \left(\partial_\tau + \boldsymbol{A}_i^0 \cdot \boldsymbol{\sigma}\right) \psi_i\right] - H, \tag{16.15}$$

where $H$ is the Hamiltonian and $\boldsymbol{A}_i^0 = (A_{ix}^0, A_{iy}^0, A_{iz}^0)$ a real, three-component vector which plays the role of a Lagrange multiplier for the three constraints of (16.13). Consider a time-dependent gauge transformation

$$\psi_i(\tau) \to W_i(\tau)\psi_i(\tau). \tag{16.16}$$

To ensure the invariance of the action (16.15), $\boldsymbol{A}_i^0$ must transform as the time component of an $SU(2)$ gauge field,

$$\boldsymbol{A}_i^0 \cdot \boldsymbol{\sigma} \to W_i(\tau)\left(\partial_\tau + \boldsymbol{A}_i^0 \cdot \boldsymbol{\sigma}\right).W_i^\dagger(\tau). \tag{16.17}$$

From (16.11), the Heisenberg interaction can be written as

$$\boldsymbol{S}_i \cdot \boldsymbol{S}_j = -\frac{1}{8}\mathrm{Tr}\left[\psi_i \psi_j^\dagger \psi_j \psi_i^\dagger\right], \tag{16.18}$$

a quartic term in fermionic operators which can be decoupled (Hubbard–Stratonovich procedure) by introducing a $2 \times 2$ complex matrix $U_{ij}$ on each bond $\langle i, j \rangle$. The corresponding contribution to the Lagrangian is

$$-H = -\frac{8}{J}\sum_{\langle i,j \rangle} \mathrm{Tr}\left[U_{ij}^\dagger U_{ij}\right] - \sum_{\langle i,j \rangle} \mathrm{Tr}\left[\psi_i^\dagger U_{ij} \psi_j + \mathrm{H.c}\right], \tag{16.19}$$

whence a Gaussian integration over $U_{ij}$ returns the spin–spin interaction of (16.18). From (16.9), it is clear that $U$ is invariant under spin rotations, and from (16.14) one observes that $U$ transforms as the spatial component of an $SU(2)$ gauge field under gauge transformation,

$$\psi_i(\tau) \to W_i(\tau)\psi_i(\tau), \qquad U_{ij}(\tau) \to W_i(\tau) U_{ij}(\tau) W_j^\dagger(\tau). \tag{16.20}$$

### 16.3.3 Mean-field (Spin-liquid) States

**Mean-field Hamiltonian** – Various mean-field approximations may be applied when the Heisenberg model is expressed in the form of (16.15) and (16.19). As we will show, these can describe a large variety of spin-liquid states, and in particular the "RVB spin liquids." The procedure is to replace the fluctuating *fields* $U_{ij}(\tau)$ and $\boldsymbol{A}_i^0(\tau)$ by time-independent, complex matrices $U_{ij}^0$ and complex vectors $\boldsymbol{a}_i^0$, because the mean-field Hamiltonian is then quadratic, and hence soluble, in the fermion operators,[14]

$$H_{\mathrm{MF}} = \frac{8}{J}\sum_{\langle i,j\rangle} \mathrm{Tr}\left[U_{ij}^{0\dagger} U_{ij}^0\right] + \sum_{\langle i,j\rangle} \mathrm{Tr}\left[\psi_i^\dagger U_{ij}^0 \psi_j + \mathrm{H.c}\right] + \sum_i \mathrm{Tr}\left[\psi_i^\dagger (\boldsymbol{a}_i \cdot \boldsymbol{\sigma}) \psi_i\right]. \quad (16.21)$$

The first term is a constant, the second describes spinon hopping and pairing, and the third term arises from the constraints. Minimizing the energy with respect to the parameters $U_{ij}^0$ and $\boldsymbol{a}_i$ gives the self-consistency conditions

$$\frac{8}{J} U_{ij}^0 = \left\langle \psi_i \psi_j^\dagger \right\rangle = \begin{bmatrix} -\chi_{ij}^{0*} & -\eta_{ij}^{0*} \\ -\eta_{ij}^0 & \chi_{ij}^0 \end{bmatrix}, \quad \left\langle \mathrm{Tr}\left[\psi_i^\dagger (\boldsymbol{a}_i \cdot \boldsymbol{\sigma}) \psi_i\right]\right\rangle = 0, \quad (16.22)$$

where we have used the notation $\chi_{ij}^0 \equiv \langle \chi_{ij} \rangle = \langle c_{i\uparrow}^\dagger c_{j\uparrow} + c_{i\downarrow}^\dagger c_{j\downarrow} \rangle$ and $\eta_{ij}^0 \equiv \langle \eta_{ij} \rangle = \langle c_{i\uparrow}^\dagger c_{j\downarrow}^\dagger - c_{i\downarrow}^\dagger c_{j\uparrow}^\dagger \rangle$. From (16.20), the parameters $U_{ij}^0$ and $\boldsymbol{a}_i$ are not gauge-invariant,[15] so different mean-field parameters may lead to the same physical quantities. This will have important consequences in Sect. 16.3.4.1.

**Ground state and excitations of $H_{\mathrm{MF}}$** – Equation (16.21) describes a system of free spinons. The ground state is obtained by calculating the spinon band structure and filling the negative-energy single-particle states. The resulting state satisfies the constraints (16.13) only on average, and is therefore not a valid spin-$\frac{1}{2}$ wave function. One way to obtain a spin state is to apply a Gutzwiller projection in order to remove configurations with empty or doubly occupied sites [35]. This can be performed numerically by using Monte Carlo methods [35–37]. Another approach is to analyze the qualitative effects of fluctuations, based on symmetry arguments.

Because $U_{ij}$ and $\boldsymbol{A}^0$ are invariant under spin rotations, $H_{\mathrm{MF}}$ does not have any preferred direction in spin space. The mean-field ground state is thus a total-spin singlet, without magnetic long-range order. It is already a spin-liquid state in the sense of definition 1 (Sect. 16.2.1). This type of mean-field approach is not

[14] This approximation is equivalent to particular large-$N$ limits of the model, obtained when the spin-rotation symmetry group $SU(2)$ is generalized to $SU(N)$ [26–28]. With bosonic operators instead of fermions, this type of mean-field approximation is closely related to "Schwinger-boson" approaches [29–34].

[15] $\left\langle \psi_i \psi_j^\dagger \right\rangle \neq 0$ is in apparent contradiction with Elitzur's theorem, which states that non-gauge-invariant quantities should average to zero. Some slight abuse of notation has been committed here, as true expectation values in the mean-field theory should be averaged over all gauge-rotated copies of a given representative state.

appropriate to describe Néel-ordered phases; bosonic representations of the spin are more appropriate, because bosons may *condense*.

In addition, $H_{\rm MF}$ contains spin-$\frac{1}{2}$ excitations, obtained by adding or removing one spinon to or from the ground state. A crucial question is whether the existence of "deconfined" (in fact, free at this crude level of approximation) spinons is merely an artefact of the mean-field approximation. In such a case, the inclusion of fluctuations (in particular of gauge-field fluctuations) would confine the spinons. The other possibility is that the spinons remain deconfined in the presence of fluctuations (or Gutzwiller projection), in which case the mean-field approximation is indeed a useful starting point for an accurate description.[16]

**Example: the "$\pi$-flux" state** – Consider the mean-field state on the square lattice introduced by Affleck and Marston [27,28], which is equivalent to the "mixed $s+id$" RVB state [38] and labeled "SU2B$n$0" in Wen's classification [39]. This state has $\boldsymbol{a}_i = 0$, $\eta_{ij} = 0$, and a modulus $|\chi_{ij}| = \chi_0$ identical on all bonds. The phases $\theta_{ij} = \arg(\chi_{ij})$ are such that $\theta_{12}+\theta_{23}+\theta_{34}+\theta_{41} = \pi \pmod{2\pi}$ on every square plaquette. As in the case of hopping amplitudes for a charged particle in the presence of a uniform magnetic field, there is no gauge in which $\theta_{ij}$ is translation-invariant. The unit cell defined by $\theta_{ij}$ contains at least *two* sites. A possible choice is $U^0_{ij} = i\chi_0 \begin{bmatrix} 1 & 0 \\ 0 & 1 \end{bmatrix}$ for bonds $i \to j$ oriented as in Fig. 16.1. The corresponding mean-field Hamiltonian is

$$H_{\rm MF} = 2\chi_0 \sum_{\langle i\to j\rangle,\sigma=\uparrow,\downarrow} \left( i c^\dagger_{i\sigma} c_{j\sigma} + \text{H.c.} \right) + \text{constant}, \tag{16.23}$$

and gives two bands of quasiparticles with dispersion relations [27,28]

$$E^{\pm}(k) = \pm 4\chi_0 \sqrt{\cos(k_x)^2 + \cos(k_y)^2} \tag{16.24}$$

(in the Brillouin zone defined by $|k_x| + |k_y| \le \pi$). The Fermi energy is at $E = 0$,[17] where the two bands meet at $\boldsymbol{k}_A = (\pi/2, \pi/2)$ and $\boldsymbol{k}_B = (\pi/2, -\pi/2)$. To describe the long-distance properties of the system, it is useful to focus on low-energy excitations and to linearize the spectrum in the vicinity of $\boldsymbol{k}_A$ and $\boldsymbol{k}_B$. The corresponding Hamiltonian is that for *four* fermion flavors (two for the spin and two for the $A$- or $B$-"valley" index) of two-component[18] Dirac fermions.

[16] A similar question arises concerning the presence of a gap in the excitation spectrum. $H_{\rm MF}$ can be gapless, as in the $\pi$-flux example below. It is then important to understand whether fluctuations beyond the mean-field approximation can act to open a gap. In some cases, the spectrum is expected to remain gapless, although fluctuations will in general change the correlation exponents. This is the case if the terms which could potentially open a gap (terms relevant in the renormalization-group sense) are actually forbidden by gauge invariance or by symmetry (Sect. 16.3.4.1). Then the mean-field approximation is again a good starting point to describe a gapless spin liquid.

[17] In agreement with (16.22), the system is half-filled.

[18] There are *two* zero-energy single-particle states when $E^+ = E^-$.

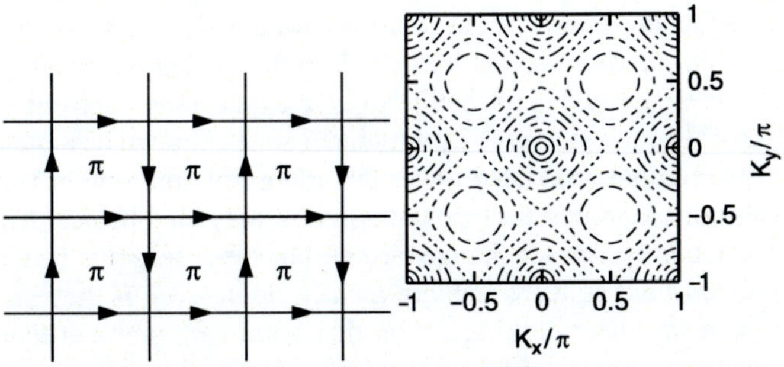


**Fig. 16.1** *Left*: bond orientations used to define the mean-field $\pi$-flux state on the square lattice ($\chi_{i\to j} = -\chi_{j\leftarrow i} = i\chi_0 \in \mathbb{R}$) [27,28]. *Right*: minimum energy for a *pair* of spinons with total momentum $\boldsymbol{k} = (k_x, k_y)$

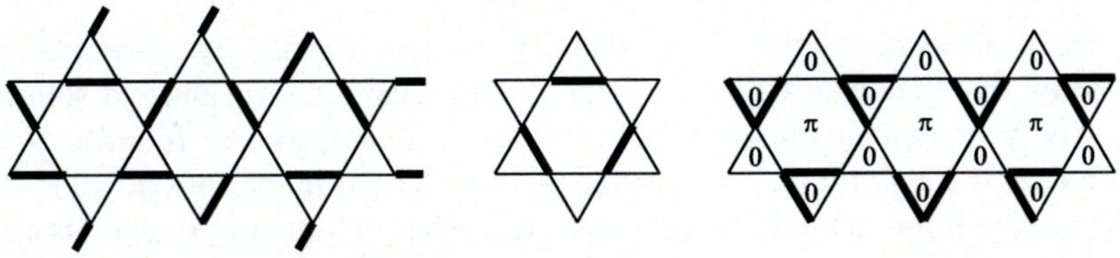


**Fig. 16.2** Examples of mean-field states on the kagomé lattice. A: dimerized state with $\chi^0_{ij} = J/2$ on the thick bonds and $\chi^0_{ij} = 0$ on all other bonds. Such states are the (degenerate) lowest-energy states at the mean-field level [41]. B: state including fluctuations $\delta\chi_{ij}$ of the bond field $\chi_{ij}$ at order $(\delta\chi_{ij})^3$ lifts this degeneracy in favor of configurations maximizing the number of hexagons with three "dimers" [42]. C: $[0,\pi]$-flux phase on the kagomé lattice [43,44]. Thin (thick) bonds have $\chi_{ij} = +\chi_0(\chi_{ij} = -\chi_0) \in \mathbb{R}$, such that the flux is 0 on *triangles* and $\pi$ on *hexagons*

Because spinons are necessarily created in pairs, we show in Fig. 16.1 the energy of the two-spinon continuum, $E^+(\boldsymbol{k}-\boldsymbol{q}) + E^+(\boldsymbol{q})$, as a function of $\boldsymbol{k}$. These excitations are gapless and linearly dispersive around the four minima (located at $\boldsymbol{k} = (0,0)$, $(\pi,0)$, $(0,\pi)$, and $(\pi,\pi)$). As we will discuss in Sect. 16.3.4, the presence of gapless fluctuation modes around this mean-field state means that the stability of the mean-field approximation is a priori not at all clear. A somewhat involved analysis suggests, however, that it could indeed represent a stable spin-liquid phase with gapless magnetic excitations (remnants of the excitations discussed above) and algebraic correlations [40].

**Dimerized mean-field states** – Among all the different self-consistent mean-field states, the "dimerized" states have the lowest energy at the mean-field level for a large class of lattices [41]. Such a mean-field solution can be viewed as a hard-core dimer covering of the lattice (Fig. 16.2): $\chi^0_{ij} = J_{\max}/2$ on the bonds occupied by a "dimer" and $\chi^0_{ij} = 0$ otherwise, while $\eta^0_{ij} = 0$ everywhere. The configuration of "dimers" is such that each site is touched by exactly one dimer, and only occupies the bonds $(ij)$ with the strongest antiferromagnetic exchange, $J_{ij} = J_{\max}$. Because the number of such dimer coverings is (usually) an exponential function of the number of sites, these mean-field solutions are massively degenerate. However, we expect that the *fluctuations* of the field $U$ will lift this degeneracy. Indeed, if these fluctuations are treated perturbatively ($1/N$ expansions), one obtains an effective model [42, 45] in the subspace of dimerized states, known as a quantum dimer model [46]. Fluctuations may also lower the energy of some other (undimerized) mean-field solutions, and certain types of solution may also be stabilized by other interactions (notably by ring-exchange terms). For these reasons, it is very important to study mean-field states which are *locally stable* even if they are not global energy minima at the mean-field level; on this point we comment that it is much more significant to compare energies *after* Gutzwiller projection.

### *16.3.4 Gauge Fluctuations*

For a given mean-field solution, its stability must be verified by investigating the low-energy fluctuation modes. In a large $N$ formalism, the mean-field solution is exact at $N = \infty$ and the question to be adressed is: does the phase found at $N = \infty$ survive at large but *finite* $N$, that is when the fluctuations modes of the fields $A_i^0(\tau)$ and $U_{ij}(\tau)$ are introduced. In other words, do we have a (quantum) phase transition between the (mean-field) state at $N = \infty$ and $1 \ll N < \infty$. If any mode or modes drive(s) the system to an instability, the mean-field approximation is not an appropriate starting point and the physical properties of the system (presence/absence of a gap, broken symmetries, excitations) cannot be those of the mean-field Hamiltonian. On the other hand, if no modes lead to a divergence of physical quantities, the sign of an instability, the mean-field solution may describe, to some extent, a real phase of the spin model (at least for large enough $N$). Of course, it is difficult to examine all possible fluctuations, as this would be equivalent to solving the original spin model. As a first approximation, gapped degrees of freedom may be integrated over (or simply ignored), as they are expected to play no qualitative role at low energies. By contrast, gapless modes are potential sources of instability, and thus are likely to influence the low-energy, long-distance physics of the system.

Fermion density fluctuations may be gapped or gapless, depending on the spectrum of $H_{\mathrm{MF}}$. When these modes are gapped, the fermions form an incompressible state at the mean-field level, and the density fluctuations are expected to have no effect on the long-distance properties of the system. There are three simple cases [4] where such a thing happens: (1) the bond parameters $U_{ij}^0$ break the translational invariance in such a way that the ground state of $H_{\mathrm{MF}}$ is a band insulator (Fermi level between a completely filled and a completely empty band), as in the dimerized solutions of [41]; (2) $H_{\mathrm{MF}}$ contains a pairing term ($\eta \neq 0$) so that its ground state is a BCS-like gapped "superconductor" [4]; (3) $H_{\mathrm{MF}}$ contains a non-trivial flux (i.e. different from 0 or $\pi$) piercing some of the plaquettes, and its ground state is analogous to a set of completely filled Landau levels, as in the integer Hall effect [47].

Gauge excitations with a continuous gauge group are also natural candidates for gapless excitations. The reason for this is that the gauge invariance forbids "naive" mass terms, such as $(A^{\mu})^2$, for the gauge field in the same way that it preserves the masslessness of ordinary photons; however, more elaborate mechanisms, such as Anderson-Higgs, may still open a gap. Gauge excitations are also important because they can mediate long-range interactions between the spinons. The importance of gauge modes in slave-boson mean-field theory was first put forward by Baskaran and Anderson [48].

The spin models of interest here have a local $SU(2)$ gauge invariance (Sect. 16.3.2). Thus, one may ask why the fluctuations are not always described by an $SU(2)$ gauge field, or alternatively, why the nature of the gauge field does actually depend on the particular mean-field state. The answer is that the mean-field parameters $U^0_{ij}$ break partially the local $SU(2)$ gauge invariance (unless all the $U^0_{ij}$ are diagonal matrices on all bonds), leaving only a lower invariance symmetry. This is in a sense analogous to global symmetries, where the number of Goldstone modes depends on the number of broken continuous symmetries. Following Wen [4, 39], we will discuss how to construct the gauge fields describing fluctuations about a given mean-field state.

#### 16.3.4.1 Projective Symmetry Group and Invariant Gauge Group

Because the mean-field parameters $U^0_{ij}$ are not gauge-invariant, two apparently different solutions may be physically identical. A set of parameters $U^0_{ij}$ which are not translationally invariant may describe a (mean-field) QSL with no broken symmetry.

As an example, consider the $\pi$-flux state defined by (16.23). Under any translation by one lattice constant in the $x$ direction, $U^0_{ij}$ is changed into $\tilde{U}^0_{ij} = -U^0_{ij}$ on all the vertical bonds, and remains unchanged on horizontal bonds. However, the gauge transformation associated with $W_i = (-1)^{i_y}\mathbb{I}$ maps $\tilde{U}^0$ back to $U^0$, and thus both $U^0$ and its translation, $\tilde{U}^0$, label the same mean-field state.[19] This result illustrates that the physical symmetries are encoded in a non-trivial way in the mean-field parameters $U^0_{ij}$. In fact, this is a fundamental property, inherent to any description of the system in terms of fractional excitations (in this case spinons): the Hamiltonian describing the hopping of the spinons requires a gauge choice and is apparently less symmetric than the original spin model. This led X.-G. Wen [39] to introduce the concept of Projective Symmetry Group (PSG).

**Definition of the PSG** – Let $T : i \mapsto T(i)$ be a lattice symmetry of the original spin model and $W$ be a (time-independent) gauge transformation (16.20). The PSG associated with the mean-field parameters $\{U^0_{ij}, \boldsymbol{a}_i\}$ is defined as the set of all the pairs $(T, W)$ satisfying

$$U^0_{ij} = W_i U^0_{T(i)T(j)} W_j^\dagger \,, \quad \boldsymbol{a}_i \cdot \sigma = W_i(\boldsymbol{a}_{T(i)} \cdot \sigma) W_i^\dagger \qquad \forall i, j. \tag{16.25}$$

[19] The two mean-field states give the same expectation values for any observable which conserves the total number of fermions on each site (therefore including physical spin observables). They also give the same spin-$\frac{1}{2}$ wave function after Gutzwiller projection.

An element of the PSG is thus a lattice symmetry followed by a gauge transformation, such that the mean-field parameters $U^0_{ij}$ and $\boldsymbol{a}_i$ are unchanged. In the "$\pi$-flux" example above, the PSG contains (among other elements) the horizontal translation by one unit cell associated with $W_i = (-1)^{i_y}\mathbb{I}$, and the vertical translations associated to the trivial gauge transformation $W_i = \mathbb{I}$.

The invariant gauge group (IGG) [39] is a special subgroup of the PSG, containing all the elements $(T, W)$ where $T$ is the identity. As we will show, *the IGG determines the gauge group, and therefore the nature of the gauge fluctuations around the mean-field solution.*

We consider a mean-field state and denote by $\mathcal{I}$ its IGG. We first assume for simplicity that $\mathcal{I}$ is isomorphic to $U(1)$. In such a case, the gauge transformations $W^\theta \in \mathcal{I}$ can be parameterized as

$$W^\theta : i \mapsto W^\theta_i = \exp(\mathrm{i}\theta\, \boldsymbol{n}_i \cdot \sigma)\,, \tag{16.26}$$

where $W^\theta_i$ is an $SU(2)$ rotation of angle $\theta \in [0, 2\pi[$ about the axis defined by the (spatially varying) unit vector $\boldsymbol{n}_i$. At each site, we rotate $\boldsymbol{n}_i$ to the $z$ axis,

$$V_i W^\theta_i V^\dagger_i = \exp(\mathrm{i}\theta\, \sigma^z)\,. \tag{16.27}$$

These elements $V_i \in SU(2)$ define a new gauge, in which $U^0$ becomes $\tilde{U}^0$, defined by

$$\tilde{U}^0_{ij} = V_i U^0_{ij} V^\dagger_j, \tag{16.28}$$

and certain types of fluctuation of $\tilde{U}$ about $\tilde{U}^0$ may be parameterized by a real field $A$,

$$\tilde{U}_{ij} = \tilde{U}^0_{ij} \mathrm{e}^{\mathrm{i}A_{ij}\sigma^z}. \tag{16.29}$$

We will now show that $A$ is the spatial component of a $U(1)$ gauge field, and is thus potentially important in describing the low-energy excitations of the system. We consider a particular family of gauge transformations, $i \mapsto \exp[\mathrm{i}\theta(i)\,\sigma^z]$, where the angle $\theta$ parameterizing the elements of the IGG (16.27) has been promoted to a *local* variable $\theta(i)$. The bond field $\tilde{U}_{ij}$ transforms according to $\tilde{U}_{ij} \to \mathrm{e}^{\mathrm{i}\theta(i)\sigma^z} \tilde{U}_{ij} \mathrm{e}^{-\mathrm{i}\theta(j)\sigma^z}$, whence only a few short algebraic manipulations are required[20] to cast $\tilde{U}_{ij}$ in the form of (16.29), with the replacement $A_{ij} \to A_{ij} + \theta(i) - \theta(j)$.

---

[20] One uses (16.27),(16.28), and (16.29) to transform the right-hand side,

$$\mathrm{e}^{\mathrm{i}\theta(i)\sigma^z} \tilde{U}_{ij} = \mathrm{e}^{\mathrm{i}\theta(i)\sigma^z} \tilde{U}^0_{ij} \mathrm{e}^{\mathrm{i}A_{ij}\sigma^z} \tag{16.30}$$

$$= \mathrm{e}^{\mathrm{i}\theta(i)\sigma^z} V_i U^0_{ij} V^\dagger_j \mathrm{e}^{\mathrm{i}A_{ij}\sigma^z} \tag{16.31}$$

$$= V_i W^{\theta(i)}_i U^0_{ij} V^\dagger_j \mathrm{e}^{\mathrm{i}A_{ij}\sigma^z}, \tag{16.32}$$

Thus these phase fluctuations of the bond variables are those of a $U(1)$ gauge field. In some cases [39], several subgroups of the IGG are isomorphic to $U(1)$. Each one can be parameterized as in (16.26), but with different directions $\boldsymbol{n}_i^1$, $\boldsymbol{n}_i^2$, .... Repeating this construction for each subgroup leads to the same number of $U(1)$ gauge fields. We note finally that the IGG always contains the group $\mathbb{Z}_2$, because the gauge transformation $W : i \mapsto -\mathbb{I}$ leaves all $U_{ij}^0$ unchanged. The construction of the associated $\mathbb{Z}_2$ gauge field, $A_{ij} \in \{0, \pi\}$, is identical to the $U(1)$ case, except for the restriction $\theta(i) \in \{0, \pi\}$.

#### 16.3.4.2 Two Simple Examples of IGG

As a first exercise, one may determine the IGG of the $\pi$-flux state on the square lattice, where $U_{ij}^0 = \pm i\chi_0$ is proportional to the identity. By definition, an element of the IGG is a set of matrices $W_i \in SU(2)$ satisfying $W_i U_{ij}^0 W_j^\dagger = U_{ij}^0$ on all bonds. From the particular form of $U^0$, this equality becomes $W_i W_j^\dagger = \mathbb{I}$, showing that $W_i$ must be the same (arbitrary) $SU(2)$ matrix on every site. We have thus shown that the IGG of this spin-liquid state is isomorphic to $SU(2)$.

As a second example, one may consider the spin liquid proposed by Hastings [43] and Ran et al. [44] for the kagomé-lattice Heisenberg model. At the mean-field level, the bond field takes the values $U_{ij}^0 = \pm\chi_0 \begin{bmatrix} -1 & 0 \\ 0 & 1 \end{bmatrix} = \pm\chi_0\sigma^z$, with the signs chosen to produce a flux 0 for each triangle and $\pi$ for each hexagon (Fig. 16.1). From this particular form of $U^0$, the condition on $W \in \mathcal{I}$ may be expressed as $W_j = \sigma^z W_i \sigma^z$. Thus, if $W_i$ is specified at any point $i$, all the other matrices are fixed. By propagating this condition around any triangle of the lattice, one finds $W_0 = (\sigma^z)^3 W_0 (\sigma^z)^3 = \sigma^z W_0 \sigma^z$, i.e. $W_0$ must commute with $\sigma^z$, and therefore has the form of (16.26) with $\boldsymbol{n} = [0, 0, 1]$. Thus $\mathcal{I} = U(1)$.

#### 16.3.4.3 PSG Beyond the mean-field Approximation

Thus far, we have defined the PSG as the symmetry of the mean-field Hamiltonian (through $U_{ij}^0$; the time component $\boldsymbol{a}$ is omitted hereafter for simplicity). The utility of the PSG is, however, that it is robust to fluctuations, at least at the perturbative

---

and then employs the fact that, for any angle $\theta$, $W^\theta$ belongs to the IGG of the mean field $U^0$. Setting $\theta = \theta(i)$ yields $W_i^{\theta(i)} U_{ij}^0 = U_{ij}^0 W_j^{\theta(i)}$, and finally

$$\tilde{U}_{ij} \to V_i U_{ij}^0 W_j^{\theta(i)} V_j^\dagger e^{iA_{ij}\sigma^z} e^{-i\theta(j)\sigma^z} \tag{16.33}$$

$$= V_i U_{ij}^0 V_j^\dagger e^{i\theta(i)\sigma^z} e^{iA_{ij}\sigma^z} e^{-i\theta(j)\sigma^z} \tag{16.34}$$

$$= \tilde{U}_{ij}^0 e^{i\theta(i)\sigma^z} e^{iA_{ij}\sigma^z} e^{-i\theta(j)\sigma^z}. \tag{16.35}$$

level [39]. We denote by $\mathcal{L}(\psi, U)$ the exact Lagrangian of the spin model, in terms of the fermions $\psi_i(\tau)$ and bond fields $U_{ij}(\tau)$ introduced in Sect. 16.3.2. $\mathcal{L}$ is invariant under any gauge transformation $W$ and lattice symmetry $T$: $\mathcal{L}(\psi, U) = \mathcal{L}(P\psi P^{-1}, PUP^{-1})$, where $P = (W, T)$ need not be in the PSG of $U_0$. However, if $P \in$ PSG, then $U_0 = PU_0P^{-1}$, and the Lagrangian $\mathcal{L}^0$ describing the fluctuations, defined by $\mathcal{L}^0(\psi, \delta U) = \mathcal{L}(\psi, U_0 + \delta U)$, is invariant: $\mathcal{L}^0(\psi, \delta U) = \mathcal{L}(\psi, U_0 + \delta U) = \mathcal{L}(P\psi P^{-1}, P(U_0 + \delta U)P^{-1}) = \mathcal{L}(P\psi P^{-1}, U_0 + P(\delta U)P^{-1}) = \mathcal{L}^0(P\psi P^{-1}, P\delta UP^{-1})$. In addition, the mean-field ground state and the mean-field Hamiltonian are also symmetric under the PSG, because the expectation values $\langle\psi_i\psi_j^\dagger\rangle = U_0$ are by definition PSG-invariant. Thus, *the theory for the fluctuations about a given mean-field $U_0$ has the symmetry group of its PSG.*

We appeal now to the general property that, unless a phase transition occurs, broken or unbroken symmetries do not change their nature under the inclusion of perturbations. The ground state will receive some corrections once fluctuations are included but, on the assumption that the mean-field state exists within a *stable phase*, the symmetries will not change. The PSG is not only a property of the mean field, but is a property of the whole *phase*. Different spin liquids may be distinguished and hence classified by the symmetry group of the Hamiltonian driving their spinon dynamics, which can be read from the PSG of the appropriate mean-field starting point, $H_{\mathrm{MF}}$.

This symmetry principle has important consequences. To obtain an effective description of $\mathcal{L}^0(\psi, \delta U)$, some high-energy modes can be integrated out formally. One may, for example, eliminate the fermionic modes which are far from the Fermi level. In the example of the $\pi$-flux state (16.23), such a procedure leads to Dirac fermions with a linear dispersion relation. One may also integrate out the amplitude fluctuations of the bond variables $U_{ij}$, retaining only the phase fluctuations (which in the example of (16.23) is equivalent to neglecting the fluctuations of $|\chi_0|$). Under the assumption that no change of symmetry occurs when including these fluctuations, any term (even if it is invariant under all lattice symmetries) which is not PSG-invariant cannot appear in the effective action $\mathcal{L}^0$.

Consider the example of a mean-field Hamiltonian where the spinons are gapless at $N_f$ points in the Brillouin zone, and are described by $2N_f$ flavors of two-component Dirac fermions after linearization (the factor of 2 given by the spin $\uparrow, \downarrow$). The corresponding fermion operators, $\Psi_i^{\alpha=1\cdots 2N_f}$, are linear combinations of the microscopic spinon operators, $c_{i,\sigma}$, and each element of the PSG is equivalent to a particular transformation of the $\Psi$ operators (a detailed example is presented in [49]). The effective action for the Dirac fermions (and their associated gauge field) is obtained by a formal integration over higher-energy degrees of freedom. During this process, terms which are quadratic in $\Psi$ may a priori be generated and open a gap (which would spoil the algebraic nature of the spin correlations). However, from the discussion above, such terms are constrained to be invariant under the PSG. In certain cases [40, 44, 49], one may show that none of the possible terms arising this way is PSG-invariant. Such terms cannot be generated by integrating out the fluctuations (particularly of $U_{ij}$) perturbatively, and the system may remain gapless.

In such cases, the PSG "protects" the gapless spectrum. This can lead to stable critical states, even when the Hamiltonian has not been tuned to a critical point. For a recent example on the kagomé lattice, see [44,49]. We note in concluding this section that the PSG analysis does not provide any information about *non-perturbative* effects caused by fluctuations. This is the case, in particular, for the proliferation of magnetic monopoles in a $U(1)$ gauge field [31], which can lead to spinon confinement (but, however, is not expected to occur if $N_f$ is sufficiently large [40]).

## 16.4 $\mathbb{Z}_2$ Spin Liquids

The simplest spin-liquid states (according to definition 3) for a two-dimensional spin-$\frac{1}{2}$ system are the $\mathbb{Z}_2$ liquid states, which have gapped spinons. The name is taken from the fact that the gauge group (IGG) relevant for describing its elementary excitations is $\mathbb{Z}_2$. All the excitations are gapped, and the spin correlations are short-ranged. The magnetic (spin-$\frac{1}{2}$) excitations are deconfined spinons, which may be fermions or bosons. In addition, the system has singlet (total spin $S = 0$, i.e. non-magnetic) excitations which are fluxes (or vortices) of the $\mathbb{Z}_2$ gauge field. These vortices were discussed in the early days of RVB theories [4,50,51] and have more recently been christened "visons" [52]. Although such spin liquids can be discussed within the slave-fermion formalism [39], below we describe the approach based on the short-range RVB framework.

### 16.4.1 Short-range RVB Description

An RVB wave function can be written as $|\phi\rangle = \sum_c \phi(c)|c\rangle$, where $c$ labels a valence-bond covering of the lattice. This state is manifestly a spin singlet, but the nature of the spin correlations depends on the weights $\phi(c)$. In particular, if this state has a sufficient weight $\phi(c)$ for configurations $c$ with long singlet bonds, $\langle\phi|\boldsymbol{S}_i \cdot \boldsymbol{S}_j|\phi\rangle$ can even be long-range-ordered (Néel order) [53]. Here we focus instead on states where the weight $\phi(c)$ can be neglected if the valence-bond exceeds a finite length $\xi \sim \mathcal{O}(1)$. In this case, spin correlations are expected to decay exponentially.

However, this condition is not sufficient to guarantee a *liquid*, as a VBC wave function can also be written using short-range valence-bonds. In a VBC, one may define "parent" configurations $c_i$ ($i = 1, \ldots, d$, where $d$ is the degeneracy) which have the spatial periodicity of the crystal. In a columnar VBC on the square lattice, the parent states would be the 4-columnar configurations. If $|\phi\rangle$ is a crystalline state, each covering $c$ can be compared to its "closest" parent, from which it will differ only by collection of small *loops*.[21] These loops represent fluctuations around the maximally ordered configurations [17].

[21] We employ the standard notion of the *transition graph* to compare different valence-bond configurations. By overlaying two configurations $c$ and $c'$, one obtains closed loops by following alternately a valence bond of $c$ and a valence bond of $c'$.

If $|\phi\rangle$ describes an RVB *liquid*, there is no parent configuration to which to compare $c$, but one may still consider the transition graphs between two *typical* configurations $c$ and $c'$. Such loops can be visualized as resulting from a process where two neighboring spinons are created out of $c$, propagate along a closed loop, and are annihilated to form again a short-range valence-bond in $c'$. On the assumption that these virtual processes ocurring within the ground state contain some information about elementary *excitations*, the characteristic size of the resonance loops in the ground state represents the typical distance between excited spinons. If this length-scale is finite, it would indicate spinon confinement. Because the short-range RVB liquids are by contrast deconfined, the associated resonance loops should be large, with their size described by a scale-invariant (critical) distribution. We find here an interesting situation where the spectrum is gapped, and local observables are short-ranged, but some critical phenomena are "hidden" in the loop distribution of the ground-state wave function. These loops are related to Wilson loop operators in a gauge-theory description.

With periodic boundary conditions, one may choose a closed loop $\Delta_1$ on the dual lattice, which winds around the torus in the direction 1. Short-range valence-bond configurations may then be sorted according to the parity $P_1 = \pm 1$ of their number of valence-bonds crossing $\Delta_1$. $P_1$ is a topological invariant, in that it cannot be changed by any local operator (for precise statements see [54]). However, moving a spinon around the system in direction 2 *does* change $P_1$, which is the analog of the operator $U\hat{F}$ discussed in Sect. 16.2.3.3. $P_1$ defines two *topological sectors*, while $P_1$ and $P_2$ together define four. These sectors are *locally* equivalent: if a valence-bond configuration is known only over a finite region of the lattice, it is not possible to decide to which sector it belongs [55]. Because conventional liquids are insensitive to their boundary conditions (compared to solids), a short-range RVB liquid, where all local observables have short-range correlations, can reach the lowest ground-state energy equally well in all sectors.[22] Thus, a $\mathbb{Z}_2$ liquid has as many ground states as it has topological sectors.

### 16.4.2 $\mathbb{Z}_2$ Gauge Theory, Spinon Deconfinement, and Visons

In a VBC, the confining potential experienced by the spinons arises from the ordered background. It is thus *plausible* that valence-bond liquids do not generate such a confinement force. To show that spinons truly are deconfined requires a deeper

[22] It is instructive to compare with the case of a VBC: for a general VBC covering, invariant under two translations $T_1$ and $T_2$, choose a lattice size and geometry such that the periodicity vectors are an even multiple of $T_1$ and an even multiple of $T_2$. The directions of $T_1$ and $T_2$ are also taken to define the cuts $\Delta_{1,2}$ required to define parity sectors. It is easy to verify that these choices guarantee that all ordered parent configurations, and thus the degenerate ground states in the VBC phase, belong to the same "even×even" topological sector. The lowest states in the other sectors will lie higher by an energy proportional to the linear system size.

analysis. One possibility is to derive an effective $\mathbb{Z}_2$ gauge theory [56], which is known to have a deconfined phase, by analyzing the structure of the gauge fluctuations about an appropriate mean-field state [4].[23] This approach has also been described in the context of an $Sp(N)$ generalization [60] of a frustrated Heisenberg model on the square lattice (large-$N$ limit) [33], and is similar to the mean-field theory of Sect. 16.3 except in that it has *bosonic* spinons. In particular, it has been shown that the phase fluctuations of the bond variables are described by a $U(1)$ or a $\mathbb{Z}_2$ gauge theory, depending on whether the short-range spin correlations are collinear or non-collinear, respectively (these two cases have different IGGs). Short-range RVB spin liquids correspond to the $\mathbb{Z}_2$ case, where the presence of competing interactions is essential to produce the non-collinear spin structures responsible for the emergence of $\mathbb{Z}_2$ gauge degrees of freedom and spinon deconfinement. The large-$N$ description of gapped $\mathbb{Z}_2$ liquids has been extended to several 2D frustrated models [61–63].

We consider a mean-field state $\{U_{ij}^0, \boldsymbol{a}_i\}$ with a gapped spinon spectrum [obtained from (16.21)] and IGG= $\mathbb{Z}_2$. From the discussion of Sect. 16.3.4, the relevant gauge modes can be parameterized with a $\mathbb{Z}_2$ gauge field $A_{ij} \in \{0, \pi\}$ by $U_{ij} = U_{ij}^0 \mathrm{e}^{\mathrm{i}A_{ij}}$. Because the spinons are gapped at the mean-field level, they can be integrated out, which generates short-range interactions for $A_{ij}$. Given that $\sigma_{ij}^z \equiv \mathrm{e}^{\mathrm{i}A_{ij}} = \pm 1$, these interactions must be invariant under the $\mathbb{Z}_2$ gauge transformations $\sigma_{ij}^z \to \eta_i \sigma_{ij}^z \eta_j$, where $\eta_i = \pm 1$ may take an arbitrary value at each lattice site. A product $B_p = \sigma_{12}^z \sigma_{23}^z \cdots \sigma_{n1}^z$ around any plaquette $p$ of the lattice is such an invariant: this is the $\mathbb{Z}_2$ "magnetic" (or gauge) flux. On each bond we denote by $\sigma_{ij}^x$ the operator which changes $\sigma_{ij}^z = 1$ to $\sigma_{ij}^z = -1$ (and vice versa); this is the "electric" field. The $\mathbb{Z}_2$ gauge invariance requires that the Hamiltonian commutes with any $G_{i_0} = \prod_{\alpha=1}^{p} \sigma_{i_0 i_\alpha}^x$, where $i_1, \ldots, i_p$ are the neighbors of site $i_0$, because $G_{i_0}$ generates the elementary gauge transformation defined by $\eta_{i_0} = -1$ and $\eta_{j \neq i_0} = 1$.[24] Thus the "electric" field is also gauge-invariant, and hence is an allowed term in the effective Hamiltonian for the gauge fluctuations.

To discuss the typical phenomenology of such a $\mathbb{Z}_2$ gauge theory, we consider the simplest Hamiltonian

$$H_{\mathbb{Z}_2} = -\Gamma \sum_{(ijkl)=\square} \sigma_{ij}^z \sigma_{jk}^z \sigma_{kl}^z \sigma_{li}^z - J \sum_{\langle i,j \rangle} \sigma_{ij}^x, \tag{16.36}$$

[23] An alternative is to describe short-range RVB liquids by effective quantum dimer models [46], such as those considered in [57, 58], which in turn can be mapped (sometimes exactly [58]) onto $\mathbb{Z}_2$ gauge theories [59].

[24] States must also be gauge-invariant. Fermions transform according to $\psi_i \to \eta_i \psi_i$, which corresponds to the gauge generator $F_i = \mathrm{e}^{\mathrm{i}\pi(c_{i\uparrow}^\dagger c_{i\uparrow} + c_{i\downarrow}^\dagger c_{i\downarrow})}$. Physical states should therefore satisfy $G_i F_i |\phi\rangle = |\phi\rangle$. However, $F_i = 1$ because of the constraint (16.6), and thus $G_i |\phi\rangle = -|\phi\rangle$, which is the origin of the term *odd $\mathbb{Z}_2$ gauge theory* [59].

where $\Gamma$ controls the "magnetic" energy term and $J$ the "electric" one, and the sums run over all plaquettes and bonds respectively. Creating a pair of (infinitely heavy) "test" spinons at sites $i_0$ and $i_n$ can be effected by the gauge-invariant operator $c^\dagger_{i_0\uparrow}\sigma^z_{i_0 i_1}\cdots\sigma^z_{i_{n-1}i_n}c_{i_n\uparrow}$ [as in (16.8)], which changes (anticommutes with) $\sigma^x$ on all the bonds along the path $i_0\cdots i_n$. For $\Gamma/J \ll 1$ we may ignore the "magnetic" term, and the ground state has $\sigma^x = 1$ everywhere. Because any bond with $\sigma^x_{ij} = -1$ introduces a high energy penalty, the spinons experience a potential which grows linearly with their separation, i.e. they are *confined*. By contrast, the spinons are essentially free in the limit $\Gamma/J \gg 1$, where the model can be studied perturbatively from the $J = 0$ limit.

At $J = 0$, the ground state has $\sigma^z_{ij} = 1$, and the elementary excitation is a gapped and localized $\mathbb{Z}_2$ flux, for example on plaquette $p_0$, corresponding to $B_{p_0} = -1$, whereas $B_p = 1$ elsewhere (we specialize the discussion to dimension $D = 2$). Such vortices ("visons" [52]) can be created in pairs by applying to the ground state a product $\prod_l \sigma^x_l$ of electric-field operators involving all the bonds $l$ cutting some path on the dual lattice (by considering the operator $G$, one observes that the resulting state is unaffected by local deformations of the path). The cores of the two visons are located in the plaquettes at the two ends of the path, and the energy is independent of their separation. Away from $J = 0$, visons acquire a finite bandwidth and non-trivial short-range interactions. However, if short-distance effects close to the vortex core are neglected, the vison creation operator is essentially the product of $\mathbb{Z}_2$ "electric-field" operators as defined above.

To see what this means in the RVB description, we note that $e^{iA_{ij}}$ can be viewed as an operator which shifts the valence bonds by one lattice constant (Sect. 16.2.3.2). Because the electric field $\sigma^x_{ij}$ anticommutes with $\sigma^z_{ij} = e^{iA_{ij}}$, it can be interpreted as an operator measuring the presence or absence of a valence-bond between sites $i$ and $j$. Thus, the vison creation operator *counts the parity of the number of valence bonds crossing a path ending at the vortex core* (the other end may be at the boundary of the system or at another vortex core). In addition to local modifications close to the core, a vison excitation is obtained from the ground state by changing the *sign* of the valence-bond amplitude $\phi(c)$ if the number of valence-bonds crossing the path is odd.

As suggested by the names "electric" and "magnetic", a spinon winding around a vison experiences a long-range Aharonov–Bohm effect, corresponding to a phase factor $-1$. In the approach discussed here, spinons are fermionic, and thus if a bound state of a spinon and a vison happens to be energetically favorable, the resulting composite spin-$\frac{1}{2}$ excitation would be a boson.

### 16.4.3 Examples

In this section, we review a (not exhaustive) selection of lattice models with a gapped, fractionalized $\mathbb{Z}_2$ phase. The Ising-like model introduced by Kitaev [64] is quite possibly the simplest example. It contains four-site interactions between

Ising spins, and can be solved exactly, but has no continuous symmetry and is not microscopically related to the RVB states expected in frustrated, Heisenberg-like magnets. Still, it provides a very simple realization of spin systems with "spinon"- and "vison"-type excitations. A related model was introduced in [65]. The bosonic models discussed by Motrunich and Senthil [66] also have a $\mathbb{Z}_2$ phase and are somewhat closer to the types of magnets discussed here, in that they possess a global $U(1)$ symmetry. The model of Balents, Fisher, and Girvin [67] (see also [68]) is a spin-$\frac{1}{2}$ model on the kagomé lattice, with easy-axis, Heisenberg interactions between 1$^{\text{st}}$, 2$^{\text{nd}}$, and 3$^{\text{rd}}$ neighbors. It is one of the simplest known Heisenberg-like models with a well-characterized $\mathbb{Z}_2$ liquid phase.

Several numerical studies have also found indications of possible gapped QSL phases in $SU(2)$-symmetric spin-$\frac{1}{2}$ models. These systems are, however, hard to simulate, and the theoretical understanding of the candidate QSL states which emerge remains rather incomplete. Here we mention also models with four-spin "ring" exchange on the triangular lattice [69], with $J_1 - J_2 - J_3$ interactions on the honeycomb lattice [70], and with $J_1 - J_2$ [71] or $J_1 - J_3$ [72] interactions on the square lattice.

Although a dimer may be viewed as a pair of nearest-neighbor spins coupled in a singlet state, quantum dimer models are not related exactly to simple $SU(2)$ magnets. However, they do provide simple realizations of $\mathbb{Z}_2$ liquids [57, 58], and can also be used to construct "by hand" $SU(2)$ spin models with QSL ground states [73, 74].

### 16.4.4 How to Detect a Gapped $\mathbb{Z}_2$ Liquid

In this section, we consider some *observables* which can be used to investigate whether a system is a gapped QSL. First, the system should not develop any SSB upon cooling. If the system fulfils the conditions of the LSM theorem (energy gap and half-odd-integer spin per unit cell, Sect. 16.2.4), the absence of SSB at $T = 0$ is in fact sufficient to guarantee the existence of fractionalized excitations. In this case, the detection of an energy continuum in the dynamical spin structure factor (accessible through inelastic neutron scattering), as opposed to the single peak characteristic of a long-lived spin-1 excitations, is a signature of spinon deconfinement. In the case of a $\mathbb{Z}_2$ liquid, short-range vison correlations are a further necessary condition (for example [68]), and a possible experimental technique for the detection of visons in a doped $\mathbb{Z}_2$ liquid was proposed in [52]. Theoretically, another test is to search for a ground-state degeneracy and to verify that the ground state cannot be distinguished by any local observable in the thermodynamic limit [22, 55]. We mention finally that the topological order can also be detected from the wave function itself, by analyzing its bipartite entanglement entropy [75, 76].

## 16.5 Gapless (Algebraic) Liquids

The mean-field theory described in Sect. 16.3 can lead to states with a gapless spinon and/or a gapless gauge-excitation spectrum. In some cases, these gapless QSLs have been argued to be stable with respect to fluctuations. The gapless excitations mean that these new states of matter are a priori quite "fragile". If the spinons are gapless at the mean-field level, there are possible "mass" terms which could be generated when including fluctuations (even if these are weak), and which could open a spin gap (i.e. cause an instability towards a gapped QSL or some type of VBC). However, as noted in Sect. 16.3.4, these terms are sometimes forbidden by the PSG. One must then consider the effect of gapless gauge modes, for example if the IGG is $U(1)$. With *gapped* matter fields, which here are the spinons, such compact lattice gauge theories are generically in a *confined phase* in 2D due to the proliferation of "magnetic monopoles" (particular space-time configurations of the gauge field [77]), and the mean-field theory is unstable to gauge fluctuations [31, 32]. However, this result does not always apply in the presence of gapless spinons with a linear dispersion relation (one or more "Dirac cones", as in the $\pi$-flux state), in which case the gaplessness of the mean-field state may survive fluctuations [40, 78, 79]. The resulting QSLs are known as "$U(1)$", "algebraic", or "long-range RVB" spin liquids. The rich physics of these critical *states* is closely related to that of "deconfined critical points" [80], because in both cases the monopoles are irrelevant for the low-energy properties. The "$\pi$-flux" state [27, 28, 40] on the square lattice and an analog on the kagomé lattice [43, 44] are two examples of mean-field states with Dirac-type spinon spectra which have been argued to survive fluctuations and to give rise to algebraic spin liquids. It should also be stressed that spinons are not free quasiparticles after fluctuations are taken into account, even at very low energy. Because of the strong interactions with the gauge modes, many correlation functions (including spin correlations) show an algebraic decay with non-trivial exponents (different from the mean-field ones) [40, 78, 79]. To our knowledge, there is as yet no lattice *spin* model for which clear evidence of such an algebraic QSL has been found.

## 16.6 Other Spin Liquids

We have included several families of QSL in the present review, but have also omitted several important ones. Here, we provide a brief list of some of these. Chiral spin liquids [47, 81], which have spontaneous breaking of time-reversal symmetry (and therefore do not obey definition 2), have deconfined spinons. The mechanism by which these systems can escape confinement is the existence of a Chern-Simons term, allowed because of the time-reversal symmetry-breaking, in the effective action, which gaps the gauge modes. We also mention possible QSL states with rich topological structures, including fractional excitations with *non-Abelian* statistics [82–85]. A further class of quantum spin liquid is the set of "algebraic vortex liquids" [86], proposed in certain frustrated models with easy-plane interactions; their description is based on a mapping of the vortices to fermionic degrees of freedom.

## 16.7 Conclusion

We have introduced some theoretical ideas for describing disordered ground states in Mott insulators (by which is meant that the total spin is a half odd integer per unit cell). Using the fermionic representation of the spin operators, we have discussed how gauge fields emerge as fluctuation modes around given mean-field solutions of the Heisenberg model. The stability and hence validity of such a mean-field approximation depends on whether the gauge field mediates a confining interaction between the spinons (instability) or whether the spinons remain deconfined (stability). In the specific case of short-range resonating-valence-bond liquids, the fluctuating $Z_2$ gauge field is simply the valence-bond background in which the spinons propagate.

This type of approach is very useful in shedding light on the low-energy properties of spin liquids. It also allows a classification of the different possible phases and the extraction of certain universal properties. Deciding whether or not a given frustrated spin model has a spin-liquid ground state remains a difficult task, because the approaches discussed here are not easy to apply as quantitatively accurate calculations for microscopic Hamiltonians. However, the concepts we have reviewed, including gauge fluctuations, fractionalization, and topological order, are crucial elements guiding the search for and characterization of these new states of matter.

**Acknowledgements**

I am grateful to Claire Lhuillier and Frédéric Mila for numerous discussions and collaborations within the topic of quantum spin liquids as well as for valuable comments concerning this manuscript. I am also indebted to Bruce Normand for his help in improving the text.

## References

1. P.W. Anderson, Mat. Res. Bull. 8, 153 (1973)
2. X.-G. Wen, Phys. Rev. B 40, 7387 (1989)
3. X.-G. Wen, Q. Niu, Phys. Rev. B 41, 9377 (1990)
4. X.-G. Wen, Phys. Rev. B 44, 2664 (1991)
5. X.-G. Wen, Quantum Field Theory of Many-body Systems from the Origin of Sound to an Origin of Light and Electrons (Oxford University Press, Oxford, 2004)
6. E.H. Lieb, T.D. Schultz, D.C. Mattis, Ann. Phys. (N.Y.) 16, 407 (1961)
7. M.B. Hastings, Phys. Rev. B 69, 104431 (2004)
8. B. Nachtergaele, R. Sims, Commun. Math. Phys. 276, 437 (2007)

9. P.W. Anderson, Science 235, 1196 (1987)
10. G. Baskaran, Z. Zou, P.W. Anderson, Solid State Commun. 88, 853 (1987)
11. I. Affleck et al., Phys. Rev. B 38, 745 (1988)
12. X.-G. Wen, P.A. Lee, Phys. Rev. Lett. 76, 503 (1996)
13. A.F. Andreev, I.A. Grishchuk, Sov. Phys. JETP, 60, 267 (1984)
14. A. Läuchli et al., Phys. Rev. Lett. 95, 137206 (2005)
15. A. Läuchli, F. Mila, K. Penc, Phys. Rev. Lett. 97, 087205 (2006)
16. N. Shannon, T. Momoi, P. Sindzingre, Phys. Rev. Lett. 96, 027213 (2006)
17. C. Lhuillier, G. Misguich, in Quantum Spin Liquids and Fractionalization, ed. by Lacroix, Introduction to Frustrated Magnetism (Springer, Heidelberg, 2010)
18. I. Affleck et al., Phys. Rev. Lett. 59, 799 (1987).
19. C.K. Majumdar, D.K. Ghosh, J. Math. Phys. 10, 1399 (1969)
20. P. Jordan, E. Wigner, Z. Phys. 47, 631 (1928)
21. P.A. Lee, N. Nagaosa, X.-G. Wen, Rev. Mod. Phys. 78, 17 (2006)
22. S. Furukawa, G. Misguich, M. Oshikawa, Phys. Rev. Lett. 96, 047211 (2006)
23. M. Oshikawa, T. Senthil, Phys. Rev. Lett. 96, 060601 (2006)
24. M. Oshikawa, Phys. Rev. Lett. 84, 1535 (2000)
25. I. Affleck, Phys. Rev. B 37, 5186 (1988)
26. I. Affleck, Phys. Rev. Lett. 54, 966 (1985)
27. I. Affleck, J.B. Marston, Phys. Rev. B 37, 3774 (1988)
28. J.B. Marston, I. Affleck, Phys. Rev. B 39, 11538 (1989)
29. D. Arovas, A. Auerbach, Phys. Rev. B 38, 316 (1988)
30. A. Auerbach, D.P. Arovas, Phys. Rev. Lett. 61, 617 (1988)
31. N. Read, S. Sachdev, Phys. Rev. Lett. 62, 1694 (1989)
32. N. Read, S. Sachdev, Phys. Rev. B 42, 4568 (1990)
33. N. Read, S. Sachdev, Phys. Rev. Lett. 66, 1773 (1991)
34. H.A. Ceccatto et al., Phys. Rev. B 47, 12329 (1993)
35. For a review see F. Becca et al., in Quantum Spin Liquids and Fractionalization, ed. by Lacroix, Introduction to Frustrated Magnetism (Springer, Heidelberg, 2010)
36. P. Horsch, T.A. Kaplan, J. Phys. C: Solid State Phys. 16, 1203 (1983)
37. C. Gros, R. Joynt, T.M. Rice, Phys. Rev. B 36, 381 (1987)
38. G. Kotliar, J. Liu, Phys. Rev. B 38, 5142 (1988)
39. X.-G. Wen, Phys. Rev. B 65, 165113 (2002)
40. M. Hermele et al., Phys. Rev. B 70, 214437 (2004)
41. D.S. Rokhsar, Phys. Rev. B 42, 2526 (1990)
42. J.B. Marston, C. Zeng, J. Appl. Phys. 69, 5962 (1991)
43. M.B. Hastings, Phys. Rev. B 63, 014413 (2000)
44. Y. Ran et al., Phys. Rev. Lett. 98, 117205 (2007)
45. N. Read, S. Sachdev, Nucl. Phys. B, 316, 609 (1989)
46. For a review of quantum dimer models see R. Moessner, in Quantum Spin Liquids and Fractionalization, ed. by Lacroix, Introduction to Frustrated Magnetism (Springer, Heidelberg, 2010)
47. X.-G. Wen, F. Wilczek, A. Zee, Phys. Rev. B 39, 11413 (1989)
48. G. Baskaran, P.W. Anderson, Phys. Rev. B 37, 580 (1988)
49. M. Hermele et al., Phys. Rev. B 77, 224413 (2008)
50. S. Kivelson, Phys. Rev. B 39, 259 (1989)
51. N. Read, B. Chakraborty, Phys. Rev. B 40, 7133 (1989)
52. T. Senthil, M.P.A. Fisher, Phys. Rev. B 62, 7850 (2000)
53. S. Liang, B. Douçot, P.W. Anderson, Phys. Rev. Lett. 61, 365 (1988)
54. N.E. Bonesteel, Phys. Rev. B 40, 8954 (1989)
55. S. Furukawa, G. Misguich, M. Oshikawa, J. Phys.: Condens. Matter 19, 145212 (2007)
56. J.B. Kogut, Rev. Mod. Phys. 51, 659 (1979)
57. R. Moessner, S. Sondhi, Phys. Rev. Lett. 86, 1881 (2001)
58. G. Misguich, D. Serban, V. Pasquier, Phys. Rev. Lett. 89, 137202 (2002)
59. R. Moessner, S.L. Sondhi, E. Fradkin, Phys. Rev. B 65, 024504 (2002)

60. For an introduction to $Sp(N)$ antiferromagnets see Sect. 4.3 of [87]
61. S. Sachdev, Phys. Rev. B 45, 12377 (1992)
62. C.H. Chung, J.B. Marston, S. Sachdev, Phys. Rev. B 64, 134407 (2001)
63. F. Wang, A. Vishwanath, Phys. Rev. B 74, 174423 (2006)
64. A. Kitaev, Ann. Phys. 303, 2 (2003)
65. X.-G. Wen, Phys. Rev. Lett. 90, 016803 (2003)
66. O.I. Motrunich, T. Senthil, Phys. Rev. Lett. 89, 277004 (2002)
67. L. Balents, M.P.A. Fisher, S.M. Girvin, Phys. Rev. B 65, 224412 (2002)
68. D.N. Sheng, L. Balents, Phys. Rev. Lett. 94, 146805 (2005)
69. G. Misguich et al., Phys. Rev. B 60, 1064 (1999)
70. J.B. Fouet, P. Sindzingre, C. Lhuillier, Eur. J. Phys. B, 20, 241 (2001)
71. L. Capriotti et al., Phys. Rev. Lett. 87, 097201 (2001)
72. L. Capriotti, D.J. Scalapino, S.R. White, Phys. Rev. Lett. 93, 177004 (2004)
73. K.S. Raman, R. Moessner, S.L. Sondhi, Phys. Rev. B 74, 064413 (2005)
74. A. Seidel, Phys. Rev. B 80, 165131 (2009)
75. A. Kitaev, J. Preskill, Phys. Rev. Lett. 96, 110404 (2006)
76. M. Levin, X.-G. Wen, Phys. Rev. Lett. 96, 110405 (2006)
77. A.M. Polyakov, Nucl. Phys. B 120, 429 (1977)
78. W. Rantner, X.-G. Wen, Phys. Rev. Lett. 86, 3871 (2001)
79. M. Hermele, T. Senthil, M.P.A. Fisher, Phys. Rev. B 72, 104404 (2005)
80. T. Senthil et al., Science 303, 1490 (2004)
81. V. Kalmeyer, R.B. Laughlin, Phys. Rev. Lett. 59, 2095 (1987)
82. M. Freedman, C. Nayak, K. Shtengel, Phys. Rev. Lett. 94, 066401 (2005)
83. A.Y. Kitaev, Ann. Phys. 321, 2 (2006)
84. H. Yao, S.A. Kivelson, Phys. Rev. Lett. 99, 247203 (2007)
85. P. Fendley, arXiv:0711.0014v1
86. J. Alicea, O.I. Motrunich, M.P.A. Fisher, Phys. Rev. Lett. 95, 241203 (2005)
87. G. Misguich, C. Lhuillier, in Frustrated spin systems, ed. by H.T. Diep (World-Scientific, Singapore, 2005) (also on cond-mat/0310405)

# Chapter 17
# Quantum Dimer Models

**Roderich Moessner and Kumar S. Raman**

**Abstract** These lecture notes aim to provide a self-contained, pedagogical introduction to the physics of local constraints, fractionalisation and topological liquids organised around the Rokhsar–Kivelson quantum dimer model. Topics and phenomena covered include emergent photons, SU(2) invariant spin liquids, valence-bond solids and Cantor deconfinement, along with an elementary introduction to the underlying theoretical models and methods.

## 17.1 Introduction

The study of simple model Hamiltonians to describe magnetic systems has a long tradition in condensed matter and statistical physics. Starting with the initial analyses of the Ising and Heisenberg models, our understanding of the collective behaviour of matter, and of countless magnetic materials, has been advanced greatly. In particular, magnetic ordering phenomena have provided a stable basis for our understanding of order and spontaneous symmetry breaking. The ordered state of the ferromagnet, and the Néel state of an antiferromagnet, are two particularly simple, and familiar, examples of orderings which occur at low temperature.

Beyond such simple examples, qualitatively different types of behaviour have been uncovered both in experiment and in theoretical studies. This chapter is devoted to an expose of the Rokhsar–Kivelson quantum dimer model [1], a model Hamiltonian which captures the physics of some of these new phenomena in a particularly simple and transparent way. In pictorial form, the RK-QDM Hamiltonian on the square lattice reads:

$$H_{QDM} = \Sigma - t(|\parallel\rangle\langle = | + \text{h.c.}) + v(|\parallel\rangle\langle\parallel| + |=\rangle\langle=|) \tag{17.1}$$

Its historical origin lies in the study of high-temperature superconductors, where it was proposed by the two scientists whose names it now bears as a description of the short-range flavour of Anderson's resonating valence bond physics [2–4] – for a few words on history, see the review [5]. In this formulation, a dimer represents an SU(2) singlet bond between two spins located at its endpoints, and the model describes the quantum dynamics of a singlet-dominated phase: the first, kinetic, term in the above Hamiltonian describes a 'resonance' between two different dimerisations of a plaquette. The analogy to the resonance between the two dimerisations of the benzene ring is the origin of the apellation RVB.

However, the RK-QDM has since graduated from this field and has increasingly been used to describe new and unusual forms of collective behaviour in a broader variety of settings. A survey of these forms the backbone of this chapter. We will encounter new types of order (topological and quantum order), unusual (resonons and visons) and fractionalised (spinons and holons) excitations, and exotic critical points.

## 17.2 How Quantum Dimer Models Arise

### *17.2.1 Link Variables and Hard Constraints*

What is the difference between a quantum dimer and a quantum spin model? While we tend to think of spins living on *sites* of a lattice, a dimer degree of freedom is more naturally associated with a *link*: a dimer living on a link can be thought of as connecting the two nearest neighbor sites which are located at its endpoints.

There is a certain arbitrariness in calling a variable a site or a link variable. However, the Hilbert space of a dimer model is *defined* by enforcing a hard *constraint* on the sites of the lattice, i.e. where the links meet. This constraint consists of demanding that each site forms a dimer with one, and only one, of its nearest neighbors. Therefore, the configurations included in the dimer Hilbert space are the set of nearest-neighbor dimer coverings of the lattice and superpositions thereof. This constraint thus prohibits monomers, i.e. sites not attached to any dimer (Fig. 17.1a); higher-order polymers (Fig. 17.1b), in which three or more sites are connected together; or long-range dimers between sites that are not nearest-neighbors (Fig. 17.1c). The connection of this idea with gauge theories will be discussed below. In this chapter, we will use the term "dimer" to mean "nearest-neighbor dimer".

Much of the new physics captured by QDMs relies on this constraint. Such constraints are a common theme in modern many-body physics. They typically arise when the Hamiltonian contains an overwhelmingly large energy scale which needs to be taken into account at the very outset. Such a constraint may be present in the form of a large onsite Coulomb repulsion (i.e. a "Hubbard-$U$"), which prohibits more than one electron to be present on a lattice site; in a frustrated magnet, it may be due to a large ground-state manifold obeying a certain local ground-state constraint, as described in the following section.

Much of the discussion in this chapter applies more generally to systems in which the (effective) Hilbert space incorporates (1) a non-trivial local constraint and (2) a

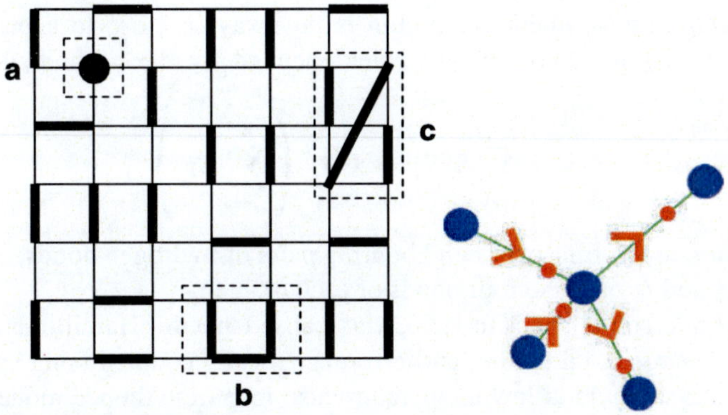


**Fig. 17.1** Constraints. *Left*: The hardcore constraint of the quantum dimer model. Also shown are defects which violate this constraint including: (**a**) monomers, (**b**) higher-order polymers, and (**c**) long-range dimers. *Right*: The local structure of ice involves protons (*red, small dots*) bonded to oxygens (*blue, large dots*)

local quantum dynamics. Let us point out at this stage that (2) tends to be much harder to achieve than (1), and we discuss point (1) first.

This definition includes, for instance, quantum vertex, ice, and coloring models, fermionic derivatives of RK-QDMs, and much more. In fact, the reader may now, with some justification, be alarmed at the broad scope of this definition, and is therefore invited to draw the somewhat arbitrary line of what to include in this class of 'locally constrained models' more restrictively. As a guide, we present a number of examples in the following.

### *17.2.2 The Origin of Constraints*

The original motivation for the quantum dimer model was high-temperature superconductivity, in particular the quest for a non-magnetic parent state of the superconducting phase. To minimise their antiferromagnetic exchange energy, neighbouring pairs of spins would form singlet bonds, denoted by a dimer. The dimer constraint reflects the fact that each spin can form a singlet bond with at most one of its neighbours. More generally, dimer models can be thought of as a simple starting point for describing phases of magnets dominated by local singlet formation. Such unusual magnetic phases are covered in this volume in the chapter by Misguich.

An even older, probably the first, example of a similar constraint is found in ice. There, oxygen atoms are connected by a network of hydrogen bonds into a fourfold coordinated lattice. The protons binding the oxygens together sit asymmetrically on each bond, but not randomly so: each oxygen has two protons sitting close to it, and two far away, so that $H_2O$ molecules effectively retain their identity. This constraint is one of the Bernal–Fowler ice rules [6]. Defining an Ising variable

$S^z = \pm 1$ according to whether a proton is far away or close to a particular set of oxygen atoms, one finds the ice rules encoded by the ground-states of the Hamiltonian

$$H = J \sum_{\langle ij \rangle} S_i^z S_j^z = J \sum_{\alpha} \left( \sum_{i_\alpha = 1}^{4} S_{i_\alpha}^z \right)^2 , \tag{17.2}$$

where the sum on $\langle ij \rangle$ runs over neighbouring pairs of hydrogen bonds, $\alpha$ runs over oxygen sites, and $i_\alpha$ over the four bonds of each oxygen.

It was in Anderson [7], back in 1956, who realised that this Hamiltonian describes a range of systems: in particular, antiferromagnetically coupled Ising spins on the pyrochlore lattice would at low temperature also map onto the ice model, with the sum on $i$ now running over the four sites of a tetrahedron, the basic building block of the ice lattice. For details, see Gingras' article on spin ice in this volume.

This avenue of investigation has more recently reappeared in the study of charged particles partially covering a lattice, where electrostatics at short distances imposes the constraint that the system be as neutral as possible [8]. For charge ordering, the Ising spins will thus encode the ionisation state of an ion.

Similarly, for particles subject to short-range repulsion, $S_i$ stands for whether a given site is empty or occupied. This Hamiltonian has thus played an important role for supersolid phases of bosons on optical lattices [9–12].

### 17.2.3 *Tunable Constraints*

A generalisation of the above Hamiltonian (17.2) is obtained by adding a field $h$ (in reduced units) pointing in the $z$-direction:

$$H = J \sum_{\alpha} \left( \sum_{i_\alpha = 1}^{4} S_{i_\alpha}^z \right)^2 - \sum_i h S_i^z = J \sum_{\alpha} \left( \sum_{i_\alpha = 1}^{4} S_{i_\alpha}^z - h/2 \right)^2 , \tag{17.3}$$

where we have dropped a constant involving $h^2$. $h$ assumes different meanings (electric field, particle density, and so on) in the different contexts mentioned above.

Crucially, $h$ allows the selection of a *sector* of the theory [13] by imposing that the lowest energy states obey a local constraint of minimising:

$$L_\alpha = \left| \sum_{i_\alpha = 1}^{4} S^z - h/2 \right| . \tag{17.4}$$

For very large $h$, this is minimised by $S^z \equiv 1$, which implies a trivial ground state manifold consisting only of the maximally polarised state. As $h$ is reduced, it becomes favourable at first to flip a spin on exactly one of the links emanating from each site $\alpha$. Denoting the link with a flipped spin by a dimer yields a hardcore dimer model. As $h$ is lowered further, it becomes advantageous to flip a second spin, which in turn gives a *loop model*: at each site $\alpha$, two links are special, which can be identified with a loop passing through the site.[1] On the square lattice, this gives the ice model. Continuing this process requires increasingly coordinated sites to yield new sectors – for the sixfold coordinated triangular lattice, the three-dimer manifold at $h = 0$ generates the basis states for the easy-axis kagomé ring exchange model [14]. Back in the language of magnetism, these sectors manifest their presence via magnetisation plateaux, as they are stable for finite ranges of field and correspond to different average local, and hence global, magnetisations.

As an aside, we remark that at the special values of the field where two sectors with different magnetisations are degenerate, the space of allowed states contains the allowed states from both sectors but also additional states[2]; such an energy crossing occurs, for example, on the kagomé lattice at zero field, $h = 0$. This crossing is an interesting feature in itself, as it can give rise to unusual signatures in the magnetothermodynamics; some examples are given in [15–17].

In model systems in statistical mechanics, these constraints can, of course, be imposed at will, and the dimer, vertex, ice, colouring, etc., models have a long history in statistical mechanics, in particular in $d = 2$ where many of them are soluble [18].

The archetypal setting in which local constraints play a central role are gauge theories [19] – the local configuration space is defined by restricting a much larger space to a 'physical' subspace by demanding that a gauge constraint be satisfied. For instance in magnetostatics, of all possible vector fields $\boldsymbol{B}$, only those satisfying the no-monopole condition $\nabla \cdot \boldsymbol{B} = 0$ are considered acceptable.

### *17.2.4 Adding Quantum Dynamics*

The most popular route of adding quantum dynamics is by *fiat*. First, the set of allowed classical configurations are elevated to basis vectors of a Hilbert space. Next, one identifies the simplest local rearrangement permissible. This does not typically involve a single link variable (e.g. removing one dimer) as this would lead to a violation of the local constraint. Rather, one finds a dynamics involving loops or plaquette flips – a number of examples are given in Fig. 17.2.

One then endows such a flip with a coherent quantum dynamics captured by a matrix element of strength $t$: for the square quantum dimer model, this is just the kinetic term in (17.1).

[1] We can identify the two special links as part of a "string" passing through the site. As the reader may verify, the constraint that every site has two such links requires the strings to form closed loops.

[2] For example, when the dimer and loop models become degenerate, the space of allowed sites will include states where some sites have one dimer emanating while others have two.

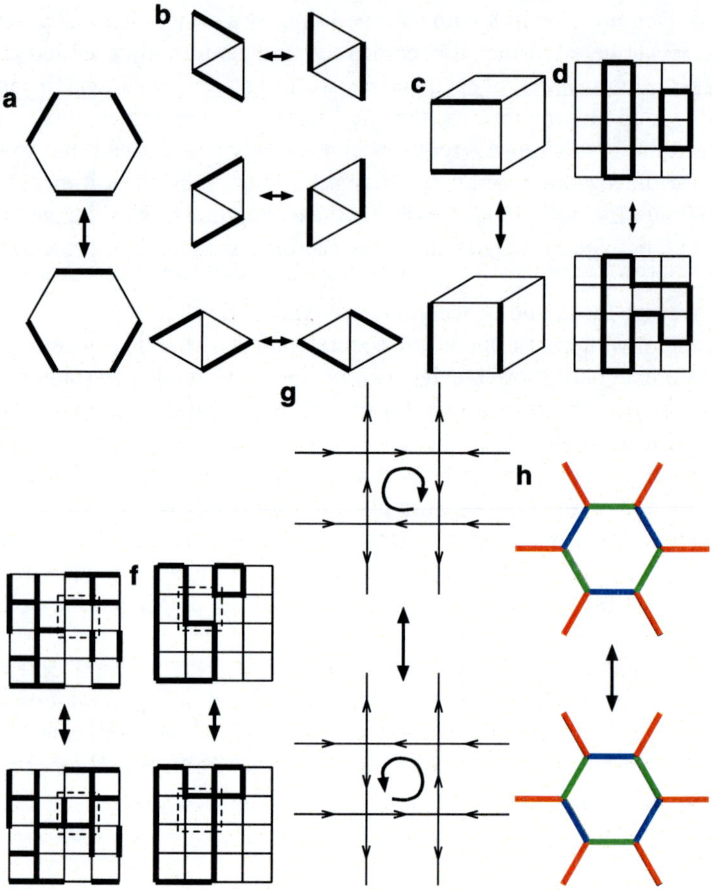


**Fig. 17.2** The analog of (17.1) can be written for any constrained model by identifying the simplest local rearrangement consistent with the constraint. For the honeycomb lattice dimer model (**a**), the simplest move involves three dimers while for the triangular lattice (**b**), there are three different (but symmetry-equivalent) types of two-dimer moves. For the cubic lattice (**c**), it is again a two-dimer move which can now occur in any plane and likewise in higher dimensions. In a loop model (**d**), each site has two dimers. The simplest move involves two dimers and results in loops fusing or separating. A $Z_2$ gauge theory may be viewed as a generalized dimer (or loop) model with a less restrictive constraint: instead of fixing the number of dimers per site, we fix the parity. This means that each site contains either an odd (**e**) or even (**f**) number of dimers and the simplest move is to exchange the empty and occupied links on an elementary plaquette as shown. Note that under this dynamics, *every* plaquette is flippable. In the six-vertex model (**g**), each site has two inward and two outward arrows. In this case, the simplest rearrangement is to reverse the direction of the *arrows* on a plaquette with circulation. In a coloring model (**h**), each site has a *red, green, and blue link*. The most basic move exchanging involves six links

In the corresponding spin model, this plaquette dynamics can be due to any term like a transverse field of strength $\Gamma$, a transverse exchange $J_{xy}$, or a 'ring-exchange' $K$ around a plaquette □:

$$H_q = -\Gamma \sum_i S_i^x - J_{xy} \sum_{\langle ij \rangle} S_i^x S_j^y - K \sum_{\square} \prod_{i \in \square} S_i^x \tag{17.5}$$

When projected onto the physical space, these terms may, in fact, lead to the same effective Hamiltonian [20].

In the original paper by RK, an estimate of the strength of the kinetic term $t$ was provided in terms of the original magnetic exchange constant $J$ of the Heisenberg model but in general, the instances in which this is underpinned by an actual microscopic calculation are few.[3] The classical potential term, $v$ in the Hamiltonian (17.1), is often added for convenience, as it allows an RK construction – this is described in detail below.

## 17.3 The Quantum Dimer Model Hilbert Space

### *17.3.1 Topological Invariants*

Before discussing the QDM Hilbert space, we first mention an important property of the classical dimer model. Because of the hardcore constraint, the number of allowed configurations is much smaller than in a spin model defined on the same lattice. For example, for a square lattice with $N$ sites, the number is $2^N$ while the number of ways to cover the square lattice with dimers grows asymptotically as $(1.3385...)^N$ [24, 25]. The price of this reduction is that it is no longer generally possible to view the system as being composed of independent degrees of freedom on sites or links.

Put another way, for spin systems this latter property implies that any spin configuration can be obtained from another spin configuration by a series of local manipulations, i.e. by sequentially flipping the spins on appropriate sites. However, because of the hard constraint, it is not possible to manipulate a dimer without also moving other dimers. On a square lattice, the simplest move is the plaquette flip mentioned above. A more general move involves interchanging the occupied and empty links along a *flippable loop* as illustrated in Fig. 17.3.

From this perspective, it is easy to see that there are quantities which remain invariant under *local* manipulations of the type just described. For example, on non-bipartite lattices such as the triangular lattice, the parity of the number of dimers crossing a non-contractible reference line spanning the lattice (Fig. 17.4a) is unaffected by local dimer rearrangements. One way to see this is to note that any flippable loop will intersect this reference line an *even* number of times so flipping

[3] Parameters in effective Hamiltonians have been estimated for frustrated magnets in [21–23].

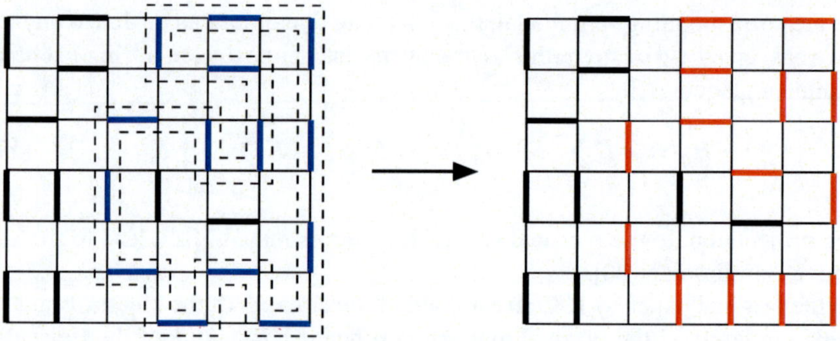

**Fig. 17.3** An example of a flippable loop. The links between the *dashed lines* in the figure on the left define a closed path where every other link is occupied by a dimer. The figure on the right is obtained from the one on the left by switching the occupied and empty links

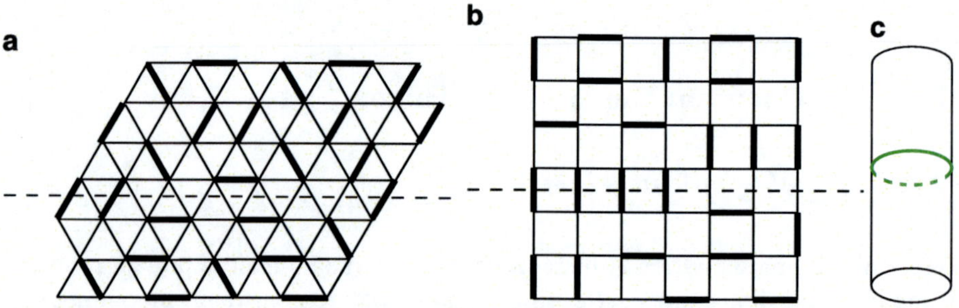


**Fig. 17.4** The *dashed line* is a reference line that fully spans the lattice, i.e. the line may end on a boundary but not in the interior of the lattice. (**a**) For non-bipartite lattices, the parity of the number of dimers crossing the reference line (i.e. whether the number is odd or even) is unaffected if the dimer pattern is changed *locally*. (**b**) For the square lattice (similar constructions apply for other bipartite lattices), one may label the *vertical lines* of the lattice alternately as $A$ and $B$ lines. Then, if $N_{A(B)}$ is the number of dimers crossing the reference line on $A(B)$ lines, the quantity $N_A - N_B$ is invariant under local dimer rearrangements. If the system is defined on a torus or cylinder, the reference line is actually a reference loop which winds around one of the toroidal directions. (**c**) shows a non-contractible loop around a cylinder; identifying the two *open circles* at the ends of the cylinder yields a torus

the dimers on this loop will not change the parity. Any local transformation involves flipping a sequence of such loops. An analogous invariant exists for bipartite lattices such as the square lattice (Fig. 17.4b).

However, if the system is defined on a torus, or on a surface of higher genus, it is possible to make a flippable loop that intersects the reference line an *odd* number of times by having the loop wind around one of the toroidal directions. Such a *nonlocal* transformation will change the value of the "invariants" just described which, in this context, are called *winding numbers*.

The winding numbers provide a natural way of partitioning the set of dimer coverings into *sectors*. On a torus, the procedure works by first choosing two reference loops that wind around the two toriodal directions. With respect to these two loops, we define two winding numbers which we call $W_x$ and $W_y$. We define the sector $(W_x, W_y)$ as the set of all dimer coverings with winding numbers $(W_x, W_y)$. Clearly, all dimer coverings in this set can be related to one another by a sequence of local dimer moves. To move from one sector to another requires a change in the winding number which involves at least one nonlocal move. For more general surfaces, the sectors are defined by specifying $2g$ winding numbers where $g$ is the genus of the surface – the genuses of a sphere and a torus are 0 and 1, respectively.

While the sector labels $(W_x, W_y)$ depend on our choice of reference loops, the reader may verify that the sectors themselves depend only on the *topology* of these loops, i.e. that they wind around the two toriodal directions. For this reason, the sectors are commonly called *topological sectors*.

### 17.3.2 Topological Order

To construct the QDM Hilbert space, the first step is to treat the set of dimer coverings as a collection of vectors. A vector space is constructed by also allowing states that are linear superpositions of dimer coverings. To obtain a Hilbert space, we need to define an inner product. The usual choice is to declare the set of dimer coverings to be an orthonormal basis for the space.[4] In this chapter, we will still refer to these quantum basis vectors as "dimer coverings".

Operators acting on this dimer Hilbert space are characterized by examining the way they act on the individual dimer coverings. The RK-QDM and other Hamiltonians considered in this chapter have the common feature of being the sum of operators which act *locally* on dimers in the sense discussed above. For such systems, the winding numbers discussed above are good quantum numbers and it is natural to subdivide the Hilbert space into dynamically independent topological sectors corresponding to different values of the winding number.

A topological sector is the subspace spanned by the set of all dimer coverings with a given winding number. The number of topological sectors depends on the lattice geometry and topology as in the classical case. However, depending on the lattice and also the terms included in the Hamiltonian, it may be possible to subdivide the topological sectors into even finer dynamical sectors. This is because quantum fluctuations only generate dimer rearrangements by repeated application of the kinetic terms included in the Hamiltonian. For the basic RK Hamiltonian (17.1) on the square lattice, the kinetic term is believed to be ergodic within each topological sector so a further subdivision of a topological sector is not possible. This is not the case for the RK Hamiltonian on a triangular lattice [27].

This Hilbert space structure makes QDMs a natural setting in which to construct and explore microscopic models of *topological order* [28]. Topological order

[4] A recent preprint by Paul Fendley argues that a more complex inner product may be a better choice in some settings [26].

describes the situation in which the ground states in the different topological sectors are exactly degenerate. To be more precise, consider the triangular lattice QDM where each of the $2g$ winding numbers can take two values corresponding to even and odd parity (Fig. 17.4a). In this case, for a system of finite linear size $L$, there is a ground-state multiplet containing $2^{2g} = 4^g$ states – one from each topological sector – separated by a gap from the other states, and split among each other by an energy which vanishes exponentially as $\exp[-cL]$, where $c$ is some constant.[5]

Topological order is like conventional 'local' order in that it leads to a low-lying multiplet of asymptotically degenerate states. However, it is fundamentally different in the following crucial respects. First, the degeneracy depends on the topology of the lattice on which the system lives. Second, there is no local order parameter which can be used to distinguish the ground states in the different topological sectors. Rather, one needs to follow a reference line all the way around the system to count the number of dimers it crosses. Whether the outcome is even or odd is uncertain until the line closes.

This is in stark contrast with an Ising ferromagnet, say, where the preference of the spins to point up or down can be detected locally. Likewise, the absence of conventional order in QDMs amounts to demanding that all local dimer correlators decay exponentially with distance.

### *17.3.3 Fractionalisation*

The concept of topological order was first discussed in the context of the fractional quantum Hall effect [28] where its most striking consequence is *fractionalisation* [30]: the phenomenon that a gas of electrons can organize so that its elementary excitations carry a fraction of the electron charge $e$ and obey fractional statistics, i.e. as if the electron has "split" into more basic constituents.

In the current context, this phenomenon can be explained in simple pictures by enlarging our Hilbert space to include monomer defects. Imagine taking one dimer out of the system – this will leave behind two monomers on the sites it occupied. These monomers can then be separated by shifting a neighbouring dimer so that it occupies the site of one of the monomers. If the monomers can be separated to large distances at finite cost in energy, they are said to be *deconfined*. They can, therefore, act as independent quasiparticles [31] – removing one dimer has given rise to two quasiparticles! The word confinement is used here in analogy to the situation in quantum chromodynamics, where it is impossible to separate a pair of isolated quarks at a finite cost in energy – they are confined.

To make contact with some commonly used vocabulary, it is useful to view the dimers as SU(2) singlets. The elementary magnetic excitation involves replacing one of the singlets by some triplet state (Fig. 17.5a – we are ignoring orthogonality

[5] More complex forms of topological order occur in different settings [29].

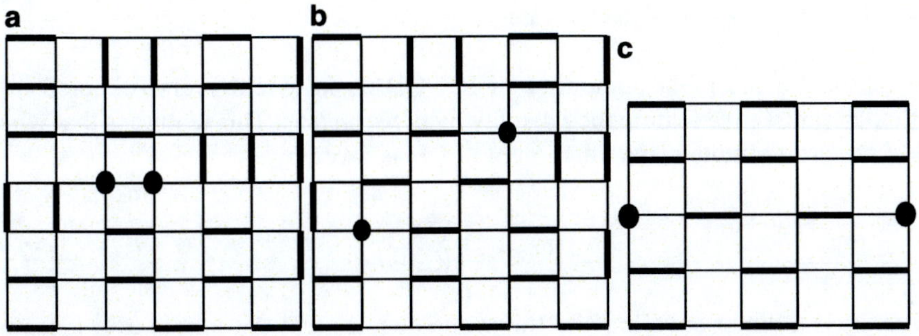


**Fig. 17.5** Monomers (spinon s or holons). **(a)** A dimer can break into two monomers. **(b)** In a resonating background, the monomers can separate and propagate as independent excitations. **(c)** Separating monomers in a crystalline phase causes a string of unfavorable bonds which requires an energy proportional to the length of separation

issues here). Fractionalisation of the triplet then means that two spin-1/2 excitations, 'spinons', can propagate independently.

Similarly, one can think of removing a *single* electron. As this electron was part of a singlet bond, it leaves behind an unpaired electron. Again, if these two defects can move apart at a finite cost in energy, the electron will have fractionalised into the spinon carrying $S = 1/2$, and another quasiparticle which is charged and spinless – known as a holon. This flavour of fractionalisation is known as spin-charge separation [2].

Note that these particles – monomers, i.e. spinons or holons – act as defects that violate the hard constraint mentioned above. This is similar to what happens in gauge theories, where Gauss's law can allow for the presence of charges: $\nabla \cdot \mathbf{E} = \rho$. monomers can similarly be treated as charges in a gauge-theoretic description of the fractionalised phases. We note that RK-QDMs provided the first theoretical examples of microscopic Hamiltonians with deconfined fractionalized phases [27].

## 17.4 QDM Phase Diagrams

### *17.4.1 General Structure of Phase Diagrams*

The presence of two terms in (17.1) immediately provides a one-parameter ($v/t$) family of models. The detailed structure of the QDM phase diagram, including which phases are present, depends strongly on the lattice geometry and dimensionality. Also, new phases may appear if we perturb (17.1) with additional interactions or consider less trivial ways of implementing the constraint (see Fig. 17.2). Nonetheless, there are a number of features common to most models that are worth noting.

#### 17.4.1.1 The Rokhsar–Kivelson Point

One reason for the great popularity of RK-QDMs in recent years is rooted in a particularly attractive feature they display by *construction*. This feature is that when $v = t$, the ground state wave function of (17.1) is given by:

$$|\Phi_{\mathrm{GS}}\rangle = \sum' |c\rangle \tag{17.6}$$

where $|c\rangle$ is a dimer covering and the prime denotes that the sum is over a sector, that is to say a set of dimer coverings that is closed under repeated action of the flip term in (17.1). In fact, we can construct such an equal amplitude superposition in every sector and these special wave functions are all degenerate ground states of (17.1) at $v = t$.

In the next section, we will derive (17.6) and discuss a number of special properties of these wave functions. At the moment, we note that often the dimer coverings contained in a sector will lead to an equal amplitude wave function without any local order. However, a nonlocal measurement will be able to detect the winding number of the topological sector in which the wave function lies. Also, if the model is generalized to include monomers, we will find that these excitations are deconfined, as shown heuristically in Fig. 17.5a, b.

Due to these unique properties, the point $v = t$ has been given a special name: the Rokhsar–Kivelson point or *RK point*. The relation of the RK point to the rest of the QDM phase diagram depends on the lattice. For a number of nonbipartite lattices in two and higher dimensions, including the $2d$ triangular and $3d$ fcc, the RK point is part of a $Z_2$ *RVB liquid phase*, which has topological order. For a number of bipartite lattices in three and higher dimensions, including the cubic lattice, the RK point is part of a *Coulomb phase*, which is also a liquid phase but with a different type of quantum order. For a number of $2d$ bipartite lattices, including the square and honeycomb, the RK point is a special critical point separating different crystalline phases. The presenting understanding is that these behaviors should be generic.

#### 17.4.1.2 Columnar Phase

For $v/t \to -\infty$, the system will seek to maximize the number of flippable plaquettes. The state which accomplishes this for the square lattice is the columnar dimer arrangement shown in Fig. 17.6a. This state is fourfold degenerate and breaks rotational symmetry and the symmetry of translation by one lattice spacing along the direction of dimer orientation (i.e. the horizontal direction in the figure).

The term *columnar state* is commonly used to denote the maximally flippable state of a lattice, though the label is especially descriptive for the square lattice. On the triangular lattice, the number of such states is exponentially large in the linear size of the system. The columnar state is literally an eigenstate only in the $v/t \to -\infty$ limit but a state with columnar correlations will generally persist up to a

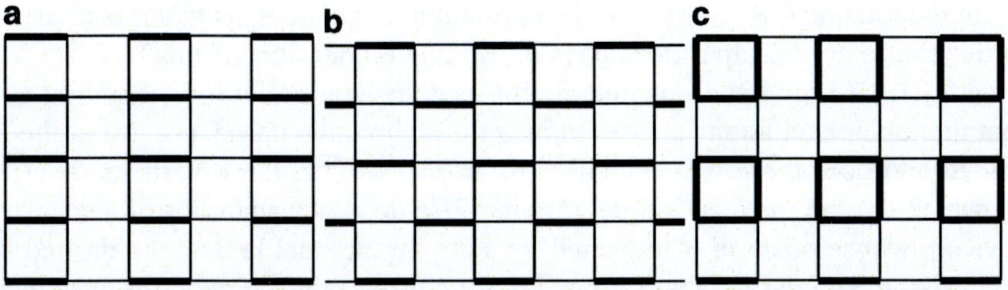


**Fig. 17.6** Examples of valence bond solids: (**a**) columnar state, (**b**) staggered state, and (**c**) Plaquette state

value of $v/t = (v/t)_c < 1$ which depends on the lattice [32–34]. Therefore, we call $-\infty < (v/t) < (v/t)_c$, the *columnar phase*. In contrast to the RK point, monomers are linearly *confined* in the columnar phase (Fig. 17.5c). Linear confinement is a generic feature of a crystalline phase.

#### 17.4.1.3 Staggered Phase

For $v/t \to \infty$. the system will seek to minimize the number of flippable plaquettes. If the lattice permits dimer arrangements with *no* flippable plaquettes, then these states will obviously be preferred. In terms of the above discussion, such states will be the only occupants of their sectors in the dimer Hilbert space. However, what is less trivial is that these states, if they are permitted, will be the ground states for $v/t > 1$. We will see this explicitly in the next section. For the moment, we point out that on the square lattice, there are, in fact, exponentially (in the linear size of the system) many such states, one of which is given in Fig. 17.6b. This arrangement is called the *staggered state* and the term is commonly used to denote a non-flippable state of a general lattice. In the triangular lattice QDM, by contrast, the number non-flippable states does not grow with the size of the lattice.

The nature of the phase diagram between the columnar phase and RK point depends strongly on the details of the lattice. We now survey a number of phases and features which have been seen in two and higher dimensions.

### *17.4.2 $Z_2$ RVB Liquid Phase*

The $Z_2$ RVB liquid phase is an example of a phase with $Z_2$ topological order [27]. This means that for a two-dimensional lattice with periodic boundary conditions, there are four degenerate gapped ground states. The ground states are "liquids" in that all dimer correlations decay exponentially. This phase was first discovered on the $2d$ triangular lattice [27] where current numerical evidence [35] shows the phase lying in the interval $0.8 < v/t < 1$. The present understanding is that this phase is a generic feature of non-bipartite QDMs in two and higher dimensions.[6]

The $Z_2$ RVB liquid phase is notable for its nontrivial excitations. The first point is that monomer excitations are *deconfined* throughout the phase, not just at the RK point. In addition, there is a second class of excitations, Ising vortices or *visons* [37], which live on the dual lattice. Figure 17.7a depicts a snapshot of a state with one vison, whose position is indicated by a dot on the dual lattice; the dashed line extends to the boundary, if there is one, or, for a torus, winds around the system and ends on one of the plaquettes neighboring the dot. At the RK point, a variational wave function describing this state is given by $|\text{vis}\rangle = \sum_c (-1)^{n_c} |c\rangle$ where $n_c$ is the number of dimers crossing the dashed line and the sum is over a sector. Note that $|\text{vis}\rangle$ is orthogonal to the liquid ground state (17.6) where the phase factor $(-1)^{n_c}$ is absent. There is numerical work which suggests the lowest lying excitation above the triangular lattice ground state is, in fact, vison-like [38].

The nontrivial feature of a vison is in its effect on the motion of monomers as depicted in Fig. 17.7b. The process of moving a monomer around a vison can be achieved by breaking a dimer into two monomers; holding one monomer fixed and moving the other around the vison until the monomers meet again; and then fusing the monomers back into a dimer. The process is equivalent to flipping the dimers along a flippable loop surrounding the vison. As the figure indicates, the number of dimers crossing the dashed line will be changed by $\pm 1$ so the wavefunction will change sign.

We now return to an issue glossed over in our preliminary discussion of fractionalisation, namely the issue of statistics. What is the relative statistics of the monomers? The answer is: "It depends", as usual in $d = 2$, where statistics is a question of energetics. Indeed, from the above discussion, it is clear that the relative statistics of monomers changes if they bind to a vison. This 'flux attachment' can be achieved by altering terms in the Hamiltonian rather than any native statistics

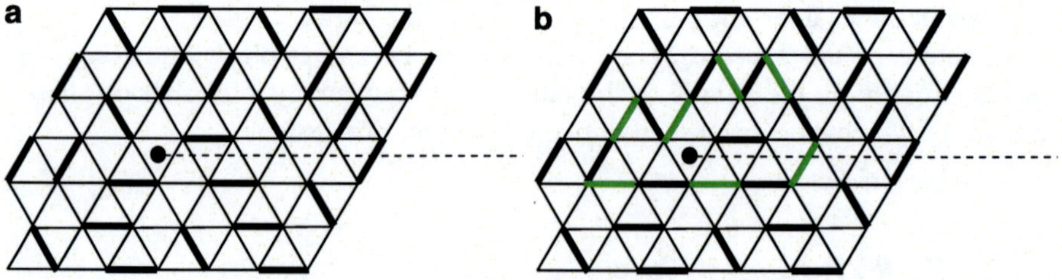


**Fig. 17.7** Visons. (**a**) visons live on the dual lattice. (**b**) Taking a monomer around a vison causes the number of dimers intersecting the *dashed line*, and hence the wave function, to change sign

[6] On the kagomé lattice, a dimer model with a $Z_2$ liquid phase can be constructed using multiple flip terms [36]. This latter construction is noteworthy because the model is exactly solvable and the resulting liquid has no correlations beyond one lattice spacing. This latter property makes the model particularly convenient for calculations.

of the monomer excitations, as has been shown explicitly in [39]. For the simple quantum dimer models, statistical transmutation between Fermionic and Bosonic monomers is thus straightforwardly possible. More complex anyonic, or even non-Abelian, statistics for defects in these kind of models are currently actively being pursued, see e.g. [26].

As mentioned earlier, the $Z_2$ RVB liquid has a higher dimensional analog. For example, on the (non-bipartite) $3d$ three-dimensional fcc lattice with periodic boundary conditions, the ground state is eightfold degenerate. The visons are now defined in terms of loops instead of lines. The interaction of a spinon with a vison involves the wave function acquiring a factor of $(-1)^n$ where $n$ is the number of times the spinon trajectory links with the vison loop. As in the two-dimensional case, the relative statistics of monomers depends on whether or not it is energetically favorable to bind a vison.

### 17.4.3 U(1) RVB Liquid Phase

The U(1) RVB liquid phase is another kind of fractionalized liquid phase. So far, it has only been observed on bipartite lattices in three and higher dimensions, including the $3d$ cubic and diamond lattices [40–42].

As with the $Z_2$ RVB liquid, the U(1) RVB liquid has a distinctly quantum type of order that is not captured by a local order parameter. However, its quantum order is different from the (also quantum) topological order of the $Z_2$ RVB liquid in a number of respects. The $Z_2$ RVB liquid is a gapped phase with exponential dimer correlations and a ground state degeneracy that depends on the system topology. This degeneracy is an essential feature of the entire phase.[7] In contrast, the U(1) RVB liquid is gapless with algebraic dimer correlations. While the ground state does have a topologically related degeneracy at the RK point (see Fig. 17.4b), this degeneracy will be *lifted* as we enter the phase.

The U(1) RVB liquid also has deconfined monomers. However, in the $Z_2$ RVB liquid, the monomers interact via a force whose range is a few lattice spacings while in the U(1) RVB phase, the monomers interact via a long-ranged inverse square force.[8]

The U(1) RVB also has a gapped excitation analogous to the vison of the $Z_2$ RVB called a *monopole* for reasons that will become clear later. These monopoles interact with each other via an inverse square force. The interaction between monomers and monopoles is directly analogous to the interaction between monomers and visons

[7] That the degeneracy, in fact, does persist away from the RK point, where the wave function is no longer known, was shown numerically in [35].

[8] This is not actually true at the RK point, which is a boundary point of the U(1) RVB phase. As we will see later, the inverse square force is like a Coulomb interaction transmitted by a photon whose speed (near the RK point) is $c = 1 - (v/t)$. At the RK point, the "speed of light" vanishes, and so does the force.

in the $Z_2$ RVB liquid. In particular, bare monomers and a monomer-monopole bound states have different statistics. Therefore, as in the $Z_2$ RVB case, the relative statistics of monomers is determined by the energetics.

The U(1) RVB liquid phase is also called a *Coulomb phase* because it has a continuum description that resembles the usual Maxwell action of the free electromagnetic field. In fact, we will later see that the gapless excitation of the U(1) RVB phase resembles a *photon*.

## 17.4.4 Deconfined Critical Points

On $2d$ bipartite lattices, it turns out that the RK point is not part of a liquid phase but rather a critical point between a plaquette phase, which is discussed in the next section, and a staggered phase. More precisely, the dimer correlations decay algebraically and the order parameter of the plaquette phases vanishes continuously. However, the staggered order appears at full strength so the ground state energy would have a derivative discontinuity. Therefore, the transition itself would be classified as first order. However, we will see below that the transition can be made continuous by weakly perturbing the model.

The plaquette and staggered phases break different symmetries so, according to the Landau theory of phase transitions, a transition between them will generically be first order. It was noted in [43] that the scale invariance (i.e. diverging correlation length) at the RK point was related to the existence of an emergent height field, which will be discussed below, that supports deconfined excitations. Critical points with this structure, now called *deconfined critical points*, have since been proposed as a generic mechanism by which such non-Landau transitions can occur in quantum systems [44–46].

## 17.4.5 Valence Bond Crystals

The most common class of phases found in QDMs are phases where the valence bonds arrange in an ordered pattern. We have already seen the columnar (Fig. 17.6a) and staggered (Fig. 17.6b) states. We now discuss some other crystalline phases which have been found.

### 17.4.5.1 Plaquette Phase

The *plaquette* state is drawn schematically in Fig. 17.6c for a square lattice. This picture should be interpreted in a mean-field sense, i.e. the thick lines indicate bonds of the lattice where the probability of having a dimer is comparatively large and the thin lines are where this probability is comparatively small. Imagine a pair of dimers on an elementary plaquette resonating between the horizontal and vertical configurations. To zeroth order, a variational wavefunction for this state may be visualized as a product of such terms.

As the figure indicates, the square lattice plaquette state is symmetric under rotation and translation by two lattice spacings in either direction. The state is fourfold degenerate as there are four possible ways to choose this doubled superlattice. Note that the plaquette state breaks different symmetries than the columnar and staggered states.[9]

The plaquette state can be generalized to other lattices and higher dimensions but so far a plaquette *phase* has been observed only on $2d$ bipartite lattices, particularly the square [32, 33, 47] and honeycomb [34] lattices, where numerical evidence suggests that the onset of the plaquette phase begins immediately to the left of the RK point and persists for some range of parameters. On the honeycomb lattice, the current picture is that the plaquette phase gives way to the columnar phase via a first order transition around $v/t = -0.2$ [34]. A similar scenario has been proposed for the square lattice [47] but there has been a very recent suggestion that, on the square lattice, a new "mixed" phase might occur between the plaquette and columnar phases [48].

#### 17.4.5.2 More Complex Crystals

A familiar difficulty in the numerical analysis of quantum many body Hamiltonians is that with available techniques only fairly small system sizes may be studied. Therefore, questions regarding thermodynamic limits are difficult to answer definitively. In the case of QDMs, the problem is compounded by the observation that crystalline phases with rather large unit cells have been discovered even for QDMs on fairly simple lattices. Indeed, for the kagomé quantum dimer model, fantastically large unit cells have been proposed [49, 50].

As a concrete example, the triangular lattice QDM exhibits a phase called the "$\sqrt{12} \times \sqrt{12}$ phase", named for its 12 site unit cell. The phase was first conjectured due to an exact mapping between the QDM at $v = 0$ and the fully frustrated transverse field Ising model on the honeycomb lattice [51]. The most current numerical evidence suggests that on the triangular lattice, the $\sqrt{12} \times \sqrt{12}$ phase is the only phase between the columnar phase and the $Z_2$ RVB liquid phase discussed above [35].

#### 17.4.5.3 Even more Complex Crystals: Cantor Deconfinement

*Cantor deconfinement* is a phenomenon predicted to occur for QDM's on $2d$ bipartite lattices when these models are perturbed near their RK points. In the unperturbed

[9] For example, on the square lattice, the staggered and columnar phases break rotational symmetry but both have higher translational symmetry than the plaquette phase.

QDM, the RK point is a critical point between a plaquette crystal, which has zero winding number, and a staggered crystal, which has the maximal winding number. It was noted in [52, 53] that for a wide class of perturbations, this picture gets modified as new crystalline phases with intermediate winding numbers appear between the plaquette and staggered states. In particular, in the limit of an infinite system, the winding number per unit length, which is called the *tilt* for reasons that will be made clear later in this chapter, will increase *continuously* from zero to its maximal value of one as we move from the plaquette phase to the staggered phase along a generic path in the phase diagram.[10]

We remind the reader of two qualitatively different ways by which a function can continuously increase from zero to one as a parameter $x$ is tuned from $x_i$ to $x_f$. There is the naive "analytic" way: place a pencil at $(x_i, 0)$ and trace out a non-decreasing curve that ends at $(x_f, 1)$ without removing the pencil from the paper. However, we can also construct a continuous function that begins at $(x_i, 0)$, ends at $(x_f, 1)$, and has *zero* derivative everywhere except on a generalized Cantor set, where the derivative does not exist. The resulting function looks like a staircase but between any two steps there are an infinite number of other steps, hence its name, *devil's staircase*. This construction is a standard exercise in analysis but, remarkably enough, is relevant in the present context as we shall now see.

Returning to physics, we note that details of these new intermediate phases will depend on specifics of the perturbation. For example, additional potential energy terms will favor states where the tilt is *commensurate* with the lattice. A commensurate crystal has a unit cell which is finite-sized and a tilt that is a rational number. However, quantum fluctuations, due to present and additional kinetic energy terms, can stabilize states where the tilt is *incommensurate* with the lattice, i.e. where the tilt is an irrational number.[11] Therefore, if the potential terms dominate, the system will prefer to "lock" into commensurate values of the tilt. The variation of the tilt along a line in the phase diagram connecting the plaquette and staggered states will resemble the devil's staircase discussed above. However, if quantum fluctuations are dominant, then the tilt will vary along this line in the conventional "analytic" fashion. Such a regime of parameter space, if it exists, would effectively be an incommensurate *phase* because rational tilts would occur only for a set of measure zero.

The central conclusion of [52, 53] was that such a fluctuation-dominant regime will exist parametrically close to the RK point for a wide class of perturbations. This would be bordered by a regime where the fluctuations compete with the locking

[10] To simplify the discussion, the reader may wish to consider a class of states where $W_x = 0$ but $W_y$ can vary. In this case, Fig. 17.6a–c are states of minimal and maximal winding $|W_y|$, respectively. We invite the reader to construct examples of states with intermediate winding or look at some of the pictures in [54]. Also, when we perturb the model, the phase diagram becomes multi-dimensional. The reader should understand that when we speak of new phases appearing between the plaquette and staggered states, we are referring to a quasi-1D slice of the actual phase diagram.

[11] Of course, this would make sense only in the limit of an infinite system.

potentials, which would eventually dominate away from the RK point.[12] Moreover, it was observed that the commensurate states were gapped with linearly confined monomers while the incommensurate states were gapless with monomers that were confined only *logarithmically*, due to the effective two-dimensional Coulomb interaction. In this sense, we can think of the incommensurate states as having monomers that are "almost" deconfined.

When the locking is strong, these "almost" deconfined points, i.e. the incommensurate states, occupy a small fraction of the phase diagram and in the limit of infinitely strong locking will exist only on the Cantor-like set defining the boundaries of the steps of the devil's staircase. However, when the locking is weak, these "almost" deconfined points occupy a finite fraction of the phase diagram and this fraction approaches unity as we approach the RK point. Hence, very close to the RK point we almost have an "almost" deconfined *phase*! This last result is what is meant by "Cantor deconfinement".

One reason why this result is noteworthy is because a famous result of Polyakov [55] states that one cannot have an actual deconfined phase in a compact $U(1)$ gauge theory in 2+1, of which the $2d$ bipartite QDM is an example. In fact, this is another explanation for why the RK point is an isolated point in these models instead of being part of a deconfined liquid phase. The Cantor deconfinement scenario provides a way around this, the fundamental property absent in the dimer models being Lorentz invariance.

The arguments summarized in this section are based on renormalization group analyses of a continuum field theory, which will be discussed more below, that is believed to describe $2d$ bipartite RK points. We encourage the reader to consult the original papers [52, 53] for details and additional caveats.[13] We would like to point that a microscopic demonstration, i.e. at the level of explicit Hamiltonian that can be viewed as a perturbed RK point, of the weak locking regime is currently lacking. However, we refer the interested reader to [54] for an explicit "proof of principle" of the strong locking regime.

### 17.4.6 Summary of Phase Diagrams

The results of this section suggest the generic RK phase diagrams shown in Fig. 17.8. However, richer phase diagrams may be obtained by considering exotic lattices or, as we have seen, by perturbing away from the prototypical RK Hamiltonian.

[12] The transition from an analytic variation of the tilt to the staircase-like variation was called a "breaking of analyticity" in analogy to a similar mechanism in classical soliton models first discussed by Aubry.

[13] For example, for the square lattice QDM it turns out that a generic perturbation will drive the transition first order, so that fine-tuning the Hamiltonian is required in order to realize the Cantor deconfinement in that case.

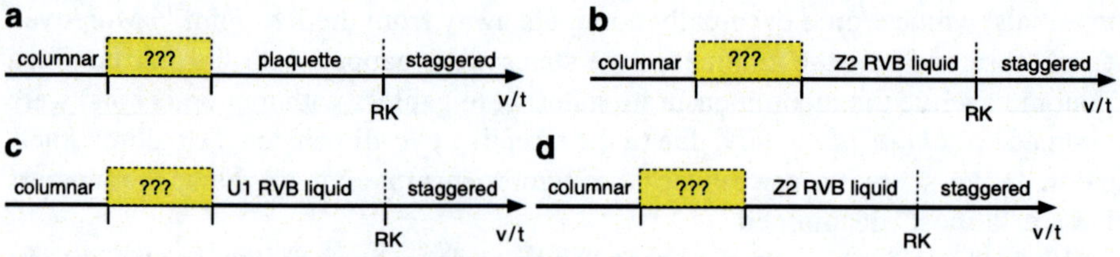


**Fig. 17.8** Generic phase diagrams. (**a**) $2d$ bipartite. For the honeycomb lattice, "???" is a first order transition between the columnar and plaquette phases. (**b**) $2d$ non-bipartite. For the triangular lattice, "???" includes the $\sqrt{12} \times \sqrt{12}$ phase and possibly others. (**c**) higher $d$ bipartite. For the cubic lattice, "???" is not known. (**d**) higher $d$ non-bipartite. For the fcc lattice, the $v/t \to -\infty$ phase, normally the columnar phase, is not characterised completely

## 17.5 The Rokhsar–Kivelson Point

In this section, we discuss a number of features of the RK point and elaborate on some of the issues discussed previously.

### 17.5.1 Ground-state Wavefunction

To derive (17.6), it is easiest to consider the explicit example of the original square lattice QDM given by (17.1). The sum is over all elementary plaquettes in the lattice. The $v$ operator acting on a dimer covering gives a potential energy if the plaquette in question has two parallel dimers, i.e. if the plaquette is *flippable*, and annihilates the state otherwise. The $t$ operator is a kinetic energy and flips the dimers if the plaquette is flippable and annihilates the state otherwise. At the point $v = t$, the Hamiltonian is a simple sum of projectors:

$$\mathrm{H}_{\mathrm{QDM}} = \Sigma(|\,\|\,\rangle - |\,=\,\rangle)(\langle\,\|\,| - \langle\,=\,|) \tag{17.7}$$

i.e. a sum of terms each of which has eigenvalues 0 or 1. Therefore, any wave function annihilated by this Hamiltonian will be a zero energy ground state. We may write a general wave function as $|\Phi\rangle = \sum_c A_c|c\rangle$, where $|c\rangle$ is a dimer covering of the lattice and the sum is over all coverings. The state $|\Phi\rangle$ will be annihilated if, and only if, the amplitude $A_c$ of any dimer covering $|c\rangle$ is the same as the amplitudes $\{A_{c'}\}$ of *all* dimer coverings $\{|c'\rangle\}$ that differ from $|c\rangle$ by one flipped plaquette. Therefore, a prototypical ground state of (17.7) will have the form of an equal amplitude superposition, i.e. (17.6).

For the square lattice, the kinetic term is believed to be ergodic in each topological sector. In this case, there is a *unique* ground state for each topological sector given by the equal amplitude superposition of all dimer coverings in that sector. Of course, because these are all degenerate, any linear combination of them will also be a ground state including the equal amplitude superposition of *all* dimer coverings.

The staggered state (Fig. 17.6b) will always be annihilated by (17.1) so hence will also be a ground state at the RK point. Of course, the staggered state wave function trivially has the form of (17.6), being the only occupant of its dynamical sector, so this is consistent with the previous discussion. When $v > t$, the RK Hamiltonian is positive semidefinite so the staggered states, which will still be zero energy eigenstates, will still be ground states.

### 17.5.2 Fractionalisation and Deconfinement

Fractionalisation and deconfinement are evident at the RK point. Imagine declaring two fixed sites as hosting monomers rather than having dimers emanating from them. Now, consider the Hamiltonian (17.7) with the two monomers held fixed. The ground state will still have zero energy and the wave function will still have the form in (17.6) where the dimers now resonate everywhere except on the sites having the monomers. As the ground state energy is independent of the separation between the monomers, the monomers are deconfined. Hence, fractionalistion generically occurs at the RK point, indepentdently of the lattice. Whenever the RK point is a critical point between two solid phases, the RK point is thence a deconfined critical point.

### 17.5.3 Spatial Correlations

The dimer–dimer correlation function $\langle \hat{n}_\sigma(r)\hat{n}_\sigma(0)\rangle$, where $\hat{n}_\sigma(r)$ is a projection operator for having a dimer of orientation $\sigma$ at site $r$, is an important quantity. The form of the equal amplitude wave function, (17.6), has an important consequence for the correlations of operators, $\hat{D}$, which are *diagonal* in the dimer basis, $\langle c|\hat{D}|c'\rangle = \delta_{c,c'}\langle c|\hat{D}|c\rangle \equiv D_c$:

$$\langle \hat{D}\rangle = \frac{\langle \Phi_{\mathrm{GS}}|\hat{D}|\Phi_{\mathrm{GS}}\rangle}{\langle \Phi_{\mathrm{GS}}|\Phi_{\mathrm{GS}}\rangle} = \frac{\sum'\langle c|\hat{D}|c\rangle}{\sum'\langle c|c\rangle} = \frac{\sum' D_c}{N_c}. \tag{17.8}$$

This expression is simply an ensemble average over the set of $N_c$ dimer coverings connected by the flip term (which are averaged over in the $\sum'$) – this is exactly the corresponding correlation function of the relevant *classical* dimer model at infinite temperature.[14]

The spatial correlations of the dimer density may be obtained by choosing $\hat{D} = n^c_\sigma(r)n^c_\sigma(0)$, where $n^c_\sigma(r) = 1(0)$ if a dimer of orientation $\sigma$ is (is not) present at site $r$ in dimer covering $|c\rangle$. For calculating spatial correlations of the analogous

[14] However, as we shall see shortly, this does not imply that the RK point is literally "connected" to the infinite temperature state in the sense of being part of the same phase in the phase diagram.

classical problem, efficient numerical and, in two dimensions, analytical [24, 25] techniques are available.

This 'dimensional reduction' from a $d+1$-dimensional (quantum) to a $d$-dimensional (classical) problem is one of the reasons RK-type models have been so popular. They enable the transpositions of relatively easily obtained classical results into a more interesting quantum setting.

### 17.5.4 Excited States

Another property of the RK point is that information about the *excited* states can be obtained by studying *temporal* correlations of the infinite temperature classical system [57]. To see this, consider a classical ensemble of dimer coverings that are connected to one another by application of the basic flip term in (17.1). At equilibrium, each dimer covering occurs with the same probability. In a classical Monte Carlo simulation, the dynamics of this equilibration is governed by the master equation:

$$\frac{dp_\alpha}{dt} = -\sum_\alpha W_{\alpha\beta}\, p_\beta \tag{17.9}$$

where $p_\alpha$ is the probability of the system being in dimer configuration $\alpha$, and $W_{\alpha\beta}$ is the rate of transition from state $\beta$ to state $\alpha$. We are interested in the case where the basic Monte Carlo move involves randomly a selecting a plaquette and flipping it if possible. In this case, the matrix element $W_{\alpha\beta}$ between two different dimer coverings $\alpha$ and $\beta$ equals $-1$ if $\alpha$ and $\beta$ are connected by a single plaquette flip and zero otherwise. The normalization condition $\sum_\alpha p_\alpha = 1$ implies that $W_{\alpha\alpha} = -\sum_{\beta\neq\alpha} W_{\alpha\beta} = n_{fl,\alpha}$ where $n_{fl,\alpha}$ is the number of flippable plaquettes in dimer covering $\alpha$. The key observation of [57] was that:

$$W_{\alpha\beta} = (H_{QDM})_{\alpha\beta} \tag{17.10}$$

i.e. the rate matrix of the classical Monte Carlo simulation is the same as the Hamiltonian matrix (17.1) of the quantum problem. Therefore, the two matrices share the same eigenvalues and eigenvectors. In other words, the relaxation modes of the master equation correspond *exactly* to the excited states of the QDM at the RK point. As we will see later in this review, this fact provides a route for constructing a continuum field theory of the RK point when the lattice is bipartite.

Moreover, any temporal correlation function of the classical problem will have the form:

$$\sum_\lambda c_\lambda e^{-\lambda t} \tag{17.11}$$

where $\{\lambda\}$ are the eigenvalues of matrix $\mathbf{W}$ (and hence of $H_{QDM}$) and are non-negative for reasons discussed above. As noted in [57], the long time behavior of such correlation functions will be dominated by the smallest nonzero eigenvalue so numerical simulations of appropriate classical time correlation functions can give nontrivial information about the energy gap and low lying spectrum at the RK point. Applications of this idea will be seen further below.

The mapping from a quantum problem in $d + 1$ to a classical $d$ dimensional one is very general and can be run in reverse: given any local set of weights for a problem in classical statistical mechanics, it becomes possible to construct a quantum Hamiltonian with a special point analogous to the RK point. At this point, the ground-state wave function can be expressed as a superposition of the classical configurations where the probability amplitudes are exactly the Boltzmann weights of the classical model. This process is known as Rokhsar–Kivelsonisation [58–60].

### 17.5.5 A Special Liquid Point or part of a Liquid Phase?

We have seen that the ground state manifold at the RK point includes wave functions with no local order and fractionalized deconfined excitations. A natural question is whether the RK point is a special point or part of a fractionalized liquid phase. We have seen that the answer depends on the geometry and dimensionality of the lattice. The present understanding is that for lattices in three and higher dimensions, as well as for $2d$ *nonbipartite* lattices, the RK point is generically part of a liquid phase. For $2d$ *bipartite* lattices, the RK point is (generically) either a special critical point or the non-critical endpoint of an ordered phase.

This classification of lattices has not been established rigorously but extrapolated from numerous specific examples. While the existence of some of these liquid phases has been independently verified through numerics, it is difficult to numerically distinguish a liquid from a crystal with a very large unit cell, as discussed in the previous section. In fact, the strongest evidence for the existence of these liquid phases is based on a careful analysis of the RK point, which is what prompted their discovery in the first place. We now review this line of reasoning.

The existence of the $Z_2$ RVB liquid phase on the triangular lattice follows from the observation that the classical dimer model at infinite temperature has exactly the same correlations as the zero temperature quantum problem at the RK point. Of course, we know the dimer–dimer correlations will match for *any* lattice by construction. We also know that for a general lattice at the RK point, monomers are deconfined with exactly zero correlation beyond one lattice spacing. The key point about the triangular lattice is that the classical (i.e. infinite temperature) monomer–monomer correlation function decays exponentially with a characteristic length of around one lattice spacing [61,62]. In other words, both the RK point and the infinite temperature state are liquids with deconfined monomers that interact ultra-locally. The simplest picture consistent with this is that the RK point is part of a zero temperature liquid phase that connects smoothly to the infinite temperature phase as the temperature is raised. Similar arguments apply for other non-bipartite lattices in two and higher dimensions.

For bipartite lattices, this argument no longer works. Its failure stems from the fact that for these lattices, the classical monomer-monomer correlation decays as a power law. This means that in the infinite temperature phase, monomers are logarithmically *confined* in $2d$ while in $3d$, they are still deconfined but now have a long range interaction. Therefore, the infinite temperature liquid is qualitatively different than the liquid occuring at the RK point so there is no obvious reason to suspect a smooth connection between the two.

Nonetheless, there are other approaches which give insight into the structure of these bipartite RK points. In the next sections, we will discuss some of these, the single mode approximation (SMA), which works for general lattices, and the height representation, which works for bipartite lattices. In those contexts, we will return to the question of why bipartite RK points in $2d$ are critical points while in $3d$, they are part of a U(1) RVB liquid phase.

## 17.6 Resonons, Photons, and Pions: Excitations in the Single mode Approximation

In the original RK paper, the authors also had a look at the excitations of the dimer model. One approach they took was via the single-mode approximation (SMA), in which one constructs a trial state with a momentum which differs from that of the ground state by $\mathbf{q}$; the momentum being a good quantum number in a translationally invariant system, the variational principle employed on the states at that momentum yields the basic result that the energy of the trial state provides an upper bound on the excitation energy at that momentum.

Therefore, as a matter of principle, one can use the SMA to prove gaplessness. In order to demonstrate the presence of a gap, a different method is needed [63].

Let us denote the ground state of the dimer model by $|0\rangle$, and let $\sigma^x_{\hat{\tau}}(\mathbf{r})$ be the Pauli spin operator, its eigenvalues $\pm 1$ corresponding to the presence or absence of a dimer on the link at location $\mathbf{r}$. $\hat{\tau}$ encodes the direction the dimer points in, i.e. its "polarisation".

Fourier transforming the dimer density operator,

$$\tilde{\sigma}^x_{\hat{\tau}}(\mathbf{q}) \equiv \sum_{\mathbf{r}} \sigma^x_{\hat{\tau}}(\mathbf{r}) \exp(i\mathbf{q}\cdot\mathbf{r}). \tag{17.12}$$

enables us to define our trial state, which is orthogonal to $|0\rangle$ for $\mathbf{q} \neq 0$:

$$|\mathbf{q}, \hat{\tau}\rangle \equiv \tilde{\sigma}^x_{\hat{\tau}}(\mathbf{q})\,|0\rangle \tag{17.13}$$

One next needs to check that one has, in fact, constructed a state rather than just annihilated the ground state, i.e. that

$$\langle \mathbf{q}, \hat{\tau}\;|\mathbf{q}, \hat{\tau}\rangle \neq 0. \tag{17.14}$$

One then obtains a variational energy of

$$E(\mathbf{q},\hat{\tau}) \leq \frac{\langle 0|\,[\tilde{\sigma}^x_{\hat{\tau}}(-\mathbf{q}),[H_{QDM},\tilde{\sigma}^x_{\hat{\tau}}(\mathbf{q})]]\,|0\rangle}{\langle 0|\,\tilde{\sigma}^x_{\hat{\tau}}(-\mathbf{q})\tilde{\sigma}^x_{\hat{\tau}}(\mathbf{q})\,|0\rangle} \equiv \frac{f(\mathbf{q})}{s(\mathbf{q})}, \tag{17.15}$$

where $f(\mathbf{q})$ is known as the oscillator strength, and $s(\mathbf{q})$ as the structure factor. Crucially, these can be evaluated as expectation values *in the ground state*, whose correlations therefore encode information on the excitation spectrum.

The utility of the SMA derives in large part from the fact that there are situations in which gaplessness is present generically. For instance, if the density $\tilde{\sigma}^x_{\hat{\tau}}(\mathbf{q}_0)$ is a conserved quantity, then $[H_{QDM},\tilde{\sigma}^x_{\hat{\tau}}(\mathbf{q}_0)] = 0$, whence $E(\mathbf{q}_0,\hat{\tau}) = 0$. The behaviour of $f(\mathbf{q})$ near $\mathbf{q}_0$ can then be used to determine a bound on the dispersion of the soft excitations. Finally, a finite $f(\mathbf{q})$ accompanied by a divergence of $s(\mathbf{q})$ can be used to infer gapless excitations. Indeed, such a "soft mode" is a classic signature of incipient order.

On the square lattice, RK identified the density of dimers pointing in a given direction ($\hat{\tau} = \hat{x}$, say) at wavevector $\mathbf{q}_0 = (q_x,\pi)$ as a conserved quantity. For the cubic lattice, the analogous density is simply that at $\mathbf{q}_0 = (q_x,\pi,\pi)$. The way to see this is to observe that dimers are always created and destroyed in pairs on opposite sides of a plaquette. This implies that the oscillator strength, $f(\mathbf{q}_0)$, vanishes at $\mathbf{q}_0$; at wavevector $\mathbf{q}_0 + \mathbf{k}$, the oscillator strength is given by [41]

$$f(\mathbf{k}) \propto (\mathbf{k}\times\hat{\tau})^2, \tag{17.16}$$

This is true not just at the RK point, but for all values of $-\infty < v/t \leq 1$.

Contrary to appearances, this does not imply a line of zero energy excitations because along with the oscillator strength, the structure factor also vanishes for all $\mathbf{q}_0$ with $q_x \neq \pi$, i.e. (17.14) is not satisfied unless $q_x = \pi$. Indeed, at the RK point itself, for momentum $\mathbf{q} = (\pi,\pi[,\pi]) + \mathbf{k}$, the structure factor is given by a transverse projector:

$$s_{\hat{x}}(\mathbf{k}) \propto \frac{k_y^2[+k_z^2]}{k^2} \equiv \frac{k_\perp^2}{k^2}. \tag{17.17}$$

Consequently, only transverse excitations are generated [64] by $\tilde{\sigma}^x_{\hat{\tau}}$, a fact that traces back to the defining constraint of the dimer model, which takes a form like $\nabla\cdot\mathbf{B} = 0$ for bipartite lattices.

The soft excitations near $\mathbf{q}_0 = (\pi,\pi[,\pi])$ were called *resonons* by RK, while in three dimensions, they are typically called *photons* [41,42], as they arise from a Maxwell theory in the standard way as discussed below. It is important to note that the RK point is in fact untypical: the photons are anomalously soft, with a disperson $\omega \propto k^2$. This is remedied upon entering the Coulomb phase to the left of the RK point, where the photons become linearly dispersing.

These photons turn out to be the *only* gapless excitations for the cubic lattice. In contrast, on the square lattice one also finds gapless excitations of the 'soft-mode' type near $(\pi, 0)$ and $(0, \pi)$, where there is a divergence of, respectively, $s_{\hat{x}}$ and $s_{\hat{y}}$. These soft excitations were called *pi0ns* [41] due to their location in the square lattice reciprocal space. This divergence is a signature of an incipient crystalline phase having order at $(\pi, 0)$ and confirms the earlier assertion that the square lattice RK point is an isolated critical point. In contrast, the absence of pi0ns at the cubic lattice RK point is consistent with the RK point being part of an extended liquid phase.

It turns out that the trial wavefunctions used in the SMA do a much better job at constructing resonons and photons than pions [65, 66]. Indeed, if the critical exponent of the correlations is such that the structure factor remains infrared convergent, the SMA does not even show the presence of a soft mode.

Finally, for the RK points of the simple non-bipartite lattices (triangular and face-centered cubic), the calculations presented above do not yield gapless excitations, in keeping with the expectation that the relevant $Z_2$ RVB liquids are gapped [41].

## 17.7 Dualities and Gauge Theories

Link variables, together with a constraint defined on sites, are defining features of lattice gauge theories. There, the gauge fields live on links of a lattice, while Gauss' laws define the physical sector of the theory, such as the familiar

$$\nabla \cdot \mathbf{E} = 0 \tag{17.18}$$

from electrostatics. In the following paragraphs, we address the relationship between QDMs and gauge theories.

This is most crisply done using the transverse-field Ising model (TFIM) as an example. It will turn out that the quantum dimer model is the strong-coupling dual partner of this simple spin model in $d = 2 + 1$. The Ising spins $S^z = \pm 1$ live on the sites of a lattice $\Lambda$:

$$H_{TFIM} = -\sum_{\langle ij \rangle} J_{ij} S_i^z S_j^z - \Gamma \sum_i S_i^x . \tag{17.19}$$

Here, $J_{ij}$ denotes the exchange constant of a given pair of spins and $\Gamma$ is the strength of the transverse field.

The mapping (see Fig. 17.9) now proceeds by identifying the link variable $\sigma_{ij}^x$ on the dual lattice with the bond energy of the spin model:

$$-|J_{ij}|\sigma_{ij}^x = J_{ij} S_i^z S_j^z , \tag{17.20}$$

so that $\sigma_{ij}^x = -1$ for a frustrated bond.

The action of the transverse field is to flip a spin, i.e. to toggle between $S_i^z = \pm 1$; this corresponds to exchanging the sign of the energy of the bonds emanating from

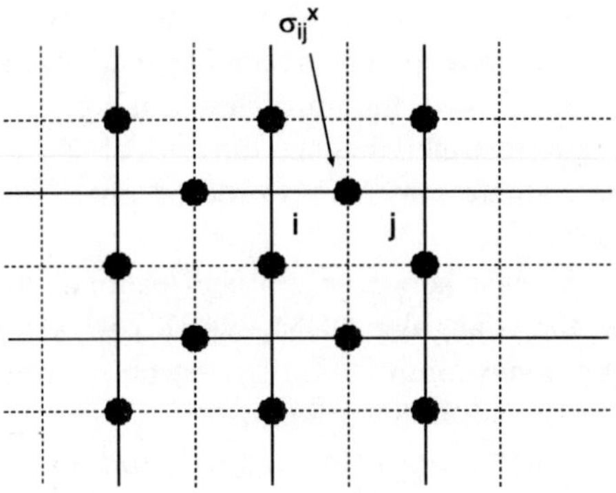


**Fig. 17.9** The *dark lines* denote the direct lattice; the *dashed lines* the dual lattice; and the *dots* are where the $\sigma$ variables are located. $i$ and $j$ are sites on the direct lattice

site $i$. These bonds form an elementary plaquette of the dual lattice, $\Box$, at the centre of which direct lattice site $i$ is located. The dual lattice Hamiltonian thus reads:

$$H_{gauge} = -\sum_{-} |J_{ij}| \sigma^x_{ij} - \Gamma \sum_{\Box} \prod_{ij \in \Box} \sigma^z_{ij}, \tag{17.21}$$

where the first sum runs over bonds and the second over plaquettes.

It would at first sight seem that (17.21) does not know whether the original spin model was frustrated or not, as the signs of the exchange constants have disappeared. Indeed, the information on exchange constants is stored in a gauge invariant fashion in a Gauss' law constraint on the physical sector of the theory, i.e. for every site of the *dual* lattice, we have:

$$\hat{G}|phys\rangle \equiv \prod_{+} \sigma^x_{ij} |phys\rangle = \pm 1 |phys\rangle. \tag{17.22}$$

Here, the product is over links emanating from a site of the *dual* lattice, which corresponds to the links forming a plaquette of the direct lattice. If the plaquette is frustrated, i.e. the product of $J_{ij}$ around its bonds is negative, the minus sign in (17.22) is chosen, otherwise the plus sign applies.

### *17.7.1 Emergence of the QDM*

In the limit $\Gamma/J \to 0$, we have to minimise the number of frustrated bonds subject to the constraint (17.22). For an unfrustrated model, this is done by choosing $\sigma^x \equiv 1$ throughout – the system is free of dimers.

However, for a frustrated model, we have to have at least one frustrated bond emanating from each site. Equating a frustrated bond with a dimer present on it thus leads us to a hardcore dimer model: each site of the dual lattice has one, and only one, dimer emanating from it in the limit $\Gamma/J \to 0$. Degenerate perturbation theory in the $\Gamma$ term leads to the quantum dynamics of the QDM. The RK-potential term in turn corresponds to an additional multi-spin interaction.

At $T = 0$, the unfrustrated Ising magnet has an ordering transition at $(\Gamma/J)_c$. For a frustrated magnet, the critical $(\Gamma/J)_c$ may be suppressed; in the most extreme case, there may be no ordered phase at all, so that only the paramagnetic phase exists.

In dual language, this corresponds to a quantum liquid. In the particular case of the triangular QDM, which can be obtained from a fully frustrated TFIM on the honeycomb lattice by the above route, the appropriate quantum liquid is the fractionalised RVB liquid.

What does the topological order of the RVB liquid mean in the spin model? In fact, working backwards, one finds that the topological degeneracy indicates that the ground-state energy is independent of the choice of [(anti-)periodic] boundary conditions for the spin model: the topological sectors correspond to having an even or odd number of frustrated bonds as one goes around the system once. This makes sense as absence of long-range order in the paramagnet implies insensitivity of the energy to the boundary conditions. Indeed, in a perturbative expansion of the ground state energy around $J = 0$, one will need to keep track of contributions involving a product of $\sim L$ terms, corresponding to a loop winding around the system, in order to discover the nature of the boundary conditions – this is a simple explanation for the origin of the $\exp(-cL)$ splitting of the topological sectors in the topologically ordered phase.

### *17.7.2 Continuum Limit of the Gauge Theory*

On the lattice, the constraint states that the (integer) number of dimers emanating from each site is always the same (namely, 1, for a hardcore dimer model, but this number can be chosen freely as we have seen above). A constraint on a *number* leads naturally to a U(1) gauge theory. It is a constraint on a parity, such as in the preceeding example (17.22), that yields a $Z_2$ gauge theory. Do all hardcore dimer models thus yield U(1) gauge theories in the continuum?

In fact, some do (e.g. the cubic lattice) but others do not (i.e. the triangular lattice). If we want to write down a U(1) Gauss' law such as $\nabla \cdot \mathbf{B} = 0$ in the continuum, we need to interpret the dimers as fluxes. To do this, they need to be oriented. This can straightforwardly be done on a bipartite lattice – one chooses to orient them to point from one sublattice to the other. This is clearly not possible on a non-bipartite lattice, as can most easliy be seen by thinking of the triangular lattice as a square lattice with a diagonal linking the bottom left to the top right of each plaquette. Dimers on this diagonal cannot thus be oriented.

The presence of these diagonal bonds spoils the $U(1)$ gauge symmetry in the continuum limit, and breaks it down to $Z_2$, as described e.g. in [67]. In more elaborate schemes, other continuum gauge theories (e.g. $Z_3$ [68]) become possible, all from models which look very similar on the level of the constraint on a single lattice site.

Indeed, a good deal of effort has been concentrated on building lattice models realising other types of constraints, and thereby further types of continuum theories. For instance, Xu has written down a theory in which spin-2 excitations ('gravitons') occur in three dimensions, and other authors have looked at a range of effective plaquette models [69–72].

## 17.8 Height Representation

For two-dimensional, *bipartite* lattices, the *height representation* provides one route for constructing a continuum theory of the QDM. Consider, for example, a square lattice which we divide into $A$ and $B$ sublattices.[15] The idea is to construct a *height field* on the plaquettes of the lattice. Choose an arbitrary plaquette to be assigned height zero. We then assign integer values to the other plaquettes through the following rule: moving clockwise around a site of the $A$ sublattice, the height increases by one if a dimer is not crossed and the height decreases by three if a dimer is crossed. The constraint of one dimer per site ensures that the mapping is consistent and unique up to the choice of the arbitrary height zero plaquette.[16]

Figure 17.10 shows some sample dimer coverings along with their respective height representations. These pictures may be interpreted by considering each plaquette as the base of a block which extends out of the page by an amount given by the integer in the center of the plaquette. From this perspective, a dimer covering may be viewed as a two-dimensional representation of the surface of a three-dimensional crystal. A typical height profile will be fluctuating on microscopic scales (Fig. 17.10a). However, if we coarse-grain this height field by averaging over small but macroscopic regions, there will be regions of the lattice such as in Fig. 17.10b where the (averaged) surface is essentially *flat* and also regions such

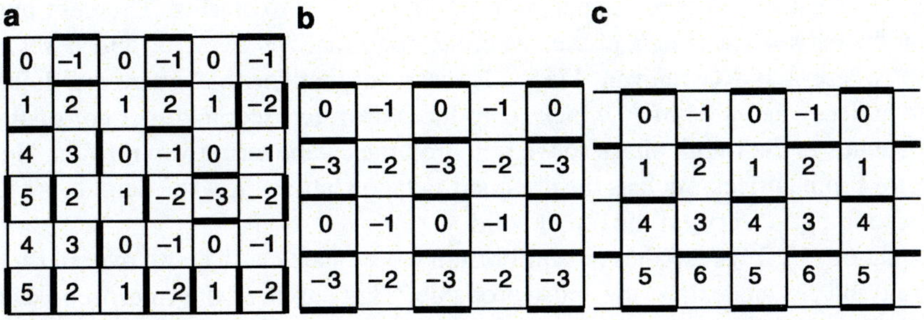


**Fig. 17.10** Examples of height mappings. In each case, the upper left plaquette is chosen to have height zero

[15] As usual, each $A$ site is surrounded by four $B$ sites and vice versa.

[16] Of course, we are also free to uniformly shift all the plaquettes by the same fixed amount.

as in Fig. 17.10c where the (averaged) surface is *tilted*. As these figures suggest, a dimer configuration can have an *overall* tilt. The columnar state is an example of a flat configuration while the staggered state is the state with maximal tilt.

The effect of flipping the dimers on a flippable plaquette is simply to increase or decrease the height of that plaquette by 4. Therefore, the overall tilt of a configuration will not be affected by any local rearrangement of the dimers. One can verify that the overall tilt of a configuration is simply its winding number per unit length.

So far, the height field is just a different way to represent a dimer covering. The usefulness of this mapping is that it provides a route for constructing continuum field theories for RK points of (bipartite) QDMs. A purely deductive argument, where the field theory is obtained by systematically coarse-graining the microscopic Hamiltonian, is currently not available. The present intuitive construction, due to Henley [57], is based on the relation between RK point of the QDM and the relaxation modes of the master equation for the classical problem discussed earlier in Sect. 17.5.4.

The starting point is the nontrivial observation that the long distance properties of the *classical* dimer problem can be captured by a continuum theory of the sine-Gordon type [76]:

$$S[h] = \frac{K}{2}\int d^2r|\nabla h(\mathbf{r})|^2 - \lambda \int d^2r \cos[2\pi h(\mathbf{r})] \qquad (17.23)$$

Here, $h(\mathbf{r})$ is a coarse-grained version of the height field so the theory implicitly assumes a bipartite lattice. The origin of the second term can be traced back to the fact that the microscopic height field can only take integer values. The route is a bit involved but it is carefully described in [73] in the setting of triangular Ising magnets, where the appropriate operator identifications are also derived. $K$ is determined by requiring that correlation functions computed with this action have the same long distance behavior as the corresponding correlations of the microscopic system.

A $2d$ sine-Gordon theory, such as (17.23), shows a Kosterlitz–Thouless phase transition between a *rough* phase, where $\lambda$ renormalizes to zero, and a *smooth* phase, where $\lambda$ is RG relevant [76].[17] It turns out that for the square and honeycomb lattices, the value of $K$ which reproduces the microscopic dimer correlations corresponds to the rough phase so we will drop the cosine term from now on. However, for some lattices, such as the diamond-octagon lattice [75], the heights are flat, corresponding to a crystalline dimer solid.

Equation (17.23) describes an equilibrium where states with small overall tilt are favored and flat states are the most probable. This captures the microscopic fact that the low winding number sectors contain more dimer coverings than the high winding number sectors so, in the absence of interactions, will be favored due to entropy. The strategy for constructing the quantum field theory of the RK point is to

[17] In the language of the 2D Coulomb gas, the rough phase is where opposite charges exist as bound pairs while the smooth phase corresponds to a plasma.

first construct the continuum analog of (17.9). This will give a continuum version of a rate matrix, which we then identify, by analogy to (17.10), as the continuum version of the QDM Hamiltonian at the RK point.

Recall that (17.9) determines how the probability $p_\alpha$ of a classical dimer system being in configuration $\alpha$ changes with time during a Monte Carlo simulation where the only allowed dynamics is the plaquette flip and where the equilibrium distribution is the one where each dimer covering occurs with equal probability. How does this appear from a continuum standpoint? The variation of the microscopic configuration with time will appear as a time dependence in the height field $h(\mathbf{r},t)$. The quantity corresponding to $p_\alpha$ is then $P[h(\mathbf{r},t)]$, which is the probability distribution function of the field $h(\mathbf{r},t)$. The presence of the Monte Carlo dynamics will appear as random, sudden, and ultra-local fluctuations of $h(\mathbf{r},t)$. At equilibrium, $P[h] = P_0[h]$ will be proportional to $e^{-S[h]}$, $S[h]$ being given by (17.23).

Of course, this picture is qualitative. In order to proceed with the derivation, we need to determine the correct way to generalize the field $h(\mathbf{r})$, which we were discussing above, to incorporate dynamics. As a guiding principle, we note that for conventional critical points described by a Landau-Ginzburg type theory of a local order parameter the way to do this is to model the dynamics via a Langevin type equation [76]:

$$\frac{dh(\mathbf{r})}{dt} = -\frac{\delta S[h]}{\delta h(\mathbf{r})} + \zeta(\mathbf{r},t) \tag{17.24}$$

where the first term is a generalized damping force that drives $P[h]$ to its equilibrium distribution and $\zeta(\mathbf{r},t)$ is a random noise source, with zero mean and uncorrelated in space and time. A more detailed discussion of the phenomenology behind (17.24) may be found in an introductory text such as [76].

We remind the reader that the RK point is a very *unconventional* critical point and $h(\mathbf{r})$ is not a local order parameter so this procedure is somewhat non-rigorous. Forging ahead, we note that if a field $h$ obeys a Langevin equation, then its probability distribution function $P[h]$ will obey a Fokker–Planck equation.[18] For technical reasons discussed in [57], it is convenient to write the equation in momentum space and in terms of the variable $\Psi[h] = P[h]/(P_0[h])^{1/2}$:

$$\frac{d\Psi[h]}{dt} = -W_h\Psi[h] = -\left[\sum_{\mathbf{q}}\left(-\frac{d}{d\hat{h}_{\mathbf{q}}} + \frac{1}{2}K|\mathbf{q}|^2\hat{h}_{-\mathbf{q}}\right)\left(\frac{d}{d\hat{h}_{-\mathbf{q}}} + \frac{1}{2}K|\mathbf{q}|^2\hat{h}_{\mathbf{q}}\right)\right]\Psi[h] \tag{17.25}$$

Identifying $W_h$ with $H_{RK}$ (see (17.10)) and changing to spatial coordinates, we obtain:

$$H_{RK} = \int d^2r\left[\frac{1}{2}\Pi^2 + \frac{K^2}{2}(\nabla^2 h)^2\right] \tag{17.26}$$

[18] Again, we recommend [76] if this material is not familiar.

where we have rescaled the field $h \to \frac{h}{\sqrt{2}}$; dropped an (infinite) additive constant term; and identified $\Pi = i\frac{d}{dh(\mathbf{r})}$ as the momentum conjugate to $h(\mathbf{r})$. In order to write this as an action, we note that $\Pi = \partial_t h$ so that:

$$S_{RK}[h] = \int d^2r\Big[\frac{1}{2}(\partial_t h)^2 + \frac{K^2}{2}(\nabla^2 h)^2\Big] \tag{17.27}$$

Equation 17.27 is the continuum theory of the RK point. As the theory is Gaussian, the correlations are critical and $K$ can be chosen so the theory reproduces the microscopic correlations, which will depend on the (bipartite) lattice in question. For example, on the square lattice, $K = \frac{\pi}{18}$ while on the honeycomb lattice $K = \frac{\pi}{32}$. Another feature of (17.27) is that there is no penalty if the state has an overall tilt, i.e. if $\nabla h = \boldsymbol{C_0} + \boldsymbol{C_1}(\mathbf{r}, \boldsymbol{t})$ where $\int d^2r\boldsymbol{C_1}(\mathbf{r}, \boldsymbol{t}) = \mathbf{0}$. This corresponds to the extensive winding number degeneracy of bipartite RK points.

A third feature of RK points captured by this action is monomer deconfinement. A monomer is a site without a dimer and in terms of the height mapping, is a site about which the height field increases (or decreases) by a multiple of 4 (for a square lattice) upon a clockwise traverse. In the continuum theory, monomers may be viewed as vortices of the height field. More formally, if $\sigma(\mathbf{r})$ is the density of vortices, then for any closed curve $C$ enclosing an area $A$, we have the relation:

$$\oint_C \nabla h \cdot d\mathbf{r} = 4\int_A d^2r\sigma(\mathbf{r}) \tag{17.28}$$

which in differential form, translates to $\nabla^2 h = 4\sigma(\mathbf{r})$. This is a $2d$ Poisson equation which may be solved to give: $h(\mathbf{r}) = 4\int d^2r'\sigma(\mathbf{r}')\ln|\mathbf{r}-\mathbf{r}'|$. If the action were a conventional Gaussian model, such as (17.23), then this fact would imply that $S[h] \sim \int(\nabla h)^2 \sim \int\int d^2rd^2r'\sigma(\mathbf{r})\sigma(\mathbf{r}')\ln|\mathbf{r}-\mathbf{r}'|$, i.e. the usual logarithmic interaction between vortices. In contrast, the RK action of (17.27) implies:

$$S[h] \sim \int(\nabla^2 h)^2 \sim \int\int d^2rd^2r'\sigma(\mathbf{r})\sigma(\mathbf{r}')\delta(\mathbf{r}-\mathbf{r}') \tag{17.29}$$

which means that at the RK point, vortices, i.e. monomers, are deconfined as required.

The RK point is a critical point separating valence bond crystals living in different topological sectors. This fact motivates the following modification of (17.27) to describe the system *near* the RK point:

$$S_{RK}[h] = \int d^2rdt\Big[\frac{1}{2}(\partial_t h)^2 + \frac{\rho_2}{2}(\nabla h)^2 + \frac{\rho_4}{2}(\nabla^2 h)^2 + \lambda\cos(2\pi h) + \cdots\Big] \tag{17.30}$$

where $\rho_4 = K^2$ and we have explicitly written the cosine term which enforces the discreteness of heights. Here, $\rho_2 = 1-(v/t)$ controls the phase transition. If $v/t = 1$, we recover the RK action (17.27) If $v/t \neq 1$, the $\rho_4$ term is higher order so, at the crudest level, may be ignored. If $v/t > 1$, this action favors a state of maximal tilt, namely the staggered state. If $v/t < 1$, the action prefers a state of minimal tilt as required but there is an additional subtlety which we will return to in a moment.

A nice feature of this construction is that it can be generalized to higher dimensions. To see this, we first view the microscopic problem from yet another perspective. Divide the bipartite lattice into $A$ and $B$ sublattices in the usual way, i.e. each link has one $A$ site and one $B$ site. A dimer covering of the lattice can be interpreted as a lattice magnetic field in the following way. The links containing dimers are vectors with magnitude $z-1$ that point from the $A$ to the $B$ sublattice, where $z$ is the coordination of the lattice. The links without dimers are vectors with unit magnitude that point from the $B$ to the $A$ sublattice.

In this way, the hard core dimer constraint is exactly the condition $\nabla \cdot \mathbf{B} = 0$ as discussed in the previous section. This constraint can be solved by writing $\mathbf{B}$ as the (lattice) curl of a potential. In $2d$, $\mathbf{B} = \nabla \times h$[19] where $h$ is a scalar field: it is precisely the height field we have been working with so far! In higher dimensions, $\mathbf{B} = \nabla \times \mathbf{A}$ where $\mathbf{A}$ is a vector potential defined on the links of the dual lattice.

We learned that the $2d$ winding numbers corresponded to the overall tilts of the height field. In this magnetic analogy, the winding numbers correspond to the magnetic flux through a line or surface.[20]

Equation (17.23) describes a situation where small tilts are favored. By analogy [41,42,74], we may conjecture that the classical dimer problem in three dimensions should be described by a continuum action that favors small magnetic flux:

$$S_C = \frac{K}{2} \int d^3\mathbf{r} (\nabla \times \mathbf{A})^2 \tag{17.31}$$

Using the same procedure as we did to derive (17.30), we can obtain the following action:

$$S = \int d^3x dt \Big[ (\partial_t \mathbf{A})^2 + \rho_2 (\nabla \times \mathbf{A})^2 - \rho_4 (\nabla \times \nabla \times \mathbf{A})^2 \Big] \tag{17.32}$$

$$= \int d^3x dt \Big[ (\mathbf{E})^2 + \rho_2 (\mathbf{B})^2 - \rho_4 (\nabla \times \mathbf{B})^2 \Big] \tag{17.33}$$

where we have used the gauge $\nabla \cdot \mathbf{A} = 0$ and $\rho_2 = 1 - v/t$. Here, $\mathbf{E} = \partial_t \mathbf{A} - \nabla A_0$ and $\mathbf{B} = \nabla \times \mathbf{A}$. In going from (17.32) to (17.33), it is not literally an equality as the field $A_0$ has been included to obtain the most general expression.

[19] To make sense of the curl of a scalar, you can think of it as a vector pointing in the $\hat{z}$ direction, yielding a $\mathbf{B}$ in the 2d XY plane. As $\mathbf{B}$ has two components and there is one constraint, $\nabla \times \mathbf{B} = 0$, a one-component (scalar) field is sufficient to encode all the information. In 3d, one needs two degrees of freedom, which are encoded by the vector potential $\mathbf{A}$ (3 components) minus a local gauge transformation.

[20] Draw a line, or surface, passing through the links of the direct lattice. The magnetic flux is defined as the net magnetic field through the surface, i.e. the sum of the fields on the pierced links.

The reader is invited to verify that (17.33) captures the salient features of a bipartite RK point including the extensive degeneracy and deconfinement. Equation (17.33) is precisely the action of quantum electrodynamics except that $\mathbf{B}$ is restricted in the range of values it can take. For this reason, (17.33) is referred to as a theory of compact QED in $3+1$ dimensions and similarly (17.30) may be viewed as a compact QED in $2+1$.

We now return to the issues raised in our preliminary discussion of bipartite RK points. Once again, if $v/t \neq 1$, we may ignore $\rho_4$ in (17.33). When $v/t < 1$, (17.33) becomes exactly the Maxwell action with a photon described by the dispersion $\omega = (1 - v/t)k$. This is the origin of the $1/r$ interaction between monomers in the U(1) RVB phase. The force is carried by a "photon" whose speed vanishes at the RK point. Thus, at the RK point, the force itself vanishes and the dispersion becomes quadratic. In contrast, in two dimensions, when $v/t < 1$, the $\lambda$ term in (17.30) becomes relevant. Therefore, when $v/t < 1$, the system is not described by a Gaussian theory but is driven into a crystalline phase. Therefore, unlike in 2d, the RK is the endpoint of a phase in three dimensions.

Finally, just as a monomer can be understood as a local violation of the $\nabla \cdot \mathbf{B} = 0$ constraint, in the $3d$ case, we can also consider violations of $\nabla \cdot \mathbf{E} = 0$. These excitations are called electric monopoles and are analogous to visons in the $Z_2$ RVB case. Just like visons, these excitations involve only dimers.

## 17.9 Numerical Methods

In common with other models of quantum matter, quantum dimer models are in principle amenable to study by perhaps the most general method, namely exact diagonalisation of finite-size systems, a method described in the contribution by Laeuchli in this volume.

An asset in the present context is the somewhat slower – yet still exponential – growth of the Hilbert space with system size. One can think of this roughly as follows: whereas a system of spins $S = 1/2$ has a Hilbert space of dimension $2^N$, the imposition of a local constraint effectively reduces that dimensionality. Therefore, whereas the current upper limit on the size of spin systems hovers around 40 sites, for constrained models such as the quantum six-vertex model [77, 78], significantly larger systems can be studied.

A further very significant help is the existence of the RK point, at which the properties of the ground state wavefunction can be studied by resorting to *classical* methods, in particular standard Monte Carlo simulations where analytical treatments are not available. These enable the study of systems very large compared to those accessible to quantum Monte Carlo studies. In particular, this allows the study of higher-dimensional models: for instance, the RK point of the cubic lattice QDM has been studied for a system with over 32000 sites [40, 41].

Given the RK-QDMs are in general constructed not to exhibit a sign problem, they are also in principle amenable to efficient numerical study using quantum Monte Carlo (rather than exact diagonalisation) methods away from the RK point.

In some cases, obtaining information on ground-state properties can be surprisingly straightforward: for instance, for the hexagonal QDM, a duality mapping onto the triangular Ising model permits a very simple quantum Monte Carlo study.

More generally, however, the constrained nature of the degrees of freedom can be hard to deal with in the numerics. The problem arises as follows. When sampling the partition function stochastically,

$$Z = \mathrm{Tr}\ \exp(-\beta H) \tag{17.34}$$

the presence of the Trace requires the imposition of periodic boundary conditions in imaginary time on the degrees of freedom in a Monte Carlo simulations.

If the quantum dynamics involves flipping single spins, such as in a transverse field Ising model, this translates into the simple and maximally local constraint that each spin be flipped an even number of times. This constraint is naturally observed, e.g., by an algorithm flipping clusters of spins in the imaginary time direction.

However, for plaquette moves as typically arise in QDMs, this is not usually possible – demanding that each plaquette flip an even number of times is more restrictive than just imposing periodic boundary conditions on the microscopic degrees of freedom.

One recent piece of progress has been the realisation that diffusion Monte Carlo (also known as Green function Monte Carlo) can be used to resolve this issue [79]. A description of this is beyond the scope of the current article, and we refer the reader to the relevant articles in the literature [80].

## 17.10 Dimer Phases in SU(2) Invariant Models

In this section, we show how QDMs can arise from local, SU(2)-invariant spin Hamiltonians where the physics is dominated by nearest-neighbor valence bonds. The resulting dimer phases may then be interpreted as valence bond crystals and liquids. The derivation involves two steps. The first is to provide a mechanism that justifies the truncation of the full Hilbert space to the much smaller nearest-neighbor valence bond manifold. The second step is to show how the Hamiltonian in (17.1) can arise as the effective description of a spin Hamiltonian in this truncated Hilbert space. We consider the second step first.

The transcription of QDM results into spin language is not an entirely trivial matter because of two essential differences between dimers and the singlets they represent. The first point is that while a dimer connecting sites 1 and 2 has no orientation, specifying a singlet bond between the spins requires a choice of sign: $\pm\frac{1}{\sqrt{2}}(1_\uparrow 2_\downarrow - 1_\downarrow 2_\uparrow)$. The second point is that while dimer coverings of the lattice were taken as orthonormal basis vectors *by construction*, any two nearest-neighbor valence bond coverings will *always* have nonzero overlap. In fact, whether the collection of nearest-neighbor valence bond states is linearly independent depends on the lattice geometry. The current understanding is that nearest-neighbor valence

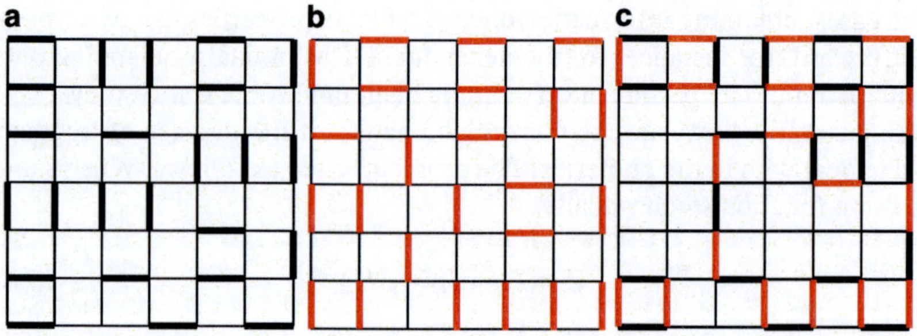


**Fig. 17.11** The transition graph (**c**) of valence bond coverings (**a**) and (**b**). In this example, the magnitude of the overlap between the two states is $|S_{ab}| = 2^3 x^{38} = \frac{1}{2^{16}}$

bond coverings of the square, honeycomb, triangular, and kagomé lattices are linearly independent for sufficiently large lattices.[21] However, linear independence will break down for highly interconnected lattices.[22]

The non-orthogonality of two valence bond coverings is most conveniently discussed in terms of their *transition graph*. As shown in Fig. 17.11, this construction involves overlaying two valence bond coverings $|a\rangle$ and $|b\rangle$ and eliminating the shared bonds. The resulting figure contains closed loops of varying even lengths. It can be shown that the magnitude of the overlap matrix element $S_{ab} = \langle a|b\rangle$ is given by $|S_{ab}| = 2^{N_l} x^{L_l}$ where $N_l$ is the number of loops, $L_l$ is the sum of the lengths of the loops, and $x = \frac{1}{\sqrt{2}}$.[23] For a large lattice, a typical transition graph will involve many long loops so the overlap between arbitrary states, though never zero, will usually be very small.

### *17.10.1 Overlap Expansion*

The *overlap expansion* is motivated by this latter observation that the states are "almost orthogonal". The idea is to treat the overlap factor $x$, which is actually $\frac{1}{\sqrt{2}}$, as if it were a small expansion parameter. For the square lattice, we may choose a sign convention for singlets so that the overlap between any two states differing by only a single (minimal) loop of length 4 is always $-2x^4$. In terms of the overlap expansion, the overlap matrix for a square lattice may then be written as:

$$S_{ab} = \delta_{ab} - 2x^4 \Box_{ab} + O(x^6) \tag{17.35}$$

[21] However, rigorous proofs currently exist only for finite-sized square and honeycomb lattices with open boundary conditions.

[22] For example, consider four spins on the corners of a square plaquette with two additional links connecting opposite corners. This lattice has two linearly independent nearest-neighbor valence bond states but three dimer coverings.

[23] The sign of $S_{ab}$ depends on the sign convention for labeling singlets discussed earlier.

where $\Box_{ab}$ is unity if the states $|a\rangle$ and $|b\rangle$ differ by a single (minimal) loop of length 4 and zero otherwise.

Now consider the following spin Hamiltonian:

$$\delta H = J\sum_{\langle ij\rangle} \boldsymbol{s}_i \cdot \boldsymbol{s}_j + v\sum_{\Box}\Big((\boldsymbol{s}_1\cdot\boldsymbol{s}_2)(\boldsymbol{s}_3\cdot\boldsymbol{s}_4) + (\boldsymbol{s}_1\cdot\boldsymbol{s}_3)(\boldsymbol{s}_2\cdot\boldsymbol{s}_4)\Big) \qquad (17.36)$$

where the first sum is over nearest neighbors and the second sum over all square plaquettes. We would like to write this as an effective operator that acts on the nearest-neighbor valence bond manifold. The first step is to form an orthogonal basis: $|\alpha\rangle = \sum_i S^{-1/2}_{\alpha i}|i\rangle$. In terms of the overlap expansion, $(S^{-1/2})_{ab} = \delta_{ab} + x^4\Box_{ab} + O(x^6)$ so the state $|\alpha\rangle$ can be labelled in terms of its order unity component. In terms of the basis $\{|\alpha\rangle\}$, (17.36) is:

$$\begin{aligned} H_{\alpha\beta} &= (S^{-1/2}\delta H S^{-1/2})_{\alpha\beta} = \sum_{ij}(S^{-1/2})_{\alpha i}\langle i|\delta H|j\rangle(S^{-1/2})_{j\beta} \\ &= -t\Box_{\alpha\beta} + vn_{fl,\alpha}\delta_{\alpha\beta} + O(vx^4 + tx^2) \end{aligned} \qquad (17.37)$$

where $t = Jx^4$ and $n_{fl,\alpha}$ is the number of flippable plaquettes contained in (the order unity component of) state $|\alpha\rangle$. Equation (17.37) is just the RK-QDM!

Of course, these are formal manipulations and in reality, $x = \frac{1}{\sqrt{2}}$ so the error terms are not small compared to the leading terms. As discussed in [81], the exponent of the error term $vx^4$ comes from the length of the minimal loop, which for the square lattice has length 4, while the exponent of the error term $tx^2$ comes from the difference in lengths of the minimal and next minimal loops, which for the square lattice gives $6 - 4 = 2$. Therefore, one may expect the overlap expansion to work comparatively well for lattice architectures that give larger values for these exponents.

### *17.10.2 Decoration*

The most straightforward way to fix this problem is to modify the problem by considering a *decorated* square lattice where an even number $N$ of sites have been added to each link (Fig. 17.12) [81]. Equation 17.36 is also slightly modified:

$$\delta H = J\sum_{\langle ij\rangle} \boldsymbol{s}_i \cdot \boldsymbol{s}_j + v\sum_{\Box}\Big((\boldsymbol{s}_1\cdot\boldsymbol{s}_{b_1})(\boldsymbol{s}_2\cdot\boldsymbol{s}_{b_2}) + (\boldsymbol{s}_1\cdot\boldsymbol{s}_{a_1})(\boldsymbol{s}_3\cdot\boldsymbol{s}_{a_3})\Big) \qquad (17.38)$$

where the labels refer to Fig. 17.12. The dimer coverings of the decorated lattice correspond exactly to those of the undecorated lattice where an occupied (empty) link in the latter case corresponds to a chain of dimers where the endpoints are (are

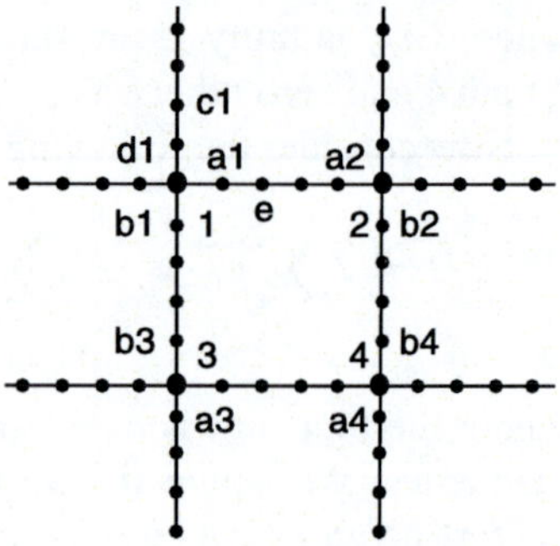


**Fig. 17.12** Decorated square lattice for the case $N = 4$

not) included. Therefore, the overlap expansion will give the same effective dimer model at leading order.[24] However, by decorating the lattice the length of a minimal loop has increased from 4 to $4(N + 1)$ so the error term is reduced from $O(x^2)$ to $O(x^{2N+1})$.

Therefore, by decorating sufficiently, the QDMs can be realized to arbitrary accuracy and the procedure can be adapted to any lattice and also higher dimensions. While fine-tuned features, such as critical points, will only be captured in the limit of infinite decoration, a finite decoration should be sufficient to realize the gapped phases, including the RVB liquid of the triangular lattice.[25]

### *17.10.3 Large-N*

An alternative way of suppressing the overlap between dimer states is to endow each dimer with an additional internal flavour variable, which can take on integer values between 1 and $N$. This can be achieved by following a line of investigation which was initiated by Arovas and Auerbach with their study of Schwinger Boson mean-field theory [83], which was then formalised in a series of papers by Read and Sachdev, where the small parameter justifying a mean-field treatment was provided by the inverse of the number of flavours, $1/N$, see e.g. [84].

This route provides a simple quantum dimer model with only a kinetic term at leading order in $1/N$. An analysis of this model for the pyrochlore lattice has found only a partially ordered dimer crystal [85], the final ordering pattern of which at higher order is at present not known, although a large-unit cell solid is a plausible outcome [86].

[24] In decorated case, $t$ is now related to $J$ in (17.38) by $t = Jx^{4(N+1)}$.

[25] We direct the interested reader to [82], where a different but complementary approach was used to construct SU(2) invariant analogs of the square and honeycomb lattice RK points. We point out that the procedure of [82] should work for any lattice where a Klein model is known to exist, which includes the decorated lattices just discussed.

### 17.10.4 Klein Models: SU(2) Invariant Spin Liquids

Having discussed how (17.1) can arise as an effective model in the nearest-neighbor valence bond subspace, we return to the more fundamental question of how such an effective subspace can arise in an SU(2)-invariant, local, spin model. The following construction, originally due to Klein, is one such route [81, 82, 87–90]. For an arbitrary lattice $\Lambda$, consider the following Hamiltonian:

$$H_0 = \sum_{i \in \Lambda} \alpha_i \hat{h}_i \tag{17.39}$$

where $\alpha_i$ is a positive constant that may, in principle, vary with $i$. $\hat{h}_i$ is an operator that projects the cluster formed by spin $i$ and its nearest-neighbors onto its highest spin sector. For example, on the square lattice, $\hat{h}_i$ projects the five spin cluster of site $i$ and its four neighbors onto the $S = 5/2$ state. On the decorated square lattice (Fig. 17.12), (17.39) includes $S = 3/2$ projectors for each link site and $S = 5/2$ projectors for each corner site. The SU(2)-invariance of the $\hat{h}$ operators may be seen by writing them out explicitly. For example, referring to the figure, $\hat{h}_1 = [S^2 - (\frac{1}{2})(\frac{3}{2})][S^2 - (\frac{3}{2})(\frac{5}{2})]$ where $\boldsymbol{S} = \boldsymbol{S}_1 + \boldsymbol{S}_{a1} + \boldsymbol{S}_{b1} + \boldsymbol{S}_{c1} + \boldsymbol{S}_{d1}$ and so on.

Being a sum of projection operators, (17.39) only has non-negative eigenvalues. If spin $i$ is in a singlet with one of its neighbors, then the cluster of spin $i$ and its neighbors has zero projection in its highest spin sector. Hence $\hat{h}_i$ will annihilate such a state any nearest-neighbor valence bond covering of the lattice will be a zero energy ground state of (17.39). Depending on the lattice, (17.39) may also have ground states outside of the nearest-neighbor valence bond manifold.[26] Another issue is whether there is a spin gap separating the nearest-neighbor valence bond manifold from magnetic states. These issues are not trivial to answer [92] but for a sufficiently decorated lattice, we may appeal to the well known [93] result that a Majumdar–Ghosh chain has a twofold degenerate, spin-gapped ground state. The size of the spin gap in the case is determined by the smallest of the $\alpha_i$.

The Hamiltonians made up of perturbed Klein models can thus be used to obtain – in a controlled way – SU(2) invariant gapped spin liquids, the existence of which had been in doubt for many years.

## 17.11 Outlook

The field of quantum dimer models is now so rich that it has become impossible to give a comprehensive and comprehensible summary even in an extended set of notes such as this one. In closing, we would like to provide a few pointers to interesting developments not covered so far.

[26] For a finite size honeycomb lattice with open boundary conditions, there is a rigorous proof that the nearest-neighbor valence bond manifold spans the ground state space [88]. Also, we direct the reader to [91], where this question is explored for a class of generalized Klein models.

### *17.11.1 Hopping Fermions . . .*

As mentioned above, the natural quantum dynamics chosen involves the smallest clusters compatible with the local constraints. However, if one thinks of a dimer as representing a Fermionic particle, there is a simple selection rule: plaquette moves involving an even number of Fermions (with identical internal quantum numbers) have zero matrix elements because they can be achieved by the Fermions hopping clockwise or counterclockwise. These two differ by an overall sign as one results in a final state which is an odd permutation of the other [94].

The natural quantum dynamics thus involves three Fermions (one being excluded by the local constraint). Interestingly, the matrix elements can sometimes still be chosen to avoid a sign problem, [95] so that the generic features of the resulting models are rather like those of the bosonic ones.

Additional structure appears, however, when the Fermions exhibit in addition a non-trivial internal degree of freedom, such as spin: for Fermions of opposite spin, the abovementioned cancellation does not take place, and one thus obtains a spin-dependent quantum dynamics (not entirely unlike Anderson superexchange) [96]. Detailed phase diagrams of this class of models are still being worked out at present.

### *17.11.2 . . . and much more*

Dimer models with dynamical defects (such as holons and spinons) have e.g. been studied in [97, 98]. Supersymmetric examples of quantum dimer models have been proposed in [99]. Dimer-type models with non-Abelian properties are also attracting a great of interest at the moment, such as in loop models [26] or in the 'Golden Chain' [100], which amusingly enough is related [101] to a simple two-leg dimer ladder [20]. An intriguing dimer model with a huge ground-state degeneracy has been uncovered in [102] Finally, there is a burgeoning literature on 'non-Landau' phase transition, or to be more precise phase transitions out of the new liquid phases discussed above [103–106]: there is still much to look forward to in this field.

**Acknowledgements**

The authors are grateful to innumerable members of the community for many useful and enlightening discussions; several of these are authors of other chapters of this book. We are of course particularly indebted to our collaborators, too numerous to list here, together with who we have developed our understanding of this field, in particular of course Shivaji Sondhi, with who both of us have extensively collaborated.

## References


1. D.S. Rokhsar, S.A. Kivelson, Phys. Rev. Lett. 61, 2376 (1988)
2. P.W. Anderson, Science 235, 1196 (1987)
3. P. Fazekas, P.W. Anderson, Philos. Mag. 30, 23 (1974)
4. S.A. Kivelson, D.S. Rokhsar, J.P. Sethna, Phys. Rev. B. 35, R8865 (1987)
5. R. Moessner, S.L. Sondhi, arXiv:cond-mat/0205029
6. J.D. Bernal, R.H. Fowler, J. Chem. Phys. 1, 515 (1933)
7. P.W. Anderson, Phys. Rev. 102, 1008 (1956)
8. P. Fulde, N. Shannon, K. Penc, Annalen der Physik 11, 892 (2002)
9. G. Murthy, D. Arovas, A. Auerbach, Phys. Rev. B 55, 3104 (1997)
10. S. Wessel, M. Troyer, Phys. Rev. Lett. 95, 127205 (2005)
11. D. Heidarian, K. Damle, Phys. Rev. Lett. 95, 127206 (2005)
12. R.G. Melko, A. Paramekanti, A.A. Burkov, A. Vishwanath, D.N. Sheng, L. Balents, Phys. Rev. Lett. 95, 127207 (2005)
13. S.R. Hassan, R. Moessner, Phys. Rev. B 73, 094443 (2006)
14. L. Balents, M.P.A. Fisher, S.M. Girvin, Phys. Rev. B 65, 224412 (2002)
15. M.E. Zhitomirsky, H. Tsunetsugu, Phys. Rev. B 70, 100403 (2004)
16. O. Derzhko, J. Richter, Phys. Rev. B 70, 104415 (2004)
17. S.V. Isakov, K.S. Raman, R. Moessner, S.L. Sondhi, Phys. Rev. B 70, 104418 (2004)
18. R.J. Baxter, Exactly Solved Models in Statistical Mechanics (Academic Press, London, New York, 1982)
19. J.B. Kogut, Rev. Mod. Phys. 51, 659 (1979)
20. R. Moessner, S.L. Sondhi, Phys. Rev. B 63, 224401 (2001)
21. J. von Delft, C.L. Henley, Phys. Rev. B 48, 965 (1993)
22. D. Bergman, R. Shindou, G.A. Fiete, L. Balents, Phys. Rev. B 75, 094403 (2007)
23. A. Sen, K. Damle, A. Vishwanath, Phys. Rev. Lett. 100, 097202 (2008)
24. P.W. Kasteleyn, Physica 27, 1209 (1961)
25. M.E. Fisher, Phys. Rev. 124, 1664 (1961)
26. P. Fendley, arXiv:0711.0014
27. R. Moessner, S.L. Sondhi, Phys. Rev. Lett. 86, 1881 (2001)
28. X.G. Wen, Q. Niu, Phys. Rev. B 41, 9377 (1990)
29. G. Moore, N. Read, Nucl. Phys. B 360, 362 (1991)
30. R.B. Laughlin, Rev. Mod. Phys. 71, 863 (1999)
31. R. Rajaraman, arXiv:cond-mat/0103366
32. S. Sachdev, Phys. Rev. B 40, 5204 (1989)
33. P.W. Leung, K.C. Chiu, K.J. Runge, Phys. Rev. B 54, 12938 (1996)
34. R. Moessner, S.L. Sondhi, P. Chandra, Phys. Rev. B. 64, 144416 (2001)
35. A. Ralko, M. Ferrero, F. Becca, D. Ivanov, F. Mila, Phys. Rev. B 71, 224109 (2005)
36. G. Misguich, D. Serban, V. Pasquier, Phys. Rev. Lett. 89, 137202 (2002)
37. T. Senthil, M.P.A. Fisher, Phys. Rev. B 62, 7850 (2000)
38. D. Ivanov, Phys. Rev. B. 70, 094430 (2004)
39. S. Kivelson, Phys. Rev. B 39, 259 (1989)
40. D.A. Huse, W. Krauth, R. Moessner, S.L. Sondhi, Phys. Rev. Lett. 91, 167004 (2003)
41. R. Moessner, S.L. Sondhi, Phys. Rev. B 68, 184512 (2003)
42. M. Hermele, M.P.A. Fisher, L. Balents, Phys. Rev. B 69, 064404 (2004)
43. R. Moessner, S.L. Sondhi, E. Fradkin, Phys. Rev. B 65, 024504 (2002)
44. T. Senthil, A. Vishwanath, L. Balents, S. Sachdev, M.P.A. Fisher, Science 303, 1490 (2004)
45. A. Sandvok, Phys. Rev. Lett. 98, 227202 (2007)
46. R.G. Melko, R.K. Kaul, Phys. Rev. Lett. 100, 017203 (2008)
47. O.F. Syljuasen, Phys. Rev. B. 73, 245105 (2006)
48. A. Ralko, D. Poilblanc, R. Moessner, Phys. Rev. Lett. 100, 037201 (2008)
49. C. Zeng, V. Elser, Phys. Rev. B 51, 8318 (1995)
50. P. Nikolic, T. Senthil, Phys. Rev B 68, 214415 (2003)

51. R. Moessner, S.L. Sondhi, P. Chandra, Phys. Rev. Lett. 84, 4457 (2000)
52. E. Fradkin, D.A. Huse, R. Moessner, V. Oganesyan, S.L. Sondhi, Phys. Rev. B 69, 224415 (2004)
53. A. Vishwanath, L. Balents, T. Senthil, Phys. Rev. B 69, 224416 (2004)
54. S. Papanikolaou, K.S. Raman, E. Fradkin, Phys. Rev. B 75, 094406 (2007)
55. A.M. Polyakov, Nucl. Phys. B 120, 429 (1977)
56. S. Aubry, in Solitons in Condensed Matter Physics, eds. A. Bishop, T. Schneider, vol. 264, (Springer, Berlin, 1978)
57. C.L. Henley, J. Stat. Phys. 89, 483 (1997)
58. C.L. Henley, J. Phys.: Condens. Matter 16, S891 (2004)
59. E. Ardonne, P. Fendley, E. Fradkin, Ann. Phys. 310, 493 (2004)
60. C. Castelnovo, C. Chamon, C. Mudry, P. Pujol, Ann. Phys. 318, 316 (2005)
61. P. Fendley, R. Moessner, S.L. Sondhi, Phys. Rev. B 66, 214513 (2002)
62. A. Ioselevich, D.A. Ivanov, M.V. Feigelman, Phys. Rev. B 66, 174405 (2002)
63. M. Freedman, C. Nayak, K. Shtengel, Phys. Rev. B 78, 174411 (2008)
64. M.B. Hastings, arXiv:cond-mat/0011125
65. R. Moessner, S.L. Sondhi, Phys. Rev. B 68, 064411 (2003)
66. A. Laeuchli, S. Capponi, F.F. Assaad, J. Stat. Mech. (2008) P01010
67. S. Sachdev, Phys. Rev. B 45, 12377 (1992)
68. O. Motrunich, Phys. Rev. B 67, 115108 (2003)
69. C. Xu, Phys. Rev. B 74, 224433 (2006)
70. S. Pankov, R. Moessner, S.L. Sondhi, Phys. Rev. B 76, 104436 (2007)
71. K. Penc, M. Mambrini, P. Fazekas, F. Mila, Phys. Rev. B 68, 012408 (2003)
72. D.P. Arovas, unpublished.
73. C. Zeng, C.L. Henley, Phys. Rev. B 55, 14935 (1997)
74. C.L. Henley, Phys. Rev. B 71, 014424 (2005)
75. K. Shtengel, private communication
76. P. Chaikin, T. Lubensky, Principles of Condensed Matter Physics (Cambridge University Press, Cambridge, 1995)
77. R. Moessner, O. Tchernyshyov, S.L. Sondhi, J. Stat. Phys. 116, 755 (2004)
78. N. Shannon, K. Penc, G. Misguich, Phys. Rev. B 69, 220403(R) (2004)
79. M. Troyer, private communication (2004)
80. See [35] and the references therein
81. K.S. Raman, R. Moessner, S.L. Sondhi, Phys. Rev. B 72, 064413 (2005)
82. S. Fujimoto, Phys. Rev. B 72, 024429 (2005)
83. A. Auerbach, D.P. Arovas, Phys. Rev. Lett. 61, 617 (1988)
84. N. Read, S. Sachdev, Nucl. Phys. B 316, 609 (1989)
85. R. Moessner, S.L. Sondhi, M.O. Goerbig, Phys. Rev. B 73, 094430 (2006)
86. E. Berg, E. Altman, A. Auerbach, Phys. Rev. Lett. 90, 147204 (2003)
87. D.J. Klein, J. Phys. A. Math. Gen. 15, 661 (1982)
88. J.T. Chayes, L. Chayes, S.A. Kivelson, Commun. Math. Phys. 123, 53 (1989)
89. C. Batista, S.A. Trugman, Phys. Rev. Lett. 93, 217202 (2004)
90. Z. Nussinov, C.D. Batista, B. Normand, S.A. Trugman, Phys. Rev. B 75, 094411 (2007)
91. Z. Nussinov, arXiv:cond-mat/0606075
92. I. Affleck, T. Kennedy, E.H. Lieb, H. Tasaki, Phys. Rev. Lett. 59, 799 (1987)
93. B.S. Shastry, B. Sutherland, Phys. Rev. Lett. 47, 964 (1981)
94. F. Pollmann, P. Fulde, E. Runge, Phys. Rev. B 73, 125121 (2006)
95. F. Pollmann, J.J. Betouras, K. Shtengel, P. Fulde, Phys. Rev. Lett. 97, 170407 (2006)
96. D. Poilblanc, K. Penc, N. Shannon, Phys. Rev. B 75, 220503(R) (2007)
97. L. Balents, L. Bartosch, A. Burkov, S. Sachdev, K. Sengupta, Phys. Rev. B 71, 144509 (2005)
98. D. Poilblanc, F. Alet, F. Becca, A. Ralko, F. Trousselet, F. Mila, Phys. Rev. B 74, 014437 (2006)
99. P. Fendley, K. Schoutens, J. de Boer, Phys. Rev. Lett. 90, 120402 (2003)
100. A. Feiguin, S. Trebst, A.W.W. Ludwig, M. Troyer, A. Kitaev, Z. Wang, M.H. Freedman, Phys. Rev. Lett. 98, 160409 (2007)

101. A. Laeuchli, private communication
102. G. Misguich, D. Serban, V. Pasquier, Phys. Rev. B 67, 214413 (2003)
103. O. Motrunich, A. Vishwanath, Phys. Rev. B 70, 075104 (2004)
104. F. Alet, G. Misguich, V. Pasquier, R. Moessner, J.L. Jacobsen, Phys. Rev. Lett. 97, 030403 (2006)
105. T.E. Saunders, J.T. Chalker, Phys. Rev. Lett. 98, 157201 (2007)
106. O. Motrunich, T. Senthil, Phys. Rev. B 71, 125102 (2005)

# Chapter 18
# Numerical Simulations of Frustrated Systems

**Andreas M. Läuchli**

**Abstract** In many fields of science, computational approaches have become one of the central cornerstones alongside experimental and theoretical approaches. This is also the case in the field of frustrated magnetic systems, where theory, simulations, and experiments drive each other mutually, thus advancing our overall understanding. This chapter aims to provide an overview of the most common numerical techniques available for strongly correlated lattice models, with an emphasis on frustrated spin systems. While it is not possible to treat all methods in sufficient depth to provide a comprehensive introduction, here the key ideas are presented and special issues regarding their application to frustrated systems are discussed. The power of these techniques is illustrated by examples taken from the literature,[1] and specific references to appropriate detailed presentations are included where possible.

## 18.1 Overview of Methods

In this chapter, I present on a general level both classical and quantum Monte Carlo methods, series-expansion techniques, and the density-matrix renormalization-group approach, and illustrate their utility by selected applications. I then treat in rather more detail the exact-diagonalization technique, as this is not documented so extensively in the literature. I close by touching briefly on a small number of methods which are not so widespread and by providing some links to available source codes.

## 18.2 Classical Monte Carlo

The Monte Carlo technique is a probabilistic approach applicable to a large class of statistical physics problems. The method performs a stochastic evaluation of the expectation values of observables $\mathcal{O}$, based for example on the canonical ensemble

[1] The choice of examples here is subjective, and unfortunately cannot be exhaustive.

for a classical system,

$$\langle \mathcal{O} \rangle_\beta = \sum_i p_i(\beta) \mathcal{O}_i, \tag{18.1}$$

with probability distribution

$$p_i(\beta) = \exp(-\beta E_i)/Z(\beta) \tag{18.2}$$

and

$$Z(\beta) = \sum_i \exp(-\beta E_i), \tag{18.3}$$

where $\beta = 1/T$ ($k_B \equiv 1$) and the sum $\sum_i$ represents the summation or integration over all configurations (with energy $E_i$) of the system. A direct summation is in general impossible due to the exponentially large number of terms as a function of system size.[2] However, the exponential suppression of the probabilities $p_i(\beta)$ with increasing energy suggests that not all configurations among the enormous number of possible states are really important. The basis of the Monte Carlo approach is to perform a stochastic sampling of system configurations based on a "random walk" in the configuration space. More precisely, one generates a Markov process of configurations with limiting distribution $p_i(\beta)$.

The Metropolis [1] algorithm provides a simple but very powerful prescription for performing a Markov process with the desired properties:[3]

1. Start with a configuration $c_n = i$ with energy $E_i$.
2. Based on some prescription $M_{i \to j}$ (to be discussed below), propose a new configuration $j$.
3. Accept the configuration $j$ as $c_{n+1}$ with probability

$$P_{\text{accept}} = \min\left\{1, \frac{M_{j \to i}}{M_{i \to j}} \frac{p_j(\beta)}{p_i(\beta)}\right\}, \tag{18.4}$$

   otherwise keep $i$ as $c_{n+1}$.
4. Increase $n$ by one.
5. If the system is thermalized, measure observable $\mathcal{O}_{n+1}$ in the new configuration $c_{n+1}$.
6. If more statistics are needed then go back to step (2), otherwise stop.
7. The average value of the observable is $\langle \mathcal{O} \rangle_\beta \approx \frac{1}{N_{\text{meas}}} \sum_n \mathcal{O}_n$.

One of the key factors responsible for the success of the Metropolis algorithm is the fact that the a priori unknown partition function $Z(\beta)$ is eliminated from the problem by considering the probability ratio $p_j(\beta)/p_i(\beta)$. In this form the algorithm

[2] Even for a simple $10 \times 10$ Ising model one would have to sum 1 267 650 600 228 229 401 496 703 205 376 terms.

[3] Here presented in the Metropolis–Hastings form.

(known as "Metropolis–Hastings") is rather general, although some of its complexity is concealed in the prescription for the process $M_{i \to j}$. In many cases one uses a symmetric trial prescription, $M_{i \to j} = M_{j \to i}$, which simplifies the acceptance step in that one always accepts trial configurations which lower the energy, while configurations with higher energy are accepted with probability $\exp[-\beta(E_j - E_i)]$. Other choices sometimes used are the heat-bath algorithm (for discrete states) and overrelaxation (for continuous variables) [2]. There is substantial freedom in the choice of $M_{i \to j}$ in the algorithm, but the full transition probability $W_{i \to j}$ must satisfy elementary requirements such as detailed balance ($p_i \times W_{i \to j} = p_j \times W_{j \to i}$) and ergodicity (each configuration $j$ is reachable from $i$ in a finite number of steps) in order to be physically valid.

Probably the most famous prescription is the single-spin-flip algorithm for the classical Ising model. In this context, a configuration $i$ consists of the state of the spins on each lattice site,

$$i = \{\sigma_1, \ldots, \sigma_{N_{\text{sites}}}\}.$$

From this one selects a site $s$ (randomly or sequentially) and flips its spin to give the new trial configuration

$$j = \{\sigma_1, \ldots, \hat{\sigma}_s = -\sigma_s, \ldots, \sigma_{N_{\text{sites}}}\}.$$

This is accepted as a new configuration of the Markov chain according to the simplified probability $\min\{1, \exp(-\beta(E_j - E_i)]\}$ (above). By using this simple spin-flip algorithm, one may already explore a broad range of classical spin systems, including those with competing interactions [3–6].

Close to second-order phase transitions, the phenomenon known as "critical slowing-down" – where large blocks of the system become correlated – sets in, and this increases severely the numerical effort required to obtain high-quality Monte Carlo data using single-spin-flip algorithms. This problem was solved for a certain class of models with the invention of powerful cluster algorithms in the formulations of Swendsen and Wang [7] and of Wolff [8]. These algorithms act to eliminate the slowing-down by flipping clusters of spins which are generated dynamically and whose sizes track the divergence of the correlation length at the transition. Cluster algorithms have enabled high-precision MC studies of critical phenomena in ferromagnetic Ising, Heisenberg, and related models which deliver impressive accuracies.

No general cluster algorithms are currently known for frustrated systems. The cluster algorithms mentioned above are in fact formally applicable also in this case [9], but are often unable to track the true, physical correlation length – which can, for example, be hidden in emergent degrees of freedom – and thus perform poorly when applied to frustrated systems. However, for specific frustrated models some very powerful methods exploiting nonlocal updates have been found. As one example, in the physics of classical interacting dimer models at finite temperature one finds both interesting phases and unexpected phase transitions [10–13].

A second challenge for local-update algorithms is posed by first-order transitions where the tunneling times between the two phases at the transition become prohibitively large (due to the extensive tunneling barrier) as the system size increases, or when the low-temperature energy landscape is very complicated, as is typical in frustrated systems. In recent years this problem has almost been solved by extended-ensemble simulation techniques, where one goes beyond the simulation of a single system at a given temperature; ways to effect this include simulating directly in energy space (for example the Wang-Landau algorithm [14]) or simulating many systems at different temperatures in parallel and allowing swaps of configurations between the temperatures. This latter is known as the "parallel tempering" or "exchange Monte Carlo" method, and was pioneered in the context of spin glasses [15]. The key idea is that for a finite system the energy distribution at a given temperature $T$ has a finite width, and if two temperatures are sufficiently close that the distributions overlap, it can be energetically beneficial to exchange two configurations at adjacent temperatures. Generalizing to more temperatures, the configurations can diffuse in temperature space, and measurements performed at a given temperature are then both much better equilibrated and have lower autocorrelation times. In initial studies by this approach, the temperatures set in the simulations were chosen heuristically, but a feedback-optimized variant of parallel tempering has been put forward more recently [16] which determines automatically a given number of temperatures in order to minimize the round-trip time for a configuration from high to low temperature and back. Over the last five years, the parallel-tempering technique has become increasingly popular for frustrated spin systems because of its improved equilibration properties at low temperature. Applications studied include vector spins on the FCC and $J_1 - J_2$ diamond lattices [17, 18].

Finally, it is important to stress that each and every Monte Carlo simulation requires a complete error analysis. First, a reasonably long thermalization phase is required, during which the Markov process should lose any memory of the initial state. Then, the autocorrelation times of all observables must be tracked carefully, because these enter the estimator of the error bars. One method to analyze the errors and autocorrelation times of a time series is the binning technique, while for more complicated quantities, such as the ratio of two random variables (as used, for example, in the Binder cumulant in studies of critical phenomena), a jackknife analysis is useful. I refer here to the ALPS library, where the binning and jackknife tools are already implemented and can be included easily in the reader's own codes.

## Further Literature


1. D.P. Landau and K. Binder, *A Guide to Monte Carlo Simulations in Statistical Physics*, (Cambridge University Press, Cambridge, 2000).
2. W. Krauth, *Statistical Mechanics: Algorithms and Computations*, (Oxford University Press, Oxford, 2006).

3. W. Janke, *Monte Carlo Methods in Classical Statistical Physics*, Lect. Notes Phys. 739, 79 (2008).
4. S. Trebst, D.A. Huse, E. Gull, H.G. Katzgraber, U.H.E. Hansmann, and M. Troyer, *Ensemble Optimization Techniques for the Simulation of Slowly Equilibrating Systems*, Springer Proceedings in Physics, Volume 115, Eds. D.P. Landau, S.P. Lewis, and H.-B. Schuettler (2007).

## 18.3 Quantum Monte Carlo

For classical systems, the Boltzmann weight provides a direct probabilistic interpretation and a Monte Carlo approach is conceptually simple. In quantum statistical mechanics, the partition function is

$$Z(\beta) = \mathrm{Tr}\exp[-\beta H] \tag{18.5}$$

and, because the eigenstates of $H$ are not known in general, it is not at first sight obvious how to treat quantum mechanical systems at finite temperature by Monte Carlo methods. It is, therefore, necessary to map the quantum mechanical problem to a representation with a probabilistic interpretation. Several such mappings are known (including the Suzuki–Trotter decomposition and path-integral formulation) and here I will discuss the Stochastic-Series-Expansion (SSE) approach, which has become very popular in recent years.

### *18.3.1 Stochastic Series Expansion (SSE)*

In the SSE method, pioneered by Sandvik, this mapping is achieved by expanding the partition function (18.5) in powers of $\beta$ and in operator strings $S_n$ labeling the action of different bond parts of the Hamiltonian,

$$Z(\beta) = \sum_{\alpha}\sum_{n=0}^{\infty}\sum_{S_n} \frac{(-\beta)^n}{n!} \left\langle \alpha \left| \prod_{i\in S_n} h^{b_i}_{(s_i,t_i)} \right| \alpha \right\rangle, \tag{18.6}$$

where $\alpha$ is the index for the trace over a specified basis set, $n$ is the current expansion order, $S_n$ specifies an operator string $[b_1,(s_1,t_1)],\ldots,[b_n,(s_n,t_n)]$, and $h^{b_i}_{(s_i,t_i)}$ is a bond term in the Hamiltonian which operates in a non-branching way on the sites $(s_i,t_i)$ in the chosen basis. The action is either diagonal or off-diagonal, as specified by $b_i$. A graphical illustration of a single term (for fixed $\alpha$, $n$, and $S_n$) in this infinite sum is displayed in Fig. 18.1. If the total weight $\frac{\beta^n}{n!}\left\langle \alpha \left| \prod_{i\in S_n} -h^{b_i}_{(s_i,t_i)} \right| \alpha \right\rangle$ is non-negative for all $\alpha$, $n$, and $S_n$, one has found a configuration space with a probabilistic interpretation, therefore enabling Monte Carlo simulations.

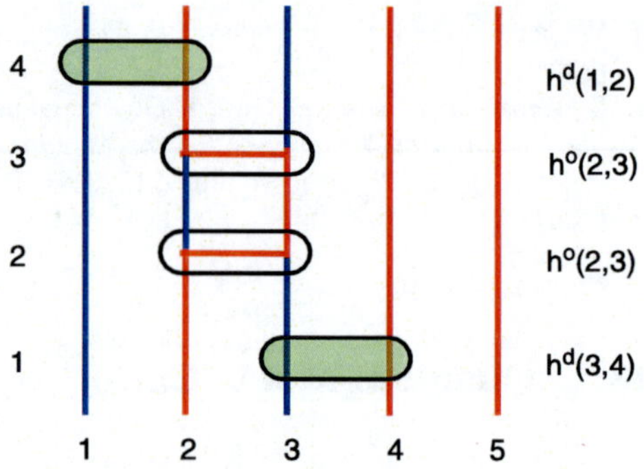


**Fig. 18.1** Schematic representation of a configuration in the SSE technique (18.6). Here the starting state is $\alpha = |\downarrow\uparrow\downarrow\uparrow\uparrow\rangle$, the expansion order is $n = 4$, and the operator string is $S_4 = [d, (3,4)], [o, (2,3)], [o, (2,3)], [d, (1,2)]$

Here I do not enter into a discussion of the Monte Carlo updates themselves, and mention only that "diagonal" update terms either insert or remove diagonal bonds in the operator string $S_n$, thereby changing the current expansion order $n$. More involved are "off-diagonal updates," which can modify the state $|\alpha\rangle$ and exchange diagonal into off-diagonal bond terms while conserving $n$. Different off-diagonal update schemes have been proposed, under the names of operator loop, directed loop, or worm updates [19–21].

To be more specific on the requirements as to when a model is amenable to QMC simulations, because at order $n$ one obtains a factor $(-\beta)^n$, the matrix elements must be negative if the total weight is to be positive for all $n$. It is clear that positive diagonal matrix elements of the Hamiltonian can in principle always be made negative by subtracting a suitable constant. The situation becomes more difficult if there are positive off-diagonal matrix elements: in the example of a $S = 1/2$ XXZ model with antiferromagnetic in-plane (XY components) exchange interactions on a bipartite lattice, then it is possible to gauge the positive sign to a negative one, again making simulations possible. However, antiferromagnetic in-plane interactions on a non-bipartite lattice cannot be gauged away, and this leads to a version of the "sign problem." Fermions are another source of negative weights due to their anticommutation relations. The sign problem can be circumvented formally by assigning the sign of a configuration to the observables calculated, and the true expectation value of the observable recovered by dividing by the average sign. The fundamental limitation of this procedure is that the average sign becomes exponentially small in $\beta$ times the simulated volume $V$, leading to exponentially large errors in the observables and therefore rendering the approach practically useless for low temperatures or large system sizes.

While on the one hand it has been shown that the sign problem is an extremely hard (and perhaps even impossible) problem to solve in general [22], on the other hand some progress towards an alleviation or even the complete elimination of the sign problem has been achieved in special cases [23, 24]. This demonstrates that it remains worthwhile to continue the search for solutions to the sign problem in particular applications.

Thus, although QMC methods cannot be applied to all quantum mechanical problems in frustrated systems, an interesting type of problem amenable to this class of simulations has been explored in recent years with much success [25–30]. These are XXZ-type spin models in which the Ising interactions ($z$-component) are frustrated while the transverse (XY) part of the interaction is chosen to be ferromagnetic (unfrustrated). As mentioned above, the diagonal frustration does not give rise to a sign problem. As in classical frustrated systems, the absence of the sign problem is not *per se* equivalent to "easy" problems. Good equilibration techniques and efficient update schemes are still required in order to obtain high-quality MC data (see for example [31]).

## Further Literature

1. A.W. Sandvik and O. F. Syljuasen, *The Directed-Loop Algorithm*, AIP Conf. Proc. 690, 299 (2003), [also available at arXiv:cond-mat/0306542v1].
2. M. Troyer, F. Alet, S. Trebst, and S. Wessel, *Non-local updates for quantum Monte Carlo simulations*, AIP Conf. Proc. 690, 156 (2003), [also available at arXiv:physics/0306128v1].
3. N. Kawashima and K. Harada, *Recent Developments of World-Line Monte Carlo Methods*, J. Phys. Soc. Jpn. 73, 1379 (2004).

### *18.3.2 Green-function Monte Carlo*

The Green-function technique has received renewed interest in recent years, because it is at present the only technique which is able to simulate quantum dimer models (QDMs) on rather large lattices at $T = 0$. See the chapter by R. Moessner and K. Raman for a detailed discussion of the physics of QDMs. Formally one could also apply the SSE approach (preceding paragraph) directly to QDMs, but this is challenging in that, due to the constrained nature of the dimer configurations, no powerful and ergodic update scheme is currently known.

The foundation of this technique is to start with a trial wave function $|\psi_T\rangle$ and to project onto the ground state by applying the operator $\exp[-\tau H]$ in the approximate form $\exp[-\tau H] \approx 1 - \tau H$ repeatedly on $|\psi_T\rangle$. The Green-function technique implements the nodeless wave function as a probability distribution and considers the stochastic dynamics of a population of walkers which are attached to certain configurations in the Hilbert space [32–35].

A number of problems in the context of QDMs have been treated using the Green-function Monte Carlo technique, including the topological degeneracy of the RVB phase on the triangular lattice, the nature of the symmetry-broken phases in the square lattice QDM, and the characterization of doped QDMs [36–38]. At the Rokhsar–Kivelson point of a QDM, Green-function Monte Carlo becomes a simple, classical Monte Carlo process. Henley has shown [39] that in this case the classical, continuous-time Monte Carlo process corresponds to the imaginary-time evolution of the corresponding QDM. By exploiting this correspondence, dimer and vison gaps on the triangular lattice have been obtained [40, 41]. The full dynamical dimer excitation spectrum at the Rokhsar–Kivelson point has also been obtained on several lattices by performing a stochastic analytical continuation [42].

The Green-function method has also been used in the past to improve variational wave functions, such as projected BCS wave functions, in the context of the fixed-node approximation. While this approach is not exact, it can still lead to valuable insight for problems which have a severe sign problem, such as the triangular-lattice Heisenberg antiferromagnet or the $J_1 - J_2$ model on the square lattice [43, 44]. The chapter by Becca et al. contains a detailed discussion of projected BCS wavefunctions for frustrated quantum spin systems.

## 18.4 Series Expansions

Series-expansion techniques come in different types, many of which are important in the field of fustrated systems. This utility has rendered series expansions quite popular. Here, I discuss two different approaches, high-temperature expansions for thermodynamic quantities and $T = 0$ perturbation series around simple ground states.

### 18.4.1 High-temperature Series

High-temperature series for physical quantities such as the susceptibility or the specific heat are very useful for comparison with experimental results. Such series can be obtained in symbolic form sometimes up to 20th order in $\beta = 1/T$, enabling a rapid fit of model parameters to experimental data. A disadvantage encountered in frustrated systems is that high-temperature series often do not extend to temperatures much lower than the exchange coupling $J$. By applying series-extrapolation techniques (such as Padé approximants) or by imposing physical constraints on the series, it is sometimes possible to go somewhat lower in temperature [45, 130, 131].

Technically, the series are obtained as a linked-cluster expansion. The series are calculated directly in the thermodynamic limit, and because the desired quantities (susceptibility, specific heat) are intensive quantities, only connected ("linked") clusters contribute to a given order in $\beta$. The computer is required here for the symbolic calculation of the quantum mechanical traces over a cluster and to enumerate the graphs and their embeddings on the lattice.

Recently, Rigol and Singh [46] have proposed a somewhat different approach, termed "numerical linked-cluster expansion," where they still perform a linked-cluster expansion, but for each cluster the complete temperature-dependence is obtained by full numerical diagonalization, thereby summing effectively the contributions of all powers of $\beta$ confined to a given cluster. This method allows one to reach slightly lower temperatures than conventional high-temperature series expansions, but at the price of losing the flexibility of the symbolic series.

### 18.4.2 $T = 0$ *Perturbative Expansions for Ground- and Excited-state Properties*

In the quest to understand the ground-state phase diagrams of frustrated spin models, perturbative series expansions are very useful. Often, one has a clear understanding of the physics in certain simple limits, and would like to know how far these phases extend in parameter space. By using high-order series expansions for the ground-state energy, one may estimate the convergence radius of the series, thereby obtaining an estimate for the extent of the phase under consideration. In the same way, one may track the excitation gaps, and their closing can indicate a continuous transition to another phase. A recent application of this technique to the $S = 1/2$ kagomé-lattice antiferromagnet supports the existence of a valence-bond crystal with a rather large unit cell [47]. Technically, these series expansions are performed via linked-cluster expansions [48]. A somewhat different approach is the continuous unitary transformation (CUT), which also yields a series expansion in the perturbative formulation. For a discussion of the CUT method, the reader may refer to the chapter of Mila and Schmidt.

Recent advances in series-expansion techniques [49] have enabled the calculation of excited-state properties, leading for example to the discovery of the highly renormalized spin-wave dispersion in the triangular-lattice antiferromagnet [50]. It has also become possible to calculate spectral properties such as dynamical spin structure factors, for example in the Shastry–Sutherland model [51].

## Further Literature

1. J. Oitmaa, C. Hamer, and W.-H. Zheng, *Series Expansion Methods for Strongly Interacting Lattice Models*, (Cambridge University Press, Cambridge, 2006).

## 18.5 Density-Matrix Renormalization Group (DMRG)

The density-matrix renormalization group is a relatively new and highly successful numerical technique applicable to one-dimensional quantum lattice models. It is based on finding an appropriate truncation of the Hilbert space which enables calculations to be performed on systems much larger than those which are accessible without truncation in Exact Diagonalization (see next section). The method was invented by White in 1992 [52, 53] while investigating the limitations of previous real-space renormalization schemes inspired by Wilson's numerical renormalization group (which is, however, very successful for the Kondo problem). This latter type of scheme truncates the Hilbert space of a block $B$ based on the low-energy eigenstates of the Hamiltonian when restricted to $B$. As can be seen by studying a simple, one-dimensional, tight-binding problem [54], the ground state of a composite block $BB'$ is not given simply as a linear combination of the product of low-energy states of the individual blocks $B$ and $B'$, rendering purely energy-based truncation schemes inaccurate. The key concept behind the DMRG algorithm is to use information from the reduced density matrix $\rho_B$ of the block $B$ as a criterion for truncating the Hilbert space of the block. The reduced density matrix $\rho_B$ is obtained by forming a superblock $BB'$, restricting the Hamiltonian to the superblock ($H_{BB'}$), solving for the ground state $|\psi_{BB'}\rangle$ of the superblock Hamiltonian, and taking the trace over the degrees of freedom in $B'$ to obtain the reduced density matrix $\rho_B = \mathrm{Tr}_{B'}|\psi_{BB'}\rangle\langle\psi_{BB'}|$. Finally, one retains those states which correspond to the $m$ largest eigenvalues of the reduced density matrix. A detailed description of the algorithm is beyond the scope of this chapter and the reader is referred to the original literature [52, 53] and to the review articles listed at the end of this section for an in-depth discussion of the method.

The potential of the DMRG method was shown shortly after its invention by the determination of the Haldane gap in the $S = 1$ Heisenberg chain to unprecedented accuracy. The ground-state properties of frustrated quantum spin systems were also studied at an early stage, starting with the frustrated $J_1 - J_2$ chain [55–57], the $CaV_4O_9$ lattice [58], and the kagomé strip [59], to name but a few examples. More recently, the method has also been applied to cases including frustrated ladders in a magnetic field [60–62] and two-dimensional clusters of considerable size, such as triangular and kagomé systems [63, 64].

### *18.5.1 Finite T*

Ideas from the DMRG technique have also been used in finding new algorithms to calculate the thermodynamic properties of one-dimensional quantum systems. One class of approaches is to apply density-matrix notions to the calculation of the transfer matrix (TMRG), based on which many thermodynamic quantities can be computed from high $T$ to moderately low temperatures. A different approach is based on recently developed algorithms for real- or imaginary-time evolution in order to evolve the trivial infinite-temperature density matrix in $\beta$ to the desired temperatures (T-DMRG). Many frustrated systems have been studied using these techniques [65–67].

### *18.5.2 Dynamical Response Functions*

For comparison with experiments one is often interested in dynamical correlation functions, and these are also accessible by DMRG. One class of methods works directly in frequency space and these are called "continued fraction," "correction vector," or "dynamical" DMRG [68–70]. A second class is based on the measurement of real-space, time-displaced correlation functions, and became possible with the advent of very efficient time-evolution algorithms for DMRG [71]. The frequency method calculates the response for a given wave vector and energy at a time and is very accurate but rather time-consuming. The real-time approach, by contrast, gives an overview of the full momentum and frequency range in a single calculation [72], but the quality depends on the accessible time window and the temporal Fourier transform requires some care.

### *18.5.3 DMRG in two and more Dimensions*

For reasons which have become completely clear through a stimulating interaction with the field of quantum information theory, DMRG works very well for one-dimensional systems, but cannot perform equally well for large two-dimensional systems. This is dictated by the fact that DMRG is effectively determining the best variational state in the class of matrix-product states (MPS) with a given matrix size. These states are well suited to model the entanglement scaling in one-dimensional gapped and critical systems, but are not ideal to model the entanglement properties in two or more dimensions (due to the area law). New classes of states such as tensor-product states (e.g. projected entangled pair states (PEPS) [73]) have been proposed, and a preliminary application to the Shastry-Sutherland lattice [74] provides a valuable proof of concept. Currently these new methods show an unfavorable computational scaling with the tensor index dimension, so that they do not yet achieve an accuracy comparable to DMRG in one dimension, but in the event that these novel methods can be made more efficient, the simulation of strongly correlated systems has an even brighter future.

## Further Literature

1. U. Schollwöck, *The density-matrix renormalization group*, Rev. Mod. Phys. 77, 259 (2005).
2. K. Hallberg, *New Trends in Density Matrix Renormalization*, Adv. Phys. 55, 477 (2006).
3. R.M. Noack and S.R. Manmana, *Diagonalization- and Numerical Renormalization-Group-Based Methods for Interacting Quantum Systems*, AIP Conf. Proc. 789, 93 (2005).

## 18.6 Exact Diagonalization (ED)

The aim of this section is to provide an introduction to the exact-diagonalization technique, which is a versatile and powerful numerical technique applicable to almost any quantum lattice model. The name "Exact Diagonalization" (ED) stems from the fact that one solves the time-independent Schrödinger equation,

$$H|\psi\rangle = E|\psi\rangle, \tag{18.7}$$

of a finite quantum many body system, which is an eigenvalue problem, by numerical diagonalization of $H$ with no approximations. In practice one often works at zero temperature and is, therefore, interested mostly in the low-energy physics of the model. In this case, only information about a small number of extremal eigenvalues and eigenvectors of the eigenproblem is required. For this specific task, many numerical algorithms with fast convergence properties exist, such as the Lanczos [75] or the Davidson Algorithm [76]. The Lanczos algorithm will be discussed later in this section. However, it should be stressed that the complexity of a modern ED code goes far beyond the numerical eigensolver part.

The primary advantage of the ED method is its flexibility: it can be used to treat large classes of spin systems, frustrated or unfrustrated, fermionic systems ($t$-$J$ model, Hubbard model, Quantum Hall problems, and many others) and bosonic models in low dimensions. Furthermore, one may calculate the expectation values of almost any observable, including dynamical correlation functions, which are often hard to obtain by other numerical methods. The major drawback of the ED method is the exponential scaling of the Hilbert space as a function of the system size: for a quantum mechanical spin-$S$ system with $N_s$ sites, one has a priori a Hilbert space $\mathcal{H}_S$ with $\dim \mathcal{H}_S = (2S+1)^{N_s}$. In the following subsection I will describe some general strategies for implementing an Exact Diagonalization program in such a way that this "exponential hard wall" can be pushed as far as possible. The present-day limits for $S = 1/2$ spin systems are 40–42 sites (for $S^z = 0$), 32 sites for a $t$-$J$ model with 4 holes, and 20 sites for a square-lattice Hubbard model at half filling. The corresponding dimensions of the Hilbert spaces range between some hundreds of millions and a few billion ($10^9$). To the best of my knowledge, the largest eigenvalue problem which has been solved in strongly correlated systems has a dimension of approximately 160 billion states [77]. This solution was possible because it is actually easier to parallelize and to solve a large eigenvalue problem when spatial symmetries are discarded. However, the largest physical systems can only be reached by including symmetries. It is also more challenging to reach large Hilbert-space sizes for spin problems than it is for $t$-$J$ or Hubbard models due to the (approximate) factorization of the Hilbert space in the latter two models, simplifying the look-up procedure.

An ED program generally consists of four parts:

1. A numerical representation of the basis states in the (symmetrized) Hilbert space $\mathcal{H}$,
2. A numerical (or virtual) representation of the Hamiltonian matrix $H$,
3. An algorithm which calculates the desired eigenvalues and eigenvectors,
4. A set of observables whose expectation values are calculated.

In the following subsections, we discuss each of these four building blocks in turn.

### 18.6.1 Basis Construction

The Hamiltonian of the system often possesses certain symmetries, and therefore some conserved quantities. Common symmetries or conservation laws encountered in quantum lattice-model systems are

1. Charge conservation: $N_e$ constant,
2. Magnetization conservation: $S^z_{\text{tot}}$ constant,
3. Translational symmetry: momentum conserved,
4. Point-group symmetry: parity and angular momentum conserved, and
5. Full SU(2) spin symmetry: $S^2_{\text{tot}}$ conserved.

Among these symmetries, the first four are implemented routinely in an ED program. The full SU(2) symmetry is in general rather hard to implement in combination with spatial symmetries. However, for the special case $S^z_{\text{tot}} = 0$, a global spin inversion which maps $S^z_i \to -S^z_i$ can be exploited. This symmetry splits the representations with even and odd total spin. The advantages of implementing and using symmetries are twofold: first, they lead to a dramatic reduction in the dimension of the Hilbert-space sector to be diagonalized and second, one may as a result resolve the computed properties as functions of the corresponding quantum numbers, for example to calculate dispersion curves for the elementary excitations (i.e. energy as a function of momentum) or to uncover symmetry-breaking patterns by investigating the tower of states in the spectrum.

The process of Hilbert-space reduction can be illustrated with the example of a $S = 1/2$ Heisenberg model on a 40-site square lattice.[4] The unrestricted Hilbert space has dimension $2^{40}$ ($\approx 10^{12}$). Restricting oneself to the $S^z_{\text{tot}} = 0$ sector reduces the number of states to $\frac{40!}{20!\cdot 20!}$ ($\approx 138 \cdot 10^9$). The space group of the tilted square lattice combines 40 translations with a four-fold symmetry axis (yielding 160 symmetry elements), and implementing these space-group symmetries (in the one-dimensional, fully symmetric representation) yields a reduction by a factor 160 to approximately 861 million. Finally, exploiting spin-inversion symmetry leads to a final result of 430909650 basis states. For Heisenberg-type models, this is among the largest problems solvable today.

## 18.6.2 Coding of Basis States

For simplicity I discuss the case of a $S = 1/2$ model on a $N_s$-site lattice. An intuitive way of representing the Hilbert space is given by mapping the basis states of the $S^z$ product basis to $N_s$-bit integers according to

$$|c\rangle \equiv |\sigma_0, \sigma_1, \sigma_2, \sigma_3, \ldots, \sigma_{N_s-1}\rangle, \quad \sigma_i \in \{\uparrow, \downarrow\}$$

$$\text{Int}(|c\rangle) \equiv \sum_{i=0}^{N_s-1} c_i 2^i, \quad c_i \in \{0(\downarrow), 1(\uparrow)\}; \tag{18.8}$$

as an example,

---

[4] This cluster is generated by the spanning vectors $\mathbf{T}_1 = (6, 2)$ and $\mathbf{T}_2 = (-2, 6)$.

$$|\uparrow_0, \uparrow_1, \downarrow_2, \uparrow_3, \downarrow_4, \ldots, \uparrow_{N_s-1}\rangle \Longleftrightarrow 11010\ldots1_2 \\ = 2^0 + 2^1 + 2^3 + \ldots + 2^{N_s-1}. \tag{18.9}$$

With this choice of basis it is particularly easy to implement the conservation of $S^z_{\text{tot}}$, which corresponds to a set with a fixed number of bits. For models with a different local state structure (such as the $t$-$J$ model, with three states, or the Hubbard model, which has four states per site), analogous mappings are easy to find. For the $t$-$J$ model, there is an alternative representation where one stores the positions of the holes in one integer and the spin states of the occupied sites in a different integer. For the Hubbard model, a convenient representation is to separate the positions of the up-spin electrons in one integer and of the down-spin electrons in a second integer. The idea behind all of these representations is to allow a rapid evaluation of the terms in the Hamiltonian at a later stage of the calculation.

The enumeration of states with fixed $S^z$ is quite straightforward. One approach is to first enumerate and label all configurations of a half system with their $S^z$ quantum number. Then one sorts this list, so that configurations with the same $S^z$ are adjacent. Finally, one loops over all the configurations of the second half of the system and combines them with the complementary $S^z$ states of the first half in order to satisfy the total-$S^z$ constraint. This approach is most effective when working at total $S^z = 0$. For large systems with $S^z$ close to saturation, a recursive enumeration of states with fixed $S^z$ is more appropriate, and is easy to implement for $S = 1/2$ by simple bit-shuffling.

### 18.6.3 Symmetrized Basis States

Consider a system with a spatial symmetry group $G$ which is the direct product of a set of translations $TG$ and a set of point-group elements $PG$,

$$G = \text{TG} \otimes \text{PG}.$$

The translational subgroup $TG$ is an abelian group and therefore has only one-dimensional irreducible representations, which can be labeled by a wavevector $\mathbf{K}$.[5] The character of this representation is

$$\chi_{\mathbf{K}}(t) = \exp(\mathrm{i}\mathbf{K} \cdot \mathbf{T}(t)), \quad \forall t \in \text{TG},$$

where $\mathbf{T}(t)$ denotes the translation vector associated with the element $t$. For a given wave vector $\mathbf{K}$, not all point-group elements are allowed: these are restricted to the

[5] The allowed set of discrete wave vectors for a finite sample is generated by the reciprocal lattice vectors of the superlattice.

"little group" of **K**, $\mathrm{PG}_{\mathbf{K}} = \{p \in \mathrm{PG} \mid p(\mathbf{K}) = \mathbf{K}\}$, i.e. to the stabiliser of **K**. The actual symmetry group is then $G'_{\mathbf{K}} := \mathrm{TG} \otimes \mathrm{PG}_{\mathbf{K}}$. The point group PG and also the little group $\mathrm{PG}_{\mathbf{K}}$ are in general nonabelian groups and their irreducible representations are not always easy to determine, but can be found tabulated in works on group theory. I assume in the following that the selected representation $\rho$ of the little group is one dimensional[6] and the character $\chi_\rho(p)$ is known $\forall p \in \mathrm{PG}_{\mathbf{K}}$. The configurations $|c\rangle$ in the Hilbert space are symmetrized by taking

$$\mathrm{symm}_{\mathbf{K},\rho}(|c\rangle) = \frac{1}{\mathcal{N}_{c,\mathbf{K},\rho}} \sum_{t\in T,\ p\in PG_{\mathbf{K}}} \chi_{\mathbf{K}}(t)\chi_\rho(p)|(p \circ t)(c)\rangle. \tag{18.10}$$

These states are invariant under the symmetry operations and can be considered as generalized Bloch waves. The normalization factor $1/\mathcal{N}_{c,\mathbf{K},\rho}$ must be determined explicitly for each state $\mathrm{symm}_{\mathbf{K},\rho}(|c\rangle)$. It is not given simply by $1/\sqrt{|G'_{\mathbf{K}}|}$ because for certain configurations the application of some operators $g \in G'_{\mathbf{K}}$ is equivalent to the identity operation. Due to cancellations it is also possible for a state to vanish completely, and such states should not be kept in the Hilbert space.

This point can be illustrated on a four-site $S=1/2$ spin chain with variable total magnetization. The symmetry group is assumed to be $C_4$ (the cyclic shift corresponding to one-step translation).

- $S^z = 2$
  The configuration $|\uparrow\uparrow\uparrow\uparrow\rangle$ is a one-state orbit with momentum $\mathbf{K} = 0$. Its norm is 1. States with other momenta do not exist.
- $S^z = 1$
  The four states $|\downarrow\uparrow\uparrow\uparrow\rangle$, $|\uparrow\downarrow\uparrow\uparrow\rangle$, $|\uparrow\uparrow\downarrow\uparrow\rangle$, and $|\uparrow\uparrow\uparrow\downarrow\rangle$ span a single orbit and can be combined to give the four symmetric states with momenta $\mathbf{K} = 0, \pi/2, \pi$, and $3\pi/2$. The normalization factors are $1/2$.
- $S^z = 0$
  The six states $|\downarrow\downarrow\uparrow\uparrow\rangle$, $|\downarrow\uparrow\downarrow\uparrow\rangle$, $|\downarrow\uparrow\uparrow\downarrow\rangle$, $|\uparrow\downarrow\downarrow\uparrow\rangle$, $|\uparrow\downarrow\uparrow\downarrow\rangle$, and $|\uparrow\uparrow\downarrow\downarrow\rangle$ separate into two orbits,
  - $|\downarrow\downarrow\uparrow\uparrow\rangle$, $|\uparrow\downarrow\downarrow\uparrow\rangle$, $|\uparrow\uparrow\downarrow\downarrow\rangle$, and $|\downarrow\uparrow\uparrow\downarrow\rangle$, which span states with momenta $\mathbf{K} = 0, \pi/2, \pi$, and $3\pi/2$ and have a norm of $1/2$;
  - $|\downarrow\uparrow\downarrow\uparrow\rangle$ and $|\uparrow\downarrow\uparrow\downarrow\rangle$, which yield two states with momenta 0 and $\pi$ and have norm $1/\sqrt{2}$.
- $S^z = -1$ and $S^z = -2$
  are analogous to $S^z = 1$ and $S^z = 2$.

[6] In some cases, a higher-dimensional representation of $PG$ can be expressed as a one-dimensional representation of a group $PG'$ with fewer elements. For example, the two-dimensional $E$ representation of $C_{4v}$ can be decomposed into the one-dimensional $(+1, -1)$ and $(-1, +1)$ representations of $C_2 \otimes C_2 \subset C_{4v}$

As will be demonstrated in the next section, it is sufficient to store only one representative state per orbit, together with the normalization factor of each orbit. Usually the configuration $|c\rangle$ with minimal Int($|c\rangle$) is taken as the representative.

For construction of the basis, one loops over all configurations $|c\rangle$ with the correct $S^z_{\text{tot}}$. Inside this loop, one generates orbit($|c\rangle$) $= \{g(|c\rangle) \mid \forall g \in G'_{\mathbf{K}}\}$. If Int($|c\rangle$) $=$ min{Int(orbit($|c\rangle$))}, $|c\rangle$ is registered together with the calculated normalization factor (given a nonzero norm). The dimension of the targeted subspace is equivalent to the number of registered states. If memory is available, one can also store for all $|c\rangle$ the index of the representative in the common orbit, together with the symmetry operation which maps $|c\rangle$ onto the representative configuration, as this can significantly accelerate the calculation of the matrix elements.

### 18.6.4 Hamiltonian

Now that the symmetry-reduced basis of the Hilbert space has been constructed, I explain how to apply the Hamiltonian matrix $H$. Assume for simplicity that the action of the Hamiltonian on a given representative $|r\rangle$ is to generate one new configuration $|n\rangle$ with some amplitude $h$,[7]

$$H|r\rangle = h|n\rangle. \tag{18.11}$$

Note that $|n\rangle$ is not in general a representative. Let $|r'\rangle$ be the representative of $|n\rangle$ and $g_n$ one of the symmetry operations which map $|n\rangle$ onto $|r'\rangle$. The matrix element is then given as

$$\langle r'|H|r\rangle = \sqrt{\frac{\mathcal{N}_{r',\mathbf{K},\rho}}{\mathcal{N}_{r,\mathbf{K},\rho}}}\, \chi_{\mathbf{K},\rho}(g_n)\, h, \tag{18.12}$$

a relation which can be proven by inserting the definitions (18.10) of $\langle r'|$ and $|r\rangle$ on the left-hand side while exploiting $[H, g] = 0\ \forall\ g \in G'_{\mathbf{K}}$ together with the property (18.11).

In the computer implementation, the challenge is to find strategies suitable for obtaining $|r'\rangle$ and its index in the list of representatives in the most efficient manner possible. For certain problems, a simple two-stage look-up table can be implemented. The first step is to map Int($|n\rangle$) to its index in the (virtual) list of all configurations with the selected $S^z_{\text{tot}}$. This can be done using what are known as Lin tables [78]. This index is then used in addressing a large table containing the index of $|r'\rangle$ and the phase factor of the operation relating $|n\rangle$ to $|r'\rangle$. This approach is probably one of the fastest look-up procedures, but is restricted to approximately 36 sites for spin-1/2 problems on present-day computers. For larger systems, an approach by two-sublattice decomposition [79] followed by a binary search or a hash-table look-up proves to be quite useful. This technique is, however, difficult to

[7] This is not a restriction because the Hamiltonian can be decomposed into a sum of terms with this property.

implement in a generic code as the symmetry group must be factorized by hand. For Hubbard models, the two-sublattice look-up idea can be extended to the spin-up and spin-down electrons. For the $t$-$J$ model, a decomposition into a hole configuration and a spin configuration is also useful at low hole doping. For fermionic models in general, care should be taken regarding the Fermi sign. Each time a symmetry operation or a term in the Hamiltonian is applied, the resulting Fermi sign must be calculated and taken into account.

## *18.6.5 Eigensolvers*

### 18.6.5.1 Lanczos Algorithm

After the computer implementation of the basis states and the Hamiltonian operator, the next requirement concerns the algorithms to solve the eigenproblem (18.7). In practice, the Lanczos algorithm is used for this task [75]. The Lanczos algorithm builds a special basis in the Krylov space

$$\mathcal{K}_n = \{|\phi_1\rangle, H|\phi_1\rangle, H^2|\phi_1\rangle, \dots, H^n|\phi_1\rangle\}$$

in which the operator $H$ takes a tridiagonal form. The recursive sequence is defined by

$$\text{choose a random normalized starting vector } |\phi_1\rangle, \tag{18.13}$$

$$\beta_1|\phi_2\rangle = H|\phi_1\rangle - \alpha_1|\phi_1\rangle, \tag{18.14}$$

$$\beta_n|\phi_n + 1\rangle = H|\phi_n\rangle - \alpha_n|\phi_n\rangle - \beta_n|\phi_{n-1}\rangle, \tag{18.15}$$

$$\text{where } \alpha_n = \langle\phi_n|H|\phi_n\rangle \text{ and } \beta_n = \| \langle\phi_{n+1}|A|\phi_n\rangle \| \,. \tag{18.16}$$

After $n$ steps, the Hamiltonian in the Krylov space takes the matrix form

$$T_n = \begin{pmatrix} \alpha_1 & \beta_1 & & & \\ \beta_1 & \alpha_2 & \beta_2 & & \\ & \ddots & \ddots & \ddots & \\ & & \ddots & \ddots & \beta_{n-1} \\ & & & \beta_{n-1} & \alpha_n \end{pmatrix}, \tag{18.17}$$

which is known as the $T$-matrix. The defining strength of the Lanczos tridiagonalisation procedure lies in the fact that the extreme eigenvalues of the $T$-matrix, $T_n$, converge rapidly towards the extreme eigenvalue of $H$. In practice $n \sim 100$

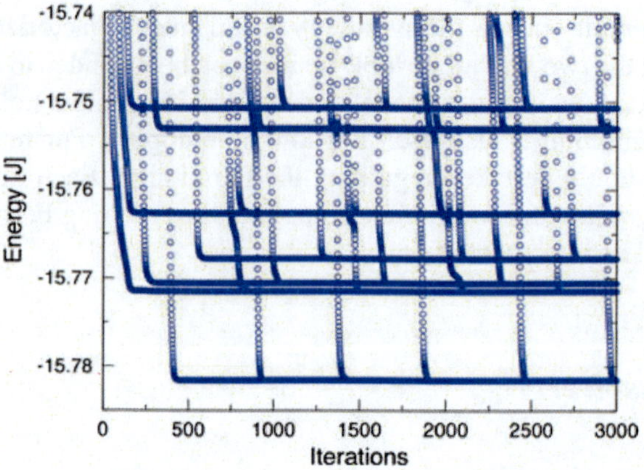


**Fig. 18.2** Convergence of $T$-matrix eigenvalues as a function of the number of iterations. The system is a 36-site kagomé lattice with a starting vector orthogonal to the ground state. Still, the Lanczos algorithm converges finally to the true ground state and accumulates fake degeneracies approximately every 500 iterations. The points aligned on horizontal lines correspond to physical eigenvalues

is usually enough to reach a tolerance of $10^{-14}$ for the ground-state energy.[8] The evolution of the lower end of the $T$-matrix spectrum as a function of $n$ for a realistic application is shown in Fig. 18.2.

Once convergence for the targeted energy has been reached, the corresponding eigenvector of the $T$-matrix can be calculated. The Lanczos recursion is restarted with exactly the same initial vector $|\phi_1\rangle$. The eigenvector of $H$ can then be constructed based on the coefficient of $|\phi_i\rangle$ in the $T$-matrix eigenvector. Restarting the Lanczos process is necessary because for large problems the recursion vectors cannot be kept in memory or on disk.

The rapid convergence is one of the primary advantages of this algorithm. In a large class of quantum many-body models, the Hamiltonian matrix is sparse, i.e. the number of non-zero matrix elements is proportional to the dimension of the Hilbert space, $\dim \mathcal{H}$.[9] For sparse matrices the Lanczos algorithm has a time complexity $O(\dim \mathcal{H})$, compared to eigensolvers for dense matrices which scale as $O(\dim^3 \mathcal{H})$. Memory complexity also plays an important role in a large-scale application: in a matrix-free implementation it is possible to limit the space requirements to two or three Lanczos vectors to be kept simultaneously in memory. On the other hand, one may trade more memory to store the Hamiltonian as well, gaining a faster $H|\psi\rangle$ iteration.

[8] Still, one should terminate the recursion based on a convergence criterion, and not on a fixed number of iterations.

[9] For real-space problems with short-range interactions, the number of matrix elements per row or column is proportional to $N_{\text{sites}}$. Even the $k$-space version with $N_k^3$ matrix elements yields a sparse matrix due to the exponential growth of $\dim \mathcal{H}$ with $N_k$.

I would like to point out one peculiar problem of the Lanczos algorithm, related to finite-precision arithmetic in real-world computers: the appearance of spurious eigenvalues (a.k.a. "ghosts") in the spectrum of $T_n$. In practice, these ghosts appear for example if one attempts to achieve convergence for excited states (cf. Fig. 18.2). The problem is caused by a loss of orthogonality among the recursion vectors once the ground state has converged. It is also worth noting that the Lanczos algorithm in its simplest form, as presented here, is not able to resolve the true degeneracy of an eigenvalue, irrespective of the loss of orthogonality. Readers interested in heuristic methods for dealing with the detection of ghosts or techniques to resolve the degeneracy of multiple eigenvalues may wish to consult the excellent books by Cullum and Willoughby [80] and by Bai et al. [81].

### 18.6.6 Implementation Details and Performance Aspects

A modern ED code is written in modular form, meaning among other things that the core of the program is independent of the lattice and its symmetries. One may then easily add a new lattice structure by enumerating the elementary bonds of the Hamiltonian and the symmetry elements with their phase factors in a given representation. The back end of the code generates automatically the symmetrized basis states and reduces the dimension of the Hilbert space accordingly.

Regarding the efficiency of matrix-vector multiplication – which is the most time-consuming part of the calculation – modern approaches in general adopt a matrix-free strategy. This means that all matrix elements of the Hamiltonian are recalculated in each iteration without hard-disk or main-memory storage. This approach parallelizes well on current shared-memory machines, and depending on the structure of the problem even on distributed-memory machines. The matrix-free technique is definitely the method of choice when dealing with the largest system sizes on a given machine. It might, however, be faster to store the Hamiltonian when the memory is available. The parallelization of a Lanczos code is straighforward, as each thread is responsible for calculating a certain part of the new Lanczos vector $|\phi_{n+1}\rangle$, while having full read access to the complete vector $|\phi_n\rangle$.[10] An outstanding challenge in the near future will be the development of ED codes able to run efficiently on petaflop machines, which have extremely large numbers of computing cores.

[10] Concurrent read access is no problem and is treated by the hardware, whereas concurrent write access would require explicit synchronization and locking in the code, thereby reducing the performance very significantly.

## 18.6.7 *Observables*

An important ingredient of an ED code is the set of observables which can be calculated. In ED, the implementation of observables follows closely the above discussion of the Hamiltonian, meaning that one has to symmetrize the observables when using a symmetry-reduced basis, and for complicated correlators parallelization is the key to rapid evaluation. Depending on the physical system, typical correlation functions to consider include spin, dimer, plaquette, scalar chirality, vector chirality, quadrupolar, and octupolar correlations. Plaquette correlations provide an example equivalent to eight-spin correlators, which are very difficult to determine in other numerical approaches.

In the following, I discuss briefly two particularly useful tools to investigate frustrated quantum spin systems: first the tower-of-states analysis is illustrated by an application to the frustrated nuclear magnetism of bcc $^3$He, and then recent proposals for the systematic detection of ordering tendencies and dominant correlation functions in ED simulations are outlined.

### 18.6.7.1 Tower-of-States Analysis – the Nuclear Magnetism of bcc $^3$He

The tower of states is the finite-size manifestation of the breaking of a continuous symmetry, which arises frequently in U(1)- or SU(2)-symmetric spin models. The method was pioneered in an application to the Heisenberg antiferromagnet on the triangular lattice [82, 83], and was later used to show the absence of magnetic order on the $S = 1/2$ kagomé lattice [84]. The key idea is that symmetry-breaking is manifest already in finite systems by a specific arrangement of the low-lying eigenstates as a function of total spin (or total $S^z$), as shown in the left part of Fig. 18.3. These levels have specific quantum numbers (such as momentum, parity, angular momentum, etc.) and degeneracies dictated by the structure of the order parameter. The method is not restricted to magnetic order, and can also be applied to spin nematics, i.e. to states which do not have a magnetic moment, but a instead possess an order parameter of higher rank. The reader is referred to the review by Misguich and Sindzingre listed at the end of this section for a detailed discussion of the method, and to the chapter by Penc and Läuchli for examples of towers of states in magnetically ordered and spin-nematic states on the $S = 1$ triangular lattice.

As an illustration that the method can also be applied to three dimensional systems, I present some new ED results on an old problem, the nuclear magnetism of solid $^3$He in the body-centered-cubic (bcc) phase. This system is reviewed in [85] and [86]. The magnetism is provided by the nuclear spins rather than by the electrons of a conventional magnetic system. These spins are now believed to interact through cyclic exchange processes of various loop lengths. The experimental phase diagram shows a low-field phase whose structure is compatible with an up-up-down-down (UUDD) magnetic structure (based on NMR experiments). In a small magnetic field, the nuclear spins undergo a metamagnetic transition to a canted normal antiferromagnetic (NAF) phase at high fields, which is stable between $m/m_{\text{sat}} \sim 0.6$

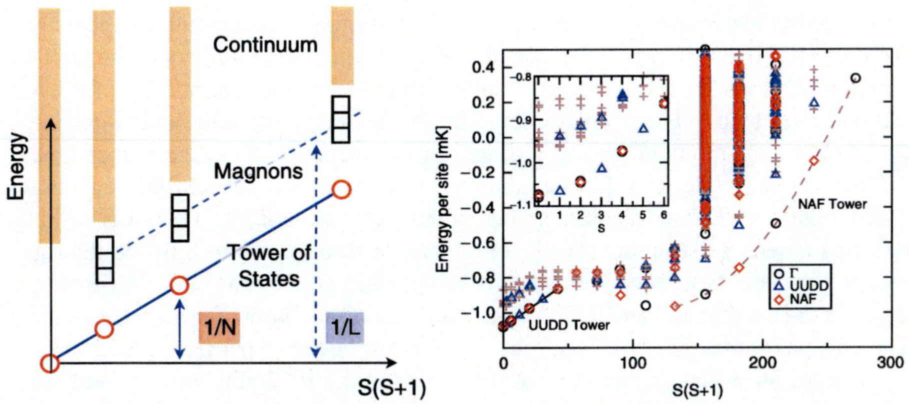


**Fig. 18.3** *Left:* schematic representation of the tower-of-states structure in a finite-size spectrum. The levels denoted by the *circles* correspond to states forming the ground-state manifold in the thermodynamic limit and its quantum numbers are severely constrained by the type of symmetry-breaking. *Right:* spin-resolved spectrum of the minimal $^3$He model on a 32-site bcc sample. The occurrence of towers of states for the UUDD state at low magnetization and the NAF state at high magnetization is confirmed clearly. Inset: zoom of the low-energy, low-spin region of the same spectrum

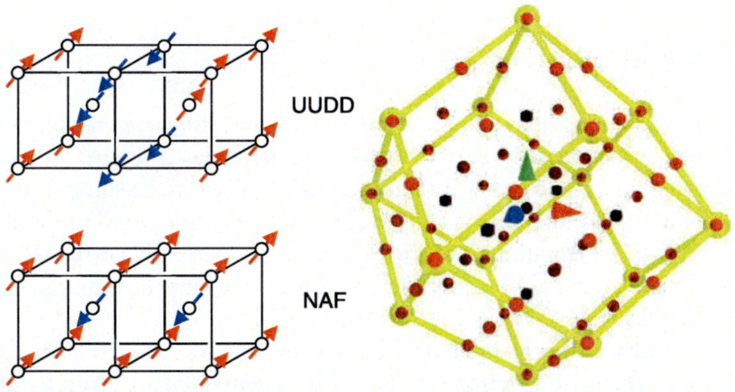


**Fig. 18.4** *Left:* schematic representation of the UUDD (up-up-down-down) and the NAF (normal antiferromagnet) magnetic structures of bcc $^3$He. *Right:* Brillouin zone of the 32-site bcc sample. *Red spheres* denote all allowed $k$-points, *blue boxes* the six wave vectors relevant for the UUDD phase. The wave vector for the NAF corresponds to the *yellow spheres* at the corners of the Brillouin zone

and the full saturation magnetization. The two magnetic structures are illustrated in the left part of Fig. 18.4.

A minimal ab initio model is based on the cyclic exchange frequencies calculated by Ceperley and Jacucci [87–89], where only the nearest-neighbor two-body exchanges, the three-body exchanges, and the planar four-body exchanges are retained. The coupling constants are $J = 0.46$ mK, $J_t = -0.19$ mK, and $K_p = 0.27$ mK [87–89]. The ED simulations are performed on a 32-site sample.

I now apply the tower-of-states analysis to the two magnetic structures expected to occur in solid bcc $^3$He (Fig. 18.4). On deriving the quantum numbers of the UUDD state, one obtains six levels per total spin with distinct quantum numbers for even and odd total spin.[11] For even spin, one finds one singly and one doubly degenerate level at zero momentum and a threefold-degenerate level at the corner of the Brillouin zone, while for odd total spin one finds a level at the sixfold-degenerate characteristic UUDD momentum shown in the right panel of Fig. 18.4. It is striking that the low-energy structure revealed in the finite-size spectrum shown in the right part of Fig. 18.3 is in excellent agreement with the predictions, and thus strongly supports the presence of an UUDD phase in the minimal model for bcc $^3$He. A similar analysis for the high-field NAF phase yields a single level per spin, a level at zero momentum for even spin, and one at the corner of the Brillouin zone for odd spin. (This is the bcc analog of the well-known $(\pi, \pi)$ Néel state in a magnetic field on the square lattice, *cf.* [90]). Again these levels are found in the exact spectra. The scaling with $(S^z)^2$ is not visible here, because for this one would need to include the magnetic field in the graphical presentation. In conclusion, a tower-of-states analysis of ED simulations for a minimal model of the nuclear magnetism in three-dimensional bcc $^3$He confirms the presence of an UUDD phase at low magnetization and a NAF phase at high magnetization, in agreement with experiments.

#### 18.6.7.2 Detection of Exotic Order

In numerical simulations of quantum lattice models, detecting and characterizing the important correlations in different phases is a fundamental problem. This question has usually been addressed either by intuition or by successive testing of candidate correlation functions. While this approach clearly works for most (and for all familiar) models [91–95], the hallmark of truly complex and strongly correlated systems is that different and non-intuitive types of correlation can become important and will compete in the physical state.

A first approach to the systematic detection of local order parameters was put forward in [96]. The method compares the reduced density matrices on a small area obtained from two states which become degenerate in the thermodynamic limit, and yields a local operator which can serve as an order parameter. In [96], this approach was shown to work well in the case of discrete symmetry-breaking, and the method can also be extended to continuous symmetry-breaking.

A second promising approach, devised recently by Cheong and Henley [97, 98], is based on the correlation density matrix and its decomposition into the correlation content. Starting with a wave function of interest, one selects two distinct subsystems $A$ and $B$, in whose mutual correlations one is interested. One then defines the correlation density matrix, based on standard reduced density matrices, as

[11] The fact that the number of levels is independent of the total spin is a consequence of the collinear spin structure

$$\rho^c_{AB} = \rho_{AB} - \rho_A \otimes \rho_B; \tag{18.18}$$

this central object contains all information on any connected correlation function between $A$ and $B$,

$$\langle \mathcal{O}_A \mathcal{Q}_B \rangle - \langle \mathcal{O}_A \rangle \langle \mathcal{Q}_B \rangle = \mathrm{Tr}\left[\mathcal{O}_A \mathcal{Q}_B \rho^c_{AB}\right]. \tag{18.19}$$

The key step in the analysis of this quantity is to perform a singular-value decomposition of the matrix $\rho^c_{AB}$. The resulting singular values give the strengths of the various normalized correlations, while the singular vectors give the corresponding operators acting in the Hilbert spaces of either block $A$ or block $B$. Thus, this decomposition yields direct information on which correlations between the two subsystems $A$ and $B$ are important. This type of unbiased analysis of correlation functions may shed additional light on enigmatic frustrated models such as the kagomé antiferromagnet or the $J_1 - J_2$ square-lattice Heisenberg model.

### *18.6.8 Dynamical Response Functions*

Another important class of observables contains the dynamical response functions, such as the dynamical spin structure factor or the single-particle spectral functions. Their generic form is

$$I(\omega) = \lim_{\eta \to 0+} -\frac{1}{\pi} \mathrm{Im}[\langle \Psi_0 | \mathcal{O}^\dagger \frac{1}{\omega + E_0 + i\eta - H} \mathcal{O} | \Psi_0 \rangle], \tag{18.20}$$

where $\mathcal{O}$ is the operator under consideration. Examples important for correlated systems include $\mathcal{O} = S^z(q)$ for the dynamical spin structure factor and $\mathcal{O} = c^\dagger_{k,\sigma}$ for the single-particle spectral function. $|\Psi_0\rangle$ denotes the ground state and $E_0$ the ground-state energy. Equation (18.20) can be reexpressed as

$$I(\omega) = \sum_n |\langle \Psi_n | \mathcal{O} | \Psi_0 \rangle|^2 \delta\big(\omega - (E_n - E_0)\big), \tag{18.21}$$

a form in which the spectral decomposition is more obvious. $|\Psi_n\rangle$ and $E_n$ denote the eigenvectors and energies of the Hamiltonian $H$. The spectral representation (18.21) can be calculated by a Lanczos recursion starting with the specific vector

$$|\phi_1\rangle = \frac{\mathcal{O}|\Psi_0\rangle}{\sqrt{\|\mathcal{O}|\Psi_0\rangle\|^2}} \tag{18.22}$$

rather than with an arbitrary starting vector [one is constructing a Krylov space around $\mathcal{O}|\Psi_0\rangle$ for an optimal approximation of $(\omega + E_0 + i\eta - H)^{-1}$]. Iterating $n$ times gives a new $T$-matrix with new elements $\alpha$ and $\beta$. One may prove that the continued-fraction expansion

$$I(\omega) = -\frac{1}{\pi}\mathrm{Im}\left[\frac{\|\mathcal{O}|\Psi_0\rangle\|^2}{z-\alpha_1-\frac{\beta_1^2}{z-\alpha_2-\frac{\beta_2^2}{z-\alpha_3-\cdots}}}\right] \tag{18.23}$$

is a representation of the response function, with $z := \omega + E_0 + i\eta$ and $\alpha_i$ and $\beta_i$ determined in the Lanczos recursion. This identity follows from Cramer's rule when applied to the first matrix element of the inverse of a tridiagonal matrix. An equivalent representation is given through (18.21) with the identity

$$|\langle\Psi_n|\mathcal{O}|\Psi_0\rangle|^2 = |c_0^n|^2\ \|\mathcal{O}|\Psi_0\rangle\|^2, \tag{18.24}$$

where $c_0^n$ denotes the overlap of the $n-1$ eigenvector of the $T$-matrix with the starting vector. A comment regarding the practical application is in order here: the representation (18.24) faces problems similar to the ghosts mentioned earlier. The fake multiplicities of the $T$-matrix can lead to a very small[12] splitting of the poles and a redistribution of the spectral weight among these poles. However, the spurious eigenvalues do not in my experience seem to carry any spectral weight. A recent application of this formalism predicts the absence of a quasiparticle peak in the photoemission spectral functions of a doped kagomé antiferromagnet [99].

In recent years, a somewhat different method, the "Kernel Polynomial Method" (KPM) has been advocated for the calculation of spectral functions [100], and appears to be free of the numerical instability discussed above. For many purposes the two approaches are of equal quality. One advantage of the continued-fraction method is that the individual pole weights are readily accessible, with a rapid convergence from the extremal poles towards the inner part of the spectrum, whereas in the KPM approach this information is not simple to obtain. One advantage of the KPM method is that it is possible to calculate off-diagonal dynamical correlation functions, $I(\omega) = -\frac{1}{\pi}\mathrm{Im}[\langle\Psi_0|\mathcal{A}\frac{1}{\omega+E_0+i\eta-H}\mathcal{B}|\Psi_0\rangle]$, where $\mathcal{A}$ and $\mathcal{B}$ are not adjoint operators, whereas this is rather cumbersome within the continued-fraction framework.

### 18.6.9 Time Evolution

An obvious way to perform real-time evolution in ED is to diagonalize fully the Hamiltonian $H$ so that the propagator $U(t) := \exp[-itH]$ is known for all initial states and times $t$. One major drawback is that the full diagonalization limits the applicability to rather small systems ($\dim\mathcal{H} \approx 10^4 \sim 10^5$). If, however, one is only

---

[12] of order the machine precision.

interested in computing the time evolution of a specific initial state $|\psi(0)\rangle$, then very powerful Krylov methods are available [101]. The idea of these is to find a very good approximation for $|\psi(\Delta t)\rangle = U(\Delta t)|\psi(0)\rangle$ (but not for other initial states) by constructing the Krylov space $\mathcal{K}_n$ =span$\{|\psi(0)\rangle, H|\psi(0)\rangle H^2|\psi(0)\rangle, \ldots, H^n|\psi(0)\rangle\}$ using a Lanczos recursion. In this basis the Hamiltonian is tridiagonal and can easily be exponentiated. For a given (and not necessarily small) time interval $\Delta t$, the method converges exponentially in the Krylov space dimension $n$, analogous to the conventional Lanczos algorithm for eigenvalues. This type of method will certainly be very useful for the study of non-equilibrium dynamics and decoherence in frustrated quantum systems in the future.

### *18.6.10 Finite Temperatures*

In this section, I have focused on ED tools at $T = 0$, but the ED method is also suitable for the calculation of finite-temperature properties. At intermediate to high temperatures the results are essentially exact, even on the small systems which can be handled, because the correlation lengths are very short at these temperatures. The first way to proceed is simply by diagonalizing the Hamiltonian completely using a numerical linear-algebra routine (e.g. from the `LAPACK` or `SCALAPACK` libraries). From the full spectrum it is straightforward to calculate thermodynamic quantities such as the specific heat or the uniform susceptibility (if $S^z$ is conserved). If one obtains also all of the eigenfunctions, then any other observable including dynamical correlation functions becomes accessible. Presently, the largest spin systems to be diagonalized completely are up to 24 $S = 1/2$ spins using all spatial symmetries and $S^z$ conservation (corresponding to a Hilbert space of approximate dimension 100,000) [102]. Codes employing a second approach to this type of calculation, known as "finite-temperature-Lanczos" techniques [103–106], combine a stochastic sampling of the Hilbert space with Lanczos recursions on the randomly chosen states in order to approximate the full density of states. The key advantage is that only operations of the type $|\phi\rangle = H|\psi\rangle$ are needed, and the sparseness of the Hamiltonian enables studies of somewhat larger systems than are possible with full-diagonalization techniques.

## Further Literature

1. N. Laflorencie and D. Poilblanc, *Simulations of pure and doped low-dimensional spin-1/2 gapped systems*, Lect. Notes Phys 645, 227 (2004).
2. R.M. Noack and S.R. Manmana *Diagonalization- and Numerical Renormalization-Group-Based Methods for Interacting Quantum Systems*, AIP Conf. Proc. 789, 93 (2005).
3. A. Weisse and H. Fehske, *Exact Diagonalization Techniques*, Lect. Notes Phys. 739, 529 (2008).
4. G. Misguich and P. Sindzingre, *Detecting spontaneous symmetry breaking in finite-size spectra of frustrated quantum antiferromagnets*, J. Phys.: Condens. Matter 19, 145202 (2007).

## 18.7 Miscellaneous Further Methods

As stated in the introduction, the field of numerical simulations is too vast for a chapter of this type to give a detailed overview of all the existing methods. Selected methods which presently have something of a "niche" status in the field of frustrated systems, but which have the potential to become more widely used in the future, are thus collected here for a brief mention.

### 18.7.1 *Classical Spin Dynamics (Molecular Dynamics)*

The Monte Carlo approach to classical spin systems is discussed in Sect. 18.2. For classical spin systems which are endowed with a non-trival Hamiltonian dynamics (such as $O(3)$ vector spins with Heisenberg interactions) it is also possible to integrate the classical equations of motion in time to obtain thermodynamic and even dynamical properties [107], in a spirit similar to the molecular-dynamics methods applied for classical particles in the continuum. For a classical Heisenberg system the equations of motion are

$$\frac{\mathrm{d}\boldsymbol{S}_i}{\mathrm{d}t} = \boldsymbol{S}_i \times \boldsymbol{H}_i(t), \tag{18.25}$$

where $\boldsymbol{H}_i$ is the exchange field acting on site $i$. Conventional Runge-Kutta schemes can be used to integrate the equations of motion, although they suffer from a drift in energy at long times (symplectic algorithms have been proposed to eliminate this problem). The method has been applied to a number of classical frustrated systems, some examples being the kagomé Heisenberg antiferromagnet [108–110], a study of the low-temperature properties of the classical pyrochlore antiferromagnet [111], and more recently also highly frustrated magnetic molecules [112].

### 18.7.2 *Coupled-Cluster Method*

The coupled-cluster method is a technique used widely in quantum chemistry and nuclear physics, which expands the ground state being sought by starting from a reference wave function as a vacuum. The expansion is performed in terms of the contributions of certain clusters of operators. Reviews of the method may be found in [113, 114]. In the field of frustrated systems, the method has been used to determine the region of stability of magnetically ordered phases, for example in the $J_1 - J_2$ square-lattice model and the Shastry–Sutherland lattice [114]. The method is in principle also applicable to non-magnetic ordered states, but has not to date received much exploration in this context.

### 18.7.3 Dynamical Mean-Field Theory (DMFT)

The dynamical mean-field theory has become a very popular technique in the field of strongly correlated metals, providing a significant improvement in our understanding of the interaction- and doping-driven Mott transitions in real materials. Extensions of the method to larger interacting clusters coupled to a self-consistent bath (a review may be found in [115]) has also enabled its first applications to model systems for itinerant charge carriers on frustrated geometries such as the kagomé lattice [116]. These studies showed that the Mott transition in frustrated systems is very rich and that many questions remain to be addressed, especially in the light of possible spin-liquid phases close to the Mott transition in triangular-lattice organic conductors [117].

### 18.7.4 Contractor Renormalization (CORE)

The contractor renormalization technique [118,119] is a real-space block-decimation technique providing an effective Hamiltonian for describing the chosen degrees of freedom of a building block of a system. The reader is referred to the chapter by Mila and Schmidt for a short description of the method. Depending on the problem at hand, the procedure can sometimes be repeated until a fixed point is reached [120]. Another approach is to perform a certain number of CORE steps and then to apply a mean-field-type analysis on the renormalized Hamiltonian to obtain a solution of the problem [121, 122]. Yet another approach is to simulate the effective Hamiltonians numerically [123], which then corresponds to solving the Hamiltonian on lattices larger than would have been possible for the initial Hamiltonian. For example, by this approach it is possible to simulate a 48-site $S = 1/2$ kagomé cluster, a size which remains beyond the reach of present-day ED codes.

### 18.7.5 SR-RVB Calculations

An approximate numerical approach which has turned out to be rather useful for some systems is the projection of a frustrated $S = 1/2$ Heisenberg Hamiltonian onto the subspace of short-range singlet coverings of the lattice under consideration. Because these coverings are composed of SU(2) singlets, they are non-orthogonal. The numerical problem to be solved is therefore a generalized eigenvalue problem consisting of the Hamiltonian and the overlap matrix. Lattices up to 60 sites can be diagonalized using a SCALAPACK parallelized version of the code. Applications of this approach first appeared and became popular for the kagomé lattice [124–127], but other frustrated models on square lattices have also been studied, highlighting the presence of valence-bond crystal phases [128, 129].

## 18.8 Source Code Availability

- ALPS project
  The "Algorithms and Libraries for Physics Simulations" project provides a set of libraries to facilitate simulations of quantum lattice models, such as a lattice and model library, a library for the evaluation of Monte Carlo data, and a parallel scheduler. These libraries can easily be included in the reader's own codes. Furthermore, the ALPS project provides a number of applications which build on these libraries. Currently, an exact diagonalization code (sparsediag, fulldiag), a density-matrix renormalization-group code, different Quantum Monte Carlo codes (SSE, Loop, Worm, and Quantum Wang-Landau) and a Monte Carlo code for classical spin problems are included and available as `C++` source code.
  For more information, tutorials, and the license condition ("cite-me" licence), consult the following website:

  `http://alps.comp-phys.org`

- TITPACK
  Hidetoshi Nishimori distributes his exact diagonalization package TITPACK under a cite-me license. Registration for download at:

  `http://www.stat.phys.titech.ac.jp/nishimori/titpack2/index-e.html`

- Spinpack
  Jörg Schulenburg developed a C/C++ implementation of an exact diagonalization code for spin models which is freely available. It is optimized for speed and also determines the spatial symmetry automatically.

  `http://www-e.uni-magdeburg.de/jschulen/spin/`

- Stochastic Series Expansion
  Anders Sandvik has a sample code of a SSE program for the square lattice Heisenberg antiferromagnet available for download at the following location:

  `http://physics.bu.edu/~sandvik/programs/index.html`

- Coupled Cluster Method
  Jörg Schulenburg also hosts a page with a coupled-cluster code package available for download at:

  `http://www-e.uni-magdeburg.de/jschulen/ccm/index.html`

**Acknowledgements**

In my research on numerical methods and frustrated systems I have profited from many stimulating and insightful collaborations and discussions with the following colleagues: F. Alet, R.N. Bhatt, S. Capponi, J.-B. Fouet, C.L. Henley, A. Honecker,

C. Kollath, N. Laflorencie, C. Lhuillier, A. Lüscher, M. Mambrini, S.R. Manmana, I. McCulloch, F. Mila, G. Misguich, R. Moessner, R.M. Noack, K. Penc, D. Poilblanc, T.M. Rice, J. Richter, I. Rousochatzakis A. Sandvik, K.P. Schmidt, U. Schollwöck, J. Schulenburg, P. Sindzingre, R. R. P. Singh, J. Sudan, S. Trebst, M. Troyer, C. Weber, P. Werner, S. Wessel, and S.R. White.

## References

1. N. Metropolis, A.W. Rosenbluth, M.N. Rosenbluth, A.M. Teller, E. Teller, J. Chem. Phys. 21 1087 (1953)
2. M. Creutz, Phys. Rev. D 36, 515 (1987)
3. J.T. Chalker, P.C. Holdsworth, E.F. Shender, Phys. Rev. Lett. 68, 855 (1992)
4. R. Moessner, J.T. Chalker, Phys. Rev. Lett. 80, 2929 (1998)
5. M.E. Zhitomirsky, Phys. Rev. Lett. 88, 057204 (2002)
6. C. Weber, L. Capriotti, G. Misguich, F. Becca, M. Elhajal, F. Mila, Phys. Rev. Lett. 91, 177202 (2003)
7. R.H. Swendsen, J.-S. Wang, Phys. Rev. Lett. 58, 86 (1987)
8. U. Wolff, Phys. Rev, Lett. 62, 361 (1989)
9. P.-W. Leung, C.L. Henley, Phys. Rev. B 43, 752 (1991)
10. D.A. Huse, W. Krauth, R. Moessner, S.L. Sondhi, Phys. Rev. Lett. 91, 167004 (2003)
11. A.W. Sandvik, R. Moessner, Phys. Rev. B 73, 144504 (2006)
12. F. Alet, J.L. Jacobsen, G. Misguich, V. Pasquier, F. Mila, M. Troyer, Phys. Rev. Lett. 94, 235702 (2005)
13. F. Alet, G. Misguich, V. Pasquier, R. Moessner, J.L. Jacobsen, Phys. Rev. Lett. 97, 030403 (2006)
14. F. Wang, D.P. Landau, Phys. Rev. Lett. 86, 2050 (2001)
15. K. Hukushima, K. Nemoto, J. Phys. Soc. Jpn. 65, 1604 (1996)
16. H.G. Katzgraber, S. Trebst, D.A. Huse, M. Troyer, J. Stat. Mech. P03018 (2006)
17. M.V. Gvozdikova, M.E. Zhitomirsky, JETP Lett. 81, 236 (2005)
18. D. Bergman, J. Alicea, E. Gull, S. Trebst, L. Balents, Nat. Phys. 3, 487 (2007)
19. A.W. Sandvik, Phys. Rev. B 59, 14157 (1999)
20. O.F. Syljuåsen, A.W. Sandvik, Phys. Rev. E 66, 046701 (2002)
21. F. Alet, S. Wessel, M. Troyer, Phys. Rev. E 71, 036706 (2005)
22. M. Troyer, U.-J. Wiese, Phys. Rev. Lett. 94, 170201 (2005)
23. S. Chandrasekharan, U.-J. Wiese, Phys. Rev. Lett. 83, 3116 (1999)
24. M. Nyfeler, F.-J. Jiang, F. Kämpfer, U.-J. Wiese, Phys. Rev. Lett. 100, 247206 (2008)
25. T. Roscilde, A. Feiguin, A.L. Chernyshev, S. Liu, S. Haas, Phys. Rev. Lett. **93**, 017203 (2004)
26. S. Wessel, M. Troyer, Phys. Rev. Lett. 95, 127205 (2005)
27. D. Heidarian, K. Damle, Phys. Rev. Lett. 95, 127206 (2005)
28. R. Melko et al., Phys. Rev. Lett. 95, 127207 (2005)
29. S.V. Isakov, S. Wessel, R.G. Melko, K. Sengupta, Y.-B. Kim, Phys. Rev. Lett. 97, 147202 (2006)
30. A. Banerjee, S.V. Isakov, K. Damle, Y.-B. Kim, Phys. Rev. Lett. 100, 047208 (2008)
31. R.G. Melko, J. Phys.: Condens. Matter 19, 145203 (2007)
32. N. Trivedi, D.M. Ceperley, Phys. Rev. B 41, 4552 (1990)
33. M.C. Buonaura, S. Sorella, Phys. Rev. B 57, 11446 (1998)
34. O.F. Syljuåsen, Phys. Rev. B 71, 020401(R) (2005)
35. O.F. Syljuåsen, Int. J. Mod. Phys. B 19, 1973 (2005)
36. A. Ralko, M. Ferrero, F. Becca, D. Ivanov, F. Mila, Phys. Rev. B 71, 224109 (2005)
37. O.F. Syljuåsen, Phys. Rev. B 73, 245105 (2006)
38. A. Ralko, F. Mila, D. Poilblanc, Phys. Rev. Lett. 99, 127202 (2007)

39. C.L. Henley, J. Phys.: Condens. Matter 16, S891 (2004)
40. D.A. Ivanov, Phys. Rev. B 70, 094430 (2004)
41. A. Ralko, M. Ferrero, F. Becca, D. Ivanov, F. Mila, Phys. Rev. B 74, 134301 (2006)
42. A. Läuchli, S. Capponi, F.F. Assaad, J. Stat. Mech. P01010 (2008)
43. L. Capriotti, A.E. Trumper, S. Sorella, Phys. Rev. Lett. 82, 3899 (1999)
44. L. Capriotti, S. Sorella, Phys. Rev. Lett. 84, 3173 (2000)
45. B. Bernu, G. Misguich, Phys. Rev. B 63, 134409 (2001)
46. M. Rigol, T. Bryant, R.R.P. Singh, Phys. Rev. Lett. 97, 187202 (2006)
47. R.R.P. Singh, D.A. Huse, Phys. Rev. B 76, 180407 (2007)
48. M.P. Gelfand, R.R.P. Singh, Adv. Phys. 49, 93 (2000)
49. S. Trebst, H. Monien, C.J. Hamer, Z. Weihong, R.R.P. Singh, Phys. Rev. Lett. 85, 4373 (2000)
50. W. Zheng, J.O. Fjærestad, R.R.P. Singh, R.H. McKenzie, R. Coldea, Phys. Rev. Lett. 96, 057201 (2006)
51. C. Knetter, G.S. Uhrig, Phys. Rev. Lett. 92, 027204 (2004)
52. S.R. White, Phys. Rev. Lett. 69, 2863 (1992)
53. S.R. White, Phys. Rev. B 48, 10345 (1993)
54. S.R. White, R.M. Noack, Phys. Rev. Lett. 68, 3487 (1992)
55. R.J. Bursill, G.A. Gehring, D.J.J. Farnell, J.B. Parkinson, C. Zeng, T. Xiang, J. Phys. Condens. Matter 7, 8605 (1995)
56. R. Chitra, S. Pati, H.R. Krishnamurthy, D. Sen, S. Ramasesha, Phys. Rev. B 52, 6581 (1995)
57. S.R. White, I. Affleck, Phys. Rev. B 54, 9862 (1996)
58. S.R. White, Phys. Rev. Lett. 77, 3633 (1996)
59. S.R. White, R.R.P. Singh, Phys. Rev. Lett. 85, 3330 (2000)
60. J.-B. Fouet, A. Läuchli, S. Pilgram, R.M. Noack, F. Mila, Phys. Rev. B 73, 014409 (2006)
61. J.-B. Fouet, F. Mila, D. Clarke, H. Youk, O. Tchernyshyov, P. Fendley, R.M. Noack, Phys. Rev. B 73, 214405 (2006)
62. K. Penc, J.-B. Fouet, S. Miyahara, O. Tchernyshyov, F. Mila, Phys. Rev. Lett. 99, 117201 (2007)
63. S.R. White, A.L. Chernyshev, Phys. Rev. Lett. 99, 127004 (2007)
64. H.C. Jiang, Z.Y. Weng, D.N. Sheng, Phys. Rev. Lett. 101, 117203 (2008)
65. K. Maisinger, U. Schollwöck, Phys. Rev. Lett. 81, 445 (1998)
66. Ch. Waldtmann, H. Kreutzmann, U. Schollwöck, K. Maisinger, H.-U. Everts, Phys. Rev. B 62, 9472 (2000)
67. A.E. Feiguin, S.R. White, Phys. Rev. B 72, 220401 (2005)
68. K.A. Hallberg, Phys. Rev. B 52, 9827 (1995)
69. T.D. Kühner, S.R. White, Phys. Rev. B 60, 335 (1999)
70. E. Jeckelmann, Phys. Rev. B 66, 045114 (2002)
71. U. Schollwöck, S.R. White, in Methods for Time Dependence in DMRG, eds. by G.G. Batrouni, D. Poilblanc, Effective Models for Low-dimensional Strongly Correlated Systems, (AIP, Melville, New York, 2006) p. 155; arXiv:cond-mat/0606018
72. S.R. White, I. Affleck, Phys. Rev. B 77, 134437 (2008)
73. F. Verstraete, J.I. Cirac, arXiv:cond-mat/0407066
74. A. Isacsson, O.F. Syljuåsen, Phys. Rev. E 74, 026701 (2006)
75. C. Lanczos, J. Res. Nat. Bur. Stand. 45, 255 (1950)
76. E. Davidson, J. Comput. Phys. 17, 87 (1975); see also Comput. Phys. 7, 519 (1993)
77. S. Yamada, T. Imamura, M. Machida, Lect. Notes Comput. Sci. 4759, 402 (2008)
78. H.Q. Lin, Phys. Rev. B 42, 6561 (1990)
79. H.J. Schulz, T.A.L. Ziman, D. Poilblanc, J. Phys. I France 6, 675 (1996)
80. J.K. Cullum, R.A. Willoughby, Lanczos Algorithms for Large Symmetric Eigenvalue Computations (Birkhäuser, Boston, 1985)
81. Z. Bai, et al.(eds.), Templates for the solution of Algebraic Eigenvalue Problems: A Practical Guide (SIAM, Philadelphia, 2000). Online book available at http://www.cs.ucdavis.edu/ bai/ET/contents.html
82. B. Bernu, C. Lhuillier, L. Pierre, Phys. Rev. Lett. 69, 2590 (1992)
83. B. Bernu, P. Lecheminant, C. Lhuillier, L. Pierre, Phys. Rev. B 50, 10048 (1994)

84. P. Lecheminant, B. Bernu, C. Lhuillier, L. Pierre, P. Sindzingre, Phys. Rev. B 56, 2521 (1997)
85. M. Roger, J.H. Hetherington, J.M. Delrieu, Rev. Mod. Phys. 55, 1 (1983)
86. M.C. Cross, D.S. Fisher, Rev. Mod. Phys. 57, 881 (1985)
87. D.M. Ceperley, G. Jacucci, Phys. Rev. Lett. 58, 1648 (1987)
88. D.M. Ceperley, G. Jacucci, Phys. Rev. Lett. 59, 380 (1987)
89. D.M. Ceperley, Rev. Mod. Phys. 67, 279 (1995)
90. A. Lüscher, A. Läuchli, Phys. Rev. B 79, 195102 (2009)
91. A. Läuchli, G. Schmid, M. Troyer, Phys. Rev. B 67, 100409(R) (2003)
92. A. Läuchli, J.-C. Domenge, C. Lhuillier, P. Sindzingre, M. Troyer, Phys. Rev. Lett. 95, 137206 (2005)
93. N. Shannon, T. Momoi, P. Sindzingre, Phys. Rev. Lett. 96, 027213 (2006)
94. T. Momoi, P. Sindzingre, N. Shannon, Phys. Rev. Lett. 97, 257204 (2006)
95. A. Läuchli, F. Mila, K. Penc, Phys. Rev. Lett. 97, 087205 (2006)
96. S. Furukawa, G. Misguich, M. Oshikawa, Phys. Rev. Lett. 96, 047211 (2006)
97. S.-A. Cheong, C.L. Henley, Phys. Rev. B 79, 212402 (2009)
98. J. Sudan, C.L. Henley, A.M. Läuchli, unpublished.
99. A. Läuchli, D. Poilblanc, Phys. Rev. Lett. 92, 236404 (2004)
100. A. Weisse, G. Wellein, A. Alvermann, H. Fehske, Rev. Mod. Phys. 78, 275 (2006)
101. T.J. Park, J.C. Light, J. Chem. Phys. 85, 5870 (1986)
102. F. Heidrich-Meisner, A. Honecker, T. Vekua, Phys. Rev. B 74, 020403(R) (2006)
103. J. Jaklic, P. Prelovsek, Phys. Rev. B 49, 5065 (1994)
104. J. Jaklic, P. Prelovsek, Adv. Phys. 49, 1 (2000)
105. M. Aichhorn, M. Daghofer, H.-G. Evertz, W. von der Linden, Phys. Rev. B 67, 161103 (2003)
106. N. Shannon, B. Schmidt, K. Penc, P. Thalmeier, Eur. Phys. J. B 38, 599 (2004)
107. D.P. Landau, M. Krech, J. Phys.: Condens. Matter 11, R179 (1999)
108. A. Keren, Phys. Rev. Lett. 72, 3254 (1994)
109. J. Lee, J. Magn. Magn. Mater. 292, 462 (2005)
110. J. Robert, B. Canals, V. Simonet, R. Ballou, Phys. Rev. Lett. 101, 117207 (2008)
111. R. Moessner, J.T. Chalker, Phys. Rev. B 58, 12049 (1998)
112. C. Schröder, H.-J. Schmidt, J. Schnack, M. Luban, Phys. Rev. Lett. 94, 207203 (2005)
113. D.J.J. Farnell, R.F. Bishop, Lect. Notes Phys. 645, 307 (2004)
114. J. Richter, R. Darradi, R. Zinke, R.F. Bishop, Int. J. Mod. Phys. B 21, 2273 (2007)
115. T. Maier, M. Jarrell, T. Pruschke, M.H. Hettler, Rev. Mod. Phys. 77, 1027 (2005)
116. T. Ohashi, N. Kawakami, H. Tsunetsugu, Phys. Rev. Lett. 97, 066401 (2006)
117. Y. Shimizu, K. Miyagawa, K. Kanoda, M. Maesato, G. Saito, Phys. Rev. Lett. 91, 107001 (2003)
118. C.J. Morningstar, M. Weinstein, Phys. Rev. Lett. 73, 1873 (1994)
119. C.J. Morningstar, M. Weinstein, Phys. Rev. D 54, 4131 (1996)
120. M. Weinstein, Phys. Rev. B 63, 174421 (2001)
121. E. Berg, E. Altman, A. Auerbach, Phys. Rev. Lett. 90, 147204 (2003)
122. R. Budnik, A. Auerbach, Phys. Rev. Lett. 93, 187205 (2004)
123. S. Capponi, A. Läuchli, M. Mambrini, Phys. Rev. B 70, 104424 (2004)
124. C. Zeng, V. Elser, Phys. Rev. B 51, 8318 (1995)
125. M. Mambrini, F. Mila, Eur. Phys. J. B 17, 651 (2000)
126. G. Misguich, D. Serban, V. Pasquier, Phys. Rev. B 67, 214413 (2003)
127. S. Dommange, M. Mambrini, B. Normand, F. Mila, Phys. Rev. B 68, 224416 (2003)
128. J.-B. Fouet, M. Mambrini, P. Sindzingre, C. Lhuillier, Phys. Rev. B 67, 054411 (2003)
129. M. Mambrini, A. Läuchli, D. Poilblanc, F. Mila, Phys. Rev. B 74, 144422 (2006)
130. G. Misguich, B. Bernu, Phys. Rev. B 71, 014417 (2005)
131. G. Misguich, B. Bernu, L. Pierre, Phys. Rev. B 68, 113409 (2003)

# Chapter 19
# Exact Results in Frustrated Quantum Magnetism

**Shin Miyahara**

**Abstract** Most of the exact results in frustrated spin systems have for a long time been regarded as of purely academic interest, being realized only due to the special geometry of the lattices concerned. However, recent developments in material design offer the genuine possibility of producing such exact states in real materials. In fact, the exact dimer singlet state of the two-dimensional Shastry–Sutherland model has already been found as the ground state of the quasi-two-dimensional material $SrCu_2(BO_3)_2$. The cooperation between experimentalists and theorists in investigating this material has caused rapid development in the understanding of low-dimensional frustrated spin systems in general, due to the extreme utility of cases where the ground state is known exactly. This fact provides information essential to recognizing novel magnetic behavior in external magnetic fields, at finite temperatures, and in other regimes. In this chapter, we introduce spin-$1/2$ models which have an exact ground state, considering first exactly solvable spin-$1/2$ Heisenberg models, exemplified by the sawtooth–chain model, the Majumdar–Ghosh model, the two-dimensional Shastry–Sutherland model, and a frustrated ladder model. Such exact states can be realized due to special symmetries on geometrically frustrated lattices. As a second class of examples, we introduce also some exact ground states in spin-1/2 models with multiple-spin interactions.

## 19.1 Introduction

In some branches of theoretical physics, finding an exact solution for a Hamiltonian can be the highlight of a research career. Such interests were for a long time limited only to theorists in more abstract pursuits, because most models with exact results have a complex structure. However, recent developments in material design have changed the situation dramatically, to the point where finding an exact solution in some materials, rather than merely in models, can be one of the highlights for an experimentalist. Indeed, the number of spin-1/2 models previously considered as objects of purely mathematical curiosity because of their unique structures, but now realized as materials, is not small. Examples include $SrCu_2O_3$ [1] for a spin ladder system, $CaV_4O_9$ [2] for a 1/5-depleted Heisenberg square-lattice system, and $Cu_3V_2O_7(OH)_2 \cdot 2H_2O$ [3] and $ZnCu_3(OH)_6Cl_2$ [4,5] for kagomé lattices. Although an exact ground state is not obtained in any of these examples, the very fact that such models can be synthesized suggests the possibility of realizing exact ground states in real materials.

Indeed, the exact ground state of the two-dimensional Shastry–Sutherland model [6] is found in the orthogonal-dimer material $SrCu_2(BO_3)_2$ [7,8]. This is a spectacularly successful example of collaboration between experimentalists and theorists leading to a full understanding of the properties of a quantum spin system. In this material, experiments including magnetization in external fields, inelastic neutron scattering, electron spin resonance, (ESR), Raman scattering, nuclear magnetic resonance (NMR), and others, have been performed in a short period, and because of the exactly known ground state, most of the experimental results can be understood rather well [9]. At the same time, experimental results act to promote theoretical development, for example in the understanding of magnetization plateaus in two-dimensional lattices (as discussed in chap. 10). In fact, many properties of such strongly correlated materials, including excited states and the response to external magnetic fields, can be interpreted much more easily than in regular cases. The obvious advances in the field of quantum magnetism made possible by the realization of a material with an exact ground state makes their synthesis a goal of primary importance. Thus it is the aim of this chapter to present for experimentalists, as well, needless to say, as for theorists, an overview of the exactly solvable models already known, as a basis for the further development of this type of research.

### 19.1.1 Dimer Model

We begin with the simplest example which has an exact ground state, a single-dimer model [10, 11]. Knowledge of the ground state of this system is of key importance because, as will emerge shortly, most exactly known states consist of the products of dimer singlets.

The Hamiltonian for the Heisenberg model on a dimer cluster is written as

$$h^d = J\boldsymbol{S}_1 \cdot \boldsymbol{S}_2, \tag{19.1}$$

where $\boldsymbol{S}_i$ is an $S = 1/2$ spin operator and we represent its spin states by using the eigenstates for the $S^z$ operator, $\uparrow_i$ for $S_i^z = 1/2$ and $\downarrow_i$ for $S_i^z = -1/2$. In terms of the raising and lowering operators $S_i^+ = S_i^x + iS_i^y$ and $S_i^- = S_i^x - iS_i^y$, the dimer Hamiltonian may be reexpressed as

$$h^d = \frac{J}{2}\left(S_1^+ S_2^- + S_1^- S_2^+\right) + JS_1^z S_2^z. \tag{19.2}$$

Using the basis

$$|\uparrow_1\uparrow_2\rangle,\quad |\uparrow_1\downarrow_2\rangle,\quad |\downarrow_1\uparrow_2\rangle,\quad |\downarrow_1\downarrow_2\rangle, \tag{19.3}$$

the matrix for $h_d$ is represented by

$$\begin{pmatrix} \frac{J}{4} & 0 & 0 & 0 \\ 0 & -\frac{J}{4} & \frac{J}{2} & 0 \\ 0 & \frac{J}{2} & -\frac{J}{4} & 0 \\ 0 & 0 & 0 & \frac{J}{4} \end{pmatrix}. \tag{19.4}$$

This matrix is easy to solve, yielding a singlet state, $|s\rangle$, with eigenenergy $e_{\mathrm{s}}^d = -3J/4$ and threefold-degenerate triplet states, $|t_\mu\rangle$ ($\mu = 1, 0, -1$), with eigenenergies $e_{\mathrm{t}}^d = J/4$, whose wave functions are

$$|s\rangle = \frac{1}{\sqrt{2}}\left(|\uparrow_1\downarrow_2\rangle - |\downarrow_1\uparrow_2\rangle\right), \tag{19.5}$$

$$|t_1\rangle = |\uparrow_1\uparrow_2\rangle, \tag{19.6}$$

$$|t_0\rangle = \frac{1}{\sqrt{2}}\left(|\uparrow_1\downarrow_2\rangle + |\downarrow_1\uparrow_2\rangle\right), \tag{19.7}$$

$$|t_{-1}\rangle = |\downarrow_1\downarrow_2\rangle. \tag{19.8}$$

The degeneracy of each spin manifold is $2S + 1$, whence, again, the $S = 0$ state is a singlet and the $S = 1$ state a triplet. Here, $S$ is simply the quantum number for total spin, $\boldsymbol{S}^{\mathrm{tot}} = \boldsymbol{S}_1 + \boldsymbol{S}_2$, the eigenvalues of $(\boldsymbol{S}^{\mathrm{tot}})^2$ being $S(S+1)$. These results can be seen in a straightforward manner on rewriting the Hamiltonian as

$$h^d = \frac{J}{2}\left\{(\boldsymbol{S}^{\mathrm{tot}})^2 - (\boldsymbol{S}_1)^2 - (\boldsymbol{S}_2)^2\right\}, \tag{19.9}$$

$$= \frac{J}{2}\left(\boldsymbol{S}^{\mathrm{tot}}\right)^2 - \frac{3J}{4}, \tag{19.10}$$

whence an $S = 0$ state gives an energy of $-3J/4$ and $S = 1$ an energy $J/4$. From the point of view of symmetry, the singlet is antisymmetric with respect to the exchange of spins, while the triplet states are symmetric.

For $J > 0$, the singlet is the ground state and the triplet states are excited. The energy difference $J$ between the singlet and the triplets can be considered as a spin gap. On application of an external magnetic field, the excited states split into three by the Zeeman effect. By extension from the one-dimer singlet state, a ground state represented by the product of singlet states also has a spin gap.

## 19.2 Exact Results in Spin-1/2 Heisenberg Models

In frustrated spin systems, a particular special geometry of the lattice may occasionally produce an exact ground state. In this chapter, several spin-1/2 Heisenberg models with exact ground states for certain sets of parameters are introduced. In most of them, the Hamiltonian $H$ can be written as a sum of cluster Hamiltonians,

$$H = \sum_{\alpha} h_{\alpha}. \tag{19.11}$$

The cluster Hamiltonian $h_{\alpha}$ is such that its ground state is easy to obtain. If the whole space can be tiled by using the ground-state spin configurations of the cluster, the resulting state is an exact ground state of the full Hamiltonian $H$ [12]. If the ground state of the cluster is non-degenerate, as in the dimer model, it is rather difficult to tile the whole space using only the cluster ground state. However, the ground state of many frustrated clusters is degenerate, which enhances the possibility of tiling the whole space by using some combination of these cluster ground states. In fact, the exact ground state can be constructed by filling the whole space with one of the degenerate ground states of the frustrated cluster: this is the procedure which generates the exact Affleck–Kennedy–Lieb–Tasaki ground state of the spin-1 chain model[13] (although in this chapter we discuss only spin-1/2 systems). From a complementary point of view, the ground state of the cluster is an eigenstate for some symmetry operators in most cases, and thus a state tiled by it can be considered as a symmetrical ordered state of a set of cluster states.

### *19.2.1 Exact Ground States in Coupled Triangular Cluster Models*

In this section, we treat models which consist of triangular spin-1/2 antiferromagnetic Heisenberg clusters,

$$H^{\mathrm{tri}} = \sum_{\alpha} h_{\alpha}^{\mathrm{tri}}. \tag{19.12}$$

Here, $h_{\alpha}^{\mathrm{tri}}$ is a spin-1/2 Heisenberg model on a triangular cluster, and is written as

$$h^{\mathrm{tri}} = J\left(\boldsymbol{S}_1 \cdot \boldsymbol{S}_2 + \boldsymbol{S}_2 \cdot \boldsymbol{S}_3 + \boldsymbol{S}_3 \cdot \boldsymbol{S}_1\right) \tag{19.13}$$

$$= \frac{J}{2}\left(\boldsymbol{S}_1 + \boldsymbol{S}_2 + \boldsymbol{S}_3\right)^2 - \frac{9}{8}J, \tag{19.14}$$

where $\boldsymbol{S}$ is a spin-1/2 operator and the spin labels are shown in Fig. 19.1a. Clearly, from (19.14), the ground state of the triangular cluster is a state of total spin $S^{\mathrm{tot}} = 1/2$, with an eigenenergy per cluster of $e^{\mathrm{tri}} = -3J/4$. As in the dimer case, the matrix can be constructed from eight basis states of the form $|\uparrow_1\uparrow_2\uparrow_3\rangle$ and $|\uparrow_1\uparrow_2\downarrow_3\rangle$, and solving it yields two eigenenergies, $-3J/4$ and $3J/4$. Here, both energy states have a degeneracy of four, and the four-fold-degenerate ground state can, for example, be written as

$$|\phi_1\rangle = [1,2] \times |\uparrow_3\rangle, \tag{19.15}$$

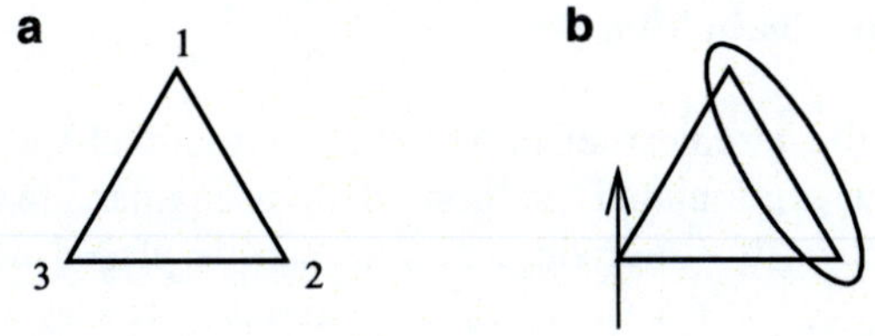


**Fig. 19.1** (**a**) Labeling of triangular cluster for spin-1/2 Heisenberg model. (**b**) Sketch of one of the ground states on the triangular cluster. The ellipse represents the dimer singlet state; the spin state on site 3 is arbitrary

$$|\phi_2\rangle = \frac{1}{\sqrt{3}}([2,3]\times|\uparrow_1\rangle - [3,1]\times|\uparrow_2\rangle)$$
$$= \frac{1}{\sqrt{6}}\{2|\uparrow_1\uparrow_2\rangle\times|\downarrow_3\rangle - (|\uparrow_1\downarrow_2\rangle + |\downarrow_1\uparrow_2\rangle)\times|\uparrow_3\rangle\}, \quad (19.16)$$
$$|\phi_3\rangle = [1,2]\times|\downarrow_3\rangle, \quad (19.17)$$
$$|\phi_4\rangle = \frac{1}{\sqrt{3}}([2,3]\times|\downarrow_1\rangle - [3,1]\times|\downarrow_2\rangle)$$
$$= \frac{1}{\sqrt{6}}\{-2|\downarrow_1\downarrow_2\rangle\times|\uparrow_3\rangle + (|\uparrow_1\downarrow_2\rangle + |\downarrow_1\uparrow_2\rangle)\times|\downarrow_3\rangle\}, \quad (19.18)$$

where

$$[i,j] \equiv \frac{1}{\sqrt{2}}(|\uparrow_i\downarrow_j\rangle - |\downarrow_i\uparrow_j\rangle) \quad (19.19)$$

represents a dimer singlet state on the dimer bond $ij$. This basis is selected to diagonalize the parity operation with respect to the bond 12. The states in (19.15) and (19.17) have odd parity, while the others have even parity. This is obvious from the fact that in the former the spin state on the bond 12 is a singlet and in the latter a triplet. For the states in both (19.15) and (19.17), the fact that a dimer singlet lies on the bond 12 means that the spin state for an arbitrary spin at site 3 is already a ground state. A sketch of the ground state of (19.15) is shown in Fig. 19.1 b. In this way, any state having a dimer singlet on one of the bonds is a ground state of $h^{\text{tri}}$, consistent with the fact that the ground state is a state of $S^{\text{tot}} = 1/2$. The state shown in Fig. 19.1b is a useful starting point for seeking an exact ground state of the full Hamiltonian (19.12). If it is possible to fulfill the condition of having a singlet dimer on every triangle for the full Hamiltonian $H^{\text{tri}}$, such a state is an exact ground state.

As examples of the realization of exactly solvable one-dimensional models with triangular motifs, we discuss a sawtooth chain and the Majumdar–Ghosh model. The Shastry–Sutherland model is a case with two-dimensional geometry, and is discussed in the end of this section.

#### 19.2.1.1 Sawtooth–Chain Model

Let us start with the simplest example, the sawtooth–chain model shown in Fig. 19.2a, whose Hamiltonian is composed of triangular clusters as shown in Fig. 19.2b. From above, if dimer singlet states can be placed on every triangular unit within the lattice, such a state has the minimum energy obtainable for the Hamiltonian. In the case with periodic boundary conditions, such states can be constructed as shown in Fig. 19.2c, d, i.e., the ground state is described as the product of singlet dimers [14]. The two-fold degenerate ground states are

$$|\Psi_1^{\mathrm{saw}}\rangle = \prod_{i=1}^{N/2} [2i-1, 2i], \tag{19.20}$$

$$|\Psi_2^{\mathrm{saw}}\rangle = \prod_{i=1}^{N/2} [2i, 2i+1], \tag{19.21}$$

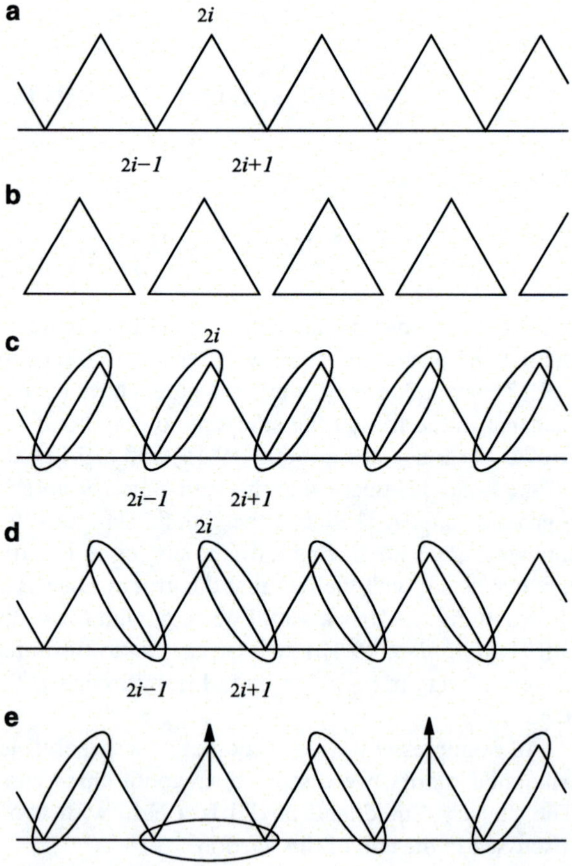


**Fig. 19.2** (**a**) The one-dimensional sawtooth-chain Heisenberg model. (**b**) The Hamiltonian is described as a sum of triangular clusters. (**c**) One of the exact ground states. Ellipses represent the dimer singlet state. (**d**) The other exact ground state; the ground state is two-fold degenerate. (**e**) An excited state: the free spin with a dimer singlet pair in its triangle is called a "kink" and that which does not is an "antikink"

and sketches representing (19.20) and (19.21) are shown in Fig. 19.2c, d, respectively. For the ground state in (19.20) and (19.21), the odd-parity state is realized between spins $2i - 1$ and $2i$ ($2i$ and $2i + 1$).

For the system with open boundary conditions, it can be shown that the ground state is $2(N + 1)$-fold degenerate, with $N$ singlet dimers and one free spin. The degeneracy arises from the freedom in position and orientation of the free spin.

The existence of an excitation energy gap above the ground state was shown rigorously in [14]. The low-energy excited state can be described as a domain wall of "kink"-"antikink" type, based on the excitations shown in Fig. 19.2e [15,16]. A kink has no excitation energy and is localized, while an antikink has a finite excitation energy and propagates. The spin gap in this system can be explained very well by the excitation energy of a kink-antikink pair: a finite energy gap plus a dispersive kinetic energy contribution due to the free motion of the antikink.

#### 19.2.1.2 Majumdar–Ghosh Model

We turn now to perhaps the best known example of exactly solvable frustrated models, the Majumdar–Ghosh chain [17]. This model is a special point in the parameter space of the one-dimensional spin-1/2 $J_1$–$J_2$ Heisenberg model, or zig–zag chain model, shown in Fig. 19.3a. The $J_1$–$J_2$ Heisenberg model has been studied intensively due to the existence over much of its phase diagram of a non-magnetic ground state with a spin excitation gap. Specifically, the ground state of the model is either a gapless state or this non-magnetic, spin-gapped state, and the quantum phase-transition point separating the two is estimated to be $(J_2/J_1)_c = 0.24$ [18]. The non-magnetic ground state is stable for $J_2/J_1 > 0.24$ and the gapless, quasi-long-range-ordered state for $J_2/J_1 < 0.24$.

At the parameter value $J_2/J_1 = 0.5$ in the spin-gap phase, which is the Majumdar–Ghosh point, the ground-state wave function is exact. The Hamiltonian of the $J_1$-$J_2$ model at this point can be written again as a sum of triangular clusters, as shown in Fig. 19.3b. Here, the $J_1$ bonds are shared by two triangles, which is the origin of the fact that the special point occurs when the magnitude of the interaction for $J_1$ bonds is twice that of the $J_2$ bonds. The energies of each triangle can be minimized by placing a dimer singlet on every one. There are only two possibilities which minimize the energies of all triangles simultaneously, and the ground states are

$$|\Psi_1^{\mathrm{MG}}\rangle = \prod_{i=1}^{N/2} [2i - 1, 2i], \tag{19.22}$$

$$|\Psi_2^{\mathrm{MG}}\rangle = \prod_{i=1}^{N/2} [2i, 2i + 1], \tag{19.23}$$

where we have assumed periodic boundary conditions and $N + 1 \equiv 1$. Once again, $[i, j]$ indicates the dimer singlet state defined in (19.19). Sketches of these ground

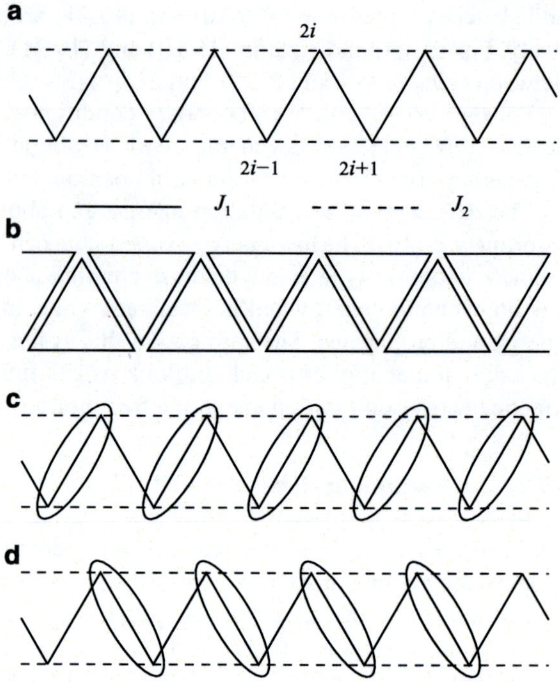


**Fig. 19.3** (**a**) The one-dimensional $J_1$–$J_2$ Heisenberg model, or zig–zag chain model. (**b**) At the Majumdar–Ghosh point, $J_1 = 2J_2$, the Hamiltonian is considered as coupled triangular clusters. (**c**) One of the exact ground states for the Majumdar–Ghosh model. Ellipses represent the dimer singlet state. (**d**) The other exact ground state; the ground state is two-fold degenerate

states are shown in Fig. 19.3c, d. A proof that the energy $E^{\mathrm{MG}} = -3J_1N/8$ is the ground-state energy is provided by van den Broek and by Shastry and Sutherland [19, 20]. Affleck, Kennedy, Lieb, and Tasaki proved further that only the two states (19.22) and (19.23) can form the ground-state manifold, and that there is a spin excitation gap between these ground states and the first excited state [21].

#### 19.2.1.3 Two-Dimensional Shastry–Sutherland Model

The spin-1/2 Heisenberg Hamiltonian

$$H^{\mathrm{ssm}} = J' \sum_{n.n.} \boldsymbol{S}_i \cdot \boldsymbol{S}_j + J \sum_{n.n.n.} \boldsymbol{S}_i \cdot \boldsymbol{S}_j \tag{19.24}$$

on the Shastry–Sutherland lattice has been put forward as an example of a two-dimensional model with an exact ground state [6]. There are two types of interactions, a nearest-neighbor bond $J'$ and a next-nearest-neighbor bond $J$ (Fig. 19.4a). For $J'/J = 0.5$, the model is a sum of isotropic triangular clusters as shown in Fig. 19.4b. As in the Majumdar–Ghosh model, the ground state is described as the product of dimer singlet states on all of the $J$ bonds, which makes it possible to place dimer singlet states on every triangle simultaneously. The remarkable property of Shastry–Sutherland model is that the dimer-product state is an eigenstate for

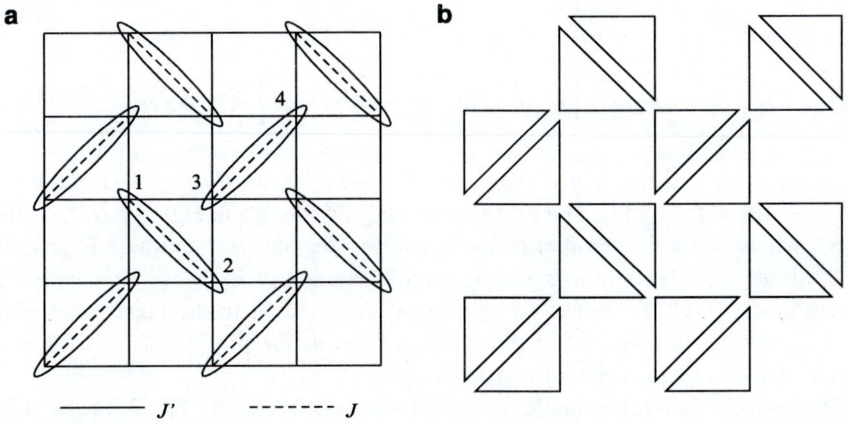


**Fig. 19.4** (**a**) The two-dimensional spin-1/2 Shastry–Sutherland Heisenberg model, where an exact dimer-product ground state is realized. Ellipses represent the dimer singlet bonds. (**b**) For $J'/J = 0.5$, the Hamiltonian is described as a sum of triangular clusters

all values of $J'/J$. Thus this state is the ground state in some region of parameter space. Further differences between this case and the Majumdar–Ghosh model are that translational symmetry is broken in the latter case but preserved here, and that the ground state of the Shastry–Sutherland model is non-degenerate.

The proof for the exact eigenstate is simple. Let us consider two coupled dimers,

$$h^{\text{od}} = J'\boldsymbol{S}_2 \cdot \boldsymbol{S}_3 + J'\boldsymbol{S}_1 \cdot \boldsymbol{S}_3 + J\boldsymbol{S}_1 \cdot \boldsymbol{S}_2 + J\boldsymbol{S}_3 \cdot \boldsymbol{S}_4, \tag{19.25}$$

$$= J'(\boldsymbol{S}_1 + \boldsymbol{S}_2) \cdot \boldsymbol{S}_3 + J\boldsymbol{S}_1 \cdot \boldsymbol{S}_2 + J\boldsymbol{S}_3 \cdot \boldsymbol{S}_4, \tag{19.26}$$

where the indices are shown in Fig. 19.4a. It is obvious from (19.26) that the $J'$ term vanishes if one considers a state in which the total spin for sites 1 and 2 is zero, $\boldsymbol{S}_1 + \boldsymbol{S}_2 = 0$, i.e., one with a singlet state on the $J$ bond. Thus, also for the Hamiltonian $H^{\text{ssm}}$ of the full system, the state which is a product of dimer singlets on the $J$ bonds is an eigenstate (and, in addition, is actually the ground state at least for $J'/J \ll 1$). From the symmetry point of view, this can be understood by considering the dimer bonds: all bonds in the ground state are odd under the parity operation with respect to the $J$ bond. For orthogonal configurations, the Hamiltonian conserves the parity. On the other hand, when a spin operator is applied to a singlet, a finite matrix element is possible only for some triplet components. These requirements are contradictory, and therefore all matrix elements should vanish. This restriction due to the parity leads to an almost complete localization of the triplet excitations [8].

We prove next that the dimer-product state is the ground state in some parameter range, by following the variational method discussed by Shastry and Sutherland [6]. The Hamiltonian (19.24) can be considered as a sum of isosceles triangles also for $J'/J \neq 0.5$, whence

$$H^{\mathrm{ssm}} = \sum h^{\mathrm{ist}} \tag{19.27}$$

$$= \sum \left\{ J' (\boldsymbol{S}_2 \cdot \boldsymbol{S}_3 + \boldsymbol{S}_3 \cdot \boldsymbol{S}_1) + \frac{J}{2} \boldsymbol{S}_1 \cdot \boldsymbol{S}_2 \right\}. \tag{19.28}$$

The ground-state energy per cluster is $e^{\mathrm{ist}} = -3J/8$ for $J'/J \leq 0.5$ and $e^{\mathrm{ist}} = J/8 - J'$ for $J'/J \geq 0.5$. The estimate of the ground-state energy for $H^{\mathrm{ssm}}$ is then $e^{\mathrm{ist}} N_t$, where $N_t$ is the number of the triangles. Because this estimate is given by a variational calculation, the actual ground state energy, $E_{\mathrm{g}}$, of $H^{\mathrm{ssm}}$ satisfies the condition $E_{\mathrm{g}} \geq e^{\mathrm{ist}} N_t$. For $J'/J \leq 0.5$, $e^{\mathrm{ist}} N_t$ is equal to the eigenvalue of the exact eigenstate, the product of dimer singlets. Thus the product state is the exact ground state in this parameter range.

On the other hand, the model has not been solved exactly for $J'/J > 0.5$, a fact which indicates that the real phase-transition point should lie above 0.5. In fact, several calculations indicate that the phase transition occurs at $(J'/J)_c \sim 0.7$ [9]. In the limit $J' \gg J$, the model is equivalent to the square lattice, and therefore the antiferromagnetically ordered ground state is expected for small values of $J/J'$. It has been proposed that a further, intermediate phase lies between the dimer-product state and the antiferromagnetically ordered state. While its existence is regarded as probable, there remain open questions regarding its nature, with both helical ordered states [22, 23] and plaquette resonating-valence-bond states [24–26] as viable candidates.

### *19.2.2 Exact Ground States in Coupled Tetrahedral Cluster Models*

We consider now a tetrahedral cluster,

$$h^{\mathrm{tet}} = J\,(\boldsymbol{S}_1 \cdot \boldsymbol{S}_2 + \boldsymbol{S}_1 \cdot \boldsymbol{S}_3 + \boldsymbol{S}_1 \cdot \boldsymbol{S}_4 + \boldsymbol{S}_2 \cdot \boldsymbol{S}_3 + \boldsymbol{S}_2 \cdot \boldsymbol{S}_4 + \boldsymbol{S}_3 \cdot \boldsymbol{S}_4) \tag{19.29}$$

$$= \frac{J}{2} (\boldsymbol{S}_1 + \boldsymbol{S}_2 + \boldsymbol{S}_3 + \boldsymbol{S}_4)^2 - \frac{3}{2} J, \tag{19.30}$$

where the spin indices are shown in Fig. 19.5a, and which is topologically equivalent to the square cluster with diagonal bonds (Fig. 19.5b). The ground state of the cluster has total spin $S^{\mathrm{tot}} = 0$, with two-fold degeneracy, and its energy is $e^{\mathrm{tet}} = -3J/2$. One way of writing the two ground states is

$$|\phi_1^{\mathrm{tet}}\rangle = [1, 2] \times [3, 4], \tag{19.31}$$

$$|\phi_2^{\mathrm{tet}}\rangle = \frac{1}{\sqrt{3}} \left( [1, 3] \times [2, 4] + [1, 4] \times [2, 3] \right), \tag{19.32}$$

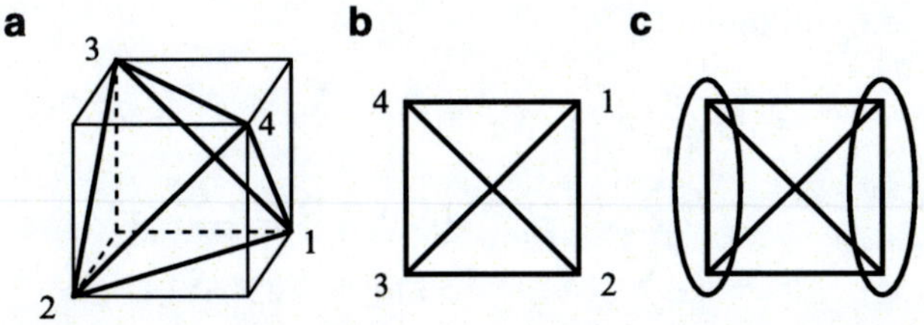


**Fig. 19.5** (**a**) Tetrahedral cluster. (**b**) Square cluster with diagonal interactions, which is topologically equivalent to the tetrahedral cluster. (**c**) Schematic representation of one of the ground states of the Heisenberg model on the tetrahedral cluster. The ellipses represent dimer singlets on individual bonds

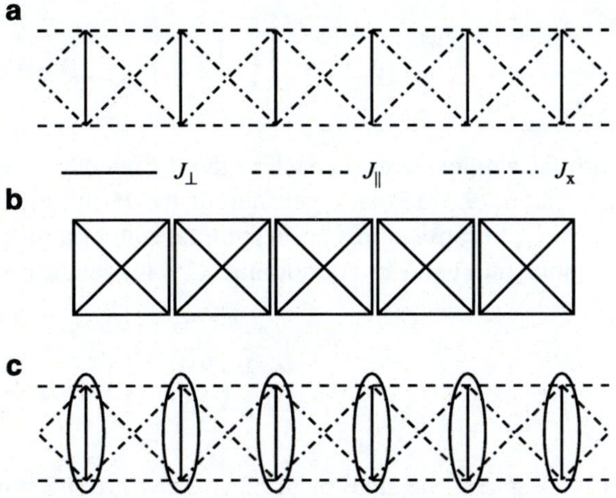


**Fig. 19.6** (**a**) The spin-1/2 frustrated-ladder Heisenberg model. There are three types of antiferromagnetic interaction, $J_\perp$, $J_\parallel$, and $J_x$. (**b**) At $J_\perp = 2J_\parallel = 2J_x$, the Hamiltonian can be considered in terms of coupled tetrahedral clusters. (**c**) The exact dimer-product ground state. Ellipses represent the singlet state on a dimer bond $J_\perp$

where $[i, j]$ is again the dimer singlet state on the bond $ij$ (19.19). The state in (19.31) satisfies the conditions $s_1 + s_2 = 0$ and $s_3 + s_4 = 0$, whereas in (19.32) the spin states on bonds 12 and 34 are triplet. These two states belong to the doubly degenerate $E$ representation of the $T_\mathrm{d}$ group. For the purpose of filling every tetrahedron in a Hamiltonian of the form

$$H^{\mathrm{tet}} = \sum h^{\mathrm{tet}} \qquad (19.33)$$

by ground states of $h^{\mathrm{tet}}$, the state with singlet dimers (19.19) on two of the six bonds can be expected to play an important role. However, because it is necessary to place two singlets on a single cluster simultaneously, but each site is shared between clusters, examples in which one may construct an exact state are limited.

#### 19.2.2.1 Frustrated Ladder Model

The most straightforward model in this category has only Heisenberg spin interactions and is shown in Fig. 19.6a [27]. There are three types of antiferromagnetic interaction in the Hamiltonian

$$H^{\text{fru}} = J_\perp \sum \boldsymbol{S}_{1,i} \cdot \boldsymbol{S}_{2,i} + J_\parallel \sum (\boldsymbol{S}_{1,i} \cdot \boldsymbol{S}_{1,i+1} + \boldsymbol{S}_{2,i} \cdot \boldsymbol{S}_{2,i+1}) + J_x \sum (\boldsymbol{S}_{1,i} \cdot \boldsymbol{S}_{2,i+1} + \boldsymbol{S}_{2,i} \cdot \boldsymbol{S}_{1,i+1}), \quad (19.34)$$

where indices 1 and 2 refer to the two ladder legs and $i$ indicates the rungs. When the condition $J_\perp = 2J_\parallel = 2J_x$ is satisfied, the model may be considered as a set of coupled tetrahedral clusters (Fig. 19.6b), and by placing singlets on every tetrahedron, as shown in Fig. 19.5c, it is possible to minimize the energy of the Hamiltonian (19.34). This ground state has the form

$$|\Psi^{\text{fru}}\rangle = \prod_i \frac{1}{\sqrt{2}} (|\uparrow_{1,i}\downarrow_{2,i}\rangle - |\downarrow_{1,i}\uparrow_{2,i}\rangle), \quad (19.35)$$

with the singlet states on each of the $J_\perp$ bonds naturally of odd parity (Fig. 19.6c). The state (19.35) is an eigenstate of the Hamiltonian (19.34) under the condition $J_\parallel = J_x$, because at this point the total spin of a rung, $\boldsymbol{T}_i = \boldsymbol{S}_{1,i} + \boldsymbol{S}_{2,i}$, is a good quantum number. The Hamiltonian (19.34) can be rewritten in terms of the operator $\boldsymbol{T}_i$ as

$$H^{\text{fru}} = J_\perp \sum_i \frac{1}{2}\left(\boldsymbol{T}_i^2 - \frac{3}{2}\right) + J_\parallel \sum_i (\boldsymbol{T}_i \cdot \boldsymbol{T}_{i+1}). \quad (19.36)$$

It is clear that the state of (19.35) is the ground state for small values of $J_\parallel/J_\perp$. The phase transition is estimated to occur at $(J_\parallel/J_\perp)_c = 0.71$ [28, 29], and the ground state is the dimer-product state of (19.35) for all $J_\parallel/J_\perp < (J_\parallel/J_\perp)_c$. In the opposite limit, $J_\parallel/J_\perp \gg 1$, the model is equivalent to a spin-1 chain model whose ground state is a Haldane state. This Haldane-type gapped phase is realized for all $J_\parallel/J_\perp > (J_\parallel/J_\perp)_c$.

Not only the ground state but also some of the excited states can be constructed exactly in the dimer singlet phase. The total spin on each rung, $\boldsymbol{T}_i$, can be 0 or 1. If it is 0, there is no coupling with the neighboring rungs, so any state with separated triplets on the rungs is an eigenstate of the Hamiltonian. This kind of excited state has an important role in the understanding of magnetization plateaus observed in the model [30] (as discussed in chap. 10).

### *19.2.3 Realization of Exact Ground States*

It seems difficult to find a material in which the exact ground states proposed in Sects. 19.2.1 and 19.2.2 are realized. In fact, materials with exact ground states are quite rare, but once they are realized, their magnetic properties can be explained rather elegantly. One real example is the quasi-two-dimensional material $SrCu_2(BO_3)_2$, which is a realization of two-dimensional Shastry–Sutherland lattice discussed above [7,8]. A sketch of a $CuBO_3$ layer is shown in Fig. 19.7a. The magnetic properties of this compound are due solely to the $Cu^{2+}$ ions, which are well represented by localized spins of $S = 1/2$. The two-dimensional linkage of the $Cu^{2+}$ ions is illustrated in Fig. 19.7b, and is topologically equivalent to the two-dimensional Shastry–Sutherland lattice (Fig. 19.7c). The estimated parameter value $J'/J = 0.635$ indicates that the exact product state of dimer singlets is realized in $SrCu_2(BO_3)_2$ [31]. This material shows a number of unique features as a consequence of strong frustration effects: (a) spin-gapped behavior, (b) the almost localized nature of magnetic excitations, and (c) magnetization plateaus. All of these features are explained well by an analysis in terms of the two-dimensional Shastry–Sutherland model [9].

Indeed, the exact dimer-product ground state is realized even in the full three-dimensional model for $SrCu_2(BO_3)_2$ [32]. The three-dimensional $Cu^{2+}$ structure is shown in Fig. 19.8a: in this model, the product of dimer singlet states on the $J$ bonds is still an eigenstate, a result which becomes clear on applying the operators for inter-layer coupling, $J''$, to the dimer singlet states. One finds

$$J''(\mathbf{s}_1 + \mathbf{s}_2) \cdot (\mathbf{s}_3 + \mathbf{s}_4)|s\rangle_a |s\rangle_b = 0, \tag{19.37}$$

where the site indices are shown in Fig. 19.8b. It is further obvious that the dimer-product state is the ground state for small $J'/J$ and $J''/J$. A phase diagram for the three-dimensional model, obtained by series expansion methods, is presented in [33].

The interlayer interactions in $SrCu_2(BO_3)_2$ are thought to be considerably smaller than $J$ and $J'$, so that the two-dimensional orthogonal dimer model is a very good starting point for a theoretical description. However, the exactness of the

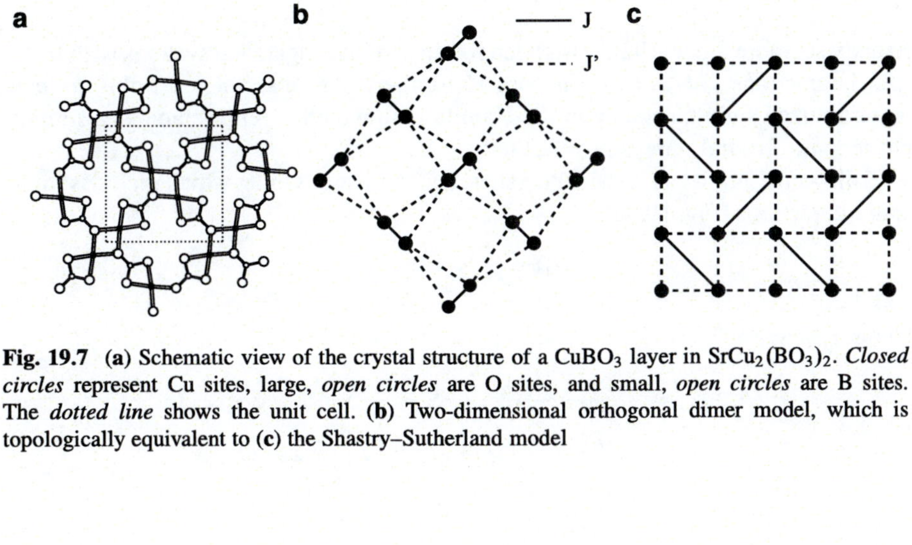


**Fig. 19.7** (**a**) Schematic view of the crystal structure of a $CuBO_3$ layer in $SrCu_2(BO_3)_2$. *Closed circles* represent Cu sites, large, *open circles* are O sites, and small, *open circles* are B sites. The *dotted line* shows the unit cell. (**b**) Two-dimensional orthogonal dimer model, which is topologically equivalent to (**c**) the Shastry–Sutherland model

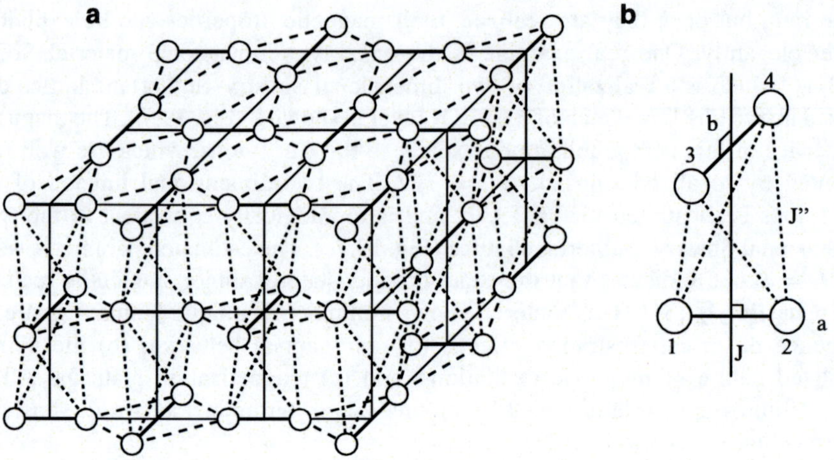


**Fig. 19.8** (**a**) Three-dimensional orthogonal dimer model given by $Cu^{2+}$ ion network in $SrCu_2(BO_3)_2$. (**b**) Configuration of orthogonal dimers for inter-layer coupling $J''$

dimer-product ground state for the three-dimensional model is one of the reasons why the magnetic properties of $SrCu_2(BO_3)_2$ are so well described by the two-dimensional model. The coupling constants estimated for $SrCu_2(BO_3)_2$ from the model of Fig. 19.8 are $J'/J = 0.635$ (above) and $J''/J = 0.09$ [31], which indicates that the treatment is fully consistent with a dimer-product phase.

## 19.3 Exact Results in Frustrated Spin-1/2 Models with Four-Spin Interactions

### *19.3.1 General Ladder Model with Four-Spin Interactions*

Exactly solvable models have also been found and investigated in systems with four-spin interactions [34–38]. As in pure Heisenberg systems, when all clusters can be minimized simultaneously in the Hamiltonian which is represented as a sum of clusters, the ground state is given exactly.

Following [34], let us consider a general ladder Hamiltonian which includes four-spin interactions (Fig. 19.9a),

$$H^{\text{fou}} = \sum h^{\text{fou}}_{i,i+1}, \tag{19.38}$$

where

$$\begin{aligned} h^{\text{fou}}_{i,i+1}, = & -E_0 + \frac{J}{2}(\boldsymbol{S}_{1,i} \cdot \boldsymbol{S}_{2,i} + \boldsymbol{S}_{1,i+1} \cdot \boldsymbol{S}_{2,i+1}) \\ & + J_{\parallel}(\boldsymbol{S}_{1,i} \cdot \boldsymbol{S}_{1,i+1} + \boldsymbol{S}_{2,i} \cdot \boldsymbol{S}_{2,i+1}) \end{aligned}$$

Fig. 19.9 (a) The spin-1/2 general ladder model with four-spin interactions. (b) The exact dimer-product ground states in the one-dimensional Shastry–Sutherland model. Ellipses represent the dimer singlet state

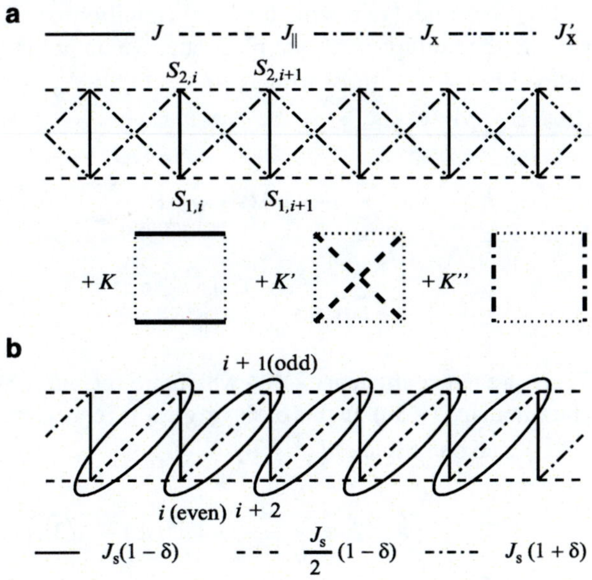


$$
\begin{aligned}
&+J_x \boldsymbol{S}_{1,i} \cdot \boldsymbol{S}_{2,i+1} + J'_x \boldsymbol{S}_{2,i} \cdot \boldsymbol{S}_{1,i+1} \\
&+K\,(\boldsymbol{S}_{1,i} \cdot \boldsymbol{S}_{1,i+1})\,(\boldsymbol{S}_{2,i} \cdot \boldsymbol{S}_{2,i+1}) \\
&+K'\,(\boldsymbol{S}_{1,i} \cdot \boldsymbol{S}_{2,i+1})\,(\boldsymbol{S}_{2,i} \cdot \boldsymbol{S}_{1,i+1}) \\
&+K''\,(\boldsymbol{S}_{1,i} \cdot \boldsymbol{S}_{2,i})\,(\boldsymbol{S}_{1,i+1} \cdot \boldsymbol{S}_{2,i+1}) \,. 
\end{aligned} \tag{19.39}
$$

As in the preceding section, the indices 1 and 2 distinguish the ladder legs while $i$ labels the rungs. For the model with periodic boundary conditions, exact ground states can be written as a trace of matrix-product wave functions,

$$
|\Psi_0\rangle = \mathrm{tr}\left(\prod_i g_i\right), \tag{19.40}
$$

where $g_i$ is defined as

$$
g_i = \begin{pmatrix} b|s\rangle_i + a|t_0\rangle_i & -\sqrt{2}a|t_1\rangle_i \\ \sqrt{2}a|t_{-1}\rangle_i & b|s\rangle_i - a|t_0\rangle_i \end{pmatrix}. \tag{19.41}
$$

Here, the singlet $|s\rangle$ and triplets $|t_\mu\rangle$ are states of a single rung. The exact ground state of (19.40), with energy density $E_0$ per rung, is obtained when the parameters in (19.39) and the free parameter $u = b/a$ in $g_i$ [(19.41)] satisfy the conditions (a) $|\Psi_0\rangle$ is annihilated by the Hamiltonian, i.e.

$$
h^{\mathrm{fou}}_{i,i+1} g_i g_{i+1} = 0, \tag{19.42}
$$

and (b) all other states of $H$ have energy $\epsilon > 0$.

It is convenient to rewrite the local Hamiltonian $h^{\text{fou}}$ in terms of projection operators to the multiplets of one plaquette, which are in turn one quintuplet ($J = 2$), defined as $|\Psi_{2,M}\rangle$, three triplets ($J = 1$), $|\Psi^{(k)}_{1,M}\rangle$ ($k = 1, 2, 3$), and two singlets ($J = 0$), $|\Psi^{(k)}_{0,0}\rangle$ ($k = 1, 2$),

$$\hat{h}^{\text{fou}} = \lambda_2 \sum_M |\Psi_{2,M}\rangle\langle\Psi_{2,M}| + \sum_{k,l=1,2,3} \lambda_1^{(k,l)} \sum_M |\Psi^{(k)}_{1,M}\rangle\langle\Psi^{(l)}_{1,M}| + \sum_{k,l=1,2} \lambda_0^{(k,l)} |\Psi^{(k)}_{0,0}\rangle\langle\Psi^{(l)}_{0,0}|. \tag{19.43}$$

There are ten coefficients $\lambda_J^{(k,l)}$ which are related to the parameters of the original Hamiltonian (19.39). In this basis, $g_i g_{i+1}$ is represented in terms of only one singlet, $|\Psi_0^{(1)}\rangle$, and one triplet, $|\Psi_1^{(1)}\rangle$, by

$$|\Psi_0^{(1)}\rangle \equiv \frac{1}{\sqrt{3+u^4}}\left(u^2|ss\rangle - \sqrt{3}|tt\rangle_{J=0}\right), \tag{19.44}$$

$$|\Psi_1^{(1)}\rangle \equiv \frac{1}{\sqrt{1+u^2}}\left(\frac{u}{\sqrt{2}}\left(|ts\rangle + |st\rangle\right) - |tt\rangle_{J=1}\right), \tag{19.45}$$

where the multiplets specify states with singlets $|s\rangle$ or triplets $|t\rangle$ on dimers $i, i+1$.

To satisfy the conditions (a), the parameters $\lambda_J^{(k,l)}$ must obey the equalities

$$\lambda_0^{(1,1)} = \lambda_0^{(1,2)} = \lambda_1^{(1,1)} = \lambda_1^{(1,2)} = \lambda_1^{(1,3)} = 0. \tag{19.46}$$

For the eigenvalues on the remaining multiplets, i.e., one single $\epsilon_0^{(2)}$, two triplets $\epsilon_1^{(\alpha)}$ ($\alpha = 2, 3$), and one quintuplet $\epsilon_2$, condition (b) results in the inequalities

$$\epsilon_0^{(2)} > 0, \ \epsilon_1^{(\alpha)} > 0\,(\alpha = 2, 3), \ \epsilon_2 > 0. \tag{19.47}$$

Further, we assume the restricted parameter sets $K - K' = K'' = 0$, in which case there are six coefficients, $E_0$, $J$, $J_\parallel$, $J_x$, $J_x'$, and $K$. (A more detailed treatment is discussed in [36].) From the condition (a), one obtains four linear equations whose general solution contains two arbitrary constants when the parameter $u = b/a$ is fixed. Absorbing one of the two constants into the energy scale,

$$1 \equiv -\frac{4}{3}E_0 + \frac{1}{3}J - \frac{5}{6}K, \tag{19.48}$$

and denoting the remaining one by $x$ yields

$$x \equiv -\frac{4}{3}E_0 - J + \frac{1}{2}K, \tag{19.49}$$

whence by defining $u^2 = F$ one obtains the family of Hamiltonians with an exact ground state, which is specified by

$$E_0 = -3[9 + 2x - 3(1+x)F + 3xF^2]/32, \tag{19.50}$$
$$J = 3[2 - x - (1+x)F + xF^2]/4, \tag{19.51}$$
$$2J_\parallel + J_x + J'_x = [15 - 9(1+x)F + 7xF^2]/4, \tag{19.52}$$
$$2J_\parallel - J_x - J'_x = -xF, \tag{19.53}$$
$$2J_x - 2J'_x = \sqrt{F}[3(1+x)/2 - xF], \tag{19.54}$$
$$K = 3(F-1)(xF-1)/4, \tag{19.55}$$

and has eigenvalues

$$\epsilon_0^{(2)} = x[3 + F^2(x)]/4 > 0, \tag{19.56}$$
$$\epsilon_1^{(2)} = [3(1+x) + 2xF(x)]/8 > 0, \tag{19.57}$$
$$\epsilon_1^{(3)} = [1 + F(x)][3(1+x) - 2xF(x)]/8 > 0, \tag{19.58}$$
$$\epsilon_2 = [18 + 8xF^2(x) - 9(1+x)F(x)]/8 > 0. \tag{19.59}$$

In this way, one may determine the Hamiltonians which have a matrix-product wave function (19.40) as an exact ground state. Several known models with exact ground states can be obtained as particular cases from the general solution of this family [36]. In the following, we show that the one-dimensional Shastry–Sutherland model and the generalized AKLT model are particular members of this general family of Hamiltonians.

#### 19.3.1.1 One-Dimensional Shastry–Sutherland Model

The one-dimensional Shastry–Sutherland model (Fig. 19.9b) is a generalization of the Majumdar–Ghosh model [20]. The Hamiltonian,

$$H = J_s \sum_i \left( (1 + (-1)^i \delta) \mathbf{S}_i \cdot \mathbf{S}_{i+1} + \frac{1}{2}(1-\delta) \mathbf{S}_i \cdot \mathbf{S}_{i+2} \right), \tag{19.60}$$

includes the Majumdar–Ghosh model ($\delta = 0$) and isolated dimers ($\delta = 1$) as special cases.

The one-dimensional Shastry–Sutherland model is obtained when the condition

$$F = 1, \quad x = \frac{3\delta}{2+\delta} \quad (0 < \delta < 1) \tag{19.61}$$

is satisfied. For these parameters, the interactions can be written as

$$E_0 = -\frac{9(1+\delta)}{8(2+\delta)}, \tag{19.62}$$

$$J = \frac{3(1-\delta)}{2(2+\delta)} \equiv J_s(1-\delta), \tag{19.63}$$

$$J_\parallel = \frac{3(1-\delta)}{4(2+\delta)} \equiv \frac{J_s}{2}(1-\delta), \tag{19.64}$$

$$J_x = \frac{3(1+\delta)}{2(2+\delta)} \equiv J_s(1+\delta), \tag{19.65}$$

$$J'_x = 0, \tag{19.66}$$

$$K = K' = K'' = 0. \tag{19.67}$$

The wave function (19.40) with $F = 1$, i.e. $a = b$, corresponds to singlets on one type of diagonal bond (here $J_x$) (Fig. 19.9b). The case $a = 0$ is included by a translation of one of the ladder legs. The eigenvalue $\epsilon_0^{(2)}$ vanishes for $x = 0$, which is to be expected because the ground state is two-fold degenerate in the Majumdar–Ghosh limit.

#### 19.3.1.2 Generalized AKLT Model

Setting

$$F = 0, \quad x \text{ finite}, \tag{19.68}$$

i.e. fixing $b = 0$, ensures that only triplets may occur on the rung bonds. By choosing a convenient scale factor to render $J_{\text{eff}} = (J_\parallel + J_x)/2 + K/4 = 1$, one obtains the Hamiltonian with parameters

$$E_0 = -\frac{3}{4} - \frac{x}{3}, \tag{19.69}$$

$$J = \frac{4}{3} - \frac{2x}{3}, \tag{19.70}$$

$$J_\parallel = J_x = J'_x = 5/6, \tag{19.71}$$

$$K = K' = \frac{2}{3}, \tag{19.72}$$

$$K'' = 0. \tag{19.73}$$

This is essentially equivalent to the AKLT Hamiltonian, but without the explicit requirement that the two $S = 1/2$ spins be coupled into a triplet. The AKLT model,

$$H = \sum_i \boldsymbol{S}_i \cdot \boldsymbol{S}_{i+1} - \frac{1}{3}\left(\boldsymbol{S}_i \cdot \boldsymbol{S}_{i+1}\right)^2, \tag{19.74}$$

is obtained in the limits $x \to \infty$ and $xF(x) \to 0$; note that the operator $\boldsymbol{S}$ in (19.74) is an $S = 1$ operator.

Finally, we note that the ground state of the Majumdar–Ghosh model belongs to the same phase as that of the AKLT model for $S = 1$. Let us consider the family of models with

$$F(x) = \frac{1}{(1+x)^2}. \tag{19.75}$$

Because both the Majumdar–Ghosh model and the AKLT limit are included in this family, (19.75) demonstrates the possibility of transforming the Hamiltonian of the Majumdar–Ghosh model to the AKLT limit, by changing $x$ from 0 to $+\infty$, without encountering any singularities in the ground state.

## 19.3.2 Two-Dimensional Model with Four-Spin Interactions

The antiferromagnetic, spin-1/2 $J_1$-$J_2$ model on the square lattice is a well-known example of a frustrated spin system. Unfortunately, an exact ground state is not known for this model, because it is not possible to minimize the energy on all plaquettes simultaneously by using the states in (19.31) and (19.32), even at the highly frustrated point $J_1 = J_2/2$. However, the inclusion of a four-spin interaction can produce exactly solvable models even in two-dimensional lattices [39, 40]. In [39], the Hamiltonian is written as a sum of projection operators which project the spin state of the plaquette onto the subspace with total spin $S = 2$. In [40], the spin projectors onto the quartet state are used. In the former, the ground state is highly (macroscopically) degenerate, while in the latter it is much lower, with a ground state of four-fold degeneracy.

### 19.3.2.1 Generalized Two-Dimensional $J_1$–$J_2$ Model

Let us consider the spin-1/2 Hamiltonian with four-spin interactions on a square lattice,

$$\begin{aligned} H^{2df} &= J_1 \sum_{n.n.} \boldsymbol{S}_i \cdot \boldsymbol{S}_j + J_2 \sum_{n.n.n.} \boldsymbol{S}_i \cdot \boldsymbol{S}_j \\ &\quad + K \sum_{i,j,k,l} \{ (\boldsymbol{S}_i \cdot \boldsymbol{S}_j)(\boldsymbol{S}_k \cdot \boldsymbol{S}_l) + (\boldsymbol{S}_j \cdot \boldsymbol{S}_k)(\boldsymbol{S}_i \cdot \boldsymbol{S}_l) \\ &\quad + (\boldsymbol{S}_i \cdot \boldsymbol{S}_k)(\boldsymbol{S}_j \cdot \boldsymbol{S}_l) \}. \end{aligned} \tag{19.76}$$

At the fully frustrated point, $J_2 = J_1/2$ and $K = J_1/8$,

$$H^{2df} = \frac{3J}{2} \sum_i P_i, \tag{19.77}$$

where $P_i$ is a projection operator onto the state of total spin $S = 2$. It is clear that any state having at least one dimer singlet per plaquette is a ground state of the Hamiltonian (19.77), because this implies that the total spin on each plaquette may be only 0 or 1.

There are two types of ground state. One is a state which is a product of local singlet dimers, as shown in Fig. 19.10a. It is important that any state obtained by rotating all dimers about one of their sites successively along a diagonal direction (Fig. 19.10a) is also a ground state. As a consequence of this freedom, the degeneracy of this type of state is $2^{N_d+1}$, where $N_d \propto \sqrt{N}$ is the number of diagonal chains and $N$ is the total number of sites.

The second type of ground state is one with interfaces between vertical and horizontal configurations along the two diagonal directions in an infinite system, two examples of which are shown in Fig. 19.10b, c. There is a defect at the intersection between the two diagonal interfaces, and it is possible to have a free but localized spin-1/2 object as a defect (Fig. 19.10c).

We conclude this section with a brief mention of the excited states. If one of the singlet dimers is excited to a triplet state, denoted by two parallel spins, the two spins can propagate along the diagonal directions to arbitrary separations with no energy cost (Fig. 19.10d). Thus, the effective dimensionality of the low-energy spectrum is reduced dynamically from two to one.

#### 19.3.2.2 Two-Dimensional $J_1$–$J_2$ Model with Products of Three-Spin Projectors

Models containing products of three-spin projectors can also be constructed in terms of two- and four-spin interactions. Such a model can be expressed using 6-site rectangular plaquettes, as shown in Fig. 19.11a, where the two sublattices $A$ and $B$ are distinguished. The Hamiltonian takes the form

$$H^{\text{pro}} = \sum_{[i,j,k,l,m,n]} \frac{1}{4} P^A_{i,j,k} P^B_{l,m,n}, \tag{19.78}$$

where the spin projectors on the quartet state are defined as

$$P^A_{i,j,k} = |\boldsymbol{S}_i + \boldsymbol{S}_j + \boldsymbol{S}_k|^2 - \frac{3}{4}, \tag{19.79}$$

$$P^B_{l,m,n} = |\boldsymbol{S}_l + \boldsymbol{S}_m + \boldsymbol{S}_n|^2 - \frac{3}{4}. \tag{19.80}$$

The indices $i, j, k, l, m, n$ are defined in Fig. 19.11a, and the sum in (19.78) runs over all horizontal and vertical 6-site plaquettes.

In the model (19.78), the ground state is given by a product of dimer singlets, as illustrated in Fig. 19.11b. This is reminiscent of the two-dimensional Shastry–Sutherland ground state (Fig. 19.4) [6], and is thus known as a Shastry–Sutherland

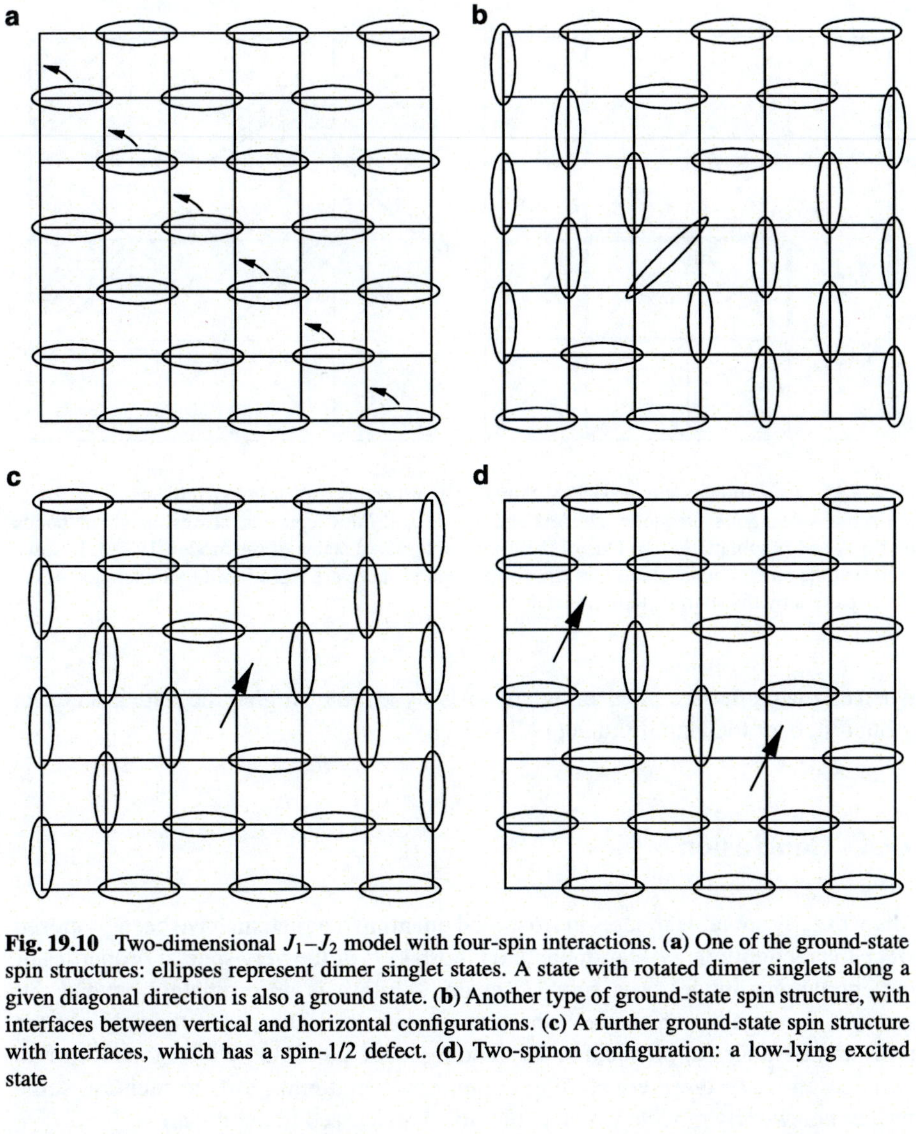


**Fig. 19.10** Two-dimensional $J_1$–$J_2$ model with four-spin interactions. (**a**) One of the ground-state spin structures: ellipses represent dimer singlet states. A state with rotated dimer singlets along a given diagonal direction is also a ground state. (**b**) Another type of ground-state spin structure, with interfaces between vertical and horizontal configurations. (**c**) A further ground-state spin structure with interfaces, which has a spin-1/2 defect. (**d**) Two-spinon configuration: a low-lying excited state

valence-bond solid. A key difference from the Shastry–Sutherland model is that the ground state of (19.78) presents a spontaneous symmetry-breaking, and is four-fold degenerate. The proof that the Shastry–Sutherland valence-bond solid is an exact ground state is simple: the values of the projectors (19.79) and (19.80) are zero when the three spins are in a doublet state, and has positive values when they constitute a quartet. The doublet state emerges whenever any two of the three spins form a singlet. Note that these two spins could be either second or third neighbors. Because $H^{\text{pro}}$ is a product of the two projectors, the presence of only one dimer on the 6-site plaquette is already sufficient to minimize the plaquette energy. Thus,

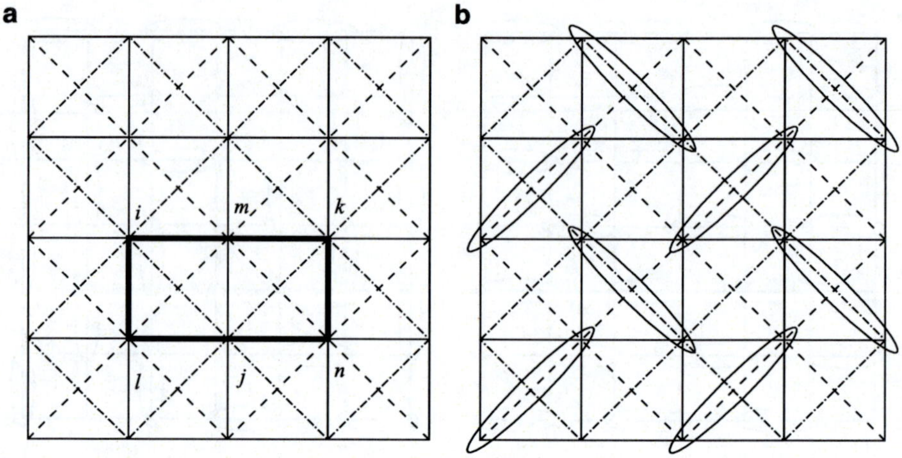


**Fig. 19.11** $J_1$–$J_2$ model with products of three-spin projectors. (**a**) *Bold line* indicates a $3 \times 2$-site horizontal rectangular plaquette. *Dashed* and *dot-dashed lines* represent nearest-neighbor bonds on the A and B sublattices. (**b**) One of the four exact ground states of the model (19.78). Ellipses represent the dimer singlet state. The three other ground states are reproduced by translation of the dimer pattern by the shortest lattice vectors

when all plaquettes are filled simultaneously by a dimer singlet, the state is an exact ground state of the Hamiltonian (19.78).

## 19.4 Conclusion

Many exactly solvable models in frustrated quantum magnetism have been proposed since the recognition by Majumdar and Ghosh [17] of the very special properties of the eponymous model on a spin-1/2 chain. While it is unfortunately beyond our scope to cover all such models, the aim of this article has been to highlight the profound advances in physical understanding made possible by the existence of an exact solution. In these concluding comments, we attempt also to mention some further noteworthy models which could not be discussed in depth here.

In spin-1/2 Heisenberg systems composed of frustrated clusters, when the ground state of the cluster is degenerate due to the frustration, there is a freedom to select one state from among the ground states on each cluster. If all such clusters can be filled at the same time by their cluster ground states, an exact ground state is obtained, i.e., such a state has the minimum energy because of the variational principle. Most of the well known examples of such ground states are products of dimer singlet states, and in this article we have reviewed the sawtooth–chain model, the Majumdar–Ghosh model, the two-dimensional Shastry–Sutherland model, and the frustrated ladder model. In addition to these, many other exact ground states represented as products of dimer singlet states have been proposed, including in the one-dimensional Shastry–Sutherland model [20], the distorted diamond chain model [41], the one-dimensional orthogonal dimer model [42], and the generalized Shastry–Sutherland models in three dimensions [43]. The general procedure for constructing a model which has an exact dimer-product ground state is proposed in [44]. Of particular interest in this context is also the distorted diamond chain model which is equivalent to a trimer chain system [41,45,46]. This model is realized in $Cu_3Cl_6(H_2O)_2 \cdot 2H_8C_4SO_2$ [47] and $Cu_3(CO_3)_2(OH)_2$ [48], although it seems unlikely that the exact ground state is stabilized in these materials.

Exact ground states are realized also in some spin-1/2 Heisenberg models with four- (or higher-)spin interactions. In this chapter, we have reviewed only a small number of examples, specifically a general ladder model and generalized two-dimensional $J_1$–$J_2$ models both with four-spin interactions and with products of three-spin projectors. In these examples, the procedure for constructing the model is similar to that for pure Heisenberg models, namely minimization of the cluster energy. For models including multiple-spin interaction terms, Klein has proposed a general scheme for constructing the exact ground state [49,50], the basic idea of the Klein model being the following. The Klein Hamiltonian is given by a sum of projectors onto the highest total-spin state of the clusters constituting the lattice, which have non-negative eigenvalues in cases such as generalized two-dimensional $J_1$–$J_2$ models. Any state in which two of the spins in a given cluster form a singlet bond prevents this cluster from having maximal total spin, and thus the valence-bond state can be a ground state with zero energy. Several types of model have been constructed based on the Klein approach, all (to date) spin-1/2 Heisenberg models with multiple-spin interactions on various lattices, including honeycomb, square, checkerboard, and pyrochlore geometries [51–53]. Some of these cases have very close links to quantum dimer models [54], which are discussed in chap. 17.

We conclude by expressing the hope that the research motivated by this review may give birth to new models and new materials which have exact ground states.

**Acknowledgements**

We thank Kenn Kubo, Kazuo Ueda, Bruce Normand, and Frédéric Mila for many helpful suggestions and comments concerning this article.

## References

1. Z. Hiroi, M. Azuma, M. Takano, Y. Bando, J. Solid State Chem. **95**, 230 (1991)
2. S. Taniguchi, T. Nishikawa, Y. Yasui, Y. Kobayashi, M. Sato, T. Nishioka, M. Kontani, K. Sano, J. Phys. Soc. Jpn. **64**, 2758 (1995)
3. Z. Hiroi, M. Hanawa, N. Kobayashi, M. Nohara, H. Takagi, Y. Kato, M. Takigawa, J. Phys. Soc. Jpn. **70**, 3377 (2001)
4. P. Mendels, F. Bert, M.A. de Vries, A. Olariu, A. Harrison, F. Duc, J.-C. Trombe, J.S. Lord, A. Amato, C. Baines, Phys. Rev. Lett. **98**, 077204 (2007)

5. J.S. Helton, K. Matan, M.P. Shores, E.A. Nytko, B.M. Bartlett, Y. Yoshida, Y. Takano, A. Suslov, Y. Qiu, J.-H. Chung, D.G. Nocera, Y.S. Lee, Phys. Rev. Lett. **98**, 107204 (2007)
6. B.S. Shastry, B. Sutherland, Physica **108B&C**, 1069 (1981)
7. H. Kageyama, K. Yoshimura, R. Stern, N.V. Mushnikov, K. Onizuka, M. Kato, K. Kosuge, C. P. Slichter, T. Goto, Y. Ueda, Phys. Rev. Lett. **82**, 3168 (1999)
8. S. Miyahara, K. Ueda, Phys. Rev. Lett. **82**, 3701 (1999)
9. S. Miyahara, K. Ueda, J. Phys. Condens. Matter **15**, R327 (2003)
10. R.P. Feynman, R.B. Leighton, M. Sands, *The Feynman Lectures on Physics Vol. III Quantum Mechanics* (Addison-Wesley Publishing Company, Massachusetts, 1965), pp. 12-1–12-16
11. S. Blundell, *Magnetism in Condensed Matter*. (Oxford University Press, New York, 2001), pp. 9–15
12. B. Sutherland, B.S. Shastry, J. Stat. Phys. **33**, 477 (1983)
13. I. Affleck, T. Kennedy, E.H. Lieb, H. Tasaki, Phys. Rev. Lett. **59**, 799 (1987)
14. F. Monti, A. Sütő, Phys. Lett. A **156**, 197 (1991)
15. K. Kubo, Phys. Rev. B **48**, 10552 (1993)
16. T. Nakamura, K. Kubo, Phys. Rev. B **53**, 6393 (1996)
17. C.K. Majumdar, D.K. Ghosh, J. Math. Phys. **10**, 1388 (1969)
18. K. Okamoto, K. Nomura, Phys. Lett. A **169**, 433 (1992)
19. P.M. van den Broek, Phys. Lett. A **77**, 261 (1980)
20. B.S. Shastry, B. Sutherland, Phys. Rev. Lett. **47**, 964 (1981)
21. I. Affleck, T. Kennedy, E.H. Lieb, H. Tasaki, Commun. Math. Phys. **115**, 477 (1988)
22. M. Albrecht, F. Mila, Europhys. Lett. **34**, 145 (1996)
23. C.-H. Chung, J.B. Marston, S. Sachdev, Phys. Rev. B **64**, 134407 (2001)
24. A. Koga, N. Kawakami, Phys. Rev. Lett. **84**, 4461 (2000)
25. D. Carpentier, L. Balents, Phys. Rev. B **65**, 024427 (2002)
26. A. Läuchli, S. Wessel, M. Sigrist, Phys. Rev. B **66**, 014401 (2002)
27. I. Bose, S. Gayen, Phys. Rev. B **48**, 10653 (1993)
28. Y. Xian, Phys. Rev. B **52**, 12485 (1995)
29. H. Kitatani, T. Oguchi, J. Phys. Soc. Jpn. **65**, 1387 (1996)
30. A. Honecker, F. Mila, M. Troyer, Eur. Phys. J. B **15**, 227 (2000)
31. S. Miyahara, K. Ueda, J. Phys. Soc. Jpn. Suppl. B **69**, 72 (2000)
32. K. Ueda, S. Miyahara, J. Phys. Condens. Matter **11**, L175 (1999)
33. A. Koga, J. Phys. Soc. Jpn. **69**, 3509 (2000)
34. A.K. Kolezhuk, H.-J. Mikeska, Phy. Rev. B **56**, 11380 (1997)
35. A.A. Nersesyan, A.M. Tsvelik, Phy. Rev. Lett. **78**, 3939 (1997)
36. A.K. Kolezhuk, H.-J. Mikeska, Phy. Rev. Lett. **80**, 2709 (1998)
37. A.K. Kolezhuk, H.-J. Mikeska, Int. J. Mod. Phys. B **12**, 2325 (1998)
38. V. Gritsev, B. Normand, D. Baeriswyl, Phy. Rev. B **69**, 094431 (2004)
39. C.D. Batista, S.A. Trugman, Phy. Rev. Lett. **93**, 217202 (2004)
40. A. Gellé, A. Läuchli, B. Kumar, F. Mila, Phy. Rev. B **77**, 014419 (2008)
41. K. Okamoto, T. Tonegawa, Y. Takahashi, M. Kaburagi, J. Phys. Condens. Matter **11**, 10485 (1999)
42. M.P. Gelfand, Phys. Rev. B **43**, 8644 (1991)
43. N. Surendran, R. Shankar, Phys. Rev. B **66**, 024415 (2002)
44. B. Kumar, Phys. Rev. B **66**, 024406 (2002)
45. K. Takano, K. Kubo, H. Sakamoto, J. Phys. Condens. Matter **8**, 6405 (1996)
46. A. Honecker, A. Läuchli, Phys. Rev. B **63**, 174407 (2001)
47. M. Ishii, H. Tanaka, M. Mori, H. Uekusa, Y. Ohashi, K. Tatani, Y. Narumi, K. Kindo, J. Phys. Soc. Jpn. **69**, 340 (2000)
48. H. Kikuchi, Y. Fujii, M. Chiba, S. Mitsudo, T. Idehara, Phys. B **329**, 967 (2003)
49. D.J. Klein, J. Phys. A. Math. Gen. **15**, 661 (1982)
50. J.T. Chayes, L. Chayes, S.A. Kivelson, Commun. Math. Phys. **123**, 53 (1989)
51. S. Fujimoto, Phys. Rev. B **72**, 024429 (2005)
52. K.S. Raman, R. Moessner, S.L. Sondhi, Phys. Rev. B **72**, 064413 (2005)
53. Z. Nussinov, C.D. Batista, B. Normand, S.A. Trugman, Phys. Rev. B **75**, 094411 (2007)
54. D.S. Rokhsar, S.A. Kivelson, Phys. Rev. Lett. **61**, 2376 (1988)

# Chapter 20
# Strong-Coupling Expansion and Effective Hamiltonians

**Frédéric Mila and Kai Phillip Schmidt**

**Abstract** When looking for analytical approaches to treat frustrated quantum magnets, it is often very useful to start from a limit where the ground state is highly degenerate. This chapter discusses several ways of deriving effective Hamiltonians around such limits, starting from standard degenerate perturbation theory and proceeding to modern approaches more appropriate for the derivation of high-order effective Hamiltonians, such as the perturbative continuous unitary transformations (CUTs) or contractor renormalization (CORE). In the course of this exposition, a number of examples taken from the recent literature are discussed, including frustrated ladders and other dimer-based Heisenberg models in a field, as well as the mapping between frustrated Ising models in a transverse field and quantum dimer models (QDMs).

## 20.1 Introduction

As emphasized several times throughout this book, frustrated magnets often have a highly degenerate ground state in the classical limit. This is sometimes even taken as a definition of highly frustrated magnets. This degeneracy makes the semi-classical expansion in $1/S$ effectively uncontrolled (if it does not already fail simply because of divergent quantum fluctuations), because usually one cannot perform and thus compare the expansions around all classical ground states. An infinite degeneracy is also often present in other limiting cases such as decoupled local units (such as triangles in the $S = 1/2$ trimerized kagomé lattice) or the Ising limit (for systems such as the antiferromagnetic Heisenberg model on the triangular or the kagomé lattice). In such limits, which preserve the quantum nature of the problem, this degeneracy is not the end of the story, but rather the starting point of a systematic expansion, namely degenerate perturbation theory, which leads to an effective Hamiltonian. In the context of strongly correlated systems, this type of method usually goes by the name 'strong-coupling expansion,' because the starting point of the perturbative expansion is a Hamiltonian where only the interaction terms are kept, the kinetic terms being treated as the perturbation.

There are several ways to perform this expansion, or more generally to derive an effective Hamiltonian. There are in fact two types of effective Hamiltonian: those which act only in the degenerate subspace of a non-perturbed Hamiltonian, and those which act in the full Hilbert space but, through a canonical transformation, are rewritten as a series in terms of the ratio of two parameters. While the first type can always be written down explicitly, the second type can be derived in a simple way only provided that one can find a suitable generator. In the following, we will discuss both types of effective Hamiltonian, starting with the expansion in the degenerate subspace because this is more standard.

The derivation of an effective Hamiltonian is extremely useful in isolating the relevant degrees of freedom. However, the problem is usually not solved once the effective Hamiltonian has been derived. Indeed, the new Hamiltonian often poses a problem as difficult as the original one. The primary advantage of the effective Hamiltonian is that, because the relevant degrees of freedom have been selected, simple approximations to the problem defined by the effective Hamiltonian often give deep insight into the physics of the problem. We will illustrate this point with several examples.

This chapter is organized as follows. In Sect. 2, we review briefly the degenerate perturbation theory approach to effective Hamiltonians, with a concise but self-contained discussion of the second-order result, and a description of the form the expansion takes at higher orders. In Sect. 3, we discuss three examples taken from the field of frustrated magnetism where this approach has proven very useful: coupled dimers in a magnetic field, the Ising model in a transverse field, and the trimerized, spin-1/2 kagomé antiferomagnet. In Sect. 4, we review more sophisticated approaches based on the same foundation: canonical transformations, continuous unitary transformations (CUTs), and the contractor renormalization group approach (CORE). Note that linked cluster expansions are discussed in Chap. 18 of this book (that by Läuchli). We conclude in Sect. 5 with a discussion that includes a comparison of the various approaches.

## 20.2 Strong-Coupling Expansion

Let us consider a system described by a Hamiltonian

$$H = H_0 + V$$

acting in a Hilbert space $\mathcal{H}$ such that the ground state of $H_0$ is degenerate. We denote by $\mathcal{H}_0$ the Hilbert space of the ground-state manifold. The goal is to find an effective Hamiltonian $H_{\text{eff}}$ acting in $\mathcal{H}_0$ such that

$$H_{\text{eff}}|\phi\rangle = E|\phi\rangle \Rightarrow H|\psi\rangle = E|\psi\rangle, \quad |\phi\rangle \in \mathcal{H}_0, \quad |\psi\rangle \in \mathcal{H}.$$

### 20.2.1 Second-Order Perturbation Theory

Up to second order, the relation to be derived below is a standard result of quantum mechanics (See for example [1]). Denoting by $E_0$ the ground-state energy of $H_0$ and by $E_m$ the other eigenenergies, $H_0|m\rangle = E_m|m\rangle$ for $E_m \neq E_0$. Thus for two vectors $|\phi\rangle$, $|\phi'\rangle \in \mathcal{H}_0$, and up to second order in $V$,

$$\langle\phi|H_{\text{eff}}|\phi'\rangle = \langle\phi|H_0|\phi'\rangle + \langle\phi|V|\phi'\rangle + \sum_{|m\rangle\notin\mathcal{H}_0} \frac{\langle\phi|V|m\rangle\langle m|V|\phi'\rangle}{E_0 - E_m}.$$

This result can be reformulated as an operator identity (See for example [2]). If one denotes by $P$ the projector on $\mathcal{H}_0$, and defines $Q = 1 - P$, then to second order in $V$

$$PH_{\text{eff}}P = PH_0P + PVP + PVQ\frac{1}{E_0 - QH_0Q}QVP.$$

*Proof.* Suppose $H|\psi\rangle = E|\psi\rangle$. Because $P + Q = 1$, this can be written as

$$(P + Q)H(P + Q)|\psi\rangle = E|\psi\rangle.$$

Projecting onto $\mathcal{H}_0$ and $\mathcal{H} - \mathcal{H}_0$ gives

$$PHP|\psi\rangle + PHQ|\psi\rangle = EP|\psi\rangle, \quad (1)$$

$$QHP|\psi\rangle + QHQ|\psi\rangle = EQ|\psi\rangle, \quad (2)$$

whence

$$(2) \Rightarrow Q|\psi\rangle = (E - QHQ)^{-1}QHP|\psi\rangle,$$

$$(1) \Rightarrow PHP|\psi\rangle + PHQ\frac{1}{E - QHQ}QHP|\psi\rangle = EP|\psi\rangle.$$

Expansion of $(E - QHQ)^{-1}$ using $(A - B)^{-1} = A^{-1}\sum_{n=0}^{\infty}\left(BA^{-1}\right)^n$, with $A = E_0 - QH_0Q$ and $B = QVQ - E + E_0$, leads to

$$\begin{aligned} PH_{\text{eff}}P &= PH_0P + PVP + PVQ(E_0 - QH_0Q)^{-1} \\ &\quad \times \sum_{n=0}^{\infty}\left((QVQ - E + E_0)(E_0 - QH_0Q)^{-1}\right)^n QVP. \end{aligned} \quad (20.1)$$

The expansion is then truncated at $n = 0$. □

### 20.2.2 High-Order Perturbation Theory

This form of the expansion is not well suited to the derivation of higher-order expansions, because beyond second order it contains explicitly the exact eigenenergy $E$.

An expansion only in terms of the unperturbed eigenenergies can nevertheless be derived. The first systematic method for this dates to the work of Kato [3]. Here we follow the formulation of Takahashi [4], in which the expansion takes the form

$$PH_{\mathrm{eff}}P = \Gamma^\dagger H \Gamma,$$

$$\Gamma = \bar{P}P(P\bar{P}P)^{-1/2},$$

$$(\bar{P}PP)^{-1/2} = P + \sum_{n=1}^{\infty} \frac{(2n-1)!!}{(2n)!!}[P(P-\bar{P})P]^n,$$

$$\bar{P} = P - \sum_{n=1}^{\infty} \sum_{k_1+\ldots+k_{n+1}=n, k_i\geq 0} S^{k_1}VS^{k_2}V\ldots VS^{k_{n+1}},$$

$$S^0 = -P, \; S^k = \left(\frac{Q}{E_0 - QH_0Q}\right)^k.$$

The true eigenstates $\psi$ are related to the eigenstates $\phi$ of $H_{\mathrm{eff}}$ by

$$|\psi\rangle = \Gamma|\phi\rangle,$$

and likewise the observables transform according to

$$O \to \Gamma^\dagger O \Gamma.$$

Thus the $n$th order term of $H_{\mathrm{eff}}$ has the form

$$H_{\mathrm{eff}}^{(n)} = \sum_{k_1+\cdots+k_{n-1}=n-1, k_i\geq 0} f(k_1, k_2, \ldots, k_{n-1})VS^{k_1}VS^{k_2}V\ldots S^{k_{n-1}}V,$$

where the coefficients $f(k_1, \ldots, k_{n-1})$ are deduced by appropriate bookkeeping from the previous expansions. The number of terms in such a strong-coupling expansion grows exponentially with $n$. In practice, beyond the fourth order it can generally be carried out only with the help of a computer. An alternative formulation based on CUTs, which is simpler when applicable, will be discussed in the next section. In the remainder of this section, we discuss a number of selected examples where low-order degenerate perturbation theory provides considerable additional insight into the problem.

### 20.2.3 Examples

In the field of quantum magnetism, the best known example is the derivation of the Heisenberg model starting from the half-filled Hubbard model. The Hubbard model is defined by

$$H = V + H_0 = -t \sum_{\langle i,j\rangle,\sigma} (c_{i\sigma}^{\dagger} c_{j\sigma} + h.c.) + U \sum_i n_{i\uparrow} n_{i\downarrow}.$$

At half-filling, the ground state of the interaction term $H_0$ is $2^N$-fold degenerate, where $N$ is the number of sites. Treating the kinetic term $V$ as a perturbation leads at second order (up to a constant) to the effective Hamiltonian

$$H_{\text{eff}} = J \sum_{\langle i,j\rangle} \boldsymbol{S}_i \cdot \boldsymbol{S}_j,$$

with $J = 4t^2/U$ [5]. As noted in the introduction, this is a case where the effective Hamiltonian is itself very difficult to solve, which is indeed true for the Heisenberg model on most lattices. As we shall see in the following, it is often useful to go one step further, and to derive a further effective Hamiltonian starting from one physically relevant limit.

### Example 1: Frustrated Spin-1/2 Ladder in a Magnetic Field

We consider the Heisenberg model for a frustrated spin-1/2 ladder in a magnetic field defined by the Hamiltonian

$$H = \sum_n J_\perp \boldsymbol{S}_{1n} \cdot \boldsymbol{S}_{2n} - B \sum_n (S_{1n}^z + S_{2n}^z) + \sum_n \sum_{i,j=1,2} J_{ij} \boldsymbol{S}_{in} \cdot \boldsymbol{S}_{jn+1}. \quad (20.2)$$

In the spin operators $\boldsymbol{S}_{in}$, the index $i$ refers to the leg and the index $n$ to the rung (Fig. 20.1). The goal is to derive an effective Hamiltonian that describes the magnetization process in the limit $J_\perp \gg J_{ij}$ [6, 7].

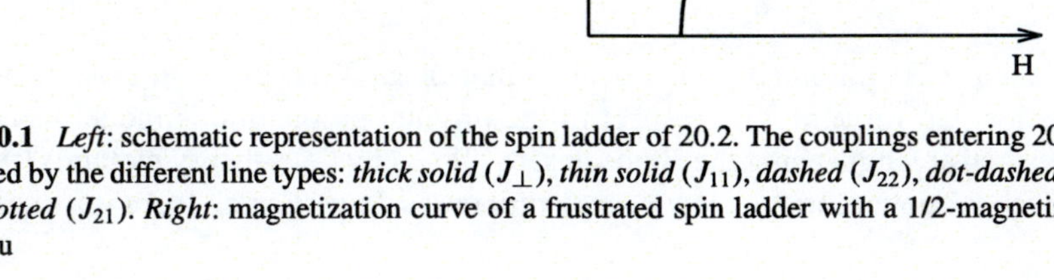


**Fig. 20.1** *Left*: schematic representation of the spin ladder of 20.2. The couplings entering 20.2 are denoted by the different line types: *thick solid* ($J_\perp$), *thin solid* ($J_{11}$), *dashed* ($J_{22}$), *dot-dashed* ($J_{12}$) and *dotted* ($J_{21}$). *Right*: magnetization curve of a frustrated spin ladder with a 1/2-magnetization plateau

The starting point of the perturbative expansion is the Hamiltonian of isolated dimers at the critical field $B_c = J_\perp$ where, for one dimer, the triplet polarized in the field direction crosses the singlet,

$$H_0 = \sum_n J_\perp \boldsymbol{S}_{1n} \cdot \boldsymbol{S}_{2n} - B_c \sum_n (S^z_{1n} + S^z_{2n}). \tag{20.3}$$

The ground state of this Hamiltonian is $2^N$-fold degenerate, where $N$ is the number of ladder rungs, and the full Hamiltonian can be treated within degenerate perturbation theory, the perturbation being given by

$$V = \sum_{\langle nm\rangle} \sum_{i,j=1,2} J_{ij} \boldsymbol{S}_{in} \cdot \boldsymbol{S}_{jm} - (B - B_c) \sum_n (S^z_{1n} + S^z_{2n}). \tag{20.4}$$

The sum over $\langle nm\rangle$ refers to the nearest-neighbor rungs. A perturbative expansion can be performed under the condition that the matrix elements of $V$ are small compared to the excited states of $H_0$, i.e. as long as $J_{ij}$, $(B - B_c) \ll J_\perp$. The condition on $B - B_c$ might suggest that such a calculation cannot give access to the full magnetization curve, but in fact the magnetization is rigorously equal to zero or to the full saturation value outside a window whose width is of order $J_{ij}$. Thus, this type of perturbation theory can indeed give the full magnetization curve.

To write down the effective Hamiltonian, one needs a description of the Hilbert space. Because there are two states per rung, one simple choice is to introduce the Pauli matrices $\boldsymbol{\sigma}_n$ at each rung such that the singlet corresponds to $|\sigma^z_n = -1/2\rangle$ and the triplet polarized along the field to $|\sigma^z_n = 1/2\rangle$. In this basis, and up to first order in $V$, the effective Hamiltonian is given by

$$H_{\text{eff}} = J^{xy} \sum_n (\sigma^x_n \sigma^x_{n+1} + \sigma^y_n \sigma^y_{n+1}) + J^z \sum_n \sigma^z_n \sigma^z_{n+1} - B_{\text{eff}} \sum_n (\sigma^z_n) + C, \tag{20.5}$$

with

$$J^{xy} = \frac{1}{2}(J_{11} + J_{22} - J_{12} - J_{21}), \tag{20.6}$$

$$J^z = \frac{1}{4}(J_{11} + J_{22} + J_{12} + J_{21}), \tag{20.7}$$

$$B_{\text{eff}} = B - B_c - \frac{1}{4}(J_{11} + J_{22} + J_{12} + J_{21}), \tag{20.8}$$

$$C = \frac{1}{8}(J_{11} + J_{22} + J_{12} + J_{21}) - (B - B_c). \tag{20.9}$$

The effective Hamiltonian is identical to that of an XXZ chain in a field, which is a major step forward with respect to the original problem: this model has been investigated at length using the Bethe ansatz [8] and by field-theory methods [9, 10], and much is known about its low-energy properties. In particular, there is a quantum phase transition in zero field at $J^z = J^{xy}$ between a gapless phase at small $J^z$ and a gapped phase at large $J^z$.

To understand the physics of this phase transition for the original problem, it is expedient to perform a Jordan–Wigner transformation (See for example [11]) of the effective model. In this language, the Hilbert space is that of spinless fermions on a chain, and the elementary operators are fermion creation ($c_i^\dagger$), annihilation ($c_i$) and density ($n_i = c_i^\dagger c_i$) operators at site $i$. An empty site corresponds to a rung singlet, an occupied one to a rung triplet, and the effective Hamiltonian becomes

$$H_{\text{eff}} = -t \sum_i (c_i^\dagger c_{i+1} + c_{i+1}^\dagger c_i) + v \sum_i n_i n_{i+1} - \mu \sum_n n_i, \tag{20.10}$$

with $t = J^{xy}/2$, $v = J^z$, and $\mu = B_{\text{eff}} + J^z$. In this model, the gapped phase of the ladder is a half-filled insulating phase of the fermionic chain, while the gapless one is a metallic phase (a Luttinger liquid in this one-dimensional (1D) system). Thus the density as a function of the chemical potential has a plateau, the width of which is equal to the gap of the insulating phase. In the original model, this implies that the magnetization can have a plateau for certain parameters, a result which has been confirmed by density-matrix renormalization-group (DMRG) calculations [12]. The physics of the plateau state is discussed elsewhere in this volume (the Chap. 10 by Takigawa and Mila). For the purposes of the present chapter, we note only how powerful the effective-Hamiltonian method can be: a very simple, first-order calculation can basically solve the problem by mapping it onto another non-trivial but well-understood one.

### Example 2: Expansion Around the Ising Limit

In the previous example, as in the case of the Hubbard model, the unperturbed Hamiltonian is a sum of local terms, and the macroscopic ground-state degeneracy is given simply from the ground-state degeneracy of each term. This is not the only case where the effective-Hamiltonian approach is useful. Another important example is the ground state of the antiferromagnetic Ising model, which is degenerate on non-bipartite lattices such as the triangular and kagomé geometries. Again this can be the starting point of a degenerate perturbation theory towards the Heisenberg model on the same lattice if the transverse exchange is treated as a perturbation. In the same spirit, degenerate perturbation theory can be used to treat the effect of a transverse magnetic field applied to a frustrated Ising model. In this section we concentrate on the Ising case, which is of direct relevance to Quantum Dimer Models (QDMs, discussed in the Chap. 17 by Moessner and Raman).

Consider the fully frustrated Ising model in a transverse field on the honeycomb lattice, defined by the Hamiltonian

$$H = H_0 + V = -J \sum_{\langle i,j \rangle} M_{ij} \sigma_i^z \sigma_j^z - \Gamma \sum_i \sigma_i^x. \tag{20.11}$$

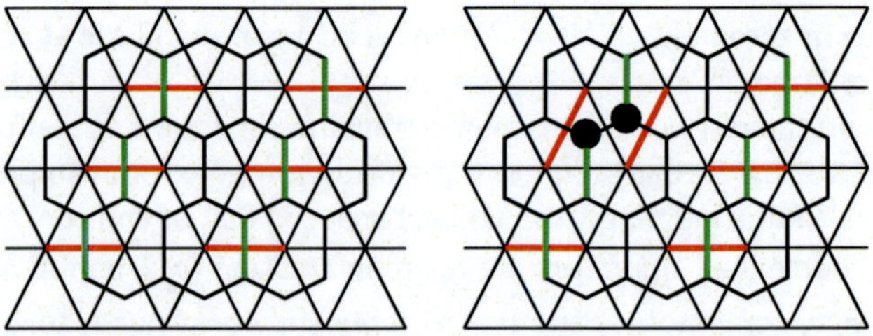

**Fig. 20.2** *Left*: representation of the model of 20.11. The *thick vertical lines* on the honeycomb lattice correspond to $M_{ij} = -1$, all others to $M_{ij} = 1$. The *thick horizontal lines* on the triangular lattice correspond to dimers in the ferromagnetic ground state. *Right*: dimer covering obtained from the ferromagnetic ground state after flipping the two spins shown as *black dots*

In this expression, the parameters $M_{ij} = \pm 1$ are chosen in such a way that their product around each hexagonal plaquette is equal to $-1$. All models in this class are then equivalent up to a gauge transformation, $\sigma_i^z \to \epsilon_i \sigma_i^z$, $M_{ij} \to \epsilon_i \epsilon_j M_{ij}$, with $\epsilon_i = \pm 1$. One possible choice is shown in Fig. 20.2. For this particular choice, the two ferromagnetic states, with all spins either up or down, are ground states of $H_0$. Indeed, they satisfy five bonds out of every six, and it is clearly impossible to do better for a case with one antiferromagnetic coupling out of six. From a ferromagnetic ground state, one may then construct more ground states, and it is easy to verify that flipping the spins at the ends of a satisfied bond that connects two unsatisfied bonds leads to another ground state (right panel, Fig. 20.2).

The structure of the Hilbert space is best understood by considering the dual lattice, which is the triangular lattice of sites at the centers of the hexagonal plaquettes, and to draw a line between two neighboring sites of the dual lattice if the bond (of the original lattice) which it crosses is unsatisfied. If one imposes the constraint that each site be connected to exactly one of its neighbors, this set of lines defines a dimer covering of the triangular lattice, and there is a one-to-one correspondence between these dimer coverings and the ground states of $H_0$ which have a given spin orientation for each given site. Because there are two possible spin orientations, the degenerate manifold of $H_0$ is twice as large as the Hilbert space of dimer coverings, the two configurations leading to the same dimer covering being related by a global flip of all spins.

For simplicity, we focus on the problem defined by the Ising Hamiltonian acting in the Hilbert space where two configurations related by a global spin-flip are identified. Then the degenerate subspace can be described by the set of dimer coverings, and the effective Hamiltonian obtained by degenerate perturbation theory takes the form of a QDM. To deduce the form of this QDM, we first note that the effective Hamiltonian vanishes to first order in the transverse field. Indeed, flipping a single spin changes the nature of the three bonds connected to it, which increases the number of unsatisfied bonds by one or by three, depending on whether or not one of the bonds connected to this site was unsatisfied. To second order, the situation is still more complex unless the two sites are nearest neighbors, in which case the process does not result in a state outside the ground-state manifold provided the two sites are on a satisfied bond connected to two unsatisfied ones. The resulting process leads to a dimer flip around a square plaquette of the dual lattice, and the effective Hamiltonian is a special case of the Rokhsar–Kivelson model, defined by the Hamiltonian

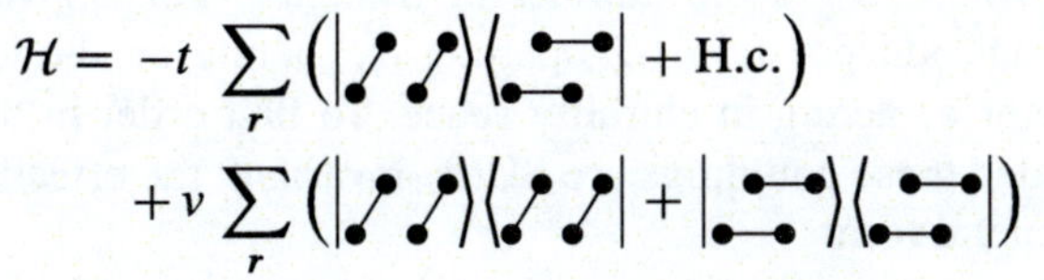


(20.12)

with $t = \Gamma^2/J$ and $v = 0$.

In this example, it is the effective model which is of direct interest in the context of frustrated magnets (as discussed in the Chap. 17 by Moessner and Raman). The connection with the Ising model turns out to be helpful in identifying a possible phase transition, as first noted in [13], and led more recently to an analytical description of the fractional excitations which exist in the Resonating-Valence-Bond phase of the QDM on the square lattice [14].

As mentioned at the beginning of this section, an expansion around the Ising limit can also be performed for the XXZ version of the Heisenberg model, using the ratio $J^{xy}/J^z$ as a small parameter. This approach has been used for the 1/3-plateau of the spin-1/2 kagomé antiferromagnet [15], for the half-magnetization plateau phase of the spin-1/2 pyrochlore Heisenberg antiferromagnet [16], and for the 1/3-plateau phase of molecular analogs of the spin 1/2 kagomé antiferromagnet [17].

### Example 3: Weakly Coupled Triangles

Frustration naturally appears in antiferromagnets in which the exchange paths create loops of odd length, the simplest of these loops being the triangle. There are several types of lattice geometry which can be considered as coupled triangles, and for which a perturbation theory starting from non-interacting triangles has proven to be useful. The effective model takes a special form due to the pecularity of the ground-state manifold of a triangle, which is four-fold degenerate. This is easily seen by rewriting the Hamiltonian in terms of the total spin of the triangle, $\boldsymbol{S}_{\text{tot}} = \boldsymbol{S}_1 + \boldsymbol{S}_2 + \boldsymbol{S}_3$, to obtain

$$H = J(\boldsymbol{S}_1 \cdot \boldsymbol{S}_2 + \boldsymbol{S}_2 \cdot \boldsymbol{S}_3 + \boldsymbol{S}_3 \cdot \boldsymbol{S}_1) = \frac{J}{2}\left[(\boldsymbol{S}_{\text{tot}})^2 - \frac{9}{4}\right], \tag{20.13}$$

a result emerging because it is possible to construct two doublets, i.e. four states, using three spin-1/2 entities. A convenient basis is provided by the simultaneous eigenstates of the scalar chirality, $\boldsymbol{S}_1.(\boldsymbol{S}_2 \times \boldsymbol{S}_3)$, and of the projection $S^z_{\text{tot}}$ of the total spin,

$$|R,\sigma\rangle = (|-\sigma\sigma\sigma\rangle + \omega|\sigma-\sigma\sigma\rangle + \omega^2|\sigma\sigma-\sigma\rangle)/\sqrt{3},$$
$$|L,\sigma\rangle = (|-\sigma\sigma\sigma\rangle + \omega^2|\sigma-\sigma\sigma\rangle + \omega|\sigma\sigma-\sigma\rangle)/\sqrt{3},$$

where $\omega = \exp(2i\pi/3)$, $\sigma = \pm 1/2$ refers to $S^z_{\mathrm{tot}}$, and $L$ and $R$ represent left- and right-handed chirality.

In a system of weakly coupled triangles, treating the inter-triangle coupling as a perturbation leads to an effective Hamiltonian which acts in a Hilbert space of dimension $4^{N_t}$, where $N_t$ is the number of triangles. For a given triangle $i$, it is convenient to introduce a spin-1/2 operator $\boldsymbol{\sigma}_i$, acting on the total spin, and a Pauli-matrix vector $\boldsymbol{\tau}_i$ acting in chirality space. To first order in the inter-triangle couplings, provided these couplings are SU(2)-invariant, the effective Hamiltonian then takes the general form

$$H_{\mathrm{eff}} = J_{\mathrm{eff}} \sum_{i,j}{}' \boldsymbol{\sigma}_i \cdot \boldsymbol{\sigma}_j H^{\tau}_{ij}, \tag{20.14}$$

where the sum is over the lattice of sites representing the arrangement of the triangles, $J_{\mathrm{eff}}$ is linear in the inter-triangle couplings, and the operator $H^{\tau}_{ij}$ depends on which sites of the triangles $i$ and $j$ are coupled by the inter-triangle interaction.

In the case of a three-leg 'spin tube,' the effective model takes the explicit form [18]

$$H_{\mathrm{eff}} = \frac{J'}{3} \sum_{\langle ij \rangle}{}' \boldsymbol{\sigma}_i \cdot \boldsymbol{\sigma}_j [1 + 2(\tau^x_i \tau^x_{i+1} + \tau^x_i \tau^x_{i+1})], \tag{20.15}$$

where $J'$ is the rung coupling of the real model. For the effective model, field-theory arguments based on bosonization show that the spectrum must be gapped in all sectors, a non-trivial prediction to be contrasted with the gapless spectrum of the three-leg ladder [18].

Another example where this type of effective Hamiltonian has provided additional valuable insight is on the kagomé lattice. Considering the trimerized version of this model, where the exchange constant within up-pointing triangles is taken to be $J$ and in down-pointing triangles $J'$, the effective Hamiltonian in the limit $J' \ll J$ can be expressed as [19, 20]

$$H_{\mathrm{eff}} = \frac{J'}{9} \sum_{\langle ij \rangle}{}' \boldsymbol{\sigma}_i \cdot \boldsymbol{\sigma}_j (1 - 4\boldsymbol{e}_{ij} \cdot \boldsymbol{\tau}_i)(1 - 4\boldsymbol{e}_{ij} \cdot \boldsymbol{\tau}_j), \tag{20.16}$$

where the vectors $\boldsymbol{e}_{ij}$ are taken from among $\boldsymbol{e}_1 = (1, 0)$, $\boldsymbol{e}_2 = (-1/2, -\sqrt{3}/2)$, and $\boldsymbol{e}_3 = (-1/2, \sqrt{3}/2)$, the choice depending on the labelling of the spins inside each triangle (For a detailed discussion of this point, see [21]). For this effective Hamiltonian, a mean-field decoupling of spin and chirality leads to a highly degenerate ground state. Each ground state can be associated with a dimer covering of the triangular lattice of up- (or down-)pointing triangles, and the number of states accordingly grows as $1.5351^{N_t}$ in terms of the number of triangles, or as $1.1536^N$ in terms of the number of spins [20]. This compares well with the number of low-lying singlets observed in numerical simulations of the spin-1/2 kagomé antiferromagnet [22].

This effective Hamiltonian is also a very useful starting point to discuss the physics in a magnetic field. Antiferromagnets composed of weakly coupled triangles exhibit a plateau at magnetization 1/3, in which all triangles have a total spin equal to 1/2 (in a $S = 1/2$ system) and oriented in the field direction. Inside the plateau, $\boldsymbol{\sigma}_i \cdot \boldsymbol{\sigma}_j = 1/4$ is a constant, and the effective Hamiltonian is a pure chirality model. This plateau has been investigated for the kagomé lattice in [23].

Finally, similar ideas based on weakly coupled tetrahedra have provided equally helpful insight into the properties of the spin-1/2 Heisenberg model on the pyrochlore lattice and in related systems [24–26].

## 20.3 Alternative Approaches Yielding Effective Hamiltonians

There are several ways of deriving effective Hamiltonians based on techniques other than direct strong-coupling expansions. The aim of this section is to provide a review of the physics underlying these approaches, with appropriate references for further reading concerning their more detailed implementations.

### *20.3.1 Canonical Transformation*

The canonical transformation of a Hamiltonian is defined by $H \to \tilde{H} = e^{\eta} H e^{-\eta}$, where $\eta$ is antihermitian, so that $e^{\eta}$ is unitary. If $|\psi\rangle$ is an eigenstate of $H$, then $|\tilde{\psi}\rangle = e^{\eta}|\psi\rangle$ is an eigenstate of $\tilde{H}$ with the same eigenvalue. If the operators are transformed simultaneously according to $A \to \tilde{A} = e^{\eta} A e^{-\eta}$, then $\langle\tilde{\psi}|\tilde{A}|\tilde{\psi}\rangle = \langle\psi|A|\psi\rangle$.

The foundation for using a canonical transformation to derive an effective Hamiltonian is the identity

$$\begin{aligned} e^{\eta} H e^{-\eta} &= H + [\eta, H] + \frac{1}{2!}[\eta, [\eta, H]] + \cdots \\ &= H + \sum_{n=1}^{\infty} \frac{1}{n!}[\eta, [\eta, \ldots [\eta, H] \ldots]], \end{aligned}$$

which is simply the Taylor expansion of $H(\lambda) = e^{\lambda\eta} H e^{-\lambda\eta}$ for $\lambda = 1$.

Considering the case where $H = H_0 + \lambda V$, if one can find an operator $\eta$ such that $[\eta, H_0] = -V$, then using $\lambda\eta$ as a generator leads to

$$\tilde{H} = H_0 + \sum_{n=1}^{\infty} \frac{n\lambda^{n+1}}{(n+1)!}[\eta, [\eta, \ldots [\eta, V] \ldots]].$$

This operator is a series in powers of $\lambda$, and hence of the perturbation $V$. While its structure is reminiscent of the results of high-order perturbation theory, there is an important difference: $\tilde{H}$ acts in the full Hilbert space of $H$, whereas $H_{\text{eff}}$ acts in the ground-state manifold of $H_0$. Depending on the problem, this may or may not be an advantage. If one is interested in high-energy states which may be detected in a particular experiment, the canonical-transformation approach has distinct advantages, because it gives all of the eigenstates up to a certain order, and not only those which have evolved from the ground-state manifold of $H_0$ under the perturbation. By contrast, if the goal is to reduce the Hilbert space to study the low-energy sector using, for example, exact-diagonalization calculations on finite clusters, then a degenerate perturbation theory is sufficient.

### *20.3.2 Continuous Unitary Transformation*

The canonical transformation introduced in the previous section is by no means the only possibility for obtaining an effective model by a unitary transformation. In fact there are many ways to do this, even in low orders of a perturbative approach, and it is therefore an obvious question to ask whether an optimal choice of transformation exists.

This question led both Wegner [27] and Głazek and Wilson [28, 29] to introduce independently of each other the concept of continuous unitary transformations (CUTs) [30]. In contrast to the one-step transformation discussed in the last section, here the unitary transformation is constructed as an infinite product of infinitesimal transformations. Although measurable (on-shell) quantities, such as energies, have to be the same independent of which kind of transformation has been chosen, off-shell quantities such as effective interactions can differ strongly. This has been demonstrated in a quite impressive manner for the case of the Fröhlich Hamiltonian, which describes conventional superconductivity mediated by the electron–phonon interaction [31]. Here, the effective electron–electron interaction at second order in the electron–phonon coupling shows divergences for the case of a one-step transformation (previous section), whereas in the continuous treatment the attractive interaction is smooth.

Another respect in which the approaches differ is that the one-step canonical transformation can be applied practically only at low orders in the perturbation. However, there are physical situations where one is interested in the quantitative determination of an effective Hamiltonian for a given parameter set in the original model. One example is the case of strongly frustrated networks of coupled dimers, as in the Shastry–Sutherland model, where processes relevant to the magnetization of the system appear only at high orders.

In the method of CUTs, a continuous parameter $l$ is introduced such that $l = 0$ refers to the initial system $H$ and $l = \infty$ corresponds to the final effective system, which should correspond to a simplified physical picture. The transformation can be constructed such that processes at higher energies are treated before those at lower energies. This renormalizing property is similar to Wilson's renormalization-group approach [32].

Let $U$ be the unitary transformation which diagonalizes the Hamiltonian $H$ and let $H(l) = U^\dagger(l)HU(l)$. This unitary transformation is then equivalent to performing an infinite sequence of unitary transformations, $e^{-\eta(l)dl}$, with the antihermitian generators

$$\eta(l) = -U^\dagger(l)\partial_l U(l). \tag{20.17}$$

Taking the derivative with respect to $l$ results in the "flow equation"

$$\partial_l H(l) = [\eta(l), H(l)], \tag{20.18}$$

which defines the change of the Hamiltonian during the flow. Note that 20.18 represents an infinite hierarchy of coupled differential equations, because for the general case an increasing number of terms is generated on the right-hand side at each order. In practice, one has therefore to perform a truncation (below).

The properties of the effective Hamiltonian depend strongly on the choice of the generator. There are in essence two different modern variants. The first one uses the generator introduced originally by Wegner [27], which aims to eliminate interaction matrix elements with the goal of obtaining an energy-diagonal effective Hamiltonian. This approach has been applied successfully to a large class of problems, with special attention being given to determining the ground state of interacting quantum many-body problems ([27, 31]; [33] and references therein). The second variant is the quasiparticle-conserving CUT which, as its name suggests, maps the Hamiltonian $H_0$ to an effective Hamiltonian conserving the number of quasiparticles [34–37]. This approach can be used either to study the excitations of an already known quantum ground state [30, 37–41], one application for which has been to bound states, or, in analogy to the previous case, to derive effective low-energy models [42–46].

Returning to the point mentioned above, in all practical calculations it is necessary to truncate the flow equation (20.18). For this there are two options: (a) cutting the hierarchy at one level and solving the equations numerically (self-similar CUTs) or (b) using a series-expansion ansatz for $\eta$ and $H$, and solving the flow equations perturbatively to high order (perturbative CUTs).

Here we focus only on presenting one illustrative example, for which we choose the perturbative version of quasiparticle-conserving CUTs [34, 37, 47, 48]. If the problem at hand meets the two conditions:

1. The unperturbed part has an equidistant spectrum bounded from below.
2. There is an integer number $N > 0$ such that the perturbing part can be separated as $\sum_{n=-N}^{N} T_n$, where $T_n$ increments (or if $n < 0$ decrements) the number of particles by $n$.

then the CUT in its quasiparticle-conserving form can be solved to high order in the perturbation and the effective Hamiltonian is given by the general form [37]

$$H_{\text{eff}}(x) = Q + \sum_{k=1}^{\infty} x^k \sum_{\substack{|\underline{m}|=k \\ M(\underline{m})=0}} C(\underline{m}) T(\underline{m}), \tag{20.19}$$

where $Q$ is the unperturbed part of the Hamiltonian, $x$ an expansion parameter, $\underline{m} = (m_1, m_2, \ldots, m_k)$, and $M(\underline{m}) = \sum_{i=1}^{k} m_i = 0$ reflects the conservation of the number of particles. The action of $H_{\text{eff}}$ can be viewed as a weighted sum of particle-number-conserving virtual excitation processes, each of which is encoded in a monomial $T(\underline{m}) = T_{m_1} T_{m_2} \ldots T_{m_k}$. The coefficients $C(\underline{m})$ are rational numbers which can be calculated (to high order in the perturbation) exactly as the ratio of two integers. It should be emphasized that the effective Hamiltonian $H_{\text{eff}}$, which has known coefficients $C(\underline{m})$, can be used straightforwardly in all perturbative problems that meet the above conditions.

We illustrate the method first for an unfrustrated spin ladder in a magnetic field, including from (20.2) only the magnetic exchange $J_\perp$ on the rungs, the magnetic field $B$, and the unfrustrated exchange along the legs of the ladder, setting $J_{11} = J_{22} \equiv J_\parallel$. Rewriting the Hamiltonian in terms of rung triplet operators $t_\alpha$, with $\alpha = \{x, y, z\}$ [49], we obtain

$$H = J_\perp Q + \frac{J_\parallel}{2} [T_0 + T_{+2} + T_{-2}] + H_B,$$

where

$$\begin{aligned}
Q &= \sum_{i,\alpha} t^\dagger_{\alpha,i} t_{\alpha,i}, \\
T_0 &= \sum_{i,\alpha} t^\dagger_{\alpha,i} t_{\alpha,i+1} + \sum_{i,\alpha,\beta} \left[ t^\dagger_{\alpha,i} t^\dagger_{\beta,i+1} t_{\beta,i} t_{\alpha,i+1} - t^\dagger_{\alpha,i} t^\dagger_{\alpha,i+1} t_{\beta,i} t_{\beta,i+1} \right], \\
T_{+2} &= \sum_{i,\alpha} t^\dagger_{\alpha,i} t^\dagger_{\alpha,i+1}, \\
T_{-2} &= \sum_{i,\alpha} t_{\alpha,i} t_{\alpha,i+1} = T^\dagger_{+2}.
\end{aligned}$$

The operator $Q$ counts the total number of triplet excitations, while the operators $T_n$ change the triplet number by $n$, and $H_B$ denotes the magnetic-field term.

In the following we consider the limit of weakly coupled rung dimers, i.e. we set $J_\perp = 1$ and consider $J_\parallel / J_\perp \equiv x$ as the small expansion parameter. The effective Hamiltonian $H_{\text{eff}}$ obtained by a quasiparticle-conserving CUT has the property $[H_{\text{eff}}, Q] = 0$, meaning that the total number of triplons (dressed triplets which are the elementary excitations of the spin ladder) is a conserved quantity. The effective Hamiltonian at second order is

$$H^{(2)}_{\text{eff}} = Q + xT_0 + \frac{x^2}{4} [T_{+2} T_{-2} - T_{-2} T_2] + H_B.$$

The total spin $S^z_{\text{tot}}$ is a conserved quantity. The magnetic-field term is therefore not changed under the unitary transformation, and the low-energy physics is influenced solely by the local singlet $|s\rangle$ and the triplet $|t^1\rangle$ polarized parallel to the magnetic field (as discussed in Sect. 2). Identifying $|s\rangle$ with an empty site and $|t^1\rangle$ with the presence of a hard-core boson, one may deduce the effective Hamiltonian in this basis by calculating matrix elements on a finite cluster [47, 48]. The result is

$$H^{hb}_{\text{eff}} = -t_n \sum_{i,j\in\{1,2\}} \left(b_i^\dagger b_{i+j} + \text{h.c.}\right) - \mu \sum_i n_i \\ - t'_1 \sum_i \left(b_{i-1}^\dagger b_{i+1} + \text{h.c.}\right) n_i + v_j \sum_{i,j\in\{1,2\}} n_i n_{i+j}, \qquad (20.20)$$

with $t_1 = x/2$, $t_2 = x^2/4$, $t'_1 = -x^2/4$, $v_1 = x/2 - 3x^2/8$, $v_2 = 0$, and $\mu = B - 1 + 3x^2/4$.

We emphasize that a calculation such as this, based on coupled dimers in an external magnetic field, can be performed for any frustrated lattice and to high orders in the perturbation. By using appropriate extrapolations, a quantitative low-energy effective Hamiltonian may be derived, which is usually applicable over a large part of the parameter space. The high-order expansion (including extrapolation of the bare series) becomes problematic only when the correlation length of the system exceeds the spatial range covered by the maximum order treated, for example close to a quantum phase transition. The method will break down if the ground state for the parameters of interest is not unitarily connected to the ground state about which one is expanding.

As a second example illustrating the importance of a quantitative effective model, meaning one obtained with high-order accuracy, we discuss the 2D spin-1/2 Heisenberg system known as the Shastry–Sutherland model [50] in a magnetic field,

$$H = J' \sum_{\langle i,j\rangle} \mathbf{S}_i \cdot \mathbf{S}_j + J \sum_{\langle\langle i,j\rangle\rangle} \mathbf{S}_i \cdot \mathbf{S}_j - B \sum_i S^z_i. \qquad (20.21)$$

The bonds denoted $\ll i,j \gg$ represent an array of orthogonal dimers, while the bonds $\langle i,j\rangle$, which are inter-dimer couplings, form a square lattice (Fig. 20.3). This theoretical model is believed to be realized experimentally in the layered copper oxide material $SrCu_2(BO_3)_2$, where the coupling ratio is $J'/J \approx 0.63$. In the theoretical model, for $J'/J$ smaller than a critical ratio of order 0.7, the ground state of the model is given exactly by the product of dimer singlets, and the magnetization process of the system can be described in terms of hard-core bosons which, as discussed for the spin ladder above, represent polarized $|t^1\rangle$ triplons on the dimers, interacting and moving on an effective square lattice [54, 55].

A consequence of the strong frustration is the appearance of several magnetization plateaus which correspond to Mott-insulating phases of the hard-core bosons [51–53], where the translational symmetry of the system is broken and triplon excitations are frozen in the ground state as in a charge-ordered state (discussed in the Chap. 10 by Takigawa and Mila). Theoretically, all approaches to the basic

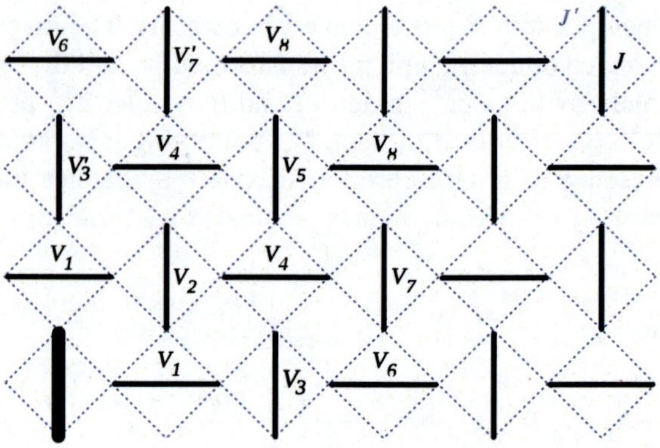


**Fig. 20.3** Shastry–Sutherland lattice and definition of the two-body interactions. $V_n$ is the coefficient of the two-body interactions between the reference dimer (*thick bond*) and the corresponding dimer. Figure courtesy of [62]

theoretical model agree on the presence of magnetization plateaus at 1/3 and 1/2 of the saturation value [54–58], in agreement with experiments [52, 59]. However, the structure below 1/3 magnetization is rather controversial. On the experimental side, the original pulsed-field data show only two anomalies which were interpreted as plateaus at 1/8 and 1/4 [52], but the presence of additional phase transitions, and of a broken translational symmetry above the 1/8-plateau have been established by recent torque and NMR measurements performed up to 31 T [60, 61]. The possibility of additional plateaus has been pointed out by Sebastian et al. [59], who interpreted their high-field torque measurements as evidence for plateaus at $1/q$, with $2 \leq q \leq 9$, and at $2/9$. On the theoretical side, the situation is also not settled. The finite clusters available for exact-diagonalization studies are not large enough to allow reliable predictions for high-commensurability plateaus, and the accuracy of the Chern–Simons mean-field approach initiated by Misguich et al. [57], and employed recently by Sebastian et al. [59] to explain their apparent additional plateaus, is not easy to assess. The essential difficulty lies in the fact that, because plateaus are a consequence of repulsive interactions between triplons, an accurate determination of the low-density, high-commensurability plateaus requires a precise knowledge of the long-range part of the interaction.

Such a precise analysis was conducted recently using perturbative CUTs [62]. The processes relevant for the physics in a finite magnetic field are those with maximum total spin and total $S_z$. Thus the general form of the effective Hamiltonian obtained by the perturbative CUT takes the form

$$H_{\text{eff}} = \sum_{n=2,4,6\ldots} \sum_{r_1,\ldots,r_n} C_{r_1,\ldots,r_n} b^\dagger_{r_1} \ldots b^\dagger_{r_{n/2}} b_{r_{n/2+1}} \ldots b_{r_n}, \qquad (20.22)$$

where $r_i$ labels the sites of the square lattice formed by the $J$ bonds, while the hard-core boson operator $b^\dagger_r$ creates a polarized $|t^1\rangle$ triplon at site $r$. The coefficients $C_{r_1,r_2,\ldots,r_n}$ are then obtained as high-order series in $J'/J$, computed up to order 15 for the two-body interactions $V_n$ to be discussed below.

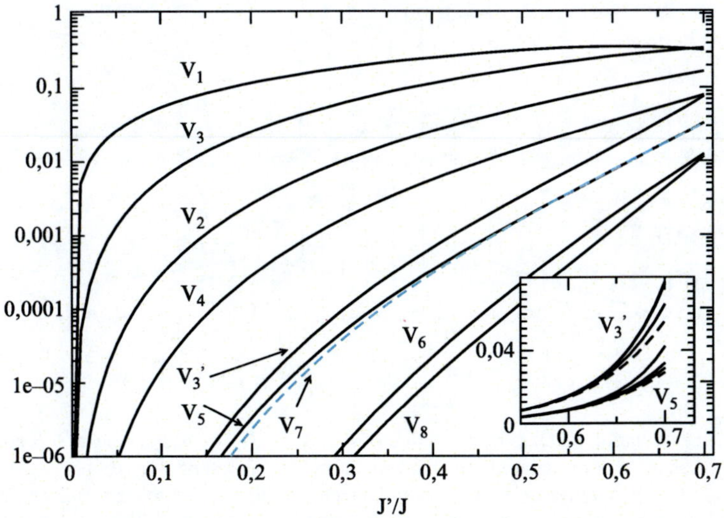


**Fig. 20.4** Coefficients of the extrapolated two-body interactions $V_n$ as function of $J'/J$. *Inset*: different extrapolations (*solid lines*) shown together with the bare series (*dashed lines*) for $V_3'$ and $V_5$. Figure courtesy of [62]

It is found that the magnitudes of all the interaction terms decrease when the separation of the sites is increased. In addition, the physics at low density is dominated by the two-body density-density interactions, while the standard two-site hopping is, as expected, strongly suppressed due to the frustration [63]. The evolution with $J'/J$ of the most relevant two-body interactions, defined in Fig. 20.3, is shown in Fig. 20.4. At small $J'/J$, interactions beyond $V_4$ are small and may be neglected, but for larger coupling ratios the higher-order terms $V_3'$, $V_5$, and $V_7$ (appearing at order 6) become important and contribute to the formation of low-density plateaus.

In general, the effective Hamiltonian $H_{\text{eff}}$ is by no means simpler than the original one, but it becomes so in the limit of low density and moderate $J'/J$. Indeed, in this limit the kinetic terms are very small, and they can be considered as a perturbation of the interaction part, which is diagonal in the local Fock basis, $|n_{r_1}, n_{r_2}, \ldots\rangle$. It is thus appropriate to use a Hartree approximation in which the variational ground state is a product of local boson wave-functions, because this approximation becomes exact in the limit of vanishing kinetic energy. This approach has the further advantage that it can be used to compare and treat rather large unit cells, which is important in the $SrCu_2(BO_3)_2$ problem, where "solid" phases with a complicated structure can be found at low magnetizations.

The resulting phase diagram in the Hartree approximation is shown in Fig. 20.5. The phase diagram is dominated by a series of plateaus, which appear at 1/3 and 1/2 (not shown) even at very small $J'/J$, and at 2/9, 1/6, 1/9, and 2/15 as $J'/J$ is increased. A plateau structure of this kind is to be expected when the kinetic terms, as here, are very small, because if these terms were completely absent, the

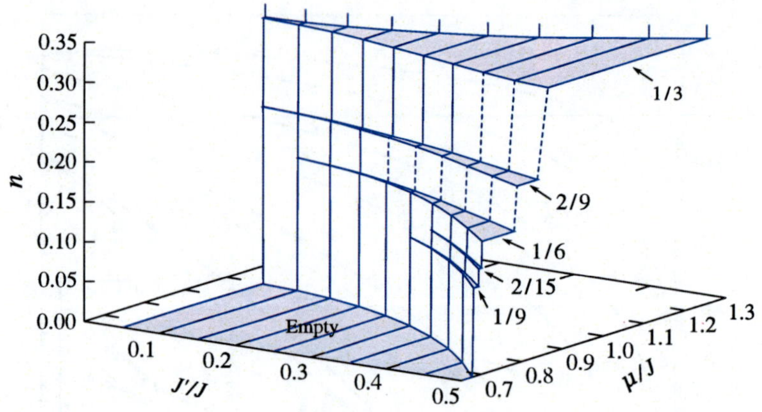


**Fig. 20.5** Magnetization plateaus as a function of $\mu$ and $J'/J$. The boson density $n$ is equal to the magnetization in units of the saturation value, and the chemical potential $\mu$ is equal to the magnetic field $B$. *Solid lines* denote results that are fully converged with respect to the terms retained in the Hamiltonian. Well-converged results are then connected by *dashed lines*. Figure courtesy of [62]

magnetization curve would consist simply of a sequence of plateaus. At $J'/J = 0.5$, the 1/6-plateau is by far the most prominent stucture below 1/3.

All of the plateaus found at low densities are actually stabilized by two-body repulsive interactions $V_n$ appearing at high orders in $J'/J$. It is, therefore, crucial to obtain the effective Hamiltonian very accurately, because the most significant features in this density regime are found to result from the competition between small interactions.

Finally, it is worth adding that the method of CUTs is also able to treat observables ([33]; and references therein [47, 62]). To this end, an observable $O$ must be transformed by the same CUT as the Hamiltonian,

$$\partial_l O(l) = [\eta(l), O(l)], \tag{20.23}$$

yielding effective observables $O_{\text{eff}} = O(l = \infty)$. Here we mention only one possible application which can be very useful for the physics of frustrated quantum magnets: a typical situation in quantum magnetism is that, in addition to the dominant nearest-neighbor Heisenberg exchange interactions, there are small coupling terms $H_{\text{add}}$, such as Dzyloshinskii–Moriya interactions, which can have a profound influence on the physics of the system. Formally

$$H_{\text{tot}} = H + H_{\text{add}},$$

and it is both elegant and efficient to perform first a CUT on $H$, which contains the dominant couplings, and then to treat $H_{\text{add}}$ as an observable

$$H_{\text{tot}}^{\text{eff}} = H^{\text{eff}} + U^\dagger H_{\text{add}} U.$$

Here, $H^{\text{eff}}$ conserves the number of quasiparticles, whereas the transformed observable $U^\dagger H_{\text{add}} U$ does not. Thus in the second step, performed after the first CUT, either ordinary perturbation theory or a second CUT can be applied to treat the term $U^\dagger H_{\text{add}} U$, which contains the small couplings. In cases where $H_{\text{add}}$ mixes low- and high-energy states, it is essential to retain access to the full Hilbert space of the problem.

### 20.3.3 Contractor Renormalization

One further method which we highlight here for the derivation of low-energy effective models is the Contractor Renormalization (CORE) technique, invented by Morningstar and Weinstein [65, 66]. The underlying idea is to derive effective Hamiltonians for a truncated local basis in such a manner that the low-energy spectrum of the model under study is reproduced exactly. As for CUTs, the CORE approach can be used either in an analytically oriented form [67–71] or as a numerical technique [72–76]. For recent reviews on CORE we refer the reader to [77,78].

The essential steps of the CORE algorithm are the following:

1. The system is divided into local subunits. One subunit is diagonalized, keeping $M$ suitable low-energy states.
2. The full Hamiltonian is diagonalized on a connected graph consisting of $N$ subunits. The low-energy eigenenergies $\epsilon_n$ and eigenstates $|n\rangle$ are calculated.
3. A basis of dimension $M^N$ is obtained by projecting the eigenstates onto the tensor product space of the retained states.
4. The effective Hamiltonian is constructed according to

$$H_N^{\text{eff}} = \sum_{n=1}^{M^N} \epsilon_n |\psi_n\rangle\langle\psi_n|.$$

5. The connected range-$N$ interactions are determined by subtracting the contributions of all connected subclusters.

Finally, the effective Hamiltonian is deduced by a cluster expansion as

$$H_{\text{CORE}}^{\text{eff}} = \sum_i H_i + \sum_{\langle i,j\rangle} H_{ij} + \cdots$$

Note that $H_{\text{CORE}}^{\text{eff}}$ reproduces exactly the low-energy physics if one considers all of the terms on the right-hand side.

In practice, it is necessary to perform a truncation. The convergence of the algorithm therefore depends both on the range of the operators taken in the cluster expansion and on the number and type of low-energy states retained for one subunit.

Hence the successful application of the CORE technique does require some physical insight concerning the problem at hand. However, once the relevant degrees of freedom have been selected, CORE represents a non-perturbative method for deriving effective low-energy Hamiltonians.

An important feature of the CORE algorithm is that it does not rely on the system being in a certain physical phase (to be contrasted with the example of quasiparticle-conserving CUTs discussed in the preceding section) and therefore does not break down even if a quantum phase transition takes place in the parameter space of the original model. As an example, we present here some illustrative CORE results for the magnetization curves of the 2D spin-1/2 Heisenberg model on the Shastry–Sutherland lattice discussed in the previous section [64]. The CORE technique is used to derive an effective Hamiltonian which is then treated by exact diagonalization (ED). The effective model obtained by CORE is found to agree very well with that obtained by perturbative CUTs. It is therefore expected that differences between the two approaches arise primarily from the method used to treat the effective model, which in the examples shown is either a classical limit [62] or ED [64].

Results obtained from CORE+ED are shown in Fig. 20.6. As in the treatment by a perturbative CUT, a rich plateau structure is found below $m = 1/3$. This again highlights the utility of an approach to the physics by first deriving a quantitative effective model which is then treated by simpler techniques. One obvious advantage of ED compared to the classical treatment is that it takes quantum fluctuations fully into account. It is, therefore, striking that the magnetization curves in both approaches are dominated by the presence of plateaus. By contrast, one drawback in using ED as a solver for the effective model (and one advantage of the classical solver) is the restriction on cluster sizes and shapes. Because the physics at low densities involves solid phases with rather large unit cells, it is a challenge within the ED approach to stabilize and to compare different solids, such as the $2/15$ phase found in [62]. Despite the differing aspects, both positive and negative, of these approaches to this problem, it should be emphasized that the advanced techniques used to derive effective Hamiltonians have proven to be crucial in the discovery and resolution of the complicated magnetization processes of strongly frustrated quantum magnets.

## 20.4 Conclusions

In this chapter, we have summarized a variety of tools which are used in the field of highly frustrated magnetism to derive effective low-energy Hamiltonians. We have aimed to capture the essential technical aspects of these different approaches and to provide examples of them for a number of physical applications. While it is not possible in a chapter of this type to cover all such methods in full detail, where relevant we have referred the interested reader to the more specialized literature.

We have shown that strong-coupling expansions and the derivation of effective models are at the same time standard techniques used by theoretical physicists for a broad range of physical questions and also a very active area of current research on

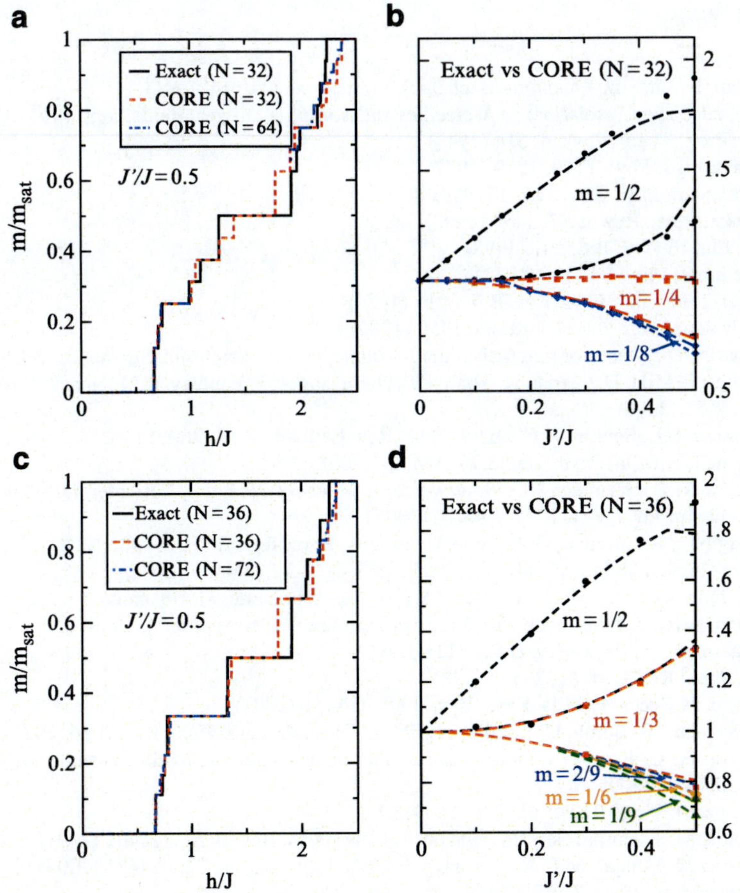


**Fig. 20.6** (**a**) Magnetization curves obtained by ED and CORE calculations on an $N = 32$ lattice and from CORE results with $N = 64$ for $J'/J = 0.5$. (**b**) Phase diagram for $N = 32$ as a function of $J'/J$ and magnetic field $h/J$: CORE results (*lines*) agree with ED (*symbols*) for locations of the $m = 1/2$, $1/4$, and $1/8$ plateaus which are allowed on this cluster. (**c,d**) same as (**a,b**) for $N = 36$ and $N = 72$. On these clusters, the $m = 1/2$, $1/3$, $2/9$, $1/6$, and $1/9$ plateaus are allowed. Figure courtesy of [64]

strongly correlated quantum systems. For the first type of application, it is often the goal to identify the relevant low-energy degrees of freedom and to use an effective model to leading order in perturbation theory to understand qualitatively the phase diagram of a given model. By contrast, the aim of the current developments is to obtain a quantitative derivation of effective models and a complete understanding of the breakdown of such derivations.

## References


1. L. Landau, E. Lifshitz, *Quantum Mechanics*, (Pergamon, Oxford, 1991)
2. P. Fulde, *Electron Correlations in Molecules and Solids*, (Springer, Heidelberg, 1991)
3. T. Kato, Prog. Theor. Phys. **4**, 514 (1949)
4. M. Takahashi, J. Phys. C **10**, 1289 (1977)
5. P.W. Anderson, Phys. Rev. **115**, 2 (1959)
6. K. Totsuka, Phys. Rev. B **57**, 3454 (1998)
7. F. Mila, Eur. Phys. J. B **6**, 201 (1998)
8. H.A. Bethe, Z. Phys. **71**, 205 (1931)
9. A. Luther, I. Peschel, Phys. Rev. B **9**, 2911 (1974)
10. F.D.M. Haldane, Phys. Rev. Lett. **45**, 1358 (1980)
11. D.C. Mattis, *The theory of magnetism made simple*, (World Scientific, Singapore, 2006)
12. J.-B. Fouet, F. Mila, D. Clarke, H. Youk, O. Tchernyshyov, P. Fendley, R.M. Noack, Phys. Rev. B **73**, 214405 (2006)
13. R. Moessner, S.L. Sondhi, P. Chandra, Phys. Rev. Lett. **84**, 4457 (2000)
14. G. Misguich, F. Mila, Phys. Rev. B **77**, 134421 (2008)
15. D.C. Cabra, M.D. Grynberg, P.C. Holdsworth, A. Honecker, P. Pujol, J. Richter, D. Schmalfuss, J. Schulenburg, Phys. Rev. B **71**, 144420 (2005)
16. D.L. Bergman, R. Shindou, G.A. Fiete, L. Balents, Phys. Rev. B **75**, 094403 (2007)
17. I. Rousochatzakis, A.M. Läuchli, F. Mila, Phys. Rev. B **77**, 094420 (2008)
18. H. J. Schulz in *Proceedings of the XXXIst Rencontres de Moriond*, eds. by T. Martin, G. Montambaux, J. Trin Thanh Vin (Editions Frontières, Gif-sur-Yvette, 1996)
19. V. Subrahmanyam, Phys. Rev. B **52**, 1133 (1995)
20. F. Mila, Phys. Rev. Lett. **81**, 2356 (1998)
21. M. Ferrero, F. Becca, F. Mila, Phys. Rev. B **68**, 214431 (2003)
22. P. Lecheminant, B. Bernu, C. Lhuillier, L. Pierre, P. Sindzingre, Phys. Rev. B **56**, 2521 (1997)
23. B. Damski, H.-U. Everts, A. Honecker, H. Fehrmann, L. Santos, M. Lewenstein, Phys. Rev. Lett. **95**, 060403 (2005)
24. H. Tsunetsugu, Phys. Rev. B **65**, 024415 (2002)
25. V.N. Kotov, M.E. Zhitomirsky, M. Elhajal, F. Mila, Phys. Rev. B **70**, 214401 (2004)
26. V.N. Kotov, M. Elhajal, M.E. Zhitomirsky, F. Mila, Phys. Rev. B **72**, 014421 (2005)
27. F.J. Wegner, Ann. Phys. **3**, 77 (1994)
28. S.D. Głazek, K.G. Wilson, Phys. Rev. D **48**, 5863 (1993)
29. S.D. Głazek, K.G. Wilson, Phys. Rev. D **49**, 4214 (1994)
30. C. Knetter, K.P. Schmidt, M. Grüninger, G.S. Uhrig, Phys. Rev. Lett. **87**, 167204 (2001)
31. P. Lenz, F. Wegner, Nucl. Phys B **482**, 693 (1996)
32. K.G. Wilson, Rev. Mod. Phys. **47**, 773 (1975)
33. S. Kehrein, *Flow-equation approach to many-particle systems*. Springer Tracts in Modern Physics, vol. 217. (Springer, Berlin, 2006)
34. J. Stein, J. Stat. Phys. **88**, 487 (1997)
35. A. Mielke, Eur. Phys. J. B **5**, 605 (1998)
36. G.S. Uhrig, B. Normand, Phys. Rev. B **58**, R14705 (1998)
37. C. Knetter, G.S. Uhrig, Eur. Phys. J. B **13**, 209 (2000)
38. C. Knetter, A. Bühler, E. Müller-Hartmann, G.S. Uhrig, Phys. Rev. Lett. **85**, 3958 (2000)
39. C.P. Heidbrink, G.S. Uhrig, Eur. Phys. J. B **30**, 443 (2002)
40. W. Brenig, Phys. Rev. B **67**, 064402 (2003)
41. K.P. Schmidt, G.S. Uhrig, Mod. Phys. Lett. B **19**, 1179 (2005)
42. A. Reischl, E. Müller-Hartmann, G.S. Uhrig, Phys. Rev. B **70**, 245124 (2004)
43. S. Dusuel, G.S. Uhrig, J. Phys. A Math. Gen. **37**, 9275 (2004)
44. S. Dusuel, J. Vidal, Phys. Rev. Lett. **93**, 237204 (2004)
45. J.N. Kriel, A.Y. Morozov, F.G. Scholtz, J. Phys. A Math. Gen. **38**, 205 (2005)
46. K.P. Schmidt, S. Dusuel, J. Vidal, Phys. Rev. Lett. **100**, 057208 (2008)
47. C. Knetter, K.P. Schmidt, G.S. Uhrig, J. Phys. Condens. Matter **36**, 7889 (2003)

48. C. Knetter, K.P. Schmidt, G.S. Uhrig, Eur. Phys. J. B **36**, 525 (2004)
49. S. Sachdev, R.N. Bhatt, Phys. Rev. B **41**, 9323 (1990)
50. B.S. Shastry, B. Sutherland, Phys. B **108B**, 1069 (1981)
51. E. Müller-Hartmann, R.R.P. Singh, C. Knetter, G.S. Uhrig, Phys. Rev. Lett. **81**, 1808 (2000)
52. K. Onizuka et al., J. Phys. Soc. Jpn. **69**, 1016 (2000)
53. K. Kodama et al., Science **298**, 395 (2002)
54. T. Momoi, K. Totsuka, Phys. Rev. B **62**, 15067 (2000)
55. S. Miyahara, K. Ueda, J. Phys. Condens. Matter **15**, R327 (2003)
56. S. Miyahara, K. Ueda, Phys. Rev. B **61**, 3417 (2000)
57. G. Misguich, T. Jolicoeur, S.M. Girvin, Phys. Rev. Lett. **87**, 097203 (2001)
58. S. Miyahara, F. Becca, F. Mila, Phys. Rev. B **68**, 024401 (2003)
59. S.E. Sebastian, N. Harrison, P. Sengupta, C.D. Batista, S. Francoual, E. Palm, T. Murphy, H.A. Dabkowska, B. D. Gaulin, Proc. Nat. Acad. Sci. **105**, 20157 (2008)
60. F. Levy, I. Sheikin, C. Berthier, M. Horvatić, M. Takigawa, H. Kageyama, T. Waki, Y. Ueda, Europhys. Lett. **81**, 67004 (2008)
61. M. Takigawa, S. Matsubara, M. Horvatić, C. Berthier, H. Kageyama, Y. Ueda, Phys. Rev. Lett. **101**, 037202 (2008)
62. J. Dorier, K.P. Schmidt, F. Mila, Phys. Rev. Lett. **101**, 250402 (2008)
63. S. Miyahara, K. Ueda, Phys. Rev. Lett. **82**, 3701 (1999)
64. A. Abendschein, S. Capponi, Phys. Rev. Lett. **101**, 227201 (2008)
65. C.J. Morningstar, M. Weinstein, Phys. Rev. Lett. **73**, 1873 (1994)
66. C.J. Morningstar, M. Weinstein, Phys. Rev. D **54**, 4131 (1996)
67. M. Weinstein, Phys. Rev. B **63**, 174421 (2001)
68. E. Altman, A. Auerbach, Phys. Rev. B **65**, 104508 (2002)
69. E. Berg, E. Altmann, A. Auerbach, Phys. Rev. Lett. **90**, 147204 (2003)
70. R. Budnik, A. Auerbach, Phys. Rev. Lett. **93**, 187205 (2004)
71. M.S. Siu, M. Weinstein, Phys. Rev. B **75**, 184403 (2007)
72. J. Piekarewicz, J.R. Shepard, Phys. Rev. B **56**, 5366 (1997)
73. J. Piekarewicz, J.R. Shepard, Phys. Rev. B **57**, 10260 (1998)
74. S. Capponi, D. Poilblanc, Phys. Rev. B **66**, 180503(R) (2002)
75. S. Capponi, A. Läuchli, M. Mambrini, Phys. Rev. B **70**, 104424 (2004)
76. A. Abendschein, S. Capponi, Phys. Rev. B **76**, 064413 (2007)
77. A. Auerbach, AIP Conf. Proc. **816**, 1 (2006)
78. S. Capponi AIP Conf. Proc. **816**, 16 (2006)

# Part VI
# Frustration, Charge Carriers and Orbital Degeneracy

# Chapter 21
# Mobile Holes in Frustrated Quantum Magnets and Itinerant Fermions on Frustrated Geometries

**Didier Poilblanc and Hirokazu Tsunetsugu**

**Abstract** In a manner similar to those of spin, charge degrees of freedom in systems with frustrated geometries often exhibit complex and exotic types of behavior. This chapter describes two classes of system with these properties. The first is the case of doped Mott insulators with magnetic frustration, where spin-charge separation and superconductivity with unconventional pairing symmetries are among the possible phenomena which emerge. The second is the case of itinerant fermions at commensurate densities, where quantum fluctuation effects, acting on top of the canonical metal–insulator transition or macroscopically degenerate classical manifold of states, can lead to novel forms of behavior, and indeed to novel phases.

## 21.1 Introduction

As discussed in several chapters of this volume, frustration leads to unconventional (insulating) ground states. On the other hand doped holes are known to have profound effects in Mott insulators. Therefore doped frustrated systems offer the prospect of novel phases with some of the most fascinating, challenging and exotic behaviour. In addition, at commensurate electron fillings and in the presence of strong (screened) Coulomb repulsion, geometrical frustration can also manifests itself as an extensive degeneracy of the classical ground-state manifold providing profound similarities with the field of quantum frustrated magnetism.

Magnetic frustration in quantum spin systems leads frequently to the formation of spin singlets (dimers). Generically, systems of fluctuating quantum dimers can often order, breaking lattice symmetries to give rise to Valence Bond Crystals (VBCs) [1], but under other circumstances they may remain in a quite unconventional quantum disordered state, the spin liquid, which breaks neither spin nor lattice symmetries. Anderson's original $d$-wave Resonating Valence Bond (RVB) state [2, 3] is a paradigm for the spin liquid (in fact, for a particular type of gapless spin liquid, while the short-range RVB state composed of only nearest-neighbor dimers is gapped spin liquid). In a number of cases, frustrated spin systems and/or dimer systems can be doped, for example by chemical substitution in a Mott insulator.

When both spin and charge degrees of freedom are present, the role of frustration becomes unclear, and to date remains only poorly explored. It is, however, clear that new and exotic phenomena emerge upon doping, including heavy-fermion behavior, spin-charge separation or quasiparticle fractionalization, unconventional superconductivity, stripe formation, bond and/or charge ordering, and many others. Such fundamental issues have motivated an increasing number of recent investigations, as well as the continuing search for new, doped materials.

In this chapter, we describe some selected topics which illustrate the richness and diversity of the field of doped, frustrated magnets. The first example concerns the dynamics of a small number of doped holes in the two-dimensional (2D) kagomé and checkerboard Heisenberg quantum antiferromagnets. Without doping, the kagom'e Heisenberg antiferromagnet is believed to be a serious candidate for spin-liquid behavior, while the checkerboard lattice is the 2D analog of the well-known and highly frustrated 3D pyrochlore structure common in real materials. With doping, issues such as particle fractionalization and pairing can be addressed. In a second example, we move to the topic of Quantum Dimer Models (QDMs), similar to those proposed by Rokhsar and Kivelson in the context of the pseudo-gap phase of high-temperature superconductors. Two classes of (weakly) doped QDMs will be discussed, which differ in the assumed statistics, bosonic or fermionic, of the bare holes. We proceed further by considering strongly correlated electrons on frustrated triangular lattices, and discuss the physics of an unconventional, reentrant metal–insulator transition. As our final example, we consider correlated fermions moving on frustrated lattices at special, commensurate densities for which exotic but once again insulating ground states (GSs) are stabilized. For systems with strong interactions (Mott insulators), we show briefly how their behavior is analogous to the physics of (undoped) QDMs.

## 21.2 Doping Holes in Frustrated Quantum Magnets

### *21.2.1 The Holon–Spinon Deconfinement Scenario*

We begin our discussion of the phenomena associated with doped holes in frustrated magnets by considering the most popular paradigm for a nonmagnetic quantum ground state, which is a dimer-based system. If we assume that a single hole is "injected," then the removal of the electron results in the breaking of one of the dimers, leaving behind an empty site (holon) and a free spin (spinon) on the same bond.

If the dimers can change their positions by quantum fluctuations, the holon and the spinon can move on the lattice, across the diagonals of the plaquettes in a square-lattice system, by exchanging with the dimers. A typical configuration is shown in Fig. 21.1. Optimization of their kinetic energies would require the holon and spin to

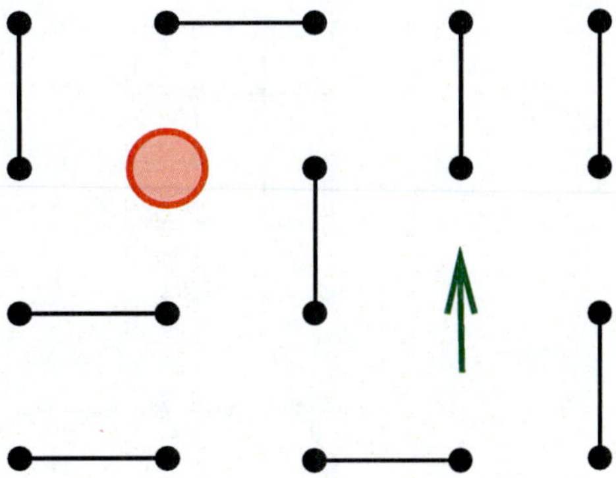

**Fig. 21.1** Schematic representation of a holon (*empty site, red*) and a spinon (*free spin, green*) embedded in a fluctuating background of bond singlets (*black*)

delocalize, and thus to become separated. In a (gapped) spin liquid, realized if the system can fluctuate through all possible dimer coverings, a complete separation, known as "deconfinement," is possible. In this situation, the Landau quasiparticle (QP) breaks apart into separate species, and an experimental technique which probes the hole Green function, such as Angle-Resolved Photoemission Spectroscopy (ARPES), would then show a broad maximum in place of the sharp QP peak. However, if dimer VBC order is present, i.e., only one specific (type of) covering lies lowest in energy, one expects an effective string potential that binds the holon and spinon: indeed, if the dimers had no internal structure, an attempt to separate these two "particles" would lead to a continuous and linear increase of their energy. In reality, this increase is bounded by the spin gap (the energy to break up a dimer), beyond which pairs of spinons would be generated spontaneously along the string.

### *21.2.2 Single Hole Doped in Frustrated Mott Insulators*

Frustrated magnets are good candidates for the observation of spin-charge separation upon doping. The checkerboard and kagomé lattices, shown in Fig. 21.2a,c, are good examples of the types of frustrated lattice on which such a phenomenon may be expected. They are composed, respectively, of strongly frustrated tetrahedra and triangles, assembled in a 2D, corner-sharing structure. While the AF Heisenberg Hamiltonian for $S = 1/2$ spins (in this chapter we consider only systems of $S = 1/2$ spins) on the checkerboard lattice is now thought to be a fully gapped system exhibiting VBC order (plaquette phase) [4–8], by contrast no sign of ordering has been found in the undoped kagomé antiferromagnet [9–13], which possesses an exponentially large number of singlet states within the (finite-size) spin gap [12–15]. It is these unconventional, low-lying excitations which open the door to new and surprising phenomena upon hole doping.

We perform exact diagonalization (ED) calculations based on the standard $t$-$J$ model Hamiltonian,

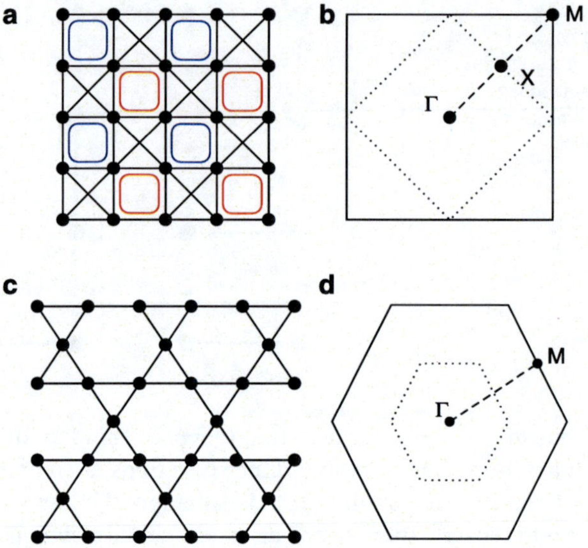


**Fig. 21.2** (**a**) Checkerboard lattice. The two degenerate VBC GSs of the nearest-neighbor Heisenberg model (*half-filling*) are represented schematically in blue and red. This figure also serves as a representation of the RSPC GSs of the quarter-filled, large-$V$ Hubbard model of Sect. 21.5.2. (**b**) Full Brillouin Zone (BZ) of the checkerboard lattice. The dotted line corresponds to the reduced BZ associated with the $\sqrt{2}\times\sqrt{2}$ super-cell reflection. The path from the zone center ($\Gamma$) to the $\mathbf{k} = (\pi, \pi)$ point ($M$) is shown as a dashed line. (**c**) kagomé lattice. (**d**) As in (**b**) but for the case of the kagomé lattice

$$H_{t\text{-}J} = -t \sum_{\langle i,j\rangle,\sigma} \mathcal{P}\left(c^{\dagger}_{i,\sigma} c_{j,\sigma} + \text{h.c.}\right)\mathcal{P} + J \sum_{\langle i,j\rangle} \left(\mathbf{S}_i \cdot \mathbf{S}_j - \frac{1}{4} n_i n_j\right), \quad (21.1)$$

where on both lattices all bonds have the same couplings $t$, describing the kinetic energy of the hopping quasiparticles, and $J$, which is the superexchange interaction between the spins. This model is believed to be a reliable description of the low-energy physics of weakly doped Mott-Hubbard insulators with large optical excitation gaps. Here and hereafter, we assume the value $J = 0.4$ (in units where $|t|$ is set to 1), which is the general order of the physical value in a number of real materials. The hole spectral functions are defined in the standard way as

$$A(\mathbf{k}, \omega) = -\frac{1}{\pi} \text{Im}\left[\left\langle \Psi_0 | c^{\dagger}_{\mathbf{k},\sigma} \frac{1}{\omega + E_0 + i\eta - H} c_{\mathbf{k},\sigma} | \Psi_0 \right\rangle\right], \quad (21.2)$$

and calculated by Lanczos ED, supplemented with a continued-fraction technique, on finite clusters with periodic boundary conditions to take advantage of the lattice translation symmetry. The reader is referred to Chap. 18 of A. Läuchli in this volume for a detailed discussion of numerical methods. Here we focus on the case of a single dynamic hole, as studied in [16]. Because of the absence of particle-hole symmetry in frustrated lattices, it is necessary to distinguish between the cases $t > 0$ and $t < 0$. Note that for $t < 0$, frustration can also appear in the hole motion: in the example of a particle with the tight-binding dispersion on an isolated triangle, the kinetic-energy gain is $|t|$, a factor of two smaller than for $t > 0$.

Typical results obtained for a 32-site cluster on the checkerboard lattice are shown in Fig. 21.3 for the line $\Gamma M$ line in the Brillouin Zone (BZ) (Fig. 21.2b). At all points, most of the spectral weight is found to be incoherent, distributed over a range of 7–9$|t|$. However, a small QP peak is visible, particularly for momenta close to the $M$ point. The region close to the $\Gamma$ point has only a very small QP peak, or possibly none at all, and the shape of the spectral function at $\Gamma$ itself is very special, probably because of its higher point-group symmetry.

The analogous spectral functions of the kagomé lattice, shown in Fig. 21.4, show definite exotic behavior; they are very broad for all momenta (widths approximately 6–8$|t|$) and, in contrast to the checkerboard lattice, show no visible QP peaks, either for $t > 0$ (left panel) or for $t < 0$ (right panel). We stress that the broad appearance of these spectra has no connection to the value of $\eta$ used in the calculation, but is an intrinsic feature of the spectral function, as can be deduced from the large number of poles carrying spectral weight (circles in Fig. 21.4). These spectral-function data support very strongly a spin-charge-separation scenario for the kagomé lattice. Indeed, this spectacular phenomenon can be observed directly in the spin-density profile in the vicinity of the hole: a repulsion between the net $S = 1/2$ moment and the mobile hole is clearly visible, providing further support to the deconfinement scenario described above in the context of dimer-based systems.

This investigation provides the first example of the observation of spin-charge separation in a 2D microscopic model. It establishes that the spin-liquid nature

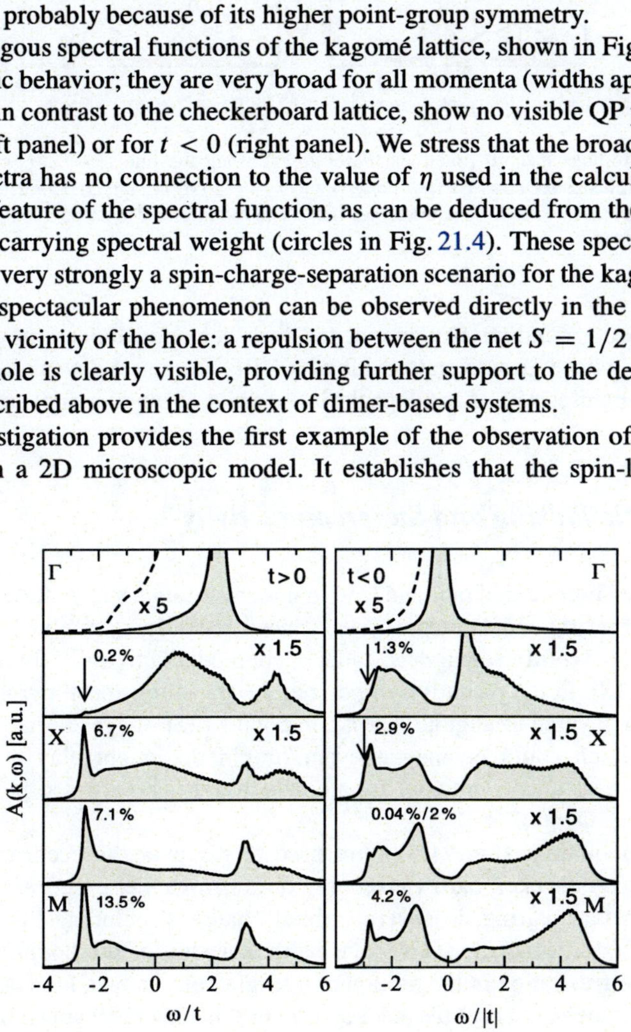


**Fig. 21.3** Single-hole spectral functions obtained on a 32-site checkerboard cluster ($\sqrt{32} \times \sqrt{32}$, tilted at 45° to the axes of Fig. 1.2) along the line $\Gamma M$. Left panel: $t = +1$; right panel: $t = -1$. In both cases, $J/|t| = 0.4$. When a quasiparticle peak is present, the corresponding weight is indicated. Magnification factors applied in some cases are as indicated. From [16].

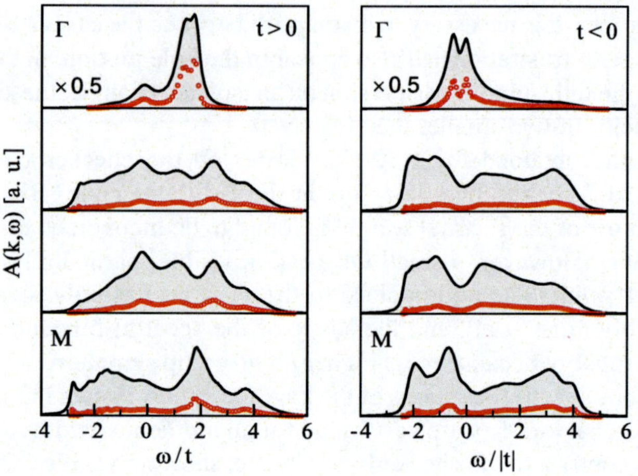


**Fig. 21.4** Single-hole spectral functions (*black lines*) along the line $\Gamma \leftrightarrow M$, computed on a 27-site kagomé cluster for $t = +1$ (*left panels*) and $t = -1$ (*right panels*). Contributions of both singlet and triplet final states are included (see text). The red circles denote pole locations and their residues. From [16]

of the undoped ground state is crucial for such behavior. Indeed, in the checkerboard lattice, whose ground state exhibits a VBC structure, a weak holon–spinon confinement manifests itself as QP peaks in the spectral function for some momenta.

## 21.2.3 Hole Pairing and Superconductivity

Whether doped holes could pair and lead to unconventional superconducting behavior is another of the fundamental issues raised recently by the new prospect of doping frustrated antiferromagnets. Indeed, superconductivity in the spinel oxide $LiTi_2O_4$ [17, 18], in the recently synthesized 5$d$ transition-metal pyrochlores [19–21], and in a layered triangular CoO compound [22] suggests that geometrical frustration, which could be magnetic and/or kinetic, might play a key role in the mechanism of unconventional superconductivity (as discussed in Chap. 22 of Z. Hiroi and M. Ogata).

Cluster calculations (Fig. 21.5) were used to discover the occurrence of pairing in the doped checkerboard Heisenberg antiferromagnet described above [23]. It was shown that pairing, in several orbital channels including $s$- and $d$-wave, appears at arbitrarily small $J/t$ for the particular sign of the hopping amplitude which leads to frustration in the motion of a single hole. In fact, hole delocalization (i.e. a gain in kinetic energy) plays a key role in this new mechanism of unconventional pairing, as also in some of the inter-layer tunneling mechanisms proposed by Anderson [24]. From these numerical data, a simple scenario might be proposed for $t > 0$: despite its suppressed coherent motion, a single hole retains a strong

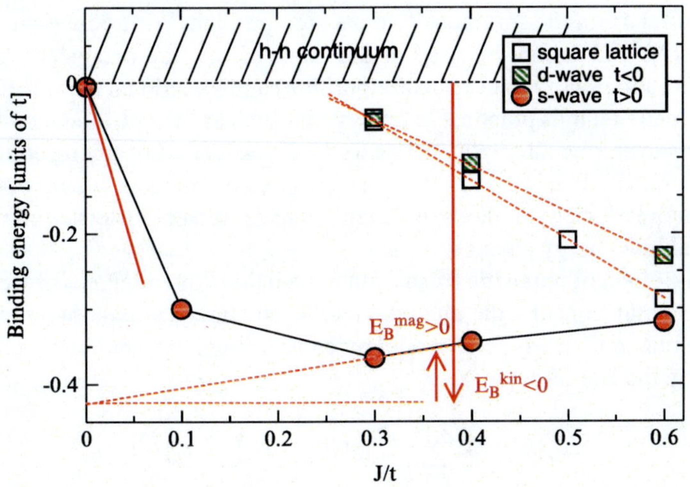


**Fig. 21.5** Binding energies $E_B$ (in units of $|t| = 1$) of two holes doped on a square, 32-site cluster ($\sqrt{32} \times \sqrt{32}$) for $t > 0$ and $t < 0$, in the two orbital-symmetry channels corresponding to the respective GSs. The thick, red line corresponds the local slope at $J/t \to 0$. Red arrows show the large kinetic-energy gain and small loss in magnetic energy whose sum gives $E_B$. $E_B < 0$ is the signature of a bound state. From [23]

incoherent motion, and thus can act to "melt" the plaquette VBC in its vicinity. This region, which may be somewhat extended in space, becomes more favorable for a second hole to gain kinetic energy, leading to correlated (or assisted) hopping. It is interesting to emphasize here the similarities with tight-binding studies of frustrated lattices, which show both localized single-particle states and interaction-induced, delocalized, two-particle bound states [25, 26].

In connection with cobaltates, superconductivity has also been investigated in the $t$-$J$ model on the triangular lattice, in particular using RVB variational Ansätze (presented in Chap. 22 of Z. Hiroi and M. Ogata). In these studies, $d_{x^2-y^2} + id_{xy}$-wave superconductivity is found to be stable near half-filling. The relationship between this phase and the three-sublattice, 120-degree AF long-range order occurring at half-filling remains at present unclear and in need of further investigation.

## 21.3 Doped Quantum Dimer Model

### 21.3.1 Origin of the Quantum Dimer Model

The conclusions obtained in Sect. 21.2.2 notwithstanding, both magnetic frustration and the introduction of fermionic variables (holes) lead, together or independently, to severe limits on the available numerical approaches. For example, quantum Monte Carlo algorithms, known to be very efficient for simple quantum spin systems, suffer from the infamous "minus-sign problem" (introduced in Chap. 18 of A. Läuchli) and cannot be used at the required low temperatures. Practically, zero-temperature ED (by the Lanczos algorithm) and variational approaches are the only controlled methods practicable for systems such as the $t$-$J$ model on frustrated lattices. However, one alternative route to the investigation of microscopic models of the $t$-$J$ and Hubbard types is to construct effective models which would allow the use of more efficient methods, or calculations on larger clusters, while retaining the essential low-energy physics.

When the effect of magnetic frustration is such that dimer degrees of freedom are relevant, one may consider the quantum hard-core dimer gas on a two-dimensional lattice. We illustrate this type of model by considering a square lattice, on which it is defined by the Hamiltonian

$$H_{\mathrm{QDM}} = V \sum_{c} N_c |c\rangle\langle c| - J \sum_{(c,c')} |c'\rangle\langle c|, \qquad (21.3)$$

where the sum over the index $(c)$ refers to all nearest-neighbor dimer coverings, $N_c$ is the number of "exchangeable" plaquettes, and the sum $(c', c)$ is over all pairs of configurations $|c\rangle$ and $|c'\rangle$ that differ by a single dimer-exchange process of the type illustrated in Fig. 21.6a. In a manner similar to the square lattice, on the triangular lattice the exchange of parallel dimer pairs can be performed on the three different types of two-triangle rhombi. This model was introduced originally by Rokhsar and Kivelson [27] in the context of the RVB theory of cuprate superconductors. The connection to the original spin formulation is, however, not completely clear: among other truncations of the spin degrees of freedom, the QDM of (21.3) deals by construction with orthogonal dimer coverings, which is not the case for the SU(2) dimer basis relevant in frustrated Heisenberg antiferromagnets. In spite of these subtleties, QDMs are expected to capture the essential physics of systems with singlet ground states, one primary reason for this being that they do possess the extreme

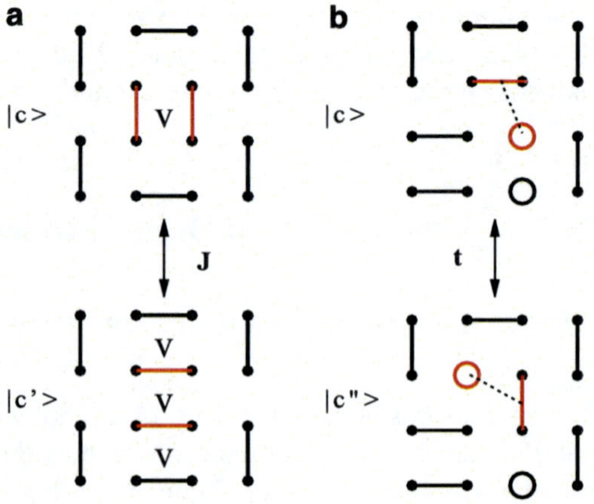


**Fig. 21.6** (**a**) Dimer-exchange process and (**b**) holon hopping process in the QDM

ground-state degeneracy of the basis manifold. More details concerning these issues may be found in Chap. 17 of R. Moessner and K.S. Raman in this volume.

It is easy to introduce doping in the QDM. Holes may be injected only in pairs (i.e. by removing dimers). However, doped holes can then move independently by hopping between nearest-neighbor (on the triangular lattice) or diagonal next-nearest-neighbor (on the square lattice) sites [27–29]. The full Hamiltonian for a doped QDM is

$$H = H_{\rm QDM} - t \sum_{(c,c'')} |c''\rangle\langle c|, \tag{21.4}$$

where the $(c'', c)$ sum involves all pairs of configurations $|c\rangle$ and $|c''\rangle$, containing a fixed number $N_h$ of vacant sites (holes), that differ by a single hole hopping along a plaquette diagonal as illustrated in Fig. 21.6b. In this formulation, bare holons, by which is meant the moving vacancies, have Bose statistics. Note that, in contrast to the triangular lattice, holes on the square lattice are constrained to remain only on one of the two sublattices.

### *21.3.2 Phase Diagrams at Zero Doping*

Somewhat remarkably, the square- and triangular-lattice QDMs have quite different phase diagrams in the undoped case. First, exactly at $V = |J|$, which is known as the "RK point," the GS takes the form of an equal superposition of all dimer coverings, and exhibits algebraic dimer correlations on the square lattice but short-ranged (exponentially decaying) correlations on the triangular lattice. Ordered VBC states appear on the square lattice immediately away from the RK point, whereas a gapped RVB liquid [30] is present over a finite region in $V/J$ on the triangular lattice. This RVB phase of the triangular lattice has also been shown to exhibit topological order [30–32], whose importance for frustrated systems is discussed in Chap. 16 of G. Misguich, and for quantum information in Chap. 23 of J. van den Brink, Z. Nussinov and A.M. Oleś. Comparative schematic phase diagrams for the two lattices are depicted in Fig. 21.7. The square lattice shows a rich variety of VBC phases [33] with, in particular, a novel mixed phase [34] which interpolates between the columnar and the plaquette phases (the blue squares in Fig. 21.7 correspond to plaquettes on which vertical and horizontal dimer pairs resonate).

### *21.3.3 Connection to the XXZ Magnet on the Checkerboard Lattice*

We have already explained the extent to which QDMs provide a natural framework to describe the dynamics of SU(2) singlets in frustrated but isotropic quantum antiferromagnets. We also mention briefly another case in which the QDM emerges

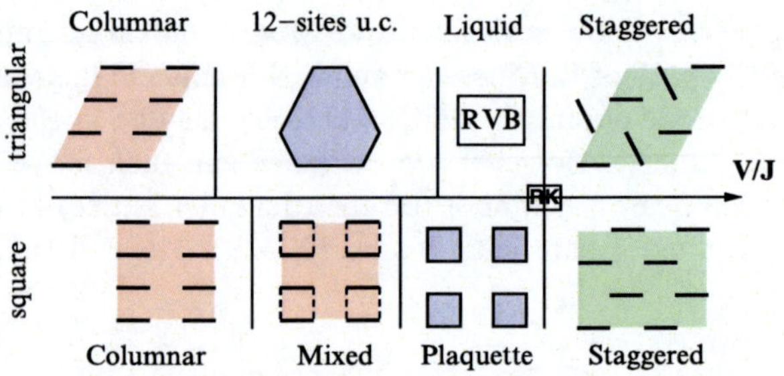


**Fig. 21.7** Schematic phase diagrams of the undoped QDM for both the triangular (*top*) and square (*bottom*) lattices. Only the triangular lattice is believed to exhibit a liquid (RVB) phase. Evidence for the mixed phase on the square lattice has been reported in [34]

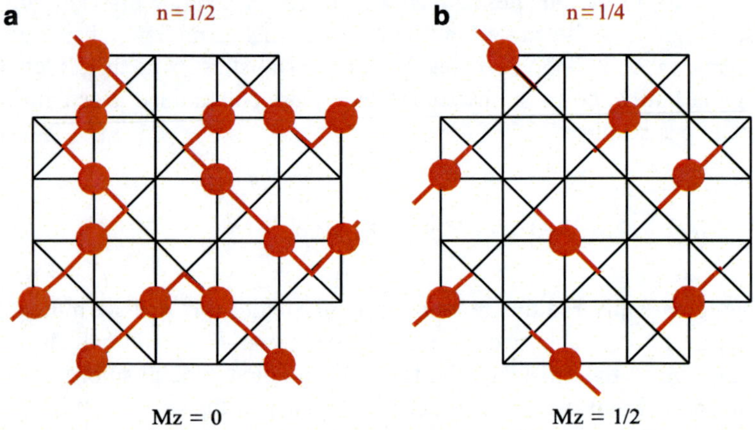


**Fig. 21.8** Representation of the Hilbert space of the large-$J_z$ XXZ Heisenberg model on the checkerboard lattice in terms of (**a**) loop and (**b**) dimer coverings, corresponding respectively to (**a**) zero magnetization and (**b**) $M_z = \pm 1/2$. The red dots represent hard-core bosons with, for example, $S_z = 1/2$ on the corresponding sites, while all other sites have $S_z = -1/2$. A dimer joining the centers of two "tetrahedra" is associated with each boson

as the model for the low-energy physics, that of the strongly anisotropic Heisenberg magnet (in the Ising limit) on the checkerboard lattice and in the presence of a magnetic field.

We begin with no magnetic field and only an Ising coupling, $J_z S_i^z S_j^z$, on the bonds of the checkerboard lattice: in this case, the (classical) ground state is highly degenerate and can be fully represented by the "loop coverings" illustrated in Fig. 21.8a. Here, an up- (down-)spin is represented by the presence (absence) of a dimer, or boson, on the bonds of an effective square lattice whose sites are in fact the centers of the "tetrahedra" (the squares including diagonal bonds). The constrained nature of the classical GS is of the "ice-rule" type: the lowest Ising energy is obtained when there are precisely two bosons on every tetrahedron.

Second-order processes in the exchange coupling $J_{xy}$ lead to the dynamics of a six-vertex model [35] or a quantum loop model [36].

By applying a magnetic field, the density of dimers (bosons) can be altered systematically. When an average of one dimer per tetrahedron is reached, again the ground-state manifold obeys an ice-rule constraint in the large-$J_z$ limit, where all states with precisely one dimer on every tetrahedron, as shown in Fig. 21.8b, are ground states. Second-order processes in $J_{xy}$ now lead to a QDM on the effective square lattice with $V = 0$ and $J = J_{xy}^2/J_z$ in (21.3).

### 21.3.4 Bosonic Doped Quantum Dimer Model

We turn now to the doped QDM and concentrate first on the case $J > 0$ in (21.4). In the mapping from SU(2) dimers, this sign of $J$ is expected for a bosonic representation of the singlet bonds. The "monomers" (holes) of the doped QDM discussed here could then be interpreted physically as entities such as unbound spinons.

For $J > 0$ and $t > 0$, the off-diagonal matrix elements of the Hamiltonian (21.4) are all non-positive, so that (from the Perron–Frobenius theorem) its GS has no node. Consequently, Green-function Monte Carlo (GFMC) techniques can be applied, particularly in the vicinity of the RK point and for small $t/J$ ratios (i.e. when the exact RK GS is still a good guiding wave function), and the phase diagrams shown in Fig. 21.9a,b can be extracted by appropriate finite-size scaling. For larger values of $t/J$, such calculations can be complemented by ED on smaller clusters [37]. The phase-separation (PS) region consists of phase coexistence between an undoped VBC and a superfluid, the latter becoming stable as a unique component above a critical doping. It is notable that this superfluid exhibits flux quantization in units of $h/2e$, in qualitative agreement with gauge theories of high-temperature superconductors [38] and recent, related $Z_2$ gauge theories [39, 40].

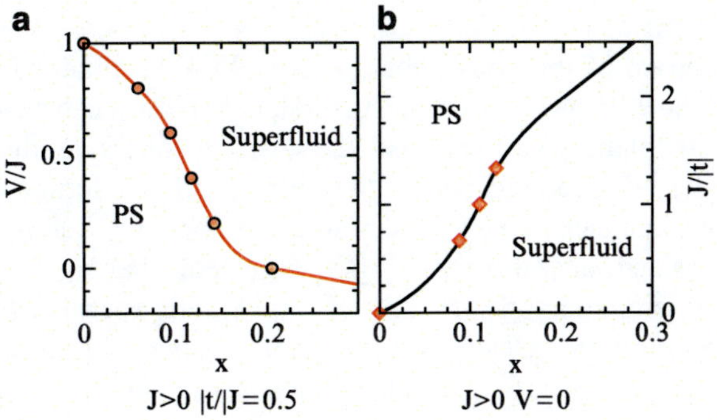


**Fig. 21.9** Schematic phase diagrams of the bosonic doped QDM as a function of dopant concentration $x$ and $V/J$ (**a**) or $J/|t|$ (**b**). Accurate GFMC data are obtained for $V/J \to 1$ and at moderate $|t|/J$ ratios (as explained in [29])

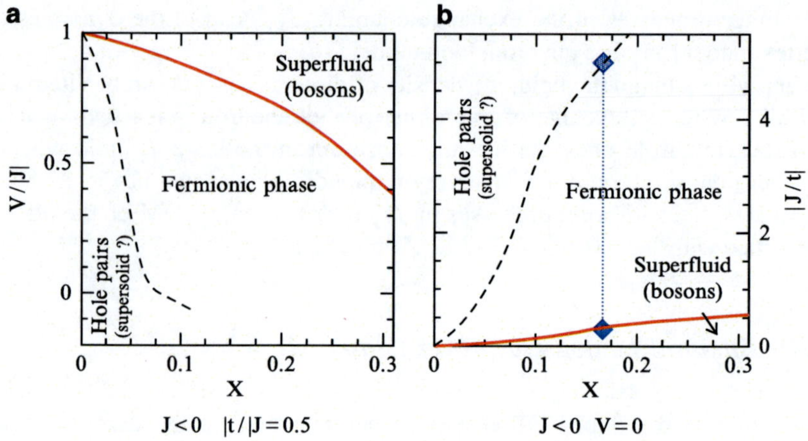


**Fig. 21.10** Schematic phase diagrams of the non-Frobenius ($J < 0$) doped QDM as a function of dopant concentration $x$ and $V/|J|$ (**a**) or $|J/t|$ (**b**), estimated by ED calculations [37]

## 21.3.5 *Non-Frobenius Doped Quantum Dimer Model on the Square Lattice*

Turning to the case of (21.4) with $J < 0$, a quite different type of behavior is expected. The "non-Frobenius" nature of the Hamiltonian (which prohibits the use of GFMC) reflects the original "Fermi sign" of the strongly correlated electrons. Indeed, if one interprets the dimers as SU(2) singlets, a dimer creation operator on the bond $ij$ can be written in the fermionic representation [41–43] as $d_{ij}^\dagger = (f_{i\uparrow}^\dagger f_{j\downarrow}^\dagger + f_{j\uparrow}^\dagger f_{i\downarrow}^\dagger)/\sqrt{2}$. In this basis, it can be verified that the effective dimer-exchange process generated by the underlying Heisenberg interaction within a plaquette occurs for the sign $J < 0$ [41, 42]. In addition, the electron-destruction operator takes the form $c_{i\sigma} = f_{i\sigma} b_i^\dagger$, where the holon (hole or monomer) -creation operator $b_i^\dagger$ is bosonic.

The phase diagram of the non-Frobenius doped QDM obtained by ED [37] is both exotic and rich, as shown in Fig. 21.10. First, bare holons can be seen binding to topological defects (namely vortices, also known as "visons"), producing fermionic composite particles; alternatively stated, the hole becomes a fermion. Secondly, in contrast to the bosonic case, no PS is seen in the immediate vicinity of the (VBC) Mott insulator. Instead, a $d$-wave pairing is expected, opening the possibility of unconventional superconductivity. Finally, at large kinetic energies, holons and vortices unbind, bosonic holes Bose-condense, and a superfluid phase is stabilized, presumably of the same type (with $2e$ charge quanta) as that obtained for $J > 0$.

## 21.4 Mott Transition on the Triangular Lattice

### 21.4.1 Frustration in Itinerant Electron Systems

In the first half of this chapter, we have discussed the dynamics of holes doped into magnetic insulators with geometrical frustration. Another important class of phenomena is driven by frustration in metallic systems. Several strongly correlated metallic systems, such as $LiV_2O_4$ [44] and (Y,Sc)$Mn_2$ [45], show unusually large entropies at temperatures much lower than their bare energy scales (band width and Coulomb repulsion), and this is thought to be related to the geometrical frustration inherent in their lattice structure [46–48]. In strongly correlated electron systems, double occupancy of a site is suppressed by the large, on-site Coulomb repulsion, and the probability of single occupancy increases. This tends to enhance the formation of a local magnetic moment at each site, which interacts with neighboring moments, and frustrated configurations may be adopted depending on the lattice geometry. The central issues for frustrated metals are the effects of exotic magnetic fluctuations on quasiparticle coherence and novel magnetic long-range order or characteristic correlations in itinerant systems.

### 21.4.2 Mott Transition in Organic Compounds with Triangular Geometry

One of the well-studied problems in the physics of frustrated metals is the Mott metal–insulator transition on a triangular lattice [49–52]. When the Coulomb repulsion is much larger than the band width ($U \gg W$), the half-filled system is described effectively by the Heisenberg spin model in the sector of energies below the Mott–Hubbard charge gap. In this case, the well-known 120° structure appears in the ground-state spin configuration. A more exotic situation can be expected when the Coulomb repulsion is comparable with the band width ($U \sim W$), when ring-exchange processes involving multiple sites become important. These processes, which are discussed in Chap. 16 by G. Misguich, open the possibility of stabilizing exotic states.

Experiments on organic compounds with a triangular lattice structure [53–55] have stimulated theoretical studies on the triangular-lattice Hubbard model. These materials are quasi-two-dimensional $\kappa$-$(ET)_2X$ systems with several possible monovalent anions $X$. ET denotes the bis(ethlylenedithio)-tetrathiafulvalene molecule, also often represented as BEDT-TTF, and dimerized pairs of ET molecules constitute a triangular lattice. Each pair provides one conduction electron, and the system is well described by a half-filled Hubbard model on the triangular lattice with nearest-neighbor hopping terms,

$$H = \sum_{\langle i,j \rangle} \sum_{\sigma} t_{ij} c_{i\sigma}^{\dagger} c_{j\sigma} - \mu \sum_{i,\sigma} n_{i\sigma} + U \sum_{i} n_{i\uparrow} n_{i\downarrow}. \tag{21.5}$$

Here, $n_{i\sigma} = c^{\dagger}_{i\sigma} c_{i\sigma}$ and, because this model lacks electron-hole symmetry, the chemical potential $\mu$ is introduced to adjust the electron density to half filling, $\sum_{\sigma} \langle n_{i\sigma} \rangle = 1$. The organic ET compounds have in fact only intermediate correlation strengths: because each site in the model represents a pair of molecules and the corresponding Wannier wave function is extended over the size of this pair, the Coulomb repulsion $U$ is smaller than in the case of typical inorganic compounds, and as a result charge fluctuations have important effects. Because of the non-spherical shape of the molecule pairs, there are two different hopping integrals between nearest-neighbor sites, $t$ and $t'$ (Fig. 21.13a). The ratio $t'/t$ depends on the anion species $X$, and is an important parameter controlling the frustration.

Extensive investigation of the ET systems [53–55] has demonstrated that their low-energy magnetic properties change dramatically for different anions $X$. Particular highlights in the series include spin-liquid-like behavior in $\kappa$-(ET)$_2$Cu$_2$CN$_3$ [54] and a reentrant metal–insulator transition with decreasing temperature at intermediate pressures in $\kappa$-(ET)$_2$Cu[N(CN)$_2$]Cl [55]. The difference between these systems lies in the different values of the frustration parameter $t'/t$. Quantum chemistry calculations estimate that $t'/t = 1.06$ for $\kappa$-(ET)$_2$Cu$_2$CN$_3$ and 0.75 for $\kappa$-(ET)$_2$Cu[N(CN)$_2$]Cl [53]. From the viewpoint of their electronic structure, the candidate spin-liquid material $\kappa$-(ET)$_2$Cu$_2$CN$_3$ is very close to being a regular triangular system, in which all the hopping integrals are the same, while the reentrant material $\kappa$-(ET)$_2$Cu[N(CN)$_2$]Cl corresponds to a triangular geometry perturbed towards an unfrustrated square lattice.

Figure 21.11 shows the temperature-pressure phase diagrams of these two compounds. In these systems, the primary effect of applying pressure is to increase the hopping integrals, and thus the region of higher pressure in experiments corresponds to smaller values of $U/W$ in the Hubbard model. The boundaries between the metallic and insulating phases in the two materials differ qualitatively in shape: in $\kappa$-(ET)$_2$Cu$_2$CN$_3$, the insulating phase appears on the high-temperature side of the boundary; in $\kappa$-(ET)$_2$Cu[N(CN)$_2$]Cl, this is only the case above approximately 30 K, while below this the insulating phase extends as the temperature decreases. The former type of behavior is beyond the naive expectation that the insulating phase appears with decreasing temperature, but it is consistent with the conventional Mott transition, as we explain in detail in the next section.

### 21.4.3 Mott Transition in the Triangular-Lattice Hubbard Model

A schematic phase diagram of the "frustrated" Hubbard model is shown in Fig. 21.12a. There is a first-order phase transition separating metallic and insulating phases. As $U/W$ is increased, spectral weight is transferred from the region around $\omega = 0$ to the lower and upper Hubbard bands at $\omega \sim \pm U$ (Fig. 21.12c). The central peak disappears at the transition point and the insulating phase is on the high-temperature side of the boundary. The model is frustrated in the sense that these results are obtained from a single-site, dynamical mean-field theory (for

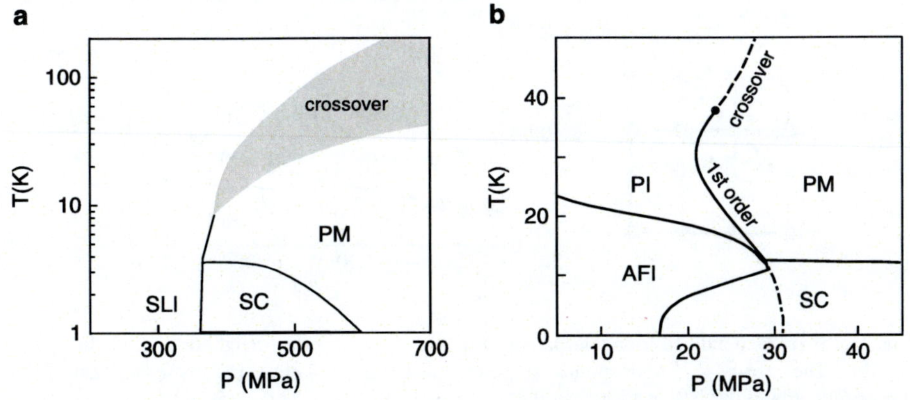


**Fig. 21.11** Temperature-pressure phase diagram of (**a**) $\kappa$-$(ET)_2Cu_2CN_3$ and (**b**) $\kappa$-$(ET)_2$ $Cu[N(CN)_2]Cl$. *PM*: paramagnetic metal, *PI*: paramagnetic insulator, *AFI*: antiferromagnetic insulator, *SLI*: spin-liquid insulator, *SC*: superconducting phase. Reproduced based on [54] and [55]

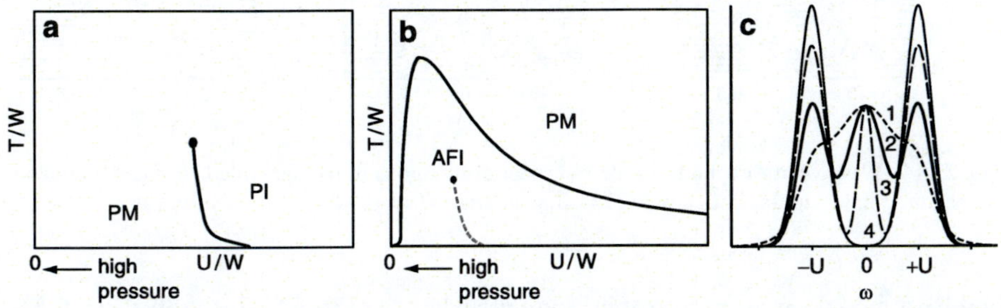


**Fig. 21.12** Phase diagram of the Hubbard model determined by dynamical mean-field theory: (**a**) single-site approximation and (**b**) results obtained when considering the possibility of AF long-range order, corresponding to an unfrustrated lattice. The dotted line marks the metal–insulator transition when magnetic order is absent. The same acronyms are used for labelling the phases as in Fig. 21.11. (**c**) Schematic illustration of the electron spectral function at the Mott transition, shown for different values of $U/W$ ($1 \rightarrow 4$)

review [56]), which assumes the absence of a magnetic instability and thus describes well the case of strong frustration; the phase diagram shows a line of Mott transitions in the original sense of this term, meaning transitions occurring with no simultaneous magnetic order. In the insulating phase, each site has a finite static magnetic moment, which is effectively decoupled from the surrounding moments, and this leads to a large spin entropy of order log 2. This is the reason that the insulating phase is stabilized at higher temperatures. In unfrustrated systems, as shown in Fig. 21.12b, the metal–insulator transition takes place simultaneously with the emergence of AF long-range order, and the genuine Mott transition does not occur.

We now return to the mysterious reentrant metal–insulator transition in $\kappa$-$(ET)_2$ $Cu[N(CN)_2]Cl$. This problem was investigated in [57] by studying the half-filled

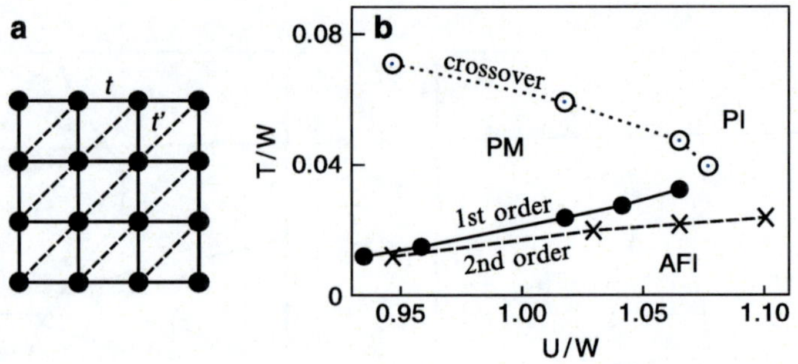


**Fig. 21.13** (**a**) Generalized triangular lattice with two types of nearest-neighbor hopping integrals, $t$ and $t'$. The case $t = t'$ corresponds to the regular triangular lattice. (**b**) Phase diagram of the anisotropic, triangular lattice with $t' = 0.8t$

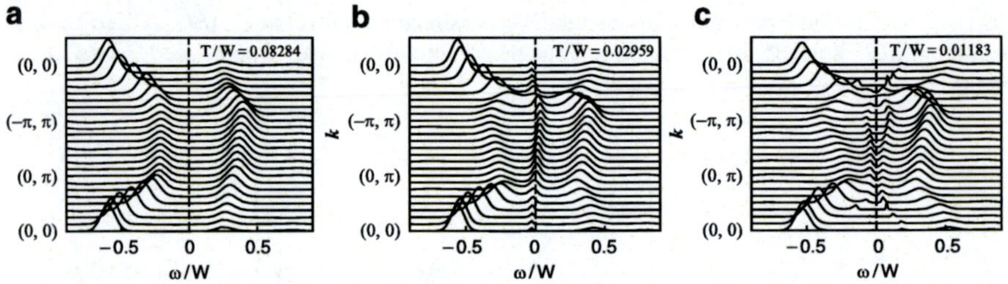


**Fig. 21.14** Wave vector-dependent electron spectral functions of the half-filled Hubbard model on an anisotropic, triangular lattice with $t' = 0.8t$ and $U/W = 0.947$

Hubbard model on an anisotropic, triangular lattice (Fig. 21.13a) with $t' = 0.8t$, to determine the $U$–$T$ phase diagram using cellular dynamical mean-field theory (CDMFT). This method is a generalization of the conventional dynamical mean-field theory which uses a cluster of multiple sites (four for this system) [58,59], and allows one to calculate electronic Green functions and different correlation functions in addition to thermodynamic quantities at finite temperatures. The CDMFT approach has the advantage that both quantum and thermal fluctuations, and thus frustration effects, are taken into account completely inside the cluster.

The phase diagram of the anisotropic, triangular-lattice Hubbard model is shown in Fig. 21.13b. The band width at $U = 0$, $W = 8.45t$ is taken as the unit of energy. The phase diagram was determined by analyzing the double occupancy $D \equiv \langle n_{i\uparrow} n_{i\downarrow} \rangle$, which is a measure of metallicity. In the metallic and insulating phases identified in this way, $D(T)$ decreases with decreasing temperature in the paramagnetic insulating (PI) phase, while it increases in the paramagnetic metallic (PM) phase. In the high-temperature regime, these two phases merge smoothly at the crossover line (dotted), which is defined by the condition $\mathrm{d}D/\mathrm{d}T = 0$. In the low-temperature regime, the two phases are separated by a first-order transition, where the double occupancy shows an abrupt jump.

It is to be noted that reentrant behavior of the metal–insulator transition/crossover is indeed found in the anisotropic, triangular-lattice Hubbard model for intermediate values, $t'/t \sim 0.8$, of the hopping ratio. The nature of this reentrant behavior is clearly visible in the wave vector-dependent electron spectral function $A_{\mathbf{k}}(\omega)$, shown in Fig. 21.14 for three representative temperatures and at the fixed value $U/W = 0.947$. In the high-temperature PI phase, there is a wide Hubbard gap in the spectrum around $\omega = 0$. In the intermediate PM phase, a heavy-quasiparticle band emerges in the Hubbard gap, a clear sign of metallic behavior and consistent with the conventional Mott transition depicted in Figs. 21.12a,c. However, the low-temperature, first-order transition line has a different character: the heavy-quasiparticle band does not disappear, splitting instead into two bands separated by a small energy gap, as shown in Fig. 21.14c. This behavior is similar to the case of a metal–insulator transition driven by magnetic instability.

This type of explanation is confirmed by calculations of the magnetic susceptibility, $\chi_{\mathbf{q}}$ [60], which has a peak at the incommensurate wave vectors $\mathbf{Q} \approx \pm(0.7\pi, 0.7\pi)$. This peak grows as the temperature decreases, and diverges at a finite temperature indicated by the crosses in the phase diagram of Fig. 21.13b. The line of magnetic instability is very close to the first-order metal–insulator-transition line, and it is reasonable to expect that the metal–insulator transition is driven by enhanced magnetic fluctuations. It should also be noted that the two lines are separate and there exists a finite region of a paramagnetic insulating phase between them.

To summarize this section, the anisotropic, triangular-lattice Hubbard model has a phase diagram showing a reentrant metal–insulator transition. This phenomenon is a direct consequence of the effects of geometrical frustration on magnetic correlations. Taking increasing pressure to reduce the ratio $U/W$, the calculated phase diagram reproduces qualitatively the essential features of the phase diagram of $\kappa$-$(ET)_2Cu[N(CN)_2]Cl$. The finite values of the magnetic transition temperature in this 2D model are a consequence of the mean-field-type approximation made for inter-cluster correlations, but may provide an estimate of the true values which would be obtained on including the 3D couplings present in real materials.

## 21.5 Ordering Phenomena at Commensurate Fermion Densities on Frustrated Geometries

In the preceding sections we have discussed only correlated systems at or near half-filling (i.e. with one electron per lattice site). However, repulsive interactions with longer range than the on-site terms considered above can also give rise to insulating behavior at different commensurate densities. Examples include quarter-filling and even 1/8-filling on the checkerboard lattice, and we illustrate this phenomenon here by discussing two scenarios occurring on frustrated geometries. One is a Bond-Order-Wave (BOW) instability is driven directly by particular nesting properties of the Fermi surface. The other concerns the effects of nearest-neighbor interactions sufficiently strong that they produce a novel type of Mott insulator exhibiting an exotic VBC order. The properties of this Mott insulator may be described by an effective QDM, hence providing a formal connection with Sect. 21.3. This latter insulator can also be doped, a point we mention briefly as a possible route towards quite new and exotic metallic and superconducting behavior.

### *21.5.1 Bond Order Waves from Nesting Properties of the Fermi surface*

Let us consider the extended Hubbard Hamiltonian, $H = H_0 + H_{\rm int}$, on the 2D frustrated kagomé and checkerboard lattices. We recall here that these lattices are composed of corner-sharing units (respectively triangles and tetrahedra) residing on an underlying bipartite lattice (respectively hexagonal and square), a point which will be important in determining their behavior. The kinetic part of the Hamiltonian is given by

$$H_0 = -t \sum_{\langle ij\rangle} \sum_{\sigma=\uparrow\downarrow} \left( c^\dagger_{i\sigma} c_{j\sigma} + \text{h.c.} \right), \tag{21.6}$$

with positive hopping matrix element $t$, and the sum $\sum_{\langle ij\rangle}$ is over all bonds on the lattice. The interaction part is given by

$$H_{\rm int} = U \sum_i n_{i\uparrow} n_{i\downarrow} + J \sum_{\langle ij\rangle} \mathbf{S}_i \cdot \mathbf{S}_j + V \sum_{\langle ij\rangle} n_i n_j, \tag{21.7}$$

with on-site repulsion $U$, nearest-neighbor spin exchange $J$, and nearest-neighbor repulsion $V$; in this section, we consider the regime of weak and intermediate couplings.

On kagomé and checkerboard lattices, the non-interacting hamiltonian $H_0$ exhibits a dispersionless (flat) band which, for $t > 0$, lies at the top of the spectrum and plays no role. For the kagomé lattice, the band structure is remarkable for the presence of "Dirac cones" positioned exactly at the Fermi level of the 1/3-filled system ($n = 2/3$) and leading to semi-metallic behavior (also relevant on the 3D pyrochlore lattice). Renormalization-group and numerical techniques [61] have been applied to demonstrate that a spontaneous symmetry-breaking occurs for arbitrarily small interactions in this system: the instability corresponds to a BOW in which the kinetic energy is staggered for neighboring triangular units on the underlying hexagonal lattice. We stress that no charge modulation is present (so that all sites remain equivalent), the BOW breaking only the spatial site-inversion (180°) symmetry, such that up- and down-pointing triangles in Fig. 21.2c become different, while translational symmetry is preserved [61].

On the checkerboard lattice, a BOW instability appears at quarter-filling ($n = 1/2$). As on the kagomé lattice, this BOW (Fig. 21.15) is characterized by two types of (tetrahedral) unit with different kinetic (and exchange) energies. However, in the

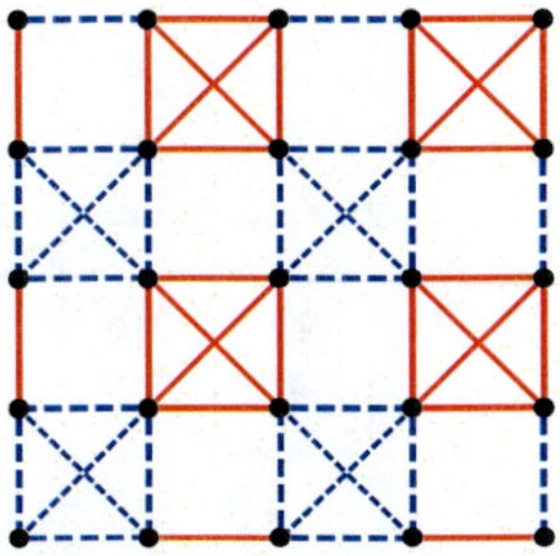

**Fig. 21.15** Bond Order Wave on the checkerboard lattice at $n = 1/2$. The four inequivalent bonds are represented by different colors/line types and by different thicknesses (diagonal and x/y bonds are different). All the sites (*black dots*) carry the same electron occupancy (1/2 electron on average)

case of the checkerboard lattice, translation symmetry is broken explicitly, although once again there is no charge order. This symmetry-breaking occurs because of perfect nesting of the square Fermi surface [62]. Alternatively, it can be physically understood as special, local (resonant) states formed on the building units (the crossed plaquettes) when the filling is such that these are preferred; the occupation states of the units can also be considered to differ. This local picture is in fact fully equivalent to the nesting instability of the Fermi surface.

### *21.5.2 Metal–Insulator Transitions and Frustrated Charge Order*

We consider next the strong-coupling limit, where at $U = \infty$ one obtains the Hamiltonian

$$H_{t\text{-}J\text{-}V} = \mathcal{P}H_0\mathcal{P} + J\sum_{\langle ij\rangle}\mathbf{S}_i\cdot\mathbf{S}_j + V\sum_{\langle ij\rangle}n_in_j, \tag{21.8}$$

$$= H_{t\text{-}J} + V'\sum_{\langle ij\rangle}n_in_j, \tag{21.9}$$

where $\mathcal{P}$ is the projection operator enforcing the single-occupancy constraint and $V' = V + J/4$. For $V' = 0$, the strong-coupling Hamiltonian reduces to the conventional $t$-$J$ model.

For simplicity, we restrict our considerations to the checkerboard lattice and state only that similar behavior can be found for the kagomé lattice. Examining first the limit $V = \infty$ for the special, commensurate filling $n = 1/4$ (1/8-filling), the minimum "classical" energy ($E = 0$) is obtained for all configurations fulfilling the "ice rule" of precisely one particle on every tetrahedron, as in Fig. 21.8b. The full Hilbert space is then obtained from all possible ways of "decorating" all dimers with a spin index, i.e. the Hilbert space at $n = 1/4$ is exactly that of a

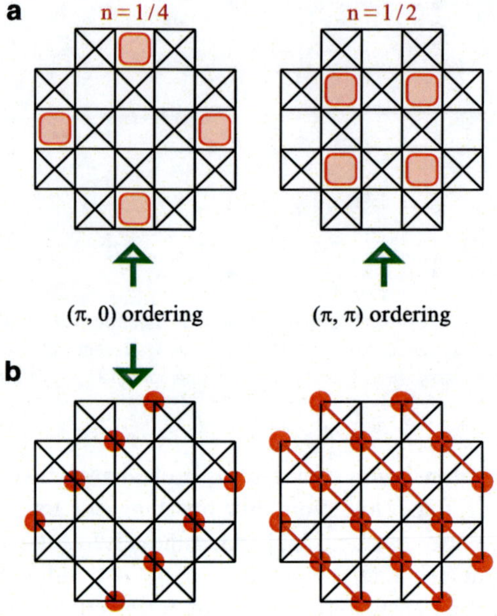


**Fig. 21.16** Schematic representation of candidate plaquette (**a**) and columnar (**b**) phases of the checkerboard lattice for electron densities $n = 1/4$, $1/2$, and $3/4$, as discussed in Sect. 21.5.2. Dots, shaded plaquettes, and thick (*red*) lines correspond respectively to electrons, singlet pairs resonating on a plaquette, and resonating four-electron plaquette singlets. From [64]

two-color QDM.[1] A similar procedure, decorating the simple loop configurations of Fig. 21.8a, can also be employed to construct the two-color loop configurations which constitute the constrained Hilbert space at quarter-filling ($n = 1/2$) in the large-$V$ limit. It is then clear that, for these special fillings, the GS should be insulating at sufficiently large $V$. The effective dimer (or loop) dynamics can be obtained by perturbation in $t/V$. We note that in the original derivation [46, 47], for spinless fermions, the lowest-order processes were of third order, whereas when spin degrees of freedom are included, terms of dimer-exchange type (below) arise at second order in $t$. Although the constrained quantum dynamics of fermions without [46, 47] and with [63, 64] spin differ, the phase diagrams of these models contain a rich variety of crystalline phases, breaking lattice translational and/or rotational symmetry, in both cases. We postpone to Sect. 21.5.3 a discussion of the properties of this type of system away from commensurate filling, and remark only that, among the novel phenomena which can arise, one of the more exotic is the fractionalization under some conditions of a single doped charge $e$ into two $e/2$ components [46, 47, 65].[2]

[1] For $n = 3/4$, empty bonds would be decorated if the dimers are considered as equivalent to the absence of electrons.

[2] In general, the two components are bound by a shallow "string" potential.

Let us now focus in more detail on the insulating phases and consider the effective Hamiltonian acting within the constrained Hilbert space as second-order processes preserving the ice rule. Here $\tilde{\mathcal{H}} = H_{\blacksquare} + H_J$ with

$$H_{\blacksquare} = -t_2 \sum_s P_{\blacksquare}(s), \tag{21.10}$$

$$\begin{aligned} P_{\blacksquare}(s) = & \left(c^\dagger_{i\uparrow}c^\dagger_{j\downarrow} - c^\dagger_{i\downarrow}c^\dagger_{j\uparrow}\right)\left(c_{k\downarrow}c_{l\uparrow} - c_{k\uparrow}c_{l\downarrow}\right) \\ & + \left(c^\dagger_{k\uparrow}c^\dagger_{l\downarrow} - c^\dagger_{k\downarrow}c^\dagger_{l\uparrow}\right)\left(c_{i\downarrow}c_{j\uparrow} - c_{i\uparrow}c_{j\downarrow}\right), \end{aligned} \tag{21.11}$$

where $t_2 = \frac{2t^2}{V}$ and the index $s$ labels the empty plaquettes of the checkerboard lattice (Fig. 21.2); the sites of a plaquette $s$ are ordered as $i, k, j, l$ in a clockwise (or anti-clockwise) direction. The operator $P_{\blacksquare}(s)$ acts on two electrons forming a singlet bond on one of the two diagonals of $s$, to rotate this bond by 90 degrees. Candidate GSs breaking the translational symmetry of the lattice are shown in Fig. 21.16. For a quarter-filled band ($n = 1/2$), the Resonating-Singlet-Pair Crystal (RSPC) of Fig. 21.16a was shown to be stable for $J/t_2 < 1$ [63,64]. In the twofold-degenerate RSPC, electron pairs resonate on every second empty plaquette, breaking translational symmetry. One therefore expects, on increasing $V/t$ and $U/t$, a first-order transition from the BOW state discussed above [62] to the RSPC. The same analysis also provides evidence [64] that the system exhibits plaquette order of the RSPC type also at $n = 1/4$ or $n = 3/4$, albeit with a quadrupling of the lattice unit cell (as opposed to the doubling found for $n = 1/2$) and a fourfold-degenerate GS. Qualitative differences between these models and their bosonic analogs, which are known for example to exhibit columnar order at $n = 1/4$ [36], emphasize the important role of the spin degrees of freedom, not least in stabilizing plaquette phases over phases breaking the rotational symmetry. However, the possibility is being investigated [66] that mixed columnar-plaquette phases, similar to one discovered recently in the square-lattice QDM [34] and which break both $\pi/2$ rotational symmetry (as does the columnar phase) and translational symmetry in two perpendicular directions (as does the plaquette phase), could be stable in some simple and natural extensions of the Hamiltonian (21.10). Note that equivalent models of spinless harcore bosons (XXZ spin systems) can also exhibit exotic (commensurate) VBC supersolid phases [67].

### *21.5.3 Away from Commensurability: Doping the Resonating-Singlet-Pair Crystal*

Whether plaquette ordering of the RSPC type can survive at sufficiently low but finite hole (electron) dopant concentrations $x = 1/2 - n$ ($x = n - 1/2$) remains unsettled. It has been shown [63] that phase separation, a generic feature of correlated systems in the vicinity of a Mott phase, is restricted to low hole kinetic energies, meaning to small $t/J$ and $t/t_2$ ratios, which leaves an extended regime over which unconventional superconducting pairing may occur. However, the phase diagram of these model for arbitrary electron densities remains largely unexplored, and can be expected to harbor further surprises.

## 21.6 Summary

The richness and diversity of systems of doped, frustrated magnets and of itinerant correlated electrons on frustrated lattices have been illustrated on selected didactic examples. The dynamics of a small number of doped holes has been investigated in the 2D kagomé and checkerboard Heisenberg quantum antiferromagnets revealing striking differences attributed to the different nature of their non-magnetic GS. Two classes of (weakly) doped QDMs have also been discussed, which differ in the assumed statistics, bosonic or fermionic, of the bare holes. We have proceeded further by considering strongly correlated electrons on frustrated triangular lattices, and discuss the physics of an unconventional, reentrant metal–insulator transition. Assuming that increasing pressure reduce on-site correlations, the calculated phase diagram reproduces qualitatively the essential features of the phase diagram of the molecular solid $\kappa$-(ET)$_2$Cu[N(CN)$_2$]Cl. As our final example, we consider correlated fermions moving on frustrated lattices at special, commensurate densities for which exotic insulating ground states (GSs) are stabilized. Interesting similarities with frustrated Heisenberg magnets showing an extensive degeneracy of the classical GS manifold are outlined and are shown to be at the heart of their fascinating properties.

**Acknowledgements**

Didier Poilblanc thanks his collaborators Andreas Läuchli, Frédéric Mila, Roderich Moessner, Karlo Penc, Arnaud Ralko and Nic Shannon who have contributed to some of the work presented here. Didier Poilblanc is also grateful to the French Research Council (ANR) for support under Grant ANR-05-BLAN-0043-01 and to IDRIS (Orsay, France) for allocation of CPU-time on supercomputers. H.T. thanks his collaborators Takuma Ohashi, Norio Kawakami, and Tsutomu Momoi for invaluable discussions. He also acknowledges support by Grants-in-Aid for Scientific Research (Nos. 17071011 and 19052003), and also by The Next-Generation Supercomputing Project, Nanoscience Program, MEXT of Japan.

## References

1. N. Read, S. Sachdev, Phys. Rev. Lett. **62**, 1694 (1989)
2. P.W. Anderson, Science **235**, 1196 (1987)
3. P.W. Anderson, G. Baskaran, Z. Zou, T. Hsu, Phys. Rev. Lett. **58**, 2790 (1987)

4. R. Moessner, O. Tchernyshyov, S.L. Sondhi, J. Stat. Phys. **116**, 755 (2004)
5. B. Canals, Phys. Rev. B **65**, 184408 (2002)
6. E. Berg, E. Altman, A. Auerbach, Phys. Rev. Lett. **90**, 147204 (2003)
7. J.-B. Fouet, M. Mambrini, P. Sindzingre, C. Lhuillier, Phys. Rev. B **67**, 054411 (2003)
8. W. Brenig, A. Honecker, Phys. Rev. B **65**, 140407(R) (2002)
9. V. Elser, Phys. Rev. Lett. **62**, 2405 (1989)
10. P.-W. Leung, V. Elser, Phys. Rev. B **47**, 5459 (1993)
11. J. T. Chalker, J.F.G. Eastmond, Phys. Rev. B **46**, 14201 (1992)
12. P. Lecheminant, B. Bernu, C. Lhuillier, L. Pierre, P. Sindzingre, Phys. Rev. B **56**, 2521 (1997)
13. C. Waldtmann, H.-U. Everts, B. Bernu, C. Lhuillier, P. Sindzingre, P. Lecheminant, L. Pierre, Eur. Phys. J. B **2**, 501 (1998)
14. F. Mila, Phys. Rev. Lett. **81**, 2356 (1998)
15. V. Subrahmanyam, Phys. Rev. B **52**, 1133 (1995)
16. A. Läuchli, D. Poilblanc, Phys. Rev. Lett. **92**, 236404 (2004)
17. D.C. Johnston, J. Low Temp. Phys. **25**, 145 (1976)
18. C.P. Sun, J.-Y. Lin, S. Mollah, P.L. Ho, H.D. Yang, F.C. Hsu, Y.C. Liao, M.K. Wu, Phys. Rev. B **70**, 054519 (2004)
19. M. Hanawa, Y. Muraoka, T. Tayama, T. Sakakibara, J. Yamaura, Z. Hiroi, Phys. Rev. Lett. **87**, 187001 (2001)
20. Z. Hiroi, S. Yonezawa, Y. Muraoka, J. Phys. Soc. Jpn. **73**, 1651 (2004)
21. S. Yonezawa, Y. Muraoka, Y. Matsushita, Z. Hiroi, J. Phys. Soc. Jpn. **73**, 819 (2004)
22. K. Takada, H. Sakurai, E. Takayama-Muromachi, F. Izumi, R.A. Dilanian, T. Sasaki, Nature (London) **422**, 53 (2003)
23. D. Poilblanc, Phys. Rev. Lett. **93**, 193204 (2004)
24. S. Chakravarty, A. Sudbo, P.W. Anderson, S. Strong, Science **261**, 337 (1993)
25. J. Vidal, B. Douçot, R. Mosseri, P. Butaud, Phys. Rev. Lett. **85**, 3906 (2000)
26. J. Vidal, B. Douçot, Phys. Rev. B **65**, 045102 (2002)
27. D.S. Rokhsar, S.A. Kivelson, Phys. Rev. Lett.**61**, 2376 (1988)
28. D. Poilblanc, F. Alet, F. Becca, A. Ralko, F. Trousselet, F. Mila, Phys. Rev. B **74**, 014437 (2006)
29. A. Ralko, F. Mila, D. Poilblanc, Phys. Rev. Lett. **99**, 127202 (2007)
30. R. Moessner, S.L. Sondhi, Phys. Rev. B **63**, 224401 (2001)
31. A. Ralko, M. Ferrero, F. Becca, D. Ivanov, F. Mila, Phys. Rev. B **71**, 224109 (2005)
32. A. Ralko, M. Ferrero, F. Becca, D. Ivanov, F. Mila, Phys. Rev. B **74**, 134301 (2006)
33. O.F. Syljuasen, Phys. Rev. B **71**, 020401(R) (2005)
34. A. Ralko, D. Poilblanc, R. Moessner, Phys. Rev. Lett. **100**, 037201 (2008)
35. E.H. Lieb, Phys. Rev. Lett. **18**, 1046 (1967). Recent attempts to quantize vertex models have been made by E. Ardonne, P. Fendley, and E. Fradkin, Ann. Phys.**310**, 493 (2004) and by O. F. Syljuasen, and S. Chakravarty, Phys. Rev. Lett. **96**, 147004 (2006)
36. N. Shannon, G. Misguich, K. Penc, Phys. Rev. B **69**, 220403(R) (2004)
37. D. Poilblanc, Phys. Rev. Lett. **100**, 157206 (2008)
38. T. Senthil, P.A. Lee, Phys. Rev. B **71**, 174515 (2005)
39. L. Balents, M.P. A Fisher, C. Nayak, Phys. Rev. B **60**, 1654 (1999)
40. T. Senthil, M.P.A. Fisher, Phys. Rev. B **62** 7850 (2000)
41. S. Kivelson, Phys. Rev. B **39**, 259 (1989)
42. N. Read and B. Chakraborty, Phys. Rev. B **40**, 7133 (1989)
43. M. Vojta, S. Sachdev, Phys. Rev. Lett. **61**, 2376 (1999)
44. S. Kondo, D.C. Johnston, C.A. Swenson, F. Borsa, A.V. Mahajan, L.L. Miller, T. Gu, A.I. Goldman, M.B. Maple, D.A. Gajewski, E.J. Freeman, N.R. Dilley, R.P. Dickey, J. Merrin, K. Kojima, G.M. Luke, Y.J. Uemura, O. Chmaissem, J.D. Jorgensen, Phys. Rev. Lett. **78**, 3729 (1997)
45. H. Wada, H. Nakamura, E. Fukami, K. Yoshimura, M. Shiga, Y. Nakamura, J. Magn. Magn. Mater. **70**, 17 (1987)
46. F. Pollmann, J.J. Betouras, K. Shtengel, P. Fulde, Phys. Rev. Lett. **97**, 170407 (2006)
47. F. Pollmann, P. Fulde, E. Runge, Phys. Rev. B **73**, 125121 (2006)
48. K. Hattori, H. Tsunetsugu, Phys. Rev. B **79**, 035115 (2009)

49. T. Kashima, M. Imada, J. Phys. Soc. Jpn. **70**, 3052 (2001)
50. H. Morita, S. Watanabe, and M. Imada, J. Phys. Soc. Jpn. **71**, 2109 (2002)
51. O. Parcollet, G. Biroli, G. Kotliar, Phys. Rev. Lett. **92**, 226402 (2004)
52. B. Kyung, A.-M.S. Tremblay, Phys. Rev. Lett. **97**, 046402 (2006)
53. Y. Shimizu, K. Miyagawa, K. Kanoda, M. Maesato, G. Saito, Phys. Rev. Lett. **91**, 107001 (2003)
54. Y. Kurosaki, Y. Shimizu, K. Miyagawa, K. Kanoda, G. Saito, Phys. Rev. Lett. **95**, 177001 (2005)
55. F. Kagawa, T. Itou, K. Miyagawa, K. Kanoda, Phys. Rev. B **69**, 064511 (2004)
56. A. Georges, G. Kotliar, W. Krauth, M.J. Rozenberg, Rev. Mod. Phys. **68**, 13 (1996)
57. T. Ohashi, T. Momoi, H. Tsunetsugu, N. Kawakami, Phys. Rev. Lett. **100**, 076402 (2008)
58. A.I. Lichtenstein, M.I. Katsnelson, Phys. Rev. B **62**, R9283 (2000)
59. G. Kotliar, S.Y. Savrasov, G. Pálsson, G. Biroli, Phys. Rev. Lett. **87**, 186401 (2001)
60. T. Ohashi, T. Momoi, H. Tsunetsugu, N. Kawakami, Prog. Theor. Phys. Suppl. **176**, 97 (2008)
61. M. Indergand, A. Läuchli, S. Capponi, M. Sigrist, Phys. Rev. B **74**, 064429 (2006)
62. M. Indergand, C. Honerkamp, A. Läuchli, D. Poilblanc, M. Sigrist, Phys. Rev. B **75**, 045105 (2007)
63. D. Poilblanc, K. Penc, N. Shannon, Phys. Rev. B **75**, 220503 (2007)
64. D. Poilblanc, Phys. Rev. B **76**, 115104 (2007)
65. P. Fulde, K. Penc, N. Shannon, Annalen der Physik, **11**, 892 (2002)
66. F. Trousselet, D. Poilblanc, R. Moessner, Phys. Rev. B **78**, 195101 (2008)
67. A. Ralko, F. Trousselet, D. Poilblanc, Phys. Rev. Lett. **104**, 127203 (2010)

# Chapter 22
# Metallic and Superconducting Materials with Frustrated Lattices

**Zenji Hiroi and Masao Ogata**

**Abstract** The nature of itinerant electrons on various frustrated lattices is discussed from both a materials and a theoretical point of view. Many metallic transition-metal oxides, intermetallic alloys, and organic compounds are reviewed, which show interesting phenomena such as superconductivity, a metal-insulator transition, anomalous magnetoresistance, and heavy-fermion-like behavior. Of particular interest in this context are the superconductors $\alpha$-$Cd_2Re_2O_7$ and $\beta$-$AOs_2O_6$, which have a pyrochlore lattice, and $Na_xCoO_2 \cdot yH_2O$ and (BEDT-TTF)$_2X$, which have triangular lattices, where some effects of frustration on the mechanisms of superconductivity may be evident. Theoretically, RVB superconductivity is discussed to be the most interesting and promising mechanism for unconventional superconductivity on frustrated lattices, although there remains as yet no direct experimental evidence for such RVB-type superconductivity in real materials. Strong- and weak-coupling theories, and spin- and charge-fluctuation mechanisms for superconductivity are reviewed. Possible mechanisms for the spin-liquid state observed in some organic compounds are also discussed.

## 22.1 Introduction

In this chapter, we describe the behavior of itinerant electrons on various geometrically frustrated lattices from the points of view of both materials and theory. The idea of geometrical frustration on triangle-based lattices is straightforward for a localized spin system, and has been studied extensively in various classes of material. In contrast, once electrons are rendered itinerant, able to move from one atom to another, the effect of frustration tends to be "diluted," and its role becomes less clear. However, at least in the case of strongly correlated electron systems, where electrons are scarcely able to hop because of the Coulomb repulsion, both spin and charge degrees of freedom survive in a form where they may potentially be affected by some of the consequences of frustration. There is in addition another degree of freedom, the orbital of the $d$ electrons, which exists for many transition-metal oxides

(TMOs) and is often influenced by frustration effects when there is a tendency to order.

It is well known that the leading physical effects of frustration are related to a macroscopic degeneracy associated with certain degrees of freedom. A finite entropy may remain down to very low temperatures, seemingly contradicting the third law of thermodynamics. This is the situation expected in simple systems composed of spins localized on frustrated lattices and interacting only with their nearest neighbors. However, in real materials, there are always, at some level, additional long-range interactions between distant spins and/or couplings to other degrees of freedom, which may lift this degeneracy partially or completely, leading respectively to less degenerate or non-degenerate ground states. For metallic systems, the introduction of a transfer integral between electrons on different sites can serve as such an additional interaction. To understand the various phenomena which can arise as a result in a range of different compounds, it is thus important to study how such a lifting of degeneracy can proceed and which ground state is selected as a result.

In many materials, because of coupling to the lattice degree of freedom, structural deformations are induced when a lifting of the degeneracy takes place. One characteristic of metallic systems is that the process is often accompanied by localization of carriers, resulting in a metal-insulator transition (MIT) or a transition showing a sudden increase in resistivity on cooling across the critical temperature. The ground state attained below the transition could be charge-ordered, and/or orbital-ordered, and may possess long-ranged magnetic order or various types of local singlet configuration, depending on the lattice and on the electronic structure of the system.

On the other hand, there is a small number of examples that remain free of lattice distortions, and in which one would expect to see the more intrinsic physics of frustration in itinerant systems, namely that associated with finite entropy. The ground state in such a case is expected to have unconventional magnetic order or to be a quantum spin liquid when the spin degree of freedom is dominant, or otherwise to be a heavy-fermion (HF) state with a high density of low-lying charge degrees of freedom [1]. It is clear that the chances of encountering such a novel state of matter are higher in fully frustrated systems.

Metallic compounds on frustrated lattices can be classified into three groups in terms of the relative sizes of their bandwidth, $W$, and electronic correlation, $U$. The first group, $W > U$, includes relatively simple metals such as $\beta$-Mn or $Y(Sc)Mn_2$ [2, 3]. Because the electrons preserve their purely itinerant character in these systems, the nature of the spin fluctuations in the metallic state may be crucial in deciding the ground state. The second group, $W \sim U$, contains most of the known organic compounds and many TMOs. These are located close to the metal-insulator (MI) boundary, and often exhibit a MIT as functions of temperature, pressure, or doping. The last category, $W < U$, consists mostly of TMOs with strong electron correlations, so that most compounds in this class with odd-integer numbers of electrons become Mott insulators. Nevertheless, some mixed-valent compounds with half-integer numbers of $d$ electrons remain metallic, exhibiting unusual metallic states or superconductivity of exotic origin.

As target compounds for frustrated metallic systems, particularly interesting would be systems where the effects of strong electronic correlations are induced by doping with carriers, as in the case of cupric oxide superconductors. It is well known that many perovskite-related TMOs are Mott insulators when they have odd-integer numbers of $d$ electrons per transition-metal (TM) ion, and readily become strongly correlated metals upon doping [4]. As for materials with frustrated lattice structures, there are also many candidate compounds for the parent insulating phase. However, few of these compounds can actually be doped with carriers to become metallic. One of the reasons for this may be a simple chemical tendency to avoid the substitution necessary for doping. Another aspect that we regard as generally important is the specific band structure near the Fermi level in frustrated compounds: it is well known that materials with kagomé or pyrochlore lattice structures possess highly degenerate flat bands near the Fermi level, which favor an insulating state with ferromagnetic interactions. If this is the case, a small concentration of doped carriers cannot contribute to conduction, even though they exist at the Fermi level. Thus, a metallic state would appear only when the carrier density becomes sufficiently large. In fact, most metallic compounds on frustrated lattices are really metals with large carrier density, and thus display fewer effects of electronic correlations in their most straightforward sense.

A further disadvantage for rendering frustrated compounds metallic by doping is a tendency towards only small electronic transfer integrals between the nearest-neighbor ions on a triangle compared with those on a square. Because the bond angle through the bridging (oxide) ions deviates very significantly from 180 degrees in the former case, one would generally expect a smaller $W$ in such frustrated systems. Moreover, the bandwidth becomes rather sensitive to subtle changes in the bond angle, so that an insulating state is easily stabilized by coupling to a lattice deformation, even when this is small. Strongly correlated electronic systems may therefore be more difficult to obtain in frustrated lattices than in square-based lattices. Stated the other way around, however, strongly correlated electron systems on frustrated lattices have remained rare to date, and would be a gold mine of interesting physics.

In this chapter, we describe various metallic and superconducting materials with frustrated lattices. We begin with an overview of the materials situation. The theoretical background important for understanding metallic frustrated systems is then discussed. We conclude by reviewing three specific metals and superconductors: the pyrochlore oxides $Cd_2Re_2O_7$ and $AOs_2O_6$ ($A$ = Cs, Rb, K), which have the pyrochlore lattice structure, sodium cobaltate, $Na_xCoO_2$, and its hydrate, which are triangular-lattice materials, and the organic compounds (BEDT-TTF)$_2X$ (for review [5, 6]), which have anisotropic triangular lattices. Although detailed studies are still in progress on these compounds, it seems that the frustrated lattices play an important role in determining both their electronic structures and the mechanism of superconductivity, as described in Chap. 21 by D. Poilblanc and H. Tsunetsugu in this volume.

## 22.2 Materials Overview

We begin by presenting a summary, in the form of Table 22.1, of the various materials possessing frustrated lattices which exhibit intriguing physical properties associated with metallic conductivity. As mentioned above, in frustrated lattice geometries there are not many metallic compounds compared to the number of insulators with localized spins. Nevertheless, several metallic compounds are found in three-dimensional frustrated lattices, and a rather small number in two-dimensional lattices. We will first describe the former and then the latter.

### *22.2.1 Pyrochlore Lattice*

Three classes of material possessing the three-dimensional pyrochlore lattice have been studied, spinel compounds containing TM ions, manganese-based metals, and pyrochlore oxides.

#### 22.2.1.1 Spinel Compounds

Spinel compounds have the general formula $AB_2O_4$ and crystallize in the cubic structure, as described in detail in Chap. 7 by H. Takagi and S. Niitaka in this volume. Because the $B$ sublattice forms a pyrochlore lattice, compounds containing TM ions with an unfilled $d$ shell at the $B$ site are of most interest. Although many TMOs crystallize in the spinel structure, most of them are Mott insulators with integer numbers of $d$ electrons, as in the examples of $ZnV_2O_4$ and $MgFe_2O_4$. However, some compounds have half-integer numbers of $d$ electrons and show metallic or nearly metallic conductivity, at least at high temperatures.

Historically, the most famous spinel compound is magnetite, $Fe_3O_4$, which shows a MIT known as the Verwey transition at $T_V \sim 120$ K [7, 8]. The crystal structure is cubic at temperatures above $T_V$, where one third of the Fe ions ($Fe^{2+}$) occupy the $A$ sites and the remaining two-thirds (on average $Fe^{2.5+}$) occupy the $B$ sites and form the pyrochlore lattice. An abrupt increase in resistivity is observed at $T_V$ on cooling, suggesting that the electrons at the $B$ sites have localized into a fixed charge-ordering pattern of $Fe^{2+}$ and $Fe^{3+}$ ions. The transition is of first order, accompanied by a lattice deformation, but the low-temperature structure appears to be very complicated, and has not yet been established experimentally. Despite very serious efforts made thus far, the origin of the transition still remains the subject of intense debate. Recent theoretical analysis proposed that orbital ordering occurs at $T_V$ [9], as opposed to the basic charge ordering, or that both charge and orbital ordering take place simultaneously [10, 11]. In any event, frustration in $Fe_3O_4$ is lifted by coupling to the lattice degree of freedom.

**Table 22.1** Metallic materials with frustrated lattices showing physical properties including superconductivity (SC), a metal-insulator transition (MIT), antiferromagnetism (AF), insulating behavior (I), large magnetoresistance (MR) in a ferromagnetic metal, and heavy-fermion-like behavior (HF). Parentheses in the fourth column indicate properties under pressure

| Lattice | Compound | Electrons | Property | $T$(K) | References |
|---|---|---|---|---|---|
| Triangular | $Na_xCoO_2 \cdot yH_2O$ | $3d^{5-6}$ | SC | 4.5 | [44] |
| | $\kappa$-(BEDT-TTF)$_2X$ | | | | |
| | $X = Cu(NCS)_2$ | $p$ | SC | 10.4 | [74] |
| | $X = Cu[N(CN)_2]Br$ | $p$ | SC | 12.6 | [75] |
| | $X = Cu[N(CN)_2]Cl$ | $p$ | AF(SC) | 27(13.2) | [75,77] |
| | $X = Cu_2(CN)_3$ | $p$ | I (SC) | – (4) | [73,79] |
| | $\alpha$-(BEDT-TTF)$_2I_3$ | $p$ | MIT(SC) | 135(7.2) | [99,103] |
| | $\theta$-(BEDT-TTF)$_2X$ | | | | |
| | $X = RbZn(SCN)_4$ | $p$ | MIT | 190 | [109] |
| | $X = CsZn(SCN)_4$ | $p$ | M | | [109] |
| | $X = I_3$ | $p$ | SC | 3 | [116] |
| | $\theta$-(BEDT-TTF)$_4X$ | | | | |
| | $X = Hg_{3-\delta}Br_8$ | $p$ | I (SC) | – (~2) | [264] |
| | $AgNiO_2$ | $3d^7$ | M | | [62] |
| | $Ag_2NiO_2$ | $3d^7$ | M | | [60] |
| | $Ag_2MnO_2$ | $3d^4$ | M | | [61] |
| Kagomé/triangular | $SrCo_6O_{11}$ | $3d^{5.33}$ | M | | [64] |
| | $NaV_6O_{11}$ | $3d^{5.5}$ | M | | [63] |
| Pyrochlore | $Fe_3O_4$ | $3d^{5.5}$ | MIT | 124 | [8] |
| (spinel oxides) | $LiTi_2O_4$ | $3d^{0.5}$ | SC | 13.7 | [12] |
| | $LiV_2O_4$ | $3d^{1.5}$ | HF | | [13] |
| | $AlV_2O_4$ | $3d^{2.5}$ | MIT | 700 | [14] |
| | $LiRh_2O_4$ | $4d^{5.5}$ | MIT | 170 | [18] |
| | $CuIr_2S_4$ | $5d^{5.5}$ | MIT | 230 | [19] |
| (Mn metals) | Y(Sc)$Mn_2$ | $3d$ | HF | | [2] |
| | $\beta$-Mn | $3d$ | HF | | [3] |
| ($\alpha$-pyrochlore oxides) | $Tl_2Mn_2O_7$ | $3d^3$ | MR | 142 | [31] |
| | $Nd_2Mo_2O_7$ | $4d^2$ | MR | 90 | [32] |
| | $Hg_2Ru_2O_7$ | $4d^3$ | MIT | 108 | [35,36] |
| | $Tl_2Ru_2O_7$ | $4d^4$ | MIT | 120 | [34] |
| | $Cd_2Re_2O_7$ | $5d^2$ | SC | 1.0 | [187–189] |
| | $Cd_2Os_2O_7$ | $5d^3$ | MIT | 225 | [37,38] |
| | $Pr_2Ir_2O_7$ | $5d^5$ | HF? | – | [41] |
| | $Sm_2Ir_2O_7$ | $5d^5$ | MIT | 117 | [42] |
| ($\beta$-pyrochlore oxides) | $CsOs_2O_6$ | $5d^{2.5}$ | SC | 3.3 | [190] |
| | $RbOs_2O_6$ | $5d^{2.5}$ | SC | 6.3 | [191–193] |
| | $KOs_2O_6$ | $5d^{2.5}$ | SC | 9.6 | [194,196–200] |

Many related spinel compounds contain mixed-valent TM ions only at the $B$ sites. A lithium spinel $LiTi_2O_4$, with $Ti^{3.5+}$ ($3d^{0.5}$), becomes superconducting at $T_c = 13.7$ K, the highest known $T_c$ for oxide materials before the discovery of cupric oxide superconductors [12]. The mechanism of superconductivity in $LiTi_2O_4$

can probably be understood in the framework of the conventional Bardeen–Cooper–Schrieffer (BCS) theory. Because no anomalous behavior in its magnetic properties has been observed, it appears that good itinerant character of the $d$ electrons and a wide band may prevail against the effects of frustration.

The situation with one more electron per TM ion added is found in $LiV_2O_4$, which has $V^{3.5+}$ ($3d^{1.5}$) and remains a good metal with no evidence for superconductivity, but shows instead unique HF-like behavior below 20 K [13]. Such a HF state is common among $f$-electron systems but unique for $d$ electrons. $LiV_2O_4$ may exist near the MI boundary, $W \sim U$, but remains on the metallic side. Full details are deferred to the next section.

Adding yet one more electron leads to the situation in $AlV_2O_4$, which has $V^{2.5+}$ ($3d^{2.5}$) and shows low resistivity at high temperatures, followed by a sudden rise at 700 K [14], as also observed in $Fe_3O_4$. A charge disproportionation and ordering take place simultaneously at $T = 700$ K, accompanied by a rhombohedral lattice distortion along the $\langle 111 \rangle$ direction of the original cubic lattice to lift the charge frustration on the pyrochlore lattice. Interestingly, it has been pointed out that the transition is characterized by the formation of V clusters called heptamers, consisting of seven V ions in a net spin-singlet state [15]. Thus, $AlV_2O_4$ clearly lies near the MI boundary, but on the insulating side.

There is in fact one exceptional compound which shows a MIT despite having an integer number of $d$ electrons per TM ion. $MgTi_2O_4$ contains $Ti^{3+}$ ($3d^2$) ions and becomes a spin-singlet insulator below 260 K [16]. The mechanism for the MIT has been ascribed to an orbitally driven Peierls transition in which a static orbital order occurs by selecting one of the $t_{2g}$ orbitals so as to form spin singlets at certain bonds [17]. Here it is an orbital degree of freedom, rather than the lattice, which appears to play an important role in lifting the frustration of spins on the pyrochlore lattice.

Compared with these $3d$ TM spinels, fewer spinel compounds are known with $4d$ and $5d$ TM elements. Of note in this category are $LiRh_2O_4$ [18] and $CuIr_2S_4$ [19, 20]. In $LiRh_2O_4$, which has $Rh^{3.5+}$ ($4d^{5.5}$) ions, the $B$ sublattice consists of low-spin (LS) $Rh^{3+}$ ($S = 0$, $4d^6$) and $Rh^{4+}$ ($S = 1/2$, $4d^5$) ions, and undergoes a MIT at 170 K. In addition, there is another transition at 230 K, which is thought to be an orbital ordering or a band Jahn-Teller transition. Thus the two degrees of freedom, charge and orbital, are frozen at two separate temperatures in $LiRh_2O_4$. In contrast, the two transitions seem to occur simultaneously in the thiospinel $CuIr_2S_4$, which has a similar electronic configuration to $LiRh_2O_4$, namely $Ir^{3+}$ ($S = 0, 5d^6$) and $Ir^{4+}$ ($S = 1/2$, $5d^5$). This system shows a MIT at 230 K, accompanied by the formation of a unique charge- and orbital-ordering pattern consisting of isomorphic octamers of $Ir_8^{3+}S_{24}$ and $Ir_8^{4+}S_{24}$ embedded in the spinel structure [20]. The origin of this phenomenon is presumably related to that of the heptamer formation found in $AlV_2O_4$ [15]. Forming such large "molecules" in the parent crystal through charge-orbital ordering is certainly a novel route to avoiding frustration.

### 22.2.1.2 Mn Metals and $LiV_2O_4$

In the previous section, we have reviewed a number of spinel compounds that exhibit MITs at low temperatures, accompanied by various kinds of symmetry-lowering which led to the loss of frustration in their ground states. On the other hand, there are compounds that remain metallic even at the lowest temperatures, and these are the Mn-based Laves-phase compounds $Y(Sc)Mn_2$ and $\beta$-Mn [2, 3], plus the more recently discovered vanadium spinel $LiV_2O_4$ [13]. The Mn sublattice in $YMn_2$ is equivalent to a pyrochlore lattice, but exhibits a complicated AF order below 100 K with a large Mn moment of $2.7\mu_B$. However, this order can be suppressed completely by substituting the smaller Sc ion for Y, even at concentrations of a few percent. In $Y_{0.97}Sc_{0.03}Mn_2$, a large enhancement of the electronic specific heat $\gamma$ is observed, up to $\gamma = 150$ mJ K$^{-2}$ mol$^{-1}$, which is 15 times the value expected from the bare DOS. Moreover, strong AF spin fluctuations leading to a reduced Mn moment of $1.3\mu_B$ are detected in neutron scattering experiments [21]. It has been suggested that the large entropy associated with the spin degree of freedom survives and is combined with the charge degree of freedom, resulting in a heavy-mass state analogous to the HF state found in many $f$-electron systems [2].

$LiV_2O_4$ presents another rare example for such a heavy-mass state on the pyrochlore lattice [13,22]. It remains metallic without showing any order down to 50 mK, and has a very large $\gamma$ value of 420 mJ K$^{-2}$ mol$^{-1}$. The low-temperature properties below 20 K are quite similar to those of $f$-electron HF compounds [22, 23]. Thus an analogous mechanism has been proposed, where electrons filling the $A_{1g}$ orbitals form localized spin-1/2 moments at each site, just as do the $f$ electrons in Ce compounds, and the remaining 0.5 electrons occupy the $E_g$ band, giving rise to Kondo screening [24]. However, alternative arguments stress the role of frustration [25–27]. It is likely that both the spins and the orbitals, or their coupled modes enhanced by the frustration, are responsible for the $d$-electron HF state [27].

It is interesting to discuss whether or not the $d$-electron HF states found in $Y(Sc)Mn_2$ and $LiV_2O_4$ have a common origin. Although it is possible that geometrical frustration on the pyrochlore lattice plays a central role in these compounds, the relevant fluctuations can be different. For example, the chemical pressure caused by Sc substitution transforms the AF order into the HF state in the former, which means that the system lies close to a magnetic instability and that enhanced spin fluctuations decide its properties [2]. By contrast, recent experiments on $LiV_2O_4$ revealed that physical pressure tends to stabilize a charge- and orbital-ordered insulating state similar to that found in $AlV_2O_4$, and not the HF state [28]. This reinforces the conclusion that $LiV_2O_4$ must be in the vicinity of the MI boundary, where electron correlations should dominate [29]. Further experimental and theoretical studies are required for a deeper understanding of the mechanisms for these $d$-electron HF states.

#### 22.2.1.3 Pyrochlore Oxides

The pyrochlore lattice is also found in the families of $\alpha$- and $\beta$-pyrochlore oxides [30], whose general formulas are respectively $A_2B_2O_7$ and $AB_2O_6$, in which $B$ is the TM element. While many TMOs are known in the former structure, the latter is found only in osmates and tungstates. Most of the $3d$ $\alpha$-pyrochlores have ferromagnetic interactions between the nearest-neighbor spins, and thus geometrical frustration is not an important topic except for some selected compounds. This occurs primarily because the superexchange pathways for the magnetic interactions are metal-oxygen-metal bonds with an angle of approximately 130 degrees, as depicted in Fig. 22.1. In contrast to the case of spinel oxides, direct exchange couplings may be less important in $3d$ $\alpha$-pyrochlore oxides with strong electron correlations. On the other hand, most but not all of the $4d$ and $5d$ pyrochlore oxides with $U$ relatively small compared to $W$ exhibit AF interactions, such that their properties may indeed be interesting from the viewpoint of frustration.

Most of the $3d$ pyrochlore oxides are ferromagnetic insulators, except for $Tl_2Mn_2O_7$ which is a ferromagnetic metal below 142 K and exhibits a large magnetoresistance (MR) [31]. The primary source for the conduction electrons in $Tl_2Mn_2O_7$ is a Tl $6s$ band crossing the Fermi level, while the $3d$ electrons from $Mn^{4+}$ ($3d^3$) are well localized to give a large magnetic moment of $3\mu_B$ on the pyrochlore lattice. The interplay between the $6s$ and $3d$ electrons may give rise to the observed MR effect. In contrast to the $3d$ compounds, many $4d$ and $5d$ pyrochlores show metallic conduction because of their extended $d$ orbitals, and sometimes have a MIT.

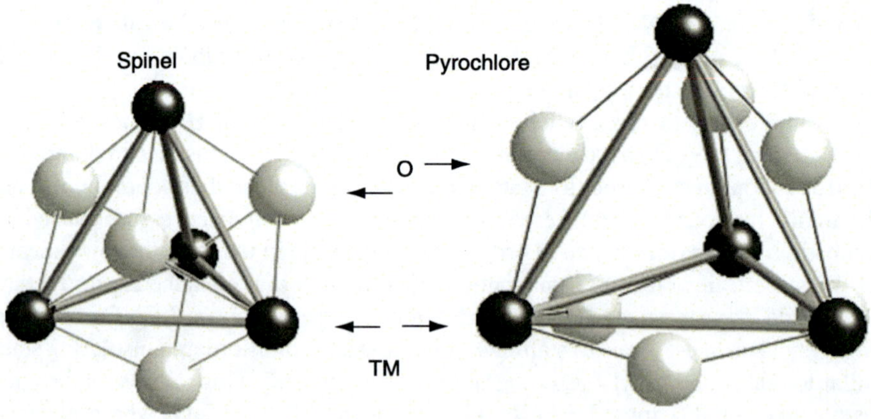


**Fig. 22.1** Local structures of the spinel (*left*) and pyrochlore geometries (*right*). One tetrahedron composed of transition metal (TM) ions and the nearest oxide ions (O), which should be primarily responsible for mediating the electron transfer and superexchange interactions, are drawn for each lattice. The relative bond length is depicted to represent the actual scale for typical compounds, indicating that the pyrochlore lattice of the spinel is much smaller than in the pyrochlore structure. The oxide ions of the former are located above the center of the triangular face, while those of the latter are above the center of the edge. Accordingly, the respective TM-O-TM bond angles are approximately 90 and 130 degrees

Some molybdenum oxides, such as $Dy_2Mo_2O_7$ ($Mo^{4+}$, $4d^2$), are ferromagnetic metals showing an anomalous Hall effect possibly associated with spin chirality on the pyrochlore lattice [32]. In contrast, $Y_2Mo_2O_7$ is an insulator showing spin-glass behavior, presumably because the smaller $A$-site cation results in a smaller bond angle and lower band width.

MITs have been reported for $Tl_2Ru_2O_7$ ($Ru^{4+}$, $4d^4$) [33,34], $Hg_2Ru_2O_7$ ($Ru^{5+}$, $4d^3$) [35,36], and $Cd_2Os_2O_7$ ($Os^{5+}$, $5d^3$) [37,38]. In these compounds, the magnetic interaction seems to be antiferromagnetic, judging from the large, negative Weiss temperatures deduced from the magnetic susceptibility at high temperature, and thus one would expect some frustration effect in these materials. $Tl_2Ru_2O_{7-\delta}$ shows a MIT at 120 K for the stoichiometric oxygen content ($\delta = 0$), but metallic behavior down to the lowest temperatures at $\delta \sim 0.3$ [34]. $Hg_2Ru_2O_7$ exhibits a similar MIT at 108 K, but this is suppressed completely by applying a pressure above 7 GPa [39]. It is worthwhile examining in detail whether these metallic states possess some unconventional features related to frustration, such as the HF states found in $LiV_2O_4$ and Y(Sc)$Mn_2$. The ground states of the ruthenates are non-magnetic, and the materials exhibit simultaneous structural transitions that are first-order, resulting in heavily distorted structures below the MIT temperature. Thus these systems realize singlet ground states accompanied by charge and/or orbital order, as is the case in some spinel compounds. By contrast, the MIT of $Cd_2Os_2O_7$ is second-order and possibly without structural deformation, suggesting that its origin is purely electronic [38]. Because the Fermi level of $Cd_2Os_2O_7$ ($Os^{5+}$, $5d^3$) is located at the center of the $t_{2g}$ manifold, it may be a magnetic instability associated with the large density of states which causes the MIT [40]. Some form of AF order sets in at the transition, but its nature has not yet been determined experimentally. The mutual alignment of the spins in order to satisfy AF interactions on the undistorted pyrochlore lattice remains a question of intense interest.

The presence or absence of MITs in the pyrochlore oxides is mysterious. It is meaningful to make a comparison among two ruthenates, $Hg_2Ru_2O_7$ and $Cd_2Ru_2$ $O_7$, and two osmates, $Hg_2Os_2O_7$ and $Cd_2Os_2O_7$, which should have similar electronic structures based on $4d^3$ and $5d^3$ electrons. A MIT is observed only in $Hg_2Ru_2O_7$ and $Cd_2Os_2O_7$, but not in $Cd_2Ru_2O_7$ and $Hg_2Os_2O_7$, which remain metallic. It is obvious that a simple discussion based on the bandwidth alone is not adequate. Band-structure calculations have revealed that stoichiometric $\alpha$-pyrochlore oxides are basically semimetals with low carrier density, because of the even number of $d$ electrons in the unit cell. Because the overall band structures of these four compounds appear to be almost the same (H. Harima, private communication), it is difficult to discern the origin of the MITs in calculations performed on the parent structure. Probably it is a subtle feature in the band structure near the Fermi level which causes an electronic instability resulting in the MIT, as in the case of $Cd_2Re_2O_7$ (discussed in Sect. 22.4.1).

It was reported recently that $Pr_2Ir_2O_7$ ($Ir^{4+}$, $5d^5$) shows good metallic conductivity without magnetic long-range order associated with either Ir $5d$ or Pr $4f$ moments [41]. Although a metallic spin liquid with a Kondo effect has been proposed, the details remain under investigation. Still more recently, the related iridium pyrochlores $A_2Ir_2O_7$ with $A$ = Nd, Sm, and Eu, which are adjacent to Pr in the periodic table, were found to exhibit MITs [42].

Superconductivity is found in $\alpha$-$Cd_2Re_2O_7$ ($Re^{5+}$, $5d^2$) at $T_c = 1.0$ K and in $\beta$-$AOs_2O_6$ with $A$ = Cs, Rb, and K at $T_c$ = 3.3, 6.3, and 9.6 K, respectively. The former is a semimetal with very low carrier density, while the latter are normal metals because of the mixed valence of $Os^{5.5+}$ ($5d^{2.5}$). The properties of these materials will be discussed in detail in Sect. 22.4.1.

## 22.2.2 *Triangular and Kagomé Lattices*

### 22.2.2.1 $Na_xCoO_2$ and Its Hydrate

The recent discoveries of an unusually large thermoelectric power in $Na_xCoO_2$ with $x \sim 2/3$ [43], and of superconductivity below 4.5 K in hydrated $Na_xCoO_2 \cdot yH_2O$ [44], stimulated a great deal of interest in this system. $Na_xCoO_2$ has a two-dimensional layered structure as in the high-$T_c$ cupric oxide superconductors, but has a triangular lattice consisting of Co atoms (Fig. 22.2a) rather than the square lattice of Cu in the $CuO_2$ plane [45].

Here, we discuss the non-hydrated samples. Because the valence of the Co ions in $Na_xCoO_2$ is $Co^{(4-x)+}$, there are $1-x$ holes per Co ion and these occupy the $t_{2g}$-orbitals due to the crystal-field splitting. Thus the material with $x = 0$ corresponds, if it exists, to half-filling. The electronic phase diagram as a function of $x$ has been proposed by several groups, a typical example being that of Foo et al.[46], where a charge-ordered magnetic insulator exists at $x = 0.5$ with a "paramagnetic metal"

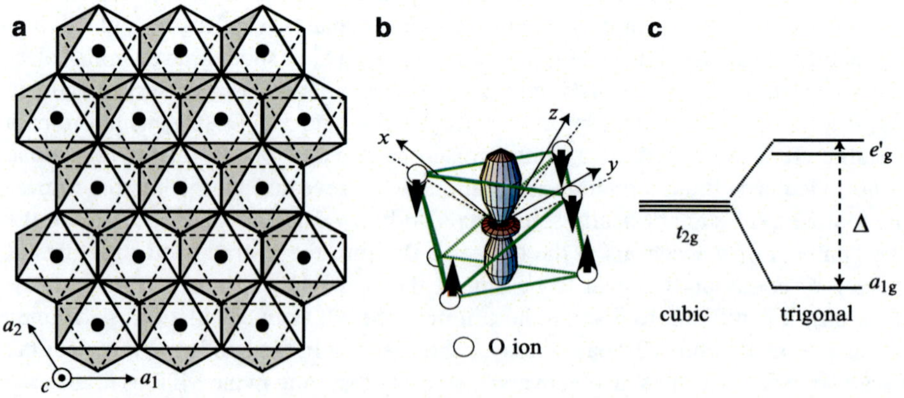


**Fig. 22.2** **(a)** $CoO_2$ layer viewed along the $c$-axis. The Co ions form a perfect triangular lattice in the layer. There are O ions on each corner of the octahedra. **(b)** Trigonal distortion of the $CoO_6$ octahedron (*arrows*) and $a_{1g}$-orbital. **(c)** $a_{1g}$-$e'_g$ splitting of the Co $t_{2g}$ levels, with energy difference $\Delta$, arising due to the trigonal crystal field induced by the distortion shown in **(b)**

for $x < 0.5$ and a "Curie–Weiss metal" for $x > 0.5$. However, Yokoi et al. [47] and recently several other groups [48, 49] have found that the boundary between the above two metals is located near $x = 0.62$ instead of at $x = 0.5$. Thus $x = 0.5$ appears as a rather singular point in the phase diagram, where Na ordering (existing already at high temperature) triggers a magnetic transition as well as a MIT [46, 47, 50, 51]. In this sense, the maximal frustration in the charge degrees of freedom at $x = 0.5$ is lifted by the Na ordering. The system with $x \neq 0.5$ is metallic probably because of the incommensurate filling.

The existence of a "Curie–Weiss metal" for $x > 0.62$ is rather surprising, because the system is well described in terms of a small number of holes in the fully occupied $t_{2g}$-band when $x \to 1$. In fact recent, detailed studies by nuclear magnetic resonance (NMR) [49, 52] have found quasi-two-dimensional ferromagnetic behavior consistent with the Curie–Weiss susceptibility. Some magnetic phases are also observed for $x \sim 0.75$–$0.82$, indicating the importance of electronic correlations in this regime [53–55]. It has been confirmed that the magnetic structure is an A-type antiferromagnetism, where ferromagnetic long-range order exists within the $CoO_2$ planes but its direction alternates between them [56–58]. Further, Na order may affect a number of possible phases appearing in the regime $x > 0.62$ [49]. In contrast, for $x < 0.62$, both the Pauli susceptibility and the Korringa behavior of the NMR relaxation rate [49] indicate weak correlation in this regime, although a weak pseudo-gap-like behavior is observed in the Knight shift.

#### 22.2.2.2 $Ag_2MO_2$ and $AgNiO_2$

The silver compounds $Ag_2MO_2$ ($M$ = Ni, Mn) are unique two-dimensional (2D) systems where the metallic $(Ag_2)^+$ layers alternate with the magnetic $M^{3+}O_2$ layers of the triangular lattice, as depicted in Fig. 22.3a [59–61]. Spins 1/2 and 2 are localized on the triangular lattice for $M = Ni^{3+}$ ($3d^7$, LS) and $Mn^{3+}$ ($3d^4$, HS), respectively. Orbital ordering associated with the $e_g$-orbital degeneracy takes place in both compounds, which tends to reduce the magnetic frustration and leads to AF order, while the metallic conductivity is preserved down to 2 K. $Ag_2MnO_2$ may be particularly interesting as a classical spin system on the triangular lattice, because it shows a peculiar magnetic phase transition at 80 K followed by a spin-glass transition at 24 K, but no long-range magnetic order down to 2 K in spite of the large spin [61]. It may be necessary to consider additional magnetic interactions mediated by the conduction electrons through the RKKY interaction in such a metallic system and, possibly as a result of these, a moderately large mass enhancement has been observed for both compounds.

A related silver nickelate, 2H-$AgNiO_2$, crystallizes in the delafossite structure and exhibits metallic conductivity down to the lowest temperatures [62]. It contains the same $Ni^{3+}$ triangular lattice as in $Ag_2NiO_2$, separated by a single Ag sheet, as depicted in Fig. 22.3b. However, a $\sqrt{3} \times \sqrt{3}$-type superstructure is already present at room temperature, and is ascribed to an unusual charge disproportionation and order expressed as $3Ni^{3+} \to Ni^{2+} + 2Ni^{3.5+}$ or $3e_g^1 \to e_g^2 + 2e_g^{0.5}$. It has been suggested

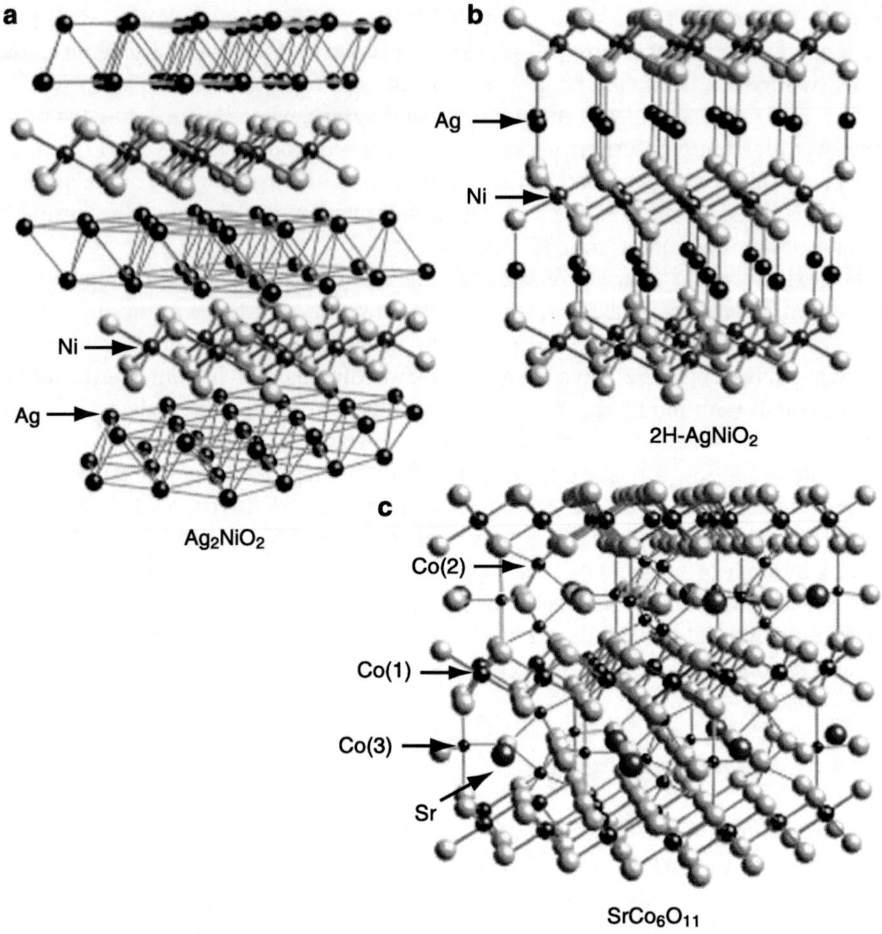


**Fig. 22.3** Crystal structures of (**a**) $Ag_2NiO_2$, (**b**) 2H-$AgNiO_2$, and (**c**) $SrCo_6O_{11}$. The $NiO_2$ layers of $Ag_2NiO_2$ and 2H-$AgNiO_2$ have the same structure as the $CoO_2$ layer in $Na_xCoO_2$, shown in Fig. 22.2. These are separated respectively by double and single Ag planes in (**a**) and (**b**). In $SrCo_6O_{11}$, the edge-sharing Co(1)$O_6$ octahedra form kagomé layers, while the dimerized, face-sharing octahedra containing Co(2) and trigonal bipyramids of Co(3) act as pillars between the kagomé layers

that the $e_g^2$ electrons are well localized at one Ni site, displaying antiferromagnetic order below 22 K, while the two $e_g^{0.5}$ electrons occupy the remaining Ni sites which form a honeycomb network and are responsible for the metallic conduction. The orbital degeneracy inherent to the $Ni^{3+}$ triangular lattice is lifted to a sufficient extent by this charge disproportionation that frustration is relieved.

#### 22.2.2.3 $AB_6O_{11}$

The family of $AB_6O_{11}$ compounds presents another example consisting of alternating metallic layers and frustrated spin systems [63, 64]. These materials crystallize in a hexagonal structure related to the magnetoplumbite structure (Fig. 22.3c). There are three TM-ion sites: $B(1)$ forms a kagomé plane, while the $B(2)$ and $B(3)$ polyhedra act as pillars between the kagomé planes. In $SrCo_6O_{11}$, electrons in the Co(1) kagomé layers are itinerant, probably due to the mixed valency, and large Ising spins exist at the Co(3) sites, which form a triangular lattice between the kagomé layers [65]. Thus one expects a unique situation where 2D conduction electrons in the kagomé lattice can interact with frustrated spins on the triangular lattice. In fact it is found that their coupling is so strong that the resistivity is affected dramatically by ordering of the moments below 20 K. Stepwise changes in resistivity are observed at low temperatures as a function of magnetic field, coupled with a spin-flip transition [64, 66].

The vanadium analogs of $SrCo_6O_{11}$, such as $NaV_6O_{11}$ and $SrV_6O_{11}$, also show metallic conductivity due to itinerant electrons in the kagomé layer [63]. However, these materials undergo successive structural transitions to lower symmetries, although the details of these processes are not yet clear. A magnetic transition takes place at 70 K but the metallic conductivity is preserved at least down to 2 K in the case of $NaV_6O_{11}$ [67].

### 22.2.3 *Organic Conductors with Triangular Lattice*

Organic conductors (for review [5, 6]) are among the most promising candidate materials for the observation of exotic quantum mechanical states. This arises because of a number of characteristic features: (1) low dimensionality, (2) relatively strong correlation, (3) variety of constituent molecules and variety of lattice structures including frustrated geometries, (4) structural softness due to the weak van der Waals bonding between molecules, (5) cleanness of the samples, and (6) simple band structure originating from the highest occupied molecular orbital (HOMO) or the lowest unoccupied molecular orbital (LUMO) [68]. Because of the softness, application of pressure can easily change the lattice parameters and hence the relative strength of interactions. A great variety of physical states have been observed and investigated intensively in organic conductors, including charge order, superconductivity, spin-Peierls states, antiferromagnetism, and spin density waves (SDWs).

Among the known organic materials, we concentrate on the quasi-two-dimensional $A_2B$ compounds, known as 1:2 salts, which are shown in Fig. 22.4. Many of these 1:2 salts show electron conduction at room temperature, where the carriers are generated as a result of a charge transfer between the $A$ molecules and the $B$ units. In many cases, the monovalent $B^-$ anion forms a closed shell and the average valence for the $A$ molecules becomes $+1/2$. The electronic band near the Fermi

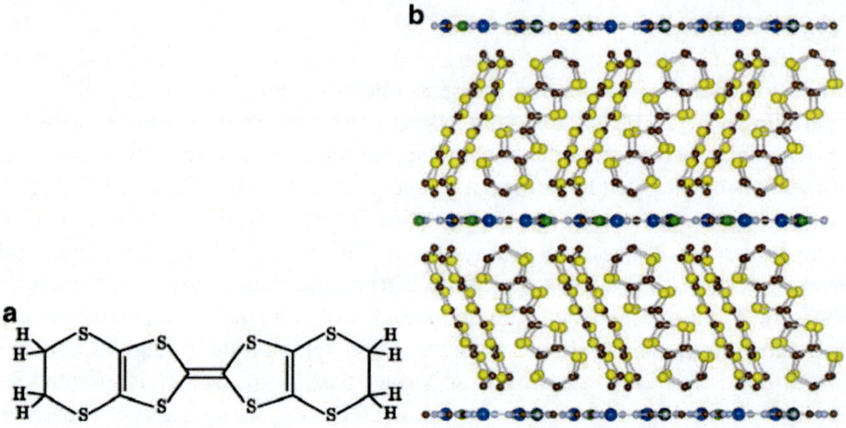


**Fig. 22.4** (**a**) BEDT-TTF molecule [72] and (**b**) layered crystal structure of $\kappa$-(BEDT-TTF)$_2$X, consisting of conducting BEDT-TTF layers and insulating anion layers. [Courtesy of Y. Hayashi.]

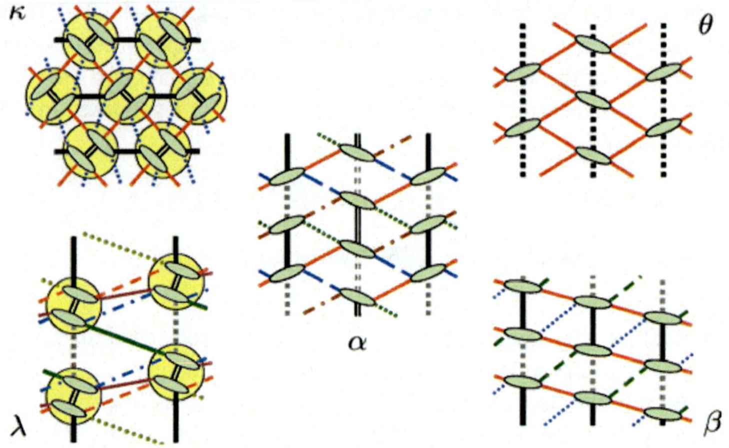


**Fig. 22.5** Polytypes in the organic conductor (BEDT-TTF)$_2X$

energy is then composed of the HOMO of the $A$ molecule, which is quarter-filled in terms of holes.

The crystal structures of $A_2B$ compounds contain alternately stacked layers of $A$ molecules and $B$ units (Fig. 22.4b), and thus realize quasi-two-dimensional electronic states; electron hopping between the layers is small due to the insulating nature of the $B$ layers. There is a variety of two-dimensional arrangements of the $A$ molecules, sometimes even for the same chemical formula, which leads to the distinguishing of polytypes, classified by the Greek letters, $\alpha, \beta, \kappa, \theta, \lambda, \ldots$ [69–71]. Some representative examples are shown in Fig. 22.5. It is apparent that these are based on triangular lattice structures, leading to possible frustration, although the degree of frustration depends on the hopping integrals between the $A$ molecules in

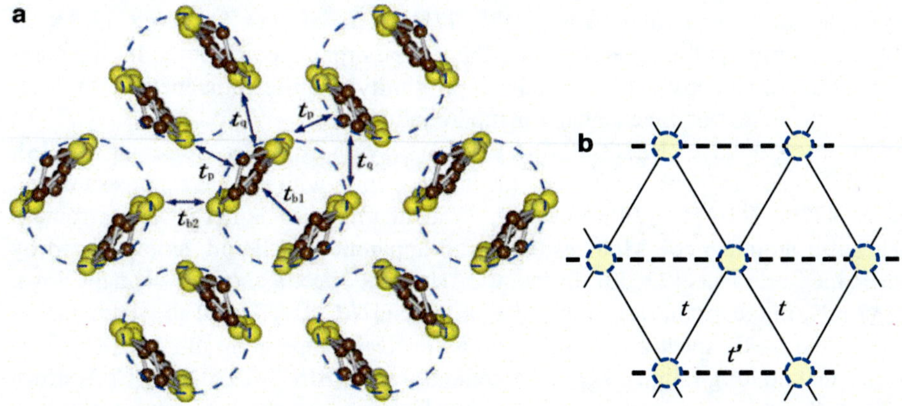


**Fig. 22.6** (**a**) Arrangement of BEDT-TTF molecules in the conducting layer of $\kappa$-(BEDT-TTF)$_2X$, viewed along the long axes of BEDT-TTF molecules. Also shown are the transfer integrals between BEDT-TTF molecules, which are estimated from the highest occupied molecular orbital (HOMO) energy and from overlap integrals calculated by the extended Hückel method. [Courtesy of Y. Hayashi.] (**b**) Schematic representation of the two-dimensional BEDT-TTF layer obtained by regarding the face-to-face BEDT-TTF dimer pair as a single unit

the different polytypes. Typical examples of interest here are $\kappa$-, $\alpha$-, and $\theta$-(BEDT-TTF)$_2X$, where BEDT-TTF is the abbreviation of bis(ethylenedithio)-tetrathiafulvalene; the structure of the BEDT-TTF molecule is shown in Fig. 22.4a.

#### 22.2.3.1 $\kappa$-(BEDT-TTF)$_2X$

The arrangement of BEDT-TTF molecules in the conducting layer of $\kappa$-(BEDT-TTF)$_2X$, viewed along the long axes of the BEDT-TTF molecules, is shown in Fig. 22.6a. In the $\kappa$-type structure, the transfer integral connecting the "face-to-face" pair of BEDT-TTF molecules, represented by $t_{b1}$ in Fig. 22.6a, is large enough to consider the pair as a dimer unit. Such dimer units form a triangular lattice, as shown in Fig. 22.6b, a simplified picture first introduced by Kino and Fukuyama [68]. The anisotropic triangular lattice shown in Fig. 22.6b will arise frequently in the discussion to follow. Band-structure calculations for $\kappa$-(BEDT-TTF)$_2X$ show that there is a distinct energy gap in the dispersion due to the strong dimerization. Because the BEDT-TTF molecules originally form a quarter-filled band, there is one hole per dimer and the resulting band becomes half-filled. The effective transfer integrals between the dimer units can be estimated as $t = (|t_p| + |t_q|)/2$ and $t' = t_{b2}/2$, where $t_p$, $t_q$, and $t_{b2}$ are defined in Fig. 22.6a [73].

Superconductivity appears in $\kappa$-(BEDT-TTF)$_2$Cu(NCS)$_2$ ($T_c = 10.4$ K) [74] and $\kappa$-(BEDT-TTF)$_2$Cu[N(CN)$_2$]Br ($T_c = 12.6$ K) [75] at ambient pressure [76]. Another compound, $\kappa$-(BEDT-TTF)$_2$Cu[N(CN)$_2$]Cl, is an antiferromagnetic insulator at ambient pressure, with a Néel temperature of 27 K and a spin moment greater than $0.4\mu_B$ per dimer. With increasing pressure, however, it undergoes an insulator-superconductor transition [75, 77], the maximum $T_c$ being 13.2 K at 300 MPa, which is the highest among the $\kappa$-type organic conductors. In this compound, superconductivity occurs in direct proximity to antiferromagnetism, suggesting a spin-fluctuation mechanism for superconductivity.

The compound $\kappa$-(BEDT-TTF)$_2$Cu$_2$(CN)$_3$, synthesized by Geiser et al. [78], also undergoes an insulator-superconductor transition with increasing pressure, and $T_c = 4$ K at 360 MPa [79]. However, its insulating state is not an antiferromagnet, but a nonmagnetic Mott insulator and apparent spin liquid, as evidenced by NMR measurements [73, 80]. In fact the $^1$H NMR spectra show no changes down to 32 mK, in clear contrast to the case of the Cu[N(CN)$_2$]Cl salt, in which a clear split of the NMR spectra is visible below the Néel temperature of 27 K [81]. This means that no long-range magnetic order exists in $\kappa$-(BEDT-TTF)$_2$Cu$_2$(CN)$_3$ down to 32 mK, which is 4 orders of magnitude smaller than the superexchange interaction, $J = 250$ K, estimated from the temperature-dependence of the magnetic susceptibility. Similar results are obtained by zero-field muon spin relaxation measurements [82]. Furthermore, the temperature-dependences of the $^1$H and $^{13}$C NMR relaxation rates [73,80], and of the static susceptibility, all show that there is no spin gap down to the lowest temperatures. More precisely, below approximately 0.4 K, the relaxation curve for $^1$H-NMR splits into two exponential functions, probably due to some inhomogeneity, and each relaxation rate, $1/T_1$, is proportional to $T$ or $T^2$. These results have led to the conclusion that a "gapless spin liquid" is realized in this system at ambient pressure. Recent specific-heat measurements support the gapless nature of the excitation [83], although a thermal conductivity measurement claims the existence of a small excitation gap [84].

There have been many theoretical proposals for explaining this quantum disordered state. Here we note only that an isotropic triangular-lattice Heisenberg model cannot explain the observed "gapless spin liquid," because its ground state would have long-range magnetic order of the three-sublattice 120-degree structure. The gapless spin liquid may be understood by taking account of either long-range interactions or anisotropy of the triangular lattice. In fact, for the $\kappa$-type structure we note that the system is slightly anisotropic [85, 86]; the anisotropies of the transfer integrals $t'/t$ in Fig. 22.6b, together with the corresponding estimates for the superexchange interactions, $J'/J = (t'/t)^2$, are summarized in Table 22.2 for various $\kappa$-(BEDT-TTF) salts [85]. Here $t'$ is the transfer integral along the one-dimensional chain and $t$ can be regarded as the interchain transfer integral, as shown in Fig. 22.6b. It should be emphasized that only $\kappa$-(BEDT-TTF)$_2$Cu$_2$(CN)$_3$, the candidate spin-liquid system, has an anisotropy $t'/t > 1$ in Table 22.2, meaning that this system has quasi-one-dimensional anisotropy [86,87]. Specifically, based on the extended Hückel method, the transfer integrals between the BEDT-TTF molecules can be evaluated as $t_{b1} = 223.6$ meV, $t_{b2} = 115.4$ meV, $t_p = 80.1$ meV, and $t_q = -29.0$ meV [88], and thus the interdimer transfer integrals become $t = 54.6$ meV and $t' = 57.7$ meV, giving a ratio $t'/t \approx 1.06$. Theoretical investigations based on this anisotropy are discussed in Sect. 22.3.1.

$\beta'$-(BEDT-TTF)$_2$ICl$_2$ also has strong dimerization of BEDT-TTF molecules, and can be regarded as a half-filled system. This salt has a pressure-temperature phase

**Table 22.2** Anisotropy of $\kappa$-(BEDT-TTF)$_2X$ salts for various anions $X$ [85, 86]. Here, $t'$ denotes the transfer integral along the one-dimensional chain and $t$ the interchain transfer integral (Fig. 22.6b). $J'/J$ is estimated from $(t'/t)^2$. Note that $t = 0$ ($J = 0$) corresponds to decoupled chains and $t' = 0$ ($J' = 0$) to the square lattice

| Anion X | $t'/t$ | $J'/J$ |
|---|---|---|
| $Cu_2(CN)_3$ | 1.06 | 1.12 |
| $Cu(NCS)_2$ | 0.84 | 0.71 |
| $Cu[N(CN)_2]Cl$ | 0.75 | 0.56 |
| $Cu[N(CN)_2]Br$ | 0.68 | 0.46 |
| $Cu(CN)[N(CN)_2]$ | 0.68 | 0.46 |
| $Ag(CN)_2 \cdot H_2O$ | 0.60 | 0.36 |
| $I_3$ | 0.58 | 0.34 |

diagram similar to that of $\kappa$-(BEDT-TTF)$_2X$ [89]. Although the critical pressure is very high (8.2 GPa), the superconducting transition temperature is the highest among all the organic conductors ($T_c = 14.2$ K).

#### 22.2.3.2 $\alpha$- and $\theta$-(BEDT-TTF)$_2X$

There are many other quasi-two-dimensional organic conductors of $A_2B$ type which show MITs without dimerization. Because these are quarter-filled, their MITs cannot be explained as a Mott transition, and instead the most common insulating state turns out to be a charge-ordered phase [90].

Charge ordering was first reported by NMR measurements in a quasi-one-dimensional organic conductor, DI-DCNQI$_2$Ag [91]. In fact a (1010)-type charge-ordered state had been predicted theoretically for this situation [92–94]. It is now known that a significant number of materials shows charge ordering, and here we focus on frustrated, two-dimensional charge-ordered systems in organic conductors. Direct observations of charge order in quasi-two-dimensional compounds were made first in $\theta$-(BEDT-TTF)$_2$RbZn(SCN)$_4$ [95, 96] and then in $\alpha$-(BEDT-TTF)$_2$I$_3$ [97] by $^{13}$C NMR experiments.

Let us discuss first the $\alpha$-type salts [98]: $\alpha$-(BEDT-TTF)$_2$I$_3$ has four BEDT-TTF molecules in a unit cell and is quarter-filled due to the absence of dimerization. At ambient pressure, there is a MIT at 135 K [99], and a spin-singlet state in the insulator is suggested by spin susceptibility measurements [100]. Seo pointed out that the insulating state should have a stripe-type charge ordering [101], which was confirmed by means of NMR [97] and synchrotron X-ray diffraction [102]. When a uniaxial pressure is applied, superconductivity with $T_c = 7.2$ K appears at 0.2 GPa [103]. At higher pressures than this, both charge order and superconductivity disappear, and, instead, a strange metal is obtained [104, 105]. In this metallic state, the resistivity is almost temperature-independent, while the Hall coefficient increases very rapidly, and by a factor of $10^6$, with decreasing temperature. This strange behavior can be understood by having a zero-gap state containing massless fermions with a linear dispersion of the Dirac-cone type [106]. These fermions are described by a tilted Weyl equation, and the Fermi surface of the system consists of two points in the Brillouin zone. It can be shown that these massless fermions induce anomalous behavior of the conductivity, Hall conductivity, and orbital susceptibility, which can explain the anomalous experimental observations [107, 108].

Moving to the $\theta$-type salts, $\theta$-(BEDT-TTF)$_2$RbZn(SCN)$_4$ undergoes a MIT at $T = 190$ K accompanied by a structural phase transition [109]. Direct observation of charge ordering was achieved by $^{13}$C NMR [95, 96]. This charge order is characterized by a diffuse rod at $\mathbf{q} = (0, k, 1/2) \equiv \mathbf{q}_1$, observed in X-ray measurements [110], which led to the term "horizontal charge order." By contrast, $\theta$-(BEDT-TTF)$_2$CsZn(SCN)$_4$ shows no phase transition: it is metallic above $T = 20$ K and in X-ray measurements a short-range, three-fold diffuse rod with $\mathbf{q} = (2/3, k, 1/3) \equiv \mathbf{q}_2$ is observed. Below $T \sim 20$ K, the resistivity increases rapidly [109], accompanied by the appearance of a short-range diffuse rod at $\mathbf{q} = \mathbf{q}_1$, which coexists with $\mathbf{q}_2$ [111, 112]. (The wave number $\mathbf{q}_1$ is same as that observed in RbZn(SCN)$_4$ salts at low temperatures, i.e. horizontal charge order.) These experiments indicate that two types of charge-order fluctuation coexist in the low-temperature regime for $\theta$-(BEDT-TTF)$_2$CsZn(SCN)$_4$, which may be a consequence of frustration in the charge degree of freedom on the triangular lattice. Probably as a result of the inhomogeneous mixture of the two charge-ordered phases, a gigantic nonlinear conductivity [113, 114] and a thyristor-like $I$–$V$ characteristic [115] are observed, which have attracted considerable attention. Related theoretical studies are discussed in Sect. 22.3.3.

Mori et al. summarized the experimental data for various $\theta$-(BEDT-TTF)$_2X$ salts in a phase diagram using the temperature and the dihedral angle between BEDT-TTF molecules [109]. The primary effect of the dihedral angle will be to change the anisotropy of the hopping integrals between the BEDT-TTF molecules. According to this phase diagram, the CsZn(SCN)$_4$ salt has a smaller dihedral angle than the RbZn(SCN)$_4$ salt and is located close to the boundary between a metallic and a charge-ordered state. $\theta$-(BEDT-TTF)$_2$I$_3$ has a dihedral angle still smaller than the CsZn(SCN)$_4$ salt, and is the only compound in the $\theta$-(BEDT-TTF)$_2X$ family which shows superconductivity ($T_c = 3$ K) [116]. In this phase diagram, the I$_3$ salt is considered to be located in the vicinity of the charge-ordered phase, suggesting a charge-fluctuation mechanism for the superconductivity.

Finally, we also mention $\beta''$-(DODHT)$_2$PF$_6$, which undergoes a MIT at $T = 225$ K accompanied by a charge ordering. When pressure is applied, superconductivity appears with $T_c = 3.1$ K at 1.65 GPa [117].

## 22.3 Theoretical Background

### *22.3.1 RVB Spin State and RVB Superconductivity*

In sufficiently frustrated quantum spin systems it is generally expected that magnetic long-range order is destroyed, and that a quantum spin liquid can be realized. As a model for such a spin liquid, Anderson proposed a resonating-valence-bond (RVB) state for the Heisenberg model on a triangular lattice [118]. This state consists of superpositions of products of singlet bonds, called valence bonds, with finite bond lengths which cover all the spins in the system.

When holes are doped into this RVB state, the pre-existing valence bonds become directly the Cooper pairs, and then a new type of superconductivity would be realized. Anderson showed that the strongly correlated BCS state can be rewritten in the form of an RVB state as [119]

$$\begin{aligned} P_G|\mathrm{BCS}\rangle &= P_G \prod_{\mathbf{k}} \left[ u_{\mathbf{k}} + v_{\mathbf{k}} c^{\dagger}_{\mathbf{k}\uparrow} c^{\dagger}_{-\mathbf{k}\downarrow} \right] |0\rangle \\ &= P_G \left(\prod_{\mathbf{k}} u_{\mathbf{k}}\right) \prod_{\mathbf{k}} \left[ 1 + \tfrac{v_{\mathbf{k}}}{u_{\mathbf{k}}} c^{\dagger}_{\mathbf{k}\uparrow} c^{\dagger}_{-\mathbf{k}\downarrow} \right] |0\rangle \\ &= P_G \left(\prod_{\mathbf{k}} u_{\mathbf{k}}\right) \prod_{\mathbf{k}} \exp\left( \tfrac{v_{\mathbf{k}}}{u_{\mathbf{k}}} c^{\dagger}_{\mathbf{k}\uparrow} c^{\dagger}_{-\mathbf{k}\downarrow} \right) |0\rangle \\ &= P_G \left(\prod_{\mathbf{k}} u_{\mathbf{k}}\right) \exp\left( \sum_{\mathbf{k}} a_{\mathbf{k}} c^{\dagger}_{\mathbf{k}\uparrow} c^{\dagger}_{-\mathbf{k}\downarrow} \right) |0\rangle \\ &= P_G \left(\prod_{\mathbf{k}} u_{\mathbf{k}}\right) \exp\left( \sum_{i,j} a_{i,j} c^{\dagger}_{i\uparrow} c^{\dagger}_{j\downarrow} \right) |0\rangle, \end{aligned} \tag{22.1}$$

where $P_G = \prod_i \left[1 - n_{i\uparrow} n_{i\downarrow}\right]$ is a projection operator which excludes the doubly occupied sites from the wave function due to the strong correlation and $a_{i,j}$ is defined as the Fourier transform of

$$a_{\mathbf{k}} = \frac{v_{\mathbf{k}}}{u_{\mathbf{k}}} = \frac{\Delta_{\mathbf{k}}}{(\varepsilon_{\mathbf{k}} - \mu) + \sqrt{(\varepsilon_{\mathbf{k}} - \mu)^2 + |\Delta_{\mathbf{k}}|^2}}. \tag{22.2}$$

By applying a projection operator, $P_N$, to the state for fixing the electron number, we obtain

$$P_N P_G|\mathrm{BCS}\rangle = \frac{1}{(N/2)!} \left(\prod_{\mathbf{k}} u_{\mathbf{k}}\right) P_G \left( \sum_{i,j} a_{i,j} c^{\dagger}_{i\uparrow} c^{\dagger}_{j\downarrow} \right)^{\frac{N}{2}} |0\rangle. \tag{22.3}$$

In the case of singlet pairing, $a_{\mathbf{k}}$ is an even function of $\mathbf{k}$, leading to the relation $a_{i,j} = a_{j,i}$. This means that the wave function $P_N P_G|\mathrm{BCS}\rangle$ is a superposition of many singlet-bond configurations, $(c^{\dagger}_{i\uparrow} c^{\dagger}_{j\downarrow} - c^{\dagger}_{i\downarrow} c^{\dagger}_{j\uparrow})/\sqrt{2}$, i.e. an RVB state with the amplitude of the singlet bond being $a_{i,j}$.

#### 22.3.1.1 Square- and Triangular-Lattice $t$-$J$ Model

The superconducting RVB state has been studied extensively in the Hamiltonian known as the $t$-$J$ model,

$$\mathcal{H}_{\text{t–J}} = -\sum_{(i,j)\sigma} P_G\left(t_{ij}c^\dagger_{i\sigma}c_{j\sigma} + \text{h.c.}\right)P_G + J\sum_{(i,j)} \mathbf{S}_i \cdot \mathbb{S}_j, \tag{22.4}$$

where the first term represents the kinetic energy for mobile doped holes and the second term is the Heisenberg superexchange interaction. The summation over $(i, j)$ denotes pairs of lattice sites depending on the lattice structure, and the parameters $t_{ij}$ can be chosen to be $t$ and $t'$ as in Fig. 22.6b. When $t' = 0$, the system is a square lattice, and when $t' = t$ it is an isotropic triangular lattice. The spin index, $\sigma$, takes the values $\uparrow$ and $\downarrow$, and $c^\dagger_{i\sigma}$ $(c_{i\sigma})$ denotes the creation (annihilation) operator for an electron with spin $\sigma$ at site $i$. The spin operator is defined as $\mathbf{S}_i = c^\dagger_{i\sigma}\sigma_{\sigma\sigma'}c_{i\sigma'}/2$.

For the two-dimensional square lattice ($t' = 0$), it has been shown that the RVB state with a $d_{x^2-y^2}$-wave superconducting order parameter, $\Delta_{\mathbf{k}}$, is a very good variational wave function [120–125]. Many of the features of high-$T_c$ superconductivity can be understood from the properties of this state [126]. At half-filling ($n = 1$), the $t$-$J$ model is equivalent to the Heisenberg model for the spin system, because the kinetic-energy term vanishes, and in this case (22.3) does not show superconductivity even if the wave function contains the superconducting order parameter, $\Delta_{\mathbf{k}}$. In fact $\langle c_{i\uparrow}c_{j\downarrow}\rangle = 0$ for the wave function of (22.3) at $n = 1$ due to the absence of holes. The variational energy of (22.3) is $E = -0.3199\,J$ per bond, which is very close to the best numerical estimate, $E = -0.3346J$ [127–129]. However, it is well known that there is antiferromagnetic (AF) long-range order for the $S = 1/2$ Heisenberg model on the square lattice, and the best variational state at half-filling studied to date accommodates coexisting RVB and AF nature which gives $E = -0.3323\,J$ [130, 131].

For the two-dimensional triangular lattice ($t' = t$), mean-field approximations for the $t$-$J$ model show that $(d_{x^2-y^2} + id_{xy})$-wave superconductivity is stable near half-filling [132–135]. $d_{x^2-y^2}$-wave and $d_{xy}$-wave superconductivity are actually degenerate here because of the $C_6$ symmetry of the triangular lattice [132, 134–136], and as a result a complex combination of $d_{x^2-y^2} + id_{xy}$ order parameters is stabilized below $T_c$, because the superconducting gap opens over all of the Fermi surface. Numerical calculations (by variational Monte Carlo simulation) based on the RVB wave function of (22.3) have confirmed the stabilization of this ground state near half-filling (less than 10% doping) [137]. High-temperature series expansion studies of the $t$-$J$ model also show a rapid growth of the $d$-wave superconducting correlation function, which indicates the appearance of RVB superconductivity [138, 139], and these studies lend weight to the speculation that a small amount of hole doping acts to destroy the magnetic long-range order and stabilize the superconducting RVB state [138].

#### 22.3.1.2 Triangular-Lattice Heisenberg Model

As discussed in Sect. 22.2.3, the spin-liquid state apparently observed in $\kappa$-(BEDT-TTF)$_2$Cu$_2$(CN)$_3$ is realized on the almost isotropic triangular lattice. The simplified model Hamiltonian for this system is thus an isotropic triangular-lattice Heisenberg spin model with $J' = J$. However, as mentioned Sect. 22.2.3, this model is believed to have a ground state with long-range magnetic order of the 120-degree type. Exact diagonalization studies of small clusters [140, 141], Ising expansions [142], high-temperature series expansions [143], and variational Monte Carlo simulations [144] all support this picture. Therefore, if one assumes finite three-dimensionality in real materials, it is difficult to explain the spin-liquid state reported in $\kappa$-(BEDT-TTF)$_2$Cu$_2$(CN)$_3$ within the simple Heisenberg model. However, if the long-range order is neglected and approaches based on a disordered ground state are adopted, one obtains an RVB state with $(d_{x^2-y^2} + id_{xy})$-wave symmetry [132–135] as in the $t$-$J$ model. Because the $(d_{x^2-y^2} + id_{xy})$-wave state should have a finite gap (the spin gap at half-filling), this state remains nevertheless inconsistent with the observed "gapless spin liquid" state.

In order to resolve this problem, the anisotropy, $J' \neq J$ [86, 87], in $\kappa$-(BEDT-TTF)$_2$Cu$_2$(CN)$_3$ (Table 22.2) has been studied. Variational Monte Carlo simulations show that the wave function continues from the one-dimensional limit ($J = 0$) until $J/J' = 0.65$ or $J/J' = 0.8$ [87], although the state considered in these studies was only a $d_{x^2-y^2}$-wave RVB state, whereas the RVB mean-field theory predicts a $(d_{x^2-y^2} + id_{xy})$-wave state in this region. Although the $(d_{x^2-y^2} + id_{xy})$-wave RVB state has a finite spin gap, the excitation spectrum obtained in the mean-field theory is almost gapless over a wide parameter range from the one-dimensional limit, probably because the interchain coupling, $J$, is frustrated [86]. Because the pure one-dimensional spin chain has no long-range order and its spin excitations are gapless (the des Cloiseau-Pearson mode), we think that this kind of one-dimensionalization [86] by frustration is a good candidate mechanism for the formation of the "gapless spin liquid".

### 22.3.2 Triangular-Lattice Hubbard Model

In contrast to the strong-coupling $t$-$J$ model, a simple model for studying superconductivity away from the strong-coupling regime is the Hubbard model,

$$\mathcal{H}_{\text{Hub}} = -\sum_{(i,j)\sigma} \left( t_{ij} c_{i\sigma}^{\dagger} c_{j\sigma} + \text{h.c.} \right) + U \sum_{i} n_{i\uparrow} n_{i\downarrow}. \tag{22.5}$$

We note that in the strong-coupling expansion of the Hubbard model in $t/U$, the $t$-$J$ model is derived, with higher terms.

In the hole-doped case, perturbation theories of the Hubbard model on a triangular lattice show that both $d$- and $p$-wave superconducting states are stable [145,146], while the renormalization-group approach finds the $d_{x^2-y^2} + id_{xy}$ state to be most stable [147]. In addition, $f$-wave pairing can be expected in the low-density region where the Fermi surface is disconnected [148]. In these weak-coupling calculations, the effects of frustration do not play an important role, and instead the superconductivity is determined from the subtle details of the Fermi surface and the low-lying spin excitations, which leads to inevitable controversy concerning the symmetry of the superconducting order parameter.

The Hubbard model at half-filling on an anisotropic lattice was used to study the superconductivity in the $\kappa$-type organic conductors. Fluctuation-exchange approximations show that the superconductivity has $d_{x^2-y^2}$-wave symmetry mediated by antiferromagnetic spin fluctuations for the $t' < t$ case [149–151]. The similarity to the high-$T_c$ cupric oxide superconductors has been discussed [152], and in fact this $d$-wave state is stable when the lattice structure is changed continuously from the triangular to the square lattice [135, 153]. Similar results were obtained in the Hubbard–Heisenberg model [154].

However, recent numerical studies do not support the appearance of superconductivity in the Hubbard model. Neither the path-integral renormalization-group method [155] nor the dynamical cluster approximation [156] find any evidence for superconductivity at half-filling. Subsequent variational Monte Carlo studies were performed [157, 158] using variational wave functions of the form

$$P_Q P_g |\mathrm{BCS}\rangle \qquad \text{or} \qquad P_Q P_g |\mathrm{AF}\rangle, \tag{22.6}$$

where $|\mathrm{BCS}\rangle$ and $|\mathrm{AF}\rangle$ are respectively mean-field BCS and antiferromagnetic wave functions. The operator $P_g$ is a Gutzwiller factor for the Hubbard model, defined as $P_g = \prod_i \left[1 - (1-g)n_{i\uparrow}n_{i\downarrow}\right]$, with $g$ being a variational parameter, and acts to reduce the probability of doubly occupied sites in the wave function by a factor of $g$. The operator $P_Q$ represents an intersite correlation factor [159, 160] which controls the binding strength between a doubly occupied site (doublon) and an unoccupied site (holon). These two projection operators represent the effects of electron correlation. The variational Monte Carlo result shows that antiferromagnetism is stabilized in a wide region of parameter space, and that superconductivity is not stabilized [158].

The stability of antiferromagnetism obtained in the variational Monte Carlo study is interesting, because the effective hopping, $t'_{\mathrm{eff}}$, in the optimized variational state becomes zero. This means that the wave function itself is acting to deform the Fermi surface so as to recover a strong nesting condition, even if the Hamiltonian contains a bare hopping term $t'$. This is similar to the "spontaneous Fermi surface deformation" studied in the $t$-$t'$-$J$ model in connection with high-$T_c$ cupric oxide superconductors [161, 162], which provides an unconventional means of reducing the frustration to stabilize an unfrustrated magnetic order.

In the search for the spin-liquid state, the Hubbard model on an anisotropic triangular lattice with intermediate interaction strengths ($U/t$) has been studied by a path-integral renormalization-group technique [155, 163] (Yoshioka et al., unpublished). It has been claimed that, as $U$ increases, the paramagnetic metallic state undergoes a first-order phase transition to a nonmagnetic insulating state, followed by another first-order phase transition to a magnetically ordered state. Although the intermediate state seems to be nonmagnetic and insulating, its nature is not yet clear. It remains to be seen whether such a state can be reproduced by other methods.

### 22.3.3 Extended Hubbard Model for Organic Conductors

For theoretical studies of organic conductors, Kino and Fukuyama [68] considered Hubbard-type models taking account of the full anisotropy in $t_{ij}$ for each polytype, and applied mean-field approximations to the on-site Coulomb interaction, $U$. In this treatment, each molecule is represented by a "site" and only the HOMO and/or LUMO closest to the Fermi level are considered. These authors provided a way to handle the variety of organic materials regardless of the apparent complexity of the crystal structures, and showed that hopping integrals between the HOMO and/or LUMO lead to a single-band model. Now it has become possible to relate various polytypes to each other and to discuss their electronic properties in a systematic manner [90, 164–166].

When one studies charge ordering at quarter-filling in the materials discussed in Sect. 22.2.3, it is important to include the longer-range Coulomb interaction terms. The effective on-site repulsive interaction is actually relatively small compared with inter-site interactions because screening effects are weak in these systems due to the extended nature of the molecular orbitals. In this case, many theoretical studies have been performed using the two-dimensional extended Hubbard model on an anisotropic triangular lattice including the full anisotropy of the materials,

$$\mathcal{H}_{\mathrm{EHM}} = -\sum_{(i,j)\sigma} \left( t_{ij} c_{i\sigma}^{\dagger} c_{j\sigma} + \mathrm{h.c.} \right) + \sum_{i} U n_{i\uparrow} n_{i\downarrow} + \sum_{(i,j)} V_{ij} n_i n_j . \tag{22.7}$$

Here $(i, j)$ denotes a pair of lattice sites, by which is meant molecules at $i$ and $j$ site, and the transfer integrals, $t_{ij}$, reflect the anisotropy resulting from the particular spatial extent of the molecular orbitals, calculated, for example, by the extended Hückel method [167, 168] or from a tight-binding fit to first-principles calculations. Actual values of the transfer integrals calculated by the extended Hückel method are given in [69–71] and [164]. Whether or not the triangular geometry leads to frustration depends on each value of $t_{ij}$ [152, 169].

In addition to the kinetic-energy terms, the inter-site Coulomb interactions, $V_{ij}$, must also be considered [170], which is crucial for properties such as charge ordering [90, 92]. The actual values for $V_{ij}/U$, estimated as 0.2–0.7, remain rather ambiguous, while $U$ is believed to be of the order of 1 eV in the BEDT-TTF compounds. Because typical values for $|t_{ij}|$ are in the approximate range 0.2–0.25 eV [171–173], these materials are strongly correlated electron systems with $U/|t_{ij}| \sim$ 4–5. We note that the degrees of anisotropy in $t_{ij}$ and $V_{ij}$ are not necessarily similar, because the parameters $t_{ij}$ are determined from the overlap integrals of the HOMO or LUMO wave functions, while $V_{ij}$ is determined largely from the distance between the molecules [173].

As discussed in Sect. 22.2.3 for the case of $\kappa$-type structures, each pair of molecules is connected by a transfer integral that is considerably larger than any others in the system, and thus can be regarded as a unit, in terms of which the effective models for such systems become half-filled Hubbard models. In this case, the on-site repulsive interaction, $U$, is determined from the strength of dimerization [164, 169, 170].

We focus here on the superconductivity which may be induced by charge fluctuations in organic conductors. Superconductivity mediated by charge fluctuations was first discussed in the extended Hubbard model on a cubic lattice [174], and the relevance of this mechanism to the organic conductors was pointed out by Merino and McKenzie [175, 176]. Theoretical studies show that the charge fluctuation associated with the nearest-neighbor Coulomb repulsion, $V$, can induce superconductivity with $d_{xy}$-wave symmetry in the square lattice [175, 177, 178]. Although the Fermi surface nesting is poor at quarter-filling, the momentum-dependence of the pairing interaction gives an attraction at wave vector $(\pi, \pi)$, leading to $d_{xy}$-wave superconductivity. This momentum-dependence is not determined by the nesting condition of the noninteracting Fermi surface, but originates from the Fourier transform of the nearest-neighbor repulsive interaction. However, when $V$ becomes larger, a checkerboard charge order appears, and as a result superconductivity is restricted to only a small parameter region in the vicinity of charge ordering. In the real-space picture, the stability of this $d_{xy}$-wave pairing can be understood simply [177]: the nearest-neighbor Coulomb repulsion repels electrons from the nearest-neighbor sites, making the amplitude of the order parameter in real space larger at the four next-nearest-neighbor sites and leading to $d_{xy}$-wave pairing. In this sense, it is reasonable that the next-nearest-neighbor superexchange interaction enhances the $d_{xy}$-wave pairing [179].

The studies mentioned above were performed on the square lattice, and the checkerboard charge-ordering pattern does not have any frustration. However, frustration in the anisotropic triangular lattice realized in some organic conductors can lead to new possibilities. The search for superconductivity close to a charge-ordering transition has been pursued by random-phase-approximation (RPA) [180], and fluctuation-exchange (FLEX) [178] calculations, and by variational Monte Carlo simulations [181]. In the anisotropic triangular lattice at quarter-filling, the nearest-neighbor repulsive interactions have strong frustration, as a result of which several charge-ordering patterns, including horizontal stripes [101], diagonal stripes, and a period-three charge order [182, 183], all compete with each other. In such a case, no one of the possible charge-ordering patterns is particularly stable, and there appears a large parameter regime in which a metallic state is realized as a consequence of the frustration. This has been called the "quantum melting" of charge order [184, 185]. It was found that spin-triplet superconductivity with $f$-wave-like symmetry is stabilized near the charge-ordering instability within the RPA [180]. This kind of superconductivity is stable, because the momentum-dependence of the pairing interaction gives an attractive force at wave vectors such as $\mathbf{Q} = (\frac{2}{3}\pi, \frac{2}{3}\pi)$. Similar to the case of the square lattice, the momentum $\mathbf{Q}$ originates from the Fourier transform of the nearest-neighbor repulsive interactions. It is worth noting that the charge fluctuation contributes equally to both the singlet and the triplet pairing channels, because it is a consequence of the charge degrees of freedom.

A recent variational Monte Carlo calculation revealed that an $f$-wave superconducting state can be stabilized in the metallic state realized due to frustration [181].

The stability of $f$-wave pairing in the anisotropic triangular lattice can also be understood from the simple real-space picture [181, 186]: because the repulsive nearest-neighbor Coulomb interaction expels electrons from the six nearest-neighbor sites, the amplitude of the Cooper pairs in real space will be larger on the six next-nearest-neighbor sites in the triangular lattice, which leads to the $f$-wave pairing.

## 22.4 Superconducting Compounds

### 22.4.1 Pyrochlore Lattice: $Cd_2Re_2O_7$ and $AOs_2O_6$

In the course of this decade, a family of pyrochlore oxide superconductors has been found and characterized. The first system to be discovered was the $\alpha$-pyrochlore $Cd_2Re_2O_7$, in which $T_c = 1.0$ K [187–189], and the second was the $\beta$-pyrochlore $AOs_2O_6$, where $T_c = 3.3, 6.3$, and 9.6 K respectively for $A =$ Cs [190], Rb [191–193], and K [194]. Figure 22.7 shows these superconducting transitions as observed in the resistivity. Perhaps surprisingly, $T_c$ varies by one order of magnitude across the series, in spite of the general similarity of the crystals and their electronic band structures, and in particular the large enhancement of $T_c$ in the $\beta$-pyrochlores from Cs to K suggests that this superconductivity is not of the conventional BCS type.

The materials all crystallize in the cubic pyrochlore structure with the same space group, $Fd$-$3m$, and possess a common 3D skeleton composed of $ReO_6$ or $OsO_6$ octahedra, as illustrated in Fig. 22.8: the Re or Os sublattice forms the pyrochlore

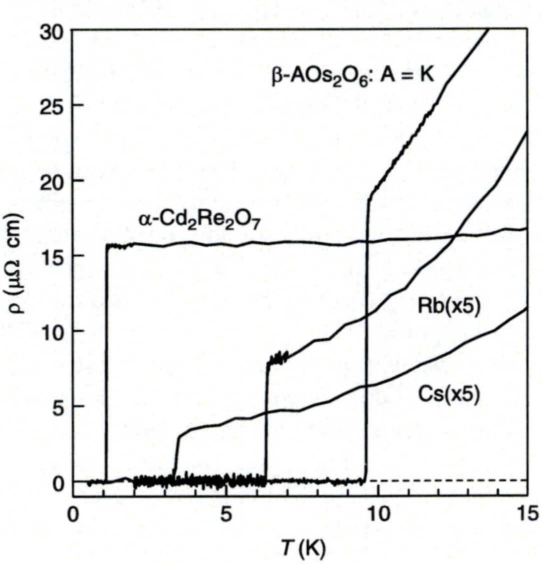


**Fig. 22.7** Resistivity measured on high-quality single crystals of the pyrochlore oxide superconductors, showing a sharp superconducting transition. The data for $RbOs_2O_6$ and $CsOs_2O_6$ are multiplied by a factor of 5 for clarity

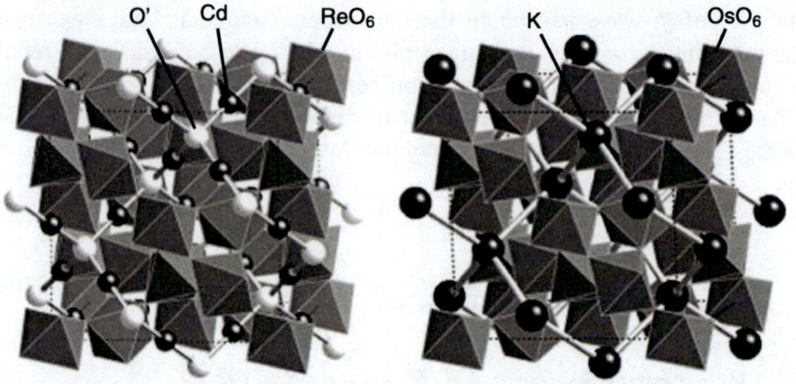


**Fig. 22.8** Comparison of crystal structures between the $\alpha$-pyrochlore $Cd_2Re_2O_6O'$ (*left*) and the $\beta$-pyrochlore $AOs_2O_6$ (right). The two structures possess an identical skeleton of $ReO_6$ or $OsO_6$, and differ in that the O' atom in the former is replaced by the $A$ atom in the latter with the nearby Cd sites vacant

lattice of interest. The difference between the two types of system comes from the fact that the O' atom at the 8$b$ site in the $\alpha$-pyrochlore $Cd_2Re_2O_6O'$ is replaced by the $A$ atom in the $\beta$-pyrochlore $AOs_2O_6$ [194–196]. The Cd 16$d$ site in the former is also vacant in the latter. Alternatively stated, a relatively large $CdO_4$ tetrahedral unit is replaced only by a single $A$ atom, which as a result suffers a large size mismatch and can "rattle" around in an oversized atomic cage. It was pointed out that this mismatch causes a peculiar, anharmonic vibration called a rattling mode, particularly for the rather small K atom [196–200].

The electronic structures of $\alpha$-$Cd_2Re_2O_7$ and $\beta$-$AOs_2O_6$ have been calculated by first-principles, density-functional methods, which reveal that metallic conduction occurs in the (Re, Os)-O network [201,202]; the electronic states near the Fermi level originate from the TM 5$d$ and O 2$p$ orbitals. Although the overall shape of the density of states (DOS) is similar for the two compounds, the difference in band filling may result in different properties; $Re^{5+}$ in $\alpha$-$Cd_2Re_2O_7$ has two 5$d$ electrons, while $Os^{5.5+}$ in $\beta$-$AOs_2O_6$ has 2.5. It is to be noted, however, that the calculated DOS at the Fermi level for all the compounds is nearly equal, and thus one must find an alternative explanation for the change in $T_c$ across the series of compounds.

The mechanism of superconductivity for the pyrochlore oxides has been studied extensively. Most of the data obtained for $Cd_2Re_2O_7$ have indicated that it is a weak-coupling, BCS-type superconductor [203, 204]. In contrast, there are several findings which suggest unconventional features for $AOs_2O_6$, including the anomalous, concave-downward curvature of the resistivity, which suggests an unusual scattering process involved in the normal state. Further, NMR experiments by Arai et al. show a very small coherence peak in the relaxation rate below $T_c$ for $RbOs_2O_6$, while no peaks were found for $KOs_2O_6$ [205]. This finding is in strong contrast to the result for $Cd_2Re_2O_7$, in which a very large coherence peak was observed [204].

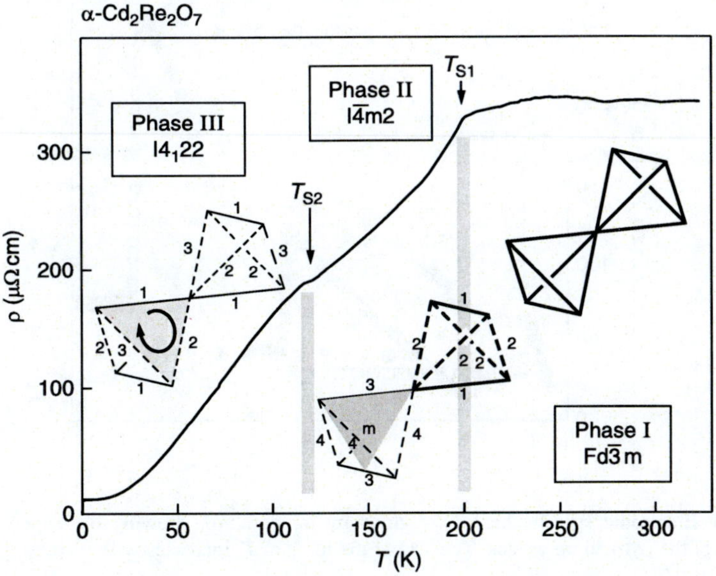


**Fig. 22.9** Temperature-dependence of the resistivity for $\alpha$-$Cd_2Re_2O_7$, showing two anomalies at $T_{s1} = 200$ K and $T_{s2} = 120$ K. These are accompanied by structural transitions between cubic and two different tetragonal phases. A pair of tetrahedra drawn for each phase shows how the pyrochlore lattice is deformed for a given space group

Although the superconductivity in $Cd_2Re_2O_7$ is conventional, two structural phase transitions were found, which represent one manifestation of the electronic instability of itinerant electrons on the pyrochlore lattice. Two symmetry-lowering structural transitions are observed at $T_{s1} = 200$ K and $T_{s2} = 120$ K, as shown in Fig. 22.9 [206–208]. The upper transition is of second order, from the ideal cubic pyrochlore structure (space group $Fd$-$3m$) to a tetragonally distorted structure ($I$-$4m2$), while the lower is of first order to another tetragonal space group, $I4_122$. It is of interest to consider how the tetrahedral lattice of Re ions can change through the successive transitions: in Phase I at high temperature, all of the tetrahedra are identical, while below $T_{s1}$ two neighboring tetrahedra become inequivalent with loss of inversion symmetry. There is also a tetragonal distortion, but it is very small. Below $T_{s2}$, the two tetrahedra become identical again, but now each has three different types of bond (Fig. 22.9). The origin of these successive transitions remains unclear at present, and is being studied in detail [209, 210].

Regarding the symmetry of the superconducting gap for the $\beta$-pyrochlores, several experiments provide evidence for $s$-wave superconductivity, which would suggest that the fundamental pairing mechanism is of phononic origin. An important question is then what kind of phonons are responsible for the occurrence of superconductivity, and an understanding of the reasons for the apparently singular behavior on approaching K in the series may well be the key to the physics involved in

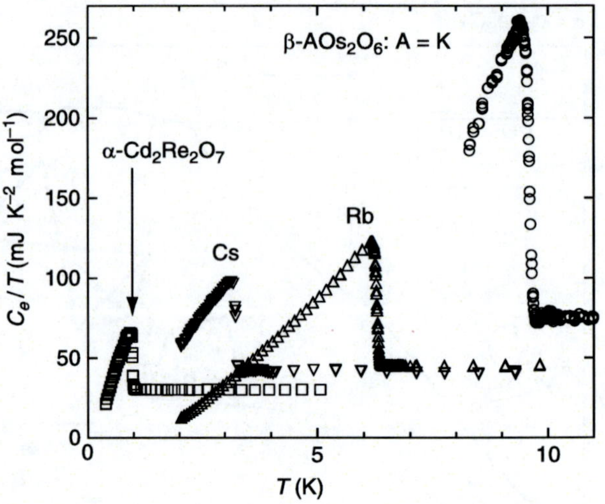


**Fig. 22.10** Electronic specific heat $C_e$ divided by temperature, showing the superconducting transitions of the pyrochlore oxides. The size of the jump at $T_c$ increases with increasing $T_c$

this system. One approach is to consider the chemical trends of various parameters in the series, which should provide insight into the pairing mechanism [211,212].

Figure 22.10 shows the temperature-dependence of the electronic specific heat for the four compounds. The magnitude of the jump at $T_c$ is a measure for the coupling strength of the Cooper pairs, the value of $\Delta C(T_c)/\gamma T_c$ being 1.43 for the weak-coupling limit in BCS theory. Here, $\Delta C(T_c)/\gamma T_c$ for $Cd_2Re_2O_7$, Cs, Rb, and $KOs_2O_6$ increases through the respective values 1.15, 1.42, 1.84, and 2.87, as shown in Fig. 22.11a. The size of the superconducting gap can be estimated from the data by taking into account the strong-coupling correction, and the result, shown in Fig. 22.11b as a function of $T_c$, demonstrates clearly that the type of superconductivity changes from weak coupling to extremely strong coupling with increasing $T_c$.

Also shown in Fig. 22.11 is the electronic specific-heat coefficient $\gamma$ in the normal state obtained from Fig. 22.10. Compared with the value deduced from band-structure calculations there is a strong enhancement, particularly for $KOs_2O_6$. The reason for this enhancement must lie not in electronic correlations but in the electron–phonon interactions, because there is no corresponding enhancement in magnetic susceptibility.

The average energy of the phonons responsible for the superconductivity, $\omega_{\ln}$, has been estimated based on a strong-coupling analysis: $\omega_{\ln}$ deceases gradually toward $KOs_2O_6$, as shown in Fig. 22.11d. This suggests strongly that a low-energy phonon is the origin of the strong-coupling superconductivity, as well as of the large mass enhancement, and that the systematic increase in the coupling strength must be ascribed to a reduction in the energy of this phonon. It is known that many strong-coupling superconductors with large coupling constants possess different low-energy phonons. As examples, the A-15 compounds show soft phonon modes

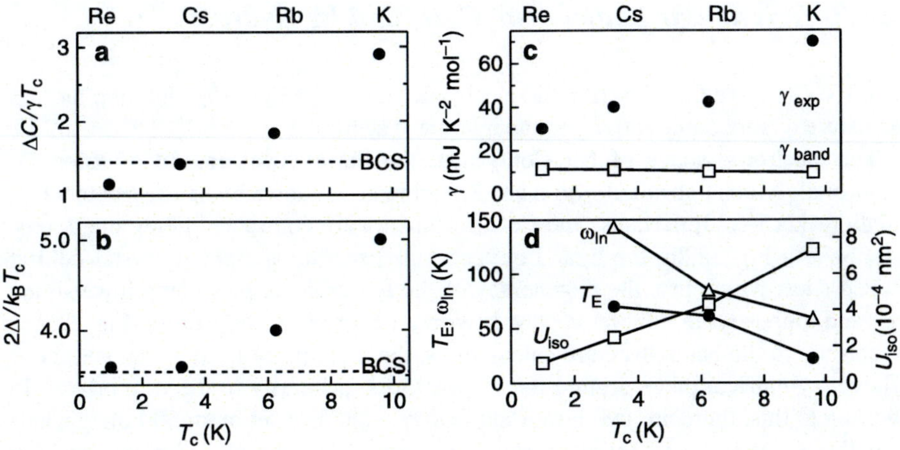


**Fig. 22.11** Chemical trends of various parameters as a function of $T_c$: the specific-heat jump at $T_c$ (**a**), the magnitude of the superconducting gap (**b**) and the Sommerfeld coefficient $\gamma$ (**c**). Shown in panel (**d**) are the atomic displacement parameters $U_{iso}$ of Cd or $A$ atoms, the Einstein temperature $T_E$ and the average phonon frequency $\omega_{ln}$ responsible for the superconductivity

associated with a cubic-to-tetragonal transition [213], while the Chevrel-phase compounds are assumed to be "molecular crystals" with low-lying phonon modes [214]. The role of such low-energy phonons in the mechanism of superconductivity has been studied extensively in these strong-coupling superconductors.

It is reasonable in the case of $\beta$-$AOs_2O_6$ to ascribe such a low-energy phonon to the rattling vibration. The energy of the rattling mode may decrease on decreasing the size of the $A$ cations, because the open space available in the (rather rigid) cage increases. In fact the isotropic atomic displacement parameter $U_{iso}$ obtained from structural analysis increases toward K, as shown in Fig. 22.11d, and it is noteworthy that the value of $U_{iso}$ for K is exceptionally large for a cation in a crystal. Moreover, a characteristic rattling energy can be estimated from the specific-heat data over a wide range of temperature by assuming that it as an Einstein mode, and the Einstein temperature, $T_E$, obtained from this procedure decreases towards K (Fig. 22.11d). It is therefore clear that the rattling mode must play a crucial role in the mechanism of superconductivity, and particularly for the strong-coupling superconductivity in $KOs_2O_6$.

It should be emphasized that in most cases the lattice itself becomes unstable when the electron-phonon coupling becomes too strong. In the $\beta$-pyrochlore oxide compounds, however, the conduction electrons are located on the hard skeleton of $OsO_6$ octahedra, while the low-energy phonons are associated with the rattling of ions that are structurally only weakly bound to the skeleton. This structural duality can give rise to such unusually strongly coupled superconductivity. It would be intriguing to investigate in further detail how this rattling, which is an essentially anharmonic vibration almost localized in a cage, can mediate the Cooper pairing. More experimental information, particularly from neutron diffraction experiments that can determine the low-energy phonon DOS, is awaited for further discussion.

### 22.4.2 *Triangular Lattice: $Na_xCoO_2$ and Its Hydrate*

In this section, we first describe the electronic states of $Na_xCoO_2$ and then discuss the nature of superconductivity in the hydrate systems.

The electronic states of $Na_xCoO_2$ near the Fermi level consist of three Co $t_{2g}$-orbitals which hybridize with the O $2p$-orbitals, as shown by band-structure calculations [215–220]. Because the $CoO_6$ octahedra are contracted along the $c$-axis, as shown in Fig. 22.2b, the ligand oxygen ions generate a trigonal crystal field at the Co sites, which lifts the degeneracy of the $t_{2g}$-orbitals, producing lower-lying $a_{1g}$ and higher-lying doublet $e'_g$ levels with an energy splitting of $\Delta$ (Fig. 22.2c). According to the band-structure calculations, the main part of the Fermi surface is a large, cylindrical piece around the $\Gamma$ point, due primarily to the $a_{1g}$ orbital. In addition to this, there are two important features: one is a set of small hole pockets near the $K$ points, originating from the $e'_g$ orbitals, to which we refer hereafter as $e'_g$ hole pockets. These are expected to appear at low fillings of order $x \sim 0.3$. (Schematic representations of the energy dispersion and the Fermi surface are shown in Fig. 22.12 (left).) The other feature is a small, concentric electron pocket around

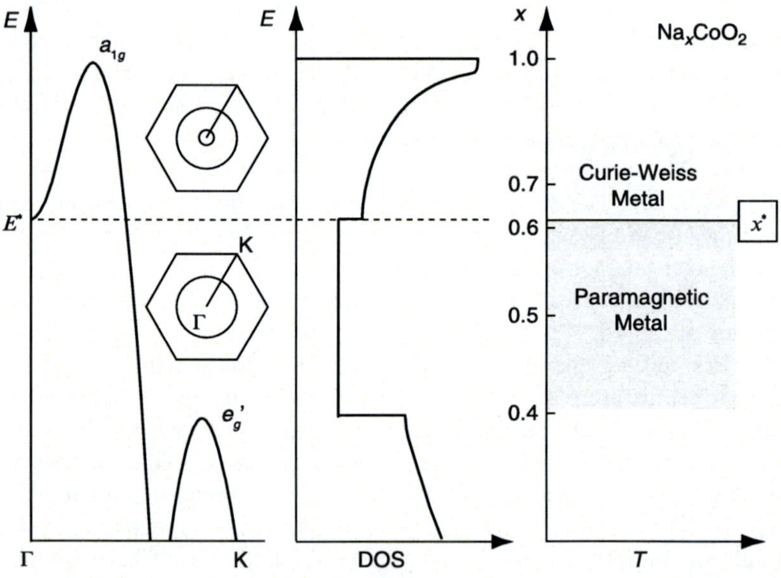


**Fig. 22.12** Schematic representation of the band structure of $Na_xCoO_2$. Depicted are band dispersions along the $\Gamma$-K line (*left*), an expected density of states (DOS) profile (*middle*), and an $x$-$T$ phase diagram (*right*). The critical Na content $x^*$ in the phase diagram corresponds to a band filling with the Fermi energy equal to $E^*$, as shown by the broken line, where the Fermi level touches the base of the dip in the $a_{1g}$ band at the $\Gamma$ point. In the left panel, two types of Fermi surface are also shown: when $E > E^*$, a small $a_{1g}$ electron pocket appears around $\Gamma$ in addition to a large hole pocket. In the phase diagram, a Curie–Weiss metal exists above $x^*$ and a Pauli paramagnetic metal below $x^*$. (Replotted from [227])

the $\Gamma$ point, whose origin is a dip in the dispersion of the $a_{1g}$ band [217, 220], as shown in Fig. 22.12 (left), and which we call the $a_{1g}$ electron pocket. This may appear at high fillings, of order $x \sim 0.7$. There has also been some discussion concerning the possibility that the band dispersions and the Fermi-surface topology are very sensitive to the thickness of the $CoO_2$ layer, which determines the trigonal crystal field [221, 222].

Angle-resolved photoemission spectroscopy (ARPES) experiments observed the large, cylindrical Fermi surface around $\Gamma$ but failed to detect either the $e'_g$ hole pockets or the small $a_{1g}$ electron pocket [223–226]. However, a recent experiment which effected very precise control over $x$ showed that dramatic changes in the electronic properties take place at $x^* \sim 0.62$ [227]. In particular, the specific heat increases rapidly for $x > x^*$, which suggests that the Fermi level touches the bottom of the $a_{1g}$ band at $\Gamma$ when $x = x^*$. A schematic illustration is shown in Fig. 22.12.

We turn next to a discussion of superconductivity in hydrated samples, which occurs when water molecules are intercalated in $Na_xCoO_2$. The superconducting properties and the symmetry of the order parameter have been studied by many experimental techniques [45]. Evidence for unconventional superconductivity has been reported in the forms of the absence of coherence peak in NMR/NQR $1/T_1T$ [228–230], the power-law behavior of $1/T_1T$ [228–231], and the specific heat [232–235], which all support anisotropic pairing. Further, a magnetic phase exists in the vicinity of the superconducting phase, and an enhancement in spin fluctuations is observed when the system approaches the magnetic phase [236–238]. These results suggest the importance of electron correlation effects, which in general lead to non-$s$-wave superconductivity. The Knight shift below $T_c$ is also observed to decrease when a magnetic field is applied in the plane [239, 240], suggesting either a singlet pairing or a triplet pairing whose $d$-vector is fixed in the plane. Recent measurements of the Knight shift in a magnetic field parallel to the $c$-axis show that it also decreases below $T_c$ [241, 242], indicating singlet pairing. However, another group has claimed a constant Knight shift in a similar experimental set-up [243].

Water intercalation is expected to have several effects: (1) enhancement of two-dimensionality, (2) smoothing the random Na potential by screening [244], and (3) enhancement of the trigonal distortion of $CoO_6$ octahedra. In fact the increase of the $c$-axis lattice constant accompanying water-intercalation pushes the $Na^+$ ions away from the $CoO_2$ layers, which makes the effective thickness of the $CoO_2$ layers smaller. As a result, the level of the $e'_g$ band becomes higher in hydrate samples than in non-hydrate ones. This change has been observed in ARPES measurements for hydrated $Na_xCoO_2$ [226], but the $e'_g$ level remains below the Fermi energy, which is not consistent with band-structure calculations [215].

In order to understand the superconductivity, there are two important parameters in this system: one is the valence of the Co ions, or the electron density, and the other is the thickness of the $CoO_2$ layers, or the energy-level splitting, $\Delta$, associated with the trigonal distortion. Regarding the valence of the Co ions, it is necessary to include the effects of the oxonium ions, $H_3O^+$ [236, 245]. Recent chemical analysis shows that $H_3O^+$ ions are intercalated simultaneously during hydration, and the compound should be actually be represented as $Na_x(H_3O)_zCoO_{2-y}H_2O$, where the

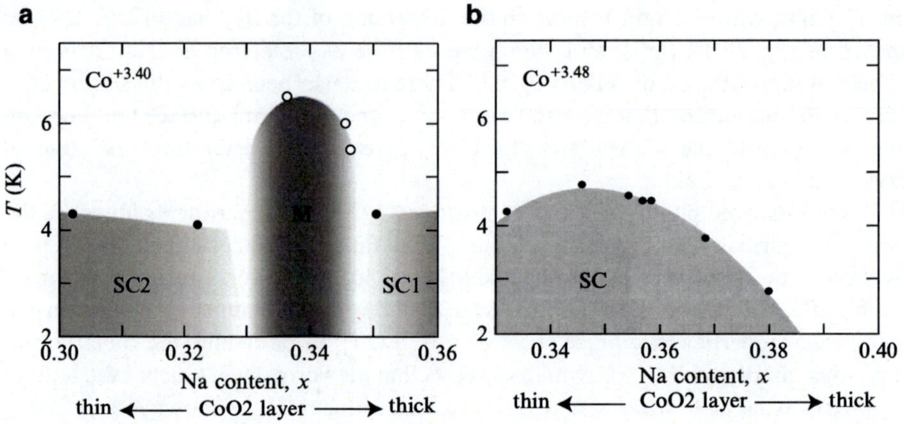


**Fig. 22.13** Experimental phase diagram of the bilayer-hydrate system in the plane of temperature and Na content $x$, as reported by Sakurai et al. [236] (Sakurai et al., unpublished). (**a**) Phase diagram for samples with approximate Co valence +3.40, which contains two superconducting phases separated by a magnetically ordered phase. (**b**) Phase diagram for samples with approximate Co valence +3.48, which contains a single superconducting phase with a dome-shaped dependence of $T_c$

Co valence is $Co^{(4-x-z)+}$. Titration analysis reveals that superconductivity appears at Co valences of +3.48 and +3.40, irrespective of $x$. These valences are different from the value +3.65, which would be the simplest expectation based on $x \sim 0.35$. However, it should be noted that one NMR result indicates that the Co valence is not modified even after hydration [246], and more careful analysis of this issue will therefore be necessary.

Based on the analysis of Co valences with oxonium ions, Sakurai et al. [236] (Sakurai et al., unpublished) proposed the phase diagrams of the type shown in Fig. 22.13; similar phase diagrams have also been proposed by other authors [231, 237, 238, 247]. When the Co valence is around +3.40, a superconducting phase (SC2), a magnetically ordered phase (MO), and then another superconducting phase (SC1) appear as $x$ is increased. By contrast, when the Co valence is close to +3.48, only one superconducting phase appears. Because the Co valence is fixed, the increase of $x$ means a decrease of $z$, and hence in the content of $H_3O^+$. This change leads to a smaller $c$-axis lattice constant, because the $H_3O^+$ ion is smaller than the $Na^+$ ion, which leads in turn to a thicker $CoO_2$ layer, meaning a smaller trigonal distortion and as a result that $\Delta$ becomes smaller [222]. This tendency has been observed in the NQR frequency, $\nu_Q$, which decreases as the trigonal distortion becomes smaller. Classifying samples on the basis of $\nu_Q$ would therefore be useful.

Before closing this section, let us discuss the implications of these results for theory. ARPES experiments show that there is only a single, cylindrical Fermi surface around the $\Gamma$ point, meaning that a single-band model should be sufficient for describing the hydrate systems. In this case, the RVB superconductivity discussed in Sect. 22.3 is one of the most probable candidates [132–135]. High-temperature series expansion studies on the $t$-$J$ model show that $d$-wave RVB superconductivity is favorable up to $x \sim 0.5$ (50% electron doping) [138, 139], which is consistent with the doping level observed in experiments. However, as discussed above and noticed in the mean-field theory, the most probable symmetry of the superconducting order parameter below $T_c$ on the triangular lattice is $(d_{x^2-y^2} + id_{xy})$-wave. If this were the case, however, time-reversal symmetry would be broken in the superconducting state, contradicting the fact that $\mu$SR experiments do not find such a symmetry-breaking [248]. Here it should be noted that pure $d$-wave superconductivity is realized in some regions of parameter space if the triangular symmetry is broken [135], but it is not clear whether this result applies in the present case. In the weak-coupling Hubbard model, on the other hand, the superconducting instability is very weak, and it would be difficult to obtain superconductivity at temperatures as high as $T_c = 4.5\,\text{K}$. As an alternative to electronic correlations, it has been proposed that a phonon mechanism can explain superconductivity with an $s$-wave order parameter [249], which would be consistent with the impurity effects [250].

Alternatively, on the assumption that the Fermi surface is not that of a single-band model, several possible mechanisms for superconductivity have been discussed. If there are $e_g'$ hole pockets near the K points, spin-triplet superconductivity due to nearly ferromagnetic spin fluctuation would be realized [251–253], for which the disconnected nature of the Fermi surface plays an important role. Another possibility, if there is a small concentric $a_{1g}$ electron pocket around $\Gamma$, as shown in Fig. 22.12, is to stabilize spin-singlet extended $s$-wave superconductivity [254].

The changes of band dispersion and Fermi surface topology have been analyzed microscopically within a multiband tight-binding model [222]. In this calculation, it is shown that several types of Fermi surface appear successively as the $CoO_6$ distortion increases, and that the phase diagrams shown in Fig. 22.13 can be explained successfully. It is interesting to note that the two superconducting phases in Fig. 22.13 have different types of pairing, SC1 being of extended $s$-wave type, due to the presence of the small $a_{1g}$ electron pocket, while SC2 is a triplet superconducting state in the presence of the small $e_g'$ hole pockets caused by the large trigonal crystal-field splitting, $\Delta$ [222, 255]. The puzzling NMR/NQR and $\mu$SR results on the character of the magnetic fluctuations can be understood by considering the strong dependence of these fluctuations on the layer thickness [255]. It has been proposed that the discrepancies in $H_{c2}$, in the specific heat, and in the superfluid density can also be resolved by considering the presence of two types of superconductivity [256]. However, these theoretical calculations are based on the presence of additional structures in the Fermi surface, which remains inconsistent with the existing ARPES data, where only a single Fermi surface is observed. Thus there is at present no reasonable theory capable of explaining all the experimental results consistently.

### 22.4.3 Anisotropic Triangular Lattice: Organic Superconductivity

The quasi-two-dimensional (BEDT-TTF)$_2X$ salts exhibit a variety of superconducting states. Because the presence of charge ordered phase next to the superconducting phase suggests a crucial role of charge fluctuations, which is not the case for high-$T_c$ cupric oxide superconductors, the issue of the mechanism for superconductivity in the organic conductors has been discussed with particular emphasis on the role of spin and charge fluctuations [90, 257, 258].

In $\kappa$-(BEDT-TTF)$_2X$, the absence of a coherence peak in the NMR relaxation rate, $1/T_1$, and its $T^3$-dependence below $T_c$ are observed, suggesting $d_{x^2-y^2}$-wave-like superconductivity [257]. Recent penetration-depth measurements [259] and specific-heat measurements [260] also support $d_{x^2-y^2}$-wave superconductivity. Because the superconducting phase is located next to SDW or antiferromagnetic phases, it has been argued that superconductivity is induced by spin fluctuations, in close analogy to many models for high-$T_c$ cupric oxide superconductors. In fact the NMR relaxation rate in deuterated $\kappa$-(BEDT-TTF)$_2$Cu[N(CN)$_2$]Br, which lies just on the border of the Mott transition [261], shows an increase on heating above $T_c$, which is similar to the pseudo-gap phenomena observed in underdoped high-$T_c$ superconductors [126, 262]. Such a Nernst effect above $T_c$, as reported in high-$T_c$ systems, is also observed in $\kappa$-(BEDT-TTF)$_2$Cu[N(CN)$_2$]Br [263]. However, as discussed in Sect. 22.3.2, it remains an open question whether or not the simple Hubbard model at half-filling contains superconductivity.

In $\kappa$-(BEDT-TTF)$_2$Cu$_2$(CN)$_3$, where the hallmarks of a spin-liquid phase are observed, superconductivity appears under pressure ($T_c$ = 4 K at 360 MPa) [79]. NMR experiments suggest the presence of line nodes in the superconducting order parameter, although no details are yet known. This superconductivity could also be discussed in the half-filled Hubbard model, but it should be stressed here that theoretical clarification of the nature of the spin-liquid state is necessary before discussing the nature of the superconductivity.

Doping into organic conductors is possible in the exceptional case of $\kappa$-(BEDT-TTF)$_4$Hg$_{3-\delta}X_8$ ($X$ = Br or Cl) [264]. The materials with Br and Cl are believed to have dopings of approximately 11% and 22%, respectively. The properties of the superconducting state observed under pressure have not yet been explored. One key issue is to compare the two superconducting states appearing when pressure is applied to the spin-liquid phase and when some holes are doped.

There are several two-dimensional $A_2B$ salts showing superconductivity in the vicinity of charge-ordered phases, such as $\theta$-(BEDT-TTF)$_2$I$_3$ [109], $\beta''$-(DODHT)$_2$PF$_6$ [117], $\theta$-(DIETS)$_2$Au(CN)$_4$ [265], and $\alpha$-(BEDT-TTF)$_2$I$_3$ [103]. In the last of these, superconductivity appears within the charge-ordered phase. RPA calculations have been performed on the basis of the reconstructed Fermi surface induced by a stripe-type charge order obtained in a mean-field approximation [266], and the results indicate that superconductivity appears in a metallic charge-ordered phase in which both small hole pockets and electron pockets exist. The pairing interaction in this case is given primarily by spin fluctuations originating from a self-doped one-dimensional spin-1/2 Heisenberg chain [266]. This implies that charge fluctuations are suppressed due to the stable charge order, and the pairing instability arises from the newly formed Fermi surface.

As discussed in Sect. 22.3.3, the charge fluctuation at quarter-filling in the extended Hubbard model can lead to $d_{xy}$-wave or $f$-wave superconductivity. The nature of the superconductivity observed under pressure in various organic conductors awaits investigation.

## 22.5 Summary

We have discussed the nature of itinerant electrons on frustrated lattices from both a materials and a theoretical point of view. In the process we have reviewed metallic transition-metal oxides and organic compounds that show interesting phenomena such as superconductivity, a metal-insulator transition, anomalous magnetoresistance, and heavy-fermion-like behavior. Of particular interest in this context are the superconductors $\alpha$-$Cd_2Re_2O_7$ and $\beta$-$AOs_2O_6$, which have a pyrochlore structure, and $Na_xCoO_2{\cdot}yH_2O$ and $(BEDT\text{-}TTF)_2X$, which have triangular lattices, where some effects of frustration on the mechanisms of superconductivity may be evident.

Theoretically, RVB superconductivity is the most interesting and promising mechanism relating the spin-liquid state of frustrated systems with unconventional superconductivity. However, there remains as yet no direct experimental evidence for such RVB-type superconductivity. In many real materials, such as $Na_xCoO_2{\cdot}yH_2O$, it seems necessary to take into account the additional effects of orbital degrees of freedom. In contrast, the organic conductors, which are considered to be single-band systems, may be good candidates for a simple RVB mechanism. Further research into both materials and theories for frustrated systems is necessary.

**Acknowledgements**

We are grateful to Junichi Yamaura, Yoshihiko Okamoto, Kazushi Kanoda, Hitoshi Seo, Masahito Mochizuki, Yoichi Yanase, Yuta Hayashi, and Emi Yukawa for various helpful discussion and comments. We also thank Bruce Normand for having corrected the manuscript.

## References

1. B. Canals, C. Lacroix, Phys. Rev. Lett. **80**, 2933 (1998)
2. M. Shiga, K. Fujisawa, H. Wada, J. Phys. Soc. Jpn. **62**, 1329 (1993)
3. H. Nakamura, K. Yoshimoto, M. Shiga, M. Nishi, K. Kakurai, J. Phys. Condens. Matter **9**, 4701 (1997)
4. M. Imada, A. Fujimori, Y. Tokura, Rev. Mod. Phys. **70**, 1039 (1998)
5. H. Fukuyama, J. Phys. Soc. Jpn. **75**, 051001 (2006)

6. For a review see P. Batail, Chem. Rev. 104, 4887 (2004), and the review articles in Chem. Rev. 104 No. 11 (2004)
7. E.J. Verwey, Nature **144**, 327 (1939)
8. P.A. Miles, W.B. Westphal, A.V. Hippel, Rev. Mod. Phys. **27**, 279 (1957)
9. H. Seo, M. Ogata, H. Fukuyama, Phys. Rev. B **65**, 085107 (2002)
10. I. Leonov et al., Phys. Rev. Lett. **93**, 146404 (2004)
11. H.-T. Jeng, G.Y. Guo, D.J. Huang, Phys. Rev. Lett. **93**, 156403 (2004)
12. D.C. Johnston, J. Low Temp. Phys. **25**, 145 (1976)
13. S. Kondo, D.C. Johnston, J.D. Jorgensen, Phys. Rev. Lett. **78**, 3729 (1997)
14. K. Matsuno et al., J. Phys. Soc. Jpn. **70**, 1456 (2001)
15. Y. Horibe et al., Phys. Rev. Lett. **96**, 086406 (2006)
16. M. Isobe, Y. Ueda, J. Phys. Soc. Jpn. **71**, 1848 (2002)
17. D.I. Khomskii, T. Mizokawa, Phys. Rev. Lett. **94**, 156402 (2005)
18. Y. Okamoto et al., Phys. Rev. Lett. **101**, 086404 (2008)
19. T. Furubayashi, T. Matsumoto, T. Hagino, S. Nagata, J. Phys. Soc. Jpn. **63**, 3333 (1994)
20. P.G. Radaelli et al., Nature **416**, 155 (2002)
21. M. Shiga et al., J. Phys. Soc. Jpn. **57**, 3141 (1988)
22. C. Urano et al., Phys. Rev. Lett. **85**, 1052 (2000)
23. A. Shimoyamada et al., Phys. Rev. Lett. **96**, 026403 (2006)
24. V.I. Anisimov et al., Phys. Rev. Lett. **83**, 364 (1999)
25. E. Runge, P. Fulde, Phys. Rev. B **70**, 245113 (2004)
26. J. Hopkinson, P. Coleman, Phys. Rev. Lett. **89**, 267201 (2002)
27. Y. Yamashita, K. Ueda, Phys. Rev. B **67**, 195107 (2003)
28. L. Pinsard-Gaudart et al., Phys. Rev. B **76**, 045119 (2007)
29. P.E. Jonsson et al., Phys. Rev. Lett. **99**, 167402 (2007)
30. M.A. Subramanian, G. Aravamudan, G.V.S. Rao, Prog. Solid State Chem. **15**, 55 (1983)
31. Y. Shimakawa, Y. Kubo, T. Manako, Nature **379**, 53 (1996)
32. Y. Taguchi et al., Science **291**, 2573 (2001)
33. H.S. Jarrett et al., in *Superconductivity*, ed. by R.D. Parks (Plenum, New York, 1977), p. 545
34. T. Takeda et al., J. Solid State Chem. **140**, 192 (1998)
35. W. Klein, R.K. Kremer, M. Jansen, J. Mater. Chem. **17**, 1356 (2007)
36. Yamamoto et al., J. Phys. Soc. Jpn. **76**, 043703 (2007)
37. A.W. Sleight, J.L. Gilson, J.F. Weiher, W. Bindloss, Solid State Commun. **14**, 357 (1974)
38. D. Mandrus et al., Phys. Rev. B **63**, 195104 (2001)
39. N. Takeshita et al., J. Phys. Soc. Jpn. **76**, 063707 (2007)
40. D.J. Singh, P. Blaha, K. Schwarz, J.O. Sofo, Phys. Rev. B **65**, 155109 (2002)
41. S. Nakatsuji et al., Phys. Rev. Lett. **96**, 087204 (2006)
42. K. Matsuhira et al., J. Phys. Soc. Jpn. **76**, 043706 (2007)
43. I. Terasaki, Y. Sasago, K. Uchinokura, Phys. Rev. B **56**, R12685 (1997)
44. K. Takada et al., Nature **422**, 53 (2003)
45. For a review see M. Ogata, J. Phys. Condens. Matt. **19**, 145282 (2007)
46. M.L. Foo et al., Phys. Rev. Lett. **92**, 247001 (2004)
47. M. Yokoi et al., J. Phys. Soc. Jpn. **74**, 3046 (2005)
48. D. Yoshizumi et al., J. Phys. Soc. Jpn. **76**, 063705 (2007)
49. G. Lang et al., Phys. Rev. B **78**, 155116 (2008)
50. P. Mendels et al., Phys. Rev. Lett. **94**, 136403 (2005)
51. G. Gaparovi et al., Phys. Rev. Lett. **96**, 046403 (2006)
52. I.R. Mukhamedshin, H. Alloul, G. Collin, N. Blanchard, Phys. Rev. Lett. **93**, 167601 (2004)
53. T. Motohashi et al., Phys. Rev. B **67**, 064406 (2003)
54. S.P. Bayrakci et al., Phys. Rev. B **69**, 100410(R) (2004)
55. J. Sugiyama et al., Phys. Rev. Lett. **92**, 017602 (2004)
56. A.T. Boothroyd et al., Phys. Rev. Lett. **92**, 197201 (2004)
57. S.P. Bayrakci et al., Phys. Rev. Lett. **94**, 157205 (2005)
58. L.M. Helme et al., Phys. Rev. Lett. **94**, 157206 (2005)
59. M. Schreyer, M. Jansen, Angew. Chem. **41**, 643 (2002)

60. H. Yoshida, Y. Muraoka, T. Soegel, M. Jansen, Z. Hiroi, Phys. Rev. B **73**, 020408 (2006)
61. H. Yoshida et al., J. Phys. Soc. Jpn. **77**, 074719 (2008)
62. E. Wawrzynska et al., Phys. Rev. Lett. **99**, 157204 (2007)
63. Y. Uchida, Y. Kanke, E. Takayama-Muromachi, K. Kato, J. Phys. Soc. Jpn. **60**, 2530 (1991)
64. S. Ishiwata, D. Wang, T. Saito, M. Takano, Chem. Mater. **17**, 2789 (2005)
65. H. Mukuda et al., J. Phys. Soc. Jpn. **75**, 094715 (2006)
66. S. Ishiwata et al., Phys. Rev. Lett. **98**, 217201 (2007)
67. Y. Kanke et al., J. Solid State Chem. **112**, 429 (1994)
68. H. Kino, H. Fukuyama, J. Phys. Soc. Jpn. **64**, 4523 (1995)
69. T. Mori, Bull. Chem. Soc. Jpn. **71**, 2509 (1998)
70. T. Mori, H. Mori, S. Tanaka, Bull. Chem. Soc. Jpn. **72**, 179 (1999)
71. T. Mori, Bull. Chem. Soc. Jpn. **72**, 2011 (1999)
72. J. Singleton, C. Mielke, Contemp. Phys. **43**, 63 (2002)
73. Y. Shimizu, K. Miyagawa, K. Kanoda, M. Maesato, G. Saito, Phys. Rev. Lett. **91**, 107001 (2003)
74. H. Urayama et al., Chem. Lett. **17**, 55 (1988)
75. U. Geiser et al., Physica C **174**, 475 (1991)
76. K. Kanoda, Physica C **282-287**, 299 (1997)
77. F. Kagawa, K. Miyagawa, K. Kanoda, Nature **436**, 534 (2005)
78. U. Geiser et al., Inorg. Chem. **30**, 2586 (1991)
79. Y. Kurosaki, Y. Shimizu, K. Miyagawa, K. Kanoda, G. Saito, Phys. Rev. Lett. **95**, 177001 (2005)
80. Y. Shimizu, K. Miyagawa, K. Kanoda, M. Maesato, G. Saito, Phys. Rev. B **73**, 140407(R) (2006)
81. K. Miyagawa, K. Kawamoto, Y. Nakazawa, K. Kanoda, Phys. Rev. Lett. **75**, 1174 (1995)
82. S. Ohira, Y. Shimizu, K. Kanoda, G. Saito, J. Low Temp. Phys. **142**, 153 (2006)
83. S. Yamashita et al., Nature Phys. **4**, 459 (2008)
84. M. Yamashita et al., Nature Phys. **5**, 44 (2008)
85. Y. Shimizu, K. Miyagawa, K. Kanoda, M. Maesato, G. Saito, Prog. Theor. Phys. Suppl. **159**, 52 (2005)
86. Y. Hayashi, M. Ogata, J. Phys. Soc. Jpn. **76**, 053705 (2007)
87. S. Yunoki, S. Sorella, Phys. Rev. B **74**, 014408 (2006)
88. T. Komatsu, N. Matsukawa, T. Inoue, G. Saito, J. Phys. Soc. Jpn. **65**, 1340 (1996)
89. H. Taniguchi et al., J. Phys. Soc. Jpn. **72**, 468 (2003)
90. For a review see H. Seo, J. Merino, H. Yoshioka, M. Ogata, J. Phys. Soc. Jpn. **75**, 051009 (2006)
91. K. Hiraki, K. Kanoda,Phys. Rev. Lett. **80**, 4737 (1998)
92. F. Mila, X. Zotos, Europhys. Lett. **24**, 133 (1993)
93. K. Penc, F. Mila, Phys. Rev. B **49**, 9670 (1994)
94. H. Seo, H. Fukuyama, J. Phys. Soc. Jpn. **66**, 1249 (1997)
95. K. Miyagawa, A. Kawamoto, K. Kanoda, Phys. Rev. B **62**, R7679 (2000)
96. R. Chiba, H.M. Yamamoto, T. Nakamura, T. Takahashi, J. Phys. Chem. Solids **62**, 389 (2001)
97. Y. Takano, K. Hiraki, H.M. Yamamoto, T. Nakamura, T. Takahashi, J. Phys. Chem. Solids **62**, 393 (2001)
98. For a review see N. Tajima, S. Sugawara, M. Tamura, Y. Nishio, K. Kajita, J. Phys. Soc. Jpn. **75**, 051010 (2006)
99. K. Bender et al., Mol. Cryst. Liq. Cryst. **108**, 359 (1984)
100. B. Rothaemel et al., Phys. Rev. B **34**, 704 (1986)
101. H. Seo, J. Phys. Soc. Jpn. **69**, 805 (2000)
102. T. Kakiuchi, Y. Wakabayashi, H. Sawa, T. Takahashi, T. Nakamura, J. Phys. Soc. Jpn. **76**, 113702 (2007)
103. N. Tajima, A. Ebina-Tajima, M. Tamura, Y. Nishio, K. Kajita, J. Phys. Soc. Jpn. **71**, 1832 (2002)
104. K. Kajita et al., J. Phys. Soc. Jpn. **61**, 23 (1992)
105. N. Tajima, M. Tamura, Y. Nishio, K. Kajita, Y. Iye, J. Phys. Soc. Jpn. **69**, 543 (2000)

106. S. Katayama, A. Kobayashi, Y. Suzumura, J. Phys. Soc. Jpn. **75**, 054705 (2006) and J. Phys. Soc. Jpn. **77**, 014710 (2008)
107. A. Kobayashi, S. Katayama, Y. Suzumura, H. Fukuyama, J. Phys. Soc. Jpn. **76**, 034711 (2007)
108. A. Kobayashi, Y. Suzumura, H. Fukuyama, J. Phys. Soc. Jpn. **77**, 064718 (2008)
109. H. Mori, S. Tanaka, T. Mori, Phys. Rev. B **57**, 12023 (1998)
110. M. Watanabe, Y. Noda, Y. Nogami, H. Mori, J. Phys. Soc. Jpn. **73**, 116 (2004) and J. Phys. Soc. Jpn. **74**, 2011 (2005)
111. M. Watanabe et al., J. Phys. Soc. Jpn. **68**, 2654 (1999)
112. Y. Nogami et al., Synth. Metals **103**, 1911 (1999)
113. K. Inagaki, I. Terasaki, H. Mori, T. Mori, J. Phys. Soc. Jpn. **73**, 3364 (2004)
114. T. Yamaguchi et al., Phys. Rev. Lett. **96**, 136602 (2006) and Phys. Rev. Lett. **98**, 116602 (2007)
115. F. Sawano et al., Nature **437**, 532 (2005)
116. H. Kobayashi et al., Synth. Metals **27**, A283 (1988)
117. H. Nishikawa et al., Phys. Rev. B **67**, 52510 (2005)
118. P.W. Anderson, Mat. Res. Bull. **8**, 153 (1973)
119. P.W. Anderson, Science **235**, 1196 (1987)
120. C. Gros, Phys. Rev. B **38**, 931 (1988)
121. C. Gros, Ann. Phys. (N.Y.) **189**, 53 (1989)
122. H. Yokoyama, H. Shiba, J. Phys. Soc. Jpn. **57**, 2482 (1988)
123. F. C Zhang, C. Gros, T.M. Rice, H. Shiba, Supercond. Sci. Technol. **1**, 36 (1988)
124. H. Yokoyama, M. Ogata, J. Phys. Soc. Jpn. **65**, 3615 (1996)
125. A. Paramekanti, M. Randeria, N. Trivedi, Phys. Rev. Lett. **87**, 217002 (2001) and Phys. Rev. B **70**, 054504 (2004)
126. For a recent review see M. Ogata, H. Fukuyama, Rep. Prog. Phys. **71**, 036501 (2008)
127. N. Trivedi, D. Ceperley, Phys. Rev. B **40**, 2737 (1989)
128. K.J. Runge, Phys. Rev. B **45**, 12292 (1992)
129. For a review see E.y Manousakis, Rev. Mod. Phys. **63**, 1 (1991)
130. T.C. Hsu, Phys. Rev. B **41**, 11379 (1990)
131. A. Himeda, M. Ogata, Phys. Rev. B **60**, R9935 (1999)
132. G. Baskaran, Phys. Rev. Lett. **91**, 097003 (2003)
133. B. Kumar, B.S. Shastry, Phys. Rev. B **68**, 104508 (2003)
134. Q.-H. Wang, D.-H. Lee, P.A. Lee, Phys. Rev. B **69**, 092504 (2004)
135. M. Ogata, J. Phys. Soc. Jpn. **72**, 1839 (2003)
136. B.J. Powell, R.H. McKenzie, Phys. Rev. Lett. **98**, 027005 (2007)
137. T. Watanabe, H. Yokoyama, Y. Tanaka, J. Inoue, M. Ogata, J. Phys. Soc. Jpn. **73**, 3404 (2004)
138. T. Koretsune, M. Ogata, Phys. Rev. Lett. **89**, 116401 (2002)
139. T. Koretsune, M. Ogata, Phys. Rev. B **72**, 134513 (2005)
140. B. Bernu, C. Lhuillier, L. Pierre, Phys. Rev. Lett. **69**, 2590 (1992)
141. P.W. Leung, K.J. Runge, Phys. Rev. B **47**, 5861 (1993)
142. R.R.P. Singh, D.A. Huse, Phys. Rev. Lett. **68**, 1766 (1992)
143. N. Elstner, R.R.P. Singh, A.P. Young, Phys. Rev. Lett. **71**, 1629 (1993)
144. L. Capriotti, A.E. Trumper, S. Sorella, Phys. Rev. Lett. **82**, 3899 (1999)
145. Y. Nisikawa, K. Yamada, J. Phys. Soc. Jpn. **71**, 2629 (2002)
146. H. Ikeda, Y. Nisikawa, K. Yamada, J. Phys. Soc. Jpn. **73**, 17 (2004)
147. C. Honerkamp, Phys. Rev. B **68**, 104510 (2003)
148. K. Kuroki, R. Arita, Phys. Rev. B **63**, 174507 (2001)
149. H. Kino, H. Kontani, J. Phys. Soc. Jpn. **67**, 3691 (1998)
150. H. Kondo, T. Moriya, J. Phys. Soc. Jpn. **67**, 3695 (1998)
151. J. Schmalian, Phys. Rev. Lett. **81**, 4232 (1998)
152. R.H. McKenzie, Science **278**, 820 (1997) and Comments Condens. Matter Phys. **18**, 309 (1998)
153. H. Kontani, Phys. Rev. B **67**, 180503(R) (2003)
154. J.Y. Gan, Y. Chen, Z.B. Su, F.C. Zhang, Phys. Rev. Lett. **94**, 067005 (2005) and J.Y. Gan, Y. Chen, F.C. Zhang, Phys. Rev. B **74**, 094515 (2006)

155. H. Morita, S. Watanabe, M. Imada, J. Phys. Soc. Jpn. **71**, 2109 (2002)
156. O. Parcollet, G. Biroli, G. Kotliar, Phys. Rev. Lett. **92**, 226402 (2004)
157. H. Yokoyama, M. Ogata, Y. Tanaka, J. Phys. Soc. Jpn. **75**, 114706 (2006)
158. T. Watanabe, H. Yokoyama, Y. Tanaka, J. Inoue, J. Phys. Soc. Jpn. **75**, 074707 (2006) and Phys. Rev. B **77**, 214505 (2008)
159. H. Yokoyama, H. Shiba, J. Phys. Soc. Jpn. **59**, 3669 (1990)
160. M. Capello, F. Becca, Y. Yunoki, S. Sorella, Phys. Rev. B **73**, 245116 (2006)
161. W.O. Putikka, M.U. Luchini, R.R.P. Singh, Phys. Rev. Lett. **81**, 2966 (1998)
162. A. Himeda, M. Ogata, Phys. Rev. Lett. **85**, 4345 (2000)
163. T. Mizusaki, M. Imada, Phys. Rev. B **74**, 014421 (2006)
164. H. Seo, C. Hotta, H. Fukuyama, Chem. Rev. **104**, 5005 (2004)
165. C. Hotta, H. Fukuyama, J. Phys. Soc. Jpn. **70**, 321 (2001)
166. C. Hotta, J. Phys. Soc. Jpn. **72**, 840 (2003)
167. R. Hoffman, J. Chem. Phys. **39**, 1397 (1963)
168. T. Mori et al., Bull. Chem. Soc. Jpn. **57**, 627 (1984)
169. H. Kino, H. Fukuyama, J. Phys. Soc. Jpn. **64**, 2726 (1995)
170. H. Kino, H. Fukuyama, J. Phys. Soc. Jpn. **65**, 2158 (1996)
171. L. Ducasse, A. Fritsch, F. Castet, Synth. Metals **85**, 1627 (1997)
172. Y. Imamura, S. Ten-no, K. Yonemitsu, Y. Tanimura, J. Chem. Phys. **111**, 5986 (1999)
173. T. Mori, Bull. Chem. Soc. Jpn. **73**, 2243 (2000)
174. D.J. Scalapino, E. Loh, Jr., J.E. Hirsch, Phys. Rev. B **35**, 6694 (1987)
175. J. Merino, R.H. McKenzie, Phys. Rev. Lett. **87**, 237002 (2001)
176. R.H. McKenzie, J. Merino, J.B. Marston, O.P. Sushkov, Phys. Rev. B **64**, 085109 (2001)
177. A. Kobayashi, Y. Tanaka, M. Ogata, Y. Suzumura, J. Phys. Soc. Jpn. **73**, 1118 (2004)
178. S. Onari, R. Arita, K. Kuroki, H. Aoki, Phys. Rev. B **70**, 094523 (2004)
179. A. Greco, J. Merino, A. Foussats, R.H. McKenzie, Phys. Rev. B **71**, 144502 (2005)
180. Y. Tanaka, Y. Yanase, M. Ogata, J. Phys. Soc. Jpn. **73**, 2053 (2004) and J. Phys. Soc. Jpn. **73**, 319 (2004)
181. H. Watanabe, M. Ogata, J. Phys. Soc. Jpn. **75**, 063702 (2006)
182. T. Mori, J. Phys. Soc. Jpn. **72**, 1469 (2003)
183. M. Kaneko, M. Ogata, J. Phys. Soc. Jpn. **75**, 014710 (2006)
184. H. Seo, M. Ogata, Phys. Rev. B **64**, 113103 (2001)
185. J. Merino, H. Seo, M. Ogata, Phys. Rev. B **71**, 125111 (2005)
186. H. Watanabe, M. Ogata, J. Phys. Soc. Jpn. **74**, 2901 (2005)
187. M. Hanawa et al., Phys. Rev. Lett. **87**, 187001 (2001)
188. H. Sakai et al., J. Phys. Condens. Matt. **13**, L785 (2001)
189. R. Jin et al., Phys. Rev. B **64**, 180503(R) (2001)
190. S. Yonezawa, Y. Muraoka, Z. Hiroi, J. Phys. Soc. Jpn. **73**, 1655 (2004)
191. S. Yonezawa, Y. Muraoka, Y. Matsushita, Z. Hiroi, J. Phys. Soc. Jpn. **73**, 819 (2004)
192. S.M. Kazakov et al., Supercond. Sci. Technol. **17**, 1169 (2004)
193. M. Brühwiler, S.M. Kazakov, N.D. Zhigadlo, J. Karpinski, B. Batlogg, Phys. Rev. B **70**, 020503(R) (2004)
194. S. Yonezawa, Y. Muraoka, Y. Matsushita, Z. Hiroi, J. Phys. Condens. Matt. **16**, L9 (2004)
195. R. Galati et al., J. Mater. Chem. **17**, 160 (2007)
196. J. Yamaura, S. Yonezawa, Y. Muraoka, Z. Hiroi, J. Solid State Chem. **179**, 336 (2006)
197. Z. Hiroi, S. Yonezawa, T. Muramatsu, J. Yamaura, Y. Muraoka, J. Phys. Soc. Jpn. **74**, 1255 (2005)
198. J. Kunes, T. Jeong, W.E. Pickett, Phys. Rev. B **70**, 174510 (2004)
199. M. Yoshida et al., Phys. Rev. Lett. **98**, 197002 (2007)
200. M. Brühwiler, S.M. Kazakov, J. Karpinski, B. Batlogg, Phys. Rev. B **73**, 094518 (2006)
201. H. Harima, J. Phys. Chem. Solids **63**, 1035 (2002)
202. D.J. Singh, P. Blaha, K. Schwarz, J.O. Sofo, Phys. Rev. B **65**, 155109 (2002)
203. Z. Hiroi, M. Hanawa, J. Phys. Chem. Solids **63**, 1021 (2002)
204. O. Vyaselev et al., J. Phys. Chem. Solids **63**, 1031 (2002)
205. K. Arai et al., Physica B **359-361**, 488 (2005)

206. J. Yamaura, Z. Hiroi, J. Phys. Soc. Jpn. **71**, 2598 (2002)
207. J.P. Castellan et al., Phys. Rev. B **66**, 134528 (2002)
208. Z. Hiroi, M. Hanawa, Y. Muraoka, H. Harima, J. Phys. Soc. Jpn. **72**, 21 (2003)
209. I.A. Sergienko et al., Phys. Rev. Lett. **92**, 065501 (2004)
210. J.C. Petersen et al., Nature Physics **2**, 605 (2006)
211. Z. Hiroi, J. Yamaura, S. Yonezawa, H. Harima, Physica C **460-462**, 20 (2007)
212. Z. Hiroi, S. Yonezawa, Y. Nagao, J. Yamaura, Phys. Rev. B **76**, 014523 (2007)
213. L.R. Testardi, Rev. Mod. Phys. **47**, 637 (1975)
214. S.D. Bader et al., Phys. Rev. Lett. **37**, 344 (1976)
215. D.J. Singh, Phys. Rev. B **61**, 13397 (2000)
216. D.J. Singh, Phys. Rev. B **68**, 020503(R) (2003)
217. M.D. Johannes, D.J. Singh, Phys. Rev. B **70**, 014507 (2004)
218. K.-W. Lee, J. Kunes, W.E. Pickett, Phys. Rev. B **70**, 045104 (2004)
219. P. Zhang, W. Luo, M.L. Cohen, S.G. Louie, Phys. Rev. Lett. **93**, 236402 (2004)
220. R. Arita, Phys. Rev. B **71**, 132503 (2005)
221. M. Mochizuki, Y. Yanase, M. Ogata, J. Phys. Soc. Jpn. **74**, 1670 (2005) and J. Phys. Soc. Jpn. **74**, 2381 (2005)(E).
222. M. Mochizuki, M. Ogata, J. Phys. Soc. Jpn. **75**, 113703 (2006)
223. M.Z. Hasan et al., Phys. Rev. Lett. **92**, 246402 (2004)
224. H.-B. Yang et al., Phys. Rev. Lett. **92**, 246403 (2004)
225. H.-B. Yang et al., Phys. Rev. Lett. **95**, 146401 (2005)
226. T. Shimojima et al., Phys. Rev. Lett. **97**, 267003 (2006)
227. D. Yoshizumi et al., J. Phys. Soc. Jpn. **76**, 063705 (2007)
228. T. Fujimoto et al., Phys. Rev. Lett. **92**, 047004 (2004)
229. K. Ishida et al., J. Phys. Soc. Jpn. **72**, 3041 (2003)
230. Y. Ihara et al., J. Phys. Soc. Jpn. **73**, 2069 (2004)
231. G.-q. Zheng et al., J. Phys.: Condens. Matter **18**, L63 (2006)
232. H.D. Yang et al., Phys. Rev. B **71**, 020504(R) (2005)
233. B. Lorenz et al., Physica C **402**, 106 (2004)
234. N. Oeschler et al., Chin. J. Phys. **43**, 574 (2005)
235. R. Jin et al., Phys. Rev. B **72**, 060512(R) (2005)
236. H. Sakurai, K. Takada, T. Sasaki, E. Takayama-Muromachi, J. Phys. Soc. Jpn. **74**, 2909 (2005)
237. Y. Ihara et al., J. Phys. Soc. Jpn. **74**, 867 (2005)
238. Y. Ihara et al., J. Phys. Soc. Jpn. **75**, 124714 (2006)
239. Y. Kobayashi, M. Yokoi, M. Sato, J. Phys. Soc. Jpn. **72**, 2453 (2003)
240. Y. Ihara et al., J. Phys. Soc. Jpn. **74**, 2177 (2005)
241. Y. Kobayashi et al., J. Phys. Soc. Jpn. **74**, 1800 (2005)
242. G.-q. Zheng, K. Matano, D.P. Chen, C.T. Lin, Phys. Rev. B **73**, 180503(R) (2006)
243. Y. Ihara et al., J. Phys. Soc. Jpn. **75**, 013708 (2006)
244. F.L. Ning, T. Imai, Phys. Rev. Lett. **94**, 227004 (2005)
245. K. Takada et al., J. Mater. Chem. **14**, 1448 (2004)
246. I.R. Mukhamedshin, H. Alloul, G. Collin, N. Blanchard, Phys. Rev. Lett. **94**, 247602 (2005)
247. C. Michioka, H. Ohta, Y. Itoh, K. Yoshimura, J. Phys. Soc. Jpn. **75**, 063701 (2006)
248. W. Higemoto et al., Phys. Rev. B **70**, 134508 (2004)
249. K. Yada, H. Kontani, J. Phys. Soc. Jpn. **75**, 033705 (2006)
250. M. Yokoi et al., J. Phys. Soc. Jpn. **73**, 1297 (2004)
251. K. Kuroki, Y. Tanaka, R. Arita, Phys. Rev. Lett. **93**, 077001 (2004) and Phys. Rev. B **71**, 024506 (2005)
252. M. Mochizuki, Y. Yanase, M. Ogata, Phys. Rev. Lett. **94**, 147005 (2005)
253. Y. Yanase, M. Mochizuki, M. Ogata, J. Phys. Soc. Jpn. **74**, 430 (2005)
254. K. Kuroki et al., Phys. Rev. B **73**, 184503 (2006)
255. M. Mochizuki, M. Ogata, J. Phys. Soc. Jpn. **76**, 013704 (2007)
256. M. Mochizuki, H.Q. Yuan, M. Ogata, J. Phys. Soc. Jpn. **76**, 023702 (2007)
257. For a review see K. Kanoda, J. Phys. Soc. Jpn. **75**, 051007 (2006)
258. For a review see K. Kuroki, J. Phys. Soc. Jpn. **75**, 051013 (2006)

259. A. Carrington et al., Phys. Rev. Lett. **83**, 4172 (1999)
260. O.J. Taylor, A. Carrington, J.A. Schlueter, Phys. Rev. Lett. **99**, 057001 (2007)
261. K. Miyagawa, A. Kawamoto, K. Kanoda, Phys. Rev. Lett. **89**, 017003 (2002)
262. See for example T. Timusk, B. Statt, Rep. Prog. Phys. **62**, 61 (1999)
263. M.-S. Nam, A. Ardavan, S.J. Blundell, J.A. Schlueter, Nature **449**, 584 (2007)
264. R.N. Lyubovskaya et al., JETP Lett. **42**, 468 (1985) and JETP Lett. **46**, 188 (1987)
265. N. Tajima, T. Imakubo, R. Kato, Y. Nishio, K. Kajita, J. Phys. Soc. Jpn. **72**, 1014 (2003)
266. A. Kobayashi, S. Katayama, Y. Suzumura, J. Phys. Soc. Jpn. **74**, 2987 (2005)

# Chapter 23
# Frustration in Systems with Orbital Degrees of Freedom

**Jeroen van den Brink, Zohar Nussinov, and Andrzej M. Oleś**

**Abstract** We review the types of frustration encountered in Mott insulators with orbital degrees of freedom and discuss the physical consequences. We first survey the driving forces for the ordering of orbital degrees of freedom and then compare the generic features of typical orbital and typical spin Hamiltonians. A primary difference between the two is caused by the sensitivity of orbital interactions to the spatial orientation of the orbitals involved. This leads in general to highly anisotropic orbital Hamiltonians, to frustration of classical ordered states on unfrustrated lattices, and to enhanced quantum fluctuations.
As a consequence of these effects, new types of symmetry can appear in orbital models, in particular in compass models. These intermediate symmetries lie midway between the extremes of global symmetries and local gauge symmetries. We discuss briefly the generic consequences of this very particular type of symmetry, and highlight in this context the relation between orbital models and the models of Kitaev for quantum computation.
As a final topic, we include spin degrees of freedom into combined spin-orbital models and consider a number of ways in which the orbital frustration in real materials is lifted by the magnetic degrees of freedom. Orbital degrees of freedom boost the tendency to form disordered states or valence-bond phases. Enhanced quantum fluctuations and spin-orbital entanglement occur in the vicinity of quantum critical points, where different types of order compete with each other. Taken together, these phenomena demonstrate that orbital and spin-orbital physics contain a number of unconventional features and peculiar symmetries which are qualitatively different from the range of properties known in pure spin models with frustration, arising either from frustrated geometries or due to longer-range magnetic interactions.

## 23.1 Introduction

The physical properties of transition-metal (TM) oxide materials are often determined not only by the charge and spin degrees of freedom of the electrons, but in many cases also by their orbital degrees of freedom. In TM oxides, the $3d$ shell of the TM ion is only partially filled, while the electrons within that shell interact very strongly through the Coulomb repulsion. As a consequence, many of these materials are Mott insulators – systems well-known for their rich variety of magnetic properties.

Due to the strong Coulomb interactions also an extra electronic degree of freedom emerges: one electron can occupy different but (quasi-) degenerate $3d$ orbitals. For extended reviews on the subject of orbital degrees of freedom in solids, and of their ordering in TM oxides, we refer the reader to [1–4]. One of the characteristics of orbital degrees of freedom is that they have a number of properties very similar to the spin degree of freedom of the electron. But, as we will discuss in this chapter, a distinct and essential feature of orbital degrees of freedom is that, quite generally, orbital–orbital interactions and the emerging orbital ordering are strongly and intrinsically frustrated.

The aim of this chapter is not only to summarize some well-established facts concerning orbital degrees of freedom and their ordering, but also to highlight some of the recent developments in the field. It should be stressed that the choice of advanced subjects is rather a reflection of the authors' interests than an appraisal of their importance *per se*. This is true in particular for the tentative relation between orbital models and Kitaev models for quantum computation which we will discuss in Sect. 23.6. One of our most important messages will be that, although orbital physics has grown into a sizable field of research, much remains unexplored.

We begin with an elementary introduction to the concepts related to orbital degrees of freedom: how they emerge (Sect. 23.2), how they interact and how they are described mathematically in terms of orbital Hamiltonians (Sect. 23.3). We then proceed in the second part by discussing in detail the types of symmetries which these orbital models possess (Sect. 23.4). We will see that the symmetry of orbital interactions is intermediate in nature between global (such as spin rotation in an antiferromagnet) and local (such as gauge invariance in electromagnetism). We also present a theorem on dimensional reduction and demonstrate how the phenomenon of "order by disorder" appears in classical orbital models (Sect. 23.5). The relationship between these orbital Hamiltonians and quantum computation is discussed in Sect. 23.6. In the final sections of this chapter, we consider the more complex situation of coupled spin-orbital superexchange models (Sect. 23.7), which are relevant in real materials where they serve to describe the physical properties of a class of TM oxides with partly filled, degenerate $3d$ orbitals. This discussion includes a brief exposition of spin-orbital entanglement (Sect. 23.8).

## 23.2 Orbital Degrees of Freedom

### *23.2.1 Orbitals and Their Energy Scales*

The orbital part of the electronic wave function gives, by definition, the angular distribution of electronic charge around the nucleus. For a free $3d$ ion, the orbital

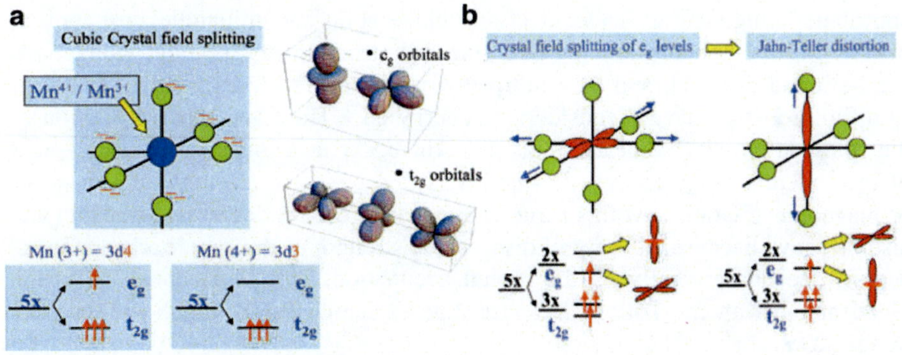


**Fig. 23.1** (**a**) Orbitals corresponding to the angular part of atomic $d$ wave functions and their splitting in a cubic crystal field. *Upper right*: the two $e_g$ orbitals, $3z^2 - r^2$ (*left*) and $x^2 - y^2$ (*right*); *lower right*: the three $t_{2g}$ orbitals, $zx$, $xy$, and $yz$. The twofold $e_g$ degeneracy remains in the case of a high-spin $d^4$ ion, such as $Mn^{3+}$ or $Cr^{2+}$, respective example materials being $LaMnO_3$ and $Rb_2CrCl_4$. By contrast, in $LaTiO_3$ or $LaVO_3$ it is the threefold-degenerate $t_{2g}$ orbitals which are relevant. (**b**) Jahn–Teller splitting of the degeneracy of $e_g$ orbitals due to deformation of the $RO_6$ octahedron, where $R$ is a TM ion. Note that different types of deformation yield the same energy-lowering

part of the wave function is given by the spherical harmonics, $Y_{2,m}$, which are fivefold-degenerate (when relativistic spin-orbit coupling is neglected), with $|m| \leq 2$. The spherical harmonic wave functions constitute a complete and orthonormal basis for the 3$d$ states. A partial filling of the 3$d$ shell therefore implies that there is a number of ways in which electrons can be distributed over the available degenerate orbitals – the orbital degree of freedom. The five different $d$ orbitals are shown in Fig. 23.1. The $e_g$ states are $x^2 - y^2 \equiv (Y_{2,-2} + Y_{2,2})/\sqrt{2}$ and $3z^2 - r^2 \equiv Y_{2,0}$, and the $t_{2g}$ states are respectively $xy \equiv (Y_{2,-2} - Y_{2,2})/\sqrt{2}$, $yz \equiv (Y_{2,-1} + Y_{2,1})/\sqrt{2}$, and $zx \equiv (Y_{2,-1} - Y_{2,1})/\sqrt{2}$. From the figure it is clear that the different orbital states correspond to quadrupole moments of the charge-density distribution. When a TM ion in a material possess an orbital degeneracy, local TM 3$d$ states with different quadrupole moments have the same energy.

The orbital-related properties of a given material depend on how exactly the 3$d$ levels are filled as electrons are introduced. Here, it should be stressed that adding a further electron to an already partially filled $d$ shell costs a Coulomb energy $U$ on the order of some electronvolts, due to the repulsion between the negatively charged electrons. This large repulsive interaction $U$ induces very strong correlations between the electrons, rendering the TM oxides insulating in spite of the partial filling of their $d$-shell [4]. In this chapter, we consider exclusively TM oxides in this Mott-insulating regime, where the 3$d$ electrons are localized and can form local magnetic moments.

The Hund rule of atomic physics states that, in order to minimize the atomic Coulomb repulsion, the electrons prefer to occupy different atomic orbitals but with parallel spin alignment. The corresponding Hund exchange element, $J_H$, also contributes a high energy scale: electrons whose spins are antiparallel cost approximately 0.5 eV per anti-aligned pair (the full structure of Coulomb interactions can be found in [5–7], and the multiplet structure of TM ions in [8]). Thus local spin-flip processes have a considerable energy cost. However, relative rotation of the spins on two different TM ions can still occur at a low energy scale – that of the superexchange energy $J$ – and it is this which is relevant for spin-wave or magnon excitations. At this same energy scale, orbital degrees of freedom can also be active because, as we will discuss, the strength of the superexchange interactions depends critically on the orbital occupation. This is encapsulated in the Goodenough-Kanamori rules (GKR) for superexchange [9,10], which will reappear in Sect. 23.8.

An essential aspect of degenerate orbitals is that the orbital degrees of freedom are in general coupled to the lattice, which leads to a Jahn–Teller distortion of the lattice structure. Here the related energy scale for static deformation is $E_{JT}$, the energy gain associated with a local Jahn–Teller distortion, while the phonon energy scale enters into the orbital dynamics.

### 23.2.2 Comparing Orbital and Spin Degrees of Freedom

In Mott insulators, both spin and orbital degrees of freedom can form local moments. This observation suggest some similarity between orbitals and spins. Let us therefore compare the physics of orbitals with the one of spins in some more detail.

On a mathematical level, the similarity between spins and orbitals can be made explicit. A *spin doublet* can be represented by the operator $\mathbf{S} = \{S^x, S^y, S^z\}$, where $S^z = \pm\frac{1}{2}$ corresponds to the spin being up/down along the $z$-axis and $S^x$, $S^y$ are linear combination of raising and lowering operators $S^+$ and $S^-$. The operator components have the well-known commutation relations $[S^\alpha, S^\beta] = i\hbar\epsilon_{\alpha\beta\gamma}S^\gamma$. An *orbital doublet* is described by exactly the same operators and the same algebra, with the only difference that $S^z = \pm\frac{1}{2}$ now corresponds to one or the other orbital being occupied.

This parallel between spins and orbitals also extends to triplets and higher multiplets. It implies that all of the techniques which have been developed to treat spin problems can be applied directly to orbital Hamiltonians. Another similarity is that both spin-spin interactions and orbital–orbital ones can be mediated via superexchange processes.

However, there are also important differences, also summarized in Table 23.1, which are related primarily to the different symmetries of spin and orbital Hamiltonians. In general, for 3$d$ systems the coupling of spins to the lattice by spin-orbit coupling (a relativistic interaction) is weak, because the dominant interaction between the spins is superexchange, which is governed by virtual charge excitations between neighboring TM ions. This leads to effective spin Hamiltonians of the Heisenberg type, with only weak anisotropies, the net effect being that the relevant

**Table 23.1** Comparing generic features of spin and orbital degrees of freedom

| Common features of spins and orbitals |
|---|
| Local moments emerge from |
| Electron-electron interactions |
| SU(2) algebra: $[S^\alpha, S^\beta] = i\hbar\epsilon_{\alpha\beta\gamma}S^\gamma, \alpha, \beta, \gamma = x, y, z$ |
| Intersite interaction due to superexchange |

Differences

| Spins | Feature | Orbitals |
|---|---|---|
| Weak | Coupling to lattice | Strong |
| High | Symmetry of Hamiltionian | Low |
| Gapless | Excitations | Gapped |
| Sometimes | Frustration of order | Almost always |

spin Hamiltonians are characterized by a high symmetry, SU(2), corresponding to rotational invariance in spin-space.

At the onset of magnetic order in a spin system, the continuous SU(2) symmetry is broken and gapless Goldstone modes appear as a consequence. These modes govern the low-energy, long-wavelength magnetic properties, at least at temperatures above the characteristic energy scale set by spin-orbit-coupling effects such as single-ion spin anisotropies and Dzyaloshinskii-Moriya contributions to the spin-spin interaction.

In contrast to spins, orbitals are spatially anisotropic degrees of freedom: as noted above, they correspond to quadrupole moments of the charge (Fig. 23.1). The interaction between two orbitals on different sites depends on the relative orientation of the lobes of the two quadrupole-moment distributions. Orbital Hamiltonians therefore depend intrinsically on the symmetry of the orbitals *and* the lattice. Shape and symmetry of the orbital lobes is determined by the local point-group symmetry of the lattice, so that the continuous symmetry which characterizes spin Hamiltonians is broken. This observation may be phrased in the form "spins exist in spin-space, orbitals exist in real space." As the symmetry which can be broken spontaneously in orbital Hamiltonians is not continuous, its breaking is not associated with the appearance of Goldstone modes.

The fact that orbital–orbital interactions are spatially very anisotropic, depending strongly on the orientation of the orbitals with respect to the direction of the bond between two neighboring lattice sites, leads in general to strong frustration of orbital ordering. As an example, if the interactions are such that it is favorable for orbitals of the $3z^2 - r^2$ type (dumbbell-shaped, Fig. 23.1a) to be aligned along the bond between each pair of nearest-neighbor sites, satisfying this tendency for a given bond implies simultaneously (in dimensions higher than one) that the orbitals are wrongly aligned on bonds in the other lattice directions.

## 23.3 Orbital Interactions and Orbital Models

### *23.3.1 Crystal-Field Splitting of Orbitals*

The degeneracy of atomic $3d$ states is without exception lifted, partially or completely, in a solid. In this section we discuss the origin of this reduction of degeneracy and which interactions become relevant in solids as a result.

In a perovskite structure, the TM ion is surrounded by oxygen ions (Fig. 23.1a). The spherical symmetry of the TM ion in free space is therefore lowered in the solid to a discrete point-group symmetry. The microscopic driving force is that the surrounding oxygen ions produce a non-spherical electrical potential at the locus of the TM ion, and this crystal-field potential changes the electronic orbital eigenstates (wave functions) and energies. This lower symmetry causes the partial or total lifting of orbital degeneracy. Whether a particular orbital degeneracy is lifted by the crystal field depends on the actual symmetry of both the orbital wave functions and the crystal-field potential.

Let us consider the case where the oxygen ions form a perfect octahedron with the TM ion in its center. In this case, all six TM–O bonds have equal length and all O–TM–O bond paths are 90-degree paths. The lobes of wave functions with $e_g$ symmetry point towards the negatively charged oxygen ions (Fig. 23.1), and the lobes of the $t_{2g}$ orbitals point precisely between two oxygen ions. The Coulomb repulsion experienced by an electron in a $t_{2g}$ state is therefore less than that in an $e_g$ state. As a consequence, the electronic $3d$ states of a TM ion located at the center of a perfect octahedron are split into threefold-degenerate (orbital triplet) $t_{2g}$ states and twofold-degenerate (orbital doublet) $e_g$ states. The $t_{2g}$ multiplet lies lower in energy than the $e_g$ multiplet and the orbital degeneracy is lifted partially.

### *23.3.2 Jahn–Teller Deformation*

Depending on the total number $n$ of electrons in an ionic configuration $d^n$, and their total spin, $S$, the $t_{2g}$ triplet or $e_g$ doublet can be partially occupied and a degeneracy of the system persists. This is the case for $t_{2g}$ orbitals when, for example, one $d$ electron is present ($d^1$ configuration), as for $Ti^{3+}$ ions in $LaTiO_3$ [11] or $V^{4+}$ ions in $SrVO_3$ [12], or in the $d^2$ configuration, as for $V^{3+}$ ions in $LiVO_2$ [13] or $LaVO_3$ [14]. However, it is not the case for $d^3$ ions, when all three electrons have the same spin (high-spin state), because then all three $t_{2g}$ orbitals are singly occupied. In this case, which is favored by the Hund rules, the Pauli principle forbids electrons with the same spin to occupy the same orbital. However, if in the $d^3$ configuration one of the electrons has a reversed spin, this electron could occupy any one of the three orbitals and there is again an orbital degree of freedom.

Orbital degeneracy in the $e_g$ doublet is also common in cubic geometry for $d^9$ configurations, relevant for $Cu^{2+}$ (in materials such as $KCuF_3$ [15]) and $Ni^{3+}$ (for example $LiNiO_2$ [16]) ions, which correspond to one hole in the $e_g$ doublet. An $e_g$ degeneracy is present also for high-spin $d^4$ ions, where the $t_{2g}$ orbitals are singly occupied and one electron is pushed into the doublet (*e.g.* $Mn^{3+}$ in $LaMnO_3$ and other manganites, $Cr^{2+}$ in $Rb_2CrCl_4$ or $KCrF_3$ [17, 18]). For other partial fillings of the $d$ shell, orbital degeneracy may be present in either the $t_{2g}$ or the $e_g$ sector, depending on the exact spin state of the ions.

The famous Jahn–Teller theorem [19] states that a local degeneracy cannot occur in the ground state: a degenerate state is unstable with respect to external perturbations, and in any physical system this degeneracy will therefore always be lifted, leaving a ground state characterized by one particular set of orbitals being occupied while the others remain empty. Note, however, that it is not *a priori* clear either which orbitals will be occupied in a solid, or whether the orbital configuration will be the same for all the TM ions in a given lattice. In order to answer questions related to the occurrence of orbital ordering, it is necessary to specify first the exact symmetry of the degenerate orbitals involved and second the nature of the interactions between orbitals on neighboring lattice sites.

As noted above, the Jahn–Teller theorem implies that a local $3d$ doublet or triplet degeneracy will be lifted. A distortion of the anion cage, for example the oxygen octahedron in a perovskite, which surrounds the central TM ion – a Jahn–Teller distortion – is very effective in achieving this, as shown in Fig. 23.2. When the Jahn–Teller distortion of an octahedron occurs, an additional crystal field is generated

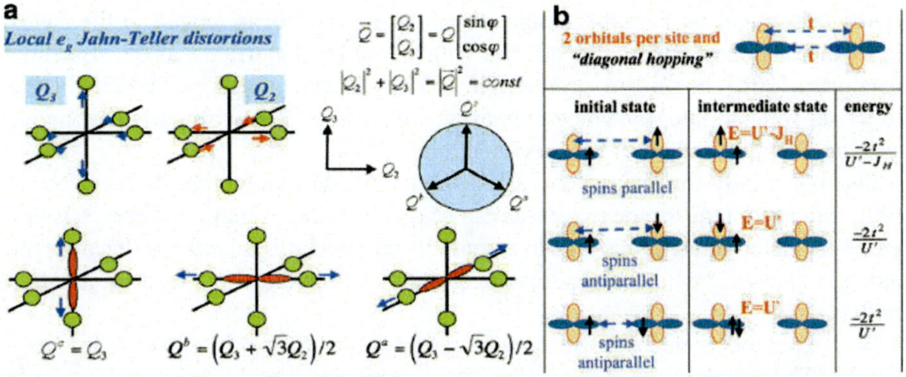


**Fig. 23.2** (**a**) Jahn–Teller distortions of different symmetry. Upper left: $Q_3$ and $Q_2$ distortion of $e_g$ symmetry; lower panel: linear combinations of these two distortions ($\theta = \pm 2\pi/3$). (**b**) Schematic representation of superexchange processes in the presence of orbital degrees of freedom. For fermions with spin, the first state of the three shown has the lowest exchange energy, because in the intermediate state the two electrons on the same ion have parallel spins (Hund rules). For spinless fermions, the lower two processes are forbidden and the first favors anti-parallel alignment of the orbital degrees of freedom (alternating orbitals). The parameter $U$ corresponds to the Kanamori intraorbital Coulomb parameter in the lowest process, while in the two upper processes it denotes the interorbital interaction $U' = U - 2J_H$

at the central TM site, whose symmetry reflects the symmetry of the distortion. The electric quadrupole moments of different shapes, corresponding to different occupied orbitals, which can be present on the central TM ion interact differently with the crystal-field potential. A local Jahn–Teller distortion thus induces a splitting of the locally degenerate states, causing particular orbitals to be occupied and others to be empty.

It is important to note that even if after a Jahn–Teller distortion the electronic degeneracy is lifted, the question *which* orbital will be occupied is still open. The reason is that when orbital $\nu$ of a certain symmetry is occupied, the lattice distorts in a manner dictated by the symmetry of $\nu$. If one views this distortion as static, one concludes that it lowers the energy of the occupied orbital $\nu$ and increases the energy of the empty orbital $\bar{\nu}$.

A Jahn–Teller distortion, however, always adopts itself to the symmetry of the particular orbital that is occupied. Thus, if orbital $\bar{\nu}$ is occupied and $\nu$ is empty, the lattice will distort in a different way, now lowering orbital $\bar{\nu}$ in energy. The total energy of the system does not depend on which particular orbital is stabilized as long as the response of the lattice is harmonic (see Sect. 23.3.3). Therefore, $\nu$ and $\bar{\nu}$ are still degenerate, provided the orbital occupation is combined with a lattice distortion of the appropriate symmetry. Non-linearities or anisotropies in the Jahn–Teller coupling and/or phonon response can in principle break this degeneracy.

### 23.3.3 Jahn–Teller-Mediated Orbital–Orbital Interactions

In a Mott insulator, interactions between orbital moments can be mediated both by the lattice and by superexchange. The first mechanism is related to the Jahn–Teller interaction of degenerate orbitals with the lattice distortions, as discussed for example in [20]. The second, proposed in 1972 by Kugel and Khomskii [21,22], is a direct generalization of the conventional superexchange [23] to orbitally degenerate systems. We discuss both mechanisms in this section.

One potentially surprising conclusion will be that the symmetries of the effective orbital Hamiltonians for the two cases can be quite similar. A major difference is that lattice-mediated interactions lead to essentially classical orbital models (because the lattice is "heavy") whereas superexchange-mediated interactions lead to quantum models.

To discuss this issue in a formal manner, we begin with an orbitally degenerate $e_g$ doublet. The doublet is spanned by the states (orbitals) $|z\rangle$ and $|x\rangle$ ($|z\rangle \equiv |3z^2 - r^2\rangle$ and $|x\rangle \equiv |x^2 - y^2\rangle$). In general, the occupied state is a linear combination of $e_g$ orbitals, $|\psi\rangle = \cos(\theta/2)|z\rangle + \sin(\theta/2)|x\rangle$, the angle $\theta$ parameterizing the ground-state wave function. We associate with this wave function a vector $\boldsymbol{T} = \{T^z, T^x\}$ whose first component is the difference between the electron densities in orbitals $|z\rangle$ and $|x\rangle$, $T^z \equiv \langle\psi|z\rangle\langle z|\psi\rangle - \langle\psi|x\rangle\langle x|\psi\rangle = \cos\theta$, and whose second component is a measure of the entanglement (mixing) between the two orbitals in the ground state, $T_x \equiv \langle\psi|z\rangle\langle x|\psi\rangle + \langle\psi|x\rangle\langle z|\psi\rangle = \sin\theta$. Thus $\boldsymbol{T} = (1, 0)$ ($\theta = 0$) corresponds to orbital $|z\rangle$ being fully occupied, and $\boldsymbol{T} = (-1, 0)$ ($\theta = \pi$) to a pure $|x\rangle$ state, while any other orbital superposition can be obtained by rotating $\boldsymbol{T}$ on the unit circle. Orbital degrees of freedom are of course quantized, because they correspond to electrons occupying particular states, and the fermionic statistics of the electrons mean that the orbital operator $\boldsymbol{T}$ obeys the spin-1/2 angular-momentum algebra, $T^z = \pm\frac{1}{2}$.

If state $|z\rangle$ is occupied, the Jahn–Teller distortion $Q_3$ is induced, while if $|x\rangle$ is occupied, the opposite distortion $-Q_3$ is induced (Fig. 23.2). A general lattice distortion can be expressed as $Q_3 \cos\phi + Q_2 \sin\phi$, and its elastic energy takes the form $E_{\text{elas}} = \omega(Q_3^2 + Q_2^2)$, or in vector notation, $\boldsymbol{Q} = (Q_3, Q_2) = |Q|(\cos\phi, \sin\phi)$ and $E_{\text{elas}} = \omega|\boldsymbol{Q}|^2$. The orbital-lattice coupling is linear, whence the total local energy is $E_{\text{local}} = -g\boldsymbol{T} \cdot \boldsymbol{Q} + \omega|\boldsymbol{Q}|^2$, where $g$ is the electron-phonon coupling constant. This local energy is clearly minimal when $\boldsymbol{T}$ and $\boldsymbol{Q}$ are parallel, *i.e.* the ground state is $\theta = \phi$. We assume henceforth that the orbital and distortion vectors are parallel. Note, however, that after fixing $\theta$ to $\phi$ the local energy is independent of $\phi$, meaning that the orbital-Jahn–Teller system in fact remains locally rotationally invariant (under the assumptions of linear electron-phonon coupling and elasticity).

This local degeneracy can of course be lifted by the interactions between neighboring orbital degrees of freedom on the lattice [20,22,24]. Such interactions occur naturally in TM oxides, where the oxygen octahedra are connected: each oxygen ion belongs to more then one octahedron [in the perovskite structure, the octahedra are corner-sharing (Fig. 23.2)]. Thus, if a given octahedron is distorted, all octahedra connected to it are automatically distorted. The orbital degeneracy can then in principle be lifted *globally*. Such a global lifting of degeneracy implies that a long-range ordering of orbitals emerges from the effective orbital–orbital interactions mediated by the lattice distortions.

To derive the effective Hamiltonian of the orbital-lattice system, let us take the example of a cubic perovskite and denote by $a$, $b$, and $c$ the crystalographic axes of the solid. To illustrate how the Jahn–Teller distortions of neighboring octahedra interact, if the orbital $3z^2 - r^2$ is occupied on site $i$, the octahedron is elongated with a $Q_3$ distortion and the octahedron connected to it along the $c$-axis is automatically compressed (distortion $-Q_3$). Thus the interaction between the distortions at nearest neighbors $\{i, j\}$ in this direction is $Q_{3,i} Q_{3,j}$. However, the orbitals can be rotated in any direction, and by choosing $\theta = 2\pi/3$ one obtains the orbital $3x^2 - r^2 = \frac{1}{2}(-|3z^2 - r^2\rangle + \sqrt{3}|x^2 - y^2\rangle)$, which is elongated along the $a$-axis. The distortion accompanying this orbital is $\frac{1}{2}(-Q_3 + \sqrt{3}\, Q_2)$, and it is this linear combination of distortions which determines the $a$-axis interaction. The situation for the $b$-axis is analogous with $\theta = -2\pi/3$. The Hamiltonian for $e_g$ orbitals on a cubic lattice with corner-sharing octahedra [25] is then

$$H_{120} = J \sum_{\langle ij \rangle \parallel \gamma} T_i^{(\gamma)} T_j^{(\gamma)}, \tag{23.1}$$

where $\langle ij \rangle$ denotes a bond between two nearest-neighbor sites and $J$ is the orbital interaction. This form emerges because the orbital vector $\boldsymbol{T} = (T^z, T^x)$ is parallel to the distortion $\boldsymbol{Q}$, and in terms of the axes $\gamma = a, b, c$ one has

$T^{(a)} = \frac{1}{2}(T^z - \sqrt{3}T^x)$, $T^{(b)} = \frac{1}{2}(T^z + \sqrt{3}T^x)$, and $T^{(c)} = T^z$ (Fig. 23.2a). We refer to this Hamiltonian as the 120-degree model because on the unit circle the three vectors $\boldsymbol{T}_i$ have relative angles of 120 degrees ($2\pi/3$). This model is essentially classical: the zero-point quantum oscillations of the heavy oxygen ions which mediate the orbital–orbital interactions (equivalent to the interactions between Jahn–Teller centers) are negligible. In the next section we discuss the proof that the ground state for this model Hamiltonian is ordered by entropic stabilization. We can define the Hamiltonian of the 90-degree model, where the three cubic directions are $T_a = T^x$, $T_b = T^y$, and $T_c = T^z$, in a similar way.

### 23.3.4 Superexchange-Mediated Orbital–Orbital Interactions

The situation for orbital degrees of freedom on different sites coupled by magnetic superexchange interactions is more complex due to the intrinsic interconnection of orbital and spin degrees of freedom. The results of considering superexchange processes in the presence of orbital degrees of freedom (Fig. 23.2b) are encapsulated in the Kugel-Khomskii (KK) spin-orbital models. These are in general quantum mechanical in nature, because only electronic degrees of freedom are present in the Hamiltonian. We will not derive these models at this point, and instead refer to the excellent review Kugel and Khomskii [1] – a classic paper in the field.

It turns out that for spinless $e_g$ electrons on a cubic lattice, the superexchange contribution to the orbital Hamiltonian is given by the same expression as in $H_{120}$ [26]. However, in contrast to the Jahn–Teller Hamiltonian, the entities $T_i$ appearing in the KK Hamiltonian are pseudospin-1/2 operators obeying the standard spin commutation relations. We postpone to the latter part of this chapter (Sect. 23.7) a more detailed discussion of KK models, where the spin degrees of freedom of the electrons are taken fully into account.

## 23.4 Symmetry and Symmetry-Breaking in Orbital Models

### 23.4.1 Types of Symmetry in Orbital Models

In terms of symmetry, orbital systems offer a richness not found elsewhere [25,27]. Before discussing these symmetries, we summarize for the present purposes the classification of types of order and their relation to symmetry.

(i) *Global symmetry-breaking order:* in many condensed matter systems there is an invariance of the basic interactions with respect to global symmetry operations (*e.g.* continuous rotations in the case of ferromagnets, uniform translations and rotations for liquids) which are performed simultaneously on all of the constituents of the system. At sufficiently low temperatures (or strong interactions), such symmetries may be broken "spontaneously." The complete symmetry is lifted and a reduced-symmetry state takes its place. Landau provided a general mean-field framework for analyzing such symmetry-breaking and symmetry-broken states.

(ii) *"Topological order"* [28]: this is a type of order which cannot be characterized by a local order parameter and may only be detected by non-local measurements. Examples of systems displaying topological order are found in gauge theories (familiar in high-energy physics) [28], whose local gauge symmetries are the corresponding symmetries to break. In these examples, indeed the only measurable quantities are non-local, pertaining to correlation functions defined on "Wilson loops," or on open contours when the interaction of matter with the fundamental gauge bosons is analyzed. The concept of topological order has been the object of some fascination in recent years. Part of this activity is stimulated by the prospect of error-tolerant quantum computation – an issue to which we return below. The key property of topological order for this purpose is the observation that, even if global symmetry-breaking cannot occur, a system may nevertheless exhibit robust order of a topological type.

The crucial point we wish to emphasize here is that many orbital systems display new, intermediate symmetries. These lie in general mid-way between the extremes of global symmetries and local gauge symmetries. Many such orbital systems can be proven to exhibit topological order. To make this statement precise, we rephrase it mathematically in the following form.

*An intermediate $d$-dimensional symmetry of a theory [29] that is characterized by a Hamiltonian $H$ is a group of symmetry transformations such that the minimal non-empty set of fields $\{\phi_{\mathrm{i}}\}$ changed by the group operations occupies a $d$-dimensional sub-volume $(\mathcal{C})$ of the full $D$-dimensional spatial volume on which the theory is defined.*

To make contact with known cases, we note that local gauge symmetries have dimension $d = 0$. Similarly, in the example of a nearest-neighbor ferromagnet on a $D$-dimensional lattice, described by $H = -J \sum_{\langle ij \rangle} \boldsymbol{S}_i \cdot \boldsymbol{S}_j$, the system is invariant under a global rotation of all spins. As the volume influenced by the symmetry operation occupies a $D$-dimensional region, in this case $d = D$. Here, we show that symmetries of intermediate dimension $0 < d < D$ arise in orbital systems.

### 23.4.2 Examples of Intermediate Symmetries in Orbital Systems

The symmetries we review are of two principal kinds: (i) symmetry operations on the electronic orbital states and (ii) symmetry operations on the electronic spin. Both types of operation appear in systems with orbital-only interactions and orbital-dependent spin exchange. We begin our discussion by focusing on symmetry operations applied to orbitals.

*Orbitals* – In TM compounds on cubic lattices, crystal fields lift the degeneracy of the five $3d$ orbitals of the TM ion to two higher-energy $e_g$ levels, $\{|3z^2 - r^2\rangle, |x^2 - y^2\rangle\}$, and to three lower-energy $t_{2g}$ states, $\{|xy\rangle, |xz\rangle, |yz\rangle\}$ (Fig. 23.1). Superexchange processes lead to the KK Hamiltonian [3, 22]

$$H = \sum_{\langle ij\rangle \parallel \gamma} H^{(\gamma)}_{\text{orb}}(ij)\left(\boldsymbol{S}_i \cdot \boldsymbol{S}_j + \frac{1}{4}\right), \tag{23.2}$$

in which $\boldsymbol{S}_i$ is the spin of the electron at site $i$ and $H^{(\gamma)}_{orb}(ij)$ are operators acting on the orbital degrees of freedom. For TM ions arranged on a cubic lattice, where each is coordinated by an octahedral cage of anions, these operators are given by

$$H^{(\gamma)}_{\text{orb}}(ij) = J\left(4\hat{\pi}^{(\gamma)}_i \hat{\pi}^{(\gamma)}_j - 2\hat{\pi}^{(\gamma)}_i - 2\hat{\pi}^{(\gamma)}_j + 1\right), \tag{23.3}$$

where $\hat{\pi}^{(\gamma)}_i$ denotes the orbital pseudospin and $\gamma = a, b, c$ is the direction of the bond $\langle ij\rangle$.

(i) In $e_g$ systems (previous section),

$$\hat{\pi}^{(a,b)}_i = T^{(a,b)}_i \equiv \frac{1}{4}(-\sigma^z_i \pm \sqrt{3}\sigma^x_i), \qquad \hat{\pi}^{(c)}_i = T^{(c)}_i \equiv \frac{1}{2}\sigma^z_i, \tag{23.4}$$

operators in terms of which the orbital-only 120-degree Hamiltonian of (23.1) for Jahn–Teller-mediated interactions in $e_g$ systems is given by

$$H_{\text{orb}} = J \sum_{\langle ij\rangle \parallel \gamma} \hat{\pi}^{(\gamma)}_i \hat{\pi}^{(\gamma)}_j. \tag{23.5}$$

The form of (23.5) with substitutions different from those of (23.4) gives rise to Hamiltonians different from the 120-degree model, including the 90-degree (or orbital compass) model mentioned in Sect. 23.3.4.

With regard to intermediate symmetries, the models contained in (23.5) are rather special. Extra symmetries emerge only in the ground-state sector of large-$S$ (classical) variants of this system [30, 31]. Taking the standard $S \to \infty$ limit results in classical analogs of the different Hamiltonians, where the quantum variables are replaced by classical two- or three-component spins [30, 31].

The classical analog of (23.5) may be specified as follows. At each site, we assign a unit-length, two-component spin [associated with the two-dimensional (2D) $e_g$ subspace] denoted by $\boldsymbol{T}$. Let **a**, **b**, and **c** be vectors evenly spaced on the unit circle, *i.e.* separated from each other by 120° angles, and **c** be set at 0° with **a** and **b** at ±120°. Next, we define $T^{(c)} = \boldsymbol{T} \cdot \boldsymbol{c}$, and similarly for $T^{(a,b)}$, whence the classical 120-degree orbital model Hamiltonian is

$$H = J \sum_i \left(T^{(a)}_i T^{(a)}_{i+e_x} + T^{(b)}_i T^{(b)}_{i+e_y} + T^{(c)}_i T^{(c)}_{i+e_z}\right). \tag{23.6}$$

As anticipated above, this large-$S$ model exhibits a high number of symmetries in its ground-state sector, because $\pm 2\pi/3$ rotations of the pseudospin in individual planes leave the energy of the system invariant (Fig. 23.3). This results in

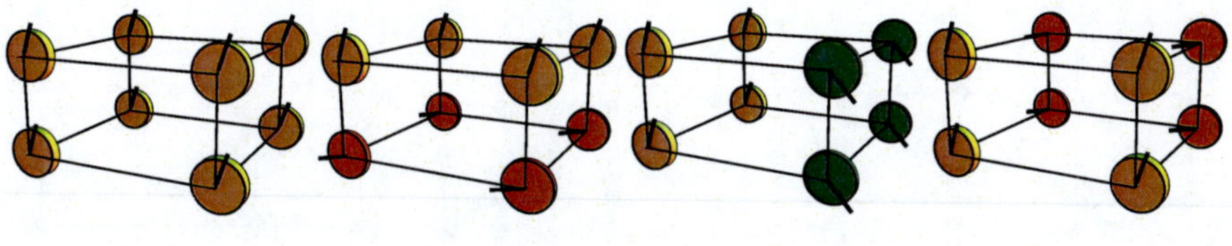

**Fig. 23.3** The symmetries of (23.7) applied to a uniform ground state (at left)

emerging $[Z_2]^{3L}$ gauge-type symmetry operators in two dimensions [27,29,32],

$$O^{(\gamma)} = \prod_{i \in P_\gamma} \mathrm{e}^{\mathrm{i}\pi \boldsymbol{T}_i \cdot \boldsymbol{e}_\gamma}, \tag{23.7}$$

where $P_\gamma$ is any plane orthogonal to the cubic $\boldsymbol{e}_\gamma$ axis. The peculiarity of the classical 120-degree orbital model is that these are not true symmetries over the entire spectrum.

(ii) For $t_{2g}$ compounds, we set in $H^\gamma_{\text{orb}}(ij)$ (23.5),

$$\hat{\pi}_i^{(\gamma)} = \tau_i^{(\gamma)} \equiv \frac{1}{2}\sigma_i^{(x,y,z)}, \tag{23.8}$$

which defines the 90-degree orbital compass model. The (exact) symmetries of this Hamiltonian are given by [27,29,32–35]

$$O^{(\gamma)} = \prod_{i \in P_\gamma} \hat{\pi}_i^{(\gamma)}, \tag{23.9}$$

the difference with the previous case being that operators of the form (23.9) commute with the Hamiltonian, $[O^{(\gamma)}, H] = 0$, *i.e.* rotations of individual planes about an orthogonal axis leave the system invariant.

Because the planes $P_\gamma$ are objects of spatial dimensionality $d = 2$, the symmetries of (23.9) are of dimension $d = 2$. The fact that different symmetry operators $O^{(\gamma)}$ of the orbital compass model do not commute implies that the ground-state sector, and indeed any other state, is at least twofold-degenerate [33]. (By Kramers' theorem, this statement also follows without the use of symmetries for any system with an odd number of spins [67–69].)

The 2D orbital compass model on the square lattice is given by (23.5,23.8) with $\gamma \in \{a, c\}$, and displays $d = 1$ $Z_2$ symmetries (where the planes $P$ of (23.9) become lines) [27,29,32–35]. Figure 23.4a, depicts the action of the $d = 1$ symmetry operations on a uniform, semi-classical ground state.

From an exact solution of the one-dimensional (1D) version of the model [36], one may conclude that the energy spectrum collapses at a quantum phase transition between two possible types of order, with short-range correlations either in $\sigma^z$ or $\sigma^x$, and is thus highly degenerate. This behavior is similar to the 2D case [37],

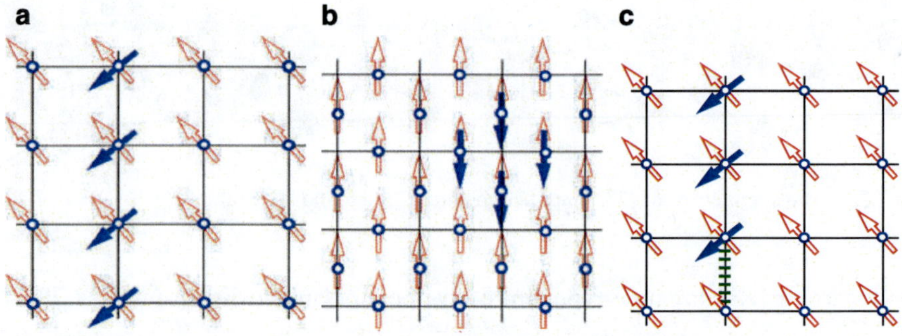


**Fig. 23.4** The two-dimensional orbital compass model ($D = 2$). (**a**) The action of the $d = 1$ symmetry operation of (23.9) when the "plane" $P$ is chosen to lie along the vertical axis. (**b**) A $d = 0$ (local) gauge symmetry. Defects within a gauge theory have a finite energy cost. Local symmetries such as this for an Ising lattice gauge theory cannot be broken. (**c**) A defect in a semi-classical ground state. Defects such as this do not allow for a finite on-site magnetization. The energy penalty for this defect is finite (there is only one unsatisfied bond, shown by the dashed line) whereas, precisely as in $d = 1$ Ising systems, the entropy associated with such defects increases monotonically with the system size (see text). Figure taken from [67–69]

where the degeneracy scales exponentially with the perimeter size ($\mathcal{O}(2^L)$) [35]. The possibility of such a high degeneracy is suggested by the $\mathcal{O}(L)$ $d = 1$ Ising symmetries $[(Z_2)^L]$ which the system possesses [27, 29, 32–35].

*Spins in TM compounds*: we label the three $t_{2g}$ orbital states by [38]

$$|a\rangle \equiv |yz\rangle, \qquad |b\rangle \equiv |xz\rangle, \qquad |c\rangle \equiv |xy\rangle. \tag{23.10}$$

In cubic $t_{2g}$ compounds, hopping through intermediate oxygen $p$ orbitals between any two electronic states of orbital flavor $|\gamma\rangle$ ($\gamma = a, b, c$) is forbidden along the $\gamma$-axis of the cubic lattice (Fig. 23.5). As a consequence [38], a uniform rotation of all spins whose electronic orbital state is $|\gamma\rangle$, in any plane ($P$) orthogonal to the $\gamma$ axis ($c^\dagger_{i\gamma\sigma} = \sum_\eta U^{(P)}_{\sigma,\eta} d^\dagger_{i\gamma\eta}$, with $\sigma$, $\eta$ the spin directions), leaves (23.2) invariant. The total spin of electrons of orbital flavor $|\gamma\rangle$ in any plane orthogonal to the cubic $\gamma$ axis is conserved. In this case, one has $d = 2$ SU(2) symmetries

$$\hat{O}_{P;\gamma} \equiv [\exp(i\boldsymbol{S}^\gamma_P \cdot \boldsymbol{\theta}^\gamma_P)/\hbar], \qquad [H, \hat{O}_{P;\gamma}] = 0, \tag{23.11}$$

where $\boldsymbol{S}^\gamma_P = \sum_{i \in P} \boldsymbol{S}^\gamma_i$ is the sum of all the spins $\boldsymbol{S}^{i,\gamma}$ in the orbital state $\gamma$ in any plane $P$ orthogonal to the direction $\gamma$ (Fig. 23.5).

### 23.4.3 A Theorem on Dimensional Reduction

The existence of intermediate symmetries has important consequences. The corresponding dimensional reduction is only with respect to expectation values of

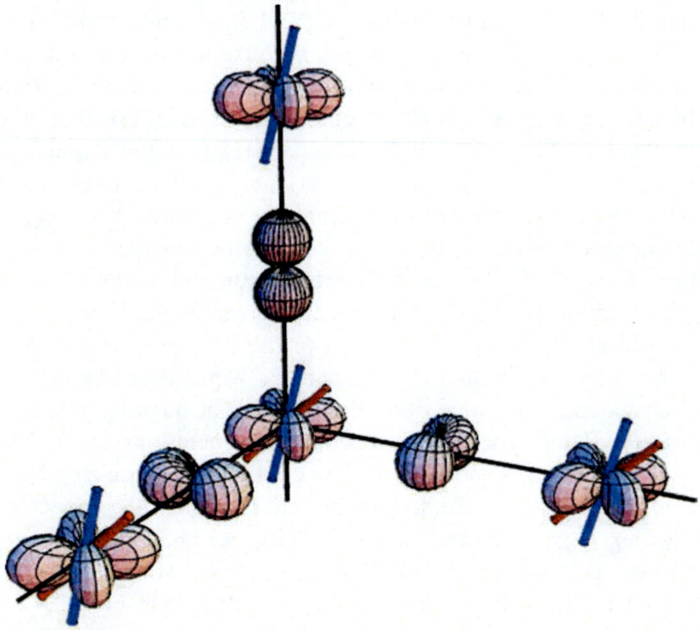

**Fig. 23.5** Representation of anisotropic hopping amplitudes leading to the KK Hamiltonian (after [29]). Spins are indicated by blue rods. Four-lobed states denote the $3d$ orbitals of a TM ion, while the intermediate $p$ states are oxygen orbitals through which the superexchange process occurs (*cf.* [38]). Orthogonality with the oxygen $p$ states forbids hopping between sites separated along the cubic $\gamma$-axis for orbital state $|\gamma\rangle$ ($|c\rangle \equiv |xy\rangle$ above). The resulting KK Hamiltonian has a $d = 2$ SU(2) symmetry corresponding to a uniform rotation of all spins whose orbital state is $|\gamma\rangle$ in any plane orthogonal to the cubic direction $\gamma$. Such a rotation in the $xy$ plane is indicated by the red spins

local quantities: the free energies of these systems and the transitions which they exhibit are generally those of systems in high dimensions [29]. More precisely, the expectation value of any quantity $\langle f \rangle$ in the original system (of dimension $D$) is bounded from above by the expectation value of the same quantity evaluated in a $d$-dimensional region,

$$|\langle f \rangle| \leq |\langle f \rangle_{H_d}|. \tag{23.12}$$

The expectation value $\langle f \rangle$ refers to that obtained in the original system (lattice), which resides in $D$ spatial dimensions. The Hamiltonian $H_d$ is defined in a $d$-dimensional subregion of the full system (lattice), with $d \leq D$. This $d$ dimensional region corresponds to the spatial region in which an intermediate $d$ dimensional symmetry is present. $H_d$ preserves the range of the interactions of the original system. It is formed by extracting from the full Hamiltonian (on the $D$-dimensional lattice) those parts which appear within the $d$-dimensional sub-region ($\mathcal{C}$) on which the symmetry operates. Fields (spins) external to $\mathcal{C}$ act as non-symmetry-breaking external fields in $H_d$. The upper bound of (23.12) becomes most powerful for quantities which are not symmetry-invariant, because in this case the expectation values $\langle f \rangle_{H_d}$ must vanish for low spatial dimensions $d$ (where no spontaneous symmetry-breaking can occur). Taken together with (23.12), this then implies that the expectation value $\langle f \rangle$ on the full $D$-dimensional lattice must vanish. By "non-invariant" is meant here that $f(\phi_\mathrm{i})$ vanishes when summed over all arguments related to each other by a $d$-dimensional symmetry operation, $\sum_k f[\mathbf{g}_{\mathrm{i}k}(\phi_\mathrm{i})] = 0$. We illustrate this result by two examples related to orbital models.

*Introductory Example I* – Consider the orbital compass model of (23.8), focusing on its 2D rendition ($D = 2$). This model possesses the $d = 1$ symmetries of (23.9), the planes $P_y$ being rays normal to the $\hat{e}_y$-direction. Let the quantity we consider be $f = \sigma_i^z$, with $i \in P_x$ and $P_x$ a plane orthogonal to the $\hat{e}_x$-direction which contains the lattice site $i$ (Fig. 23.4c). The quantity $f$ is not invariant under the discrete (Ising-type), $d = 1$ symmetry operations of (23.9), because $O^{(x)}\sigma_i^z O^{(x)} = -\sigma_i^z$. Because no discrete (nor continuous) symmetry-breaking can occur in one dimension at any finite temperature ($T > 0$), the expectation value must satisfy $\langle f \rangle_{H_{d=1}} = 0$. With the aid of (23.12), we observe that $\langle \sigma_i^z \rangle = 0$ also for the $D = 2$ orbital compass model. The physical mechanism for the loss of on-site order in $\langle \sigma_i^z \rangle$ is the proliferation of solitons, as shown in Fig. 23.4. As in one dimensional systems, the pertinent topological defects are domain walls (solitons) that entail only a finite amount of energy cost, while their entropy increases with system size. The Hamiltonian $H_{d=1}$ defined on the vertical chain of Fig. 23.4c where these operations appear is simply a one dimensional Ising Hamiltonian augmented by transverse fields generated by spins outside the chain. By virtue of their location outside the region where the symmetry of (23.9) operates, these transverse fields triggered by the spins $\sigma^x_{i \notin P_x}$ do not break the discrete $d = 1$ symmetry associated with the plane $P_x$. As in any one dimensional chain with finite range interactions (e.g., the one dimensional Ising model), this discrete $d = 1$ symmetry remains unbroken at any non-zero temperature (and, consequently, $\langle \sigma_i^z \rangle = 0$) due to the entropically favored proliferation of defects (domain walls/solitons) along the chain.

For a continuous symmetry, non-invariance means that $\int f[\mathbf{g}_\mathrm{i}(\phi_\mathrm{i})] d\mathbf{g} = 0$. Thermal fluctuations have a more dramatic effect for systems with continuous symmetries: no continuous symmetry-breaking can occur in systems with finite-range interactions in $d \leq 2$ dimensions. The bound specified by (23.12) for continuous symmetries implies even stricter results than it does for discrete symmetries: any quantity which is not invariant under $d \leq 2$ continuous symmetries must have vanishing expectation values at finite temperatures. Similarly, continuous, $d < 2$ symmetries cannot be broken even at zero temperature in systems which have an energy gap between the ground-state sector and all excited states.

*Introductory Example II* – An example of a continuous, $d = 2$ symmetry for which the bound of (23.12) prohibits local on-site order is provided by (23.11) for the KK model (23.2) of $t_{2g}$ systems (Fig. 23.5). We return to this example in Sect. 23.4.4, where we discuss in more depth the consequences of the bound on these systems.

We summarize for completeness the results of such symmetry-based analysis for general systems, reiterating our conclusions from the above examples. By choosing $f$ to be the order parameter or a two-particle correlator, one may arrive at the following general corollaries [29, 40].

*Corollary I:* any local quantity which is not invariant under local symmetries ($d = 0$) or symmetries acting on 1D regions ($d = 1$) has a vanishing expectation value $\langle f \rangle_{H_d}$ at any finite temperature. This follows because both zero- and 1D systems cannot exhibit symmetry breaking: in one- and two- dimensional systems, the expectation value of any local quantities not invariant under global symmetries are $\langle f \rangle = 0$. Physically, entropy overwhelms energetic penalties and forbids a symmetry breaking: as in zero- and one-dimensional systems, where much more entropy is gained than the energy lost on introducing defects (domain walls in discrete systems), the same energy-entropy (im)balance is replicated when these symmetries are embedded in higher dimensions. The particular case of local ($d = 0$) symmetry is the Elitzur theorem [39] familiar from gauge theories, which can be regarded more generally as a consequence of dimensional reduction.

*Corollary II:* the consequences may be extended by recalling that no symmetry-breaking occurs for continuous symmetries in two spatial dimensions. Here again, free-energy penalties are not sufficiently strong to induce order. When embedding continuous, $d = 2$ dimensional symmetries in higher dimensions, the energy-entropy balance remains the same and delivers the same result, $\langle f \rangle = 0$ at all finite temperatures for any quantity $f$ not invariant under continuous $d \leq 2$ symmetries. On noting further that order does not exist in continuous two-dimensional systems also at zero temperature in the presence of a gap between the ground and the first excited state, we find that for a $d < 2$-dimensional continuous symmetry, the expectation value of any local quantity not invariant under this symmetry vanishes strictly at zero temperature. However, although local order is forbidden, multi-particle (including topological) order can exist.

*Corollary III:* statements may be made concerning not only the absence of symmetry breaking, but also fractionalization of non-symmetry-invariant quantities in high-dimensional systems. This can occur if no (quasiparticle-type) resonant terms appear in the lower-dimensional spectral functions [40].

### 23.4.4 Consequences of the Theorem for Orbital (and Spin) Orders and Excitations

Because intermediate symmetries may be found in many types of orbital interaction, the bound of (23.12) and its corollaries reviewed above lead to a number of symmetry-based selection rules, some of which are presented below.

(a) *Selection rules for orbital and spin order in $t_{2g}$ systems* – We begin by returning to review Introductory Example II in a more general context. If the KK Hamiltonian (23.2) captures the spin physics of $t_{2g}$ compounds, then no magnetization can exist at finite temperatures [29] due to the continuous, $d = 2$ symmetries it displays (23.11) [29, 38]. Empirically, however, a low-temperature magnetization is detected in many systems. Thus the Hamiltonian relevant in these materials must be one in which (23.2) is augmented by other interactions which lift this symmetry. By relying on the theorem [29], we can propose the following experimental test of a selection rule for orbital order: if in a $t_{2g}$ system with $H = H_{\mathrm{KK}}$ one has orbital ordering in which all electrons are, for example, in states $|xy\rangle$ then, by the theorem of [29],

$$I(k_x, k_y, z, \omega) = \int \mathrm{d}k_z\ S(\boldsymbol{k}, \omega)\ \mathrm{e}^{\mathrm{i}k_z z} = 0 \tag{23.13}$$

for all $z \neq 0$. Here, $S(\boldsymbol{k}, \omega)$ is the Fourier-transformed spin-spin correlation function (the "structure factor"). Only the $z = 0$ component of $I$ will be finite. The origin of this "selection rule" is that for $z \neq 0$, one of the spins appearing in the two-spin correlation function may be rotated independently of the other with no change in energy. The quantity $I(k_x, k_y, z = 0, \omega)$ is in general always finite, and may show a Kosterlitz-Thouless-type transition. Similar considerations apply to more realistic, non-uniform orbital order. If, for example, both $|xy\rangle$ and $|yz\rangle$ orbitals order, then half of the contributions to $I$ for $z \neq 0$ will vanish by symmetry. It may be possible to test this selection rule on such 1D linear transforms of $S(\mathbf{k}, \omega)$ by considering existing data to ascertain whether $|I(z \neq 0)| \ll I(z = 0)$ for temperatures below the onset of orbital order ($T < T_{\mathrm{OO}}$).

(b) *Orbital order* – In $D = 3$ dimensions, the orbital-only Hamiltonians discussed in the preceding section exhibit a discrete, 2D $Z_2$ symmetry. The theorem of [29] allows such symmetries to be broken, and indeed, as we discuss in the next section, order already appears in these models at the classical level, a tendency which is readily enhanced by quantum fluctuations.

(c) *Nematic order in 2D orbital systems* – As discussed in Introductory Example I, the 2D orbital compass model exhibits a $d = 1$ $Z_2$ symmetry. Because these symmetries cannot be broken, no magnetization is possible, $\langle \sigma^\alpha \rangle = 0$. The simplest symmetry-allowed order parameter for spin $S > 1/2$ versions of the orbital compass model is of the nematic type. This sort of order has been shown to be realized classically [27, 32, 41].

(d) *Fractionalization in spin and orbital systems* – Corollary III allows for fractionalization in quantum systems where $d = 1$ or 2. It enables symmetry-invariant, quasiparticle excitations to coexist with non-symmetry-invariant, fractionalized excitations. Fractional excitations may propagate in $d_s = (D - d)$-dimensional regions ($D$ is the spatial dimensionality of the system). Examples can be found of frustrated spin models where spinons may propagate along lines ($d_s = 1$) on the square lattice [42] and in $d_s = D$-dimensional regions on the pyrochlore lattice [43].

## 23.5 Order by Disorder in Classical Orbital Models

When symmetry-breaking is allowed by the theorem of [29] [the bound of (23.12) and its corollaries (Sect. 23.4.3)], order often emerges by a fluctuation-driven mechanism ("order by disorder") [44, 45]. Although a number of states may appear to be equally valid candidate ground states, fluctuations act to stabilize those states which have the largest phase-space volume for low-energy fluctuations. These differences are captured in the values of the free energies for fluctuations about the contending states. Classically, fluctuations are driven by thermal effects, and quantum tunneling processes may enhance such tendencies.

When entropic contributions are omitted, the spin-wave spectrum of the classical model is gapless [15, 26, 46, 47]. This suggests that, on the classical level, these orbital systems exhibit finite-temperature disorder, and it was generally believed for some time that quantum fluctuations (tunneling between the different contending classical ground states) were mandatory for the lifting of orbital degeneracy required to account for the experimentally detected orbital order. Most studies of "quantum order by disorder" have thus focused on $1/S$ corrections ($S$ representing the spin size) to the classical spin-wave spectrum.

The classical (or large $S$ (i.e., $S \to \infty$)) renditions of the orbital models that we discussed exhibit their most extreme degeneracies within their ground state sector. In addition to the intermediate symmetries discussed earlier, within the ground state sector of their classical renditions, the orbital compass and 120°-models are invariant under a global continuous rotation [27, 29] in which all of the fields are rotated uniformly. In particular, any uniform field configuration is a ground state of the classical renditions of these models. Such an invariance does not reflect a true symmetry of the Hamiltonian (an "exact symmetry") but rather only *emerges* in the low energy limit of these classical models; such invariances are often known as *emergent symmetries*. Compounding these emergent global rotational symmetries, orbital systems such as the classical 120°-model further exhibit emergent intermediate symmetries in their ground state sector leading to an even larger degeneracy of the ground state subspace. As we discussed earlier and is depicted in Fig. 23.3, in the classical 120°-model, on starting from a given constant-field ground state, another ground state may be obtained by a symmetry operation such as reflecting all spins in the $xy$-plane through the vector $\boldsymbol{e}_c$ (see (23.6) for the definition of the unit vector $\boldsymbol{e}_c$). This operation is captured by the $d = 2$ operator of (23.7). The new ground state can be further altered by introducing more reflections of this type in other, parallel $xy$-planes. Obviously, analogous alterations of the uniform states can take place in the other two coordinate directions. These states make up the set of all possible ground states of the classical 120-degree model: there is one direction of stratification (layering); the corresponding projection of $\boldsymbol{S}_i$ is constant throughout the system, leaving two possibilities for the other projections. In planes orthogonal to the stratification direction, either of these choices can be implemented independently. This may be seen by considering all elementary cubes with a single spin fixed, and ensuring consistency in the tiling of the lattice (Fig. 23.3).

We next review in brief the effects of finite temperatures on the classical 120-degree model of (23.6) [27]. It is essential to note that the free-energy minima, not the energy minima, determine the ground states. Classical spins $\{\boldsymbol{S}_i\}$ are parameterized by their angle $\{\theta_i\}$ with the $a$-axis, and in the uniform ground states all $\theta_i = \theta^\star$. At low temperatures, fluctuations are described by the small deviations $\vartheta_i = \theta_i - \theta^\star$, and the quadratic (spin-wave) Hamiltonian corresponding to (23.6) becomes [27, 32]

$$H_{\rm SW} = \frac{1}{2} J \sum_{i,\gamma} q_\gamma(\theta^\star)\,(\vartheta_i - \vartheta_{i+e_\gamma})^2, \tag{23.14}$$

where $\gamma = a, b, c$ while $q_c(\theta^\star) = \sin^2(\theta^*)$, $q_a(\theta^\star) = \sin^2(\theta^\star + 2\pi/3)$, and $q_b(\theta^*) = \sin^2(\theta^\star - 2\pi/3)$. On a cubic lattice with periodic boundary conditions and $\theta^*$ the average of $\theta_i$ on the lattice, the partition function is given by [27, 32]

$$Z(\theta^\star) = \int \delta\Big(\sum_i \vartheta_i = 0\Big)\, \mathrm{e}^{-\beta H_{\rm SW}} \prod_i \frac{\mathrm{d}\vartheta_i}{\sqrt{2\pi}}, \tag{23.15}$$

and a Gaussian integration leads to

$$\log Z(\theta^\star) = -\frac{1}{2} \sum_{k \neq 0} \log\Big\{\sum_\gamma \beta J q_\gamma(\theta^\star)\, E_\gamma(\boldsymbol{k})\Big\}, \tag{23.16}$$

where $\boldsymbol{k} = (k_x, k_y, k_z)$ is a reciprocal-lattice vector and $E_\gamma(\boldsymbol{k}) \equiv 2 - 2\cos k_\gamma$. The spin-wave free energy $\mathcal{F}(\theta^*)$ of (23.16) has minima at $\theta^* = n\pi/3$ with integer $n$ [32].

The application of the $d = 2$ symmetry operations of (23.7) on each of these uniform configurations (Fig. 23.3) leads to an interface with an effective surface tension, a free-energy energy penalty additive in the number of operations [27, 32]. Finite-temperature fluctuations stabilize these six uniform ground states and favor them above all others. By "blocking" the lattice and employing reflection positivity bounds [27,32], it can indeed be proven that the results of the spin-wave analysis are correct: the free energy has strict minima for six uniform orientations [27,32], $\boldsymbol{S}_i = \pm S\boldsymbol{e}_a$, $\boldsymbol{S}_i = \pm S\boldsymbol{e}_b$, and $\boldsymbol{S}_i = \pm S\boldsymbol{e}_c$. Thus, out of the large number of classical ground states, only six are chosen. Orbital order already appears within the classical ($S \to \infty$) limit [27, 32] and does not rely exclusively on subtle quantum (zero-point) fluctuations of the type captured by $1/S$ calculations for its stabilization. Indeed, orbital order is detected up to relatively high temperatures [$\mathcal{O}(100K)$] in some systems [2, 48–51]. However, in a recent numerical study by Dorier et al. [35], it was shown that quantum fluctuations may not act to lift the orbital degeneracy in the simplest $S = 1/2$ system, the planar orbital compass model [the 2D version of (23.5,23.8)]. A 2D, spin-1/2 analog of the three-dimensional (3D) 120-degree model of (23.3,23.4), a model with far lower symmetry (and frustration) than the planar orbital compass model, has been shown recently to have a $S = 0$ order [32]. A mean-field analysis of the $T = 1/2$ orbital compass model on the square lattice [52] suggests that, at zero temperature, the symmetric point $J_x = J_z$ may mark a first-order quantum transition similar to the 1D case [36, 37].

We conclude this section with a brief summary of recent results on diluted (or "doped") orbital compass-type systems [53, 54]. These investigations found that the critical doping fraction ($x = 1/2$) necessary to suppress order is smaller than the doping required in the problem of diluted magnets (such as $KCu_{1-x}Zn_xF_3$) [55,56]. In typical magnetic systems, the decrease in the ordering temperature and saturation magnetization are governed by the percolation threshold (where the ordering temperature vanishes at a critical dopant concentration of $x_c = 0.69$ for the simple cubic lattice). The faster degradation of orbital order with doping compared to simple percolation physics can be attributed to the directional character of the orbital exchange interactions. Similar effects have been found in related systems [57]. The concept of an orbital-driven quantum critical point was introduced [58] by considering an exact solution of diluted 2D and 3D orbital compass models. Similar to charge- and spin-driven quantum critical fluctuations, orbital fluctuations may also drive the system to quantum criticality. In real materials, such fluctuations may be altered, and hence the system driven to criticality, by a combination of doping and uniaxial pressure or strain [58]. Such a quantum critical point was considered for spin-orbital singlets in [59]. Similarly, it can be shown that spin-glass-type behavior can arise in diluted orbital systems with random exchange constants [58], where the orbitals take the role of the spins in conventional spin-glass systems.

## 23.6 Connection with Quantum Computation

Quantum states can in principle be encoded in a fault-tolerant manner in order to be protected against decoherence [60]. If one assumes that errors are of a local nature, topological quantum memories [61, 62] would seem to be intrinsically stable because of physical fault-tolerance to weak, quasi-local perturbations. In this section, we describe two of the prototypical models which have been invented to explore the basic principles of topological quantum computing, and mention their connection with spin-orbital Hamiltonians. The first model ("Kitaev's Toric Code model") is essentially a pure Ising gauge theory. The second ("Kitaev's honeycomb model") is far richer, and enables non-Abelian quantum computation. The form of Kitaev's honeycomb model is reminescent of the orbital models of (23.5), but defined on a hexagonal lattice. In its more stable (gapped) phase, Kitaev's honeycomb model can be related to the Toric Code model).

### *23.6.1 Kitaev's Honeycomb Model*

This model is very similar to both the orbital compass model and the 120-degree model. It reduces to Kitaev's Toric Code model in the limit that one coupling constant is much larger than all of the others. The Kitaev model on the hexagonal lattice [62] is defined by the $S = 1/2$ Hamiltonian (Fig. 23.6)

$$H_{K_h} = -J_x \sum_{x\text{–bonds}} \sigma_j^x \sigma_k^x - J_y \sum_{y\text{–bonds}} \sigma_j^y \sigma_k^y - J_z \sum_{z\text{–bonds}} \sigma_j^z \sigma_k^z, \quad (23.17)$$

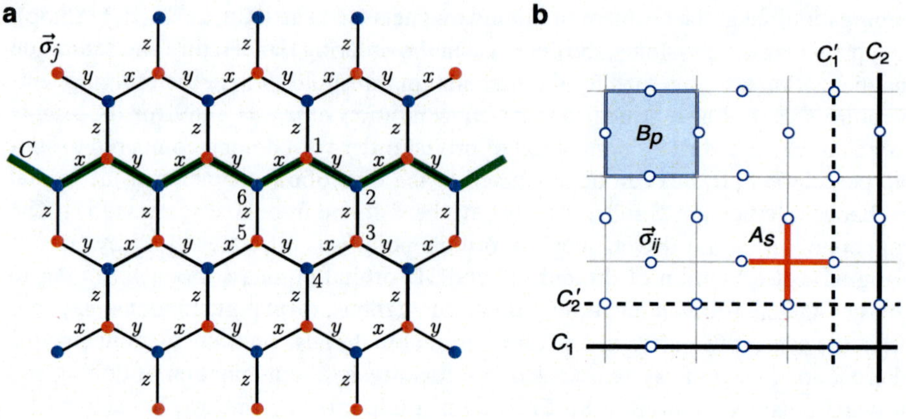


**Fig. 23.6** (**a**) Kitaev model on a honeycomb lattice with three types of bond. On each vertex there is a $S = 1/2$ degree of freedom, indicated by a Pauli matrix $\boldsymbol{\sigma}_j$ (see text). (**b**) Elementary plaquette ($B_p$) and star ($A_s$) interaction terms in Kitaev's Toric Code model. Empty circles on the bonds (links) represent a $S = 1/2$ degree of freedom, while thick (*dashed or solid*) lines represent topological ($d = 1$) symmetry operators (see text). Figure taken from [67–69]

and is exactly solvable in its ground-state sector [62,71]. Local ($d = 0$) symmetries are given by the products of spins around any hexagon (Fig. 23.6),

$$O_h = \sigma_1^x \sigma_2^y \sigma_3^z \sigma_4^x \sigma_5^y \sigma_6^z. \tag{23.18}$$

The model also has $d = 1$ symmetries given for example by $\prod_{i \in C} \sigma_i^\alpha$ for any loop $C$ that spans the system, when it is placed on a torus, in the $a$-direction. A schematic representation of such a $d = 1$ invariant is given in Fig. 23.6, where for the contour $C$ illustrated, the product of $\{\sigma_i^z\}$ is taken for $i \in C$.

A recent study [63] suggested that it may be possible to engineer, by the use of spin-orbit coupling, interactions of Kitaev- and orbital-model type in Mott insulators. The consideration of doubly degenerate orbital systems on the honeycomb lattice has led a number of authors to study of a 120-degree version of Kitaev's honeycomb model, namely (23.1) with the $a$-, $b$-, and $c$-axis directions taken parallel to the three bonds in the honeycomb lattice [64–66]. Dimer-covering overlap matrix elements [64] may be computed in a manner similar to that employed in spin systems [43], and shown to reproduce faithfully the ground states. The dimers in the 120-degree version of the Kitaev model correspond to orbital states on bonds connecting two nearest-neighbor sites that minimize the bond energy [64]. Similar to the orbital compass and 120-degree models, orbital order by disorder was found to occur in this system.

In the limit $J_z \gg J_{x(y)}$, Kitaev's honeycomb model (23.17) reduces to the Toric Code model of (23.19,23.20) [62]. In this limit, the spins on all of the bonds along the $z$-direction are so strongly correlated that they may be replaced by one effective spin. The resulting geometry is the square lattice of Kitaev's Toric Code model.

### 23.6.2 Kitaev's Toric Code model

Kitaev's Toric Code model [61] in $D = 2$ spatial dimensions is defined on a square lattice with $L \times L = N_s$ sites, where on each link $\langle ij \rangle$ there is a $S = 1/2$ degree of freedom (Fig. 23.6b). The Hamiltonian is

$$H_K = -\sum_s A_s - \sum_p B_p, \tag{23.19}$$

with operators

$$A_s = \prod_{\langle ij \rangle \in \mathrm{star}(s)} \sigma_{ij}^x, \qquad B_p = \prod_{\langle ij \rangle \in \mathrm{plaquette}(p)} \sigma_{ij}^z, \tag{23.20}$$

and Pauli matrices $\sigma_{ij}^\kappa$ ($\kappa = x, y, z$). $B_p$ and $A_s$ describe respectively the plaquette (or face) and star (or vertex) operators, with

$$[A_s, A_{s'}] = [B_p, B_{p'}] = [A_s, B_p] = 0 \tag{23.21}$$

$\forall\, s, s', p, p'$, which generate an Abelian group known as the "code stabilizer" [61]. The model has two $d = 1$ $\mathbb{Z}_2$ symmetries, given by [61]

$$\begin{gathered} Z_{1,2} = \prod_{\langle ij \rangle \in C_{1,2}} \sigma_{ij}^z, \qquad X_{1,2} = \prod_{\langle ij \rangle \in C'_{1,2}} \sigma_{ij}^x, \\ \{X_\mu, Z_\mu\} = 0, \qquad [X_\mu, Z_\nu] = 0, \qquad \mu \neq \nu, \end{gathered} \tag{23.22}$$

where $C_1$ ($C'_2$) are horizontal and $C_2$ ($C'_1$) vertical closed contours [*i.e.* loops on the lattice (dual lattice)]. The non-commutativity of the $d = 1$ symmetry operators $X_\mu$ and $Z_\mu$ implies that the ground-state sector is degenerate. It is on basis states in this ground-state manifold that schemes for topological quantum computing have been devised. The logical operators $Z_{1,2}$ and $X_{1,2}$ commute with the code stabilizer but are not part of it, thus acting non-trivially on the two "encoded" Toric Code qubits.

As shown in [67–69], $H_K$ is related to the plaquette model of Wen [70] and to a model of two Ising chains by exact duality mappings. These three models therefore share the same spectrum. The ground state (protected subspace of the code) is fourfold-degenerate (Abelian $\mathbb{Z}_2 \times \mathbb{Z}_2$ symmetry) and there is a gap to all excitations. The spectrum is essentially that of two uncoupled, circular Ising chains of length $N_s$ ($2N_s$ being the total number of links of the original $D = 2$ lattice). From this mapping it is apparent that, in thermal equilibrium, this model is disordered at all non-zero temperatures. However, if temperatures far below the spectral gap can be achieved, the autocorrelation time associated with the topological invariants (the $d = 1$ symmetries) of (23.22) can be made very large [67–69].

The elementary excitations of $H_K$ are of two types [61],

$$|\Psi_z(\Gamma)\rangle = \prod_{\langle ij\rangle\in\Gamma} \sigma_{ij}^z |\Psi_0\rangle \equiv S^z(\Gamma)|\Psi_0\rangle,$$
$$|\Psi_x(\Gamma')\rangle = \prod_{\langle ij\rangle\in\Gamma'} \sigma_{ij}^x |\Psi_0\rangle \equiv S^x(\Gamma')|\Psi_0\rangle, \tag{23.23}$$

where $\Gamma$ ($\Gamma'$) is an open string on the lattice (dual lattice) and $|\Psi_0\rangle$ is a ground state. [If $\Gamma$ ($\Gamma'$) were a closed contour circumscribing an entire Toric cycle, then the string operators $S^{x,z}$ would become the Toric symmetries of (23.22)]. In the case of the open contours of (23.23), the operators $S^{x,z}$ generate excitations at the end points of these strings (thus always coming in pairs), which have abelian fractional statistics (anyons). Excitations based on the vertices are the analogs of electric charges while those on the plaquettes are magnetic vortices. These effective electric and magnetic excitations obey fusion rules which enable abelian quantum computation. Due to the exact equivalence between Kitaev's Toric Code model and two Ising chains, no non-trivial, finite-temperature, spontaneous symmetry-breaking, or any other transition, can take place. The spectrum exhibits a multitude of low-energy states. At any (arbitrarily low) finite temperature, entropic contributions to the free energy overwhelm energy penalties, leading to a free energy which is analytic everywhere.

### *23.6.3 Recent Discussions of Quantum Computing Realizations*

There has been extensive recent activity in the study of orbital-related models as candidates for topological quantum memories. One variant was proposed by Bacon [72], while Douçot et al. [33] discussed a possible implementation of the 2D orbital compass model in a Josephson-junction array. Orbital and general frustrated models such as those of Kitaev may be implemented in atomic gases [73,74]. Another potential realization, closer to the orbital models we discuss here, was investigated by Jackeli and Khaliullin [63].

## 23.7 Spin-Orbital Frustration

### *23.7.1 General Structure of Spin-Orbital Superexchange Models*

Frustration of orbital interactions takes on a new quality in the spin-orbital superexchange models which originate from the realistic electronic structure of TM oxides with partly filled (quasi-)degenerate $e_g$ or $t_{2g}$ orbitals. As explained in Sect. 23.3, frustration of orbital interactions is intrinsic in such models, and occurs even in unfrustrated lattice geometries, such as square or cubic lattices. However, in some cases this frustration may be removed by the onset of an appropriate combination of coexisting spin and orbital order. The topic of spin-orbital models is the subject of much active research, and a rich variety of both spin and orbital order, as well as phases where one or both are disordered, has been found in these systems. Here we concentrate only on some representative physical systems to demonstrate the reasons behind spin-orbital frustration and, in some cases, the mechanisms for removing it.

We begin this section by considering models for the perovskites. Thus we take magnetic TM ions of $d^n$ configuration in a Mott (or charge-transfer) insulator, which are located on a 3D lattice with three cubic axes labeled by $\gamma = a, b, c$. As a consequence of the strong Coulomb interaction, $U \gg t$, the low-energy physics is described fully by the superexchange interaction $J = 4t^2/U$; here $t$ is an effective hopping matrix element (through the ligand orbitals), between either two directional $e_g$ orbitals [$(dd\sigma)$ element] or two identical $t_{2g}$ orbitals [$(dd\pi)$ element], belonging to the TM ions at neighboring sites. As discussed in Sect. 23.4, spin-orbital superexchange (23.2) in TM perovskites involves SU(2)-symmetric spin interactions $\mathbf{S}_i \cdot \mathbf{S}_j$ on the bond $\langle ij\rangle$, coupled to orbital operators which obey only a lower (cubic) symmetry, and its general form is [5–7]

$$\mathcal{H}_J = J \sum_{\langle ij\rangle \parallel \gamma} \left\{ (\boldsymbol{S}_i \cdot \boldsymbol{S}_j)\, \hat{J}_{ij}^{(\gamma)} + \hat{K}_{ij}^{(\gamma)} \right\}. \tag{23.24}$$

Here, the orbital operators $\hat{J}_{ij}^{(\gamma)}$ and $\hat{K}_{ij}^{(\gamma)}$ involve active orbitals on each individual bond $\langle ij\rangle \parallel \gamma$, either $e_g$ or $t_{2g}$, which participate in $d_i^n d_j^n \rightleftharpoons d_i^{n+1} d_j^{n-1}$ virtual charge excitations, so in a perovskite lattice they have a cubic symmetry. These orbital operators depend on the $T = 1/2$ pseudospin operators at sites $i$ and $j$, as introduced in Sect. 23.3 (this holds also for $t_{2g}$ orbitals, because only two of these are active in any given cubic direction $\gamma$). The form of the operators $\hat{J}_{ij}^{(\gamma)}$ and $\hat{K}_{ij}^{(\gamma)}$ depends on the system under consideration, *i.e.* on the actual valence $n$ of the TM ions in the ground state of the $d^n$ electronic configuration, and on the type ($e_g$ or $t_{2g}$) of active orbital degrees of freedom.

### 23.7.2 Spin-Orbital Models for $e_g$ Perovskites

Our first example is the KK model derived for $Cu^{2+}$ ions ($d^9$ configuration) in $KCuF_3$ [15, 47]. Its structure becomes transparent when one realizes that virtual excitations on the bond $\langle ij\rangle$ involve a pair of $e_g$ orbitals (at least one of which must allow hopping of its electron), and the spin state on each bond, which is conserved in the charge excitation, may be either a triplet or a singlet, with respective projection operators $P_{S=1} = (\boldsymbol{S}_i \cdot \boldsymbol{S}_j + \frac{3}{4})$ and $P_{S=0} = (\frac{1}{4} - \boldsymbol{S}_i \cdot \boldsymbol{S}_j)$. The orbital projections on a directional orbital $|\zeta\rangle$ [$(3z^2 - r^2)$-type] oriented along the bond (and along the cubic axis) $\gamma$, or $|\xi\rangle$ [$(x^2 - y^2)$-type] oriented orthogonal to it, are $\mathcal{P}_{i\zeta} \equiv (\frac{1}{2} - T_i^{(\gamma)})$ and $\mathcal{P}_{i\xi} \equiv (\frac{1}{2} + T_i^{(\gamma)})$, with $\mathcal{P}_{i\zeta} + \mathcal{P}_{i\xi} = 1$ and $T_i^{(\gamma)}$ defined as in (23.4).

The Hamiltonian consists of three terms,

$$\mathcal{H}(d^9) = \frac{1}{2}J \sum_{\langle ij\rangle \| \gamma} \left\{ \left[ -r_1 \left( \boldsymbol{S}_i \cdot \boldsymbol{S}_j + \frac{3}{4} \right) + r_2 \left( \boldsymbol{S}_i \cdot \boldsymbol{S}_j - \frac{1}{4} \right) \right] \left( \frac{1}{4} - T_i^{(\gamma)} T_j^{(\gamma)} \right) \right.$$
$$\left. + (r_2 + r_4) \left( \boldsymbol{S}_i \cdot \boldsymbol{S}_j - \frac{1}{4} \right) \left( \frac{1}{2} - T_i^{(\gamma)} \right) \left( \frac{1}{2} - T_j^{(\gamma)} \right) \right\}, \tag{23.25}$$

which follow from virtual excitations at sites $i$ and $j$ involving either two holes in different orbitals (first line), with singlet and triplet excited states, or two holes in $|\zeta\rangle$ orbitals (second line), leading to spin singlets in doubly occupied $|\zeta\rangle$ orbitals. In addition to the common energy scale $J$, the interactions are parameterized by the ratio of the Hund exchange $J_H$ to the intraorbital Coulomb element $U$ [5–7],

$$\eta = \frac{J_H}{U}. \tag{23.26}$$

The coefficients $r_1, r_2$, and $r_4$ in (23.25) depend on $\eta$, which typically lies in the range $0.1 < \eta < 0.2$, according to

$$r_1 = \frac{1}{1 - 3\eta}, \qquad r_2 = \frac{1}{1 - \eta}, \qquad r_4 = \frac{1}{1 + \eta}, \tag{23.27}$$

and are determined by the multiplet structure of the excited $d^8$ ions [8], where the excitation energies are $(U - 3J_H)$ for a spin triplet and $(U - J_H)$ or $(U + J_H)$ for the possible spin-singlet states (the first of which is doubly degenerate) [47].

Individual terms in the Hamiltonian $\mathcal{H}(d^9)$ (23.25) can be minimized for chosen spin and orbital configurations: when two different ($|\zeta\rangle$ and $|\xi\rangle$) orbitals are occupied, the favored spin state is a triplet (for finite Hund exchange, $\eta > 0$), while for two $|\zeta\rangle$ orbitals the favored spin state is a singlet. This behavior highlights immediately the intrinsic frustration of the spin-orbital model (23.25), and is also consistent with the GKR [9, 10]. These dictate that alternating orbital (AO) order favors ferromagnetic (FM) spin order, while ferro-orbital (FO) order favors antiferromagnetic (AF) spin order.

The interactions in (23.25) are particularly strongly frustrated at $\eta = 0$, where

$$\mathcal{H}_0(d^9) = \frac{1}{2}J \sum_{\langle ij\rangle \| \gamma} \left\{ 2 \left( \boldsymbol{S}_i \cdot \boldsymbol{S}_j - \frac{1}{4} \right) \left( \frac{1}{2} - T_i^{(\gamma)} \right) \left( \frac{1}{2} - T_j^{(\gamma)} \right) - \left( \frac{1}{4} - T_i^{(\gamma)} T_j^{(\gamma)} \right) \right\}, \tag{23.28}$$

implying that on each bond the energy is minimized by a singlet state when both orbitals (occupied by holes) are oriented along the bond. This frustration is absent only in a 1D model, where for $\gamma \equiv c$ the occupied orbitals may be taken parallel to the $c$-axis. Then both orbital operators give a constant $\langle T_{i(j)}^{(c)} \rangle = -\frac{1}{2}$, whence $\langle (\frac{1}{2} - T_i^{(c)})(\frac{1}{2} - T_j^{(c)}) \rangle = 1$. The resulting AF Heisenberg model,

$$\mathcal{H}_J = J \sum_{\langle ij \rangle \parallel c} \left( \boldsymbol{S}_i \cdot \boldsymbol{S}_j - \frac{1}{4} \right), \tag{23.29}$$

is equivalent to that derived from the nondegenerate Hubbard model. Here the orbitals play no role because their state has been fixed to maximize the energy gain from the nearest-neighbor spin correlations, similar to the situation encountered on a single bond. However, this becomes impossible when the coordination increases with the dimension of the lattice, *i.e.* on a square (2D) or a cubic lattice (3D). In these cases, for any frozen orbital configuration on the lattice, the orbitals will not be oriented along the bond axes on some (or even all) of the bonds, giving $\langle (\frac{1}{2} - T_i^{(\gamma)})(\frac{1}{2} - T_j^{(\gamma)}) \rangle < 1$, and thus the spin exchange interaction and corresponding energy gains per bond are reduced.

To appreciate further the intrinsic frustration of orbital interactions, let us compare the exact ground states of the Heisenberg (23.29) and spin-orbital (23.28) models for a plaquette in the *ab* plane. The ground state of the spin model is a spin singlet obtained as a symmetric combination of two valence-bond (VB) wave functions, with singlets either on two horizontal or on two vertical bonds. It is crucial to examine the resonance processes stabilising these VB states, which leads to the energy $E_{\square}^{(s)} = -3J$ (here and below the energy includes a constant $-\frac{1}{4}J$ per bond which is contained in the superexchange term). Indeed, this value is obtained for a symmetric (bonding) linear combination of two VB states, with two spin singlets occupying either the horizontal or the vertical bonds, so this state may be seen as a bonding state formed by these two configurations. The energy of each VB state on the plaquette is $E_{\text{VB}}^{(s)} = -2.5J$, and the energy gain due to the resonance amounts to $E_{\square}^{(s)} - E_{\text{VB}}^{(s)} = -0.5J$. When the same calculation is performed for the spin-orbital model (23.28), one finds instead much higher energies: $E_{\square}^{(so)} = -2.3744J$ for the plaquette ground state [75] and $E_{\text{VB}}^{(so)} = -2.2188J$ for each individual VB state. In the latter, the spin singlets are associated with a $|\zeta_i \zeta_j\rangle$ orbital configuration in which the orbitals are oriented along each singlet bond (the other bonds give only a small energy contribution because the orbitals are neither aligned with these bonds nor in the opposite states). Note that the energy gain due to the resonance, as well as the overlap, between the two VB states is reduced in the spin-orbital model when compared to the spin model. This illustrative example demonstrates that the orbital variables frustrate spin interactions due to their directional nature and partly suppress the resonance between different VB states, contributing to an intrinsic frustration of spin-orbital superexchange on square and cubic lattices. In fact this frustration occurs predominantly in the orbital sector (Sects. 23.4 and 23.5), but the combined spin-orbital state is always entangled (Sect. 23.8).

Frustration of spin-orbital interactions is removed partially in real materials due to the finite Hund exchange $J_H$ (and to Jahn–Teller distortions, discussed below). This acts to modify the strength of the interaction terms in (23.25), favoring FM superexchange. As a compromise between competing superexchange terms, the state known as an *A*-type AF (*A*-AF) phase, which has FM planes alternating their spin direction along the *c* axis, emerges in $KCuF_3$ below a rather low Néel temperature, $T_N \sim 38$ K [76]. This systems is best considered as AF chains along the $c$ axis, with weak FM interactions which induce FM order within the $ab$ planes, accompanied by an AO order (this order is of $C$-type, where the orbitals repeat along the $c$-axis and alternate in the $ab$-planes). This compound actually provides a good example of dimensional reduction, because the competing magnetic interactions nearly compensate each other in the $ab$-planes (when charge-transfer contributions are included [5–7]), and as a result $KCuF_3$ becomes one of the best-known realizations of a disordered AF Heisenberg chain, exhibiting a spinon excitation spectrum over a broad temperature range [76].

Another mechanism which removes frustration is the coupling to the lattice. When large, local Jahn–Teller distortions occur for $RO_6$ octahedra, where $R$ represents a TM ion, they remove the orbital degeneracy and induce an orbital splitting $E_z$ (parameterized by $\varepsilon = E_z/J$), so that the orbital dynamics is quenched and the orbital operators may be simply replaced by their expectation values. An example here is a 2D $G$-type AF order ($G$-AF, AF in both directions in each $ab$ plane) observed in $La_2CuO_4$, where the holes occupy $x^2 - y^2$ orbitals.

Whether or not some kind of joint spin-orbital order may emerge at the point $(\eta, \varepsilon) = (0, 0)$ in the model of (23.28) remains the subject of some controversy [77, 78], and this case is regarded as a good candidate for a spin-liquid state with only short-range orbital correlations. To date it has been established that the long-range order in the $G$-AF and $A$-AF phases is destabilized near this point due to enhanced quantum fluctuations [78], and wave functions with short-range VB-type orbital correlations are favored instead [15].

Superexchange interactions between the $Mn^{3+}$ ($d^4$) ions, which have $S = 2$ spins, in $LaMnO_3$ show an intrinsic frustration similar to $KCuF_3$. The dominant part of the superexchange in this system comes from $(t_{2g}^3 e_g^1)_i (t_{2g}^3 e_g^1)_j \rightleftharpoons (t_{2g}^3 e_g^2)_i (t_{2g}^3)_j$ charge excitations arising from the hopping of an $e_g$ electron, but $t_{2g}$ excitations also contribute to a superexchange interaction $J_t$ (defined in [79] and frequently called core-spin superexchange) through $(t_{2g}^3 e_g^1)_i (t_{2g}^3 e_g^1)_j \rightleftharpoons (t_{2g}^4 e_g^1)_i (t_{2g}^2 e_g^1)_j$ charge excitations. This latter term is AF ($J_t > 0$), because it does not involve any high-spin excited states for half-filled $t_{2g}$ orbitals. The spin-orbital model in this system has the form [79]

$$\mathcal{H}(d^4) = \frac{1}{2} J \sum_{\langle ij \rangle \parallel \gamma} \left\{ \left[ -\frac{1}{10} r_1 \left( \mathbf{S}_i \cdot \mathbf{S}_j + 6 \right) + \left( \frac{3}{80} r_5 + \frac{1}{16} r_6 \right) \left( \mathbf{S}_i \cdot \mathbf{S}_j - 4 \right) \right] \right.$$
$$\left. \times \left( \frac{1}{4} - T_i^{(\gamma)} T_j^{(\gamma)} \right) + \frac{1}{16} (r_6 + r_7) \left( \mathbf{S}_i \cdot \mathbf{S}_j - 4 \right) \left( \frac{1}{2} - T_i^{(\gamma)} \right) \left( \frac{1}{2} - T_j^{(\gamma)} \right) \right\}$$
$$+ J_t \sum_{\langle ij \rangle} \left( \mathbf{S}_i \cdot \mathbf{S}_j - 4 \right), \tag{23.30}$$

where the coefficients

$$r_5 = \frac{1}{1 + 3\eta/4}, \qquad r_6 = \frac{1}{1 + 5\eta/4}, \qquad r_7 = \frac{1}{1 + 13\eta/4}, \tag{23.31}$$

follow again from the multiplet structure of $Mn^{2+}$ ($d^5$) excited states [$r_1$ is as in (23.25)] [8]. The spin operators in (23.30) are for spins of total $S = 2$, and the prefactors of the different terms are determined by their spin algebra and by the respective Clebsch-Gordon coefficients. One may recognize the FM $(\boldsymbol{S}_i \cdot \boldsymbol{S}_j + 6)$ and AF $(\boldsymbol{S}_i \cdot \boldsymbol{S}_j - 4)$ terms, which are analogous to the spin projections in (23.25). The large value of the spin makes quantum effects in the spin subsystem less important in this case, the behavior of the model being considerably more classical than for $S = 1/2$ spins (23.25). Indeed, spin and orbital operators may be treated here as decoupled from each other [5–7].

In (23.30), one recognizes another mechanism which removes frustration of spin-orbital superexchange interactions in the manganites. In addition to the interactions with the lattice, which here play a role as important as in the case of $KCuF_3$ [80], the orbital frustration in $LaMnO_3$ may be removed by the spin interactions. This effect is best shown by investigating a quantum phase transition in the 1D manganite model derived from (23.30), where, as a function of the $t_{2g}$ superexchange $J_t$, the system undergoes a transition between the FM spin phase with AO order and the AF spin phase with FO order [81]. This transition can also be driven by changing the value of the orbital splitting between two AO states, which is a consequence of the Jahn–Teller potential, and in this latter case the spin state is dictated by the orbital order. Such competition between different phases plays an important role also in the more realistic situations of 2D [82] and 3D models [80, 83, 84] at finite temperatures. The orbital superexchange interactions are in fact supported by the Jahn–Teller interactions [79], and thus AO order in the *ab* planes occurs first at high temperature, $T_{OO} = 780$ K. The *A*-AF phase observed in $LaMnO_3$ below $T_N \simeq 140$ K can be thus considered to be dictated by one particular compromise between spin and orbital interactions, with the orbital frustration quenched by the Jahn–Teller effect.

### 23.7.3 Spin-Orbital Superexchange for $t_{2g}$ Perovskites

Many challenging and still not fully understood examples of spin-orbital frustration, with even more pronounced quantum effects than in the $e_g$-orbital systems, are encountered in titanate and vanadate perovskites. The relevant model for the $S = 1/2$ spins of $Ti^{3+}$ ions ($d^1$ configuration), which have only $t_{2g}$ orbital degrees of freedom, is quite different from that of (23.25). In this case the symmetry allows hopping for only two of the three $t_{2g}$ orbitals along each cubic $\gamma$ axis [24, 85], whereas the third orbital lies in the plane perpendicular to this axis, along which hopping through the intermediate $2p_\pi$ oxygen orbital is forbidden (Fig. 23.5). This motivates the convenient notation used below (23.10), where the inactive orbital along a given cubic direction $\gamma$ is labeled by its index as $|\gamma\rangle$.

The superexchange Hamiltonian for the titanate perovskites is determined by intersite charge excitations of the type $(t_{2g}^1)_i(t_{2g}^1)_j \rightleftharpoons (t_{2g}^2)_i(t_{2g}^0)_j$, which lead to a form similar to the $d^9$ configuration of $Cu^{2+}$ ions in $KCuF_3$, with either a triplet virtual state at energy $(U - 3J_H)$, or to one of three different singlet states, with the energies $(U - J_H)$ (doubly degenerate) and $(U + 2J_H)$ [8]. Once again, if these excitation energies are parameterized by $\eta$ (23.26) and one uses the coefficients $r_1$ and $r_2$ (23.27) together with

$$r_3 = \frac{1}{1+2\eta}, \tag{23.32}$$

one obtains compact expressions for the superexchange interactions between $S = 1/2$ spins in the cubic titanates [85, 86],

$$\begin{aligned}\mathcal{H}(d^1) = J \sum_{\langle ij\rangle\|\gamma} \Bigg\{ & r_1 \left(\mathbf{S}_i \cdot \mathbf{S}_j + \frac{3}{4}\right)\left(\boldsymbol{\tau}_i \cdot \boldsymbol{\tau}_j - \frac{1}{4}(n_i + n_j) + \frac{1}{4}n_i n_j\right)^{(\gamma)} \\ & + r_2 \left(\mathbf{S}_i \cdot \mathbf{S}_j - \frac{1}{4}\right)\left(\boldsymbol{\tau}_i \cdot \boldsymbol{\tau}_j + \frac{1}{4}(n_i + n_j) + \frac{1}{4}n_i n_j\right)^{(\gamma)} \\ & - \frac{2}{3}(r_2 - r_3)\left(\mathbf{S}_i \cdot \mathbf{S}_j - \frac{1}{4}\right)\left(\boldsymbol{\tau}_i \times \boldsymbol{\tau}_j + \frac{1}{4}n_i n_j\right)^{(\gamma)}\Bigg\}, \end{aligned} \tag{23.33}$$

where the term

$$\boldsymbol{\tau}_i \times \boldsymbol{\tau}_j = \tau_i^x \tau_j^x - \tau_i^y \tau_j^y + \tau_i^z \tau_j^z \tag{23.34}$$

originates from the structure of Coulomb interactions [5–7] and is responsible for nonconservation of the total orbital quantum numbers $\{\mathcal{T}, \mathcal{T}^z\}$. The operators $\boldsymbol{\tau}_i^{(\gamma)} = \{\tau_i^x, \tau_i^y, \tau_i^z\}$ represent the conventional components of a pseudospin $\tau = 1/2$ entity active along the $\gamma$ axis, and the number operators $n_i^{(\gamma)}$ represent the density of electrons in the two active $t_{2g}$ orbitals for each individual bond $\langle ij\rangle \parallel \gamma$. Thus it is only when each of both sites $i$ and $j$ contains a single electron in one of the active $t_{2g}$ orbitals that the scalar product $\boldsymbol{\tau}_i \cdot \boldsymbol{\tau}_j$ and the cross product $\boldsymbol{\tau}_i \times \boldsymbol{\tau}_j$ (23.34) contribute to the superexchange energy. Clearly, bonds which do not contribute with finite energy occur more frequently in this case than for $e_g$ orbitals.

The model for cubic vanadates with $S = 1$ spins, relevant for $V^{3+}$ ions [87, 88],

$$\begin{aligned}\mathcal{H}(d^2) = J \sum_{\langle ij\rangle\|\gamma} \Bigg\{ & \frac{1}{3} r_1 (\mathbf{S}_i \cdot \mathbf{S}_j + 2)\left(\boldsymbol{\tau}_i \cdot \boldsymbol{\tau}_j + \frac{1}{4}n_i n_j\right)^{(\gamma)} \\ & + \frac{1}{12}(\mathbf{S}_i \cdot \mathbf{S}_j - 1)\left[-3\left(\boldsymbol{\tau}_i \cdot \boldsymbol{\tau}_j + \frac{1}{4}n_i n_j\right)^{(\gamma)} + 2\right] \\ & + \frac{1}{12}(2 + 3r_3)(\mathbf{S}_i \cdot \mathbf{S}_j - 1)\left(\boldsymbol{\tau}_i \times \boldsymbol{\tau}_j + \frac{1}{4}n_i n_j\right)^{(\gamma)}\Bigg\}, \end{aligned} \tag{23.35}$$

is the consequence of virtual $(t_{2g}^2)_i (t_{2g}^2)_j \rightleftharpoons (t_{2g}^3)_i (t_{2g}^1)_j$ charge excitations and contains the same coefficients $r_1$ and $r_3$ as (23.32). The origin of the additional terms lies in two intermediate, low-spin states with energy $U$, which therefore contribute with the coefficient 1. As in (23.33), the operators $\boldsymbol{\tau}_i^{(\gamma)}$ describe orbital pseudospins $\tau = 1/2$ on the axis $\gamma$. In the known materials in this class, the intrinsic frustration of spin-orbital superexchange is removed by lattice distortions: structural data for the $R$VO$_3$ perovskites ($R$ = Lu, Yb, ..., La [14]) suggest that the $|c\rangle \equiv |xy\rangle$ orbitals are occupied at every site ($n_{ic} = 1$), and hence the cubic symmetry of (23.35) is broken. In this case there is precisely one electron in both active orbitals along the bonds $\langle ij\rangle \parallel c$ ($n_{ia} + n_{ib} = 1$ per site), as a result of which quantum fluctuations contribute only in this direction, but are blocked in the $ab$-planes.

Two different types of order may emerge from the superexchange model of (23.35), as a function of the Hund exchange $\eta$. In the regime of small $\eta$, the quantum fluctuations stabilize a dimerized 1D orbital VB (OVB) phase, with ($ab$) orbital singlets on every second bond supporting locally FM spin order, and weak AF spin interactions between these chains [88, 89], as shown in Fig. 23.7a. By contrast, at large $\eta$ the interdimer AF interactions in the $c$-direction are suppressed and quantum fluctuations along the disordered orbital chain support uniform FM spin interactions [87]. The cubic symmetry of the orbital superexchange is broken here, the full SU(2) pseudospin (orbital) interaction $\boldsymbol{\tau}_i \cdot \boldsymbol{\tau}_j$ contributing only along the $c$-axis, while the orbital interactions in the $ab$-planes are of Ising type. This explains the physical origin of the $C$-AF order observed in the $R$VO$_3$ perovskites [14], which coexists with rather weak $G$-AO order due to $\{a, b\}$ orbital fluctuations, as shown in Fig. 23.7b. Orbital order (occurring at $T < T_{\text{OO}}$) and magnetic order (at $T < T_{N1}$) set in almost simultaneously ($T_{\text{OO}} \simeq T_{N1}$) in LaVO$_3$, whereas in other $R$VO$_3$ perovskites, which have smaller ionic radii, the $G$-AO order occurs first and is followed later by $C$-AF order at a rather lower magnetic transition [14]. A recent theoretical analysis [90] demonstrates that the systematic changes observed in the phase diagram of the $R$VO$_3$ perovskites for these two types of order can be

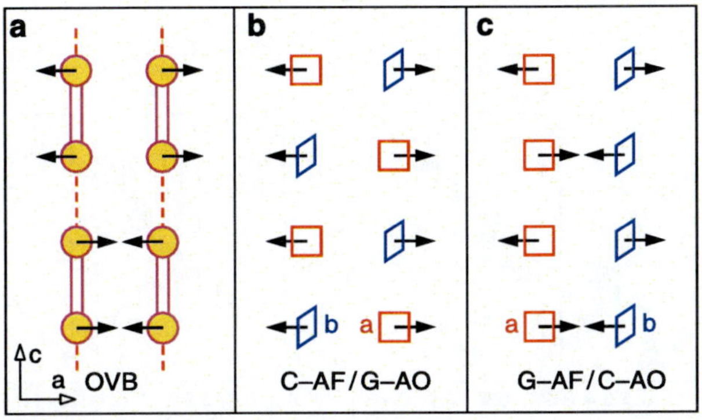


**Fig. 23.7** Possible spin and orbital order in the $ac$-planes of cubic $R$VO$_3$ perovskites [92]: (**a**) OVB phase, with alternating FM bonds occupied by ($ab$) orbital singlets (*double lines*), coupled by weak AF bonds (*dashed lines*); (**b**) $C$-AF and $G$-AO order; (**c**) $G$-AF and $C$-AO order ($yz$ and $zx$ orbitals are represented schematically by squares). In the phases with orbital order, (**b**) and (**c**), spins and orbitals alternate also along the $b$ axis (not shown); these states [but not state (**a**)] are consistent with the GKR [9, 10]

well understood by extending the microscopic model of (23.35) to include both the orbital-lattice coupling and the orbital–orbital interactions induced by Jahn–Teller and by $GdFeO_3$-type distortions [12].

In some $RVO_3$ perovskites with small rare-earth ions, such as $YVO_3$, a complementary type of spin-orbital order has also been observed [14]: this features coexisting $G$-AF and $C$-AO order, and is shown in Fig. 23.7c. This phase was first observed in $YVO_3$ [91], at a magnetic phase transition which is particularly surprising because the ordering moments change direction at $T_{N2} \simeq 77$ K from approximately parallel to the $c$-axis in the low-temperature $G$-AF phase to lying almost within the $ab$-planes in the high-temperature $C$-AF phase, with only a small, alternating $G$-AF component. The stability of the low-temperature $G$-AF phase in $YVO_3$ was addressed theoretically, and it was argued that the orbital interactions induced by lattice distortions stabilize this phase [87]. In this way, the lattice removes the spin-orbital frustration and dictates a different and more classical type of the spin-orbital order then that induced by superexchange (23.35). It has been argued further [92] that a transition to the $C$-AF phase with weak $G$-AO order could be triggered by the large gain in the free energy of the $C$-AF phase due to its large entropy. In fact, both spin and orbital excitations have lower energies in the latter phase than in $G$-AF phase.

The above entropy argument follows from the experimental observation that the two AF phases of $YVO_3$ have remarkably different magnon spectra, as shown in Fig. 23.8: these change due not only to the different type of the AF order, but also to the strong reduction in the characteristic energy scale of the magnetic excitations in the high-temperature $C$-AF phase [93]. This latter effect remains an outstanding unsolved problem in the field. Another spectacular effect, which may be considered as a consequence of competing interactions, is the dimerization observed in the magnon spectra in the $C$-AF phase: this can be described by a FM interaction $J_c(1 \pm \delta)$ alternating along the $c$ axis (Fig. 23.8b). A recent theoretical study [94]

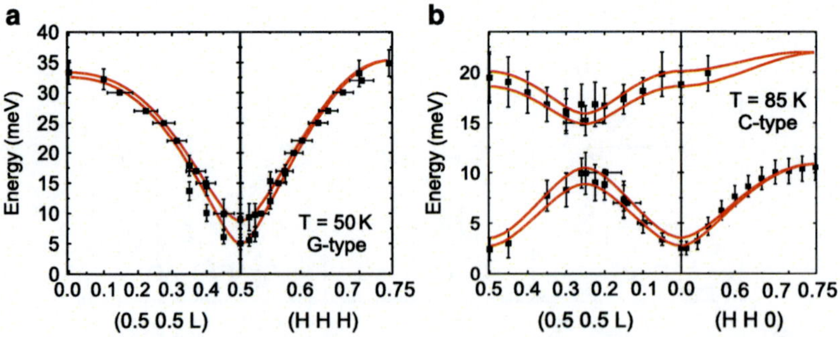


**Fig. 23.8** Magnon dispersion data obtained by neutron scattering in (**a**) the $G$-AF phase and (**b**) the $C$-AF phase of $YVO_3$. The data points are shown with effective linewidths indicated by the bars; lines show the fit to the data obtained with $J_{ab} = J_c = 5.7$ meV in the $G$-AF phase and with $J_{ab} = 2.6$, $J_c = -3.1$ meV and $\delta = 0.35$ (see text) in the $C$-AF phase. Figure reproduced from [93]

indicates that, somewhat surprisingly, a Peierls dimerization can in fact be activated by thermal fluctuations in a FM spin chain, when either spin-phonon or spin-orbital coupling is present, the latter being relevant for $YVO_3$. The measured magnon dispersion curves can be reproduced within the vanadate spin-orbital model of (23.35) after appropriate averaging over the orbital operators (see also Sect. 23.8).

We close this discussion of cubic systems by noting that, unlike the case of active $e_g$ orbitals, a relativistic spin-orbit coupling modifies the local states of strongly correlated $t_{2g}$ electrons, particularly in later TM oxides such as $Na_xCoO_2$ [95]. This coupling plays a role also in the $RTiO_3$ and $RVO_3$ perovskites, being found to modify the spin and orbital order for realistic parameters. As one example of its effects, it has been suggested [88] that the parameters of $YVO_3$ drive a system close to the regime of local entangled states induced by the spin-orbit interaction. Thus, the inclusion of a finite spin-orbit coupling could be crucial for a complete understanding of spin and orbital excitations even in $3d$ TM systems.

### *23.7.4 Spin-Orbital Frustration on a Triangular Lattice*

Spin-orbital superexchange interactions on a triangular lattice are particularly intriguing. In general, geometrical frustration enhances the overall frustration of orbital interactions in spin-orbital models. An important difference between the perovskite lattice and lattices with geometrical frustration is the angle between the oxygen-metal bonds connecting two neighboring TM ions: while this angle is close to 180° in perovskites, in layered (quasi-2D) materials with triangular lattice geometry (Fig. 23.9) and in spinels (Sect. 23.7.5) it is 90°.

We begin our consideration of the triangular lattice with the prominent example of $LiNiO_2$, which has active $e_g$ orbital degrees of freedom [15]. It is a rather remarkable and still outstanding problem in this field that the Mott insulators $LiNiO_2$ and $NaNiO_2$ behave quite differently in spite of their structural and electronic similarity: while $NaNiO_2$ undergoes a cooperative Jahn–Teller transition followed by a magnetic transition at $T_N = 20$ K, both transitions are absent in $LiNiO_2$. Possible reasons for this distinctly different behavior have been discussed in a number of papers, most notably in [15, 16] and [96–99]. We do not present here all of the theoretical concepts which have been suggested to account for the differences between $LiNiO_2$ and $NaNiO_2$, but focus instead on extracting some essential features. The triangular lattice which occurs in both compounds (and also in $NaTiO_2$, discussed below) is formed by edge-sharing $MO_6$ octahedra, shown in Fig. 23.9, where successive $\langle 111 \rangle$ planes of TM ions are well separated from each other.

There is to date no consensus concerning the simplest microscopic spin-orbital model which would capture all the essential features of both compounds. On the one hand, following the approach of the previous section and the relevant symmetry arguments [97], a general spin-orbital model for $Ni^{+3}$ ions in $d^7$ configuration,

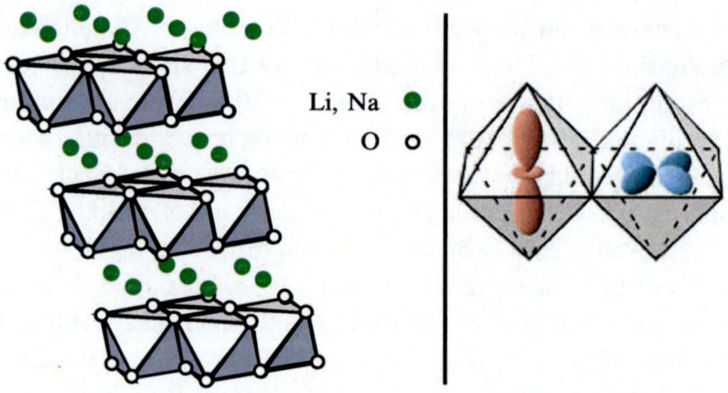


**Fig. 23.9** *Left*: $LiNiO_2$ (or $NaTiO_2$) structure. TM ions are located in the centers of the $TMO_6$ octahedra. *Right*: orbital $e_g$ states of the seventh $d$-electron of $Ni^{+3}$ ($d^7$) in two representative $NiO_6$ octahedra. Figure reproduced from [99]

$$\mathcal{H}_\Delta(d^7) = \frac{1}{2}J \sum_{\langle ij\rangle \parallel \gamma} \Big\{ r_1 \left( \boldsymbol{S}_i \cdot \boldsymbol{S}_j + \frac{3}{4} \right) \Big[ 2\lambda\, \boldsymbol{T}_i \cdot \boldsymbol{T}_j + (1-\lambda^2) T_i^{(\gamma)} T_j^{(\gamma)} - c_0 \Big]$$
$$+ r_2 \left( \boldsymbol{S}_i \cdot \boldsymbol{S}_j - \frac{1}{4} \right) \Big[ 4\lambda T_i^y T_j^y + \frac{1}{2}(1-\lambda^2) \left( T_i^{(\gamma)} + T_j^{(\gamma)} \right) + 2c_0 \Big]$$
$$+ r_4 \left( \boldsymbol{S}_i \cdot \boldsymbol{S}_j - \frac{1}{4} \right) \Big[ 2\lambda\, \boldsymbol{T}_i \cdot \boldsymbol{T}_j - 4\lambda T_i^y T_j^y + (1-\lambda^2) T_i^{(\gamma)} T_j^{(\gamma)}$$
$$+ \frac{1}{2}(1-\lambda^2) \left( T_i^{(\gamma)} + T_j^{(\gamma)} \right) + c_0 \Big] \Big\}, \qquad (23.36)$$

is characterized by rather anisotropic orbital interactions. Here, $r_m$ are the coefficients given in (23.27), $\boldsymbol{T}_i \cdot \boldsymbol{T}_j$ is a pseudospin scalar product for $T = 1/2$ pseudospins with components $T_j^\alpha$ ($\alpha = x, y, z$), the operators $T_j^{(\gamma)}$ are defined as in (23.4) and depend on the bond direction $\gamma$ in the triangular lattice, $J = 4t^2/U$, and $c_0 = \frac{1}{4}(1 + \lambda^2)$. The parameter $\lambda = t'/t$ is an unknown ratio of two effective hopping matrix elements between pairs of identical $|z\rangle$ or $|x\rangle$ orbitals in the triangular geometry (the interorbital hopping vanishes by symmetry). A schematic representation may be found in Fig. 23.9. The phase diagram of the model Hamiltonian (23.36) investigated in a mean-field (MF) approach on a 16-site cluster suggests two different realizations of resonating valence-bond (RVB) states as a possible explanation of the disordered phase in $LiNiO_2$, the "SU(4) phase" (see also Sect. 23.8) and the fluctuating dimer phases. The latter states have motivated some authors to consider approaches based on quantum dimer models [98, 99].

On the other hand, the excitations on oxygen orbitals are crucial in the geometry of $LiNiO_2$, which has 90° Ni–O–Ni bonds, and arguments have been articulated [96] that the leading term in the $LiNiO_2$ spin-orbital model is of the form

$$\mathcal{H}'_{\Delta}(d^7) = \sum_{\langle ij\rangle \in \alpha\beta} \left\{ J_T T_i^{(\alpha)} T_j^{(\beta)} - J_{ST} \mathcal{I}_{ij}^{\alpha\beta} \left( \mathbf{S}_i \cdot \mathbf{S}_j + \frac{3}{4} \right) + J_S \mathbf{S}_i \cdot \mathbf{S}_j \right\}, \tag{23.37}$$

where each Ni–Ni bond lies in the plane $\alpha\beta$, with $\alpha, \beta = x, y.z$ being the main directions in the crystal structure of $NaNiO_2$ [96]. The orbital operators which couple to the spins are

$$\mathcal{I}_{ij}^{\alpha\beta} = \left( \frac{3}{2} - T_i^{(\alpha)} \right) \left( \frac{3}{2} - T_j^{(\beta)} \right) + \left( \frac{3}{2} - T_i^{(\beta)} \right) \left( \frac{3}{2} - T_j^{(\alpha)} \right), \tag{23.38}$$

and all the interaction parameters $(J_T, J_{ST}, J_S)$ in (23.37) are positive. By construction, excited $Ni^{2+}$ $(d^8)$ states do not occur here, and terms with coefficients $r_m$ are absent. Analysis of the relevant charge-transfer model shows that $J_T \gg J_{ST}$ and $J_T \gg J_S$, so the strongly anisotropic orbital interaction $J_T$ dominates [16]. In fact, the structure of the model derived by considering 90° Ni–O–Ni bonds is even richer, because the low-spin states must also be included [16], but the general conclusion that orbital interactions of the compass-model type dominate in $LiNiO_2$ remains valid.

This discussion of the relevant energy scales suggests that a scenario in which the spin and orbital degrees of freedom are decoupled [96] could be realistic, particularly if the coupling to the lattice is also included in the spin-orbital model. Local $NiO_6$ distortions in fact favor directional $|z\rangle$-type orbitals and induce a frozen orbital order [16]. In addition, the real materials feature a non-stoichiometry of the form $Li_{1-x}Ni_{1+x}O_2$, and disorder due to $Ni^{2+}$ ions in the Li planes can prevent the onset of 3D magnetic order. Thus at present one may conclude only that $LiNiO_2$ is unlikely to be a spin-orbital liquid, and that the reasons behind the observed disordered state in $LiNiO_2$ are quite subtle [16, 97].

The properties of the spin-orbital model analogous to (23.33), derived for the $d^1$ occupancy of $t_{2g}$ orbitals but on a triangular lattice (as realized in the geometry of $NaTiO_2$), were recently discussed *in extenso* [100]. This model is somewhat involved, because the electrons change their $t_{2g}$ orbital flavor during charge excitations, in a manner similar to hopping in the $CoO_2$ layer of $Na_xCoO_2$ [101]. By considering the phase diagram as a function of the ratio between the superexchange and direct-exchange interactions, one finds that the effective orbital interactions are always frustrated, and that orbital correlations are generally dictated by the spin state, manifesting an intrinsic entanglement of these degrees of freedom. In the absence of Hund exchange, the ground state changes from a highly resonating, dimer-based, symmetry-restored spin and orbital liquid phase in the superexchange limit to one based on completely static, but highly degenerate, spin-singlet VB states in the direct-exchange limit. The latter limit, also discussed in [102], is reminiscent of the dimer singlet states in spinels, which we discuss in the next section. It is worth noting that the generic properties of frustration and entanglement survive in the triangular-lattice $t_{2g}$ model even when the spins and orbitals are nominally decoupled in the FM phases stabilized by strong Hund coupling [100].

### 23.7.5 Spin-Orbital Frustration in Spinels

It was shown recently that novel phases based on singlet dimer coverings arise in $S = 1/2$ spin models with the frustrated pyrochlore geometry, and may thus be expected on spinel lattices [43] (the reader is referred also to Chap. 7 by H. Takagi and S. Niitaka in this volume). One may expect that spin-orbital superexchange leads also to similar phases of singlet dimer states, and with particular types of symmetry-breaking in the singlet distribution on the lattice. Indeed, the frustration of superexchange obtained for $t_{2g}$ orbital degrees of freedom is further amplified when the lattice is frustrated [103], as in the example of the spinel titanate $MgTi_2O_4$. In this case, only one orbital is active for a bond in a given plane, *e.g.* the $|c\rangle \equiv |xy\rangle$ orbital in the $ab$-plane. This situation thus resembles the $d^9$ model (23.25), and the orbital variables are of Ising type. The model for the spinel titanate $MgTi_2O_4$, which has $Ti^{3+}$ ions with spins $S = 1/2$, is [103, 104]

$$\mathcal{H}_{\bowtie}(d^1) = \frac{1}{4}J\sum_{\langle ij\rangle\in\alpha\beta}\left\{-r_1\left(\boldsymbol{S}_i\cdot\boldsymbol{S}_j+\frac{3}{4}\right)\mathcal{O}_{ij,\alpha\beta}+r_2\left(\boldsymbol{S}_i\cdot\boldsymbol{S}_j-\frac{1}{4}\right)\mathcal{O}_{ij,\alpha\beta}\right.$$
$$\left.+\frac{1}{3}(2r_2+r_3)\left(\boldsymbol{S}_i\cdot\boldsymbol{S}_j-\frac{1}{4}\right)\bar{\mathcal{O}}_{ij,\alpha\beta}\right\}. \tag{23.39}$$

This expression contains the same coefficients $\{r_1, r_2, r_3\}$ as in the perovskite titanate. Its first line (terms with orbital operators $\mathcal{O}_{ij,\alpha\beta}$) describes the competing FM and AF contributions when one active and one inactive orbital are occupied on a bond $\langle ij\rangle$ in the plane $\alpha\beta$ (with $\{\alpha,\beta\} = \{x,y,z\}$), while the last term describes the AF interaction obtained for bonds $\langle ij\rangle$ with active $\alpha\beta$ orbitals at both sites of the bond. The bond projection operators

$$\mathcal{O}_{ij,\alpha\beta} = P_{i,\alpha\beta}(1-P_{j,\alpha\beta}) + P_{j,\alpha\beta}(1-P_{i,\alpha\beta}), \tag{23.40}$$
$$\bar{\mathcal{O}}_{ij,\alpha\beta} = P_{i,\alpha\beta}P_{j,\alpha\beta}, \tag{23.41}$$

include the on-site projectors $P_{i,\alpha\beta}$ onto the $\alpha\beta$ orbital at site $i$.

The superexchange model on the spinel lattice (23.39) features the now familiar intrinsically frustrated interactions. However, unlike in the $d^9$ model for $KCuF_3$ on the perovskite lattice (23.25), here the three orthogonal $t_{2g}$ orbitals correspond to the bonds in three different planes, $ab$, $bc$, and $ca$, and they contribute with finite energy only when at least one site contains an electron in an active orbital. Because only one out of the three orbitals on each site is active, the probability $\frac{4}{9}$ of obtaining a neutral bond is relatively high. In fact, the interactions favor a dimer phase consisting of spin singlets on bonds whose effective spin interaction is AF and on which both electrons are in active orbitals. *A priori*, these bonds appear with a relatively low probability of $\frac{1}{9}$, but they contribute a large energy $-J$ in the limit $\eta = 0$ [103]. This situation resembles the VB states in the $d^9$ model (23.25), but in the spinel case the spin dimer state is exact in the limit $\eta \to 0$. In this state, the bonds with one active and one inactive orbital also give a finite energy contribution independent of the spin state (spins on the interdimer bonds are disordered), and this orbital configuration favors local triplet spin states at $\eta > 0$. As a result, the ground state is a collection of spin-singlet dimers, and is highly degenerate with respect to the dimer configurations. It was shown that this "orientational" degeneracy can be lifted by a magnetoelastic interaction that optimizes the energy gain by distorting the bonds in suitable directions, leading to a tetragonal distortion and to the formation of a VB crystal through the condensation of dimers along helical chains winding around the tetragonal $c$-axis. In this way, the orbital state with FO order along the helices and AO order between them, as observed in $MgTi_2O_4$, could be explained [103, 104].

The vanadate model on the spinel lattice [105–107] is equally strongly frustrated. In this case, the superexchange involves $S = 1$ spins from the $d^2$ triplet configurations of $V^{3+}$ ions in the $AV_2O_4$ spinels, with $A =$ Zn, Mg, and Cd. Two $t_{2g}$ orbitals are now singly occupied at each site due to the finite Hund exchange ($J_H > 0$), and similar excitation energies (and coefficients $r_m$) follow from the active orbitals in this geometry as in the perovskite case of (23.35). The Hamiltonian [105–107]

$$\mathcal{H}_{\bowtie}(d^2) = \frac{1}{4}J \sum_{\langle ij\rangle \in \alpha\beta} \left\{ -\frac{1}{3} r_1 (\mathbf{S}_i \cdot \mathbf{S}_j + 2)\mathcal{O}_{ij,\alpha\beta} + \frac{1}{3}(\mathbf{S}_i \cdot \mathbf{S}_j - 1)\mathcal{O}_{ij,\alpha\beta} \right.$$
$$\left. + \frac{1}{2}(1 + r_3)(\mathbf{S}_i \cdot \mathbf{S}_j - 1)\bar{\mathcal{O}}_{ij,\alpha\beta} \right\} \tag{23.42}$$

contains the same projection operators as $\mathcal{H}_P(d^1)$, given in (23.40) and (23.41). One finds again a particularly strong frustration of the spin-orbital interactions when $\eta \to 0$. It has been shown [107] that at $\eta = 0$, all real orbital patterns are degenerate. Several factors play a role in removing this frustration in physical systems, where the observed spin structure is AF. This may be explained by considering not only a finite $\eta$ and the orbital ordering of complex orbital states of the type $(|xz\rangle \pm i|yz\rangle)/\sqrt{2}$, but also a finite spin-orbit coupling [107]. This type of orbital order also implies the coherent flattening of $VO_6$ octahedra along a unique axis, leading to the experimentally observed tetragonal distortion.

## 23.8 Spin-Orbital Entanglement

With the exception of specific cases in one dimension where the orbital operators are of Ising type [81], spin and orbital operators are in general entangled. Spin-orbital entanglement means here that the eigenstates cannot be written as product states of spin and orbital states, i.e. the eigenstates cannot be factorized. It occurs both for $e_g$ and $t_{2g}$ spin-orbital superexchange models, but is qualitatively stronger and has more profound consequences in the latter case [75]. Here, we discuss spin-orbital entanglement in the ground state of the $t_{2g}$ spin-orbital models on the perovskite lattice (23.33) and (23.35), by evaluating in exact-diagonalization calculations the composite correlation function defined for a bond $\langle ij \rangle$ as

$$C_{ij} = \frac{1}{(2S)^2} \left\{ \langle (\boldsymbol{S}_i \cdot \boldsymbol{S}_j)(\boldsymbol{T}_i \cdot \boldsymbol{T}_j) \rangle - \langle \boldsymbol{S}_i \cdot \boldsymbol{S}_j \rangle \langle \boldsymbol{T}_i \cdot \boldsymbol{T}_j \rangle \right\}. \qquad (23.43)$$

It it sufficient to investigate four-site spin-orbital chains oriented along the $c$-axis and with active $a$ and $b$ orbitals, using periodic boundary conditions. By definition, if $C_{ij} = 0$, the spin and orbital operators are disentangled and their MF decoupling is exact, while if $C_{ij} < 0$, the operators are entangled and a MF procedure cannot be justified (leading to systematic errors). Spin-orbital entanglement is evident in the titanate $d^1$ model (23.33), where one recovers the SU(4)-symmetric point [46] at $\eta = 0$. In this clear example, spin-orbital entanglement is manifest in a large, negative value of the spin-orbital correlation function, $C_{ij} = -0.25$ (Fig. 23.10a). At finite $\eta$, the SU(4) degeneracy of all intersite correlations is removed and one finds $T_{ij} < C_{ij} < S_{ij} < 0$, where

$$S_{ij} = \frac{1}{(2S)^2} \langle \boldsymbol{S}_i \cdot \boldsymbol{S}_j \rangle \quad \text{and} \quad T_{ij} = \langle \boldsymbol{T}_i \cdot \boldsymbol{T}_j \rangle \qquad (23.44)$$

are the conventional spin and orbital correlation functions, when the system is in the regime, $\eta \lesssim 0.21$, with a spin singlet ($S = 0$) ground state. The GKR, which imply that the signs of $S_{ij}$ and $T_{ij}$ should be complementary, are thus violated in this region. Instead, by analyzing the values of the intersite correlations, one finds that the ground-state wave function for each bond $\langle ij \rangle$ is close to a global SU(4) singlet, equivalent to a linear combination of (spin singlet/orbital triplet) and (spin triplet/orbital singlet) bond wavefunctions [75]. The vanadate $d^2$ model (23.35) behaves in a similar way, with all three correlation functions $S_{ij}$, $T_{ij}$, and $C_{ij}$ being negative simultaneously in the spin-singlet regime of fluctuating $a$ and $b$ orbitals, which is obtained in that case for $\eta \lesssim 0.07$ (Fig. 23.10b). Here again the composite spin-orbital correlations are significant ($C_{ij} < 0$) and their MF factorization is not allowed due to the manifest entanglement of spin and orbital variables.

To provide further evidence that the GKR do not apply to the $R\mathrm{TiO_3}$ and $R\mathrm{VO_3}$ perovskites, which have active $t_{2g}$ orbitals and low $\eta$, we compare the intersite spin correlation function $S_{ij}$ with the effective values of the spin exchange constants

$$J_{ij} \equiv \left\langle \hat{J}_{ij}^{(\gamma)} \right\rangle \qquad (23.45)$$

obtained by averaging the operators $\hat{J}_{ij}^{(\gamma)}$ in (23.24) over the orbital variables in a MF approach. Such a factorization of spin and orbital operators is usually assumed [5–7] in the analysis of results from neutron scattering experiments or optical spectroscopy within spin-orbital superexchange models. However, this procedure yields a net FM exchange interaction, $J_{ij} < 0$ (23.45), on averaging the orbital operators over an orbital-disordered phase with $T_{ij} < 0$ (Fig. 23.10c and d), and one finds that this is in fact accompanied by AF spin correlations ($S_{ij} < 0$) (Fig. 23.10a and b).

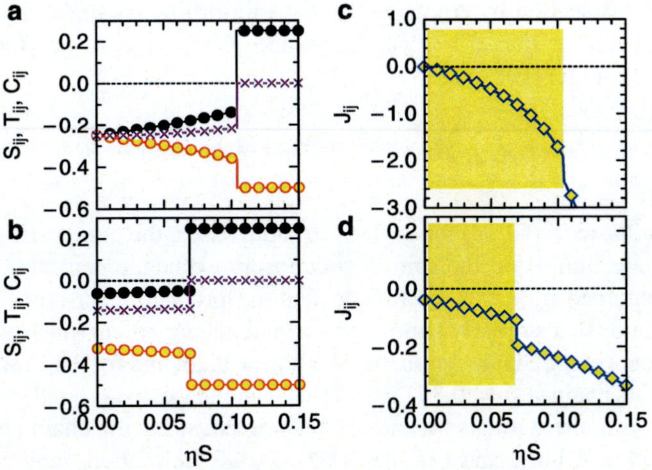


**Fig. 23.10** Spin-orbital entanglement in: (**a**), (**c**) $d^1$ model (23.33), and (**b**), (**d**) $d^2$ model (23.35), after [75]. *Left*: intersite spin, orbital, and spin-orbital correlation functions along the $c$-axis as a function of Hund exchange $\eta S$, showing respectively the quantities $S_{ij}$ (*filled circles*), $T_{ij}$ (*empty circles*), and $C_{ij}$ (*crosses*) of (23.44) and (23.43). *Right*: spin exchange constants $J_{ij}$ (23.45). In the shaded areas $J_{ij}S_{ij} > 0$, violating the classical GKR

Thus $J_{ij}S_{ij} > 0$ and the ground-state energy deduced from such a MF decoupling is unrealistically high, indicating the qualitative importance of spin-orbital entanglement [75]. In contrast, a similar analysis performed for the $d^9$ spin-orbital model (23.25), which has $e_g$ orbital degrees of freedom, gives $J_{ij}S_{ij} < 0$ in the entire parameter regime, and the GKR are satisfied [75]; however, here also one finds quantitative corrections to the ground-state energy which lie beyond the MF decoupling.

This qualitative difference between $t_{2g}$ and $e_g$ systems is a consequence of the importance of composite spin-orbital fluctuations, which are responsible for the dynamical nature of the exchange constants $J_{ij}$. By considering this quantity, one finds that its fluctuations on the bonds,

$$\delta J = \left\{ \left\langle \left( \hat{J}_{ij}^{(\gamma)} \right)^2 \right\rangle - J_{ij}^2 \right\}^{1/2}, \tag{23.46}$$

are particularly large at $\eta = 0$ in the $d^1$ (23.33) and $d^2$ models (23.35), while its average $J_{ij}$ is small in both cases ($J_{ij} \simeq 0$ for $d^1$ and $J_{ij} \simeq -0.04$ for $d^2$ [75]). In the $d^1$ case, the fluctuations between ($S = 0/T = 1$) and ($S = 1/T = 0$) wavefunctions on each bond are so large that $\delta J = 1$. Also in the $d^2$ model, the orbital bond correlations change dynamically from singlet to triplet, resulting in $\delta J > |J_{ij}|$, with $\delta J \simeq 0.25$, whereas these fluctuations are small for the $e_g$ $d^9$ model (23.25).

We close this section by emphasizing that spin-orbital entanglement effects are most pronounced for spin $S = 1/2$ and pseudospin $T = 1/2$ operators in the 1D SU(2)⊗SU(2) model [108]

$$\mathcal{H}_J = J \sum_i (\boldsymbol{S}_i \cdot \boldsymbol{S}_{i+1} + p)(\boldsymbol{T}_i \cdot \boldsymbol{T}_{i+1} + p), \tag{23.47}$$

which is similar to the SU(4) model [46]. In both cases, the composite correlations $C_{ij}$ (23.43) are finite, and the spin or pseudospin exchange constants $J_{ij}$ (23.45) cannot be obtained by using the MF decoupling procedure for any $p > -0.25$. In fact, one finds that similarly strong spin-orbital entanglement (measured again by $C_{ij} < 0$) occurs at the SU(4) symmetry point, $p = 0.25$, and for the exactly solvable 1D spin-orbital chain at $p = 0.75$ [108]. In this latter case, the ground state is doubly degenerate, with alternating spin and orbital singlets along the chain [109], similar to the Majumdar-Ghosh states of the 1D $J_1$-$J_2$ AF Heisenberg model with $J_2 = J_1/2$ [4]. This demonstrates once again the intrinsic frustration present in spin-orbital superexchange models, which occurs here in the absence of any geometrical frustration and with nearest-neighbor interactions only.

**Acknowledgements**

It is our pleasure to thank every one of our collaborators for all of the insightful discussions which contributed to our present understanding of the subject of frustration in systems with orbital degrees of freedom. We acknowledge financial support by the Nederlandse Organisatie voor Wetenschappelijk Onderzoek (NWO) and the Stichting voor Fundamenteel Onderzoek der Materie (FOM), by the National Science Foundation under Grant No. PHY05-51164, and by the Foundation for Polish Science (FNP) and the Polish Ministry of Science and Higher Education Project No. N202 069639.

## References

1. K.I. Kugel, D.I. Khomskii, Sov. Phys. Usp. **25**, 231 (1982)
2. Y. Tokura, N. Nagaosa, Science **288**, 462 (2000)
3. J. van den Brink, G. Khaliullin, D.I. Khomskii, *Orbital Effects in Manganites*, ed. by T. Chatterji *Colossal Magnetoresistive Manganites* (Kluwer, 2004)
4. P. Fazekas, *Lecture Notes on Electron Correlation and Magnetism* (World Scientific, Singapore, 1999)
5. A.M. Oleś, G. Khaliullin, P. Horsch, L.F. Feiner, Phys. Rev. B **72**, 214431 (2005)
6. A.M. Oleś, G. Stollhoff, Phys. Rev. B **29**, 314 (1984)
7. A.M. Oleś, Acta Phys. Polon. A **115**, 36 (2009)
8. J.S. Griffith, *The Theory of Transition Metal Ions* (Cambridge University Press, Cambridge, 1971)
9. J.B. Goodenough, *Magnetism and the Chemical Bond* (Interscience, New York, 1963)
10. J. Kanamori, J. Phys. Chem. Solids **10**, 87 (1959)

11. B. Keimer, D. Casa, A. Ivanov, J.W. Lynn, M. v. Zimmermann, J.P. Hill, D. Gibbs, Y. Taguchi, Y. Tokura, Phys. Rev. Lett. **85**, 3946 (2000)
12. E. Pavarini, A. Yamasaki, J. Nuss, O.K. Andersen, New J. Phys. **7**, 188 (2005)
13. H.F. Pen, J. van den Brink, D.I. Khomskii, G.A. Sawatzky, Phys. Rev. Lett. **78**, 1323 (1997)
14. S. Miyasaka, J. Fujioka, M. Iwama, Y. Okimoto, Y. Tokura, Phys. Rev. B **73**, 224436 (2006)
15. L.F. Feiner, A.M. Oleś, J. Zaanen, Phys. Rev. Lett. **78**, 2799 (1997)
16. A. Reitsma, L.F. Feiner, A.M. Oleś, New J. Phys. **7**, 121 (2005)
17. S. Margadonna, G. Karotsis, J. Am. Chem. Soc. **128**, 16436 (2006)
18. G. Giovannetti, S. Margadonna, J. van den Brink, Phys. Rev. B. **77**, 075113 (2008)
19. H.J. Jahn, E. Teller, Proc. Roy. Soc. A **161**, 220 (1937)
20. G.A. Gehring, K.A. Gehring, Rep. Progr. Phys. **38**, 1 (1975)
21. K.I. Kugel, D.I. Khomskii, JETP Lett. **15**, 446 (1972)
22. K.I. Kugel, D.I. Khomskii, Sov. Phys. JETP **37**, 725 (1973)
23. P.W. Anderson, Phys. Rev. **115**, 2 (1959)
24. J. Zaanen, A.M. Oleś, Phys. Rev. B **48**, 7197 (1993)
25. J. van den Brink, New J. Phys. **6**, 201 (2004)
26. J. van den Brink, P. Horsch, F. Mack, A.M. Oleś, Phys. Rev. B **59**, 6795 (1999)
27. Z. Nussinov, M. Biskup, L. Chayes, J. van den Brink, Europhys. Lett. **67**, 990 (2004)
28. X.-G. Wen, *Quantum Field Theory of Many-Body Systems* (Oxford University Press, Oxford, 2004), and references therein
29. C.D. Batista, Z. Nussinov, Phys. Rev. B **72**, 045137 (2005)
30. E.H. Lieb, Commun. Math. Phys. **31**, 327 (1973)
31. B. Simon, Commun. Math. Phys. **71**, 247 (1980)
32. M. Biskup, L. Chayes, Z. Nussinov, Comm. Math. Phys. **255**, 253 (2005)
33. B. Doucot, M.V. Feigel'man, L.B. Ioffe, A.S. Ioselevich, Phys. Rev. B **71**, 024505 (2005)
34. Z. Nussinov, E. Fradkin, Phys. Rev. B **71**, 195120 (2005)
35. J. Dorier, F. Becca, F. Mila, Phys. Rev. B **72**, 024448 (2005)
36. W. Brzezicki, J. Dziarmaga, A. M. Oleś, Phys. Rev. B **75**, 134415 (2007)
37. W. Brzezicki, A. M. Oleś, Phys. Rev. B **82**, 060401 (2010)
38. A.B. Harris, T. Yildirim, A. Aharony, O. Entin-Wohlman, I.Y. Korenblit, Phys. Rev. Lett. **91**, 087206 (2003)
39. S. Elitzur, Phys. Rev. D **12**, 3978 (1975)
40. Z. Nussinov, C.D. Batista, E. Fradkin, Int. J. Mod. Phys. B **20**, 5239 (2006)
41. A. Mishra, M. Ma, F.-C. Zhang, S. Guertler, L.-H. Tang, S. Wan, Phys. Rev. Lett. **93**, 207201 (2004)
42. C.D. Batista, S.A. Trugman, Phys. Rev. Lett. **93**, 217202 (2004)
43. Z. Nussinov, C.D. Batista, B. Normand, S.A. Trugman, Phys. Rev. B **75**, 094411 (2007)
44. E.F. Shender, Sov. Phys. JETP **56**, 178 (1982)
45. C.L. Henley, Phys. Rev. Lett. **62**, 2056 (1989)
46. Y.Q. Li, M. Ma, D.N. Shi, F.C. Zhang, Phys. Rev. Lett. **81**, 3527 (1998)
47. A.M. Oleś, L.F. Feiner, J. Zaanen, Phys. Rev. B **61**, 6257 (2000)
48. Y. Murakami, H. Kawada, H. Kawata, M. Tanaka, T. Arima, Y. Moritomo, Y. Tokura, Phys. Rev. Lett. **80**, 1932 (1998)
49. Y. Endoh, K. Hirota, S. Ishihara, S. Okamoto, Y. Murakami, A. Nishizawa, T. Fukuda, H. Kimura, H. Nojiri, K. Kaneko, S. Maekawa, Phys. Rev. Lett. **82**, 4328 (1999)
50. S. Ishihara, S. Maekawa, Phys. Rev. Lett. **80**, 3799 (1998)
51. I.S. Elfimov, V.I. Anisimov, G.A. Sawatzky, Phys. Rev. Lett. **82**, 4264 (1999)
52. H.-D. Chen, C. Fang, J. Hu, H. Yao, Phys. Rev. B **75**, 144401 (2007)
53. T. Tanaka, M. Matsumoto, S. Ishihara, Phys. Rev. Lett. **95**, 267204 (2005)
54. T. Tanaka, S. Ishihara, Phys. Rev. Lett. **98**, 256402 (2007)
55. R.B. Stinhcombe, in *Phase Transition and Critical Phenomena*, ed. by C. Domb, J.L. Lebowitz, vol. 7 (Academic, London, 1983)
56. D.J. Breed, K. Gilijamse, J.W.E. Sterkenburg, A.R. Miedema, J. Appl. Phys. **41**, 1267 (1970)
57. A. Honecker, D.C. Cabra, H.-U. Everts, P. Pujol, F. Stauffer, J. Phys. Condens. Matter **19**, 145249 (2007)

58. Z. Nussinov, G. Ortiz, Europhys. Lett. **84**, 36005 (2008)
59. G. Chen, L. Balents, A.P. Schnyder, Phys. Rev. Lett. **102**, 096406 (2009)
60. P. Shor, Proceedings of the *Symposium on the Foundations of Computer Science* (CA IEEE, Los Alamos, 1996)
61. A.Yu. Kitaev, Ann. Phys. **303**, 2 (2003)
62. A. Kitaev, Ann. Phys. **321**, 2 (2006)
63. G. Jackeli, G. Khaliullin, Phys. Rev. Lett. **102**, 256402 (2009)
64. J. Nasu, A. Nagano, M. Naka, S. Ishihara, Phys. Rev. B **78**, 024416 (2008)
65. C. Wu, Phys. Rev. Lett. **100**, 200406 (2008)
66. E. Zhao, W.V. Liu, Phys. Rev. Lett. **100**, 160403 (2008)
67. Z. Nussinov, G. Ortiz, Phys. Rev. B **77**, 064302 (2008)
68. Z. Nussinov, G. Ortiz, Ann. Phys. (N.Y.) **5**, 064302 (2009)
69. Z. Nussinov, G. Ortiz, Phys. Rev. B **79**, 214440 (2009)
70. X.-G. Wen, Phys. Rev. Lett. **90**, 016803 (2003)
71. H.-D. Chen, Z. Nussinov, J. Phys. A **41**, 075001 (2008)
72. D. Bacon, Phys. Rev. A **73**, 012340 (2006)
73. M. Lewenstein, A. Sanpera, V. Ahufinger, B. Damski, A. Sen De, U. Sen, Adv. Phys. **56**, 243 (2007)
74. L.M. Duan, E. Demler, M.D. Lukin, Phys. Rev. Lett. **91**, 090402 (2003)
75. A.M. Oleś, P. Horsch, L.F. Feiner, G. Khaliullin, Phys. Rev. Lett. **96**, 147205 (2006)
76. D.A. Tennant, R.A. Cowley, S.E. Nagler, A.M. Tsvelik, Phys. Rev. B **52**, 13368 (1995)
77. G. Khaliullin, V. Oudovenko, Phys. Rev. B **56**, R14243 (1998)
78. L.F. Feiner, A.M. Oleś, J. Zaanen, J. Phys.: Condens. Matter **10**, L555 (1998)
79. L.F. Feiner, A.M. Oleś, Phys. Rev. B **59**, 3295 (1999)
80. A. Weiße, H. Fehske, New J. Phys. **6**, 158 (2004)
81. M. Daghofer, A.M. Oleś, W. von der Linden, Phys. Rev. B **70**, 184430 (2004)
82. M. Daghofer, A.M. Oleś, D.R. Neuber, W. von der Linden, Phys. Rev. B **73**, 104451 (2006)
83. E. Dagotto, *Nanoscale Phase Separation and Colossal Magnetoresistance* (Springer, Heidelberg, 2003)
84. E. Dagotto, New J. Phys. **7**, 67 (2005)
85. G. Khaliullin, S. Maekawa, Phys. Rev. Lett. **85**, 3950 (2000)
86. G. Khaliullin, Phys. Rev. B **64**, 212405 (2001)
87. G. Khaliullin, P. Horsch, A.M. Oleś, Phys. Rev. Lett. **86**, 3879 (2001)
88. P. Horsch, G. Khaliullin, A.M. Oleś, Phys. Rev. Lett. **91**, 257203 (2003)
89. S.Q. Shen, X.C. Xie, F.C. Zhang, Phys. Rev. Lett. **88**, 027201 (2002)
90. P. Horsch, A.M. Oleś, L.F. Feiner, G. Khaliullin, Phys. Rev. Lett. **100**, 167205 (2008)
91. Y. Ren, T.T.M. Palstra, D.I. Khomskii, A.A. Nugroho, A.A. Menovsky, G.A. Sawatzky, Phys. Rev. B **62**, 6577 (2000)
92. A.M. Oleś, P. Horsch, G. Khaliullin, Phys. Rev. B **75**, 184434 (2007)
93. C. Ulrich, G. Khaliullin, J. Sirker, M. Reehuis, M. Ohl, S. Miyasaka, Y. Tokura, B. Keimer, Phys. Rev. Lett. **91**, 257202 (2003)
94. J. Sirker, A. Herzog, A.M. Oleś, P. Horsch, Phys. Rev. Lett. **101**, 157204 (2008)
95. G. Khaliullin, W. Koshibae, S. Maekawa, Phys. Rev. Lett. **93**, 176401 (2004)
96. M.V. Mostovoy, D.I. Khomskii, Phys. Rev. Lett. **89**, 227203 (2002)
97. F. Vernay, K. Penc, P. Fazekas, F. Mila, Phys. Rev. B **70**, 014428 (2004)
98. F. Vernay, A. Ralko, F. Becca, F. Mila, Phys. Rev. B **74**, 054402 (2006)
99. F. Mila, F. Vernay, A. Ralko, F. Becca, P. Fazekas, K. Penc, J. Phys. Condens. Matter **19**, 145201 (2007)
100. B. Normand, A.M. Oleś, Phys. Rev. B **78**, 094427 (2008)
101. W. Koshibae, S. Maekawa, Phys. Rev. Lett. **91**, 257003 (2003)
102. G. Jackeli, D.I. Ivanov, Phys. Rev. B **76**, 132407 (2007)
103. S. Di Matteo, G. Jackeli, C. Lacroix, N.B. Perkins, Phys. Rev. Lett. **93**, 077208 (2004)
104. S. Di Matteo, G. Jackeli, N.B. Perkins, Phys. Rev. B **72**, 024431 (2005)
105. H. Tsunetsugu, Y. Motome, Phys. Rev. B **68**, 060405 (2003)
106. H. Tsunetsugu, Y. Motome, Phys. Rev. B **70**, 184427 (2004)
107. S. Di Matteo, G. Jackeli, N.B. Perkins, Phys. Rev. B **72**, 020408(R) (2005)
108. A.M. Oleś, P. Horsch, G. Khaliullin, Phys. Stat. Solidi B **244**, 3478 (2007)
109. A.K. Kolezhuk, H.-J. Mikeska, Phys. Rev. Lett. **80**, 2709 (1998)

# Index